W9-ABD-044

MOLECULAR BIOLOGY: A SELECTION OF PAPERS

SEMINAL PAPERS REPRINTED FROM THE JOURNAL OF MOLECULAR BIOLOGY

Compiled by

S. Brenner

*MRC Molecular Genetics Unit,
Cambridge, UK*

ACADEMIC PRESS
Harcourt Brace Jovanovich, Publishers

London   San Diego   New York   Berkeley
Boston   Sydney   Tokyo   Toronto

ACADEMIC PRESS LIMITED
24/28 Oval Road,
London NW1 7DX

United States Edition published by
ACADEMIC PRESS INC.
San Diego, CA 92101

Copyright © 1989 by
ACADEMIC PRESS LIMITED

This book is printed on acid-free paper ∞

All rights reserved

No part of this book may be reproduced in any form by photostat, microfilm, or any other means, without written permission from the publishers.

British Library Cataloguing in Publication Data

Is available

ISBN 0-12-131200-3

Printed in Great Britain by Galliard (Printers) Ltd, Great Yarmouth

MOLECULAR BIOLOGY:
A SELECTION OF PAPERS

Preface

The Journal of Molecular Biology (JMB) was founded by John Kendrew just over thirty years ago and the first issue, a modest one of less than a hundred pages, appeared in April 1959. Few present day readers can appreciate how daring this venture seemed at the time; after all, the subject had only just begun, and the number of people willing to call themselves molecular biologists in public was still quite small. But as we now all know, the subject and, for that matter, the Journal, were doomed to success from the start. Molecular biology became pervasive and now dominates areas of biological research, and, today, nearly everybody is a molecular biologist.

This book celebrates this remarkable development and marks the thirtieth anniversary of the founding of JMB by reprinting a selection of papers from the first hundred volumes. These papers span the first half of the life of JMB, up to 1975, and cover the period when most of the fundamental discoveries of molecular biology were made. During this time the first three-dimensional structures of protein were obtained by X-ray diffraction and the molecular basis of the regulation of protein function was revealed; messenger RNA was discovered, and the mechanism of protein biosynthesis and the structure of the genetic code were elucidated; DNA replication came to be understood and a new area of the structural analysis of macromolecular assemblies in viruses and cellular components was ushered in by new technical extensions of electron microscopy and X-ray diffraction. During this time the foundations were also laid for the second and more recent phase of molecular biology based on techniques of DNA cloning and sequencing developed in the mid-1970s.

The papers in this volume have been selected to illustrate some of these developments in molecular biology, and represent the twin streams of genetics and structural biochemistry that united to form the subject. The first reprint, from Volume 1 of JMB, reports the famous 'pajamo' experiment which proved that repression was dominant and that it involved the synthesis of a cytoplasmic factor. This volume also included the paper on *E. coli* ribosomes and another that showed that bacteriophages were built up of a set of structural components. Volume 3, published in 1961, contained the classic review of Jacob and Monod on genetic regulatory mechanisms which set the scene for the great developments that were to follow in this field. There was also a short note in this volume on the Theory of Mutagenesis with the essential idea that led to the use of acridine induced mutations to analyse the genetic code. The short note in Volume 4 by Champe and Benzer describes what we would now call a fusion protein, made by a genetic deletion, $r1589$. Volume 6 was marked by a second important review on "Allosteric Proteins and Cellular Control Systems" in which Monod, Changeux and Jacob first put forward a set of important ideas on regulation.

Cairns has two papers that exploit the direct use of autoradiography to visualize the chromosomes of bacteriophages and bacteria. Volume 6 contained the remarkable demonstration that the DNA of *E. coli* was one circular molecule, 1-mm long. Included are a number of important papers on electron microscopy. There is the 1963 paper by Huxley on the structure of filaments of striated muscle and a definitive paper on the structure of viruses by Klug and Finch, which appeared in 1968. Also included is the earlier 1964 paper of Klug on optical methods for image reconstruction which laid the foundations for the correct and objective analysis of negatively stained preparations.

Volume 94, 1970, contained the important paper by Unwin and Henderson on a novel and powerful method of determining molecular structure by electron microscopy.

There are many papers on the genetic code and the mechanism of protein synthesis. In 1964 Sanger and Marcker reported a transfer RNA (then called s-RNA) with formylmethione attached to it, and with this method discovered the special mechanism for chain initiation in protein synthesis. Crick's Wobble Hypothesis of codon–anticodon recognition, appeared in Volume 19, together with the experimental confirmation of the ideas by the Khorana group in the succeeding paper. Natural mutants of a tyrosine suppressor RNA are reported in Volume 47, 1970, using sequencing methods developed by Sanger and reported in a 1965 paper reprinted here. In 1971, an entire issue of JMB was devoted to the publication of Khorana's work on the total chemical synthesis of a transfer RNA, the introduction to the series is included in the collection.

From 1970 onwards, papers appear that were important steps in the development of the modern approach in molecular biology. Many are technical, and it is to the credit of the Editors that they were willing to recognize new techniques and publish papers many of which exercised a seminal role in the development of the subject. Thus, there is the first report in Volume 53, the transfection calcium-treated E. coli by DNA, of a method which proved to be absolutely essential in the development of cloning methods using this bacterium. Restriction enzymes make their appearance; there is the first report of a recognition site by Smith in Volume 51 and we also reprint one of Nathan's papers in Volume 78 on the use of these enzymes to map a DNA, in this case, the virus SV40. Southern's paper on his method of detecting nucleic sequences by hybridization is reprinted from Volume 98, and Sanger's paper on a new method of DNA sequencing appeared in 1975. We have also included a paper by Kaiser and Lobban reprinted from Volume 78, 1973, which is an early remarkable piece of work on the beginning of DNA cloning. The important paper by Weissmann on site-directed mutagenesis, using Qβ, reprinted from Volume 89, 1974, will be recognized by readers as already very much in the modern style. And, finally, we have included two papers on the analysis of protein sequences, one being the well-known computer algorithm of Needleman and Wunsch from Volume 48, 1970.

The reader should recognize that this selection is a personal one, and I must emphasize that I accept sole responsibility, not only for the papers included here but also for all the omissions. My first list would have generated a book three or four times the present size and each successive culling became more and more difficult. The papers that remained are not necessarily "The Best of JMB", although many certainly are; and in addition to papers which reflected the achievements of the subject I have included some whose value lay in the future and could not have been fully appreciated at the time. I hope too that the reader will learn something of the development of our subject. Many ideas, which today we find simple and obvious, only became established by complex and difficult argument as people struggled to extract themselves from the rigid mould of contemporary thinking, and many experiments that today are carried out with kits ordered by telephone were than heroic acts at the limit of existing technology.

Science advances at an ever-increasing rate and molecular biology is no slouch. Anything more than 2-years old is already history and anything older than 4-year lifetime of a graduate student is prehistorical. Even so, there is something we recognize as modern about all of these papers; they represent steps on the long, uncompleted road of molecular biology, leading to understanding how the nucleotide sequences of genes specify the complex functions of protein molecules, and how these interact to produce the regulated behaviour of cells.   *Sydney Brenner*

Contents

viii CONTENTS

J. Mol. Biol. (1959) **1**, 165-178

The Genetic Control and Cytoplasmic Expression of "Inducibility" in the Synthesis of β-ġalactosidase by *E. Coli*†

Arthur B. Pardee‡, François Jacob and Jacques Monod

Institut Pasteur, Paris and University of California, Berkeley, California, U.S.A.

(*Received 16 March 1959*)

A number of extremely closely linked mutations have been found to affect the synthesis of β-galactosidase in *E. coli*. Some of these (*z* mutations) are expressed by loss of the capacity to synthesize active enzyme. Others (*i* mutations) allow the enzyme to be synthesized constitutively instead of inducibly as in the wild type. The study of galactosidase synthesis in heteromerozygotes of *E. coli* indicates that the *z* and *i* mutations belong to different cistrons. Moreover the constitutive allele of the *i* cistron is recessive over the inducible allele. The kinetics of expression of the i^+ (inducible) character suggest that the *i* gene controls the synthesis of a specific substance which represses the synthesis of β-galactosidase. The constitutive state results from loss of the capacity to synthesize active repressor.

1. Introduction

Any hypothesis on the mechanism of enzyme induction implies an interpretation of the difference between "inducible" and "constitutive" systems. Conversely, since specific, one-step mutations are known, in some cases, to convert a typical inducible into a fully constitutive system, an analysis of the genetic nature and of the biochemical effects of such a mutation should lead to an interpretation of the control mechanisms involved in induction. This is the subject of the present paper.

It should be recalled that the metabolism of lactose and other β-galactosides by intact *E. coli* requires the sequential participation of two distinct factors:

(1) The galactoside-permease, responsible for allowing the entrance of galactosides into the cell.

(2) The intracellular β-galactosidase, responsible for the hydrolysis of β-galactosides.

Both the permease and the hydrolase are inducible in wild type *E. coli*. Three main types of mutations have been found to affect this sequential system:

(1) $z^+ \rightarrow z^-$: loss of the capacity to synthesize β-galactosidase;

(2) $y^+ \rightarrow y^-$: loss of the capacity to synthesize galactoside-permease;

(3) $i^+ \rightarrow i^-$: conversion from the inducible (i^+) to the constitutive (i^-) state.

The $i^+ \rightarrow i^-$ mutation always affects *both* the permease and the hydrolase. All these mutations are extremely closely linked: so far all independent occurences of each of these types have turned out to be located in the "*Lac*" region of the *E. coli* K 12 chromosome. However, the mutations appear to be *independent* since all the different phenotypes resulting from combinations of the different alleles are observed (Rickenberg, Cohen, Buttin & Monod, 1955; Cohen & Monod, 1957; Cohn, 1957).

† This work has been aided by a grant from the Jane Coffin Childs Memorial Fund.
‡ Senior Postdoctoral Fellow of the National Science Foundation (1957–58).

It should also be recalled that conjugation in *E. coli* involves the injection of a chromosome from a ♂ (Hfr) into a ♀ (F⁻) cell, and results generally in the formation of an incomplete zygote (merozygote) (Wollman, Jacob & Hayes, 1956). Recombination between ♂ and ♀ chromosome segments does not take place until about 60 to 90 min after injection; moreover segregation of recombinants from heteromerozygotes occurs only after several hours, thus allowing ample time for experimentation.

In order to study the interaction of these factors, their expression in the cytoplasm and their dominance relationships, we have developed a technique which allows one to determine the kinetics of β-galactosidase synthesis in merozygotes of *E. coli*, formed by conjugation of ♂ (Hfr) and ♀ (F⁻) cells carrying different alleles of the factors z, y and i (Pardee, Jacob & Monod, 1958). Before discussing the results obtained with this technique, we shall summarize some preliminary observations on the genetic structure of the "*Lac*" region in *E. coli* K 12.

2. Materials and Methods

(a) *Bacterial strains*

A ♂ (Hfr) strain (no. 4,000) of *E. coli* K 12 was used in most experiments. It was derived from strain 58,161 F⁺, and was selected for early injection of the "*Lac*" marker (Jacob & Wollman, 1957). This strain is streptomycin sensitive (S^s), requires methionine for growth and carries the phage λ. A second Hfr strain (no. 3,000), isolated by Hayes (1953), was used in some experiments. This strain is S^s, requires vitamin B₁, and does not carry λ prophage. Other Hfr strains carrying mutations for galactosidase (z), inducibility-constitutivity (i), and permease (y) were isolated from the Hayes strain after u.v. irradiation. These markers were also put into ♀ (F⁻) strains, by appropriate matings and selection of the desired recombinants.

A synthetic medium (M 63) was commonly used. It contained per liter: 13·6 g KH_2PO_4, 2·0 g $(NH_4)_2SO_4$, 0·2 g $MgSO_4 . 7H_2O$, 0·5 mg $FeSO_4 . 7 H_2O$, 2·0 g glycerol, and KOH to make pH 7·0. If amino-acids were required, they were added at a concentration of 10 mg/l. of the L-form. For mating experiments, the above stock medium was adjusted to pH 6·3 and vitamin B₁ (0·5 mg/l.) was added prior to use. Aspartate (0·1 mg/ml.) was generally added at the time of mating, according to Fisher (1957).

(b) *Mating experiments*

The desired volume of fresh medium was inoculated with an overnight culture (grown in the same medium) to an initial density of approximately 2×10^7 bacteria/ml. This culture was aerated by shaking at 37°C in a water bath. Turbidity was measured from time to time; and when the density reached 1 to 2×10^8 bacteria/ml., the experiment was started. Usually small volumes of ♂ and ♀ bacteria were mixed in a large Erlenmeyer flask, with the ♀ strain in excess (e.g. 3 ml. ♂ plus 7 ml. ♀ in a 300 ml. flask). The mixed bacteria were agitated very gently so that the motion of the liquid was barely perceptible. From time to time samples were removed for enzyme assay and plating on selective media, usually lactose-B₁-streptomycin agar, for measurement of recombinants. Under these conditions, in a mating of ♂ z^+Sm^s by ♀ z^-Sm^r, up to 20 % of the ♂ population formed z^+Sm^r recombinants (as tested by selection on lactose-streptomycin agar). More often 5 to 10 % recombinants were found.

Streptomycin (Sm)† was used in many mating experiments, to block enzyme synthesis by z^+Sm^s ♂ cells. Controls showed that the synthesis of β-galactosidase was blocked in these strains immediately upon addition of 1 mg/ml. of Sm. Incorporation of ³⁵S from $^{35}SO_4^-$ as well as increase of turbidity were also suppressed by this treatment. This concentration of Sm had no effect on Sm-resistant (Sm^r) mutants. In some experiments, virulent phage (T6) was used to kill the ♂ cells, thus preventing remating.

† The following abbreviations are used in this paper:

Sm = streptomycin   ONPG = *o*-nitrophenyl-β-D-galactoside
IPTG = *iso*propyl-thio-β-D-galactoside   TMG = methyl-thio-β-D-galactoside

It should be noted that if streptomycin was added initially, it significantly reduced the number of recombinants (e.g., 75 % fewer colonies were formed on lactose-B$_1$-streptomycin plates after 80 min mating in the presence of 1 mg/ml. streptomycin) relative to mating in the absence of streptomycin; but the antibiotic had little effect on enzyme formation by zygotes if added at the commencement of the experiment or after the z^+ locus had been injected.

When galactosidase synthesis had to be induced in zygotes, isopropyl-thio-β-D-galactoside (IPTG) was used at 10^{-3}M, a concentration at which this inducer is known to be active even in the absence of permease (Rickenberg et al., 1956).

(c) Recombination studies

The blender technique of Wollman & Jacob (1955) was used to determine the times of penetration of markers into the zygotes. It should be noted that this treatment reduces enzyme-forming capacity in zygotes by 30 to 60 %. Recombinant colonies, selected on appropriate selective media, were restreaked on the selector medium and replica plating was used to determine unselected characters. Tests for galactosidase synthesis (with or without induction) were performed on maltose-synthetic agar plates with or without IPTG, using filter paper impregnated with ONPG, according to Cohen-Bazire & Jolit (1953).

Transductions were performed with phage 363, according to Jacob (1955).

(d) β-galactosidase assay

For this enzyme assay, 1 ml. aliquots of culture were pipeted into tubes containing 1 drop of toluene. The tubes were shaken vigorously and were incubated for 30 min at 37°C. They were then brought to 28°C; 0·2 ml. of a solution of M/75 o-nitrophenyl-β-D-galactoside in M/4 sodium phosphate (pH 7·0) was added, and the tubes were incubated a measured time, until the desired intensity of color had developed. The reaction was halted by addition of 0·5 ml. of 1 M-Na$_2$CO$_3$, and the optical density was measured at 420 mμ with the Beckman spectrophotometer. A correction for turbidity could be made by multiplying the optical density at 550 mμ by 1·65 and subtracting this value from the density at 420 mμ. One unit of enzyme is defined as producing 1 mμ-mole o-nitrophenol/ minute at 28°C, pH 7·0. The units of enzyme in the sample can be calculated from the fact that 1 mμ-mole/ml. o-nitrophenol has an optical density of 0·0075 under the above conditions (using 10 mm light-path).

(e) Chemicals

o-nitrophenyl-β-D-galactoside (ONPG), methyl-thio-β-D-galactoside (TMG) and iso-propyl-thio-β-D-galactoside (IPTG) were synthesized at the Institut Pasteur by Dr. D. Türk. Other chemicals were commercial products.

3. Genetic Structure of the "Lac" Region

Figure 1 presents the structure of the "Lac" region, as it can be sketched from the data available at present. This complex locus, as established long ago by Lederberg (1947) and confirmed by the blender experiments of Wollman & Jacob (1955), lies at about equal distances from the classical markers TL and Gal. The closest known markers are Proline (left) and Adenine (or T6) (right). As shown in the map, the several (about 10) occurrences of the y^- mutation all lie together probably at the left of the segment, while the different z^- mutations and the i^- mutant are packed together at the other end. No attempt has been made to establish the order of individual y^- mutations. The order of the z^- mutations relative to each other and to the i^- marker is unambiguously established, as shown, except for the z_U^- mutation, whose position is largely undetermined. Several independent occurrences of the i^- mutation have been isolated. They all appear to be closely linked to the i_3^- marker, but they have not been mapped, for lack of adequate methods of selection i^+ recombinants. The evidence for this structure is briefly as follows:

(1) The frequency of recombination between z and y mutations is very low:

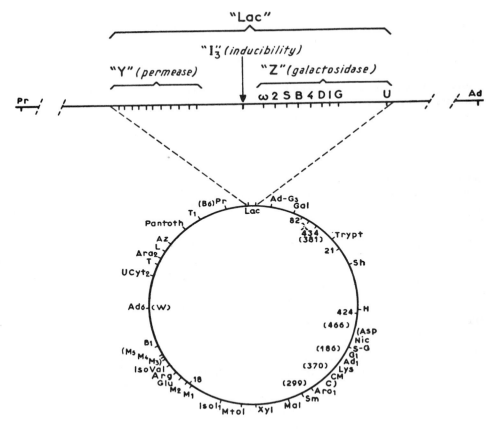

FIG. 1. Fine structure of the "*Lac*" segment.

The "*Lac*" segment is shown enlarged and positioned with respect to the rest of the *E. coli* K 12 linkage group for which the circular model (Jacob & Wollman, 1958) has been adopted.

roughly 1/100th of the frequency of recombination between TL and Gal. The frequency of recombination between individual z markers is about one order of magnitude lower.

(2) When y^+z^+ recombinants are selected (by growth on lactose-agar) in crosses of the type:

$$y^+i^-z^- \times y^-i^+z^+$$

the i^+ marker remains associated with z^+ 85 % of the time.

(3) The frequency of cotransduction of i with z (selecting for z^+ alone) is very high (> 90 %), while the frequency for i and y is also high, although definitely lower (about 70 %). (These data are somewhat ambiguous, because of the heterogeneity of the clones resulting from a transduction.)

(4) The selection of z^+ recombinants in crosses involving different z^- mutants, and i as unselected marker, invariably results in about 90 % of the progeny being either i^- or i^+, depending on the particular z^- mutants used. Assuming this result to be due to the position (left or right) of i with respect to the z group:

a linear order can be established, without contradictions, for the eight markers shown. This however leaves an ambiguity as to whether i lies *between* the y and the z groups, or outside.

Let us emphasize that this sketch of the *Lac* region is preliminary and very incomplete, and that the results concerning the relationships of certain markers are not understood. For instance, the z_U marker recombines rather freely with all the other mutants shown (both y and z) yet, by cotransduction tests, it is closely linked to i (25 % cotransduction). It should also be mentioned that certain of the z^- mutants (z_ω^- ; z_2^- ; z_G^-) have apparently lost the capacity to synthesize *both* the galactosidase *and* the permease. Yet these mutations do not seem to be deletions. We shall not attempt, here, to interpret this finding, since we shall center our attention on the interaction between the i marker and the z region.†

A question which should now be considered is whether we may regard the z region as possessing the specific structural information concerning the galactosidase molecule. The fact that so far all the independent mutations resulting in loss of the capacity to synthesize galactosidase were located in this region might not constitute sufficient evidence‡. However, it has been found by Perrin, Bussard & Monod (1959, in preparation) that several of the z^- mutants synthesize, instead of active galactosidase, an antigenically identical, or closely allied, protein. Moreover several of these mutant proteins are different from one another by antigenic and other tests. These findings appear to prove that the z region indeed corresponds to the "structural" genetic unit for β-galactosidase.

4. β-Galactosidase Synthesis by Heteromerozygotes

(a) *Preliminary experiments*

The feasibility and significance of experiments on the expression and interaction of the z, y and i factors depended primarily on whether *E. coli* merozygotes are physiologically able to synthesize significant amounts of enzyme very soon after mating. It was equally important to determine whether the mating involved any cytoplasmic mixing. These questions were investigated in a series of preliminary experiments.

Since the physical separation of *E. coli* zygotes from unmated or exconjugant parent cells cannot be achieved at present, test conditions must be set up, such that the zygotes only, but not the parents, can synthesize the enzyme. This is obtained when the following mating:

$$\text{♂} \; z^+y^+i^+Sm^s \times \text{♀} \; z_s y^+i^+Sm^r \qquad \text{(A)}$$

is performed in the presence of inducer (IPTG) and of 1 mg/ml. of streptomycin. The ♀ lack the z^+ factor; the ♂ are inhibited by streptomycin (cf. Methods); the zygotes are not, because they inherit their cytoplasm from the ♀ cells (see below

---

† Interaction of i with the y region is of course equally interesting, but since determinations of activity are much less sensitive with the galactoside-permease than with the galactosidase, we have used the latter almost exclusively.

‡ In addition to the mutants shown on Fig. 1, 20 other galactosidase-negative mutants, as yet unmapped, have been found to belong to the same segment by contransduction tests. None was found outside. Lederberg *et al.* (1951), however, have isolated some lactose-"non-fermenting" mutants (as tested on EMB-lactose agar) which are located at other points on the *E. coli* chromosome. In our hands, one of these mutants (Lac_3^-) formed normal amounts of both galactosidase and galactoside-permease (although it did form white colonies on EMB-lactose). Another one (Lac_7^-) formed reduced, but significant, amounts of both. A third (Lac_2^-) which is a galactosidase-negative, appears to belong to the "*Lac*" segment, by cotransduction tests.

pages 170 and 171), and because the type of ♂ used transfers the Sm^s gene to only a very small percentage of the cells. Under these conditions, enzyme is formed in the mated population with a time course and in amounts showing that the synthesis can be due only to zygotes having received the z^+ factor. Figure 2 shows the

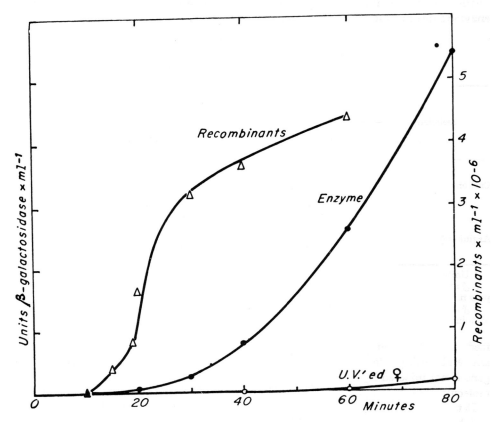

FIG. 2. Enzyme formation and appearance of recombinants in mating A.
Mating in presence of streptomycin (1 mg/ml.) and IPTG (10^{-3}M). A control with u.v.-treated ♀ cells (0·01 % survival) is shown. Recombinants (z^+Sm^r) selected by plating on Sm-lactose agar after blending (separate experiment with the same ♂ culture).

kinetics of galactosidase accumulation, compared with the appearance of z^+Sm^r recombinants, determined on aliquots of the same population (cf. Methods). The latter curve corresponds, as shown by Wollman & Jacob (1955), to the distribution of times of penetration of z^+ genes in the zygote population. It will be remarked that enzyme synthesis commences just within a few minutes after the first z^+ genes enter into zygotes. Assuming that the number of zygotes having received a z^+ gene is 4 to 5 times the number of recovered z^+Sm^r recombinants, and taking into account the fact that normal cells are on the average trinucleate (i.e., have three z^+ genes), the rate of enzyme synthesis per injected z^+ appears nearly normal.

This rapid expression of the z^+ factor poses the problem whether cytoplasmic constituents are injected from the ♂ into the zygote. This already appeared unlikely from the previous observations of Jacob & Wollman (1956). We reasoned that if there occurred any significant cytoplasmic mixing, such a mixing should allow the

♂ cells to feed the ♀ cells with any small metabolites which the ♂ had and the ♀ lacked. This condition is obtained in the following mating:

$$\text{♂ } z^{+}Sm^{s}\ maltose^{+} \times \text{♀ } z^{-}Sm^{r}\ maltose^{-}$$

if it is performed in presence of maltose as sole carbon source, using a ♂ which **virtually does not inject the** *maltose*$^{+}$ **gene.** It results in a very strong inhibition of enzyme synthesis (and recombinant formation) showing that the ♂ cannot effectively

TABLE 1

Enzyme formation in nutritionally deficient zygotes

Deficiency	Rate of enzyme formation †			Mean % inhibition of recombinant formation
	Control	Deficient	Mean % inhibition	
Carbon source ‡	1·6	0·4	73	75
	0·66	0·20		
Arginine §	0·28	0·02	96	65
	0·36	0·01		

† Units of enzyme × hr^{-1}.

‡ ♂ $z^{+}Sm^{s}$ *maltose*$^{+}$ × ♀ $z^{-}Sm^{r}$ *maltose*$^{-}$ mated in presence of inducer and Sm, with glycerol plus maltose (control) or maltose as sole carbon source.

§ ♂ $z^{+}Sm^{s}$ Arg^{+} × ♀ $z^{-}Sm^{r}$ Arg^{-} mated in presence of inducer and Sm with and without arginine (10 μg/ml.).

feed the ♀. An even stronger effect is observed when the ♀ requires arginine, the ♂ not, and mating takes place in absence of arginine (again on condition that the Ar^{+} gene is not injected by the ♂) (Table 1). These observations indicate that even small molecules do not readily pass from the ♂ into the ♀ cell during conjugation.†

It therefore appears that cytoplasmic fusion or mixing does not occur to an extent which might allow cross-feeding. That the contribution of the ♂ is exclusively genetic, and does not involve cytoplasmic constituents of a nature, or in amounts, significant for our purposes, is however only proved by the results of the opposite matings, which we shall consider in the next section.

(b) *Expression and interaction of the alleles of the z and i factors*

We should first consider which of the alleles of the z factors are dominant, and whether they all belong to a single cistron. Experiments of the type described above (mating A) were performed with each of the eight z^{-} mutants, used as ♀ cells, receiving a z^{+} from the ♂. Enzyme was synthesized to similar extents in all cases, showing that the z^{-} mutants in question were all recessive. Each of the mutants was also mated (as ♂) to a z^{-} ♀. No enzyme was synthesized by any of these double recessive heterozygotes where the mutations were in the *trans* position

† However such leakage may occur when the concentration of a compound is exceptionally high in the ♂. This happens when a ♂ with the constitution $z^{-}i^{-}y^{+}$ is used in the presence of lactose. The constitutive permease then may concentrate lactose up to 20 % of the cells' dry-weight (Cohen & Monod, 1957). Adequate tests have shown that this lactose does flow from the ♂ into a permease-less ♀ during conjugation.

$$\frac{z_\alpha^+ \, z_\beta^-}{z_\alpha^- \, z_\beta^+}$$

showing that all the (tested) z^- mutants belong to the same cistron as defined by Benzer (1957).

The next and most critical problem is whether the z and i factors also belong to the same unit of function (gene or cistron) or not. Let us recall that cells with the constitution z^+i^+ synthesize enzyme in presence of inducer only, while z^+i^- cells synthesize enzyme without induction, and z^-i^+ or z^-i^- cells do not synthesize enzyme under any condition. The extremely close linkage of z and i mutations suggests that they may belong to the same unit. If this were so, they would not be able to interact through the cytoplasm, but could act together only when in *cis* position within the same genetic unit. The heterozygote, z^+i^+/z^-i^- would then be expected not to synthesize galactosidase constitutively.

In order to test this expectation, the following mating:

$$\text{♂}\ z^+i^+ \ \times \ \text{♀}\ z_2^- i_3^- \tag{B}$$

was performed *in absence of inducer*. The ♂ cannot synthesize enzyme, because they are i^+. The ♀ cannot because they are z^-. The zygotes however do synthesize enzyme (Fig. 3): during the first hour following mating the synthesis is, if anything, even more rapid and vigorous than when both parents are i^+ and inducer is used, as in mating (A).

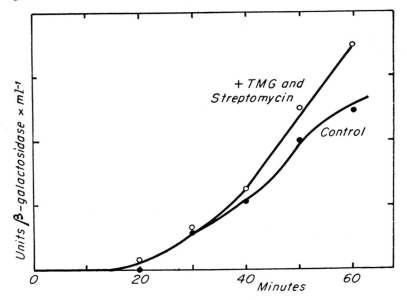

FIG. 3. Enzyme formation during first hour in mating B.
Mating under usual conditions. To an aliquot streptomycin (0·8 mg/ml.) was added at 20 minutes, and TMG at 25 minutes, to allow comparison of synthesis with and without inducer.

Such a mating therefore allows immediate and complete interaction of the z^+ from the ♂ with the i^- from the ♀. The possibility that the interaction depends upon actual recombination yielding z^+i^- in *cis* configuration is excluded because: (a) the synthesis begins virtually immediately after injection whereas genetic recombination is known (Jacob & Wollman, 1958) not to occur until 60 to 90 min after injection; (b) the factors z and i are so closely linked that recombination is an exceedingly rare event (less than 10^{-4} of the zygotes) while the rate of enzyme synthesis is of an order indicating that most or all of the zygotes participate.

The possibility should also be considered that, rather than taking place through the cytoplasm, the interaction requires actual *pairing* of the homologous chromosome segments. This is excluded by the fact that the following mating:

$$\male\ z_2^- i_3^- \times \female\ z^+ i^+ \tag{C}$$

when performed in the *absence* of inducer, yields no trace of enzyme, at any time after mixing, although conjugation and chromosome injection occur normally as shown by adequate controls involving other markers. The zygotes obtained in matings B and C are genetically identical, except that the wild type alleles ($z^+ i^+$) are in relative excess (about 3 to 1) in (B), while the mutant alleles are in similar excess in (C). This quantitative difference cannot account for the absolute contrast of the results of the reciprocal matings, one allowing vigorous constitutive synthesis, the other none at all. This can only be attributed to the fact that the cytoplasm of the zygote is entirely furnished by the $\female$ cell, with no significant contribution from the $\male$. Therefore the $i^- \rightarrow z^+$ interaction must be considered to take place through the cytoplasm.

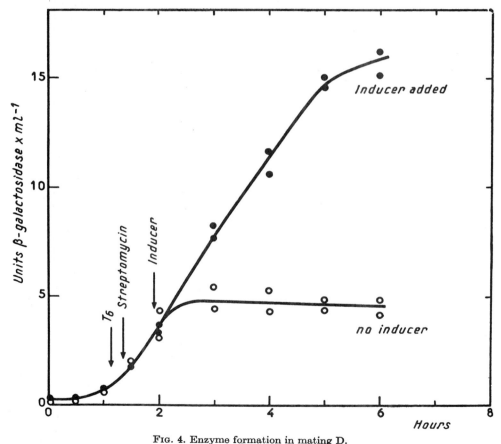

FIG. 4. Enzyme formation in mating D.
Mating performed under usual conditions in quadruplicate in absence of inducer. At times indicated, a suspension of phage T6 (20ϕ/B final concentration) and streptomycin (1 mg/ml.) were added to all of the cultures and TMG (2×10^{-3}M) was added to two of them (black circles) while the other two (white circles) received no addition.

This result may also be expressed by saying that the i factor sends out a cytoplasmic message which is picked up by the z gene, or gene products. Postulating, as we must, that this message is borne by a specific compound synthesized under the control of the i gene, we may further assume that one of the alleles of the i gene provokes the synthesis of the message, while the other one is inactive in this respect. If these assumptions are adequate, one of the alleles should be absolutely dominant over the other, but the dominance should become expressed only gradually when the cytoplasm of the zygotes came from the recessive parent, while it should be expressed immediately when the cytoplasm came from the dominant parent.

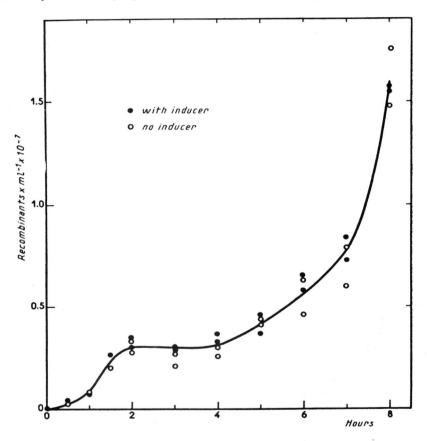

FIG. 5. Recombinant appearance in mating D.

Formation of z^+Sm^r recombinants tested by plating aliquots of the four cultures used in the experiment above (Fig. 4) on lactose-Sm agar. Portions of the culture were diluted 1000-fold and shaken vigorously at 100 minutes to prevent further mating. The increase up to the second hour is due to increasing numbers of zygotes. The increase after the fourth hour is due to *multiplication* of segregants (Wollman, Jacob & Hayes, 1956).

The fact that in matings of type (C) no enzyme is synthesized, even several hours after mating, means that the constitutive (i^-) allele from the ♂ is never expressed. This suggests that the dominant allele is the inducible (i^+). If so, the i^+ should eventually become expressed in matings of type (B)—i.e., the zygotes, initially constitutive (since their cytoplasm comes from the i^- parent), should eventually become inducible. To test this prediction, the following mating was performed:

$$\text{♂ } z^+i^+Sm^s\ T6^s \times \text{♀ } z_2^-i_3^-Sm^r\ T6^r \qquad \textbf{(D)}$$

and the synthesis of enzyme, in the absence and in the presence of inducer, was followed over several hours (in order to block induction of the ♂ and remating, a mixture of streptomycin and T6 phage was used). Figure 4 shows that, in the absence of inducer, enzyme synthesis stops about 90 min (or earlier) after entry of the z^+i^+ genes into the ♀ cells. When inducer is added at this stage, enzyme synthesis is resumed, showing that the initially constitutive z^+i^+/z^-i^- zygotes have not been inactivated, but have become inducible.

It should be asked whether this conversion to inducibility, rather than occurring in the heterozygotes, might not correspond to the segregation of homozygous $z^+i^+Sm^rT6^r$ recombinants with concomitant disappearance of the heterozygotes. This is excluded because the earliest homozygous recombinants only appear 2 hr after the time when constitutive synthesis ceases† (Fig. 5).

From these observations we may conclude that the constitutive (i^-) allele is inactive, while the i^+ is dominant, provoking the synthesis of a substance responsible specifically for the inducible behaviour of the galactosidase enzyme-forming-center.

5. Discussion and Conclusions

(1) The conclusions which can be directly drawn from the evidence presented above may be summarized as follows:

The synthesis of β-galactosidase and galactoside-permease in *E. coli* is controlled by three extremely closely linked genes (cistrons), z, i and y. The z gene determines, in part at least, the structure of the galactosidase protein molecule. The y gene probably does the same for the permease molecule, but there is no evidence on this point. The i gene in its active form controls the synthesis of a product which, when present in the cytoplasm, prevents the synthesis of β-galactosidase and galactoside-permease, unless inducer is added externally (inducible behaviour). When the i gene-product is absent or inactive as a result of mutation within the gene, no external inducer is required for β-galactosidase and galactoside-permease synthesis (constitutive behaviour). The i gene product is very highly specific, having no effect on any other known system.

(2) While proving that the interaction of the i and z factors involves a specific cytoplasmic messenger, the data presented here do not, by themselves, give any indication as to the mode of action of this compound. Two alternative models of this action should be considered.

According to one, which we shall call the "inducer" model, the activity of the galactosidase-forming system‡ requires the presence of an inducer, both in the constitutive and in the inducible organism. Such an inducer (a galactoside) is synthesized by *both* types of organisms. The i^+ gene controls the synthesis of an enzyme which destroys or inactivates the inducer : hence the requirement for external inducer in the wild type. The i^- mutation inactivates the gene (or its product, the enzyme) allowing accumulation of endogenous inducer. This model accounts for the dominance of inducibility over constitutivity, and for the kinetics of conversion of the zygotes.

† It may also be recalled that, according to Anderson & Maze (1957), heterozygosis prevails for many generations in the descendants of *E. coli* zygotes.

‡ By this term we designate the system of all cellular constituents *specifically* involved in galactosidase synthesis. This includes the z gene and its cytoplasmic products.

According to the other, or "repressor", model the activity of the galactosidase-forming system is inhibited in the wild type by a specific "repressor" (probably also involving a galactosidic residue) synthesized under the control of the i^+ gene. The inducer is required only in the wild-type as an *antagonist* of the repressor. In the constitutive (i^-), the repressor is not formed, or is inactive, hence the requirement for an inducer disappears. This model accounts equally well for the dominance of i^+ and for the kinetic relationships.

(3) The "repressor" hypothesis might appear strictly *ad hoc* and arbitrary were it not also suggested by other facts which should be briefly recalled. That the synthesis of certain constitutive enzyme systems may be specifically inhibited by certain products (or even substrates) of their action, was first observed in 1953 by Monod & Cohen-Bazire working with constitutive galactosidase (of *E. coli*) (1953a) or with tryptophan-synthetase (of *A. aerogenes*) (1953b), and by Wijesundera & Woods (1953), and Cohn, Cohen & Monod (1953) independently working with the methionine-synthase complex of *E. coli*. It was suggested at that time that this remarkable inhibitory effect could be due to the displacement of an internally-synthesized inducer, responsible for constitutive synthesis, and it was pointed out that such a mechanism could account, in part at least, for the proper adjustment of cellular syntheses (Cohn & Monod, 1953; Monod, 1955). During the past two or three years, several new examples of this effect have been observed and studied in some detail by Vogel (1957), Yates & Pardee (1957), Gorini & Maas (1957). It now appears to be a general rule, for bacteria, that the formation of sequential enzyme systems involved in the synthesis of essential metabolites is *inhibited* by their end product. The convenient term "repression" was coined by Vogel to distinguish this effect from another, equally general, phenomenon: the control of enzyme *activity* by end products of metabolism.

(4) The facts which demonstrate the existence and wide occurrence of repression effects justify the basic assumptions of the repressor model. They do not allow a choice between the two models. Further considerations make the repressor model appear much more adequate:

(a) The repressor model is simpler since it does not require an independent inducer-synthesizing system.

(b) It predicts that constitutive mutants should, as a rule, synthesize more enzyme than induced wild-type. This appears to be the case for such different systems as galactosidase, amylomaltase (Cohen-Bazire & Jolit, 1953), glucuronidase (Stoeber, 1959, unpublished data), galactokinase of *E. coli* and penicillinase of *B. cereus* (Kogut, Pollock & Tridgell, 1956).

(c) The inducer model, if generalized, implies that internally synthesized inducers (Buttin, unpublished) operate in all constitutive systems. This assumption, first suggested as an interpretation of repression effects, has not been vindicated in recent work on repressible biosynthetic systems (Vogel, 1957; Gorini & Maas, 1957; Yates & Pardee, 1957). In contrast, the synthesis of numerous inducible systems has been known for many years (Dienert, 1900; Stephenson & Yudkin, 1936; Monod, 1942) to be inhibited by glucose and other carbohydrates. The recent work of Neidhardt & Magasanik (1957) has shown this glucose effect to be comparable to a non-specific repression and these authors have suggested that glucose acts as a preferential metabolic source of internally synthesized repressors. If this is so, and if our repressor model is correct, the conversion of glucose into specific galactosidase-repressor should be blocked in the constitutives. Accordingly the galactosidase-forming system of the mutant should be largely insensitive to the glucose effect while other inducible systems should retain their sensitivity. That this is precisely the case (Cohn & Monod, 1953) is a very strong argument in favor of the repressor model.

(5) If adopted and confirmed with other systems, the repressor model may lead to a generalizable picture of the regulation of protein syntheses; according to this scheme, the basic mechanism common to all protein-synthesizing systems would be inhibition by specific repressors formed under the control of particular genes, and antagonized, in some cases, by inducers. Although the wide occurrence of repression effects is certain, the situation revealed with the present system, namely a· genetic "complex" comprising, besides the "structural" genes (z, y) a repressor-making gene (i) whose function is to block or regulate the expression of the neighboring genes is, so far, unique for enzyme systems. But the formal analogy between this situation and that which is known to exist in the control of immunity and zygotic induction of temperate bacteriophage is so complete as to suggest that the basic mechanism might be essentially the same. It should be recalled that according to Jacob & Wollman (1956), when a chromosome from a λ-lysogenic ♂ of E. coli is injected into a non-lysogenic ♀, the process of vegetative phage development is started, which involves as an essential, probably as a primary, step the synthesis of specific proteins. When the reverse mating (♂ non-lysogenic × ♀ λ-lysogenic) is performed, zygotic induction does not occur; nor does vegetative phage develop when such zygotes are superinfected with λ particles. The λ-lysogenic cell is therefore immune against manifestations of prophage or phage potentialities, *and the immunity is expressed in the cytoplasm* (Jacob, 1958–59). Moreover the immunity is strictly specific, since it does not extend to other, even closely related, phages. The formation, under the control of a phage gene, of a specific repressor, able to block synthesis of proteins determined by other genes of the phage, would account for these findings.

(6) Implicit in the repressor model are two critical questions, which for lack of evidence we have avoided discussing, but which should be explicitly stated in conclusion. These questions are:

(a) What is the chemical nature of the repressor? Should it be considered a primary or a secondary product of the gene?

(b) Does the repressor act at the level of the gene itself, or at the level of the cytoplasmic gene-product (enzyme-forming system)?

We are much indebted to Professor Leo Szilard for illuminating discussions during this work and to Mme M. Beljanski, Mme M. Jolit and Mr. R. Barrand for assistance in certain experiments.

REFERENCES

Anderson, T. F. & Maze, R. (1957). *Ann. Inst. Pasteur*, **93**, 194.
Benzer, S. (1957). "The elementary units of heredity", in *The Chemical Basis of Heredity*, ed. by W. McElroy& B. Glass, p. 70. Baltimore: Johns Hopkins Press.
Cohen, G. N. & Monod, J. (1957). *Bact. Rev.* **21**, 169.
Cohen-Bazire, G. & Jolit, M. (1953). *Ann. Inst. Pasteur*, **84**, 937.
Cohn, M. (1957). *Bact. Rev.* **21**, 140.
Cohn, M., Cohen, G. N. & Monod, J. (1953). *C.R. Acad. Sci., Paris*, **236**, 746.
Cohn, M. & Monod, J. (1953). In *Adaptation in Microorganisms*, p. 132. Cambridge: University Press.
Dienert, F. (1900). *Ann. Inst. Pasteur*, **14**, 139.
Fisher, K. W. (1957). *J. Gen. Microbiol.* **16**, 120.

Gorini, L. & Maas, W. K. (1957). *Biochim. biophys. Acta*, **25**, 208.
Hayes, W. (1953). *Cold Spr. Harb. Symp. Quant. Biol.* **18**, 75.
Jacob, F. (1955). *Virology*, **1**, 207.
Jacob, F. (1958–59). Harvey Lectures, Series **54**, in the press.
Jacob, F. & Wollman, E (1956). *Ann. Inst. Pasteur*, **91**, 486.
Jacob, F. & Wollman, E. (1957). *C.R. Acad. Sci., Paris*, **244**, 1840.
Jacob, F. & Wollman, E. (1958). *Symp. Soc. Exp. Biol.* **12**, 75. Cambridge: University Press.
Kogut, M., Pollock, M. R. & Tridgell, E. J. (1956). *Biochem. J.* **62**, 391.
Lederberg, J. (1947). *Genetics*, **32**, 505.
Lederberg, J., Lederberg, E. M., Zinder, N. D. & Lively, E. R. (1951). *Cold Spr. Harb. Symp. Quant. Biol.* **16**, 413.
Monod, J. (1942). *Recherches sur la croissance des cultures bactériennes*. Paris: Herman Edit.
Monod, J. (1955). *Exp. Ann. Biochim. Méd.*, série **17**, 195. Paris: Masson & Cie Edit.
Monod, J. & Cohen-Bazire, G. (1953a). *C.R. Acad. Sci., Paris*, **236**, 417.
Monod, J. & Cohen-Bazire, G. (1953b). *C.R. Acad. Sci., Paris*, **236**, 530.
Neidhardt, F. C. & Magasanik, B. (1957). *J. Bact.* **73**, 253.
Pardee, A. B., Jacob, F. & Monod, J. (1958). *C.R. Acad. Sci., Paris*, **246**, 3125.
Rickenberg, H. V., Cohen, G. N., Buttin, G. & Monod, J. (1956). *Ann. Inst. Pasteur*, **91**, 829.
Stephenson, M. & Yudkin, J. (1936). *Biochem. J.* **30**, 506.
Vogel, H. J. (1957). In *The Chemical Basis of Heredity*, ed. by W. D. McElroy & B. Glass, p. 276. Baltimore: Johns Hopkins Press.
Wijesundera, S. & Woods, D. D. (1953). *Biochem. J.* **55**, viii.
Wollman, E. & Jacob, F. (1955). *C.R. Acad. Sci., Paris*, **240**, 2449.
Wollman, E., Jacob, F. & Hayes, W. (1956). *Cold Spr. Harb. Symp. Quant. Biol.* **21**, 141.
Yates, R. A. & Pardee, A. B. (1957). *J. Biol. Chem.* **227**, 677.

J. Mol. Biol. (1959) **1**, 221-233.

Ribonucleoprotein Particles from *Escherichia coli*

A. Tissières, J. D. Watson, D. Schlessinger and B. R. Hollingworth

Biological Laboratories, Harvard University, Cambridge, Mass., U.S.A.

(*Received 14 May 1959*)

In exponential cultures, 25% of the dry weight of *E. coli* is accounted for by RNA. Of this about 80 to 90% is present in ribonucleoprotein particles and 10 to 20% in the "soluble" or "non-sedimentable" fraction. Magnesium stabilizes the particles, and on varying its concentration, four kinds of components are observed, with sedimentation constants of 30, 50, 70 and 100 *S*. 70 *S* is formed of one 30 *S* and one 50 *S*; 100 *S*, of two 70 *S*. Each type of particle has been isolated. They all contain about 63% RNA and 37% protein, and have the same density. The molecular weights of the 30 *S*, 50 *S* and 70 *S* particles are about 0.7×10^6, 1.8×10^6, and 2.6×10^6 respectively.

1. Introduction

Bacteria release on lysis a large number of small particles of uniform size which in electron micrographs seem to constitute the bulk of the bacterial cytoplasm (Luria, Delbrück & Anderson, 1943). These particles contain most of the ribonucleic acid (RNA) of the cells; they form sharp boundaries in the analytical centrifuge, have a molecular weight of the order of 10^6, and are found in all the bacterial species examined (Schachman, Pardee & Stanier, 1952). They resemble in chemical composition and size the ribonucleoprotein particles which have been studied extensively in animal tissues (Petermann & Hamilton, 1952, 1957; Palade, 1955; Palade & Siekewitz, 1956*a* and *b*), in plants (Tso, Bonner & Vinograd, 1956, 1958), and in yeast (Chao & Schachman, 1956; Chao, 1957), and which are believed to be an important site of protein synthesis (Littlefield, Keller, Gross & Zamecnik, 1955; Littlefield & Keller, 1956, 1957; Schweet, Lamfrom & Allen, 1958).

Bacteria in the exponential phase of growth are one of the richest sources of ribonucleoprotein particles. In *Escherichia coli* dividing every 30 min, about 30% of the dry weight of the organism is accounted for by these particles (see below). While they are often firmly attached to the reticulum in animal tissues (Palade, 1955), they are free in bacterial extracts. *E. coli* is therefore convenient for the preparation of ribonucleoprotein particles in amounts sufficient for physical and chemical studies.

Magnesium stabilizes ribonucleoprotein particles from *E. coli* (Bolton, Hoyer & Ritter, 1958) from yeast (Chao, 1957) and from plants (Tso *et al.*, 1956, 1958). In the experiments reported here, in suitable concentrations of magnesium, four kinds of ribonucleoprotein particles were observed, with sedimentation coefficients of 30 *S*, 50 *S*, 70 *S* and 100 *S*. The isolation and some properties of each component will be described, and their molecular weights, based on sedimentation, diffusion, viscosity, and partial specific volume measurements, will be presented and discussed. A preliminary account of some of this work has appeared elsewhere (Tissières & Watson, 1958).

2. Methods

E. coli strain B was grown in enriched broth at 37°C under forced aeration. The broth had the following composition: 1% bactotryptone (Difco), 1% NaCl, 0·5% yeast extract

(Difco), 0·1% glucose, in distilled water. 1 ml. 1 N-NaOH was added to 1 l. of medium to adjust the pH to about 7·0. The cells were harvested in the exponential phase of growth and washed twice in the refrigerated centrifuge with the buffer used to make the extract. They could be kept frozen. Unless otherwise mentioned, the cell-free extract was made by grinding the well packed cells by hand in a cold mortar for 2 to 3 min with 3 parts (wt/wt) of alumina powder (Norton levigated alumina from Norton Abrasives, Worcester 6, Mass.) according to McIlwain (1948), and extracting with 3 vol of buffer. All manipulations were carried out at 0° to 4°C. Deoxyribonuclease (1 μg/ml.) obtained from Worthington Biochemical Corp., Freehold, New Jersey, was added to depolymerize the viscous bacterial deoxyribonucleic acid. The mixture of alumina, broken cells and buffer was centrifuged at 6000 g for 15 min, giving a sediment composed of two layers: a lower layer of alumina and some unbroken cells and a brown upper layer of large *cell debris*. The supernatant after this centrifugation, which is referred to as *crude extract*, contained the bulk of the ribonucleic acid of the cells, mostly in the form of ribonucleoprotein particles. The crude extract was usually centrifuged once more at 6000 g for 15 min to sediment any remaining alumina and cell debris. From this the ribonucleoprotein particles were then isolated by high speed centrifugation as described below.

A Model L Spinco ultracentrifuge was used for the preparation of the various particle fractions. The centrifugal forces given here represent g average, as calculated for the middle of the centrifuge tube. Analytical ultracentrifugation was done with a model E Spinco centrifuge, with either schlieren or ultra-violet optics. The films taken with the latter were analyzed for their optical density with a Spinco Analytrol densitometer. The centrifuge cell had a light path of 12 mm. The sedimentation constants were corrected to 20°C, water and zero concentration. They are expressed in Svedberg units,

$$S \ (= S°_{20w} \times 10^{-13} \text{ cm/sec}).$$

Electrophoresis and diffusion measurements were performed with a Spinco model H apparatus, in an 11 ml. Tiselius cell at 1°C.

Viscosity was measured with Ostwald-Fenske type viscosimeters, solvent running time 350 to 370 sec, at 24·97 ± 0·02°C.

RNA was estimated by the orcinol method as described by Dische (1953), with correction, when necessary, for deoxyribonucleic acid (Schneider, 1945) by the diphenylamine reaction (Dische, 1955).

Protein was estimated by the biuret test (Gornall, Bardawill & David, 1949).

Dry weight measurements were done on aliquots dried at 110°C to constant weight. The dry weight of the buffer, in which the ribonucleoprotein particles were suspended, was measured in the same way and subtracted from the weight of the particle preparations.

Partial specific volumes were measured in a 25 ml. pycnometer.

The relative amount of each ribonucleoprotein component present in a solution was measured on plates from ultracentrifuge runs by determining (a) the area covered by each peak on the schlieren diagrams and (b) the optical density given by each component on pictures taken with the ultra-violet optics.

Crystalline pancreatic ribonuclease was obtained from Worthington Biochemical Corp.

0·01 M-phosphate buffer pH 7·0 was prepared by mixing 1 vol of 0·01 M-KH_2PO_4 with 3 vol of 0·01 M-Na_2HPO_4. The pH was then adjusted to 7·0 by addition of 0·01 M-KH_2PO_4 or Na_2HPO_4.

Tris(hydroxymethyl)aminomethane was Sigma 7–9 biochemical buffer from Sigma Chemical Company. The pH of tris solutions was adjusted by addition of 0·1 N-HCl. Unless otherwise mentioned, 0·01 M-tris, pH 7·4, was used.

The magnesium salt, added to stabilize the particles, was magnesium acetate.

3. Experimental

(a) *Amount of RNA in E. coli and in fractions from cell free extracts*

RNA and protein were estimated on *E. coli* cells and on the following three fractions: (a) *cell debris*, consisting of pieces of cell walls and cell membranes which sedimented as a brown layer on top of the alumina during the first centrifugation at 6000 g for 15 min. To obtain this fraction free of crude extract and of intact cells, which might well modify considerably the RNA content, it was washed five times in the centrifuge with the buffer used to make the extract. Each time the upper layer of the sediment was resuspended in buffer in order to eliminate the intact cells, unbroken during the grinding with alumina, which would, in view of their mass, centrifuge first to the bottom of the tube. This fraction was examined under the microscope and was found free of intact cells; (b) *crude extract*, or supernatant after centrifugation at 6000 *g* for 15 min; (c) *supernatant* obtained by carefully removing 2 ml. on top of the 11 ml. centrifuge tube, after centrifuging the crude extract at 100,000 *g* for 120 min.

TABLE 1

Amounts of RNA and protein in intact cells and fractions from cell free extract. The cells were ground with 3 parts of alumina and extracted with 0·001 M-tris pH 7·4 and 0·01 M-magnesium acetate.

(Similar results are obtained when the cells are extracted with 0·01 M-phosphate buffer pH 7·0 and 0·001 M-magnesium acetate. The *cell debris* is formed mostly of pieces from cell walls and cell membranes (see text). The *crude extract* is the fraction obtained after removing the alumina, the intact cells and the large cell debris by 15 min centrifugation at 6000 *g*. The crude extract, centrifuged for 120 min at 100,000 *g*, yields the *supernatant* and a small pellet formed essentially of ribonucleoprotein particles.)

		Protein	RNA	RNA/Protein
Intact cells (% of dry weight)		62·5	24·1	0·386
% of amount in intact cells	cell debris	15·0	0·9	0·06
	crude extract	85·0	97·0	1·17
% of amount in crude extract	} supernatant	45·0	17·0	0·378
mg/ml.	crude extract	5·2	3·44	0·66
	supernatant	2·34	0·585	0·25
Difference (mg sedimented in 120 min centrifugation at 100,000 *g* from 1 ml. crude extract)		2·86	2·855	0·99

The results given in Table 1 show that: (a) 24% of the dry weight of the cell is accounted for by RNA; (b) the bulk of the RNA of the cell is present in the crude extract; (c) the debris of cell wall and cell membrane, after washing with buffer, contain less than 1% of the amount of RNA present in the cells; (d) 17% of the RNA in the crude extract is in the form of " non-sedimentable RNA", and is found in the supernatant after centrifugation at 100,000 *g* for 120 min (this value ranged in various experiments from 15 to about 25%); (e) about equal amounts of RNA and protein are found in the unwashed fraction sedimented from the crude extract in a field of 100,000 g. All the RNA and most of the protein in this fraction belong to ribonucleoprotein particles, which are formed, as shown below, of about 63% of RNA and 37% of protein. A few per cent of the protein particles sedimenting with this fraction probably consist of fragments derived from the cell membrane.

(b) Sedimentation diagrams of crude extracts

The schlieren diagram of a crude extract made in 0·01 M-Mg^{++} and 0·01 M-tris buffer pH 7·4 showed two ribonucleoprotein peaks with sedimentation constants of 70 S and 100 S. The latter was predominant. With 0·001 M-Mg^{++}, the 70 S peak was the major one and in addition there were small amounts of 100 S, 50 S and 30 S peaks. With still lower magnesium concentrations (0·0001 to 0·0002 M), the ribonucleo-protein particles were present as 30 S and 50 S peaks. In the crude extracts the bulk of the soluble proteins of the cells appeared as a peak with a coefficient of about 5 S.

When the crude extract was made with 0·001 M-Mg^{++} and 0·01 M-phosphate buffer pH 7·0, the ribonucleoprotein particles were present as 30 S and 50 S components. This can be partly explained by the fact that some of the magnesium is bound to phosphate.

Thus four kinds of ribonucleoprotein particles could be observed in bacterial extracts, with sedimentation coefficients of 30 S, 50 S, 70 S and 100 S. The largest particles were predominant in presence of the highest concentration of magnesium. In low magnesium, only the 30 S and 50 S peaks were visible; these two components always appeared together and in the same ratio.

Schachman et al. (1952) found that extracts made by breaking the cells in different ways showed essentially similar sedimentation patterns. This was confirmed here: extracts made by sonic vibration or by lysing the protoplasts formed following the addition of penicillin to the culture medium according to Lederberg (1956) gave identical schlieren diagrams.

(c) Isolation of the 70 S component

The crude extract was made with 0·01 M-Mg^{++} and tris buffer pH 7·4. Under these conditions, the schlieren diagram showed a main 100 S component and a small amount of a 70 S peak. The 30 S and 50 S peaks were not visible. This extract was centrifuged at 78,000 g for 180 min. The supernatant was poured off and the yellow brown gelatinous pellet was resuspended in 0·005 M-Mg^{++} and 0·01 M-tris buffer by means of a teflon homogenizer used directly in the nitrocellulose centrifuge tube. The suspension was first centrifuged at 8000 g for 15 min in order to sediment and discard some aggregated material; then it was centrifuged at 78,000 g for 180 min. Washing the particles by resuspending them in buffer, followed by low and high speed centri-fugation, as described above, was repeated twice. Finally the pellet was resuspended in tris buffer containing 0·001 M-Mg^{++} and the suspension was centrifuged at 8000 g for 15 min. The sediment was discarded. In the supernatant there was a major 70 S component, with in addition 10 to 15% of a 100 S peak and smaller amounts of 30 S and 50 S peaks. In one preparation 1 g (dry weight) of particles was obtained from 70 g (wet weight) of bacteria. The yield varied somewhat from preparation to pre-paration.

The 100 S particle is obtained by increasing the magnesium concentration of a 70 S preparation to 0·005–0·01 M. Sedimentation and viscosity measurements suggest that it is formed of two 70 *S* particles. This view is supported by electron micrographs of 100 *S* (Dr. C. Hall, personal communication) showing large numbers of two particle aggregates.

All the 70 *S* and 100 *S* pellets were yellow-brown, although the intensity of this color was reduced by repeated centrifugation. When a 70 *S* pellet was examined as such in a 3 mm thick cell, with the Cary spectrophotometer, in the presence and in the absence of a reducing agent, the typical absorption spectrum of all the cytochrome pigments found in *E. coli* (Keilin & Harpley, 1941) was observed. It is likely that the ribonucleoprotein particles are contaminated at this stage by nonparticle protein, to the extent of about 5%. This view is consistent with the percentages of RNA and protein reported in Table 2.

(d) *Conversion of* 70 *S particles into* 30 *S and* 50 *S components*

The observations with the crude extracts described above suggested that the 70 *S* and 100 *S* particles were formed by the aggregation of 30 *S* and 50 *S* particles. Experiments with isolated 70 *S* particles proved this conjecture and showed that as the magnesium concentration is decreased 70 *S* breaks down to 30 *S* and 50 *S*. Two procedures were used to effect this transition.

(i) A 70 *S* preparation was centrifuged at 78,000 *g* for 180 min and the sedimented particles were resuspended in sufficient 0·00025 M-Mg^{++} and tris buffer to give a concentration of ribonucleoprotein of about 5 mg/ml. This was then dialyzed overnight in the cold with stirring against 0·00025 M-Mg^{++} and tris buffer. In some cases this treatment led to a complete conversion of 70 *S* particles into 30 *S* and 50 *S* components. However, in many instances a small amount of the 70 *S* peak was still visible on the schlieren diagram. The amount of this remaining 70 *S* component can be further reduced by lowering the magnesium concentration or by exposure to 45°C for several hours.

(ii) The pellet of particles after high speed centrifugation was resuspended in 0·001 M-Mg^{++} and 0·01 M-phosphate buffer pH 7·0. This procedure gave a 100% conversion of 70 *S* into the two smaller units, and was therefore adopted in most cases for the preparation of 30 *S* and 50 *S*. The conversion could also be obtained by dialysis against the phosphate-magnesium mixture.

(e) *Relative amounts of* 30 *S and* 50 *S arising from* 70 *S*

70 *S* was split to 30 *S* and 50 *S* and the resulting solutions were examined in the analytical ultra-centrifuge. For two different preparations, the areas of the 30 *S* and 50 *S* peaks on the schlieren plates were in the ratio of about 1 : 2. Applying the Johnston & Ogston (1946) correction would somewhat lower this ratio. Since, as shown below, the RNA and protein composition of 30 *S* and 50 *S* is the same, the optical density for each component on films taken with the ultra-violet optics is also a measure of their relative proportions. In four such experiments the ratio of 30 *S* to 50 *S* varies from 1 : 2 to 1 : 2·5. The errors in these measurements were possibly as great as 20%.

The molecular weights of the 30 *S* and 50 *S*, calculated from sedimentation and diffusion rates (see Table 3), are in the ratio of 1 to 2·5. Furthermore the molecular weight of the 70 *S* is approximately the sum of those of the 30 *S* and 50 *S*. These data show clearly that the 70 *S* particle is formed of one 30 *S* and one 50 *S*.

(f) *Isolation of the 30 S component*

The mixture of 30 S and 50 S, in concentrations of 5 to 10 mg/ml., was centrifuged at 100,000 **g** for 360 min. 7/10ths of the supernatant fluid was carefully removed without stirring and discarded while the remaining fluid was shaken gently by hand for about 5 sec in order to resuspend the upper layer of the pellet. This fluid was poured off and saved. It consisted of 30 S particles with 10 to 30 % of 50 S. With one or two more cycles of centrifugation at 100,000 **g** for 360 min and gentle resuspension of the upper layer of the pellet, the ratio of 30 S to 50 S could be brought to about 20 to 1, as judged by the area covered by each peak under the schlieren curve. The 30 S preparation could be sedimented by centrifugatio at 100,000 **g** for 360 min. The pellet thus obtained was colorless. The buffer was either tris with 0·00025 M-Mg^{++} or phosphate with 0·001 M-Mg^{++} depending on whether procedure (a) or (b) was used to make 30 S and 50 S.

(g) *Isolation of the 50 S component*

The pellet after the first centrifugation at 100,000 **g** for 360 min was resuspended in buffer and magnesium mixture (either 0·01 M-tris pH 7·4 and 0·00025 M-Mg^{++}, or 0·01 M-phosphate pH 7·0 and 0·001 M-Mg^{++}). It was then centrifuged at 15,000 **g** for 15 min, the sediment was discarded and the supernatant was centrifuged at 100,000 **g** for 90 min. This last step was repeated twice. Further purification could be achieved by lowering the pH of the preparation to 5·0 with 0·1 N-CH$_3$.COOH, centrifuging at 8000 **g** for 15 min and finally bringing the pH of the supernatant to 7·0 with 0·1 N-NH$_4$OH. This treatment removed some cell debris derived probably from the membrane and the cell wall (see below) and also small amounts of 30 S which may still be present at this stage. After this last step the preparation showed only one peak on the sedimentation diagram and upon high speed centrifugation formed a colorless pellet. In addition it moved as a single symmetrical peak on electrophoresis at pH 7·6 or 4·7.

(h) *Formation of 70 S particles from purified 30 S and 50 S*

30 S and 50 S particles (in tris buffer and 0·00025 M-Mg^{++}) were mixed together in equal number amounts and magnesium added to a final concentration of 0·001 M. No aggregation occurred. On raising magnesium to 0·005 M, 70 S particles appeared and at 0·01 M the majority of particles were in the 70 S form. Subsequent lowering of the magnesium level to 0·001 M reveals a majority of 70 S particles. Thus a higher magnesium concentration is required to form 70 S particles than to effectively stabilize them. Our data are summarized by the equation

$$\text{increasing Mg}^{++} \longrightarrow$$
$$2\,(30\,S) + 2(50\,S) \rightleftharpoons 2(70\,S) \rightleftharpoons (100\,S) \qquad (1)$$

When a mixture of peaks is observed, they are usually very sharp. This would not be true if the various particle types were in rapid equilibrium, as the peaks would blend into each other. This implies that the transitions need not occur immediately, but that an activation energy is required when 70 S splits into 30 S and 50 S. This is supported by the observation that incubation of 70 S particles in low magnesium (0·00025 M) does not lead to immediate breakdown and that the transition is hastened by incubation at 45°C. The same reason may explain why the 70 S particle is not formed at 0·001 M-Mg^{++} even though once formed it initially seems stable and can be isolated at this magnesium level. In fact, prolonged incubation (several days to a week) of 70 S particles at 45°C leads to their breakdown to 30 S and 50 S particles. It is thus probable that the 70 S particles are completely stable only at magnesium concentrations greater than 0·005 M where in fact they begin to further aggregate to 100 S particles.

(i) *Chemical composition*

The results of RNA and protein estimations, together with dry weights and extinctions at 260 mμ on a dry weight basis, are given in Table 2. The dry weights corresponded, within the limits of error, to the sum of the weights of RNA and protein found by chemical estimation. The errors in these estimations were probably 3 to 4%. Thus it appears that the particles are formed mostly, if not exclusively, of RNA and protein. It is likely that the ratio of RNA to protein is the same in the three kinds of particles, 30 S, 50 S and 70 S, and that the small variations shown in Table 2 are due to varying amounts of impurities. This view is supported by the finding (see below) that the densities of the three kinds of particles are very similar. The values for the extinction at 260 mμ suggest that the purest preparations are those of 30 S and of 50 S, the latter after treatment at pH 5·0.

TABLE 2

Dry weight, chemical composition and specific extinction coefficients at 260 mμ of 30 S, 50 S and 70 S ribonucleoprotein particle preparations.

Particle fraction	Dry weight mg/ml.	RNA mg/ml.	Protein mg/ml.	RNA + Protein mg/ml.	% RNA	% Protein	Specific extinction coefficient at 260 m$\mu = E_{1\,cm}^{1\%}$
30 S	3·72	2·38	1·50	3·88	62·0	38·0	167·0
30 S	1·58	0·95	0·60	1·55	61·3	38·7	162·0
50 S	2·26	1·39	0·83	2·22	62·5	37·5	—
50 S	1·06	0·63	0·37	1·00	63·4	36·6	182·0
50 S	2·82	1·80	1·00	2·80	64·0	36·0	—
50 S†	6·96	3·94	2·85	6·79	58·0	42·0	149·0
70 S	39·00	23·50	17·60	41·10	57·3	42·7	147·0
70 S	—	—	—	—	—	—	145·0
70 S	13·60	8·80	5·70	14·50	60·7	39·3	157·0

† 50 S: this preparation had not been treated at pH 5·0.

(j) *Absorption spectrum*

The absorption curves of the 30 S, 50 S and 70 S in the ultra-violet region were nearly identical. They all showed an absorption peak with a maximum at 259 mμ. The RNA in 30 S, 50 S and 70 S was found to react with formaldehyde, as described by Fraenkel-Conrat (1954) for several plant viruses: the absorption in the 260 mμ region increased by about 20% and the maximum was shifted about 3 mμ towards the longer wavelengths. The spectrum of a 50 S preparation and the effect of formaldehyde are shown in Fig. 1.

(k) *Relative densities*

A particle preparation containing 30 S and 50 S components was added to salt solutions to give an optical density of 1·0 at 260 mμ. The resulting samples contained

0·01 M-phosphate pH 7·0 and 0·001 M-Mg^{++} in 0·0, 0·5, 1·5, 3·5, and 5·0 molal CsCl. The sedimentation rates observed with ultra-violet optics are shown in Fig. 2. Corrected for the viscosity and density of the medium, they correspond to 30 S and 50 S. In all cases, the relative sedimentation constants of the two components were the same, indicating that they had similar if not identical partial specific volumes. Further experiments were performed in varying concentrations of sucrose and NaCl with identical results. In one experiment involving 70 S particles in NaCl solution, the sedimentation run was done before complete 70 S breakdown (see below) and the sedimentation rate of 70 S was reduced at the same rate as those of 30 S and 50 S.

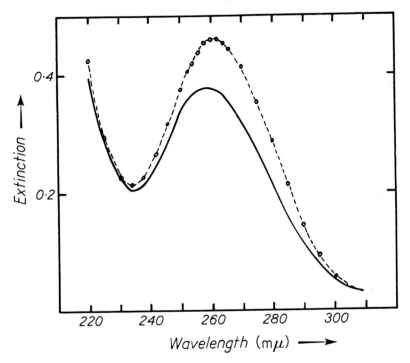

FIG. 1. Absorption spectra of 50 S particles. The solid line represents the spectrum of 50 S (20·6 μg/ml.) in 0·01 M-phosphate and 0·001 M-Mg^{++}, before and after 70 hr at 45°C. The dashed line shows the spectrum after similar incubation in 2% HCHO.

(1) Stability

(i) *Effect of magnesium.* The four kinds of particles were usually stable in the presence of magnesium for over a month in the cold in 10 μg/ml. streptomycin, although occasionally after a few weeks the particles were found to break down. Buffer was not necessary; preparations remained intact for several days in distilled water and magnesium.

Magnesium is required not only for the existence of the 70 S and 100 S, but also for the integrity of the 30 S and 50 S subunits. If 30 S or 50 S is dialyzed against buffer containing no magnesium, or against distilled water, or if 0·01 M-ethylene-diamine tetra-acetate (versene) is added, breakdown of the particles occurs.

It has been mentioned that 0·003 to 0·01 M-Mg^{++} causes aggregations to the 100 S component. The particles are stable up to 0·02 M-Mg^{++}, above which breakdown was observed.

(ii) *Effect of freezing.* The particles could be kept frozen for several months without any noticeable change of their properties.

(iii) *Effect of pH.* The 50 *S* was stable for 24 hr at pH 4·7 and 9·5. The 30 *S*, 70 *S* and 100 *S* broke down under these conditions.

If the pH of a suspension of 30 *S*, 50 *S* and 70 *S* (1·5 mg/ml.) was lowered to 4·4 with 0·01 M-ammonium acetate or 0·1 N-CH$_3$.COOH, and the resulting mixture dialyzed for 15 hr against 100 vol of 0·001 M-Mg^{++} and 0·01 M-ammonium acetate pH 4·4, a completely diffuse peak, suggesting random aggregation, was seen in the

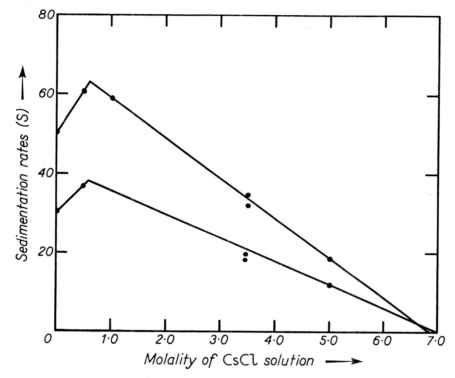

FIG. 2. Sedimentation rates in CsCl. Mixtures of 30 *S* and 50 *S* were examined in the analytical ultracentrifuge with ultra-violet optics, immediately after dilution in different concentrations of CsCl. The ratio of sedimentation rates was always 50 : 30. The graphs extrapolate to a particle density in CsCl of 1·73 ± 2% for both components. The initial rise in sedimentation rate is probably attributable to initial replacement of Na$^+$ in the particles by Cs$^+$.

analytical ultracentrifuge. If the same sample was then dialyzed for 15 hr against 0·01 M-tris, pH 7·4, a 50 *S* peak was again visible, with traces of 30 *S* and 70 *S*. If the pH was lowered further to 4·2, a white precipitate was observed. The precipitate did not revert to ribonucleoprotein components on raising the pH.

(iv) *Effect of ribonuclease.* 1 to 5 μg/ml. of pancreatic ribonuclease, at pH 7·6, had no effect on most particle preparations, even after 4 to 6 hr at room temperature, as judged by the appearance of the peak in the analytical ultracentrifuge. In some cases, however, the peak heights were reduced. The reason for these discrepancies is not clear.

(v) *Effect of NaCl and CsCl.* 30 *S*, 50 *S*, or 70 *S*, or mixtures of them, in 0·001 M-Mg^{++} solutions, were allowed to stand at 4°C in different NaCl or CsCl concentrations. After

15 hr, samples were examined in the analytical centrifuge with the ultra-violet optics. 30 S and 50 S were stable below about 0·15 and above 2·0 molal NaCl and CsCl. They broke down in the intermediate range and consequently no sedimentation rate greater than 5 S was observed. 70 S was stable in low salt concentrations up to about 0·15 molal. In higher concentrations it split to 30 S and 50 S.

Elson (1958) has shown that a latent ribonuclease is present in ribonucleoprotein particle preparations from *E. coli*. The unusual stability of the particles in salt reported here may be partly explained by the presence of this enzyme, assuming that above 0·15 molal NaCl or CsCl the particle structure is altered in such a way that the endogenous ribonuclease can degrade the particulate RNA. The stability of the particles above 2·0 molal NaCl or CsCl might be due to inhibition of ribonuclease by high salt concentrations, or else simply to salt effect on the particles.

(m) *Molecular weights*

Values shown in Table 3 are based on the following data:

(i) *Sedimentation constants.* The coefficients for 30 S, 50 S and 70 S were measured both with the ultra-violet optics at concentrations of about 60 μg/ml., and with the schlieren optics at concentrations of 0·5 to 5·0 mg/ml., in 0·01 M-tris and the suitable magnesium concentrations. The results are shown in Fig. 3. The salt concentration had no appreciable effect between 0·01 and 0·05 M-phosphate and 0·001 and 0·05 M-tris.

(ii) *Diffusion coefficients.* Free diffusion was allowed to proceed for 2 to 3 days. Measurements were made on two different preparations of 30 S (both at 0·11%, in 0·01 M-tris), three of 50 S (at 0·06%, in 0·01 M-phosphate; 0·07%, in 0·05 M-phosphate; and 0·20%, in 0·01 M-phosphate), and one of 70 S (at 0·10% and 0·20%, in 0·01 M-tris).

TABLE 3

Physical constants and molecular weights. Percentages of error are given as the average deviation from the mean. The molecular weights are consistent with equation (1).

Compo-nents	Precise S ($S^0_{20,w} \times 10^{13}$ cm/sec)	D ($D^0_{20,w} \times 10^7$ cm²/sec)	$\bar{V}$	$\eta_{int.}$ cm/dl.	Molecular weights from S & D (Svedberg & Pedersen, 1940)	from S & η (Scheraga & Mandelkern, 1953)
30 S	30·6 ± 1	2·95 ± 4 %	0·64	0·080 ± 8 %	0·7 × 10⁶ ± 8 %	1·0 × 10⁶ ± 16 %
50 S	50·0 ± 1	1·91 ± 1 %	0·64	0·054 ± 8 %	1·8 × 10⁶ ± 4 %	1·8 × 10⁶ ± 16 %
70 S	69·1 ± 1	1·83 ± 2 %	0·64	0·061 ± 10 %	2·6 × 10⁶ ± 6 %	3·1 × 10⁶ ± 18 %
100 S	100·0 ± 2		0·64	0·071 ± 12 %		5·9 × 10⁶ ± 20 %

Because of the impossibility of forming infinitely sharp initial boundaries with proteins, the calculated values of D decreased with time. Plots of D against $1/t$, which were linear, when extrapolated to infinite time, gave corrected values of D. Zero-time corrections were of the order of 20 to 50 min. The values of D were then corrected to 20°C in water. For 30 S, they were 2·81 and 3·10 × 10⁻⁷ cm²/sec; for 50 S, 1·88, 1·90, and 1·95 × 10⁻⁷; and for 70 S, 1·82 and 1·84 × 10⁻⁷. Averaged values are shown in Table 3.

(iii) *Partial specific volume.* The values obtained were 0·64 and 0·66 for two 50 S preparations, and 0·63 for one 70 S, at concentrations varying from 0·5 to 4%. Since the density and chemical composition of the three kinds of particles is probably the same (see above), an average value of 0·64 was used in all calculations.

(iv) *Viscosity.* The preparations used contained at least 85% of one ribonucleoprotein component, as measured on plates taken with the schlieren optics. For various 30 S, 50 S and 70 S preparations, and for one of 100 S, series of viscosity measurements were extrapolated to an average intrinsic viscosity (Table 3). Concentrations were calculated assuming that 67 μg/ml. corresponded to an optical density of 1·0 for 50 S, 70 S and 100 S, and 61 μg/ml. for 30 S. These values were deduced from the specific extinction coefficients at 260 mμ given in Table 2.

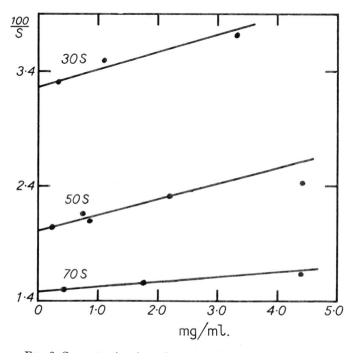

Fig. 3. Concentration dependence of sedimentation constants.

The molecular weight of the 50 S particles calculated from sedimentation, diffusion and partial specific volume data agrees well with that derived from sedimentation, viscosity and partial specific volume. As the value involving the diffusion coefficient is independent of the shape of the particle, while that using intrinsic viscosity assumes a spherical shape, this result, in agreement with the electron micrographs (Hall & Slayter, 1959), suggests a spherical shape for the 50 S particle.

In contrast the molecular weight of the 30 S particles obtained by using intrinsic viscosity seems high when compared to that based on diffusion values, and may reflect the less symmetrical form shown in electron micrographs.

4. Discussion

The construction of ribonucleoprotein particles from two unequal units which join in presence of a suitable magnesium concentration seems to be a general phenomenon.

With both pea seedlings (Tso *et al.*, 1956) and yeast (Chao, 1957) 80 S particles break down reversibly to 60 S and 40 S, and we have seen that in *E. coli* the 70 S component is made of one 30 S and one 50 S, the molecular weights of which are in the ratio of about 1 : 2·5. Viscosity measurements suggest greater asymmetry of 30 S than of 50 S and 70 S, and the electron micrographs (Hall & Slayter, 1959) show that 30 S is definitely asymmetric, while 50 S and 70 S appear nearly spherical. A simple model for the structure of 70 S would thus be a capped sphere or acorn shape.

The particles from *E. coli* seem to differ from those of other sources in at least three ways: (a) the molecular weight is lower (2·8 × 10⁶ for 70 S, while values varying between 4 and 4·5 × 10⁶ have been reported for animal (Dintzis, Borsook & Vinograd, 1958), plant (Tso *et al.*, 1956), and yeast particles (Chao & Schachman, 1956)); (b) the percentage of RNA (60 to 65%) is higher than that reported for particles from other sources (Tso *et al.*, 1956; Chao & Schachman, 1956; Petermann & Hamilton, 1957; Dintzis *et al.*, 1958) with the exception of particles from rabbit appendix (Takata & Osawa, 1957); (c) the magnesium requirement for the integrity of the *E. coli* particles appears to be greater.

If we take the weight of one *E. coli* cell to be 10⁻¹² gm, with 25% RNA in exponential cultures, there is 2·5 × 10⁻¹³ gm of RNA per cell. One 70 S particle contains 2·74 × 10⁻¹⁸ gm RNA. Thus there are about 90,000 particles in one bacterium. In the stationary phase, or upon incubation in phosphate buffer, this number decreases by a factor of 4 to 5 (Mendelsohn & Tissières, 1959), or to about 20,000 particles per cell.

We wish to thank Dr. J. T. Edsall for his interest and encouragement of this work, and Dr. D. M. Skinner and Miss C. Laumont for their assistance in some of these experiments. This work has been supported by grants from the U.S. National Science Foundation and an institutional grant from the American Cancer Society.

REFERENCES

Bolton, E. T., Hoyer, B. H. & Ritter, D. B. (1958). *Microsomal Particles and Protein Synthesis*, p. 18. New York: Pergamon Press.
Chao, F. C. (1957). *Arch. Biochem. Biophys.* **70**, 426.
Chao, F. C. & Schachman, H. K. (1956). *Arch. Biochem. Biophys.* **61**, 220.
Dintzis, H. M., Borsook, H. & Vinograd, T. (1958). *Microsomal Particles and Protein Synthesis*, p. 95. New York: Pergamon Press.
Dische, Z. (1953). *J. Biol. Chem.* **204**, 983.
Dische, Z. (1955). *The Nucleic Acids*, Vol. I, p. 287. New York: Academic Press.
Elson, D. (1958). *Biochem. biophys. Acta*, **27**, 216.
Fraenkel-Conrat, H. (1954). *Biochem. biophys. Acta*, **15**, 307.
Gornall, A. G., Bardawill, C. T. & David, M. M. (1949). *J. Biol. Chem.* **177**, 751.
Hall, C. E. & Slayter, H. S. (1959). *J. Mol. Biol.* **1**, in press.
Johnston, J. P. & Ogston, A. G. (1946). *Trans. Faraday Soc.* **42**, 789.
Keilin, D. & Harpley, C. H. (1941). *Biochem. J.* **35**, 688.
Lederberg, J. (1956). *Proc. Nat. Acad. Sci.*, *Wash.* **42**, 574.
Littlefield, J. W. & Keller, E. B. (1956). *Fed. Proc.* **15**, 302.
Littlefield, J. W. & Keller, E. B. (1957). *J. Biol. Chem.* **224**, 13.
Littlefield, J. W., Keller, E. B., Gross, J. & Zamecnik, P. C. (1955). *J. Biol. Chem.* **217**, 111.
Luria, S. E., Delbruck, M. & Anderson, T. F. (1943). *J. Bact.* **46**, 57.
McIlwain, H. (1948). *J. Gen. Microbiol.* **2**, 288.
Mendelsohn, J. & Tissières, A. (1959). *Biochem. biophys. Acta*, in press.
Palade, G. (1955). *J. Biophys. Biochem. Cytol.* **1**, 59.
Palade, G. & Siekewitz, P. (1956a). *J. Biophys. Biochem. Cytol.* **2**, 171.
Palade, G. & Siekewitz, P. (1956b). *J. Biophys. Biochem. Cytol.* **2**, 671.

Petermann, M. L. & Hamilton, M. G. (1952). *Cancer Research*, **12**, 373.
Petermann, M. L. & Hamilton, M. G. (1957). *J. Biol. Chem.* **224**, 723.
Schachman, H. K., Pardee, A. B. & Stanier, R. Y. (1952). *Arch. Biochem. Biophys.* **38**, 245.
Scheraga, H. A. & Mandelkern, L. (1953). *J. Amer. Chem. Soc.* **75**, 179.
Schneider, W. C. (1945). *J. Biol. Chem.* **161**, 293.
Svedberg & Pedersen, K. O. (1940). *The Ultracentrifuge*, p. 5. Oxford: Clarendon Press.
Schweet, R., Lamfrom, H. & Allen, E. (1958). *Proc. Nat. Acad. Sci., Wash.* **44**, 1029.
Takata, K. & Osawa, S. (1957). *Biochem. biophys. Acta*, **24**, 207.
Tissières, A. & Watson, J. D. (1958). *Nature*, **182**, 778.
Tso, P. O. P., Bonner, J. & Vinograd, J. (1956). *J. Biophys. Biochem. Cytol.* **2**, 451.
Tso, P. O. P., Bonner, J. & Vinograd, J. (1958). *Biochem. biophys. Acta*, **30**, 570.

J. Mol. Biol. (1959) **1**, 281-292

Structural Components of Bacteriophage

S. Brenner[†], G. Streisinger[†1], R. W. Horne[§], S. P. Champe[†2],
L. Barnett[†], S. Benzer[†3] and M. W. Rees[||]

[†] *Medical Research Council Unit for Molecular Biology and* [§] *Electron Microscope Group, Cavendish Laboratory, and* [||] *Department of Biochemistry, University of Cambridge, Cambridge, England*

(*Received 23 July 1959*)

A procedure is described for the efficient disjoining of the even-numbered T bacteriophages into structural components. Tail fibres and tail sheaths from T2L have been separated and purified, and sheaths can be obtained from the other T-even phages. The components have been characterized by electron microscopy using the phosphotungstate method and by chemical and physical methods. The tail of the bacteriophage consists of a sheath surrounding a core at the base of which are attached tail fibres. The sheath appears to be built of helically arranged subunits which form a hollow cylinder. We have confirmed that the sheath can contract in length, and have shown that the contraction is accompanied by an increase in diameter such that the volume of the sheath is approximately conserved. The core of the tail has been found to be a hollow cylinder with the central hole 25 Å in diameter.

Chemical studies, using fingerprinting, suggest that the sheath is composed of about 200 repeated subunits of approximately 50,000 molecular weight. No N-terminal amino acid could be detected in sheaths. Fingerprints confirm the previous finding that the head membrane of the phage is composed of a large number of repeated subunits with a molecular weight of 80,000. The tail fibres appear to have a subunit with a molecular weight not less than 100,000. These studies show that the proteins composing the head, sheath and tail fibres of the phage have different primary structures.

1. Introduction

Bacteriophages, once regarded as simple nucleoprotein particles, are now known to be amazingly sophisticated in structure. Phages of the T-even group (T2, T4, T6) are tadpole-shaped, the head having the form of a bipyramidal hexagonal prism (Anderson, 1946; Williams & Fraser, 1953), while a cylindrical tail is the organ of attachment of the phage to its host cell (Anderson, 1953; Kellenberger & Arber, 1955; Williams & Fraser, 1956). The demonstration that the head and tail contain distinct antigens suggests that these morphological parts are made of different proteins (Levinthal & Fisher, 1952; Lanni & Lanni, 1953; De Mars, Luria, Levinthal & Fisher, 1953; Anderson, Rappaport & Muscatine, 1953).

Recent work has shown that the tail itself is a complex structure. Kozloff & Henderson (1955), Kellenberger & Arber (1955) and Williams & Fraser (1956) reported structural changes of the tail produced by treatment of the phage with cadmium

[1] Fellow of the National Foundation for Infantile Paralysis on leave of absence from the Department of Genetics, Carnegie Institution of Washington, Cold Spring Harbor, Long Island, N.Y., U.S.A.

[2] Research Fellow of the Purdue Research Foundation on leave of absence from the Department of Biological Sciences, Purdue University, W. Lafayette, Indiana, U.S.A.

[3] Senior Postdoctoral Fellow of the National Science Foundation on leave of absence from the Department of Biological Sciences, Purdue University, W. Lafayette, Indiana, U.S.A.

cyanide, hydrogen peroxide and freezing and thawing respectively. The most striking alteration, as revealed in the electron microscope, is a shortening of the outer layer (the sheath) of the tail, revealing an inner core coaxial with the sheath, which displays at its distal tip a small diffuse structure and a number of outstretched fibres. Apparently some degree of fragmentation of the phage is effected by these treatments, since isolated cores and fibres are also observed.

The proteins of bacteriophage are of particular interest because they might offer favourable material for studying the genetic control of protein structure. The high resolving power of genetic analysis in phage makes it possible to construct maps of the genetic structure on the molecular level, as exemplified by the work of Benzer (1955, 1957) with the r_{II} mutants of bacteriophage T4. The relevant protein in this case has not been found, and does not appear to be an integral part of the phage structure. However, other regions of the genetic structure do affect proteins of the phage particle. The h region of T2L controls the serological specificity of the phage with respect to inactivating antiserum, and also its ability to attach to specific bacterial cells (Streisinger, 1956a,b; Streisinger & Franklin, 1956). It seems likely that this genetic region controls a protein of the tail fibres, since the fibres attach to bacterial hosts with the same specificity as the whole phage particle from which they are derived (Williams & Fraser, 1956). A particular advantage of this protein is the possibility of assaying it by its ability to block inactivating antiserum. Also promising are the c mutations of bacteriophage T4, which cause L-tryptophan to be required for attachment of the phage to the bacterium (Anderson, 1945, 1946; Delbruck, 1948; Wollman & Stent, 1950). The h and c mutations are localized within small and separate regions of the genetic map, each segment forming a functional unit or cistron (Streisinger & Franklin, 1956; Brenner, 1957). It is probable that the protein affected by the c mutations is also part of the tail (Ping-Yao Cheng, 1956). We have therefore attempted to isolate and characterize the structural components of the T-even phages in order to assess their suitability for analysis on the chemical, as well as genetical, levels.

In this paper we describe a technique for disjoining the components of the phage particle more effectively than the methods previously described; this has made possible the purification of tail fibres and tail sheaths. A study of the morphological and chemical properties of these and other phage components shows that the protein coat of the bacteriophage is a complex superstructure, built from different kinds of structural elements, each consisting of distinct proteins. In other papers we will deal with the serological and functional properties of the components and their genetic control.

2. Materials and Methods

Phages. T2, strains T2L (Luria) and T2H (Hershey), T4, strain T4B (Benzer), and T6 were used.

Bacteria. Escherichia coli, strain B was used for all assays and *E. coli* strain BB (Berkeley) for growth of phages.

Phage techniques. The methods of handling phages were those described by Adams (1959). Serum blocking power assays were carried out by the procedure described by De Mars (1955).

Hydroxyapatite columns were prepared following the directions of Tiselius, Hjertén & Levin (1956).

Enzymes were the crystalline products of Worthington Biochemical Corporation, Freehold, N.J.

(a) *Growth and purification of large amounts of phage*

Two methods were used to produce phage on a large scale. In the first, 35 l. cultures of *E. Coli* BB grown in M9 medium (Adams, 1959) to a cell count of 3×10^8/ml. were infected with 10^6 phage particles per ml. Aeration was continued for 3 hr, at which time the cells were artificially lysed by the addition of 1 l. of chloroform. Such lysates yielded a phage titre of 1 to 2×10^{11}/ml. Lysates of T2L could be concentrated by the acid precipitation technique described by Herriott & Barlow (1952). However, this technique was not successful with T4 or T6 strains and gave erratic results with T2L. The second and preferred method was to grow the phage in 5 l. batches in a simplified form of the apparatus designed by Fraser (1951). Using the medium 3XD (Fraser & Jerrel, 1953) *E. coli* BB was grown under vigorous aeration to a concentration of 2×10^9 cells/ml. and infected with 10^7 phage per ml. Artificial lysis with chloroform at $3\frac{1}{2}$ to 4 hr after infection yielded extremely viscous lysates containing from 0·6 to $1·5 \times 10^{12}$ phage/ml. After reducing the viscosity by the addition of a small amount of deoxyribonuclease (DNase) the phage was purified from the lysate by cycles of low and high speed centrifugation until clean phage pellets were obtained. To avoid ghosting, phage pellets were resuspended slowly in phosphate buffer pH 7·0 in the cold, in a total volume of about 500 ml. Each 5 l. batch generally yielded between $\frac{1}{2}$ and 1 gm of purified phage.

(b) *Electron microscopy*

Samples were prepared for electron microscopy in one of the following ways: (1) Solutions of phage or components in a volatile buffer (1% ammonium acetate or 2% ammonium bicarbonate) were sprayed onto nitrocellulose-coated grids and shadowed with chromium metal at a 60° angle. (2) Phosphotungstate (PTA) "embedded" preparations were made according to the procedure previously described (Brenner & Horne, 1959). The sample, in 1% ammonium acetate, was mixed with an equal volume of a 2% aqueous solution of phosphotungstic acid previously adjusted to pH 7·0 with N-KOH, and the mixture sprayed onto carbon-coated grids. Metal shadow casting (in method (1)) and carbon evaporation (in method (2)) were carried out in an experimental evaporator unit (Cosslett & Horne, 1957) using mercury diffusion pumps and traps to reduce possible oil vapour contamination.

A Siemens Elmiskop was used at instrumental magnifications of 8,000, 40,000 and 80,000. To minimize heating effects which were observed in the droplet patterns of PTA preparations when employing single condenser lens illumination, double condenser illumination was used, and the illuminating beam reduced to 25 to 30 microns in diameter. Through-focal series were taken during the investigation of the components to ensure that the structures observed in the micrographs were not due to electron phase contrast effects.

Ultracentrifugation was performed in a Spinco Model E centrifuge.

Electrophoresis was performed in a Perkin-Elmer Tiselius apparatus after overnight equilibration of the sample with the buffer solution by dialysis.

(c) *Preparation of samples for "fingerprinting"*

Proteins were digested with enzymes in one of two ways. When large amounts were available, the denatured protein, suspended in water, was adjusted to pH 8·0 at 30°C. An amount of enzyme (trypsin or chymotrypsin in 0·001 N-HCl) equal to 1% of the weight of the protein substrate was added, and the course of peptide bond hydrolysis was followed by recording the amount of sodium hydroxide required to maintain the pH at 8·0. Digestions with trypsin were essentially complete in 90 min, but chymotrypsin digestions required some hours. For smaller quantities the denatured protein was digested in 2% ammonium bicarbonate which buffers at pH 8·0. A drop of acetic acid was then added and the volatile salt removed by lyophilization. One-dimensional ionophoretic separation of peptides was carried out on Whatman 3MM paper using pyridine-acetic acid buffer pH 6·4 (Michl, 1951). For two-dimensional separations, or fingerprinting, Ingram's procedure was followed (Ingram, 1958).

3. Experiments and Results

(a) Purification of tail components

Kozloff, Lute & Henderson (1957) reported that cadmium cyanide, $Cd(CN_3)^-$, removed the tail fibres from T2, as judged by electron micrographs, and failure to recover all of the serum blocking power (SBP) in high speed pellets of treated phage. In our hands this technique did not yield free fibres from T2L or T2H. After treatment of the phage with 0·0062 M-$Cd(CN_3)^-$, less than 3% of the SBP remained in the supernatant after high speed centrifugation, while controls showed that the reagent inactivated the phage but did not appreciably reduce the total SBP. Electron micrographs of $Cd(CN_3)^-$ treated phage prepared by the PTA method showed that tail fibres remained attached to most of the altered particles. We have made several attempts to disjoin phage by freezing and thawing following the procedure of Williams & Fraser (1956), but very poor yields of free components were obtained. The method which was developed for efficient disjoining of phage T2L and purification of several of its components is as follows (Fig. 1). Purified phage suspensions in 0·1 M-NaCl were added to an equal volume of 0·2 M-glycine-HCl buffer with 0·1 M-NaCl at pH 2, and the pH adjusted to 2 if necessary. A precipitate of DNA and protein is obtained. This precipitate gave a very viscous suspension when reneutralized due to the liberation of DNA from the phage heads. Digestion with DNase reduced the viscosity, but did not solubilize the phage protein, which was collected by low speed centrifugation. The morphology of the altered phage in this precipitate cannot be easily studied with the electron microscope. However, the precipitate can be dispersed by brief treatment with 8 M-urea (5 min at room temperature) followed by dilution and dialysis against 1% ammonium acetate. Plate I is an electron micrograph of this material showing clumps of altered phage particles aggregated by damaged heads. Sheaths, cores, and tail fibres may be seen, indicating that the phage had been disjoined, but attempts to extract the components were unsuccessful at this stage. Digestion of the precipitate with trypsin or chymotrypsin at pH 8·0 (50 µg/ml. in 2% ammonium bicarbonate) results in rapid clearing of the suspension, leaving little insoluble material. Plate II(a) is an electron micrograph of such a digest showing that the damaged heads have been completely dissolved leaving a solution of contracted free sheaths, cores and tail fibres.

Removal and purification of the sheaths, which sediment at 100,000 g in 1 hr, was accomplished by differential centrifugation. Resuspension of the sheath pellet, in 1% ammonium acetate pH 6·5, will generally aggregate any non-sheath material, which can then be removed by low speed centrifugation. Recentrifugation of the sheaths yields almost completely translucent pellets.

Chromatographic separation of the tail fibres from the remaining mixture of fibres, cores and head peptides was achieved by chromatography on a hydroxyapatite column. The suspension was dialysed against 0·01 M-phosphate buffer (pH 6·8) and the precipitate discarded. The solution was then absorbed to a hydroxyapatite column, washed exhaustively with 0·01 M-phosphate buffer, and eluted in a stepwise manner with increasing concentrations of phosphate buffer at the same pH. Over 90% of the input protein (head peptides) was removed from the column by the initial washing. The tail fibres, detected by electron microscopy and serum blocking power assays, eluted between 0·08 and 0·12 M buffer concentrations. Alternatively, adsorption and elution can be performed as a batch process, washing the hydroxyapatite by centrifugation. An electron micrograph of purified tail fibres is shown in Plate II(b).

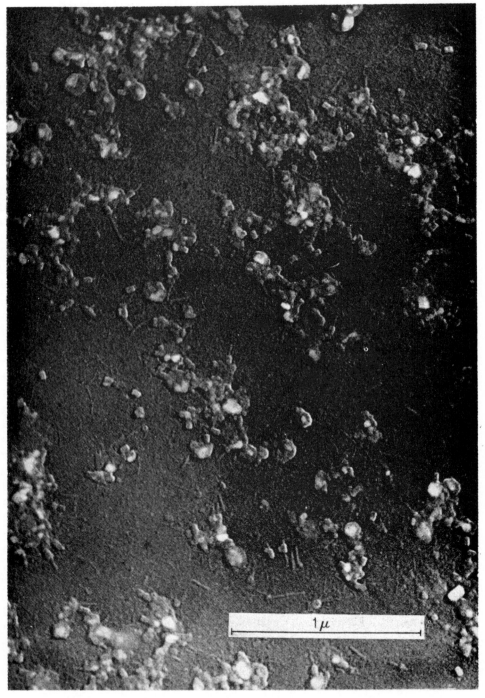

PLATE I. Electron micrograph of pH 2·0 treated T2L phage, dispersed with urea. (Chromium shadowed, × 50,000.)

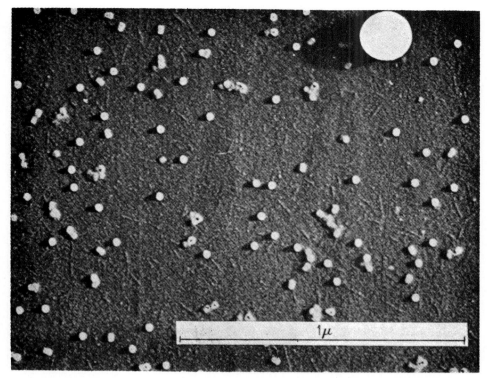

PLATE II(a). Electron micrograph of pH 2·0 treated T2L, digested with chymotrypsin. (Chromium shadowed, × 75,000.)

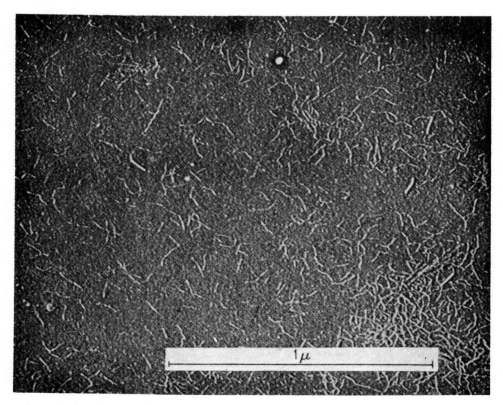

PLATE II(b). Electron micrograph of purified tail fibres. A few cores are also present. (Chromium shadowed, × 70,000.)

In the cases of T2H, T4B and T6 we were not able to use the foregoing procedure to obtain fibres because stronger acid was required to denature the heads of these

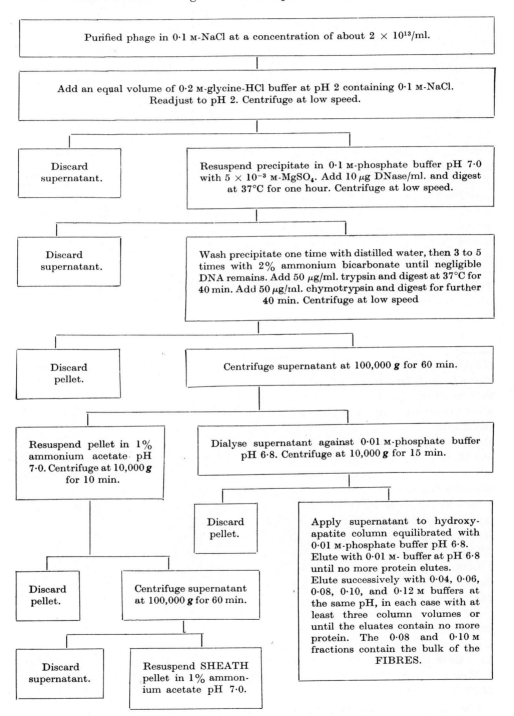

FIG. 1. Procedure for purifying sheaths and tail fibres from T2L.

strains. However, sheaths may be prepared from these strains, at the sacrifice of tail fibres and cores, by treating a suspension of phage in 0·3 M-NaCl with an equal volume of N-H₂SO₄ followed by DNase and proteolytic enzyme digestion as described for T2L. The acid treatment was usually repeated after removal of the DNA, since the presence of large quantities of DNA can interfere with the effective denaturation of the heads. The only structures which remain undigested are the sheaths, which can be purified by centrifugation.

(b) *Electron microscopy of phage components*

Kellenberger & Arber (1955) described the effects of exposure to hydrogen peroxide on the morphology of the phage. Plate III is an electron micrograph of bacteriophage treated with hydrogen peroxide to 40% survival of infectivity. Both normal and altered particles are present, aggregated into a rosette by their tail fibres. The sheaths in altered particles are contracted, and are shorter and thicker than the normal sheaths. The core is attached to the hexagonal head which is filled with DNA, but the sheaths appear unattached to the head in both normal and altered phages. The relations between the components can be more clearly seen in the isolated particle from the same preparation shown in Plate IV. At the distal end of the core there is a rather diffuse structure† to which are attached the tail fibres.

Electron micrographs of purified contracted sheaths are shown in Plates V and VI. They are hollow cylinders, 350 Å long and 250 Å in diameter, the diameter of the inner hole being 120 Å. The total volume of the contracted sheath is thus $1·3 \times 10^7$ Å³. By comparison, the diameter of the extended sheath is 165 Å in PTA preparations, and its length is probably equal to that of the core (800 Å). If it is assumed that the diameter of the cylindrical hole in the extended sheath is equal to that of the core (70 Å), then the volume of the extended sheath is computed to be $1·4 \times 10^7$ Å³. This suggests that the volume of the sheath is conserved when it contracts. Very often sheaths standing upright display a "cog wheel" appearance, and those lying obliquely show helical grooving (Plate VI) which suggests a helical arrangement of subunits in the sheath. In addition, striations can often be seen across the tail of intact phage as shown in Plate VII(a). The number of striations in the extended sheath is close to 25 and their spacing is estimated to be 30 to 40 Å. We have occasionally found structures of the type shown in Plate VII(b), which appears to be a stretched sheath revealing again its helical character.

Several tail fibres are shown in Plate VIII, where it is seen that they are long slender structures 1300 Å by 20 Å and are characteristically kinked in the middle. This structure is partially obscured when fibres from hydroxyapatite columns are examined (Plate II(b)) since the column appears to fragment them.

Cores, as can be seen from Plate VIII, are hollow cylinders 800 Å long and 70 Å in diameter, the diameter of the central hole being estimated to be 25 Å. No attempt has yet been made to obtain pure cores.

By mild acid treatment (pH 2·0) of T4, followed by DNase and trypsin digestion, preparations can be obtained that show empty heads which retain their hexagonal shape and are clearly constituted of a thin membrane, which is estimated to be about 35 Å thick (Plate IX). If it is assumed that the head to sheath volume ratio is equal to their mass ratio (8·5/1), the thickness of the head membrane should be of the order of 60 Å.

---

† Recently, it has been possible to characterize this structure more fully. It is a hexagonal plate at the base of the core to which six tail fibres are attached (Brenner & Horne, in preparation).

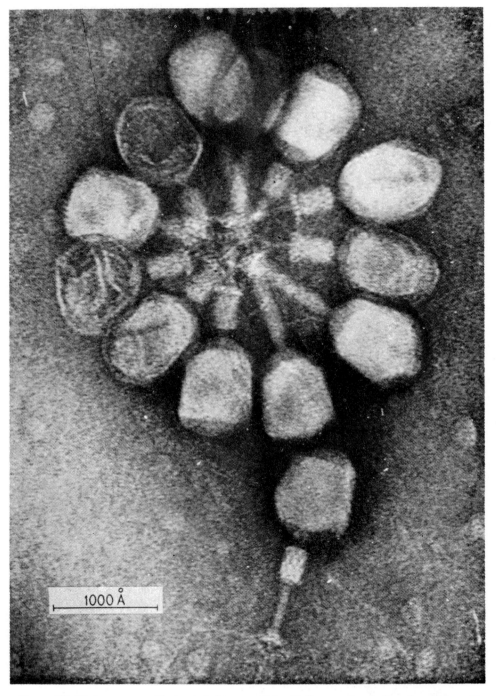

PLATE III. Hydrogen peroxide treated phage particles. Both normal and altered particles with shortened and thickened sheaths are seen in the rosette. (PTA embedded, × 280,000.)

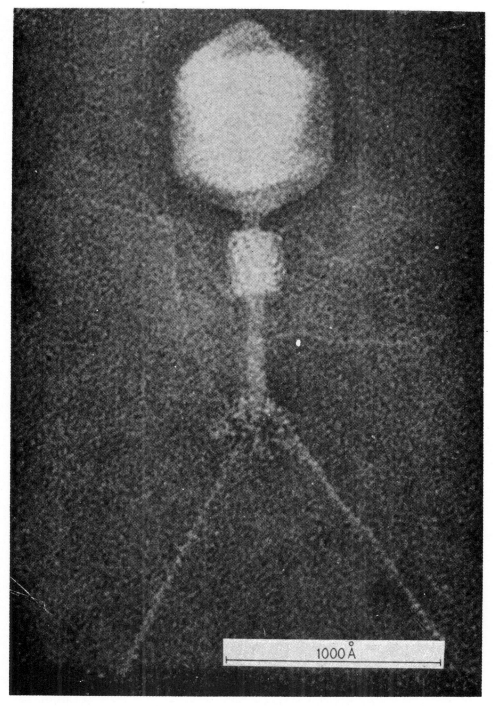

PLATE IV. Isolated hydrogen peroxide treated phage particle, showing the relations of the filled head, contracted sheath, core and tail fibres. (PTA embedded, × 570,000.)

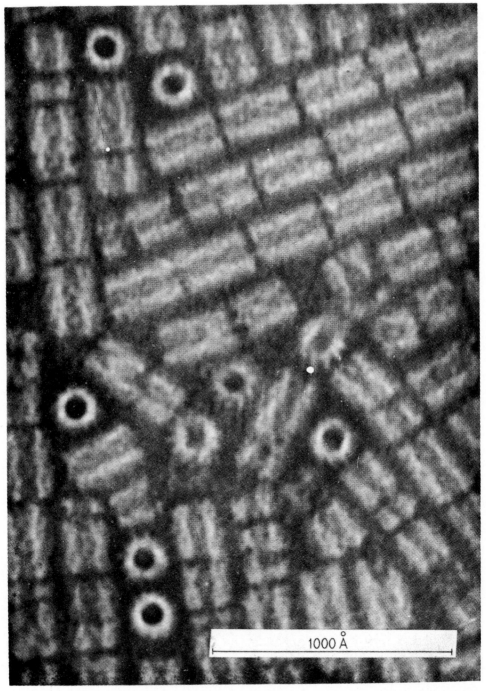

1000 Å

PLATE V. Purified contracted sheaths. A number of sheaths can be seen standing upright revealing the "cog wheel" arrangement of the subunits. (PTA embedded, × 630,000.)

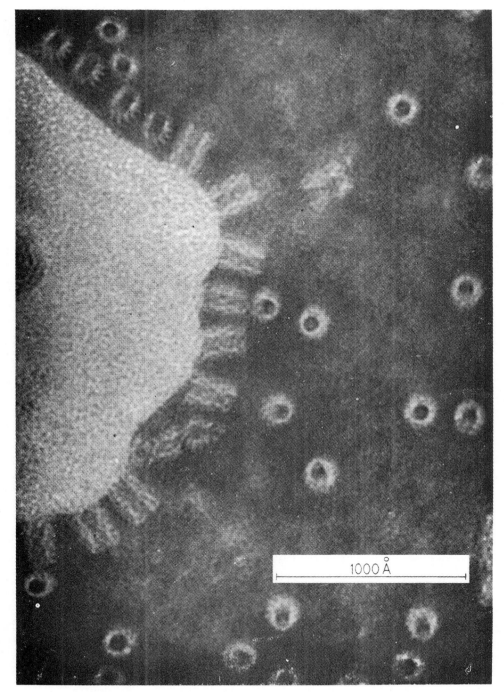

PLATE VI. Purified contracted sheaths. A number of sheaths can be seen standing upright revealing the "cog wheel" arrangement of the subunits (PTA embedded, × 500,000.)

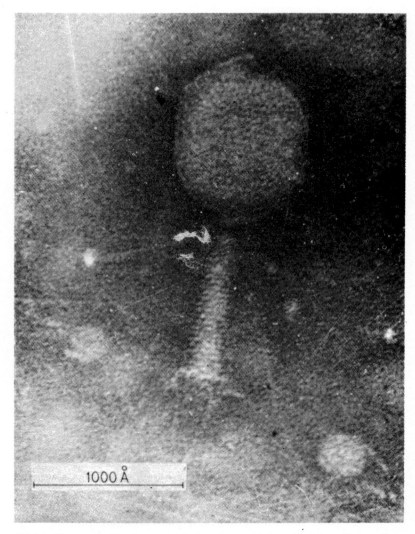

1000 Å

PLATE VII(a). Phage particle showing striations along the length of the tail spaced at approximately 30 Å. (PTA embedded, × 400,000.)

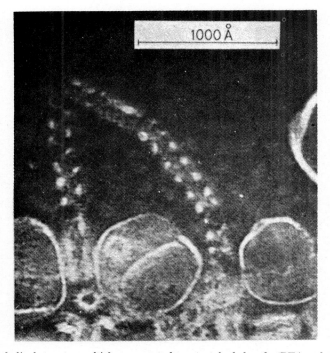

PLATE VII(b). A helical structure which appears to be a stretched sheath. (PTA embedded, × 370,000.)

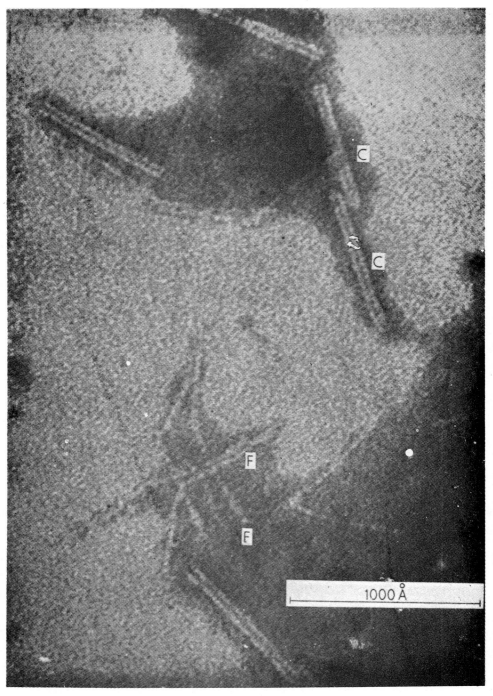

PLATE VIII. Tail fibres (F) and hollow cores (C) embedded in a PTA droplet. The full length of the tail fibres can be seen together with the characteristic kinking at the centre. (PTA embedded, × 500,000.)

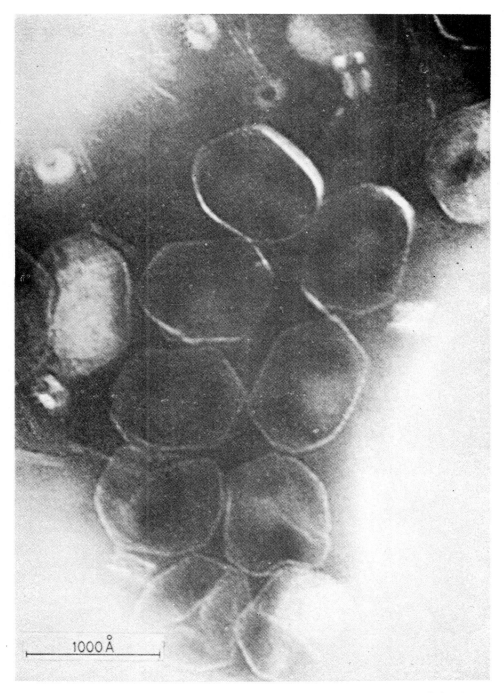

1000 Å

PLATE IX. Empty T4 head membranes with preserved hexagonal shape. Cores and sheaths are also present (PTA embedded, × 350,000.)

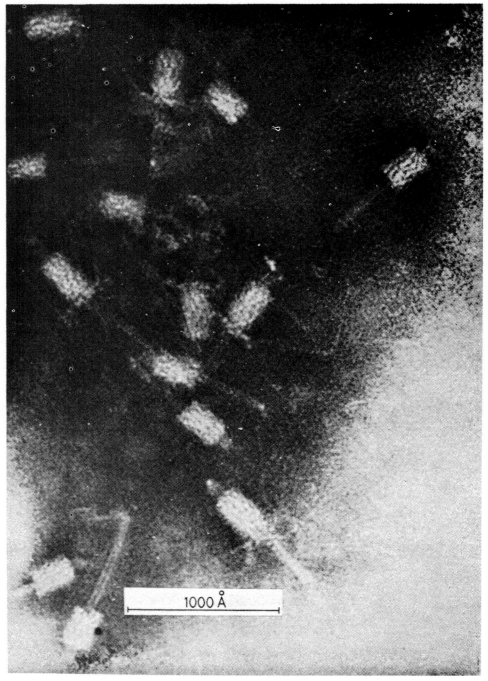

PLATE X. Structures resulting from the treatment of phage with sodium dodecyl sulphate at pH 10·0. These consist of cores surrounded by contracted sheaths. Tail fibres are also present. (PTA embedded, × 400,000.)

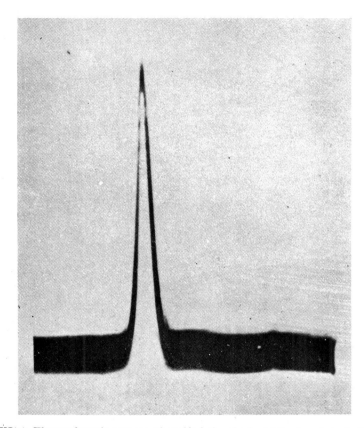

PLATE XI(a). Electrophoresis pattern of purified sheaths in 0·1 M-phosphate buffer pH 6·5.

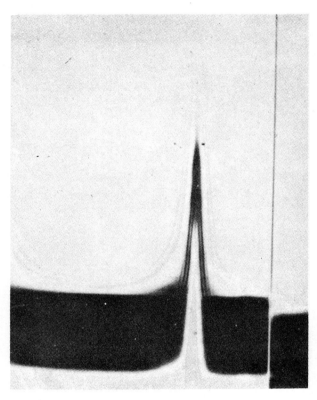

PLATE XI(b). Sedimentation pattern of purified sheaths in 0·1 M-phosphate buffer pH 6·5, 16 min after reaching 20,400 rev/min.

A characteristic structure is produced by treating phage ghosts with sodium dodecyl sulphate at pH 10 following the procedure of Van Vunakis, Baker & Brown (1958). The heads are completely solubilized leaving contracted sheaths surrounding hollow cores (Plate X). Tail fibres may also be seen. This material is presumably identical with the fast-sedimenting component A of Van Vunakis et al. which constituted roughly 10% of the structural protein of the phage.

(c) Physical and chemical properties

Only sheaths have been isolated in amounts sufficient to investigate by physical techniques. Preparations of sheaths are electrophoretically homogeneous, giving single symmetrical boundaries with a mobility of 7.9×10^{-5} cm²/volt sec in 0.1 M-phosphate buffer pH 6.5 (Plate XI(a)). Sheaths give a single symmetrical peak in the ultracentrifuge (Plate XI(b)) with a sedimentation coefficient $S_{20,\mathrm{W}} = 114\ S$ when extrapolated to zero concentration. Preliminary light scattering measurements indicate a particle weight of approximately 12,000,000 (Doty, P., personal communication), and a similar value is estimated from the electron microscopic dimensions of the sheath.

Sheaths are remarkably resistant structures, retaining their characteristic appearance after treatment with acid and alkali (N-H₂SO₄ or 0.2 N-NaOH), after boiling for 20 min, or after treatment for 24 hr with 8 M-urea, 8 M-guanidine hydrochloride, or 1% sodium dodecyl sulphate at pH 10.5. However, we have found it possible to

TABLE 1

Amino acid composition of the sheaths

Amino acid	g/100 g protein	Residues per 54,000 molecular weight
½ cystine	0.23	1
methionine	0.81	3
aspartic acid	13.0	52
threonine	7.78	35
serine	5.78	29
glutamic acid	9.94	36
proline	3.1	14
glycine	4.98	35
alanine	7.6	46
valine	6.61	30
*iso*leucine	7.5	31
leucine	7.0	28
tyrosine	6.35	19
phenylalanine	5.0	16
lysine	6.21	23
arginine	5.70	17
histidine	0.55	2
NH₃	1.94	(61)

dissociate the sheaths and sensitize them to enzyme action by the same procedure used by Fraenkel-Conrat (1957) to dissociate the subunits of tobacco mosaic virus (TMV). An aqueous suspension of sheaths, cooled to 0°C and mixed with two volumes of glacial acetic acid just above its freezing point, promptly loses its characteristic opalescence and becomes extremely viscous. Microscopic observation shows complete disappearance of the organised structure. Dialysis of the solution against distilled water or phosphate buffer pH 6·5 results in precipitation of the protein which cannot be readily redissolved even in 8 M-urea. However, if the acetic acid solution is dialysed directly into 8 M-urea buffered at pH 6·5 by 0·1 M-phosphate, most of the protein remains in solution, although an increased opalescence indicates that some reaggregation occurs. This prevented study of the physical properties of the dissociated sheaths.

Attempts to demonstrate an N-terminal amino acid using the fluorodinitrobenzene technique of Sanger (1945) have been unsuccessful. Dinitrophenylation of dissociated sheaths in 8 M-urea yielded only a trace of DNP-glutamic acid (one residue per 3×10^6 molecular weight), although one ϵ-DNP-lysine residue was recovered per 7,000 molecular weight showing that dinitrophenylation had occurred. It is possible that the N-terminal group is protected as in the subunit of TMV (Narita, 1958).

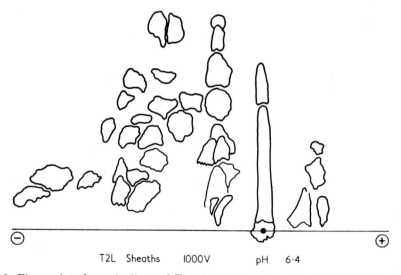

T2L Sheaths   1000V   pH   6·4

FIG. 2. Fingerprint of trypsin digest of T2L sheaths. Ionophoresis : pH 6·4, 1,000 v, 2½ hours. Chromatography : n-butanol, acetic acid, water (3 : 1 : 1), 15 hours ascending.

A quantitative amino acid analysis of sheaths was carried out and the results are given in Table 1. From these data, combined with the results of fingerprinting a tryptic digest of the protein, an estimate of the size of the repeating subunit may be made. Since trypsin cleaves specifically those bonds in which the carbonyl group is contributed by lysine and arginine residues, the fact that fingerprints of sheaths (Fig. 2) display about 37 peptides indicates that the number of lysine and arginine residues in the subunit is of this order. From the proportion of these amino acids to the total, the molecular weight indicated for the subunit would be about 50,000. Based upon the relatively rare histidine residues, the molecular weight should be an integral multiple of 27,000. Only one histidine-containing peptide has been found in the fingerprint, but some material invariably remains at the origin. It is not yet known whether this represents undigested protein or peptides insoluble in the buffer.

We have investigated the head protein of T2L by fingerprinting the tryptic peptides of phage denatured by acid. Since the sheaths remain undigested, contamination from other proteins is only a few per cent. The fingerprint, developed with ninhydrin,

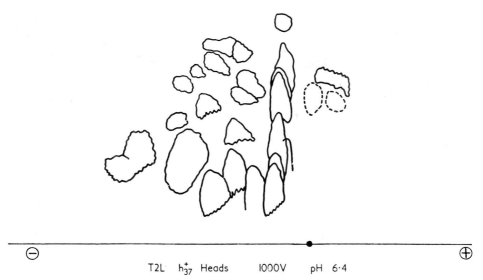

$T2L\ h^{+}_{37}$ Heads   1000V   pH 6·4

FIG 3. Fingerprint of trypsin digest of T2L head protein. Ionophoresis : pH 6·4, 1000 v, 2½ hours. Chromatography : n-butanol, acetic, acid, water (3 : 1 : 1), 15 hours ascending,

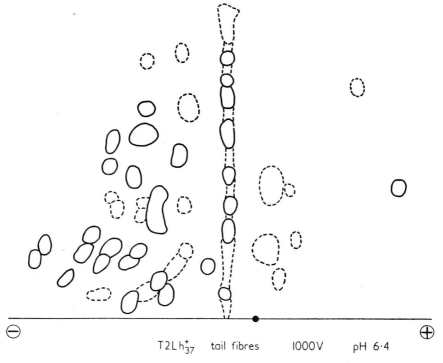

$T2Lh^{+}_{37}$ tail fibres   1000V   pH 6·4

FIG. 4. Fingerprint of trypsin digest of T2L tail fibres. Many of the peptides were faint since only a small quantity of fibres was available. Ionophoresis : pH 6·4, 1,000 v, 2½ hours. Chromatography : n-butanol, acetic acid, water (3 : 1 : 1), 15 hours ascending.

resolves approximately 28 peptides (Fig. 3). By applying specific stains for histidine, tyrosine, arginine and tryptophan, a total of some 40 peptides can be distinguished. The head subunit isolated from T2 ghosts by Van Vunakis *et al.* (1958) was reported to have a molecular weight of about 80,000 as judged from its N-terminal alanine content, and physical properties. The amino acid analysis of whole phage (Fraser, 1957) indicates 21 arginine and 32 lysine residues for a subunit of 80,000 molecular weight. Thus 53 tryptic peptides would be expected. Since the number of peptides actually resolved is a minimum value, the fingerprint supports the evidence for a repeating subunit of this molecular weight.

The quantities of tail fibres which have been isolated have been too small to permit precise physicochemical studies. Fingerprints of tail fibres are complex, showing more peptides than in the case of heads or sheaths (Fig. 4). If the fibres have about the same lysine and arginine content as the other components, a molecular weight of the order of 100,000 would be indicated for the repeating subunit. From the dimensions observed in the electron micrographs, the particle weight of an entire fibre is estimated to be of the order of 400,000.

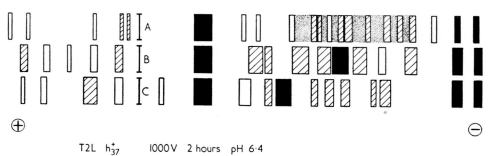

T2L h_{37}^+ 1000V 2 hours pH 6·4

FIG. 5. Ionophoretic comparison of tryptic peptides of T2L tail fibres (A), sheaths (B) and heads (C).

In addition to the fingerprints, a direct comparison of head, sheath and tail fibre proteins was made by separating the tryptic peptides of these proteins side by side, employing paper ionophoresis. Fig. 5 is a tracing of the ionophoretogram after staining with ninhydrin. The radically different patterns demonstrate that each of these morphological components is composed of a protein with primary structure distinct from the others.

4. Discussion

We have isolated two of the components of the protein coat of T-even bacteriophages, the tail sheath and the tail fibres, and have helped to characterise two more, the head and the core. To these components can be added the hexagonal plate at the base of the tail, the structure of which suggests that the number of tail fibres is six (Brenner & Horne, in preparation). Apart from these proteins and the DNA, other components have been found. Hershey (1955) described an internal protein, containing about 3% of the phage sulphur which is released by osmotic shock, while Levine, Barlow & Van Vunakis (1958) have recently shown that this fraction includes a serologically distinct internal protein, representing 4 to 6% of the total protein of the phage. Other internal constituents are the polyamines, putrescine and spermidine, and another minor polypeptide (Hershey, 1957; Ames, Dubin & Rosenthal, 1958). Koch &

Dreyer (1958) isolated a lysozyme from lysates of phage T2, and this enzyme is apparently also part of the phage structure (Koch & Weidel, 1956; Barrington & Kozloff, 1956).

The electron micrographs of the components prepared by the phosphotungstate method show a wealth of structural detail. This procedure not only provides good contrast but also can preserve three-dimensional morphology, as is particularly well demonstrated in the case of the empty head membranes which retain their hexagonal shape. The technique shows the cores to be hollow and has permitted resolution cf the helical structure of the sheaths. The dimensions of components measured in the present study differ somewhat from those found by previous authors (Kellenberger & Arber, 1955; Williams & Fraser, 1956) using shadow-cast preparations. The latter procedure is liable to overestimation due to the presence of the metal, while the phosphotungstate method is liable to underestimation because only structure impermeable to the electron dense material is visualized. Thus we do not place much reliance on the measured thickness of the head membrane of 35 Å, since calculation shows that it should be nearer to 60 Å. On the other hand, the diameter of the hole down the core cannot be greater than the measured diameter of 25 Å.

Our results support the findings of Van Vunakis et al. (1958) that the phage head is built of a large number of subunits of molecular weight about 80,000. We have also been able to show that the sheath contains a subunit of approximately 50,000 molecular weight, the N-terminal group of which appears to be unavailable to dinitrophenylation. This explains why Van Vunakis et al. (1958) found only the N-terminal alanine of the head protein in whole ghosts. From the weight of the intact sheath and that of the subunit we estimate that each sheath contains about 200 subunits. In end-on views of the contracted sheaths, about 15 "cogs" can be seen. If these represent the subunits in one turn of a helix the entire sheath should contain thirteen such turns. It is to be noted that the number of striations observed along the length of extended sheaths is about 25, and these may therefore represent the number of turns of the helix in the *extended structure*. This is consistent with the contraction of the sheaths to about one half of their extended length. The increase in diameter of the sheath accompanying its contraction can be understood for a helical structure if the contraction represents an increase in the number of subunits per turn, the total number then remaining constant. In contracted sheaths, the helical grooves run in the direction of the long axis of the structure. A helical structure built of subunits related by screw symmetry has many sets of grooves and the set which is most emphasized depends on the exact shape of the subunits. Hence the same structure may show different sets of grooves under different conditions.

The characteristic structural feature of the tail fibres is the presence of a central kink. The difficulty of isolating sufficient amounts of tail fibres has limited their characterisation. The complexity of the fingerprint shows that if a single repeating unit exists it has a molecular weight probably not less than 100,000. Conversely, if smaller subunits are present there must be a number of different ones. It is conceivable that the length of the fibre is determined by the length of a single polypeptide chain, the fibre being a bundle of such chains. An α-helix 1,300 Å long would contain about 860 amino acids and have a molecular weight of about 100,000.

The quantity, the ease of isolation, and the molecular weight of the sheath and head proteins makes their analysis possible and prompts efforts to discover their genetic control. The tail fibres, on the other hand, for which a genetic system is available, seem unsuitable for chemical analysis because of their apparently large molecular weight, and the small quantity obtainable.

The complexity of phage as a whole is to be contrasted with the relatively simple structures of plant viruses such as tobacco mosaic virus and turnip yellow mosaic virus. Although the components of phage, like the small viruses, seem to be built on a simple plan, the assembly of the parts would seem to pose a formidable problem.

REFERENCES

Adams, M. H. (1959). *Bacteriophages*. New York: Interscience Publishers.
Ames, B., Dubin, D. T. & Rosenthal, S. M. (1958). *Science*, **127**, 814.
Anderson, T. F. (1945). *J. Cell. Comp. Physiol.* **25**, 17.
Anderson, T. F. (1946). *Cold Spr. Harb. Sym. Quant. Biol.* **11**, 1.
Anderson T. F. (1953). *Cold Spr. Harb. Sym. Quant. Biol.* **18**, 197.
Anderson, T. F., Rappaport, C. & Muscatine, N. A. (1953). *Ann. Inst. Pasteur*, **84**, 5.
Barrington, L. J. & Kozloff, L. M. (1956). *J. Biol. Chem.* **223**, 615.
Benzer, S. (1955). *Proc. Nat. Acad. Sci., Wash.* **41**, 344.
Benzer, S. (1957). *The Chemical Basis of Heredity*, edited by W. D. McElroy & B. Glass. Baltimore: Johns Hopkins Press.
Brenner, S. (1957). *Virology*, **3**, 560.
Brenner, S. & Horne, R. W. (1959). *Biochim. biophys. Acta*, **34**, 103.
Cosslett, V. E. & Horne, R. W. (1957). *Vacuum*, **5**, 109.
Delbrück, M. (1948). *J. Bact.* **56**, 1.
De Mars, R. I. (1955). *Virology*, **1**, 83.
De Mars, R. I., Luria, S. E., Levinthal, C. & Fisher, H. (1953). *Ann. Inst. Pasteur*, **84**, 113.
Fraenkel-Conrat, H. (1957). *Virology*, **4**, 1.
Fraser, D. (1951). *J. Bact.* **61**, 115.
Fraser, D. (1957). *J. Biol. Chem.* **227**, 711.
Fraser, D. & Jerrel, E. A. (1953). *J. Biol. Chem.* **205**, 291.
Herriott, R. M. & Barlow, J. L. (1952). *J. Gen. Physiol.* **36**, 17.
Hershey, A. D. (1955). *Virology*, **1**, 108.
Hershey, A. D. (1957). *Virology*, **4**, 237.
Ingram, V. M. (1958). *Biochim. biophys. Acta*, **28**, 539.
Kellenberger, E. & Arber, E. (1955). *Z. Naturf.* **10b**, 698.
Koch, G. & Dreyer, W. J. (1958). *Virology*, **6**, 291.
Koch, G. & Weidel, W. (1956). *Z. Naturf.* **11b**, 345.
Kozloff, L. M. & Henderson, K. (1955). *Nature*, **176**, 1169.
Kozloff, L. M., Lute, M. & Henderson, K. (1957). *J. Biol. Chem.* **228**, 511.
Lanni, F. & Lanni, Y. T. (1953). *Cold Spr. Harb. Symp. Quant. Biol.*, **18**, 159.
Levine, L., Barlow, J. L. & Van Vunakis, H. (1958). *Virology*, **6**, 702.
Levinthal, C. & Fisher, H. (1952). *Biochim. biophys. Acta*, **9**, 419.
Michl, H. (1951). *Mh. Chem.* **82**, 489.
Narita, K. (1958). *Biochim. biophys. Acta*, **28**, 184.
Ping-Yao Cheng. (1956). *Biochim. biophys. Acta*, **22**, 433.
Sanger, F. (1945). *Biochem. J.* **39**, 507.
Streisinger, G. (1956a). *Virology*, **2**, 377.
Streisinger, G. (1956b). *Virology*, **2**, 388.
Streisinger, G. & Franklin, N. C. (1956). *Cold Spr. Harb. Symp. Quant. Biol.* **21**, 103.
Tiselius, A., Hjertén, S. & Levin, Ö. (1956). *Arch. Biochem. Biophys.* **65**, 132.
Van Vunakis, H., Baker, W. & Brown, R. (1958). *Virology*, **5**, 327.
Williams, R. C. & Fraser, D. (1953). *J. Bact.* **66**, 458.
Williams, R. C. & Fraser, D. (1956). *Virology*, **2**, 289.
Wollman, E. L. & Stent, G. S. (1950). *Biochim. biophys. Acta*, **6**, 292.

J. Mol. Biol. (1960) **2**, 143-152

Molecular Homogeneity of the Deoxyribonucleic Acid of Phage T2 †

A. D. Hershey and Elizabeth Burgi

Department of Genetics, Carnegie Institution of Washington, Cold Spring Harbor, N.Y., U.S.A.

(*Received 12 April 1960*)

Chromatographic analysis by means of a column of basic protein is used to measure the breakage of T2 DNA by stirring. Broken molecules elute from the column at lower salt concentrations than do the original molecules. At a critical low speed of stirring, one can produce single breaks near the centers of the molecules, as shown by the all-or-none character of the initial change in chromatographic behavior, and by the survival of unbroken molecules according to an exponential function of time of stirring. In an analogous way, one can produce fragments resulting from three breaks per molecule, and in general stirring for a long time at a given speed produces a moderately homogeneous collection of fragments whose mean size is smaller, the higher the speed of stirring.

The sensitivity of the DNA to breakage by stirring is strongly dependent on the concentration of DNA, being greater the lower the concentration. This self-protective action is greatly reduced when the DNA is broken by stirring.

By chromatographic refractionation of the fragments produced by initial breaks, it can be shown that the fragments are not identical in chromatographic properties, presumably because of a continuous distribution of lengths centered about the mean half-length. Unbroken DNA, on the contrary, is homogeneous in chromatographic properties, therefore presumably homogeneous in molecular length. We conclude that DNA can be extracted from T2 by phenol without manipulative breakage, and that it exists in the phage particle in the form of one or more molecules of identical length.

1. Introduction

Existing methods for the measurement of molecular weights and molecular weight distributions of deoxyribonucleic acids are of doubtful significance at best and may be inapplicable when the molecular weight becomes very high. This is so both because the methods are theoretically ambiguous or technically complicated or both, and because the nucleic acids are prone to preparative degradation. The resulting confusion is well illustrated by the status of the DNA of T2, which is variously reported to consist of a mixture of larger (molecular weight 40 million) and smaller pieces (Levinthal & Thomas, 1957); to be homogeneous at a molecular weight about 14 million (Meselson, Stahl & Vinograd, 1957); and to be composed of two fractions differing both in composition (Brown & Martin, 1955) and in molecular size (Brown & Brown, 1958).

We have re-examined the DNA of T2 at some length by chromatographic and ultracentrifugal methods that did not, at the outset, seem particularly auspicious. We now conclude, however, that both methods are exceedingly useful for the analysis

†Aided by grant C2158 from the National Cancer Institute, National Institutes of Health, United States Public Health Service.

of nucleic acids of high molecular weight; that the molecular weight of T2 DNA is indeed high; and that undegraded preparations homogeneous in molecular weight can be made.

In this paper we present evidence concerning the homogeneity of our preparations, based on the following argument. Samples of DNA that have been stirred for a long time at a given speed may be expected to consist of fairly uniform fragments whose average molecular length will be shorter the higher the speed of stirring. This expectation is borne out by the fact that such samples elute in a single band from a column of basic protein at a salt concentration that is lower the higher the speed of stirring. Nevertheless, DNA broken by stirring can easily be resolved into chromatographically distinguishable components by chromatographic separation. Unstirred DNA, on the contrary, is chromatographically homogeneous, except that the samples usually contain a small and variable fraction of broken molecules.

It will be noticed that our present argument rests solely on the application of chromatographic methods, which perhaps leave something to be desired. In a paper to follow (Burgi & Hershey, in preparation) we shall extend the analysis in other ways.

2. Methods

DNA is prepared from phage T2 by shaking the purified phage suspensions with water-saturated phenol (Mandell & Hershey, 1960). The resulting preparations contain 0·4 mg of DNA per ml. in a solution 0·1 M in NaCl and 0·1 M in sodium and potassium phosphate of pH 7·1.

Chromatographic analysis is made by means of a column of methyl-esterified serum albumin adsorbed to diatomaceous earth (Mandell & Hershey, 1960). About 0·4 mg of DNA in 50 ml. of 0·66 M-NaCl is applied to the column, from which the DNA is eluted again (at about 0·72 M) under a gradient of rising salt concentration. The salt solutions are buffered at pH 6·7 with 0·05 M-phosphate. In the illustrations shown in this paper, "tube number" means the number of 3·6 ml. samples that have passed through the column behind the gradient front under an approximately constant gradient of, usually, $3·3 \times 10^{-4}$ M/ml.

The chromatographic analyses are of two kinds. "Chromatographic separations" are made by passing material through a column and collecting fractions in the usual way. For "chromatographic comparison", two samples of DNA, one labeled with radiophosphorus and one not, are mixed in such proportions that the labeled sample does not contribute appreciably to the optical density of the mixture. The mixture is then passed through a column equipped for continuous recording of the ultraviolet absorbency and radioactivity of the effluent, which permits the labeled and unlabeled samples of DNA to be compared with great precision. The results obtained in this way are then converted into percentage distributions of DNA among the hypothetical 3·6 ml. tubes. This method has about the same accuracy as actual collection and assay of fractions.

Intentional breakage of DNA molecules is achieved by stirring for specified times at specified speeds and concentrations as follows. For high speeds (>600 rev/min) we use a Virtis Co. (Yonkers, New York) stirring blade and 5 ml. flask attached to a powerful series-wound motor. Lower speeds are obtained with a larger Virtis stirrer and 50 ml. flask attached to a gear-reduction motor capable of speeds between 100 and 600 rev/min. The speed of both motors is controlled by a variable transformer and voltmeter.

3. Results

(a) *Effects of stirring*

When solutions of T2 DNA containing 0·4 mg/ml. are stirred for 30 min at various speeds, no change is detectable by chromatographic comparison with unstirred DNA after stirring at speeds below 6000 rev/min. At 8000 rev/min a very considerable change is effected, causing the stirred DNA to elute from the column at a lower salt concentration than before. Stirring at progressively higher speeds produces DNA that elutes at progressively lower salt concentrations, but always as a single band if the stirring period has not been too short. We can produce at least five chromatographically distinguishable species of DNA by stirring at various speeds up to about 25,000 rev/min, the maximum speed of our motor. In Fig. 1 we show a chromatographic comparison of three of them, representing unstirred DNA, DNA stirred at 8000 rev/min and DNA stirred at 18,000 rev/min, respectively. In interpreting the results of Fig. 1, it must be remembered that chromatographic comparison of two DNAs is essentially a test of identity or nonidentity; the distance of separation of two different DNAs along the concentration axis depends on the relative amounts of the two kinds of DNA and other factors (Mandell & Hershey, 1960).

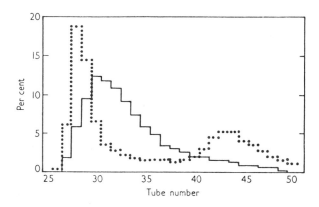

FIG. 1. Chromatographic comparison of T2 DNA stirred at 18,000 rev/min (left), 8000 rev/min (center), and unstirred (right). Solid line, D_{260} (optical density at 260 mμ); dotted line, ^{32}P.

It is important to note that no combination of speeds and times of stirring will produce DNA that elutes from the column between the positions characteristic of unstirred DNA and of DNA stirred at 8000 rev/min. This initial change is the effect of single breaks per molecule, as shown both by its all-or-none character and by the kinetics of breakage described below.

Consider first the result obtained when samples of ^{32}P-labeled DNA, diluted to contain 0·4 μg/ml. in 0·1 M-NaCl, 0·05 M-phosphate of pH 6·7, are stirred for various times at 260 rev/min, and the samples are compared chromatographically with carrier DNA that has been stirred at 0·4 mg/ml. and 8000 rev/min to produce initial breaks. Fig. 2 shows the comparison for one such sample in which 30% of the labeled DNA has been broken, and illustrates how this percentage is measured. Fig. 3 shows the kinetics of breakage, which conform to expectations if the change in chromatographic properties results from a single event per molecule. Note also in Fig. 2 that the chromatographic properties of the broken DNA in the labeled sample (30% of the total), and of the broken DNA in the unlabeled sample (60%), are the same.† Likewise, the chromatographic properties of the labeled broken DNA in the various

†Actually they are not quite the same in this instance, but for irrelevant reasons. We often find that DNA extracted from different phage preparations (not different extracts from the same preparation) are slightly different in chromatographic properties. Fig. 2 is an example of this. The same difference was seen with these two preparations of DNA when tested without stirring and when tested after stirring in mixture.

samples (always compared with the same carrier) prove to be independent of the time of stirring, at least for as long as some unbroken DNA remains. This we interpret to mean that the fragments produced by the initial breaks are not subject to further breakage at the given speed of stirring, evidently because most of the initial breaks occur near the centers of the molecules, producing fragments too short to be broken again.

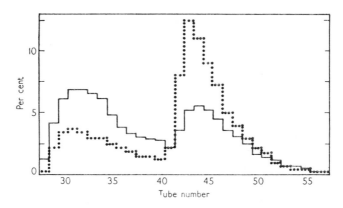

FIG. 2. Measurement of breakage. Dotted line, ^{32}P-labeled DNA stirred for 2 min at 260 rev/min and 0·4 μg/ml., 30% broken. Solid line, unlabeled DNA stirred for 15 min at 8000 rev/min and 0·4 mg/ml., 60% broken.

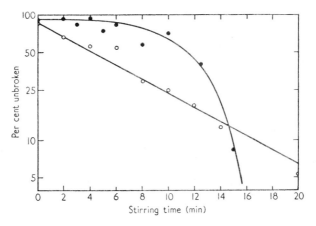

FIG. 3. Kinetics of breakage. Open circles: DNA stirred at 260 rev/min and 0·4 μg/ml. Each point is the average of 2 independent experiments. Filled circles: DNA stirred at 500 rev/min and 4 μg/ml. Measurements as in Fig. 2.

Fig. 3 also presents the contrasting result obtained when DNA is stirred at 500 rev/min and 4 μg/ml. (the same type of result is seen at 8000 rev/min and 0·4 mg/ml.). The two types of result are not inconsistent but are explained by the fact that, at appreciable concentrations, DNA exerts a strong self-protection against breakage under hydrodynamic shear, an effect that diminishes as breakage proceeds. This is proved by the experiment described in Table 1. Evidently the protective effect is sharply reduced when the molecules are broken in half, and the protective effect of unbroken DNA only disappears at some concentration between 4 and 0·4 μg/ml.

The question arises whether or not the changes in chromatographic properties produced by stirring are due solely to a reduction in molecular length. It might be supposed that the creation of unnatural ends, resulting perhaps from breaks at slightly different levels in the apposed polynucleotide chains, would also affect chromato graphic behavior. If this were a predominant factor in our results, one would not expect to be able to produce a second comparable stepwise change when half-molecules are broken into quarters, because the quarters would be even more heterogeneous with respect to end-type than the halves. We therefore sought a condition under which DNA already altered by stirring could be stirred again so as to produce a mixture of two species, one of which was further altered and one not.

TABLE 1

Self-protection by DNA against breakage under shear and destruction of protective effect by stirring

Composition of mixture (μg/ml.)			
Labeled, unstirred	Unlabeled, unstirred	Unlabeled, stirred	% of labeled DNA broken
2	0	0	76, 91
2	2	0	23, 23
2	0	2	76, 82

The mixtures shown (in which the stirred DNA had been stirred for 30 min at 500 rev/min and 4 μg/ml.) were stirred for 5 min at 500 rev/min. The resulting breakage was measured as in Fig. 2. Measurements are given for two independent experiments.

This was achieved by first isolating a relatively homogeneous sample of labeled "half-molecules" by chromatographic separation of DNA that had been subjected to initial breakage, and stirring it further under carefully selected conditions. The fraction chosen for further study (8% of the total) was collected after 34% of the DNA had passed through the column, and should correspond to molecules of half the original length (Burgi & Hershey, in preparation).

When the labeled half-molecules isolated as described above were mixed with unlabeled, unfractioned DNA that had likewise been stirred to produce initial breaks, and the mixture was then stirred for 15 min at 11,000 rev/min and 0·2 mg/ml. and passed through a column, the labeled DNA separated into two bands (Fig. 4). The unlabeled DNA did not, evidently because the unfractionated half-molecules, hence a fortiori the quarter molecules, were diverse in length.

We also verified that labeled, fractionated, half-molecules, when stirred at 1000 rev/min and 0·4 μg/ml., survived according to an exponential function of time of stirring (18%, 47% and 73% broken in 13, 28 and 90 min, respectively). This result shows that there is no peculiarly resistant fraction with respect to the second stepwise change in chromatographic properties, in contrast to what one might expect if the column were merely registering the production of fragments with two broken ends from fragments with single broken ends.

The experiments just described also show that when half-molecules are diluted from 200 μg/ml. to 0·4 μg/ml., the speed of stirring must be reduced 11-fold to produce similar effects. In either case, the speed is critical. Half-molecules exhibit self-protection.

We conclude that the changes caused by stirring that we detect by chromatographic means are due solely or mainly to reduction in length of the molecules, which can be broken into approximate halves by initial central breaks and, at a higher speed,

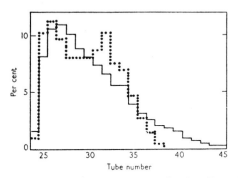

FIG. 4. Production of quarter molecules from half molecules. Dotted line, ^{32}P-labeled DNA originating from the half-molecule fraction of DNA that had been stirred for 30 min at 8000 rev/min and 0·4 mg/ml. Solid line, D_{260} of DNA previously stirred in the same way but not fractionated. The result shown was obtained after mixing the two materials and stirring the mixture again for 15 min at 11,000 rev/min and 0·2 mg/ml.

again into quarters, and so on. If this is correct, the result illustrated in Fig. 3 means that T2 DNA as we isolate it is homogeneous with respect to molecular length: otherwise progressive conversion of DNA from one length class to another by continued stirring at constant speed, without progressive alteration of either class, would be impossible. For the same reasons, it can be inferred that our DNA has not been broken during preparation by shearing phosphate ester bonds in a random manner. The possibility of breakage at specified points to produce fragments of uniform length is not, of course, excluded by our experiments.

We turn now to a more direct demonstration that our preparations of T2 DNA are chromatographically homogeneous before, and heterogeneous after, breakage by stirring.

(b) *Heterogeneity of stirred DNA*

Recall that stirring T2 DNA for 30 min at 8000 rev/min and 0·4 mg/ml. produces mainly the product of a first all-or-none change that we attribute to single breaks occurring near the centers of the molecules. If so, the product should be chromatographically heterogeneous, because the breaks could not be centered exactly. The following experiment confirms this expectation.

The starting materials were a sample of ^{32}P-labeled DNA stirred as described above, and an unlabeled sample stirred in the same manner ("stirred carrier"). The stirred, labeled DNA was submitted to a separation run through a column which yielded 15 samples of 3·6 ml. each among which the DNA was distributed in the typical manner to be illustrated below. Chromatographic comparisons showed that the labeled DNA was the same as the unlabeled DNA before stirring, and differed from it in the expected way after stirring. Additional comparisons showed that the stirred labeled DNA before separation, was identical to the stirred carrier, and that pooled 0·1 ml. samples of the separated fractions formed a mixture showing little effect of passage through the column. The latter comparison is reproduced in Fig. 5. It suggests that only a small fraction of the slower running material in the labeled specimen had been converted into faster running material, either by breakage in the column or during transfer by pipet to reconstruct the pool. Subsequent experience confirmed the second alternative, because it was found to be exceedingly difficult to pipet dilute solutions of T2 DNA of high molecular weight without breaking it (Mandell & Hershey, 1960).

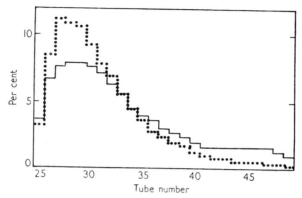

Fig. 5. Chromatographic comparison of pooled fractions from stirred DNA with the starting material. Dotted line, ³²P-labeled, pooled fractions from chromatographic separation of DNA stirred for 15 min at 8000 rev/min and 0·4 mg/ml. Solid line, D_{260} corresponding to unlabeled, unfractionated DNA similarly stirred.

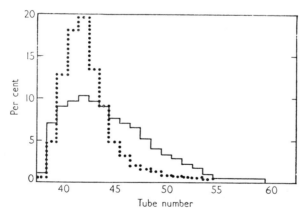

Fig. 6. Chromatographic comparison with starting material of peak fraction from stirred DNA. Dotted line, ³²P-labeled, peak fraction (9% of total) obtained by chromatographic separation of DNA stirred for 15 min at 8000 rev/min and 0·4 mg/ml. Solid line, D_{260}, unlabeled, unfractionated DNA similarly stirred.

Comparisons were next made by passing through the column mixtures of stirred carrier DNA, and labeled DNA taken from individual fractions obtained during the initial separation. In this way nearly all of the fractions were tested. In each case the fractionated material formed a much narrower band than the unfractionated carrier and eluted, in fact, very close to the position on the chromatogram from which it had been taken. Two examples are shown in Figs. 6 and 7.

We conclude that our column can resolve with considerable precision a heterogeneous collection of DNA molecules produced by stirring and, insofar as these may be expected to differ in molecular length, can discriminate molecules differing only slightly in length.

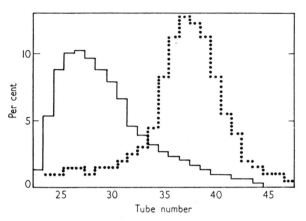

FIG. 7. Chromatographic comparison with the starting material of the 9th fraction to the right of the peak obtained with stirred DNA. Dotted line, ^{32}P-labeled DNA (3% of total) from the specified fraction obtained by chromatographic separation of DNA stirred as in Fig. 6. Solid line, D_{260}, unlabeled, unfractionated DNA stirred in the same way.

(c) Homogeneity of native T2 DNA

The following experiment is identical in other respects to the experiment just described, except that the starting material is labeled, unstirred DNA. The several fractions of labeled DNA obtained by a preliminary separation run were separately mixed with unfractionated, unlabeled carrier DNA, and the mixtures were passed

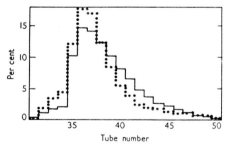

FIG. 8. Chromatographic comparison of unfractionated T2 DNA (solid line, D_{260}) with ^{32}P-labeled peak fraction (dotted line) from a previous chromatographic separation.

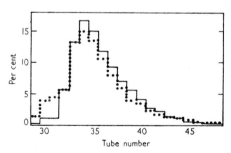

FIG 9. Chromatographic comparison of unfractionated T2 DNA (solid line D_{260}) with the 7th fraction to the right of the peak (dotted line, ^{32}P) of a previous chromatographic separation.

through additional columns. In this way nearly all of the original fractions were tested. Those comprising the small fraction of DNA found at the leading edge of the band (some 10% of the total, varying in amount in different preparations) proved indeed to differ from the remainder, resembling half-molecules. All fractions taken from the main band, including its trailing edge, however, proved to be indistinguishable from the unfractionated material. Typical results for two fractions are reproduced in Figs. 8 and 9.

4. Discussion

We find that T2 DNA subjected to stirring elutes from a column of basic protein at a lower salt concentration than does unstirred DNA. The effect is progressive and depends, when other conditions are kept constant and the time is not too short, mainly on the speed, rather than on the time, of stirring. At a minimum critical speed and after a short time of stirring, the initial change is all-or-none with respect to individual molecules. The same is true, at a higher minimum critical speed, for the transition between the product of the initial change and a second stepwise change, but only if the product of the initial change has been fractionated to render it chromatographically homogeneous. Under suitable conditions, the all-or-none character of these changes is also reflected in the survival of the original molecules according to an exponential function of time of stirring. We conclude that these stepwise changes result from transverse breaks occurring near the centers of molecular length and that the altered chromatographic behavior is due mainly or entirely to the reduction in length. If this conclusion is correct, molecules of T2 DNA can be reduced by stirring at moderate speeds to considerably less than one quarter of their original length, presumably because the original length is very great.

A point of practical importance and theoretical interest emerges from our finding that the effects of stirring at a given speed are much more severe when the DNA solution is dilute than when it is concentrated. Infinite dilution with respect to self-protection lies somewhere between 4 and 0·4 μg/ml. The protective effect (measured with respect to the production of initial breaks) is greatly reduced when the molecules are broken in half. The protection could be ascribed to a viscous effect in reducing turbulence or to a reversible molecular interaction that strengthens the molecules themselves, alternatives that we have not attempted to distinguish.

By chromatographic separation of DNA stirred to produce initial breaks, one can readily isolate DNA molecules whose mean length is apparently slightly less or slightly greater than the half-length, as identified chromatographically. By contrast, chromatographic separation of unstirred DNA does not reveal any molecules differing appreciably from the starting material, except for a very small fraction that resembles broken DNA. We conclude that DNA prepared by phenol extraction under our conditions is nearly homogeneous in molecular length and cannot have been broken in a random manner during preparation. Our main conclusion is thus at variance with evidence from radioautography (Levinthal & Thomas, 1957) that the DNA in T2 is diverse in molecular size, as well as with evidence (obtained by the use of a different column from ours) that the DNA from T2 is heterogeneous in chromatographic properties (Brown & Martin, 1955). Evidently the possibility that the earlier results were obtained with inadvertently broken preparations is not excluded. The extreme sensitivity of T2 DNA to such breakage has been pointed out before (Davison, 1959).

Our results concerning breakage by hydrodynamic shear are somewhat different from, but not inconsistent with, the results of previous work. Doty, McGill & Rice (1958) produced low molecular weight DNA by subjecting thymus DNA to sonic frequencies. They concluded that the breaks were distributed at random with respect to the original molecular lengths, yielding a very broad distribution of lengths whose mean depended on the time of treatment. We have concluded, on the contrary, that breaks caused by stirring show a strong tendency to occur near the centers of molecular length, producing a narrow distribution of lengths whose mean depends chiefly on the speed of stirring. It seems to us that these differences are readily explained as the effects, on the one hand, of strong shearing forces acting for a short time and, on the other hand, of weak shearing forces acting for a long time. Evidently the final product obtained at any given rate of shear must consist of fragments too short to be broken by that shear, which implies at the same time that the production of fragments much shorter than the maximum length resistant to breakage is prohibited. As pointed out by Cavalieri & Rosenberg (1959), one might expect a range of lengths little greater than twofold.

Cavalieri & Rosenberg (1959) subjected pneumococcal DNA to the shearing forces produced in an atomizer operated under various pressures. They found that the resulting molecular size depended primarily on the pressure, rather than on the number of passages through the orifice. Their broken preparations were more homogeneous (with respect to sedimentation coefficient) than their starting material. Our results are the other way around, which may be due to the difference in the starting materials or to the different tests for homogeneity. Cavalieri & Rosenberg did not observe any effect of DNA concentration on sensitivity to breakage, in marked contrast to our results. Their result may be attributed to the relatively low molecular weight ($3 \cdot 5 \times 10^6$) of their starting material, or possibly to the different method of producing breaks.

The homogeneity of our preparation of DNA from T2 is to be contrasted with the heterogeneity of preparations from other sources studied in the past (Doty et al., 1958). It remains to be determined whether the DNA of T2 is unique in this respect, or whether the different results merely reflect the different methods of preparation. In other respects the DNA of T4 (similar to that of T2) is unique. Doty, Marmur & Sueoka (1959) report that phage DNA, in contrast to DNA from other sources, shows an exceedingly sharp thermal denaturation temperature. According to the results of these authors, this means that the phage DNA is relatively homogeneous with respect to base composition. Presumably this method yields results that would not be affected appreciably by previous breakage of the DNA.

REFERENCES

Brown, G. L. & Brown, A. V. (1958). *Symp. Soc. Exp. Biol.* **12**, 6.
Brown, G. L. & Martin, A. V. (1955). *Nature*, **176**, 971.
Cavalieri, L. F. & Rosenberg, B. H. (1959). *J. Amer. Chem. Soc.* **81**, 5136.
Davison, P. F. (1959). *Proc. Nat. Acad. Sci., Wash.* **45**, 1560.
Doty, P., Marmur, J. & Sueoka, N. (1959). *Brookhaven Symposia in Biology*, **12**, 1.
Doty, P., McGill, B. B. & Rice, S. A. (1958). *Proc. Nat. Acad. Sci., Wash.* **44**, 432.
Levinthal, C. & Thomas, C. A., Jr. (1957). *Biochim. biophys. Acta*, **23**, 453.
Mandell, J. D. & Hershey, A. D. (1960). *Analytical Biochemistry*, in the press.
Meselson, M., Stahl, F. W. & Vinograd, J. (1957). *Proc. Nat. Acad. Sci., Wash.* **43**, 581.

J. Mol. Biol. (1961) **3**, 18–30

Structural Considerations in the Interaction of DNA and Acridines†

L. S. Lerman‡

Medical Research Council Unit for Molecular Biology, University of Cambridge, England and Department of Biophysics, University of Colorado Medical Center, Denver, Colorado, U.S.A.

(*Received 15 August 1960*)

The combination in solution of DNA with small amounts of acridine, proflavine, or acridine orange results in markedly enhanced viscosity and a diminution of the sedimentation coefficient of the DNA. These changes are contrary to those expected on the basis of aggregation or simple electrostatic effects. Characteristic changes, which suggest considerable modification of the usual helical structure of DNA, are found in the X-ray diffraction patterns of fibers of the complex with proflavine.

It is inferred that these compounds, which are potent mutagens, are intercalated between adjacent nucleotide-pair layers by extension and unwinding of the deoxyribose-phosphate backbone. The hydrodynamic changes are the consequence of the diminished bending between layers, the lengthening of the molecule, and the diminished length-specific mass. The effects are fully reversible at ordinary temperatures. The proposed structural change is compatible with the normal restrictions on bond lengths, angles, and non-bonded contacts, and maintenance of the hydrogen-bonded base pairs perpendicular to the axis of the molecule.

Another singly charged dye, pinacyanol, fails to elicit the effects of intercalation, yielding only lowered viscosity and increased sedimentation coefficient.

1. Introduction

The mode of combination of acridine derivatives with DNA is of interest because of the mutagenicity of these compounds and the carcinogenicity of the related benzacridines. While the possibility of strong electronic interactions would favor flat, face-to-face binding of the acridine to the bases of the nucleotides, this mode is not accessible in the Watson-Crick structure, where the bases are in close van der Waals contact. However, it can be supposed that untwisting of the double helix might provide spaces of suitable depth between layers, yet leave undisturbed the hydrogen-bonded pairing of the nucleotides constituting each layer. The partly extended molecule containing intercalated acridine layers would differ from native DNA in several properties measurable in solution: (a) The length of the molecule would be increased in proportion to the amount of acridine bound. (b) The further extensibility of the helix, which ordinarily permits considerable irregular coiling of the DNA molecule in solution, would be diminished. If the molecule with intercalated acridine is locally straighter

† Department of Biophysics contribution number 111. This work was supported by a research grant from the Institute of Arthritis and Metabolic Diseases, Public Health Service and received some assistance from a grant by the Damon Runyon Fund to the Department of Biophysics.

‡ Senior Research Fellow, Public Health Service.

and stiffer, the random coil will be less compact. (c) Since the additional length due to each acridine is about the same as that of a nucleotide pair, while the mass increment is less than one-half of a nucleotide pair, the average mass per unit length of the molecule would decrease. If the structure in solution is retained on drying, it may be expected to introduce substantial changes in the X-ray diffraction pattern of an oriented fiber.

The intrinsic viscosity of a coiled macromolecule depends on both the contour length along the molecule and the average breadth of the random coil. The sedimentation coefficient of a long slender molecule (which is not tightly coiled) depends only weakly on the length but is expected to be nearly proportional to the mass per unit length (Schachman, 1959). The changes found in the diffraction pattern and in these physical properties of DNA when combined with some acridine derivatives will be seen to be consistent with the expected effects of intercalated layers and incompatible with any simple alternative scheme of bonding.

2. Methods

The DNA used in viscometry and sedimentation was isolated from fresh calf thymus and chicken erythrocytes by a conventional extraction with strong NaCl and deproteinization with chloroform (but not dried or rendered salt-free) (Chargaff, 1955). Commercial dyes were used without further purification; the acridine orange contained an equimolar quantity of zinc chloride and some dextrin. Proflavine is 2,8-diaminoacridine; acridine orange is similar but each amino nitrogen carries two methyl groups. All solutions were buffered near pH 6·8 with 0·001 M-phosphate and contained, in addition, NaCl and 10^{-4} M-EDTA (ethylenediaminetetra-acetate). DNA-acridine complexes were prepared by mixing dilute solutions of both components. Mixing at high concentrations frequently produces fibrous precipitates. DNA analyses were carried out by the modified diphenylamine method (Burton, 1956).

For dissociation of the acridine orange complex the solution was equilibrated with blotted beads of Dowex 50, which strongly binds the dye, and decanted several times. The resin was first washed by repeated cycling with acid and base and equilibrated with the DNA solvent to prevent any pH disturbance.

Viscosities were measured in glass capillary shear viscometers at 25·1 °C. In solutions where the reduced specific viscosity was high a marked deviation of the measurements from linear dependence on the shear rate was noted. Since plots of the logarithm of the reduced specific viscosity against the square root of the shear were more persuasively linear and provided a more sharply defined extrapolation, this method was used in deriving the zero-shear values. The tabulation also includes, however, values obtained by extrapolation of a visually fitted straight line in Cartesian coordinates.

Sedimentation of the DNA and its complexes was observed during ultracentrifugation by means of ultraviolet absorption optics. Approximate median sedimentation coefficients were calculated from the position at which each densitometer profile reached half its plateau value. This procedure, while simple and rapid, introduces a slight, but probably significant, underestimate of the actual median value; but the error should be similar for the complexes and controls. A calculation, based on an assumed extreme case having a uniform distribution of sedimentation coefficients extending from 0·5 to 1·5 of the median, indicates that the true median of this distribution would be underestimated by 3%. In all cases the first derivative of the absorption profiles shows only a single peak, in some cases slightly sharper in the complex than in pure DNA.

Because of the complexity of a dissociating system, the sedimentation coefficients have not been extrapolated to zero concentration, nor are all of the viscosities extrapolated to the intrinsic viscosity. The difference between the measured sedimentation coefficient and its extrapolated limit is negligible for DNA at these concentrations in the presence of 0·1 M salt, but because of the lower salt concentration in the present experiments, the measured values will be smaller than the values at infinite dilution.

For the preparation of fibers, solutions containing about 4 mg/ml. lithium DNA (kindly provided by Dr. M. H. F. Wilkins) were dialysed to equilibrium against a suitable proflavine solution. Centrifugation of the resulting solution at 10^5 g yielded a stiff gel. Slices of the gel were allowed to dry while stretched by small attached weights (paper clips) to their maximum extension, and equilibrated over salt solutions of the appropriate relative humidity. Diffraction patterns were taken in a Philips microcamera (lent by Dr. D. Davies) with CuKα radiation from a rotating anode tube. Hydrogen bubbled through a saturated salt solution was circulated through the camera for an hour preceding and during each exposure. Distances between symmetrical spots were measured on microdensitometer tracings drawn at 5- to 20-fold magnification. The fibers contained roughly one proflavine per three nucleotide pairs. The dichroism of the fibers was easily observed under the polarizing microscope.

3. Results and Discussion

While an accurate general theoretical description of the hydrodynamic properties of DNA has not yet been achieved, the molecule appears to conform roughly to the properties of an open random coil of limited flexibility through which solvent can pass more or less freely. Although it is convenient to discuss the results in terms of this model, the main questions are answered by the direction in which the properties change, and the interpretation should not be sensitive to the details of the model.

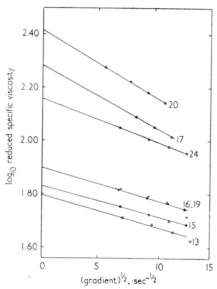

FIG. 1. The effect of salt, pinacyanol and some acridines on the reduced specific viscosity of calf thymus DNA solutions. For details of the conditions, see Table 1.

    16—no dye
    15—no dye, added salt
    13—pinacyanol
    20—proflavine
    24—acridine
    17—acridine orange
    19—(X), solution 17 after ion exchange treatment to remove the dye

(a) Viscosities

A set of measurements carried out at uniform concentration of calf thymus DNA, showing reduced specific viscosity plotted against shear gradient, is presented in Fig. 1.

The extrapolated intercepts at zero shear for these and similar experiments under varying conditions are presented in Table 1. The viscosity contribution of the DNA is

TABLE 1

Reduced specific viscosities at zero shear for DNA-acridine complexes, DNA preparation S

Index number	Dye	Nominal dye concentration $M \times 10^5$	DNA concentration $\mu g/ml.$	Sodium concentration $M \times 10^3$	Zero shear reduced specific viscosity (g/dl.)$^{-1}$ (a)†	(b)†
16	—	—	68	3·7	68·5	78
7	—	—	60·8	3·7	68·7	82·5
13	Pinacyanol	2	66	3·7	55·5	61
9	Acridine orange	2	50	3·7	110	140
10	Acridine orange	3	52	3·7	115	153
14	Acridine orange	8	58·6	3·7	124	157
17	Acridine orange	8	65	3·7	141	196
19	(17 after ion exchange)	—	65	3·8	68·5	79
20	Proflavine	8	63	3·7	203	257
11	Acridine orange	3	30·4	2·7	94·5	115
24	Acridine	10	38	3·7	122	155
21	—	—	35	12·7	47	56·7
22	Proflavine	10	18·6	12·7	199	247
15	—	—	68·6	13·7	60·5	67
18	Acridine orange	8	76	13·7	86·1	100

† (a) linear extrapolation.
 (b) log *vs.* (shear)$^{1/2}$ extrapolation.

somewhat depressed, in the expected way (Cavalieri, Rosoff & Rosenberg, 1956) on increasing the sodium chloride concentration (comparing curve 16, the control under standard conditions, with curve 15, that obtained at a higher salt concentration).

For comparison with the supposed geometrical suitability of the acridine for inter-calary binding, a supposedly unsuitable singly charged dye, pinacyanol, was studied;

it could be expected neither to be fully accommodated in the internucleotide space nor to fill it completely if partially admitted. Curve 13 shows that pinacyanol, like other cations, diminishes the viscosity of the DNA solution.

The acridines, on the other hand, produce an astonishing enhancement of viscosity. The maximum effect here is given by proflavine, which yields a more than threefold higher value.

The complete reversibility of the effect is shown by the return of the viscosity of the acridine orange complex to the value for pure DNA after removal of the dye by ion exchange. A more sensitive test of the reversibility of the effect was carried out by measurement of the transforming activity of genetically marked pneumococcal DNA which had been dissociated in the same way from a complex with acridine orange. No change in the transforming activity was detected (Lerman & Tolmach, in preparation).

(b) The apparent intrinsic viscosity

In the simple theory of the relation of the viscosity of a polymer solution to its molecular parameters, the contribution of intermolecular interactions to the viscosity of the solution is neglected. It is necessary, therefore, to infer from the experimental measurements the limiting value of the viscosity contribution per molecule as the concentration approaches zero. In dealing with a dissociating multicomponent system, as in the present experiments, direct measurement of the concentration dependence of a complex of fixed composition is impracticable. Instead, a mixture of proflavine and

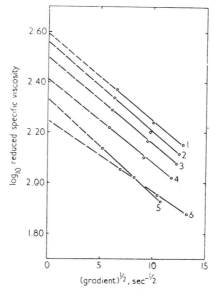

FIG. 2. The effect of proflavine on the reduced specific viscosity of chicken erythrocyte DNA solutions with 3×10^{-3} M-Na. Nearly superposed curves for some DNA concentrations have been omitted for clarity.

1, 2, 3, and 4—$2 \cdot 0 \times 10^{-5}$ M-proflavine; 41, 27·3, 20·5, and 14·7 μg/ml. DNA, respectively
5 and 6—no dye, 82 and 29·5 μg/ml. DNA, respectively

DNA was further diluted after each set of viscosity measurements with solvent containing the same total proflavine concentration. There are thus two major contributions to the slope of the curve which is to be extrapolated to infinite dilution, (a) the concentration dependence of the reduced specific viscosity of the complex, and (b) the changing composition of the complex as more proflavine is provided. In the equilibrium composition at the extrapolated limit, the concentration of *free* proflavine is the same as the total concentration. An extrapolation of this kind, labeled curve P, is shown in Fig. 3 together with a conventional dilution curve for pure DNA in the same solvent without proflavine, curve D. The points in Fig. 3 are taken from the extrapolation to zero shear at each dilution, shown in Fig. 2 (chicken erythrocyte

DNA). It will be seen that pure DNA shows the usual slight positive concentration dependence; while with proflavine η_{sp}/c rises strongly as infinite dilution is approached, apparently linear with concentration. The interesting feature of the measurements, the very high values in the presence of an acridine, is retained in the extrapolation. Similar results were obtained with calf thymus DNA (preparation Z).

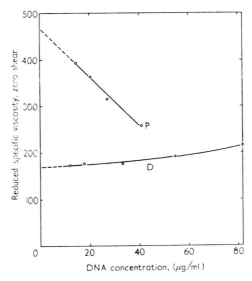

Fig. 3: The concentration dependence of the reduced specific viscosity of chicken erythrocyte DNA at zero shear, alone and in the presence of proflavine. The points are the extrapolated values given by Fig. 2.

D—no dye
P—$2{\cdot}0 \times 10^{-5}$ M-proflavine

(c) *Sedimentation*

Since aggregation would lead to an increased viscosity, it is useful to anticipate what changes it would introduce into the sedimentation properties of DNA. Although trimers or larger aggregates must be invoked to provide random coil dimensions corresponding to the observed viscosity changes, a discussion of the properties of dimers will be sufficient. The effect of the variable length of overlapping in the dimer can be ignored by considering only the limiting cases in which two molecules are joined either end-to-end or are in contact along their entire length. The sedimentation coefficient of a DNA molecule of twice the normal length is best estimated from the empirical relation derived from measurements on sonic fragments (Doty, McGill & Rice, 1958), for which S^0 varies as the 0·37 power of the molecular weight. The sedimentation coefficient of the other extreme hypothetical dimer, a cable of four polynucleotide strands, is estimated from the mass per unit length as roughly twice that of the monomer (Schachman, 1959). Thus, the dimer can be expected to have a sedimentation coefficient between 1·29 and 2 times that of the original DNA. Sedimentation coefficients for these DNA preparations and their complexes with proflavine, acridine, and pinacyanol are presented in Table 2. It will be seen that the positive increments in Table 2 fall below the lower limit for dimers, and the largest value is given by the pinacyanol complex.

The changes in the sedimentation coefficient seem to be accountable in terms of the changes in the mass per unit length of the DNA molecule according to the mode of binding of the attached dye. Binding to the outer surface increases the effective mass while intercalated binding decreases the effective mass of a given length of the molecule. Where a molecule of molecular weight 210 (proflavine, free base) replaces by intercalation a nucleotide pair of mean molecular weight 670, while maintaining constant length, density and frictional properties, the sedimentation coefficient will be expected to decline about 0·69% for each substituted segment in a hundred (neglecting possible changes in sodium binding). With an upper limit of 0·22 proflavine molecules strongly bound per nucleotide (Peacocke & Skerrett, 1956), the fractional replacement is 0·305 and hence the sedimentation coefficient would be 0·79 that of

TABLE 2

Sedimentation coefficients for DNA-acridine complexes

Index	Preparation	Dye	Total dye concentration $\text{M} \times 10^5$	Sodium concentration $\text{M} \times 10^3$	$S_{20,\,w}$ s	Ratio to control
869	Z	—	—	18	15·4	1·0
871	Z	Proflavine	2	18	13·5	0·88
870	Z	Proflavine	5	18	12·9	0·84
844	S	—	—	3·7	13·2	1·0
854	S	Proflavine	10	3·7	15·6	1·18
850	S	Acridine	3·3	3·7	14·3	1·08
851	S	Pinacyanol	2·0	3·7	16·4	1·24

Two separate DNA preparations are represented. All measurements were carried out with the DNA concentration near 33 μg/ml.

unmodified DNA. Since the length does not remain constant, but increases in direct proportion to the number of proflavine molecules bound, the effect of the diminished length-specific mass would be in some part compensated by the increased size. Although a simple estimate of the net effect is not possible, it appears unlikely that the sedimentation is more sensitive than the 0·37 power to the extra length due to intercalation; the net change in S at proflavine saturation would then be to 0·90 of pure DNA.

External binding, also assuming constant length, density, sodium binding and frictional properties, would effect an increase of 0·31% for each proflavine bound per hundred nucleotide pairs, or 0·51% for each pinacyanol (mol wt 353). At 44 dye molecules per hundred nucleotide pairs (an arbitrary figure), S would be increased by factors of 1·14 or 1·22 respectively.

Table 2 shows a depression in the sedimentation coefficient in the region of that predicted for intercalation when moderate amounts of proflavine are bound. At higher dye concentrations the net effect suggests a balance between the consequences of intercalation and some further external binding (see Section (h)) such that the sedimentation coefficient increases slightly. The gain is largest with pinacyanol where the external binding is not counterbalanced by intercalation.

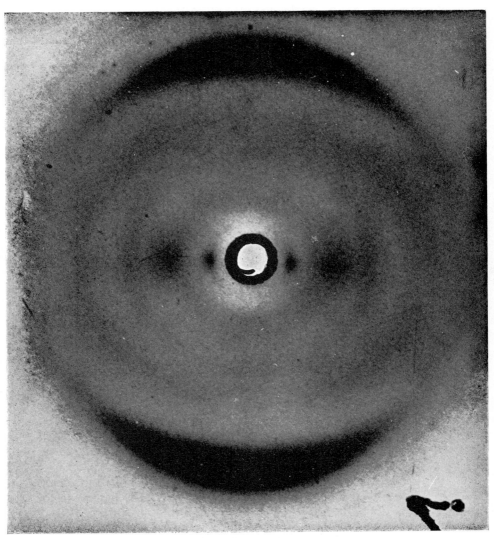

PLATE I. X-ray diffraction pattern given by the complex of Li DNA with proflavine at 75% relative humidity.

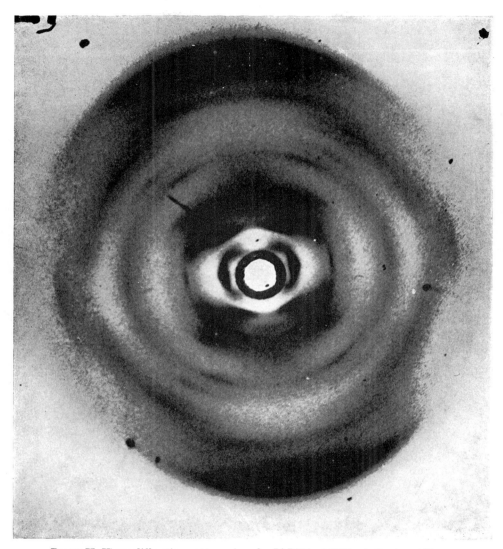

PLATE II. X-ray diffraction pattern given by Li DNA at 95% relative humidity.

It may be noted that two additional kinds of structural alterations which might be supposed to result from the binding of acridines would also have significant effects on the mass per unit length. If a space for the insertion of the acridine into the DNA molecule were provided by the rotation of each member of a base pair to the outside of the helix (see Fresco & Alberts, 1960), the same increment in mass per unit length as from simple external binding would be expected. If it is supposed that external attachment of acridines (but not other dyes) induces the structural modification which has been described for stretched DNA fibers (Wilkins, Gosling & Seeds, 1951), a decreased sedimentation coefficient can be expected. In this structure, where the bases are tilted about 45° from the fiber axis, the molecule is about 50% longer, with a corresponding diminution in the mass per unit length. Using the same mass increment for proflavine binding as above and the 0·37 power dependence on total length, the calculated net sedimentation coefficient at 0·44 proflavines per nucleotide pair is 0·89 of that for pure DNA. On the basis of sedimentation alone this model is indistinguishable from intercalation.

(d) Fiber diffraction patterns

Three main features may be discerned in the X-ray diffraction patterns of DNA-proflavine fibers: (a) the 3·4 Å meridional spots, which are prominent in the patterns of the B configuration of pure DNA, remain very strong in the complex and no new meridional reflexions appear. (b) The peak of the strong equatorial reflexion near the center, corresponding to the 110 reflexion of pure DNA, is found to lie slightly further from the center than the 110 reflexion of pure DNA B in its closest packing. (c) The characteristic layer-line pattern is abolished. Typical patterns of the proflavine complex and a comparably poorly oriented thick lithium DNA fiber are shown in Plates I and II.

The retention of the 3·4 Å spot as the sole meridional reflexion indicates that in the complex the purine and pyrimidine rings are perpendicular, or nearly so, to the helix axis. If the bases are replaced by intercalated acridines, the same 3·4 Å spacing obtains between the acridine and the bases above and below.

The spacing between DNA molecules, as indicated by the equatorial spots, appears to be incompatible with models based on binding of the acridines to the surface of the DNA helix. In fibers containing about one proflavine per three nucleotide pairs, the first equatorial spot indicates a spacing of 17·4 Å at relative humidity of 75% and 100%. For comparison, the closest spacing observed for pure DNA in the B configuration is 18·4 Å; the molecules are then in van der Waals contact through the penetration of the ridges of each helix in the grooves of four of its six neighbors (Langridge, Wilson, Hooper, Wilkins & Hamilton, 1960). The exterior attachment of one proflavine per two or three nucleotide pairs appears to just fill the large groove of the helix and would be expected, therefore, to prevent intermolecular separations closer than roughly 22 Å. Unlike pure DNA, which swells in both directions perpendicular to the c axis, the proflavine complex does not swell at higher humidities. This suggests that the 17·4 Å spacing is measured in the immediate vicinity of proflavine, rather than in regions from which it might be excluded.

In the C configuration of lithium DNA (see Langridge et al., 1960b), which is found at 44% relative humidity, the closest intermolecular spacing is 17·0 Å. The slight tilt of the bases which is responsible for the smaller helix diameter would not be detectable in the meridional spots of the complexes within the precision of the present work.

However, the same considerations given for the B form apply, and the exterior attachment of proflavine would be expected to increase the spacing by 3 Å.

The requirements for additional intermolecular space are absent, of course, from the intercalated model, and it will be seen below (Section (f)) that because of the increased average pitch, further interpenetration of the helices is possible.

The disappearance of the layer lines, while both the meridional and equatorial spots are retained, suggests a disordering of the helical backbone, as would be expected if proflavine were intercalated aperiodically. It is reasonable to suppose that insertion of the acridines reflects the nucleotide sequence to some extent; the intercalation would then appear crudely random. Where there is an average of one proflavine per three nucleotide pairs the occurrence of large regions of well-ordered structure as statistical fluctuations in the proflavine distribution will be exceedingly rare.

(e) Fiber dichroism

The maximum light absorption by proflavine in fibers of the complex occurs when the electric vector of the incident light is perpendicular to the fiber axis. It may be inferred (see Seeds, 1953) that the normal to the plane of the proflavine molecules lies less than 55° from the fiber axis. While the ease of visual observation suggests that the angle is considerably smaller, no quantitative estimate has been made.

(f) Extension of the DNA molecule

The possibility of extending a double helix to accommodate an intercalated acridine has been tested with brass rod atomic models. The following conditions were imposed

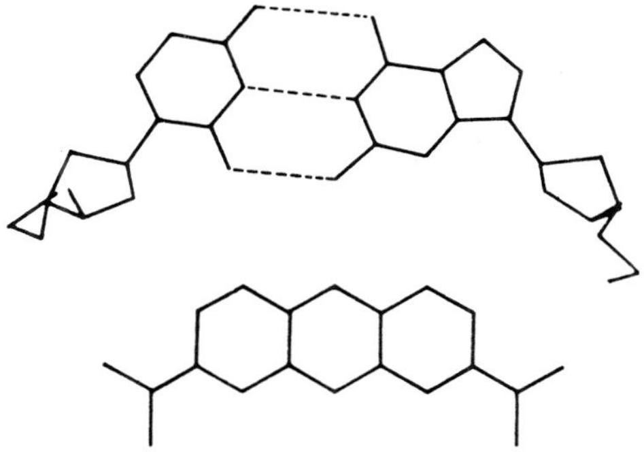

Fig. 4. A nucleotide pair at the site of intercalation, according to the model described in the text, projected on the plane perpendicular to the helix axis. A molecule of acridine orange is shown below to the same scale; hydrogen atoms are not indicated. The drawing shows the 5′ O—P—O attached to one base (the purine) and the 3′ O—P—O of the other as the extended sections of the backbone passing the intercalated acridine. The − 9° rotation of the next nucleotide pair would place it turned slightly clockwise above the pair shown, such that the ring oxygen of the left upper sugar is almost above the 1′ carbon of the lower. The most central obstructions to the placement of the intercalated molecule are the ring oxygens of the sugars of the right-hand nucleotide in the pair below and the left-hand nucleotide of the pair above.

arbitrarily: the purine-pyrimidine pairs should remain (1) at their normal, hydrogen-bonded configuration, (2) at their normal displacement from the helix axis, (3) perpendicular or nearly perpendicular to the helix axis, and (4) the backbone must continue demonstrably distortion-free to the next nucleotide above and below the separated pair at the normal internucleotide spacing.

Four bases of one chain were positioned with respect to the axis according to the co-ordinates of Langridge *et al.* (1960a) with vertical separations of 3·36, 6·72 and 3·36 Å, respectively, and a rotation of 36° between the normally spaced pairs. The deoxy-ribose-phosphate backbone, constructed with the bond angles selected for poly-adenylic acid (Crick, personal communication, 1960) could be attached with very nearly equivalent sugar orientations and without unacceptably short contacts when the rotation between the extended pair was − 9° (left-handed, opposite to the normal direction of helix rotation). Little or no further extension of the backbone was possible without bond distortion. One intercalation per two nucleotide pairs would then extend the pitch of the helix from 33·6 Å to an average of 135 Å. While a limited search failed to disclose any other acceptable configurations, it is not impossible that others are equally appropriate.

The aperture between the backbone chains provides ample space for the acridines when suitably oriented, and can also comfortably accommodate the broader carcino-genic acridine, dibenz[a, j]acridine, found by Booth & Boyland (1953) to form soluble complexes with DNA.

A projection of a nucleotide pair with the backbones in the extended configuration is shown in Fig. 4, together with the skeleton of acridine orange to the same scale.

(g) *A model for the viscosity effects*

A polymer is described as a "worm-like chain" if bending between monomer seg-ments is restricted to a small angle β, such that $\beta << \pi/2$ (see Tompa, 1956). Only for a fairly long string of segments can the direction of one end be considered to be randomly related to the direction of the other. The correlation between the directions of any two segments of the chain declines as the interval between them increases. The rate of decline is characterized by a length, a, known as the persistence length, given by the expression

$$a = t/(1 - \cos \beta) \tag{1}$$

if β and t are small. The mean-square end-to-end length $\langle l^2 \rangle$ is then

$$\langle l^2 \rangle = 2a(L - a + ae^{-L/a}) \tag{2}$$

where L is the total contour length of the polymer. For some purposes the properties of a complex chain model can be approximated by an "equivalent random chain" selected to have the same contour length and the same mean square end-to-end length. For the worm-like chain a fit of this kind is achieved when the length of the links of the equivalent random chain is made equal to twice the persistence length.

For a DNA molecule consisting of r nucleotide pairs, the persistence length is given approximately by equation (1), and can be inserted in equation (2) so that

$$\langle l^2 \rangle_0 = \frac{2rt^2}{1 - \cos \beta} \tag{3}$$

using only the first term, since L is much larger than a. Here t is the thickness of a nucleotide pair.

If n additional molecules with the same thickness, t, are intercalated into the chain, the new contour length is now

$$L_n = (r + n)t \tag{4}$$

If it is supposed that the insertion of the new molecule abolishes the flexibility of the backbone between the two nucleotide pairs that it separates, the number of flexible links will be diminished and there will be a larger effective mean link length, $\bar{t}$. This will be the arithmetic mean of n units of length $3t$ where a molecule is intercalated, and the $(r - 2n)$ units of length t;

$$\bar{t} = (r + n)t/(r - n) \tag{5}$$

A new persistence length can be calculated from $\bar{t}$ by means of equation (1) and entered in the first term of equation (2), together with the new contour length, to determine the mean square end-to-end length of the polymer containing n intercalated layers. Thus

$$\langle l^2 \rangle_n = \frac{2t^2(r + n)^2}{(1 - \cos \beta)(r - n)} \tag{6}$$

Compared with the original polymer

$$\frac{\langle l^2 \rangle_n}{\langle l^2 \rangle_0} = \frac{(1 + n/r)^2}{1 - n/r} \tag{7}$$

Since the intrinsic viscosity of a free-draining random coil (the equivalent random chain) will be proportional to $\langle l^2 \rangle$, equation (7) represents also, approximately, the viscosity enhancement at infinite dilution due to the formation of the intercalated complex.

For the limit of strong binding in herring sperm DNA, where $n/r = 0.44$ (Peacocke & Skerrett, 1956), the viscosity enhancement calculated from equation (7) is 3·7-fold; where 90% of the sites are occupied, the calculated enhancement factor is 3·2.

In the calf thymus DNA experiments, numbers 20 and 22, where the observed enhancement factors (not extrapolated) are 3·2 and 3·3, the fractional occupancy is estimated (from Peacocke & Skerrett's data on herring sperm DNA) to be near 0·95. The lower proflavine concentration in the experiments shown in Fig. 3, with chicken erythrocyte DNA, allows the estimated occupancy to vary from 0·60 at the highest DNA concentration to 0·95 at the extrapolated limit. Here, the observed limiting enhancement factor is 2·9.

The assumption of the stiffness of the link containing the intercalated layer is critical to the expectation of large viscosity changes. The assumption is supported by the impossibility of further extension of the backbone in the atomic model of the structure, and by consideration of the local electrostatic interactions. It may be noted that when $n/r = 0.44$, the viscosity enhancement factor due only to the increment in contour length is 1·44, while the factor due to diminished curvature alone is 2·57. The latter would be reduced if flexibility is allowed around the intercalated layers.

(h) *The binding of dye aggregates*

It was shown by Peacocke & Skerrett (1956) that proflavine combines with DNA in at least two different ways. Regarding the DNA concentration as fixed, there is a strong, first order combination which saturates at one dye bound per four or five nucleotides, and a much weaker, higher order process with a limit of one dye per nucleotide. At the highest proflavine concentration of the present experiments, a small part would be expected to be bound by the second process.

The higher order process, which has been interpreted as the binding of dye aggregates, has been studied by Bradley & Wolf (1959) and Bradley & Felsenfeld (1959) with respect to other polyanions as well as DNA. Since aggregates cannot be accommodated by intercalation, and external binding is incompatible with the observed physical changes, intercalation is presumably to be identified with the mode of strong binding.

(i) *Structure of the complex*

The conclusions with respect to the mode of binding can be summarized by reference to the molecular hypotheses that have been considered and the inferences from the observed physical properties. The experimental results indicate that:

(i) The mass per unit length is diminished.

(ii) The molecules in fibers of the complex are more closely packed than DNA *B*.

(iii) The 3·4 Å spacing along the helix is retained.

(iv) The molecule is straighter in solution, in opposition to the electrostatic effect. Structurally dissimilar substances fail to elicit the straightening.

(v) The regular helical structure is lost.

It can be supposed that an acridine binds according to one of the following schemes:

(i) surface attachment to DNA in the *B* configuration.

(ii) surface attachment to a stretched form of DNA.

(iii) insertion into a DNA helix by separation and displacement of a base pair.

(iv) intercalation into the helix by extension of the backbone.

Of these, only the last is fully satisfactory.

(j) *Interactions* in vivo

Acridine, proflavine and acridine orange, as well as a variety of other amino-substituted acridines, have been shown to be strongly mutagenic in inducing reversion to wild type among a selected set of *r*II mutants of the bacteriophage T4 (Orgel & Brenner, in preparation), yielding reversion rates as much as 10^3 to 10^4 times greater than the spontaneous rates.

It is tempting to consider the well-known requirements of planarity for the carcinogenic activity of polycyclic aromatic hydrocarbons (see Greenstein, 1954) in terms of an intercalated complex with DNA. The formation of water-soluble complexes between certain polycyclic hydrocarbons and purines is second order in purine concentration (Weil-Malherbe, 1946) also suggesting a sandwich configuration.

I should like to thank Dr. F. H. C. Crick and Dr. J. R. Cann for their helpful discussions. I am indebted to Mrs. Catherine Fogg and Miss Christine Froud for assistance in viscometry, and to Mr. Stanley Fisher for operation of the ultracentrifuge.

REFERENCES

Booth, J. & Boyland, E. (1953). *Biochim. biophys. Acta*, **12**, 75

Bradley, D. F. & Felsenfeld, G. (1959). *Nature*, **184**, 1920.

Bradley, D. F. & Wolf, M. K. (1959). *Proc. Nat. Acad. Sci., Wash.* **45**, 944.

Burton, K. (1956). *Biochem. J.* **52**, 315.

Cavalieri, L. F., Rosoff, M. & Rosenberg, B. H. (1956). *J. Amer. Chem. Soc.* **78**, 5239.

Chargaff, E. (1955). In *The Nucleic Acids*, ed. by E. Chargaff & J. N. Davidson, Vol. I. p. 307. New York: Academic Press.

Doty, P., McGill, B. B. & Rice, S. A. (1958). *Proc. Nat. Acad. Sci., Wash.* **44**, 432.

Fresco, J. R. & Alberts, B. M. (1960). *Proc. Nat. Acad. Sci., Wash.* **46**, 311.

Greenstein, J. P. (1954). In *Biochemistry of Cancer*. New York: Academic Press.

Langridge, R., Marvin, D. A., Seeds, W. E., Wilson, H. R., Hooper, C. W., Wilkins, M. H. F. & Hamilton, L. D. (1960a). *J. Mol. Biol.* **2**, 38.

Langridge, R., Wilson, H. R., Hooper, C. W., Wilkins, M. H. F. & Hamilton, L. D. (1960b). *J. Mol. Biol.* **2**, 19.

Peacocke, A. R. & Skerrett, J. N. H. (1956). *Trans. Faraday Soc.* **52**, 261.

Schachman, H. K. (1959). In *Ultracentrifugation in Biochemistry*. New York: Academic Press.

Seeds, W. E. (1953). In *Progress in Biophysics and Biophysical Chemistry*, ed. by J. A. V. Butler & J. T. Randall, Vol. 3, p. 27. London: Pergamon Press.

Tompa, H. (1956). In *Polymer Solutions*. London: Butterworths.

Weil-Malherbe, H. (1946). *Biochem. J.* **40**, 351.

Wilkins, M. H. F., Gosling, R. G. & Seeds, W. E. (1951). *Nature*, **167**, 759.

J. Mol. Biol. (1961) **3**, 121–124

The Theory of Mutagenesis

In this preliminary note we wish to express our doubts about the detailed theory of mutagenesis put forward by Freese (1959*b*), and to suggest an alternative.

Freese (1959*b*) has produced evidence that shows that for the r_{II} locus of phage T4 there are two mutually exclusive classes of mutation and we have confirmed and extended his work (Orgel & Brenner, in manuscript). The technique used is to start with a standard wild type and make a series of mutants from it with a particular mutagen. Each mutant is then tested with various mutagens to see which of them will back-mutate it to wild type.

It is found that the mutations fall into two classes. The first, which we shall call the base analogue class, is typically produced by 5-bromodeoxyuridine (BD) and the second, which we shall call the acridine class, is typically produced by proflavin (PF). In general a mutant made with BD can be reverted by BD, and a mutant made with PF can be reverted by PF. A few of the PF mutants do not appear to revert with either mutagen, but the strong result is that no mutant has been found which reverts identically with both classes of mutagens, and that (with a few possible exceptions) mutants produced by one class cannot be reverted by the other.

Freese also showed that 2-aminopurine falls into the base analogue class, and that most (85%) spontaneous mutants at the r_{II} locus were not of the base analogue type. We have confirmed this and shown that they are in fact revertible by acridines. We have also shown that a number of other acridines, and in particular 5-aminoacridine, act like proflavin (Orgel & Brenner, in manuscript).

Freese has produced an ingenious explanation of these results, which should be consulted in the original for fuller details. In brief he postulated that the base analogue class of mutagens act by altering an A—T base-pair on the DNA (A = adenine, T = thymine) into a G—C pair, or *vice versa* (G = guanine, C = cytosine, or, in the T even phages, hydroxymethylcytosine). The fact that BD, which replaces thymine, could act both ways (from A—T to G—C or from G—C to A—T) was accounted for (Freese, 1959*a*) by assuming that in the latter case there was an error in pairing of the BD (such that it accidentally paired with guanine) while *entering* the DNA, and in the former case after it was already in the DNA.

Such alterations only change a purine into another purine, or a pyrimidine into another pyrimidine. Freese (1959*b*) has called these "transitions." He suggested that other conceivable changes, which he called "transversions" (such as, for example, from A—T to C—G) which change a purine into a pyrimidine and *vice versa*, occurred during mutagenesis by proflavin. This would neatly account for the two mutually exclusive classes of mutagens, since it is easy to see that a transition cannot be reversed by a transversion, and *vice versa*.

We have been led to doubt this explanation for the following reasons.

Our suspicions were first aroused by the curious fact that a comparison between the *sites* of mutation for one set of mutants made with BD and another set made with PF (Brenner, Benzer & Barnett, 1958) showed there were no sites in the r_{II} gene, among the samples studied, common to both groups.

Now this result alone need not be incompatible with Freese's theory of mutagenesis, since we have no good explanation for "hot spots" and this confuses quantitative argument. However it led us to the following hypothesis:

> that acridines act as mutagens because they cause the insertion or the deletion of a base-pair.

This idea springs rather naturally from the views of Lerman (1960) and Luzzati (in preparation) that acridines are bound to DNA by sliding *between* adjacent base-pairs, thus forcing them 6·8 Å apart, rather than 3·4 Å. If this occasionally happened between the bases on *one* chain of the DNA, but not the other, during replication, it might easily lead to the addition or subtraction of a base.

Such a possible mechanism leads to a prediction. We know practically nothing about coding (Crick, 1959) but on most theories (except overlapping codes which are discredited because of criticism by Brenner (1957)) the deletion or the addition of a base-pair is likely to cause not the substitution of just one amino acid for another, but a much more substantial alteration, such as a break in the polypeptide chain, a considerable alteration of the amino acid sequence, or the production of no protein at all.

Thus one would not be surprised to find on these ideas that mutants produced by acridines were not capable of producing a slightly modified protein, but usually produced either no protein at all or a grossly altered one.

Somewhat to our surprise we find we already have data from two separate genes supporting this hypothesis.

(1) The *o* locus of phage T4 (resistance to osmotic shock) is believed to control a protein of the finished phage, possibly the head protein, because it shows phenotypic mixing (Brenner, unpublished). Using various base analogues we have produced mutants of this gene, though these map at only a small number of sites. We have failed on several occasions to produce any *o* mutants with proflavin. On another occasion two mutants were produced; one never reverted to wild type, while the other corresponded in position and spontaneous reversion rate to a base analogue site. We suspect therefore that these two mutants were not really produced by proflavin, but were the rarer sort of spontaneous mutant (Brenner & Barnett, unpublished).

(2) We have also studied mutation at the *h* locus in T2L, which controls a protein of the finished phage concerned with attachment to the host (Streisinger & Franklin, 1956).

Of the six different spontaneous h^+ mutants tested, all were easily induced to revert to *h* with 5-bromouracil (BU)†. This is especially significant when it is recalled that 85% of the spontaneous r_{II} mutants could not be reverted with base analogues (Freese, 1959*b*).

We have also shown (Brenner & Barnett, unpublished) that it is difficult to produce h^+ mutants from *h* by proflavin, though relatively easy with BU. The production of *r* mutants was used as a control.

It can be seen from Table 1 that if the production of h^+ mutants by BU and proflavin were similar to the production of *r* mutants we would expect to have obtained $\frac{57 \times 26}{108} = 13h^+$ mutants with proflavin, whereas in fact we only found 1, and this may be spontaneous background.

---

† (Added in proof.) Five of these have now been tested and have been shown not to revert with proflavin.

Let us underline the difference between the *r* loci and the *o* and *h* loci. The former appear to produce proteins which are probably *not* part of the finished phage. For both the *o* and the *h* locus, however, the protein concerned forms part of the finished phage, which presumably would not be viable without it, so that a mutant can be picked up only if it forms an *altered* protein. A mutant which deleted the protein could not be studied.

TABLE 1

	r	h^+
BU	108	57
Proflavin	26	1

It is clear that further work must be done before our generalization—that acridine mutants usually give no protein, rather than a slightly modified one—can be accepted. But if it turns out to be true it would support our hypothesis of the mutagenic action of the acridines, and this may have serious consequences for the naïve theory of mutagenesis, for the following reason.

It has always been a theoretical possibility that the reversions to wild type were not true reversions but were due to the action of "suppressors" (within the gene), possibly very closely linked suppressors. The most telling evidence against this was the existence of the two mutually exclusive classes of mutagens, together with Freese's explanation.

For clearly if the forward mutation could be made at one base-pair and the reverse one at a different base-pair, we should expect, on Freese's hypothesis, exceptions to the rule about the two classes of mutagens. Since these were not found it was concluded that even close suppressors were very rare.

Unfortunately our new hypothesis for the action of acridines destroys this argument. Under this new theory an alteration of a base-pair at one place *could* be reversed by an alteration at a different base-pair, and indeed from what we know (or guess) of the structure of proteins and the dependence of structure on amino acid sequence, we should be surprised if this did not occur.

It is all too easy to conceive, for example, that at a certain point on the polypeptide chain at which there is a glutamic residue in the wild type, and at which the mutation substituted a proline, a further mutation might alter the proline to aspartic acid and that this might appear to restore the wild phenotype, at least as far as could be judged by the rather crude biological tests available. If several base-pairs are needed to code for one amino acid the reverse mutation might occur at a base-pair close to but not identical with the one originally changed.

On our hypothesis this could happen, and yet one would still obtain the two classes of mutagens. The one, typified by base analogues, would produce the substitution of one base for another, and the other, typically produced by acridines, would lead to the addition or subtraction of a base-pair. Consequently the mutants produced by one class could not be easily reversed by the mutagens of the other class.

Thus our new hypothesis reopens in an acute form the question: which back-mutations to wild type are truly to the original wild type, and which only appear to be so? And on the answers to this question depend our interpretation of all experiments on back-mutation.

We suspect that this problem can most easily be approached by work on systems for which the amino acid sequence of the protein can be studied, such as the phage lysozyme of Dreyer, Anfinsen & Streisinger (personal communications) or the phosphatase from *E. coli* of Levinthal, Garen & Rothman (Garen, 1960). Meanwhile we are continuing our genetic studies to fill out and extend the preliminary results reported here.

Medical Research Council Unit
for Molecular Biology
Cavendish Laboratory

Pathology Laboratory
both of Cambridge University
England

S. BRENNER
LESLIE BARNETT
F. H. C. CRICK

ALICE ORGEL

Received 16 December 1960

REFERENCES

Brenner, S. (1957). *Proc. Nat. Acad. Sci., Wash.* **43**, 687.
Brenner, S., Benzer, S. & Barnett, L. (1958). *Nature*, **182**, 983.
Crick, F. H. C. (1959). In *Brookhaven Symposia in Biology*, **12**, 35.
Freese, E. (1959a). *J. Mol. Biol.* **1**, 87.
Freese, E. (1959b). *Proc. Nat. Acad. Sci., Wash.* **45**, 622.
Garen, A. (1960). 10th Symposium *Soc. Gen. Microbiol.*, London, 239.
Lerman, L. (1961). *J. Mol. Biol.* **3**, 18.
Streisinger, G. & Franklin, N. C. (1956). In *Cold Spr. Harb. Sym. Quant. Biol.* **21**, 103.

J. Mol. Biol. (1961) **3**, 318–356

REVIEW ARTICLE

Genetic Regulatory Mechanisms in the Synthesis of Proteins †

FRANÇOIS JACOB AND JACQUES MONOD

Services de Génétique Microbienne et de Biochimie Cellulaire,
Institut Pasteur, Paris

(*Received 28 December 1960*)

The synthesis of enzymes in bacteria follows a double genetic control. The so-called structural genes determine the molecular organization of the proteins. Other, functionally specialized, genetic determinants, called regulator and operator genes, control the rate of protein synthesis through the intermediacy of cytoplasmic components or repressors. The repressors can be either inactivated (induction) or activated (repression) by certain specific metabolites. This system of regulation appears to operate directly at the level of the synthesis by the gene of a short-lived intermediate, or messenger, which becomes associated with the ribosomes where protein synthesis takes place.

1. Introduction

According to its most widely accepted modern connotation, the word "gene" designates a DNA molecule whose specific self-replicating structure can, through mechanisms unknown, become translated into the specific structure of a polypeptide chain.

This concept of the "structural gene" accounts for the multiplicity, specificity and genetic stability of protein structures, and it implies that such structures are not controlled by environmental conditions or agents. It has been known for a long time, however, that the synthesis of individual proteins may be provoked or suppressed within a cell, under the influence of specific external agents, and more generally that the relative rates at which different proteins are synthesized may be profoundly altered, depending on external conditions. Moreover, it is evident from the study of many such effects that their operation is absolutely essential to the survival of the cell.

It has been suggested in the past that these effects might result from, and testify to, complementary contributions of genes on the one hand, and some chemical factors on the other in determining the final structure of proteins. This view, which contradicts at least partially the "structural gene" hypothesis, has found as yet no experimental support, and in the present paper we shall have occasion to consider briefly some of this negative evidence. Taking, at least provisionally, the structural gene hypothesis in its strictest form, let us assume that the DNA message contained within a gene is both necessary and sufficient to define the structure of a protein. The elective effects of agents other than the structural gene itself in promoting or suppressing the synthesis of a protein must then be described as operations which control the rate of transfer of structural information from gene to protein. Since it seems to be established

† This work has been aided by grants from the National Science Foundation, the Jane Coffin Childs Memorial Fund for Medical Research and the Commissariat à l'Energie Atomique.

that proteins are synthesized in the cytoplasm, rather than directly at the genetic level, this transfer of structural information must involve a chemical intermediate synthesized by the genes. This hypothetical intermediate we shall call the structural messenger. The rate of information transfer, i.e. of protein synthesis, may then depend either upon the activity of the gene in synthesizing the messenger, or upon the activity of the messenger in synthesizing the protein. This simple picture helps to state the two problems with which we shall be concerned in the present paper. If a given agent specifically alters, positively or negatively, the rate of synthesis of a protein, we must ask:

(a) Whether the agent acts at the cytoplasmic level, by controlling the activity of the messenger, or at the genetic level, by controlling the synthesis of the messenger.

(b) Whether the specificity of the effect depends upon some feature of the information transferred from structural gene to protein, or upon some specialized controlling element, not represented in the structure of the protein, gene or messenger.

The first question is easy to state, if difficult to answer. The second may not appear so straightforward. It may be stated in a more general way, by asking whether the genome is composed exclusively of structural genes, or whether it also involves determinants which may control the rates of synthesis of proteins according to a given set of conditions, without determining the structure of any individual protein. Again it may not be evident that these two statements are equivalent. We hope to make their meaning clear and to show that they are indeed equivalent, when we consider experimental examples.

The best defined systems wherein the synthesis of a protein is seen to be controlled by specific agents are examples of enzymatic adaptation, this term being taken here to cover both enzyme induction, i.e. the formation of enzyme electively provoked by a substrate, and enzyme repression, i.e. the specific inhibition of enzyme formation brought about by a metabolite. Only a few inducible and repressible systems have been identified both biochemically and genetically to an extent which allows discussion of the questions in which we are interested here. In attempting to generalize, we will have to extrapolate from these few systems. Such generalization is greatly encouraged, however, by the fact that lysogenic systems, where phage protein synthesis might be presumed to obey entirely different rules, turn out to be analysable in closely similar terms. We shall therefore consider in succession certain inducible and repressible enzyme systems and lysogenic systems.

It might be best to state at the outset some of the main conclusions which we shall arrive at. These are:

(a) That the mechanisms of control in all these systems are negative, in the sense that they operate by inhibition rather than activation of protein synthesis.

(b) That in addition to the classical structural genes, these systems involve two other types of genetic determinants (regulator and operator) fulfilling specific functions in the control mechanisms.

(c) That the control mechanisms operate at the genetic level, i.e. by regulating the activity of structural genes.

2. Inducible and Repressible Enzyme Systems

(a) *The phenomenon of enzyme induction. General remarks*

It has been known for over 60 years (Duclaux, 1899; Dienert, 1900; Went, 1901) that certain enzymes of micro-organisms are formed only in the presence of their specific substrate. This effect, later named "enzymatic adaptation" by Karstrom (1938), has been the subject of a great deal of experimentation and speculation. For a long time, "enzymatic adaptation" was not clearly distinguished from the selection of spontaneous variants in growing populations, or it was suggested that enzymatic adaptation and selection represented *alternative* mechanisms for the acquisition of a "new" enzymatic property. Not until 1946 were adaptive enzyme systems shown to be controlled in bacteria by discrete, specific, stable, i.e. genetic, determinants (Monod & Audureau, 1946). A large number of inducible systems has been discovered and studied in bacteria. In fact, enzymes which attack exogeneous substrates are, as a general rule, inducible in these organisms. The phenomenon is far more difficult to study in tissues or cells of higher organisms, but its existence has been established quite clearly in many instances. Very often, if not again as a rule, the presence of a substrate induces the formation not of a single but of several enzymes, sequentially involved in its metabolism (Stanier, 1951).

Most of the fundamental characteristics of the induction effect have been established in the study of the "lactose" system of *Escherichia coli* (Monod & Cohn, 1952; Cohn, 1957; Monod, 1959) and may be summarized in a brief discussion of this system from the biochemical and physiological point of view. We shall return later to the genetic analysis of this system.

(b) *The lactose system of* Escherichia coli

Lactose and other β-galactosides are metabolized in *E. coli* (and certain other enteric bacteria) by the hydrolytic transglucosylase β-galactosidase. This enzyme was isolated from *E. coli* and later crystallized. Its specificity, activation by ions and transglucosylase *vs* hydrolase activity have been studied in great detail (*cf.* Cohn, 1957). We need only mention the properties that are significant for the present discussion. The enzyme is active exclusively on β-galactosides unsubstituted on the galactose ring. Activity and affinity are influenced by the nature of the aglycone moiety both being maximum when this radical is a relatively large, hydrophobic group. Substitution of sulfur for oxygen in the galactosidic linkage of the substrate abolishes hydrolytic activity completely, but the thiogalactosides retain about the same affinity for the enzyme site as the homologous oxygen compounds.

As isolated by present methods, β-galactosidase appears to form various polymers (mostly hexamers) of a fundamental unit with a molecular weight of 135,000. There is one end group (threonine) and also one enzyme site (as determined by equilibrium dialysis against thiogalactosides) per unit. It is uncertain whether the monomer is active as such, or exists *in vivo*. The hexameric molecule has a turnover number of 240,000 mol $\times$ min^{-1} at 28°C, pH 7·0 with *o*-nitrophenyl-β-D-galactoside as substrate and Na$^+$ (0·01 M) as activator.

There seems to exist only a single homogeneous β-galactosidase in *E. coli*, and this organism apparently cannot form any other enzyme capable of metabolizing lactose, as indicated by the fact that mutants that have lost β-galactosidase activity cannot grow on lactose as sole carbon source.

However, the possession of β-galactosidase activity is not sufficient to allow utilization of lactose by *intact E. coli* cells. Another component, distinct from β-galactosidase, is required to allow penetration of the substrate into the cell (Monod, 1956; Rickenberg, Cohen, Buttin & Monod, 1956; Cohen & Monod, 1957; Pardee, 1957; Képès, 1960). The presence and activity of this component is determined by measuring the rate of entry and/or the level of accumulation of radioactive thiogalactosides into intact cells. Analysis of this active permeation process shows that it obeys classical enzyme kinetics allowing determination of K_m and V_{max}. The specificity is high since the system is active only with galactosides (β or α), or thiogalactosides. The spectrum of apparent affinities ($1/K_m$) is very different from that of β-galactosidase. Since the permeation system, like β-galactosidase, is inducible (see below) its formation can be studied *in vivo*, and shown to be invariably associated with protein synthesis. By these criteria, there appears to be little doubt that this specific permeation system involves a specific protein (or proteins), formed upon induction, which has been called galactoside-permease. That this protein is distinct from and independent of β-galacto-sidase is shown by the fact that mutants that have lost β-galactosidase retain the capacity to concentrate galactosides, while mutants that have lost this capacity retain the power to synthesize galactosidase. The latter mutants (called cryptic) cannot however use lactose, since the intracellular galactosidase is apparently accessible exclusively *via* the specific permeation system.

Until quite recently, it had not proved possible to identify *in vitro* the inducible protein (or proteins) presumably responsible for galactoside-permease activity. During the past year, a protein characterized by the ability to carry out the reaction:

Ac. Coenzyme A + Thiogalactoside → 6-Acetylthiogalactoside + Coenzyme A

has been identified, and extensively purified from extracts of *E. coli* grown in presence of galactosides (Zabin, Képès & Monod, 1959). The function of this enzyme in the system is far from clear, since formation of a free covalent acetyl-compound is almost certainly not involved in the permeation process *in vivo*. On the other hand:

(a) mutants that have lost β-galactosidase and retained galactoside-permease, retain galactoside-acetylase;

(b) most mutants that have lost permease cannot form acetylase;

(c) permeaseless acetylaseless mutants which revert to the permease-positive condition simultaneously regain the ability to form acetylase.

These correlations strongly suggest that galactoside-acetylase is somehow involved in the permeation process, although its function *in vivo* is obscure, and it seems almost certain that other proteins (specific or not for this system) are involved. In any case, we are interested here not in the mechanisms of permeation, but in the control mechanisms which operate with β-galactosidase, galactoside-permease and galactoside-acetylase. The important point therefore is that, as we shall see, galactoside-acetylase invariably obeys the same controls as galactosidase.†

(c) *Enzyme induction and protein synthesis*

Wild type *E. coli* cells grown in the absence of a galactoside contain about 1 to 10 units of galactosidase per mg dry weight, that is, an average of 0·5 to 5 active molecules

† For reasons which will become apparent later it is important to consider whether there is any justification for the assumption that galactosidase and acetylase activities might be associated with the same fundamental protein unit. We should therefore point to the following observations:

(a) There are mutants which form galactosidase and no acetylase, and *vice versa*.

(b) Purified acetylase is devoid of any detectable galactosidase activity.

(c) The specificity of the two enzymes is very different.

(d) The two enzymes are easily and completely separated by fractional precipitation.

(e) Acetylase is highly heat-resistant, under conditions where galactosidase is very labile.

(f) Anti-galactosidase serum does not precipitate acetylase; nor does anti-acetylase serum precipitate galactosidase.

There is therefore no ground for the contention that galactosidase and acetylase activities are associated with the same protein.

per cell or 0·15 to 1·5 molecules per nucleus. Bacteria grown in the presence of a suitable inducer contain an average of 10,000 units per mg dry weight. This is the induction effect.

A primary problem, to which much experimental work has been devoted, is whether this considerable increase in specific activity corresponds to the synthesis of entirely "new" enzyme molecules, or to the activation or conversion of pre-existing protein precursors. It has been established by a combination of immunological and isotopic methods that the enzyme formed upon induction:

(a) is distinct, as an antigen, from all the proteins present in uninduced cells (Cohn & Torriani, 1952);

(b) does not derive any significant fraction of its sulfur (Monod & Cohn, 1953; Hogness, Cohn & Monod, 1955) or carbon (Rotman & Spiegelman, 1954) from pre-existing proteins.

The inducer, therefore, brings about the complete *de novo* synthesis of enzyme molecules which are new by their specific structure as well as by the origin of their elements. The study of several other induced systems has fully confirmed this conclusion, which may by now be considered as part of the *definition* of the effect. We will use the term "induction" here as meaning "activation by inducer of enzyme-protein synthesis."

(d) *Kinetics of induction*

Accepting (still provisionally) the structural gene hypothesis, we may therefore consider that the inducer somehow accelerates the rate of information transfer from gene to protein. This it could do either by provoking the synthesis of the messenger or by activating the messenger. If the messenger were a *stable* structure, functioning as a catalytic template in protein synthesis, one would expect different kinetics of induction, depending on whether the inducer acted at the genetic or at the cytoplasmic level.

The kinetics of galactosidase induction turn out to be remarkably simple when determined under proper experimental conditions (Monod, Pappenheimer & Cohen-Bazire, 1952; Herzenberg, 1959). Upon addition of a suitable inducer to a growing culture, enzyme activity increases at a rate proportional to the increase in total protein within the culture; i.e. a linear relation is obtained (Fig. 1) when total enzyme activity is plotted against mass of the culture. The slope of this line:

$$P = \frac{\Delta z}{\Delta M}$$

is the "differential rate of synthesis," which is taken by definition as the measure of the effect. Extrapolation to the origin indicates that enzyme formation begins about three minutes (at 37 °C) after addition of inducer (Pardee & Prestidge, 1961). Removal of the inducer (or addition of a specific anti-inducer, see below) results in cessation of enzyme synthesis within the same short time. The differential rate of synthesis varies with the concentration of inducer reaching a different saturation value for different inducers. The inducer therefore acts in a manner which is (kinetically) similar to that of a dissociable activator in an enzyme system: activation and inactivation follow very rapidly upon addition or removal of the activator.

The conclusion which can be drawn from these kinetics is a negative one: the inducer does not appear to activate the synthesis of a stable intermediate able to accumulate in the cell (Monod, 1956).

Similar kinetics of induction have been observed with most or all other systems which have been adequately studied (Halvorson, 1960) with the exception of penicillinase of *Bacillus cereus*. The well-known work of Pollock has shown that the synthesis of this enzyme continues for a long time, at a decreasing rate, after removal of inducer (penicillin) from the medium. This effect is apparently related to the fact that minute amounts of penicillin are retained irreversibly by the cells after transient exposure to the drug (Pollock, 1950). The unique behavior of this system therefore does not contradict the rule that induced synthesis stops when the inducer is removed from the cells. Using this system, Pollock & Perret (1951) were able to show that the inducer acts catalytically, in the sense that a cell may synthesize many more enzyme molecules than it has retained inducer molecules.

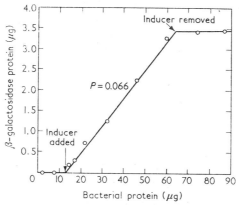

FIG. 1. Kinetics of induced enzyme synthesis. Differential plot expressing accumulation of β-galactosidase as a function of increase of mass of cells in a growing culture of *E. coli*. Since abscissa and ordinates are expressed in the same units (micrograms of protein) the slope of the straight line gives galactosidase as the fraction (P) of total protein synthesized in the presence of inducer. (After Cohn, 1957.)

(e) *Specificity of induction*

One of the most conspicuous features of the induction effect is its extreme specificity. As a general rule, only the substrate of an enzyme, or substances very closely allied to the normal substrate, are endowed with inducer activity towards this enzyme. This evidently suggests that a correlation between the molecular structure of the inducer and the structure of the catalytic center on the enzymes is *inherently* involved in the mechanism of induction. Two main types of hypotheses have been proposed to account for this correlation, and thereby for the mechanism of action of the inducer:

(a) The inducer serves as "partial template" in enzyme synthesis, molding as it were the catalytic center.

(b) The inducer acts by combining specifically with preformed enzyme (or "preenzyme"), thereby somehow accelerating the synthesis of further enzyme molecules.

It is not necessary to discuss these "classical" hypotheses in detail, because it seems to be established now that the correlation in question is in fact *not* inherent to the mechanism of induction.

Table 1 lists a number of compounds tested as inducers of galactosidase, and as substrates (or specific inhibitors) of the enzyme. It will be noted that:

(a) no compound that does not possess an intact unsubstituted galactosidic residue induces;

TABLE 1

Induction of galactosidase and galactoside-transacetylase by various galactosides

Compound	Concentrations	β-galactosidase			Galactoside-transacetylase	
		Induction value	V	$1/K_m$	Induction value	V/K_m
β-D-thiogalactosides						
(*iso*propyl)	10^{-4} M	100	0	140	100	80
(methyl)	10^{-4} M	78	0	7	74	30
(methyl)	10^{-5} M	7·5	—	—	10	—
(phenyl)	10^{-3} M	<0·1	0	100	<1	100
(phenylethyl)	10^{-3} M	5	0	10,000	3	—
β-D-galactosides						
(lactose)	10^{-3} M	17	30	14	12	35
(phenyl)	10^{-3} M	15	100	100	11	—
α-D-galactoside						
(melibiose)	10^{-3} M	35	0	<0·1	37	<1
β-D-glucoside						
(phenyl)	10^{-3} M	<0·1	0	0	<1	50
(galactose)	10^{-3} M	<0·1	—	4	<1	<1
Methyl-β-D-thiogalactoside (10^{-4} M) + phenyl-β-D-thiogalactoside (10^{-3} M)		52	—	—	63	—

Columns "induction value" refer to specific activities developed by cultures of wild type *E. coli* K12 grown on glycerol as carbon source with each galactoside added at molar concentration stated. Values are given in percent of values obtained with *iso*propyl-thiogalactoside at 10^{-4} M (for which actual units were about 7,500 units of β-galactosidase and 300 units of galactoside-transacetylase per mg of bacteria). Column V refers to maximal substrate activity of each compound with respect to galactosidase. Values are given in percent of activity obtained with phenyl-galactoside. Column $1/K_m$ expresses affinity of each compound with respect to galactosidase. Values are given in percent of that observed with phenylgalactoside. In case of galactoside-transacetylase, only the relative values V/K_m are given since low affinity of this enzyme prevents independent determination of the constants. (Computed from Monod & Cohn, 1952; Monod *et al.*, 1952; Buttin, 1956; Zabin *et al.*, 1959; Képès *et al.*, unpublished results.)

(b) many compounds which are not substrates (such as the thiogalactosides) are excellent inducers (for instance *iso*propyl thiogalactoside);

(c) there is no correlation between affinity for the enzyme and capacity to induce (*cf.* thiophenylgalactoside and melibiose).

The possibility that the enzyme formed in response to different inducers may have somewhat different specific properties should also be considered, and has been rather thoroughly tested, with entirely negative results (Monod & Cohn, 1952).

There is therefore no quantitative correlation whatever between inducing capacity and the substrate activity or affinity parameters of the various galactosides tested. The fact remains, however, that only galactosides will induce galactosidase, whose binding site is complementary for the galactose ring-structure. The possibility that this correlation is a necessary requisite, or consequence, of the induction mechanism was therefore not completely excluded by the former results.

As we shall see later, certain mutants of the galactosidase structural gene (*z*) have been found to synthesize, in place of the normal enzyme, a protein which is identical to it by its immunological properties, while being completely devoid of any enzymatic activity. When tested by equilibrium dialysis, this inactive protein proved to have no measurable affinity for galactosides. In other words, it has lost the specific binding site. In diploids carrying both the normal and the mutated gene, both normal galactosidase and the inactive protein are formed, to a quantitatively similar extent, in the presence of different concentrations of inducer (Perrin, Jacob & Monod, 1960).

This finding, added to the sum of the preceding observations, appears to prove beyond reasonable doubt that the mechanism of induction does not imply any inherent correlation between the molecular structure of the inducer and the structure of the binding site of the enzyme.

On the other hand, there is complete correlation in the induction of galactosidase and acetylase. This is illustrated by Table 1 which shows not only that the same compounds are active or inactive as inducers of either enzyme, but that the relative amounts of galactosidase and acetylase synthesized in the presence of different inducers or at different concentrations of the same inducer are constant, even though the absolute amounts vary greatly. The remarkable qualitative and quantitative correlation in the induction of these two widely different enzyme proteins strongly suggests that the synthesis of both is directly governed by a common controlling element with which the inducer interacts. This interaction must, at some point, involve stereospecific binding of the inducer, since induction is sterically specific, and since certain galactosides which are devoid of any inducing activity act as competitive inhibitors of induction in the presence of active inducers (Monod, 1956; Herzenberg, 1959). This suggests that an enzyme, or some other protein, distinct from either galactosidase or acetylase, acts as "receptor" of the inducer. We shall return later to the difficult problem raised by the identification of this "induction receptor."

(f) *Enzyme repression*

While positive enzymatic adaptation, i.e. induction, has been known for over sixty years, negative adaptation, i.e. specific inhibition of enzyme synthesis, was discovered only in 1953, when it was found that the formation of the enzyme tryptophan-synthetase was inhibited selectively by tryptophan and certain tryptophan analogs (Monod & Cohen-Bazire, 1953). Soon afterwards, other examples of this effect were observed (Cohn, Cohen & Monod, 1953; Adelberg & Umbarger, 1953; Wijesundera &

Woods, 1953), and several systems were studied in detail in subsequent years (Gorini & Maas, 1957; Vogel, 1957a,b; Yates & Pardee, 1957; Magasanik, Magasanik & Neidhardt, 1959). These studies have revealed that the "repression" effect, as it was later named by Vogel (1957a,b), is very closely analogous, albeit symmetrically opposed, to the induction effect.

Enzyme repression, like induction, generally involves not a single but a sequence of enzymes active in successive metabolic steps. While inducibility is the rule for catabolic enzyme sequences responsible for the degradation of exogeneous substances, repressibility is the rule for anabolic enzymes, involved in the synthesis of essential metabolites such as amino acids or nucleotides.† Repression, like induction, is highly specific, but while inducers generally are substrates (or analogs of substrates) of the sequence, the repressing metabolites generally are the product (or analogs of the product) of the sequence.

That the effect involves inhibition of enzyme *synthesis*, and not inhibition (directly or indirectly) of enzyme *activity* was apparent already in the first example studied (Monod & Cohen-Bazire, 1953), and has been proved conclusively by isotope incorporation experiments (Yates & Pardee, 1957). It is important to emphasize this point, because enzyme repression must not be confused with another effect variously called "feedback inhibition" or "retro-inhibition" which is equally frequent, and may occur in the same systems. This last effect, discovered by Novick & Szilard (in Novick, 1955), involves the inhibition of activity of an early enzyme in an anabolic sequence, by the ultimate product of the sequence (Yates & Pardee, 1956; Umbarger, 1956). We shall use "repression" exclusively to designate specific inhibition of enzyme *synthesis*.‡

(g) *Kinetics and specificity of repression*

The kinetics of enzyme synthesis provoked by "de-repression" are identical to the kinetics of induction (see Fig. 2). When wild type *E. coli* is grown in the presence of arginine, only traces of ornithine-carbamyltransferase are formed. As soon as arginine is removed from the growth medium, the differential rate of enzyme synthesis increases about 1,000 times and remains constant, until arginine is added again, when it immediately falls back to the repressed level. The repressing metabolite here acts (kinetically) as would a dissociable inhibitor in an enzyme system.

The specificity of repression poses some particularly significant problems. As a rule, the repressing metabolite of an anabolic sequence is the ultimate product of this sequence. For instance, L-arginine, to the exclusion of any other amino acid, represses the enzymes of the sequence involved in the biosynthesis of arginine. Arginine shows no specific affinity for the early enzymes in the sequence, such as, in particular, ornithine-carbamyltransferase. In this sense, arginine is a "gratuitous" repressing metabolite for this protein, just as galactosides are "gratuitous inducers" for the mutated (inactive) galactosidase. The possibility must be considered however that arginine may be converted back, through the sequence itself, to an intermediate product

---

† Certain enzymes which attack exogeneous substrates are controlled by repression. Alkaline phosphatase (*E. coli*) is not induced by phosphate esters, but it is repressed by orthophosphate. Urease (*Pseudomonas*) is repressed by ammonia.

‡ We should perhaps recall the well-known fact that glucose and other carbohydrates inhibit the synthesis of many *inducible* enzymes, attacking a variety of substrates (Dienert, 1900; Gale, 1943; Monod, 1942; Cohn & Horibata, 1959). It is probable that this non-specific "glucose effect" bears some relation to the repressive effect of specific metabolites, but the relationship is not clear (Neidhardt & Magasanik, 1956a,b). We shall not discuss the glucose effect in this paper.

or substrate of the enzyme. This has been excluded by Gorini & Maas (1957) who showed that, in mutants lacking one of the enzymes involved in later steps of the sequence, ornithine transcarbamylase is repressed by arginine to the same extent as in the wild type. Moreover, neither ornithine nor any other intermediate of the sequence is endowed with repressing activity in mutants which cannot convert the intermediate into arginine. It is quite clear therefore that the specificity of action of the repressing metabolite does not depend upon the specific configuration of the enzyme site.

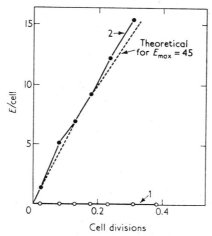

FIG. 2. Repression of ornithine-transcarbamylase by arginine. *E. coli* requiring both histidine and arginine were grown in a chemostat with 1 μg/ml. histidine + 6 μg/ml. arginine (curve 1) or with 10 μg/ml. histidine + 5 μg/ml. arginine (curve 2). Cultures are inoculated with washed cells taken from cultures growing exponentially in excess of arginine. The theoretical curve was calculated from the constant enzyme/cell value reached after 4 cell divisions. (After Gorini & Maas, 1958.)

The same conclusion is applicable to the enzymes of the histidine synthesizing pathway which are repressed in the presence of histidine, both in the wild type and in different mutants lacking one of the enzymes. The work of Ames & Garry (1959) has shown that the rates of synthesis of different enzymes in this sequence vary in *quantitatively* constant ratios under any set of medium conditions, and that the ratios are the same in various mutants lacking one of the enzymes and in the wild type. Here again, as in the case of the lactose system, the synthesis of widely different, albeit functionally related, enzymes appears to be controlled by a single common mechanism, with which the repressing metabolite specifically interacts.

In summary, repression and induction appear as closely similar effects, even if opposed in their results. Both control the rate of synthesis of enzyme proteins. Both are highly specific, but in neither case is the specificity related to the specificity of action (or binding) of the controlled enzyme. The kinetics of induction and repression are the same. Different functionally related enzymes are frequently co-induced or co-repressed, quantitatively to the same extent, by a single substrate or metabolite.

The remarkable similarity of induction and repression suggests that the two effects represent different manifestations of fundamentally similar mechanisms (Cohn & Monod, 1953; Monod, 1955; Vogel, 1957a, b; Pardee, Jacob & Monod, 1959; Szilard,

1960). This would imply either that in inducible systems the inducer acts as an antagonist of an internal repressor or that in repressible systems the repressing metabolite acts as an antagonist of an internal inducer. This is not an esoteric dilemma since it poses a very pertinent question, namely what would happen in an adaptive system of either type, when *both* the inducer and the repressor were eliminated? This, in fact, is the main question which we shall try to answer in the next section.

3. Regulator Genes

Since the specificity of induction or repression is not related to the structural specificity of the controlled enzymes, and since the rate of synthesis of different enzymes appears to be governed by a common element, this element is presumably not controlled or represented by the structural genes themselves. This inference, as we shall now see, is confirmed by the study of certain mutations which convert inducible or repressible systems into constitutive systems.

(a) *Phenotypes and genotypes in the lactose systems*

If this inference is correct, mutations which affect the controlling system should not behave as alleles of the structural genes. In order to test this prediction, the structural genes themselves must be identified. The most thoroughly investigated case is the lactose system of *E. coli*, to which we shall now return. Six phenotypically different classes of mutants have been observed in this system. For the time being, we shall consider only three of them which will be symbolized and defined as follows:

(1) Galactosidase mutations: $z^+ \rightleftharpoons z^-$ expressed as the loss of the capacity to synthesize active galactosidase (with or without induction).

(2) Permease mutations: $y^+ \rightleftharpoons y^-$ expressed as the loss of the capacity to form galactoside-permease. Most, but not all, mutants of this class simultaneously lose the capacity to synthesize active acetylase. We shall confine our discussion to the acetylaseless subclass.

(3) Constitutive mutations: $i^+ \rightleftharpoons i^-$ expressed as the ability to synthesize large amounts of galactosidase *and* acetylase in the absence of inducer (Monod, 1956; Rickenberg *et al.*, 1956; Pardee *et al.*, 1959).

The first two classes are specific for either galactosidase or acetylase: the galactosidaseless mutants form normal amounts of acetylase; conversely the acetylaseless mutants form normal amounts of galactosidase. In contrast, the constitutive mutations, of which over one hundred recurrences have been observed, invariably affect both the galactosidase and the permease (acetylase).[†] There are eight possible combinations of these phenotypes, and they have all been observed both in *E. coli* ML and K12.

The loci corresponding to a number of recurrences of each of the three mutant types have been mapped by recombination in *E. coli* K12. The map (Fig. 3) also

---

[†] The significance of this finding could be questioned since, in order to isolate constitutive mutants, one must of course use selective media, and this procedure might be supposed to favour double mutants, where the constitutivity of galactosidase and permease had arisen independently. It is possible, however, to select for $i^+ \rightarrow i^-$ mutants in organisms of type $i^+z^+y^-$, i.e. permeaseless. Fifty such mutants were isolated, giving rise to "constitutive cryptic" types $i^-z^+y^-$ from which, by reversion of y^-, fifty clones of constitutive $i^-z^+y^+$ were obtained. It was verified that in each of these fifty clones the permease was constitutive.

indicates the location of certain other mutations (o mutations) which will be discussed later. As may be seen, all these loci are confined to a very small segment of the chromosome, the *Lac* region. The extreme proximity of all these mutations raises the question whether they belong to a single or to several independent functional units. Such functional analysis requires that the biochemical expression of the various genetic structures be studied in heterozygous diploids. Until quite recently, only transient diploids were available in *E. coli*; the recent discovery of a new type of gene transfer in these bacteria (sexduction) has opened the possibility of obtaining stable clones which are diploid (or polyploid) for different small segments of the chromosome.

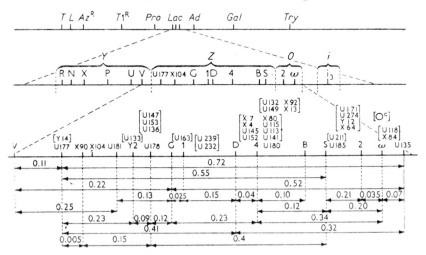

Fig. 3. Diagrammatic map of the lactose region of *E. coli* K12. The upper line represents the position of the *Lac* region with respect to other known markers. The middle line represents an enlargement of the *Lac* region with the four loci y, z, o and i. The lower line represents an enlargement of the z and o loci. Recombination frequencies (given at the bottom) are obtained in two factor crosses of the type $Hfr\ Lac_A^-ad^+S^s \times F^-Lac_B^-ad^-S^r$, from the ratios "recombinants $Lac^+ad^+S^r$/ recombinants ad^+S^r." The total length of the z gene may be estimated to be 0·7 map units, i.e. about 3,500 nucleotide pairs for about 1,000 amino acids in the monomer of β-galactosidase.

In this process, small fragments of the bacterial chromosome are incorporated into the sex factor, F. This new unit of replication is transmissible by conjugation, and is then added to the normal genome of the recipient bacterium which becomes diploid for the small chromosomal fragment. Among the units thus isolated, one carries the whole *Lac* region (Jacob & Adelberg, 1959; Jacob, Perrin, Sanchez & Monod, 1960). To symbolize the genetic structure of these diploids, the chromosomal alleles are written in the usual manner, while the alleles attached to the sex factor are preceded by the letter F.

Turning our attention to the behaviour of z and y mutant types, we may first note that diploids of structure z^+y^-/Fz^-y^+ or z^-y^+/Fz^+y^- are wild type, being able to ferment lactose, and forming normal amounts of both galactosidase and acetylase. This complete complementation between z^- and y^- mutants indicates that they belong to independent cistrons. Conversely, no complementation is observed between different y^- mutants, indicating that they all belong to a single cistron. No complementation is observed between most z^- mutants. Certain diploids of structure $z_a^-z_b^+/Fz_a^+z_b^-$ synthesize galactosidase in reduced amounts, but pairs of mutually non-complementing mutants overlap mutually complementing mutants, suggesting again that a single cistron is involved, as one might expect, since the monomer of galactosidase has a single N-terminal group. It should be recalled that intracistronic partial complementation has been observed in several cases (Giles, 1958), and has (tentatively) been explained as related to a polymeric state of the protein.

Mutations in the z gene affect the structure of galactosidase. This is shown by the fact that most of the z^- mutants synthesize, in place of active enzyme, a protein which is able to displace authentic (wild type) galactosidase from its combination with specific antibody (Perrin, Bussard & Monod, 1959). Among proteins synthesized by different z^- mutants (symbolized Cz_1, Cz_2, etc.) some give complete cross reactions (i.e. precipitate 100% of the specific antigalactosidase antibodies) with the serum used, while others give incomplete reactions. The different Cz proteins differ therefore, not only from wild type galactosidase, but also one from the other. Finally, as we already mentioned, diploids of constitution z^+/z_1^- synthesize wild type galactosidase and the modified protein simultaneously, and at similar rates (Perrin *et al.*, 1960). These observations justify the conclusions that the z region or cistron contains the structural information for β-galactosidase. Proof that mutations in the y region not only suppress but may in some cases modify the structure of acetylase has not been obtained as yet, but the assumption that the y region does represent, in part at least, the structural gene for the acetylase protein appears quite safe in view of the properties of the y mutants.

(b) *The i^+ gene and its cytoplasmic product*

We now turn our attention to the constitutive (i^-) mutations. The most significant feature of these mutations is that they invariably affect simultaneously two different enzyme-proteins, each independently determined, as we have just seen, by different structural genes. In fact, most i^- mutants synthesize more galactosidase and acetylase than induced wild type cells, but it is quite remarkable that the *ratio* of galactosidase to acetylase is the same in the constitutive cells as in the induced wild type, strongly suggesting that the mechanism controlled by the i gene is the same as that with which the inducer interacts.

The study of double heterozygotes of structures: i^+z^-/Fi^-z^+ or i^-y^+/Fi^+y^- shows (Table 2, lines 4 and 5) that the inducible i^+ allele is dominant over the constitutive and that it is active in the *trans* position, with respect to both y^+ and z^+.

Therefore the i mutations belong to an independent cistron, governing the expression of y and z *via* a cytoplasmic component. The dominance of the inducible over the constitutive allele means that the former corresponds to the active form of the i gene. This is confirmed by the fact that strains carrying a *deletion* of the izy region behave like i^- in diploids (Table 2, line 7). However, two different interpretations of the function of the i^+ gene must be considered.

(a) The i^+ gene determines the synthesis of a repressor, inactive or absent in the i^- alleles.

(b) The i^+ gene determines the synthesis of an enzyme which destroys an inducer, produced by an independent pathway.

The first interpretation is the most straightforward, and it presents the great interest of implying that the fundamental mechanisms of control may be the same in inducible and repressible systems. Several lines of evidence indicate that it is the correct interpretation.

First, we may mention the fact that constitutive synthesis of β-galactosidase by $i^-z^+y^+$ types is not inhibited by thiophenyl-galactoside which has been shown (Cohn & Monod, 1953) to be a competitive inhibitor of induction by exogenous galactosides (see p. 325).

(see p. 325).

TABLE 2

Synthesis of galactosidase and galactoside-transacetylase by haploids and heterozygous diploids of regulator mutants

Strain No.	Genotype	Galactosidase		Galactoside-transacetylase	
		Non-induced	Induced	Non-induced	Induced
1	$i^+z^+y^+$	<0.1	100	<1	100
2	$i_6^- z^+y^+$	100	100	90	90
3	$i_3^- z^+y^+$	140	130	130	120
4	$i^+z_1^-y^+/F i_3^- z^+y^+$	<1	240	1	270
5	$i_3^- z_1^- y^+/F i^+ z^+ y\overline{U}$	<1	280	<1	120
6	$i_3^- z_1^- y^+/F i^- z^+ y^+$	195	190	200	180
7	$\varDelta_{izy}/F i^- z^+ y^+$	130	150	150	170
8	$i^s z^+ y^+$	<0.1	<1	<1	<1
9	$i^s z^+ y^+/F i^+ z^+ y^+$	<0.1	2	<1	3

Bacteria are grown in glycerol as carbon source and induced, when stated, by *iso*propyl-thio-galactoside, 10^{-4} M. Values are given as a percentage of those observed with induced wild type (for absolute values, see legend of Table 1). $\varDelta_{izy}$ refers to a deletion of the whole *Lac* region. It will be noted that organisms carrying the wild allele of one of the structural genes (z or y) on the F factor form more of the corresponding enzyme than the haploid. This is presumably due to the fact that several copies of the *F-Lac* unit are present per chromosome. In i^+/i^- heterozygotes, values observed with uninduced cells are sometimes higher than in the haploid control. This is due to the presence of a significant fraction of i^-/i^- homozygous recombinants in the population.

A direct and specific argument comes from the study of one particular mutant of the lactose system. This mutant (i^s) has lost the capacity to synthesize *both* galacto-sidase and permease. It is not a deletion because it recombines, giving *Lac*+ types, with all the z^- and y^- mutants. In crosses with z^-i^- organisms the progeny is *ex-clusively* i^- while in crosses with z^-i^+ it is *exclusively* i^+, indicating exceedingly close linkage of this mutation with the i region. Finally, in diploids of constitution i^s/i^+, i^s turns out to be *dominant*: the diploids cannot synthesize either galactosidase or acetylase (see Table 2, lines 8 and 9).

These unique properties appear exceedingly difficult to account for, except by the admittedly very specific hypothesis that mutant i^s is an allele of i where the *structure* of the repressor is such that it cannot be antagonized by the inducer any more. If this hypothesis is correct, one would expect that the i^s mutant could regain the ability to metabolize lactose, not only by reversion to wild type ($i^s \rightarrow i^+$) but also, and pro-bably more frequently, by inactivation of the i gene, that is to say by achieving the constitutive condition ($i^s \rightarrow i^-$). Actually, Lac^+ "revertants" are very frequent in populations of mutant i^s, and 50% of these "revertants" are indeed constitutives of the i^- (recessive) type. (The other revertants are also constitutives, but of the o^c class which we shall mention later.) The properties of this remarkable mutant could evidently not be understood under the assumption that the i gene governs the synthesis of an inducer-destroying enzyme (Willson, Perrin, Jacob & Monod, 1961).

Accepting tentatively the conclusion that the i^+ gene governs the synthesis of an intracellular repressor, we may now consider the question of the presence of this substance in the cytoplasm, and of its chemical nature.

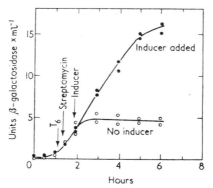

Fig. 4. Synthesis of β-galactosidase by merozygotes formed by conjugation between inducible, galactosidase-positive males and constitutive, galactosidase-negative females. Male ($Hfr\ i^+z^+T6^sS^s$) and female ($F^-\ i^-z^-T6^rS^r$) bacteria grown in a synthetic medium containing glycerol as carbon source are mixed in the same medium (time O) in the absence of inducer. In such a cross, the first zygotes which receive the Lac region from the males are formed from the 20th min. The rate of enzyme synthesis is determined from enzyme activity measurement on the whole population, to which streptomycin and phage T6 are added at times indicated by arrows to block further formation of recombinants and induction of the male parents. It may be seen that in the absence of inducer enzyme synthesis stops about 60 to 80 min after penetration of the first z^+i^+ segment but is resumed by addition of inducer (From Pardee et al, 1959).

Important indications on this question have been obtained by studying the kinetics and conditions of expression of the i^+ and z^+ genes when they are introduced into the cytoplasm of cells bearing the inactive (z^- and i^-) alleles. The sexual transfer of the Lac segment from male to female cells provides an adequate experimental system for such studies. It should be recalled that conjugation in $E.\ coli$ involves essentially the transfer of a male chromosome (or chromosome segment) to the female cell. This transfer is oriented, always beginning at one extremity of the chromosome, and it is progressive, each chromosome segment entering into the recipient cell at a fairly precise time following inception of conjugation in a given mating pair (Wollman & Jacob, 1959). The conjugation does not appear to involve any significant cytoplasmic mixing, so that the zygotes inherit virtually all their cytoplasm from the female cell, receiving only a chromosome or chromosome segment from the male. In order to study galactosidase synthesis by the zygotes, conditions must be set up such that the unmated parents cannot form the enzymes. This is the case when mating between inducible galactosidase-positive, streptomycin-sensitive males ($\male\ z^+i^+Sm^s$) and constitutive, galactosidase-negative, streptomycin-resistant females ($\female z^-i^-Sm^r$) is performed in presence of streptomycin (Sm), since: (i) the male cells which are sensitive to Sm cannot synthesize enzyme in its presence; (ii) the female cells are genetically incompetent; (iii) the vast majority of the zygotes which receive the z^+ gene, do not become streptomycin sensitive (because the Sm^s gene is transferred only to a small proportion of them, and at a very late time). The results of such an experiment, performed in the absence of inducer, are shown in Fig. 4. It is seen that galactosidase synthesis starts almost immediately following actual entry of the z^+ gene. We shall return later to a more precise analysis of the expression of the z^+ gene. The important point to be stressed here is that during this initial period the zygotes behave like *constitutive* cells, synthesizing enzyme in the *absence* of inducer. Approximately sixty minutes later, however, the rate of galactosidase synthesis falls off to zero. If at that time inducer is added, the maximum rate of enzyme synthesis is resumed. We are, in other words, witnessing the conversion of the originally i^- phenotype of the zygote cell, into an i^+ phenotype. And this experiment clearly shows that the "inducible" state is associated with the presence, at a sufficient level, of a *cytoplasmic* substance synthesized under the control of the i^+ gene. (It may be pointed out that the use of a female strain carrying a *deletion* of the *Lac* region instead of the $i^- z^-$ alleles gives the same results (Pardee *et al.*, 1959).)

If now 5-methyltryptophan is added to the mated cells a few minutes before entry of the z^+ gene, no galactosidase is formed because, as is well known, this compound inhibits tryptophan synthesis by retro-inhibition, and therefore blocks protein synthesis. If the repressor is a protein, or if it is formed by a specific enzyme, the synthesis of which is governed by the i^+ gene, its accumulation should also be blocked. If on the other hand the repressor is not a protein, and if its synthesis does not require the preliminary synthesis of a specific enzyme controlled by the i^+ gene, it may accumulate in presence of 5-methyltryptophan which is known (Gros, unpublished results) *not* to inhibit energy transfer or the synthesis of nucleic acids.

The results of Pardee & Prestidge (1959) show that the repressor *does* accumulate under these conditions, since the addition of tryptophan 60 min after 5-methyltryptophan allows immediate and complete resumption of enzyme synthesis, *but only in the presence of inducer*; in other words, the cytoplasm of the zygote cells has been converted from the constitutive to the inducible state during the time that protein synthesis was blocked. This result has also been obtained using chloramphenicol as the agent for blocking protein synthesis, and it has been repeated using another system of gene transfer (Luria *et al.*, unpublished results).

This experiment leads to the conclusion that the repressor is not a protein, and this again excludes the hypothesis that the i^+ gene controls an inducer-destroying enzyme. We should like to stress the point that this conclusion does not imply that no enzyme is involved in the synthesis of the repressor, but that the enzymes which may be involved are *not* controlled by the i^+ gene. The experiments are negative, as far as the chemical nature of the repressor itself is concerned, since they only eliminate protein as a candidate. They do, however, invite the speculation that the repressor may be the primary product of the i^+ gene, and the further speculation that such a primary product may be a polyribonucleotide.

Before concluding this section, it should be pointed out that constitutive mutations have been found in several inducible systems; in fact wherever they have been searched for by adequate selective techniques (amylomaltase of *E. coli* (Cohen-Bazire & Jolit, 1953), penicillinase of *B. cereus* (Kogut, Pollock & Tridgell, 1956), glucuronidase of *E. coli* (F. Stoeber, unpublished results), galactokinase and galactose-transferase (Buttin, unpublished results)). That *any* inducible system should be potentially capable of giving rise to constitutive mutants, strongly indicates that such mutations occur, or at least can always occur, by a loss of function. In the case of the "galactose" system of *E. coli*, it has been found that the constitutive mutation is pleiotropic, affecting a sequence of three different enzymes (galactokinase, galactose-transferase, UDP-galactose epimerase), and occurs at a locus distinct from that of the corresponding structural genes (Buttin, unpublished results).

The main conclusions from the observations reviewed in this section may be summarized as defining a new type of gene, which we shall call a "regulator gene" (Jacob & Monod, 1959). A regulator gene does not contribute structural information to the proteins which it controls. The specific product of a regulator gene is a cytoplasmic substance, which inhibits information transfer from a structural gene (or genes) to protein. In contrast to the classical structural gene, a regulator gene may control the synthesis of several different proteins: the one-gene one-protein rule does not apply to it.

We have already pointed out the profound similarities between induction and repression which suggest that the two effects represent different manifestations of the same fundamental mechanism. If this is true, and if the above conclusions are valid, one expects to find that the genetic control of repressible systems also involves regulator genes.

(c) *Regulator genes in repressible systems*

The identification of constitutive or "de-repressed" mutants of several repressible systems has fulfilled this expectation. For the selection of such mutants, certain analogs of the normal repressing metabolite may be used as specific selective agents, because they cannot substitute for the metabolite, except as repressing metabolites. For instance, 5-methyltryptophan does not substitute for tryptophan in protein synthesis (Munier, unpublished results), but it represses the enzymes of the tryptophan-synthesizing sequence (Monod & Cohen-Bazire, 1953). Normal wild type *E. coli* does not grow in the presence of 5-methyltryptophan. Fully resistant stable mutants arise, however, a large fraction of which turn out to be constitutive for the tryptophan system.† The properties of these organisms indicate that they arise by mutation of a regulator gene R_T (Cohen & Jacob, 1959). In these mutants tryptophan-synthetase as well as at least two of the enzymes involved in previous steps of the sequence are formed at the same rate irrespective of the presence of tryptophan, while in the wild type all these enzymes are strongly repressed. Actually the mutants form more of the enzymes in the presence of tryptophan, than does the wild type in its absence (just as i^-z^+ mutants form more galactosidase in the absence of inducer than the wild type does at saturating concentration of inducer). The capacity of the mutants to concentrate tryptophan from the medium is not impaired, nor is their tryptophanase activity increased. The loss of sensitivity to tryptophan as repressing metabolite cannot therefore be attributed to its destruction by, or exclusion from, the cells, and can only reflect the breakdown of the control system itself. Several recurrences of the R_T mutation have been mapped. They are all located in the same small section of the chromosome, at a large distance from the cluster of genes which was shown by Yanofsky & Lennox (1959) to synthesize the different enzymes of the sequence. One of these genes (comprising two cistrons) has been very clearly identified by the work of Yanofsky (1960) as the structural gene for tryptophan synthetase, and it is a safe assumption that the other genes in this cluster determine the structure of the preceding

† Resistance to 5-methyltryptophan may also arise by other mechanisms in which we are not interested here.

enzymes in the sequence. The R_T gene therefore controls the rate of synthesis of several different proteins without, however, determining their structure. It can only do so *via* a cytoplasmic intermediate, since it is located quite far from the structural genes. To complete its characterization as a regulator gene, it should be verified that the constitutive (R_T^-) allele corresponds to the inactive state of the gene (or gene product), i.e. is recessive. Stable heterozygotes have not been available in this case, but the transient (sexual) heterozygotes of a cross $\male\, R_T^- \times\, \female R_T^+$ are sensitive to 5-methyltryptophan, indicating that the repressible allele is dominant (Cohen & Jacob, 1959).

In the arginine-synthesizing sequence there are some seven enzymes, simultaneously repressible by arginine (Vogel, 1957a,b; Gorini & Maas, 1958). The specific (i.e. probably structural) genes which control these enzymes are dispersed at various loci on the chromosome. Mutants resistant to canavanine have been obtained, in which several (perhaps all) of these enzymes are simultaneously de-repressed. These mutations occur at a locus (near Sm^r) which is widely separated from the loci corresponding (probably) to the structural genes. The dominance relationships have not been analysed (Gorini, unpublished results; Maas, Lavallé, Wiame & Jacob, unpublished results).

The case of alkaline phosphatase is particularly interesting because the structural gene corresponding to this protein is well identified by the demonstration that various mutations at this locus result in the synthesis of altered phosphatase (Levinthal, 1959). The synthesis of this enzyme is repressed by orthophosphate (Torriani, 1960). Constitutive mutants which synthesize large amounts of enzyme in the presence of orthophosphate have been isolated. They occur at two loci, neither of which is allelic to the structural gene, and the constitutive enzyme is identical, by all tests, to the wild type (repressible) enzyme. The constitutive alleles for both of the two loci have been shown to be recessive with respect to wild type. Conversely, mutations in the structural (P) gene do not affect the regulatory mechanism, since the altered (inactive) enzyme formed by mutants of the P gene is repressed in the presence of orthophosphate to the same extent as the wild type enzyme (Echols, Garen, Garen & Torriani, 1961).

(d) *The interaction of repressors, inducers and co-repressors*

The sum of these observations leaves little doubt that repression, like induction, is controlled by specialized regulator genes, which operate by a basically similar mechanism in both types of systems, namely by governing the synthesis of an intracellular substance which inhibits information transfer from structural genes to protein.

It is evident therefore that the metabolites (such as tryptophan, arginine, orthophosphate) which inhibit enzyme synthesis in repressible systems are not active by themselves, but only by virtue of an interaction with a repressor synthesized under the control of a regulator gene. Their action is best described as an activation of the genetically controlled repression system. In order to avoid confusion of words, we shall speak of repressing metabolites as "co-repressors" reserving the name "repressors" (or apo-repressors) for the cytoplasmic products of the regulator genes.

The nature of the interaction between repressor and co-repressor (in repressible systems) or inducer (in inducible systems) poses a particularly difficult problem. As a purely formal description, one may think of inducers as antagonists, and of co-repressors as activators, of the repressor. A variety of chemical models can be imagined to account for such antagonistic or activating interactions. We shall not go into these speculations since there is at present no evidence to support or eliminate any particular model. But it must be pointed out that, in any model, the structural specificity of inducers or co-repressors must be accounted for, and can be accounted for, only by the assumption that a stereospecific receptor is involved in the interaction. The fact that the repressor is apparently not a protein then raises a serious difficulty since the capacity to form stereospecific complexes with small molecules appears to be a privilege of proteins. If a protein, perhaps an enzyme, is responsible for the specificity, the structure of this protein is presumably determined by a structual gene and muta-tion in this gene would result in loss of the capacity to be induced (or repressed). Such mutants, which would have precisely predictable properties (they would be pleio-tropic, recessive, and they would be complemented by mutants of the other structural genes) have not been encountered in the lactose system, while the possibility that the controlled enzymes themselves (galactosidase or acetylase) play the role of "induction enzyme" is excluded.

It is conceivable that, in the repressible systems which synthesize amino acids, this role is played by enzymes simultaneously responsible for essential functions (e.g. the activating enzymes) whose loss would be lethal, but this seems hardly conceivable in the case of most inducible systems. One possibility which is not excluded by these observations is that the repressor itself synthesizes the "induction protein" and remains thereafter associated with it. Genetic inactivation of the induction enzyme would then be associated with structural alterations of the repressor itself and would generally be expressed as constitutive mutations of the regulator gene.†
This possibility is mentioned here only as an illustration of the dilemma which we have briefly analysed, and whose solution will depend upon the chemical identification of the repressor.

(e) Regulator genes and immunity in temperate phage systems

One of the most conspicuous examples of the fact that certain genes may be either allowed to express their potentialities, or specifically prohibited from doing so, is the phenomenon of immunity in temperate phage systems (cf. Lwoff, 1953; Jacob, 1954; Jacob & Wollman, 1957; Bertani, 1958; Jacob, 1960).

The genetic material of the so-called temperate phages can exist in one of two states within the host cell:

(1) In the *vegetative state*, the phage genome multiplies autonomously. This process, during which all the phage components are synthesized, culminates in the production of infectious phage particles which are released by lysis of the host cell.

(2) In the *prophage state*, the genetic material of the phage is attached to a specific site of the bacterial chromosome in such a way that both genetic elements replicate as a single unit. The host cell is said to be "lysogenic." As long as the phage genome remains in the prophage state, phage particles are not produced. For lysogenic bacteria to produce phage, the genetic material of the phage must undergo a transition from the prophage to the vegetative state. During normal growth of lysogenic bacteria, this event is exceedingly rare. With certain types of prophages, however, the transition can be induced in the whole population by exposure of the culture to u.v. light,

† Such a model could account for the properties of the i^s (dominant) mutant of the regulator gene in the lactose system, by the assumption that in this mutant the repressor remains active, while having lost the capacity to form its associated induction protein.

X-rays or various compounds known to alter DNA metabolism (Lwoff, Siminovitch & Kjeldgaard, 1950; Lwoff, 1953; Jacob, 1954).

The study of "defective" phage genomes, in which a mutation has altered one of the steps required for the production of phage particles, indicates the existence of at least two distinct groups of viral functions, both of which are related to the capacity of synthesizing specific proteins (Jacob, Fuerst & Wollman, 1957). Some "early" functions appear as a pre-requisite for the vegetative multiplication of the phage genome and, at least in virulent phages of the T-even series, it is now known that they correspond to the synthesis of a series of new enzymes (Flaks & Cohen, 1959; Kornberg, Zimmerman, Kornberg & Josse, 1959). A group of "late" functions correspond to the synthesis of the structural proteins which constitute the phage coat. The expression of these different viral functions appears to be in some way co-ordinated by a sequential process, since defective mutations affecting some of the early functions may also result in the loss of the capacity to perform several later steps of phage multiplication (Jacob et al., 1957).

In contrast, the viral functions are not expressed in the prophage state and the protein constituents of the phage coat cannot be detected within lysogenic bacteria. In addition, lysogenic bacteria exhibit the remarkable property of being specifically immune to the very type of phage particles whose genome is already present in the cell as prophage. When lysogenic cells are infected with homologous phage particles, these particles absorb onto the cells and inject their genetic material, but the cell survives. The injected genetic material does not express its viral functions: it is unable to initiate the synthesis of the protein components of the coat and to multiply vegetatively. It remains inert and is diluted out in the course of bacterial multiplication (Bertani, 1953; Jacob, 1954).

The inhibition of phage-gene functions in lysogenic bacteria therefore applies not only to the prophage, but also to additional homologous phage genomes. It depends only upon the presence of the prophage (and not upon a permanent alteration, provoked by the prophage, of bacterial genes) since loss of the prophage is both necessary and sufficient to make the bacteria sensitive again.

Two kinds of interpretation may be considered to account for these "immunity" relationships:

(a) The prophage occupies and blocks a *chromosomal* site of the host, specifically required in some way for the vegetative multiplication of the homologous phage.

(b) The prophage produces a *cytoplasmic* inhibitor preventing the completion of some reactions (presumably the synthesis of a particular protein) necessary for the initiation of vegetative multiplication.

A decision between these alternative hypotheses may be reached through the study of persistent diploids, heterozygous for the character lysogeny. A sex factor has been isolated which has incorporated a segment of the bacterial chromosome carrying the genes which control galactose fermentation, *Gal*, and the site of attachment of prophage, λ. Diploid heterozygotes with the structure *Gal⁻* λ⁻/*F Gal⁺* λ⁺ or *Gal⁻* λ⁺/*F Gal⁺* λ⁻ are immune against superinfection with phage λ, a result which shows that "immunity" is dominant over "non-immunity" and has a cytoplasmic expression (Jacob, Schaeffer & Wollman, 1960).

The study of transient zygotes formed during conjugation between lysogenic (λ⁺) and non-lysogenic (λ⁻) cells leads to the same conclusion. In crosses ♂λ⁺ × ♀λ⁻, the transfer of the prophage carried by the male chromosome into the non-immune recipient results in transition to the vegetative state: multiplication of the phage occurs in the zygotes, which are lysed and release phage particles. This phenomenon is known as "zygotic induction" (Jacob & Wollman, 1956). In the *reverse* cross $\male\lambda^- \times \female\lambda^+$, however, *no zygotic induction occurs*. The transfer of the "non-lysogenic" character carried by the male chromosome into the immune recipient does not bring about the development of the prophage and the zygotes are immune against superinfection with phage λ.

The opposite results obtained in reciprocal crosses of lysogenic by non-lysogenic male and female cells are entirely analogous to the observations made with the lactose system in reciprocal crosses of inducible by non-inducible cells. In both cases, it is evident that the decisive factor is the origin of the *cytoplasm* of the zygote, and the conclusion is inescapable, that the immunity of lysogenic bacteria is due to a cytoplasmic constituent, in the presence of which the viral genes cannot become expressed (Jacob, 1960).

The same two hypotheses which we have already considered for the interpretation of the product of the regulator gene in the lactose system, apply to the cytoplasmic inhibitor insuring immunity in lysogenic bacteria.

(a) The inhibitor is a specific repressor which prevents the synthesis of some early protein(s) required for the initiation of vegetative multiplication.

(b) The inhibitor is an enzyme which destroys a metabolite, normally synthesized by the non-lysogenic cell and specifically required for the vegetative multiplication of the phage.

Several lines of evidence argue against the second hypothesis (Jacob & Campbell, 1959; Jacob, 1960). First, for a given strain of bacteria, many temperate phages are known, each of which exhibits a different immunity pattern. According to the second hypothesis, each of these phages would specifically require for vegetative multiplication a different metabolite normally produced by the non-lysogenic cells, an assumption which appears extremely unlikely. The second argument stems from the fact that, like the repressor of the lactose system, the inhibitor responsible for immunity is synthesized in the presence of chloramphenicol, i.e. in the absence of protein synthesis: when crosses $\male\lambda^+ \times \female\lambda^-$ are performed in the presence of chloramphenicol, no zygotic induction occurs and the prophage is found to segregate normally among recombinants.

In order to explain immunity in lysogenic bacteria, we are led therefore to the same type of interpretation as in the case of adaptive enzyme systems. According to this interpretation, the prophage controls a cytoplasmic repressor, which inhibits specifically the synthesis of one (or several) protein(s) necessary for the initiation of vegetative multiplication. In this model, the introduction of the genetic material of the phage into a non-lysogenic cell, whether by infection or by conjugation, results in a "race" between the synthesis of the specific repressor and that of the early proteins required for vegetative multiplication. The fate of the host-cell, survival with lysogenization or lysis as a result of phage multiplication, depends upon whether the synthesis of the repressor or that of the protein is favoured. Changes in the cultural conditions favoring the synthesis of the repressor such as infection at low temperature, or in the presence of chloramphenicol, would favor lysogenization and *vice versa*. The phenomenon of induction by u.v. light could then be understood, for instance, in the following way: exposure of inducible lysogenic bacteria to u.v. light or X-rays would transiently disturb the regulation system, for example by preventing further synthesis of the repressor. If the repressor is unstable, its concentration inside the cell would decrease and reach a level low enough to allow the synthesis of the early proteins. Thus the vegetative multiplication would be irreversibly initiated.

The similarity between lysogenic systems and adaptive systems is further strengthened by the genetic analysis of immunity. Schematically, the genome of phage λ appears to involve two parts (see Fig. 5): a small central segment, the C region, contains a few determinants which control various functions involved in lysogenization (Kaiser, 1957); the rest of the linkage group contains determinants which govern the "viral functions," i.e. presumably the structural genes corresponding to the different phage proteins. Certain strains of temperate phages which exhibit different immunity patterns are able nevertheless to undergo genetic recombination. The specific immunity pattern segregates in such crosses, proving to be controlled by a small segment "im" of the C region (Kaiser & Jacob, 1957). In other words, a prophage contains in its C region a small segment "im" which controls the synthesis of a specific repressor, active on the phage genome carrying a homologous "im" segment.

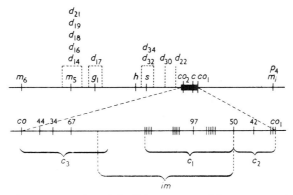

Fig. 5. Diagrammatic representation of the linkage group of the temperate bacteriophage λ. The upper diagram represents the linear arrangement of markers. Symbols refer to various plaque size, plaque type and host-range markers. Symbols d refer to various defective mutations. The C region represented by a thicker line is enlarged in the lower diagram. The figures correspond to various C mutations. The C region can be subdivided into three functional units, C_1, C_2 and C_3; the segment controlling immunity is designated im.

In the "im" region, two types of mutations arise, whose properties are extremely similar to those of the different mutations affecting the regulator genes of adaptive enzyme systems.

(1) Some mutations ($C_I^+ \rightarrow C_I$) result in the complete loss of the capacity for lysogenization in single infection. All the C_I mutations are located in a cluster, in a small part of the "im" segment, and they behave as belonging to a single cistron in complementation tests.

In mixed infections with both C_I and C_I^+ phages, double lysogenic clones carrying both C_I and C_I^+ prophages can be recovered. In such clones, single lysogenic cells segregate, which carry the C^+ type alone but never the C_I type alone. These findings indicate that the wild allele is dominant over the mutant C_I alleles and is cytoplasmically expressed, repressing the mutant genome into the prophage state. The properties of the C_I mutations are therefore similar to those of the recessive constitutive mutations of adaptive systems. The evidence suggests that the C_I locus controls the synthesis of the repressor responsible for immunity, and that the C_I mutations correspond to inactivation of this locus, or of its product.

(2) A mutation ($ind^+\rightarrow ind^-$) has been found which results in the loss of the inducible property of the prophage, i.e. of its capacity to multiply vegetatively upon exposure of lysogenic bacteria to u.v. light, X-rays or chemical inducers. This mutation is located in the C_I segment. The mutant allele ind^- is dominant over the wild allele ind^+ since double lysogenic $\lambda ind^+/\lambda ind^-$ or diploid heterozygotes of structure Gal^- $\lambda ind^+/F$ $Gal^+\lambda ind^-$ or $Gal^-\lambda ind^-/F$ $Gal^+\lambda ind^+$ are all non-inducible. In addition, the mutant λind^- exhibits a unique property. If lysogenic bacteria K12 (λ^+) carrying a wild type prophage are exposed to u.v. light, the whole population lyses and releases phage. Infection of such cells with λind^- mutants, either before or immediately after irradiation, completely inhibits phage production and lysis.

The properties of the ind^- mutant appear in every respect similar to those of the previously described mutant i^s of the lactose system. The unique properties of the ind^- mutants can be explained only by the same type of hypothesis, namely that the mutation ind^- affects, quantitatively or qualitatively, the synthesis of the repressor in such a way that more repressor or a more efficient repressor is produced. If this assumption as well as the hypothesis that the C_I mutation results in the loss of the capacity to produce an active repressor, are correct, the double mutants $C_I ind^-$ should have lost the capacity of inhibiting phage multiplication upon infection of wild type lysogenic cells. This is actually what is observed. It is evident that the properties of the ind^- mutant cannot be accounted for by the assumption that the C_I locus controls the synthesis of a metabolite-destroying enzyme (Jacob & Campbell, 1959).

In summary, the analysis of lysogenic systems reveals that the expression of the viral genes in these systems is controlled by a cytoplasmic repressor substance, whose synthesis is governed by one particular "regulator" gene, belonging to the viral genome. The identity of the proteins whose synthesis is thus repressed is not established, but it seems highly probable that they are "early" enzymes which initiate the whole process of vegetative multiplication. With the (important) limitation that they are sensitive to entirely different types of inducing conditions, the phage repression systems appear entirely comparable to the systems involved in enzymatic adaptation.

4. The Operator and the Operon

(a) *The operator as site of action of the repressor*

In the preceding section we have discussed the evidence which shows that the transfer of information from structural genes to protein is controlled by specific repressors synthesized by specialized regulator genes. We must now consider the next problem, which is the site and mode of action of the repressor.

In regard to this problem, the most important property of the repressor is its characteristic pleiotropic specificity of action. In the lactose system of *E. coli*, the repressor is both *highly specific* since mutations of the i gene do not affect any other system, and *pleiotropic* since both galactosidase and acetylase are affected simultaneously and quantitatively to the same extent, by such mutations.

The specificity of operation of the repressor implies that it acts by forming a stereospecific combination with a constituent of the system possessing the proper (complementary) molecular configuration. Furthermore, it must be assumed that the flow of information from gene to protein is interrupted when this element is combined with the repressor. This controlling element we shall call the *"operator"* (Jacob & Monod, 1959). We should perhaps call attention to the fact that, once the existence of a specific repressor is considered as established, the existence of an operator element defined as above follows necessarily. Our problem, therefore, is not whether an operator exists, but where (and how) it intervenes in the system of information transfer.

An important prediction follows immediately from the preceding considerations. Under any hypothesis concerning the nature of the operator, its specific complementary configuration must be genetically determined; therefore it could be affected by mutations which would alter or abolish its specific affinity for the repressor, without necessarily impairing its activity as initiator of information-transfer. Such mutations would result in *constitutive* synthesis of the protein or proteins. These mutations would define an "operator locus" which should be genetically distinct from the regulator gene (i.e. its mutations should not behave as alleles of the regulator); the most distinctive predictable property of such mutants would be that the constitutive allele should be *dominant* over the wild type since, again under virtually any hypothesis, the presence in a diploid cell of repressor-sensitive operators would not prevent the operation of repressor-insensitive operators.

(b) *Constitutive operator mutations*

Constitutive mutants possessing the properties predicted above have so far been found in two repressor-controlled systems, namely the phage λ and *Lac* system of *E. coli*.

In the case of phage λ, these mutants are characterized, and can be easily selected, by the fact that they develop vegetatively in immune bacteria, lysogenic for the wild type. This characteristic property means that these mutants (v) are *insensitive* to the repressor present in lysogenic cells. When, in fact, lysogenic cells are infected with these mutant particles, the development of the wild type prophage is induced, and the resulting phage population is a mixture of v and v^+ particles. This is expected, since presumably the initiation of prophage development depends only on the formation of one or a few "early" enzyme-proteins, which are supplied by the virulent particle (Jacob & Wollman, 1953).

In the *Lac* system, dominant constitutive (o^c) mutants have been isolated by selecting for constitutivity in cells diploid for the *Lac* region, thus virtually eliminating the recessive (i^-) constitutive mutants (Jacob *et al.*, 1960a). By recombination, the o^c mutations can be mapped in the *Lac* region, between the i and the z loci, the order being (*Pro*) *yzoi* (*Ad*) (see Fig. 3). Some of the properties of these mutants are summarized in Table 3. To begin with, let us consider only the effects of this mutation on galactosidase synthesis. It will be noted that in the absence of inducer, these organisms synthesize 10 to 20% of the amount of galactosidase synthesized by i^- mutants, i.e. about 100 to 200 times more than uninduced wild type cells (Table 3, lines 3 and 7). In the presence of inducer, they synthesize maximal amounts of enzyme. They are therefore only partially constitutive (except however under conditions of starvation, when they form maximum amounts of galactosidase in the absence of inducer (Brown, unpublished results)). The essential point however is that the enzyme is synthesized constitutively by diploid cells of constitution o^c/o^+ (see Table 3). The o^c allele therefore is "dominant."

If the constitutivity of the o^c mutant results from a loss of sensitivity of the operator to the repressor, the o^c organisms should also be insensitive to the presence of the altered repressor synthesized by the i^s (dominant) allele of the i^+ gene (see page 331). That this is indeed the case, as shown by the constitutive behavior of diploids with the constitution i^so^+/Fi^+o^c (see Table 3, line 12), is a very strong confirmation of the interpretation of the effects of *both* mutations (i^s and o^c). In addition, and as one would expect according to this interpretation, o^c mutants frequently arise as lactose positive "revertants" in populations of i^s cells (see p. 332).

TABLE 3

Synthesis of galactosidase, cross-reacting material (CRM), and galactoside-transacetylase by haploid and heterozygous diploid operator mutants

Strain No.	Genotype	Galactosidase		Cross-reacting material	
		Non-induced	Induced	Non-induced	Induced
1	o^+z^+	<0·1	100	—	—
2	$o^+z^+/Fo^+z_1^-$	<0·1	105	<1	310
3	o^cz^+	15	90	—	—
4	o^+z^+/Fo^cz_1	<0·1	90	30	180
5	$o^+z_1^-/Fo^cz^+$	90	250	<1	85

Strain No.	Genotype	Galactosidase		Galactoside-transacetylase	
		Non-Induced	Induced	Non-induced	Induced
6	$o^+z^+y^+$	<0·1	100	<1	100
7	$o^cz^+y^+$	25	95	15	110
8	$o^+z^+y_U^-/Fo^cz^+y^+$	70	220	50	160
9	$o^+z_1^-y^+/Fo^cz^+y_U^-$	180	440	<1	220
10	$i^+o_{84}^sz^+y^+$	<0·1	<0·1	<1	<1
11	$i^+o_{84}^sz^+y^+/Fi^-o^+z^+y^+$	1	260	2	240
12	$i^so^+z^+y^+/Fi^+o^cz^+y^+$	190	210	150	200

Bacteria are grown in glycerol as carbon source and induced when stated, with *iso*propylthiogalactoside, 10^{-4} M. Values of galactosidase and acetylase are given as a percentage of those observed with induced wild type. Values of CRM are expressed as antigenic equivalents of galactosidase. Note that the proteins corresponding to the alleles carried by the sex factor are often produced in greater amount than that observed with induced haploid wild type. This is presumably due to the existence of several copies of the *F-Lac* factor per chromosome. In o^c mutants, haploid or diploid, the absolute values of enzymes produced, especially in the non-induced cultures varies greatly from day to day depending on the conditions of the cultures.

We therefore conclude that the $o^+ \rightarrow o^c$ mutations correspond to a modification of the specific, repressor-accepting, structure of the operator. This identifies the operator locus, i.e. the genetic segment responsible for the structure of the operator, but not the operator itself.

(c) *The operon*

Turning now to this problem, we note that the o^c mutation (like the i^- mutation) is pleiotropic: it affects simultaneously and quantitatively to the same extent, the synthesis of galactosidase and acetylase (see Table 3, lines 7 and 8). The structure of the operator, or operators, which controls the synthesis of the two proteins, therefore, is controlled by a single determinant.†

Two alternative interpretations of this situation must be considered:

(a) A single operator controls an *integral* property of the z-y genetic segment, or of its cytoplasmic product.

(b) The specific product of the operator locus is able to associate in the cytoplasm, with the products of the z and y cistrons, and thereby governs the expression of both structural genes.

The second interpretation implies that mutations of the operator locus should behave as belonging to a cistron *independent* of both the z and y cistrons. The first interpretation requires, on the contrary, that these mutations behave functionally as if they *belonged to both cistrons simultaneously*. These alternative interpretations can therefore be distinguished without reference to any particular physical model of operator action by testing for the *trans* effect of o alleles, that is to say for the constitutive *vs* inducible expression of the two structural genes in o^+/o^c diploids, heterozygous for one or both of these structural genes.

The results obtained with diploids of various structures are shown in Table 3. We may first note that in diploids of constitution $o^+z^+/Fo^cz_1^-$ or $o^+z_1^-/Fo^cz^+$ (lines 4 and 5), both the normal galactosidase produced by the z^+ allele and the altered protein (CRM) produced by the z_1^- allele are formed in the presence of inducer, while in the *absence* of inducer, *only the protein corresponding to the z allele in position cis to the o^c is produced*. The o^c therefore has no effect on the z allele in position *trans*. Or putting it otherwise: the expression of the z allele attached to an o^+ remains fully repressor-sensitive even in the presence of an o^c in position *trans*. The o locus might be said to behave as belonging to the same cistron as the z markers. But as we know already, the o^c mutation is equally effective towards the acetylase which belongs to a cistron independent of z, and not adjacent to the operator locus. The results shown in Table 3, lines 8 and 9, confirm that the $o{\to}y$ relationship is the same as the $o{\to}z$ relationship, that is, the effect of the o^c allele extends *exclusively* to the y allele in the *cis* position. For instance, in the diploid $o^+z^-y^+/Fo^cz^+y_U^-$ the galactosidase is constitutive and the acetylase is inducible, while in the diploid $o^+z^+y_U^-/Fo^cz^+y^+$ both enzymes are constitutive.

These observations, predicted by the first interpretation, are incompatible with the second and lead to the conclusion that the operator governs an integral property of the genetic segment ozy, or of its cytoplasmic product (Jacob *et al.*, 1960a; Képès, Monod & Jacob, 1961).

This leads to another prediction. Certain mutations of the o segment could modify the operator in such a way as to inactivate the whole ozy segment resulting in the loss of the capacity to synthesize *both* galactosidase and permease.

These "o^o" mutants would be *recessive* to o^+ or o^c, and they would *not* be complemented either by $o^+z^+y^-$ or by $o^+z^-y^+$ mutants. Several point-mutants, possessing

† Let us recall again that no *non-pleiotropic* constitutive mutants of any type have been isolated in this system, in spite of systematic screening for such mutants.

precisely these properties, have been isolated (Jacob *et al.*, 1960*a*). They all map very closely to *o*^c, as expected (see Fig. 3). It is interesting to note that in these mutants the *i*⁺ gene is functional (Table 3, line 11), which shows clearly, not only that the *i* and *o* mutants are not alleles, but that the *o* segment, while governing the expression of the *z* and *y* genes, does not affect the expression of the regulator gene.

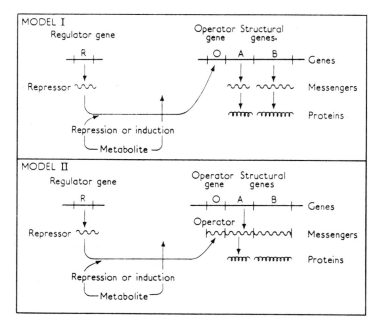

Fɪɢ. 6. Models of the regulation of protein synthesis.

In conclusion, the integral or *co-ordinate* expression of the *ozy* genetic segment signifies that the operator, which controls this expression, is and remains attached (see Fig. 6):

(a) either to the genes themselves (Fig. 6, I),

(b) or to the cytoplasmic messenger of the linked *z* and *y* genes which must then be assumed to form a single, integral, particle corresponding to the structure of the whole *ozy* segment, and functioning as a whole (Fig. 6, II).

In the former case, *the operator would in fact be identical with the o locus* and it would govern directly the activity of the genes, i.e. the synthesis of the structural messengers.

Both of these models are compatible with the observations which we have discussed so far. We shall return in the next section to the question whether the operator, i.e. the site of specific interaction with the repressor, is genetic or cytoplasmic. In either case, the *ozy* segment, although containing at least two independent structural genes, governing two independent proteins, behaves as a *unit* in the transfer of information. This *genetic unit of co-ordinate expression* we shall call the "operon" (Jacob *et al.*, 1960*a*).

The existence of such a unit of genetic expression is proved so far only in the case of the *Lac* segment. As we have already seen, the *v* mutants of phage λ, while illustrating the existence of an operator in this system, do not define an operon (because the number and the functions of the structural genes controlled by this operator are unknown). However, many observations hitherto unexplained by or even conflicting with classical genetic theory, are immediately accounted for by the operon theory. It is well known that, in bacteria, the genes governing the synthesis of different enzymes sequentially involved in a metabolic pathway are often found to be extremely closely linked, forming a cluster (Demerec, 1956). Various not very convincing speculations have been advanced to account for this obvious correlation of genetic structure and biochemical function (see Pontecorvo, 1958). Since it is now established that simultaneous induction or repression also generally prevails in such metabolic sequences, it seems very likely that the gene clusters represent units of co-ordinate expression, i.e. operons.

We have already mentioned the fact that two inducible enzymes sequentially involved in the metabolism of galactose by *E. coli*, galactokinase and UDP-galactose-transferase, are simultaneously induced by galactose, or by the gratuitous inducer D-fucose (Buttin, 1961). The genes which control specifically the synthesis of these enzymes, i.e. presumably the structural genes, are closely linked, forming a cluster on the *E. coli* chromosome. (Kalckar, Kurahashi & Jordan, 1959; Lederberg, 1960; Yarmolinsky & Wiesmeyer, 1960; Adler, unpublished results.) Certain point-mutations which occur in this chromosome segment abolish the capacity to synthesize both enzymes. These pleiotropic loss mutations are not complemented by any one of the specific (structural) loss mutations, an observation which is in apparent direct conflict with the one-gene one-enzyme hypothesis. These relationships are explained and the conflict is resolved if it is assumed that the linked structural genes constitute an operon controlled by a single operator and that the pleiotropic mutations are mutations of the operator locus.

We have also already discussed the system of simultaneous repression which controls the synthesis of the enzymes involved in histidine synthesis in *Salmonella*. This system involves eight or nine reaction steps. The enzymes which catalyse five of these reactions have been identified. The genes which individually determine these enzymes form a closely linked cluster on the *Salmonella* chromosome. Mutations in each of these genes result in a loss of capacity to synthesize a single enzyme; however, certain mutations at one end of the cluster abolish the capacity to synthesize all the enzymes simultaneously, and these mutations are not complemented by any one of the specific mutations (Ames, Garry & Herzenberg, 1960; Hartman, Loper & Serman, 1960). It will be recalled that the relative rates of synthesis of different enzymes in this sequence are constant under any set of conditions (see p. 327). All these remarkable findings are explained if it is assumed that this cluster of genes constitutes an operon, controlled by an operator associated with the *g* cistron.

The rule that genes controlling metabolically sequential enzymes constitute genetic clusters does not apply, in general, to organisms other than bacteria (Pontecorvo, 1958). Nor does it apply to all bacterial systems, even where simultaneous repression is known to occur and to be controlled by a single regulator gene, as is apparently the case for the enzymes of arginine biosynthesis. In such cases, it must be supposed that several identical or similar operator loci are responsible for sensitivity to repressor of each of the independent information-transfer systems.

It is clear that when an operator controls the expression of only a single structural cistron, the concept of the operon does not apply, and in fact there are no conceivable genetic-biochemical tests which could identify the operator-controlling genetic segment as distinct from the structural cistron itself.† One may therefore wonder whether it will be possible experimentally to extend this concept to dispersed (as opposed to clustered) genetic systems. It should be remarked at this point that many enzyme proteins are apparently made up of two (or more) different polypeptide chains. It is tempting to predict that such proteins will often be found to be controlled by two (or more) adjacent and co-ordinated structural cistrons, forming an operon.

5. The Kinetics of Expression of Structural Genes, and the Nature of the Structural Message

The problem we want to discuss in this section is whether the repressor-operator system functions at the genetic level by governing the *synthesis* of the structural message or at the cytoplasmic level, by controlling the protein-synthesizing *activity* of the messenger (see Fig. 6). These two conceivable models we shall designate respectively as the "genetic operator model" and the "cytoplasmic operator model."

The existence of units of co-ordinate expression involving several structural genes appears in fact difficult to reconcile with the cytoplasmic operator model, if only because of the size that the cytoplasmic unit would have to attain. If we assume that the message is a polyribonucleotide and take a coding ratio of 3, the "unit message" corresponding to an operon governing the synthesis of three proteins of average (monomeric) molecular weight 60,000 would have a molecular weight about 1.8×10^6; we have seen that operons including up to 8 structural cistrons may in fact exist. On the other hand, RNA fractions of *E. coli* and other cells do not appear to include polyribonucleotide molecules of molecular weight exceeding 10^6.

This difficulty is probably not insuperable; and this type of argument, given the present state of our knowledge, cannot be considered to eliminate the cytoplasmic operator model, even less to establish the validity of the genetic model. However, it seems more profitable tentatively to adopt the genetic model and to see whether some of the more specific predictions which it implies are experimentally verified.

The most immediate and also perhaps the most striking of these implications is that the structural message must be carried by a very short-lived intermediate both rapidly formed and rapidly destroyed during the process of information transfer. This is required by the kinetics of induction. As we have seen, the addition of inducer, or the removal of co-repressor, provokes the synthesis of enzyme at maximum rate within a matter of a few minutes, while the removal of inducer, or the addition of co-repressor interrupts the synthesis within an equally short time. Such kinetics are incompatible with the assumption that the repressor-operator interaction controls the rate of synthesis of *stable* enzyme-forming templates (Monod, 1956, 1958). Therefore, if the genetic operator model is valid, one should expect the kinetics of structural gene expression to be *essentially the same* as the kinetics of induction: injection of a "new" gene into an otherwise competent cell should result in virtually immediate synthesis of the corresponding protein at maximum rate; while removal of the gene should be attended by concomitant cessation of synthesis.

---

† It should be pointed out that the operational distinction between the operator locus and the structural cistron to which it is directly adjacent rests exclusively on the fact that the operator mutations affect the synthesis of several proteins governed by linked cistrons. This does not exclude the possibility that the operator locus is actually *part* of the structural cistron to which it is "adjacent." If it were so, one might expect certain constitutive operator mutations to involve an alteration of the structure of the protein governed by the "adjacent" cistron. The evidence available at present is insufficient to confirm or eliminate this assumption.

(a) *Kinetics of expression of the galactosidase structural gene*

Additions and removals of genes to and from cells are somewhat more difficult to perform than additions or removals of inducer. However, it can be done. Gene injection without cytoplasmic mixing occurs in the conjugation of *Hfr* male and *F⁻* female *E. coli*. In a mixed male and female population the individual pairs do not all mate at the same time, but the distribution of times of injection of a *given* gene can be rather accurately determined by proper genetic methods. The injection of the z^+ (galactosidase) gene from male cells into galactosidase-negative (z^-) female cells is rapidly followed by enzyme synthesis within zygotes (cf. p. 332). When the rate of enzyme synthesis in the population is expressed as a function of time, taking into

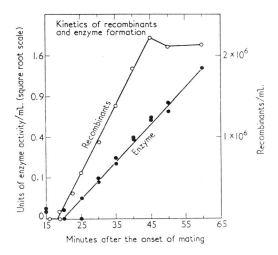

FIG. 7. Kinetics of enzyme production by merozygotes formed by conjugation between inducible galactosidase-positive males and constitutive galactosidase-negative females. Conditions are such that only the zygotes can form enzyme. Increase in the number of z^+ containing zygotes is determined by counting recombinants on adequate selective medium. Formation of enzyme is followed by enzyme activity measurements on the total population. It is seen that the enzyme increases linearly with the square of time. Since the zygote population increases linearly with time, it is apparent that the rate of enzyme synthesis per zygote is constant from the time of penetration of the z^+ gene. (From Riley *et al.*, 1960.)

account the increase with time of the number of z^+ containing zygotes, it is found (see Fig. 7):

(1) that enzyme synthesis begins within two minutes of the penetration of the z^+ gene;

(2) that the rate per zygote is constant and maximum over at least the first 40 min following penetration (Riley, Pardee, Jacob & Monod, 1960).

These observations indicate that the structural messenger is very rapidly formed by the z^+ gene, and does not accumulate. This could be interpreted in one of two ways:

(a) the structural messenger is a short-lived intermediate;

(b) the structural messenger is stable, but the gene rapidly forms a limited number of messenger molecules, and thereafter stops functioning.

If the second assumption is correct, removal of the gene after the inception of enzyme synthesis should not prevent the synthesis from continuing. This possibility is tested by the "removal" experiment, which is performed by loading the male chromosome with ^{32}P before injection. Following injection (into unlabelled female cells), ample time (25 min) is allowed for expression of the z^+ gene, before the zygotes are frozen to allow ^{32}P decay for various lengths of time. The rate of galactosidase synthesis by the population is determined immediately after thawing. It is found to decrease sharply as a function of the fraction of ^{32}P atoms decayed. If a longer period of time (110 min) is allowed for expression before freezing, no decrease in either enzyme-forming capacity or in viability of the z^+ marker are observed. This is to be expected, since by that time most of the z^+ genes would have replicated, and this observation provides an internal control showing that no indirect effects of ^{32}P disintegrations are involved.

This experiment therefore indicates that even after the z^+ gene has become expressed its integrity is required for enzyme synthesis to continue, as expected if the messenger molecule is a short-lived intermediate (Riley et al., 1960).

The interpretation of both the injection and the removal experiment rests on the assumption that the observed effects are not due to (stable) cytoplasmic messenger molecules introduced with the genetic material, during conjugation. As we have already noted, there is strong evidence that no cytoplasmic transfer, even of small molecules, occurs during conjugation. Furthermore, if the assumption were made that enzyme synthesis in the zygotes is due to pre-formed messenger molecules rather than to the activity of the gene, it would be exceedingly difficult to account for both (a) the very precise coincidence in time between inception of enzyme synthesis and entry of the gene (in the injection experiment) and (b) the parallel behaviour of enzyme-forming capacity and genetic viability of the z^+ gene (in the removal experiment).

These experiments therefore appear to show that the kinetics of expression of a structural gene are entirely similar to the kinetics of induction-repression, as expected if the operator controls the activity of the gene in the synthesis of a short-lived messenger, rather than the activity of a ready-made (stable) messenger molecule in synthesizing protein.

It is interesting at this point to recall the fact that infection of E. coli with virulent (ϕII, T2, T4) phage is attended within 2 to 4 minutes by inhibition of bacterial protein synthesis, including in particular β-galactosidase (Cohen, 1949; Monod & Wollman, 1947; Benzer, 1953). It is known on the other hand that phage-infection results in rapid visible lysis of bacterial nuclei, while no major destruction of pre-formed bacterial RNA appears to occur (Luria & Human, 1950). It seems very probable that the inhibition of specific bacterial protein synthesis by virulent phage is due essentially to the depolymerization of bacterial DNA, and this conclusion also implies that the integrity of bacterial genes is required for continued synthesis of bacterial protein. In confirmation of this interpretation, it may be noted that infection of E. coli by phage λ, which does not result in destruction of bacterial nuclei, allows β-galactosidase synthesis to continue almost to the time of lysis (Siminovitch & Jacob, 1952).

(b) Structural effects of base analogs

An entirely different type of experiment also leads to the conclusion that the structural messenger is a short-lived intermediate and suggests, furthermore, that this intermediate is a ribonucleotide. It is known that certain purine and pyrimidine analogs are incorporated by bacterial cells into ribo- and deoxyribonucleotides, and it has been found that the synthesis of protein, or of some proteins, may be inhibited in the presence of certain of these analogs. One of the mechanisms by which these effects could be explained may be that certain analogs are incorporated into the structural messenger. If so, one might hope to observe that the molecular structure of specific proteins formed in the presence of an analog is modified. It has in fact been found that the molecular properties of β-galactosidase and of alkaline phosphatase synthesized by *E. coli* in the presence of 5-fluorouracil (5FU) are strikingly altered. In the case of β-galactosidase, the ratio of enzyme activity to antigenic valency is decreased by 80%. In the case of alkaline phosphatase, the rate of thermal inactivation (of this normally highly heat-resistant protein) is greatly increased (Naono & Gros, 1960*a*,*b*; Bussard, Naono, Gros & Monod, 1960).

It can safely be assumed that such an effect cannot result from the mere presence of 5FU in the cells, and must reflect incorporation of the analog into a constituent involved in some way in the information transfer system. Whatever the identity of this constituent may be, the kinetics of the effect must in turn reflect the kinetics of 5FU incorporation into this constituent. The most remarkable feature of the 5FU effect is that it is almost immediate, in the sense that abnormal enzyme is synthesized almost from the time of addition of the analog, and that the degree of abnormality of the molecular population thereafter synthesized does not increase with time. For instance, in the case of galactosidase abnormal enzyme is synthesized within 5 min of addition of the analog, and the ratio of enzyme activity to antigenic valency remains constant thereafter. In the case of alkaline phosphatase, the thermal inactivation curve of the abnormal protein synthesized in the presence of 5FU is monomolecular, showing the molecular population to be *homogeneously* abnormal rather than made up of a mixture of normal and abnormal molecules. It is clear that if the constituent responsible for this effect were stable, one would expect the population of molecules made in the presence of 5FU to be heterogeneous, and the fraction of abnormal molecules to increase progressively. It follows that the responsible constituent must be formed, and also must decay, very rapidly.

Now it should be noted that, besides the structural gene-synthesized messenger, the information transfer system probably involves other constituents responsible for the correct translation of the message, such as for instance the RNA fractions involved in amino acid transfer. The 5FU effect could be due to incorporation into one of these fractions rather than to incorporation into the messenger itself. However, the convergence of the results of the different experiments discussed above strongly suggests that the 5FU effect does reflect a high rate of turnover of the messenger itself.

(c) *Messenger RNA*

Accepting tentatively these conclusions, let us then consider what properties would be required of a cellular constituent, to allow its identification with the structural messenger. These qualifications based on general assumptions, and on the results discussed above, would be as follows:

(1) The "candidate" should be a polynucleotide.

(2) The fraction would presumably be very heterogeneous with respect to molecular weight. However, assuming a coding ratio of 3, the average molecular weight would not be lower than 5×10^5.

(3) It should have a base composition reflecting the base composition of DNA.

(4) It should, at least temporarily or under certain conditions, be found associated with ribosomes, since there are good reasons to believe that ribosomes are the seat of protein synthesis.

(5) It should have a very high rate of turnover and in particular it should saturate with 5FU within less than about 3 min.

It is immediately evident that none of the more classically recognized cellular RNA fractions meets these very restrictive qualifications. Ribosomal RNA, frequently assumed to represent the "template" in protein synthesis, is remarkably homogeneous in molecular weight. Its base composition is similar in different species, and does not reflect the variations in base ratios found in DNA. Moreover it appears to be entirely stable in growing cells (Davern & Meselson, 1960). It incorporates 5FU only in proportion to net increase.

Transfer RNA, or (sRNA) does not reflect DNA in base composition. Its average molecular weight is much lower than the 5×10^5 required for the messenger. Except perhaps for the terminal adenine and cytidine, its rate of incorporation of bases, including in particular 5FU, is not higher than that of ribosomal RNA.

However, a small fraction of RNA, first observed by Volkin & Astrachan (1957) in phage infected *E. coli*, and recently found to exist also in normal yeasts (Yčas & Vincent, 1960) and coli (Gros, *et al.*, 1961), does seem to meet all the qualifications listed above.

This fraction (which we shall designate "messenger RNA" or M-RNA) amounts to only about 3% of the total RNA; it can be separated from other RNA fractions by column fractionation or sedimentation (Fig. 8). Its average sedimentation velocity coefficient is 13, corresponding to a minimum molecular weight of 3×10^5, but since the molecules are presumably far from spherical, the molecular weight is probably much higher. The rate of incorporation of ^{32}P, uracil or 5FU into this fraction is extremely rapid: half saturation is observed in less than 30 sec, indicating a rate of synthesis several hundred times faster than any other RNA fraction. Its half life is also very short, as shown by the disappearance of radioactivity from this fraction in pre-labelled cells. At high concentrations of Mg^{2+} (0·005 M) the fraction tends to associate with the 70s ribosomal particles, while at lower Mg^{2+} concentrations it sediments independently of the ribosomal particles (Gros *et al.*, 1961).

The striking fact, discovered by Volkin & Astrachan, that the base-composition of this fraction in T2-infected cells reflects the base composition of *phage* (rather than bacterial) DNA, had led to the suggestion that it served as a precursor of phage DNA. The agreement between the properties of this fraction and the properties of a short-lived structural messenger suggests that, in phage infected cells as well as in normal cells, this fraction served in fact in the transfer of genetic information from phage DNA to the protein synthesizing centers. This assumption implies that the same protein-forming centers which, in uninfected cells, synthesize bacterial protein, also serve in infected cells to synthesize phage protein according to the new structural information provided by phage DNA, *via* M-RNA. This interpretation is strongly supported by recent observations made with T4 infected *E. coli*. (Brenner, Jacob & Meselson, 1961).

Uninfected cells of *E. coli* were grown in the presence of ^{15}N. They were then infected and resuspended in ^{14}N medium. Following infection, they were exposed to short pulses of ^{32}P or ^{35}S, and the ribosomes were analysed in density gradients. It was found:

(1) that no detectable amounts of ribosomal RNA were synthesized after infection;

(2) labelled M-RNA formed *after* infection became associated with unlabelled ribosomal particles formed *before* infection;

(3) newly formed (i.e. phage-determined) protein, identified by its ^{35}S content, was found associated with the 70s particles before it appeared in the soluble protein fraction.

These observations strongly suggest that phage protein is synthesized by *bacterial* ribosomes formed before infection and associated with *phage-determined* M-RNA. Since the structural information for phage protein could not reside in the bacterial ribosomes, it must be provided by the M-RNA fraction.

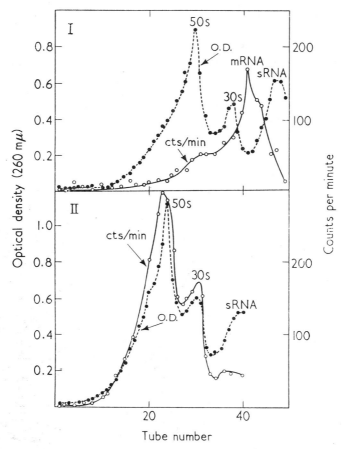

Fig. 8. Incorporation and turnover of uracil in messenger RNA. *E. coli* growing exponentially in broth were incubated for 5 sec with [^{14}C]-uracil. The bacteria were centrifuged, washed and resuspended in the original volume of the same medium containing 100-fold excess of [^{12}C]-uracil. Half the bacteria were then harvested and frozen (I) and the remainder were incubated for 15 min at 37°C (II) prior to harvesting and freezing. The frozen samples were ground with alumina and extracted with tris buffer (2-amino-2 hydroxymethylpropane-1:3-diol) containing 10^{-4}M-Mg, treated with DNase and applied to a sucrose gradient. After 3 hr, sequential samples were taken for determination of radioactivity and absorption at 260 mμ. It may be seen (part I) that after 5 sec, M-RNA is the only labelled fraction, and that subsequently (part II) uracil incorporated into M-RNA is entirely renewed. (From Gros *et al.*, 1961.)

Finally, the recent experiments of Lamfrom (1961) independently repeated by Kruh, Rosa, Dreyfus & Schapira (1961) have shown directly that species specificity in the synthesis of haemoglobin is determined by a "soluble" RNA-containing fraction rather than by the ribosomal fraction. Lamfrom used reconstructed systems, containing ribosomes from one species (rabbit) and soluble fractions from another (sheep) and found that the haemoglobin formed *in vitro* by these systems belonged in part to the type characteristic of the species used to prepare the *soluble* fraction. It is not, of course, positively proved that *inter-specific* differences in haemoglobin structure are gene-determined rather than cytoplasmic, but the assumption seems safe enough. In any case, Lamfrom's experiment proves beyond doubt that the ribosomes cannot be considered to determine entirely (if at all) the specific structure of proteins.

We had stated the problem to be discussed in this section as the choice between the genetic operator model and the cytoplasmic operator model. The adoption of the genetic operator model implies, as we have seen, some very distinctive and specific predictions concerning the behaviour of the intermediate responsible for the transfer of information from gene to protein. These predictions appear to be borne out by a considerable body of evidence which leads actually to a tentative identification of the intermediate in question with one particular RNA fraction. Even if this identification is confirmed by direct experiments, it will remain to be proved, also by direct experiments, that the synthesis of this "M-RNA" fraction is controlled at the genetic level by the repressor-operator interaction.

6. Conclusion

A convenient method of summarizing the conclusions derived in the preceding sections of this paper will be to organize them into a model designed to embody the main elements which we were led to recognize as playing a specific role in the control of protein synthesis; namely, the structural, regulator and operator genes, the operon, and the cytoplasmic repressor. Such a model could be as follows:

The molecular structure of proteins is determined by specific elements, the *structural genes*. These act by forming a cytoplasmic "transcript" of themselves, the structural messenger, which in turn synthesizes the protein. The synthesis of the messenger by the structural gene is a sequential replicative process, which can be initiated only at certain points on the DNA strand, and the cytoplasmic transcription of several, linked, structural genes may depend upon a single initiating point or *operator*. The genes whose activity is thus co-ordinated form an *operon*.

The operator tends to combine (by virtue of possessing a particular base sequence) specifically and reversibly with a certain (RNA) fraction possessing the proper (complementary) sequence. This combination blocks the initiation of cytoplasmic transcription and therefore the formation of the messenger by the structural genes in the whole operon. The specific "repressor" (RNA?), acting with a given operator, is synthesized by a *regulator gene*.

The repressor in certain systems (inducible enzyme systems) tends to combine specifically with certain specific small molecules. The combined repressor has no affinity for the operator, and the combination therefore results in *activation of the operon*.

In other systems (repressible enzyme systems) the repressor by itself is inactive (i.e. it has no affinity for the operator) and is activated only by combining with certain specific small molecules. The combination therefore leads to *inhibition of the operon*.

The structural messenger is an unstable molecule, which is destroyed in the process of information transfer. The rate of messenger synthesis, therefore, in turn controls the rate of protein synthesis.

This model was meant to summarize and express conveniently the properties of the different factors which play a specific role in the control of protein synthesis. In order concretely to represent the functions of these different factors, we have had to introduce some purely speculative assumptions. Let us clearly discriminate the experimentally established conclusions from the speculations:

(1) The most firmly grounded of these conclusions is the existence of *regulator* genes, which control the rate of information-transfer from *structural* genes to proteins, without contributing any information to the proteins themselves. Let us briefly recall the evidence on this point: mutations in the structural gene, which are reflected as alterations of the protein, do not alter the regulatory mechanism. Mutations that alter the regulatory mechanism do not alter the protein and do not map in the structural genes. Structural genes obey the one-gene one-protein principle, while regulator genes may affect the synthesis of several different proteins.

(2) That the regulator gene acts *via* a specific cytoplasmic substance whose effect is to *inhibit* the expression of the structural genes, is equally clearly established by the *trans* effect of the gene, by the different properties exhibited by genetically identical zygotes depending upon the origin of their cytoplasm, and by the fact that absence of the regulator gene, or of its product, results in uncontrolled synthesis of the protein at maximum rates.

(3) That the product of the regulator gene acts directly as a *repressor* (rather than indirectly, as antagonist of an endogenous inducer or other activator) is proved in the case of the *Lac* system (and of the λ lysogenic systems) by the properties of the dominant mutants of the regulator.

(4) The chemical identification of the repressor as an RNA fraction is a logical assumption based only on the *negative* evidence which indicates that it is not a protein.

(5) The existence of an operator, defined as the site of action of the repressor, is deduced from the existence and specificity of action of the repressor. The identification of the operator with the genetic segment which controls sensitivity to the repressor, is strongly suggested by the observation that a *single* operator gene may control the expression of *several adjacent structural genes*, that is to say, by the demonstration of the *operon* as a co-ordinated unit of genetic expression.

The assumption that the operator represents an initiating point for the cytoplasmic transcription of several structural genes is a pure speculation, meant only as an illustration of the fact that the operator controls an integral property of the group of linked genes which form an operon. There is at present no evidence on which to base any assumption on the molecular mechanisms of the operator.

(6) The assumptions made regarding the interaction of the repressor with inducers or co-repressors are among the weakest and vaguest in the model. The idea that specific coupling of inducers to the repressor could result in inactivation of the repressor appears reasonable enough, but it raises a difficulty which we have already pointed out. Since this reaction between repressor and inducer must be stereospecific (for both) it should presumably require a specific enzyme; yet no evidence, genetic or biochemical, has been found for such an enzyme.

(7) The property attributed to the structural messenger of being an unstable intermediate is one of the most specific and novel implications of this scheme; it is required, let us recall, by the kinetics of induction, once the assumption is made that the control systems operate at the genetic level. This leads to a new concept of the mechanism of information transfer, where the protein synthesizing centers (ribosomes) play the role of non-specific constituents which can synthesize different proteins, according to specific instructions which they receive from the genes through M-RNA. The already fairly impressive body of evidence, kinetic and analytical, which supports

this new interpretation of information transfer, is of great interest in itself, even if some of the other assumptions included in the scheme turn out to be incorrect.

These conclusions apply strictly to the bacterial systems from which they were derived; but the fact that adaptive enzyme systems of both types (inducible and repressible) and phage systems appear to obey the same fundamental mechanisms of control, involving the same essential elements, argues strongly for the generality of what may be called "repressive genetic regulation" of protein synthesis.

One is led to wonder whether all or most structural genes (i.e. the synthesis of most proteins) are submitted to repressive regulation. In bacteria, virtually all the enzyme systems which have been adequately studied have proved sensitive to inductive or repressive effects. The old idea that such effects are characteristic only of "non-essential" enzymes is certainly incorrect (although, of course, these effects can be detected only under conditions, natural or artificial, such that the system under study is at least partially non-essential (gratuitous). The results of mutations which abolish the control (such as constitutive mutations) illustrate its physiological importance. Constitutive mutants of the lactose system synthesize 6 to 7% of all their proteins as β-galactosidase. In constitutive mutants of the phosphatase system, 5 to 6% of the total protein is phosphatase. Similar figures have been obtained with other constitutive mutants. It is clear that the cells could not survive the breakdown of more than two or three of the control systems which keep in pace the synthesis of enzyme proteins.

The occurrence of inductive and repressive effects in tissues of higher organisms has been observed in many instances, although it has not proved possible so far to analyse any of these systems in detail (the main difficulty being the creation of controlled conditions of gratuity). It has repeatedly been pointed out that enzymatic adaptation, as studied in micro-organisms, offers a valuable model for the interpretation of biochemical co-ordination within tissues and between organs in higher organisms. The demonstration that adaptive effects in micro-organisms are primarily negative (repressive), that they are controlled by functionally specialized genes and operate at the genetic level, would seem greatly to widen the possibilities of interpretation. The fundamental problem of chemical physiology and of embryology is to understand why tissue cells do not all express, all the time, all the potentialities inherent in their genome. The survival of the organism requires that many, and, in some tissues most, of these potentialities be unexpressed, that is to say *repressed*. Malignancy is adequately described as a breakdown of one or several growth controlling systems, and the genetic origin of this breakdown can hardly be doubted.

According to the strictly structural concept, the genome is considered as a mosaic of independent molecular blue-prints for the building of individual cellular constituents. In the execution of these plans, however, co-ordination is evidently of absolute survival value. The discovery of regulator and operator genes, and of repressive regulation of the activity of structural genes, reveals that the genome contains not only a series of blue-prints, but a co-ordinated program of protein synthesis and the means of controlling its execution.

REFERENCES

Adelberg, E. A. & Umbarger, H. E. (1953). *J. Biol. Chem.* **205**, 475.
Ames, B. N. & Garry, B. (1959). *Proc. Nat. Acad. Sci., Wash.* **45**, 1453.
Ames, B. N., Garry, B. & Herzenberg, L. A. (1960). *J. Gen. Microbiol.* **22**, 369.
Benzer, S. (1953). *Biochim. biophys. Acta,* **11**, 383.
Bertani, G. (1953). *Cold. Spr. Harb. Symp. Quant. Biol.* **18**, 65.

Bertani, G. (1958). *Advanc. Virus Res.* **5**, 151.

Brenner, S., Jacob, F. & Meselson, M. (1961). *Nature*, **190**, 576.

Bussard, A., Naono, S., Gros, F. & Monod, J. (1960). *C. R. Acad. Sci., Paris*, **250**, 4049.

Buttin, G. (1956). Diplôme Et. Sup., Paris.

Buttin, G. (1961). *C. R. Acad. Sci., Paris*, in the press.

Cohen, G. N. & Jacob, F. (1959). *C. R. Acad. Sci., Paris*, **248**, 3490.

Cohen, G. N. & Monod, J. (1957). *Bact. Rev.* **21**, 169.

Cohen, S. S. (1949). *Bact. Rev.* **13**, 1.

Cohen-Bazire, G. & Jolit, M. (1953). *Ann. Inst. Pasteur*, **84**, 1.

Cohn, M. (1957). *Bact. Rev.* **21**, 140.

Cohn, M., Cohen, G. N. & Monod, J. (1953). *C. R. Acad. Sci., Paris*, **236**, 746.

Cohn, M. & Horibata, K. (1959). *J. Bact.* **78**, 624.

Cohn, M. & Monod, J. (1953). In *Adaptation in Micro-organisms*, p. 132. Cambridge University Press.

Cohn, M. & Torriani, A. M. (1952). *J. Immunol.* **69**, 471.

Davern, C. I. & Meselson, M. (1960). *J. Mol. Biol.* **2**, 153.

Demerec, M. (1956). *Cold Spr. Harb. Symp. Quant. Biol.* **21**, 113.

Dienert, F. (1900). *Ann. Inst. Pasteur*, **14**, 139.

Duclaux, E. (1899). *Traité de Microbiologie.* Paris: Masson et Cie.

Echols, H., Garen, A., Garen, S. & Torriani, A. M. (1961). *J. Mol. Biol.*, in the press.

Flaks, J. G. & Cohen, S. S. (1959). *J. Biol. Chem.* **234**, 1501.

Gale, E. F. (1943). *Bact. Rev.* **7**, 139.

Giles, N. H. (1958). *Proc. Xth Intern. Cong. Genetics*, Montreal, **1**, 261.

Gorini, L. & Maas, W. K. (1957). *Biochim. biophys. Acta*, **25**, 208.

Gorini, L. & Maas, W. K. (1958). In *The Chemical Basis of Development*, p. 469. Baltimore: Johns Hopkins Press.

Gros, F., Hiatt, H., Gilbert, W., Kurland, C. G., Risebrough, R. W. & Watson, J. D. (1961). *Nature*, **190**, 581.

Halvorson, H. O. (1960). *Advanc. Enzymol.* in the press.

Hartman, P. E., Loper, J. C. & Serman, D. (1960). *J. Gen. Microbiol.* **22**, 323.

Herzenberg, L. (1959). *Biochim. biophys. Acta*, **31**, 525.

Hogness, D. S., Cohn, M. & Monod, J. (1955). *Biochim. biophys. Acta*, **16**, 99.

Jacob, F. (1954). *Les Bactéries Lysogènes et la Notion de Provirus.* Paris: Masson et Cie.

Jacob, F. (1960). *Harvey Lectures*, 1958–1959, series **54**, 1.

Jacob, F. & Adelberg, E. A. (1959). *C.R. Acad. Sci., Paris*, **249**, 189.

Jacob, F. & Campbell, A. (1959). *C.R. Acad. Sci., Paris*, **248**, 3219.

Jacob, F., Fuerst, C. R. & Wollman, E. L. (1957). *Ann. Inst. Pasteur*, **93**, 724.

Jacob, F. & Monod, J. (1959). *C.R. Acad. Sci., Paris*, **249**, 1282.

Jacob, F., Perrin, D., Sanchez, C. & Monod, J. (1960a). *C.R. Acad. Sci., Paris*, **250**, 1727.

Jacob, F., Schaeffer, P. & Wollman, E. L. (1960b). In *Microbial Genetics*, Xth Symposium of the Society for General Microbiology, p. 67.

Jacob, F. & Wollman, E. L. (1953). *Cold Spr. Harb. Symp. Quant. Biol.* **18**, 101.

Jacob, F. & Wollman, E. L. (1956). *Ann. Inst. Pasteur*, **91**, 486.

Jacob, F. & Wollman, E. L. (1957). In *The Chemical Basis of Heredity*, p. 468. Baltimore: Johns Hopkins Press.

Kaiser, A. D. (1957). *Virology*, **3**, 42.

Kaiser, A. D. & Jacob, F. (1957). *Virology*, **4**, 509.

Kalckar, H. M., Kurahashi, K. & Jordan, E. (1959). *Proc. Nat. Acad. Sci., Wash.* **45**, 1776.

Karstrom, H. (1938). *Ergebn. Enzymforsch.* **7**, 350.

Képès, A. (1960). *Biochim. biophys. Acta*, **40**, 70.

Képès, A., Monod, J. & Jacob, F. (1961). In preparation.

Kogut, M., Pollock, M. & Tridgell, E. J. (1956). *Biochem. J.* **62**, 391.

Kornberg, A., Zimmerman, S. B., Kornberg, S. R. & Josse, J. (1959). *Proc. Nat. Acad. Sci., Wash.* **45**, 772.

Kruh, J., Rosa, J., Dreyfus, J.-C. & Schapira, G. (1961). *Biochim. biophys. Acta*, in the press.

Lamfrom, H. (1961). *J. Mol. Biol.* **3**, 241.

Lederberg, E. (1960). In *Microbial Genetics*, The Xth Symposium of the Society of General Microbiology, p. 115.

Levinthal, C. (1959). In *Structure and Function of Genetic Elements*, Brookhaven Symposia in Biology, p. 76.

Luria, S. E. & Human, M. L. (1950). *J. Bact.* **59**, 551.

Lwoff, A. (1953). *Bact. Rev.* **17**, 269.

Lwoff, A., Siminovitch, L. & Kjeldgaard, N. (1950). *Ann. Inst. Pasteur*, **79**, 815.

Magasanik, B., Magasanik, A. K. & Neidhardt, F. C. (1959). In *A Ciba Symposium on the Regulation of Cell Metabolism*, p. 334. London: Churchill.

Monod, J. (1942). *Recherches sur la Croissance des Cultures Bactériennes*. Paris: Hermann.

Monod, J. (1955). *Exp. Ann. Biochim. Méd.* série XVII, p. 195. Paris: Masson et Cie.

Monod, J. (1956). In *Units of Biological Structure and Function*, p. 7. New York: Academic Press.

Monod, J. (1958). *Rec. Trav. Chim. des Pays-Bas*, **77**, 569.

Monod, J. (1959). *Angew. Chem.* **71**, 685.

Monod, J. & Audureau, A. (1946). *Ann. Inst. Pasteur*, **72**, 868.

Monod, J. & Cohen-Bazire, G. (1953). *C.R. Acad. Sci., Paris*, **236**, 530.

Monod, J. & Cohn, M. (1952). *Advanc. Enzymol.* **13**, 67.

Monod, J. & Cohn, M. (1953). In *Symposium on Microbial Metabolism*. VIth Intern. Cong. of Microbiol., Rome, p. 42.

Monod, J., Pappenheimer, A. M. & Cohen-Bazire, G. (1952), *Biochim. biophys. Acta*, **9**, 648.

Monod, J. & Wollman, E. L. (1947). *Ann. Inst. Pasteur*, **73**, 937.

Naono, S. & Gros, F. (1960a). *C.R. Acad. Sci., Paris*, **250**, 3527.

Naono, S. & Gros, F. (1960b). *C.R. Acad. Sci., Paris*, **250**, 3889.

Neidhardt, F. C. & Magasanik, B. (1956a). *Nature*, **178**, 801.

Neidhardt, F. C. & Magasanik, B. (1956b). *Biochim. biophys. Acta*, **21**, 324.

Novick, A. & Szilard, L., in Novick, A. (1955). *Ann. Rev. Microbiol.* **9**, 97.

Pardee, A. B. (1957). *J. Bact.* **73**, 376.

Pardee, A. B., Jacob, F. & Monod, J. (1959). *J. Mol. Biol.* **1**, 165.

Pardee, A. B. & Prestidge, L. S. (1959). *Biochim. biophys. Acta*, **36**, 545.

Pardee, A. B. & Prestidge, L. S. (1961). In preparation.

Perrin, D., Bussard, A. & Monod, J. (1959). *C.R. Acad. Sci., Paris*, **249**, 778.

Perrin, D., Jacob, F. & Monod, J. (1960). *C.R. Acad. Sci., Paris*, **250**, 155.

Pollock, M. (1950). *Brit. J. Exp. Pathol.* **4**, 739.

Pollock, M. & Perret, J. C. (1951). *Brit. J. Exp. Pathol.* **5**, 387.

Pontecorvo, G. (1958). *Trends in Genetic Analysis*. New York: Columbia University Press.

Rickenberg, H. V., Cohen, G. N., Buttin, G. & Monod, J. (1956). *Ann. Inst. Pasteur*, **91**, 829.

Riley, M., Pardee, A. B., Jacob, F. & Monod, J. (1960). *J. Mol. Biol.* **2**, 216.

Rotman, B. & Spiegelman, S. (1954). *J. Bact.* **68**, 419.

Siminovitch, L. & Jacob, F. (1952). *Ann. Inst. Pasteur*, **83**, 745.

Stanier, R. Y. (1951). *Ann. Rev. Microbiol.* **5**, 35.

Szilard, L. (1960). *Proc. Nat. Acad. Sci., Wash.* **46**, 277.

Torriani, A. M. (1960). *Biochim. biophys. Acta*, **38**, 460.

Umbarger, H. E. (1956). *Science*, **123**, 848.

Vogel, H. J. (1957a). *Proc. Nat. Acad. Sci., Wash.* **43**, 491.

Vogel, H. J. (1957b). In *The Chemical Basis of Heredity*, p. 276. Baltimore: Johns Hopkins Press.

Volkin, E. & Astrachan, L. (1957). In *The Chemical Basis of Heredity*, p. 686. Baltimore: Johns Hopkins Press.

Went, F. C. (1901). *J. Wiss. Bot.* **36**, 611.

Wijesundera, S. & Woods, D. D. (1953). *Biochem. J.* **55**, viii.

Willson, C., Perrin, D., Jacob, F. & Monod, J. (1961). In preparation.

Wollman, E. L. & Jacob, F. (1959). *La Sexualité des Bactéries*. Paris: Masson et Cie.

Yanofsky, C. (1960). *Bact. Rev.* **24**, 221.

Yanofsky, C. & Lennox, E. S. (1959). *Virology*, **8**, 425.

Yarmolinsky, M. B. & Wiesmeyer, H. (1960). *Proc. Nat. Acad. Sci., Wash.* in the press.

Yates, R. A. & Pardee, A. B. (1956). *J. Biol. Chem.* **221**, 757.

Yates, R. A. & Pardee, A. B. (1957). *J. Biol. Chem.* **227**, 677.

Yčas, M. & Vincent, W. S. (1960). *Proc. Nat. Acad. Sci., Wash.* **46**, 804.

Zabin, I., Képès, A. & Monod, J. (1959). *Biochem. Biophys. Res. Comm.* **1**, 289.

J. Mol. Biol. (1961) **3**, 595–617

The Formation of Hybrid DNA Molecules and their use in Studies of DNA Homologies

Carl L. Schildkraut†, Julius Marmur‡ and Paul Doty

*Department of Chemistry, Harvard University, Cambridge 38,
Massachusetts, U.S.A.*

(*Received 29 March 1961*)

Heavy-isotope-labeled DNA has been used to study strand separation and recombination. The rate at which ^{14}N–^{15}N biologically half-labeled DNA can be made to separate into subunits corresponds very closely to what has been predicted for the rate of unwinding of the strands of a double helix.

The use of a phosphodiesterase from *E. coli* (Lehman, 1960) which selectively attacks single-stranded DNA has made it possible to remove unmatched single chain ends from renatured DNA and reduce the remaining differences between renatured and native DNA.

A mixture of heavy-isotope-labeled and normal bacterial DNA was taken through a heating and annealing cycle, treated with the phosphodiesterase and examined by cesium chloride density-gradient centrifugation. Three bands were observed, corresponding to heavy renatured, hybrid, and light renatured DNA. As would be expected for random pairing of complementary strands, the amount of the hybrid was double that of either the heavy or the light component. It has thus been demonstrated that the strands which unite in renaturation are not the same strands that were united in the native DNA but instead are complementary strands originating in different bacterial cells.

The formation of hybrids has been possible only where the heavy and normal DNA samples have a similar overall base composition. It has also been shown for DNA samples isolated from bacteria of different genera in a case where genetic exchange by conjugation has been demonstrated. The evidence for the parallelism between genetic compatibility and the formation of DNA hybrids *in vitro* has led to the proposal that organisms yielding DNA which forms hybrid molecules are genetically and taxonomically related.

1. Introduction

It has been shown recently that it is possible to separate the two strands of DNA molecules and then to reunite these strands so as to restore to a large degree the original helical structure and biological activity (Marmur & Lane, 1960; Doty, Marmur, Eigner & Schildkraut, 1960). The optimum conditions for bringing about maximum restoration, that is, renaturation, have been presented in the preceding paper (Marmur & Doty, 1961).

The work presented in this paper is aimed at the detailed examination of the process of strand separation and recombination, the formation of hybrid DNA molecules with

† Present address: Department of Biochemistry, Stanford University, Palo Alto, California.
‡ Present address: Graduate Department of Biochemistry, Brandeis University, Waltham, Massachusetts.

each of the strands coming from a different source, and the use of such hybrids in detecting overlap in nucleotide sequences in the two samples. If complementarity is the condition for maximum renaturation then it is proper to enquire to what extent deviations from complementarity are required to interfere with the renaturation. The most modest deviation is probably that which occurs between a point mutant and wild-type DNA. In a following paper (Marmur, Lane & Doty, 1961) this kind of recombination is studied with bacterial transformation as the criterion and it is concluded that such small differences do not interfere with renaturation.

The method of study in the present report was density-gradient centrifugation (Meselson, Stahl & Vinograd, 1957) coupled with the use of heavy-isotope-labeled DNA. This proves to be the best technique for following the interaction between homologous and heterologous DNA. Native DNA samples of different GC (guanine-cytosine) contents can be observed separately when banded in the density gradient (Sueoka, Marmur & Doty, 1959; Rolfe & Meselson, 1959). Two native samples having the same GC content can also be observed and their interaction studied if one of them is labeled with heavy isotopes such as ^{15}N, ^{13}C and deuterium (Meselson & Stahl, 1958; Davern & Meselson, 1960; Marmur & Schildkraut, 1961a).

A double helix resulting from the union of a heavy-labeled and non-labeled strand will have an intermediate density. It can be clearly distinguished from the original molecules if they are separated by a sufficient distance in the density gradient. The amount of hybrid molecules formed between different pairs of DNA samples expected to be homologous should vary with the degree of homology.

In order to examine the possibilities of renaturation occurring between DNA from different species that may be genetically related, more quantitative work was necessary to obtain a proper base line for comparison. The removal of non-renatured regions by means of a new phosphodiesterase, specific to single-stranded DNA (Lehman, 1960) provided a means of eliminating artifacts and improving resolution so that hybrid DNA molecules could be clearly and quantitatively resolved. As a result it was possible to examine in some detail the renaturation that was possible between pairs of DNA samples from different sources. This appears to offer a new means of estimating the extent of overlap in DNA sequence between DNA samples from genetically related species.

2. Materials and Methods

(a) *Bacterial strains*. The organisms used in this study together with the base composition and buoyant density of their DNA are listed in Table 1.

(b) *Preparation of DNA*. The isolation procedure of Marmur (1961) was used to extract DNA from cells grown in the exponential phase in Difco brain heart infusion medium. When heavy-isotope-labeled DNA was required the cells were grown in a synthetic medium containing ^{15}NH$_4$Cl as the only nitrogen source and D$_2$O of greater than 99% purity (Marmur & Schildkraut, 1961a).

(c) *Solvents*. In order to eliminate spurious effects due to divalent metal ion contamination, and to inhibit possible attack by nucleases, a chelating agent was always present in the DNA solutions. Citrate ions have proved most convenient in this respect and the standard saline solution (0·15 M-NaCl was used most often) contained 0·015 M Na-citrate. This solvent, standard saline-citrate, will be designated by the abbreviation SSC. Reference will also be made to various multiples of SSC. For example, 2 × SSC is 0·30 M-NaCl and 0·030 M-Na citrate.

(d) *Heating and Annealing*. Solutions containing equal amounts (by weight) of heavy-isotope-labeled and normal DNA in 1·9 × SSC were prepared. In the usual preparation of renatured DNA, 1 ml. was placed in a 2 ml. glass stoppered volumetric flask, immersed in a bath of boiling water for 10 min and immediately transferred to a bath thermostated at 68 °C. After 2 hr the temperature was lowered in approximately 5 °C steps at intervals of 15 min until 25 °C was reached. Although all results reported here were obtained by this cooling procedure, it has been found that the salt concentration and the rate of cooling from 68 °C to room temperature can be varied somewhat without noticeable change in the final results.

TABLE 1

Bacteria from which DNA was isolated for use in studies of hybrid formation

Species	Strain	Source	%GC†	Density (g/cm³)
Bacillus brevis	9999	ATCC‡		1·704
Bacillus macerans	7069	ATCC		1·713
Bacillus megaterium		Univ. of Penn.	38	1·697
Bacillus natto	MB-275	A Demain		1·703
Bacillus subtilis	168	Yale University	42	1·703
Clostridium perfringens	87b	M. Mandel	31	1·691
Diplococcus pneumoniae	R-36A	R. Hotchkiss	39	1·701
Erwinia carotovora	8061	ATCC	54	1·709
Escherichia coli	B	S. Luria	50	1·710
Escherichia coli	44 B	M. Mandel		1·710
Escherichia coli	C-600	R. Appleyard		
Escherichia coli	K12 (W678)	J. Lederberg	50	1·710
Escherichia coli	TAU⁻	S. Cohen	50	1·710
Escherichia coli	W-3110	M. Yarmolinsky		1·710
Escherichia coli	I	A. N. Belozersky	52	
Escherichia coli	II-IV-4	A. N. Belozersky	67	
Escherichia freundii	17	H. Blechman		1·710
Escherichia freundii	5610-52	M. Mandel		
Salmonella arizona	PCl45	Walter Reed Hosp.		1·712
Salmonella ballerup	ETS107	Walter Reed Hosp.		
Salmonella typhimurium	LT-2	M. Demerec	50, 54	1·712
Salmonella typhimurium	ETS9	Walter Reed Hosp.		
Salmonella typhosa	643	Walter Reed Hosp.	53	1·711
Shigella dysenteriae	15	S. Luria	53	1·710

† The GC contents of the DNA samples were all obtained from Belozersky & Spirin (1960).
‡ American Type Culture Collection.

(e) *CsCl.* Optical grade CsCl was obtained from the Maywood Chemical Co., Maywood, N.J. A concentrated stock solution was prepared by dissolving 130 g CsCl in 70 ml. of 0·02 M-tris buffer (2-amino-2-hydroxymethylpropane-1:3-diol) pH 8·5. The final solution was passed through a medium grade sintered glass filter to remove large amounts of solid material that seemed to contaminate the solid CsCl. When solutions in which the DNA had been annealed at low concentrations had to be brought up to the correct density for centrifugation, the solid CsCl was added directly. These solutions were not subsequently filtered.

(f) *Density-gradient centrifugation.* The technique described by Meselson *et al.* (1957) was followed. CsCl was used to bring the density of the DNA solution being examined to values between 1·71 g/cm³ and 1·75 g/cm³, depending on the specific sample involved. This could be done most easily by mixing the DNA solution with the concentrated CsCl stock solution. If the DNA solution was not concentrated enough, as was generally the case, this procedure would diminish the final concentration of the DNA below that required for accurate observation. To avoid this problem it was possible to use solid CsCl to give the proper density. To 1·03 g

CsCl was added 0·80 ml. of the slow cooled DNA solution (still in the same saline citrate solution) and 0·01 ml. of a stock solution of 50 μg/ml. of a DNA sample whose density is well established and can be used as a standard. DNA from *Cl. perfringens* is generally used since its density is the lowest observed thus far. This places it in a position where its band cannot coincide with any heavy-isotope-labeled DNA samples. The final adjustment of density was made by the addition of a small amount of water or solid CsCl. The measurement of the density at various stages of the adjustment is facilitated by the use of the linear relation between refractive index and density (Meselson, 1958, personal communication; Ifft, Voet & Vinograd, 1961)

$$\rho^{25\cdot0°C} = 10\cdot8601\ n_D^{25\cdot0°C} - 13\cdot4974.$$

Approximately 0·75 ml. of the final CsCl solution was placed in a cell containing a plastic (Kel-F) centerpiece and centrifuged in a Spinco model E analytical ultracentrifuge at 44,770 rev/min at 25°C. After 20 hr of centrifugation, ultraviolet absorption photographs were taken on Kodak commercial film. It was clear that equilibrium had been closely approached in 20 hr since there was no difference in calculated densities when photographs taken after 48 hr were used. Moreover, the variances did not decrease by more than 5% between 20 and 48 hr. Tracings were made with a Joyce-Loebl double-beam recording microdensitometer with an effective slit width of 50 microns in the film dimension.

Densities were calculated by using the position of the standard DNA as a reference. The CsCl density gradient was obtained from the data of Ifft *et al.* (1961).

(g) E. Coli *phosphodiesterase.* This enzyme, which preferentially hydrolyses single-stranded DNA, was kindly supplied by Dr. L. Grossman and had been prepared in highly purified form by the method of Lehman (1961, personal communication). The enzymatic hydrolysis was carried out by the following modification (L. Grossman, 1961, personal communication) of the method of Lehman (1960). The annealed sample, usually at a total DNA concentration of 10 μg/ml., was dialysed against two changes of 0·067 M-glycine buffer, pH 9·2. To 0·7 ml. of the dialysed DNA solution was added 1·6 μmoles of MgCl$_2$, 2·4 μmoles of 2-mercaptoethanol, and approximately 25 to 75 units of crystallized *E. coli* phosphodiesterase. The final mixture (about 0·8 ml.) was incubated at 37°C for 3 hr. Solid CsCl and the standard DNA were then added and the solution was made up to the proper density.

3. Strand Separation

In order to form hybrid molecules by renaturation it is first necessary to examine in some detail the conditions necessary for strand separation. A very useful material for these studies has been the ^{14}N-^{15}N-half-labeled or biological "hybrid" DNA molecules first described by Meselson & Stahl (1958). DNA having one strand ^{15}N-labeled and the other strand normal was isolated from *E. coli* B according to the method of Marmur (1961). It was found that the biologically formed "hybrid" DNA banded at a density of 1·717 g/cm³ in the CsCl gradient, exactly between that of fully ^{15}N-labeled and normal DNA from *E. coli*. The band profile is shown in the top tracing of Fig. 1.

(a) *Methods for inducing strand separation*

In terms of the Watson-Crick model (1953) for DNA, the dissociation of the two strands obviously requires first the rupture of the hydrogen bonds uniting the base pairs and second the uncoiling and diffusing apart of the two strands. The rupture of the hydrogen bonds has been shown to occur in a relatively narrow temperature range at elevated temperatures but the diffusion apart of the strands did not automatically follow (Rice & Doty, 1957). Later it was found that lower ionic strengths and more dilute solutions did permit the diffusing apart to occur (Doty *et al.*, 1960;

Eigner & Doty, 1961) and prevented the non-specific re-association from occurring at lower temperatures where numerous hydrogen bonds re-form.

Meanwhile it had been found by Meselson & Stahl (1958) that the DNAs containing ^{14}N and ^{15}N subunits, each of which replicate in a conservative manner, produced two bands on heating in CsCl for 30 min. Thus they demonstrated that the two molecular subunits dissociated on heating. Meselson & Stahl left the question of the molecular structure of the subunits open to further investigation. This prompted us to isolate DNA from cells grown according to the procedure described by Meselson & Stahl (1958), and to see how the hybrid DNA reacted to the various methods that produce strand separation. The evidence that these methods do produce strand separation can be found in the papers to which we will refer. We are concerned here only with their effect on the biological hybrids and the detailed investigation of this effect in order to learn more about strand separation.

First we sought to reproduce the Meselson-Stahl results. They had observed the two-band pattern after heating a cell lysate in CsCl for 30 min. We found a similar pattern was produced by heating a sample of our purified biological hybrid at 100°C for only 10 min at a concentration of 20 μg/ml. in SSC. The tracings of the bands before and after heat treatment can be seen at the top and bottom frames of Fig. 1. The apparent density of the native, hybrid DNA is 1·717 g/cm^3; that of the two denatured bands is greater by 0·007 and 0·023 g/cm^3 respectively. The average of these is, of course, greater than that of the native hybrid DNA by the 0·015 g/cm^3 characteristic of denatured DNA. As a check, it was demonstrated that the same two-band pattern results when ^{14}N and ^{15}N DNA samples are separately denatured and then mixed.

Three other methods which have recently been shown to bring about strand separation have also caused the disappearance of the native hybrid band and the formation of two heavier bands. This additional information, combined with the demonstration of Doty et al. (1960) that the above heating procedure produces strand separation in bacterial DNA samples, now leads us to conclude that Meselson & Stahl were indeed observing the separation of single strands. We shall return to this point in the discussion, and now proceed to summarize the experimental observations on the biological hybrids.

By the addition of 8 M-urea the temperature for thermal denaturation can be reduced by nearly 20°C (Rice & Doty, 1957). Consequently, by using 0·01 M-salt and 8 M-urea it was found possible to bring about strand separation by heating DNA only to 65°C (Eigner & Doty, 1961). When this procedure was applied to the biological hybrid DNA, two bands formed in the CsCl density gradient, exactly as was the case after heating at 100°C for 30 min in CsCl or for 10 min in SSC.

Formamide, being a more potent hydrogen bonding agent, and being miscible with water in all proportions, offers a substantial improvement over urea. Marmur & Ts'o (1961) have shown that in 95% formamide strand separation readily occurs at room temperature. Again, when this is applied to the biological hybrids the typical two-band pattern is produced.

Acid and base titration, by virtue of their ability to disrupt hydrogen bonding, offer another route to strand separation. It has now been shown (Cox, Marmur & Doty, 1961) that strand separation occurs when the pH is lowered to 2·5 or raised to 12·0. These two pH values lie just outside the region in which the hypochromic shift (at 260 mμ) indicates that the helix-coil transition is complete. At pH values on the other side of the hypochromic shift, strand separation has not occurred. When the biological hybrids are exposed to conditions in which the pH is below 2·5 or above 12·0, two bands again result in the CsCl gradient.

Thus it is seen that several different procedures lead to the formation of the same two-band pattern in the CsCl density gradient, and that all of them are a consequence of the prior breaking of hydrogen bonds uniting the two DNA strands. While the present work is not aimed at establishing how many subunits make up DNA molecules, it does appear that the extension to the biological hybrid DNA of the consistent finding that strand separation occurs whenever hydrogen bonds are cooperatively broken, shows that only one type of bonding is uniting the subunits. Since this can only be identified with that assumed in the Watson-Crick structure, there seems to be no alternative to equating the subunits, observed as two bands, with the two strands in the native DNA molecule.

(b) *Temperature dependence of strand separation*

With the demonstration that the two-band pattern of the biological hybrid DNA is indicative of strand separation we now attempt to use this material to obtain further details about the process of strand separation. The studies of Marmur & Doty (1959) demonstrated that it should be possible to melt out the molecules having a high AT (adenine-thymine) content leaving GC (guanine-cytosine)-rich molecules undenatured. This can be observed to occur with the biological hybrids by exposing them to temperatures only slightly above the temperature, T_m, of the midpoint of the absorbance rise. T_m is the same for all *E. coli* DNA samples, including the biological hybrids (Marmur & Doty, 1961). The band profile obtained after heating the latter at T_m for 20 min is shown as the second tracing in Fig. 1. Since there is no difference from the profile of the native sample, it is concluded that all the molecules remain in the helical form. To melt selectively the AT-rich molecules, it is necessary to expose them to temperatures a few degrees above T_m. The results of doing this are shown as the third tracing of Fig. 1, where it is seen that approximately 25% of the molecules undergo strand separation. The molecules showing the greater resistance to heat denaturation are seen to have a higher density, in agreement with the hypothesis that they should be GC-rich.

(c) *Kinetics of strand separation*

Since a small but detectable thermal depolymerization always accompanies strand separation it is desirable to know the minimum time at the elevated temperature required to obtain complete separation, and thus avoid this undesirable side effect. In addition, since a few theoretical calculations of the rate of uncoiling of the DNA molecule have been reported (Kuhn, 1957; Longuet-Higgins & Zimm, 1960), some comparison with experimental findings seems to be in order. The rate of separation into subunits can be observed experimentally by following the disappearance of the hybrid band or the appearance of the two heavier bands. The results presented in Plate I and Fig. 2 show that the complete disappearance of material of density 1·717 g/cm³ takes slightly over 60 sec at 100 °C. This is very close indeed to the time calculated by Kuhn (1957) for the unwinding of a double helix of comparable length in which no bonds are holding the strands together. As is evident from the absorbance-temperature profile, the rise in absorbance is complete at 95 °C. When the temperature of a DNA solution

PLATE I. Ultraviolet absorption photographs of ^{14}N-^{15}N-labeled DNA showing different stages of the thermally induced separation into subunits. The band at the far right has been used as a standard and is DNA isolated from *D. pneumoniae*. The other band in the top photograph is the biologically formed hybrid DNA. The second photograph shows the stability of the hybrid to a 20 min exposure at 93·8°C. At 100°C the number of molecules separating increases rapidly with time of exposure as shown in the next 3 photographs. The samples were heated in SSC at 20 μg/ml. for 30 sec, 1 min and 10 min, respectively.

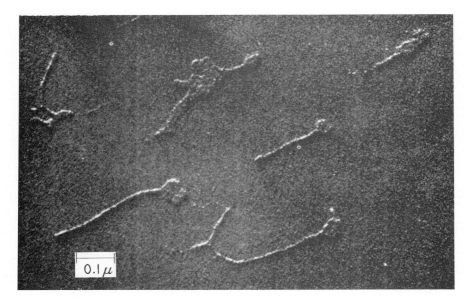

PLATE II. Electron micrograph of renatured *D. pneumoniae* DNA showing an unusually large concentration of renatured molecules with circular regions at one or both ends. Magnification × 100,000.

is raised to 100°C few hydrogen bonds should exist between strands and Kuhn's model seems applicable to such a molecule.

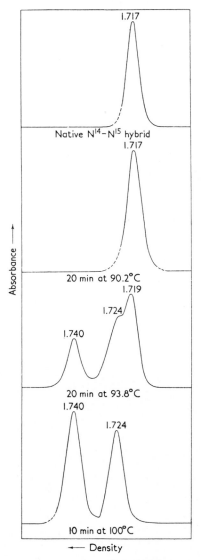

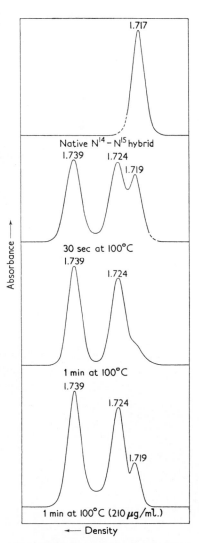

FIG. 1. The temperature dependence of strand separation. Microdensitometer tracings of ultraviolet absorption photographs of samples equilibrated in a CsCl density gradient. Each sample was heated in SSC at a concentration of 20 μg/ml.

FIG. 2. Kinetics of strand separation. The results shown in the second and third tracings were obtained by heating at a concentration of 20 μg/ml. in SSC.

(d) *Dependence of the time of strand separation on viscosity*

In the theoretical considerations just mentioned in connection with the kinetics of strand separation, the variation of the rate of unwinding with the viscosity of the medium is also considered (Longuet-Higgins & Zimm, 1960). An equation is derived which predicts that strand separation will take longer as the viscosity of the medium is increased. In the tracing shown at the bottom of Fig. 2 the time necessary for separation into subunits is shown to have increased with increased DNA concentration. Similar results have also been obtained by using concentrated sucrose solutions to produce a medium of higher viscosity (Schildkraut, Wierzchowski & Doty, 1961).

(e) *DNA from bacteriophage ΦX174*

The results of the studies with the biological hybrids indicate that the most convenient way to obtain single-stranded DNA is to heat for 10 min in SSC at a concentration of from 10 to 20 μg/ml. and quickly cool. Since one kind of DNA, that from bacteriophage ΦX174[†], is known to be single-stranded (Sinsheimer, 1959), it is expected that our thermal treatment would produce no density change in this material. This has, indeed, been found to be the case.

4. Renaturation

The general features of the specific recombination of single DNA strands to form the native helical structure have been described in the first publications (Marmur & Lane, 1960; Doty *et al.*, 1960), and methods of optimizing conditions to ensure the maximum renaturation were described in the preceding paper (Marmur & Doty, 1961). Here we have only one point to add to this: it has to do with the reduction or elimination of the remaining differences between renatured and native DNA. In the earlier work it was evident that renaturation was not complete. For example, the density had returned only about 75% of the way from the denatured to the native density. It was thought that this was due to the inequality in length of the recombined strands. DNA molecules are apt to be broken during isolation and the single strands suffer some hydrolysis during the thermal treatment. Using the measurements on rate of bond scission by Eigner, Boedtker & Michaels (1961), we would estimate that strands originally in molecules of 10,000,000 molecular weight would undergo about three scissions each as a result of a typical heat treatment: 10 min at 100°C and 100 min at 68°C. This would mean that two strands meeting in a complementary region may differ in length by as much as one quarter to one half, on the average. If they renatured completely, there would still remain 20 to 33% of the weight unrenatured and the incomplete recovery of the characteristics of the native DNA would be explained.

In the course of electron microscope studies with Professor C. E. Hall we frequently saw renatured molecules with circular regions at one or both ends that would be consistent with the protrusion of a single chain end beyond the re-formed helical region just described. One electron micrograph showing an unusually large concentration of these in a single photograph is shown in Plate II. While this interpretation lacks proof, the observation is nevertheless strikingly similar to what was expected from the point of view outlined.

From this evidence it appeared that in order to improve the extent of renaturation the unmatched single chain ends would have to be removed. The possibility of doing this suddenly became available with the discovery by Lehman (1960) of a phosphodiesterase from *E. coli* that selectively attacked single-stranded DNA. This was tested

† This material was kindly provided by Professor Robert L. Sinsheimer.

on renatured DNA from *B. subtilis* as shown in Fig. 3. The denatured DNA showed the characteristic increase in density. Upon renaturation at a concentration low enough to ensure that some denatured DNA remained, the result shown in the third frame was obtained. The density of the renatured DNA is, as expected, 0·004 g/cm³ higher than that of the native DNA. After treatment of this sample with the phosphodiesterase it is seen, in the bottom frame, that the denatured shoulder has been

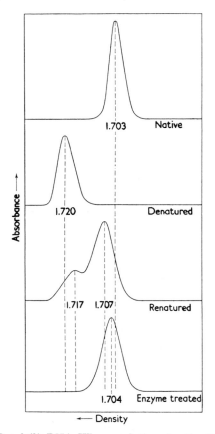

FIG. 3. Renaturation of *B. subtilis* DNA. When a solution of native DNA (band profile shown in top tracing) is heated for 10 min at 10 μg/ml. at 100°C in 1·9 × SSC and quickly cooled, the density increases 0·017 g/cm³ (second tracing). A portion of this solution was annealed and, as is evident from the third tracing, about 80% of the DNA renatures. This concentration (10 μg/ml.) was chosen so that some denatured material would still be present. Treatment with the *E. coli* phosphodiesterase causes the complete disappearance of the denatured band and a decrease in the buoyant density of the renatured band (bottom tracing).

completely removed and the density of the renatured band has been reduced to within 0·001 g/cm³ of the native DNA density. Thus the possibility of removing the denatured unmatched ends from renatured molecules has been demonstrated. As will be seen, this procedure greatly improves the resolution of bands encountered in the study of artificially produced hybrid molecules.

It may be of interest to mention at this point our failure thus far to produce renaturation of DNA from bacterial sources without thermal treatment of the type described here and in the preceding papers. We have employed urea and moderate temperatures, formamide, and low pH to produce strand separation; and then, by gradual withdrawal of the hydrogen-bond breaking agent attempted to renature the strands. Such attempts have thus far failed. While our experiments have not been exhaustive, it appears quite possible that renaturation cannot be achieved in this way. If so, a likely explanation would be that at room temperature the decreased Brownian motion of the chain segments is no longer sufficient to provide the mobility required for the exploration necessary to create nuclei, that is, for complementary regions to find each other rather than become frozen in mismatched pairings.

In connection with renaturation, we might return just briefly to the case of ΦX DNA and mention that the density does not decrease but remains exactly 1·723 g/cm³ after heating and annealing. This is as should be expected for single-stranded material whose complementary strands are not present during the annealing process.

5. Hybrid DNA Molecules produced by Renaturation

We now turn to the problem of demonstrating that the strands which unite in renaturation are not the same strands that were united in the native DNA but instead are complementary strands originating in different cells. This requires that renaturation be studied in a solution of two homologous DNA samples, one of which carries a distinctive label. With density gradient ultracentrifugation offering such good resolution it was natural to turn to the introduction of heavy atoms or to heavy isotope substitution. If the pairing of DNA strands is restrained only by the condition of complementarity one would expect that renaturation would lead to three bands, one heavy, one light and one intermediate, corresponding in density to a hybrid composed of one normal and one heavy strand. Moreover, the amount of the intermediate would be expected to be double that of either the heavy or light component provided that equal amounts of the two DNA samples had been mixed originally.

If this situation is found, then it will be of interest to attempt to form hybrid molecules from other pairs of DNA samples in which some differences exist. In order to have sufficient resolution to exploit this means of searching for such effects it is necessary to have a considerable density difference between the two samples. Practical considerations indicate that it should be at least 0·040 g/cm³. Thus the use of ¹⁵N label in one sample, such as first used by Meselson & Stahl (1958), is not sufficient since this gives a separation of only about 0·015 g/cm³.

(a) Preliminary experiments

Our first attempts to produce DNA with substantially higher density were with 5-bromouracil substitution for thymine inasmuch as this had been accomplished in both bacteria and bacteriophage (Dunn & Smith, 1954; Zamenhof & Griboff, 1954). A sample of such DNA from *B. subtilis* (Ephrati-Elizur & Zamenhof, 1959) was kindly supplied to us by Professor W. Szybalski. Unfortunately, when this sample was carried through the heating and annealing cycle it displayed a quite broad band with little indication of renaturation. The molecular weight had evidently been substantially reduced either by enzymatic attack during isolation or as a result of the greater heat sensitivity of this substituted DNA.

Nevertheless this material was mixed with high molecular weight, normal *B. subtilis* DNA, and put through the heating and annealing cycle. The banding of the mixture before and after heating is shown in Fig. 4. It is seen that the thermal treatment has

produced a broad band without any resolution of the expected hybrid. Indeed the
result is such as to indicate that the smaller heavy-labeled chains have been bound
in a random manner to the larger, normal density chains producing a complete
spectrum of densities which hides what little renaturation may have occurred.

Initial attempts to study 5-bromodeoxyuridine-labeled DNA from T4r+ bacteriophage led to similar disappointing results.

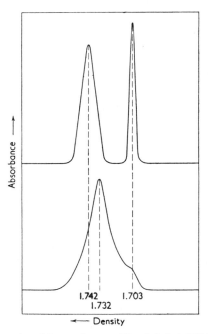

FIG. 4. Hybrid formation using 5-bromodeoxyuridine-labeled DNA. The upper tracing shows
native normal and 5-bromodeoxyuridine-labeled DNA isolated from *B. subtilis*. The properties of
the latter DNA have been described in detail elsewhere (Szybalski, Opara-Kubinska, Lorkiewicz,
Ephrati-Elizur & Zamenhof, 1960). Portions of these two samples were heated together in 1·9 × SSC
at a concentration of 20 μg/ml. each for 5 min at 100°C. The solution was then allowed to cool to
room temperature over a period of about 5 hr and CsCl was added to obtain the proper density.
The lower tracing suggests that the expected renatured species are present in solution, but the
broadening of bands due to degradation and aggregation obscures the results.

(b) ¹⁵N-deuterated DNA

Under these circumstances we turned to a combination of ¹⁵N and deuterium to
provide the density increase. After considerable manipulation this combined labeling
has been shown to be successful in producing DNA with a density increase of about
0·040 g/cm³ and the details have now been published elsewhere (Marmur & Schildkraut, 1961a).

We were, therefore, in a position to proceed with the experiment that would
demonstrate whether or not hybrid DNA molecules form from random mating of
complementary strands. The steps are shown for *B. subtilis* DNA in Fig. 5. In the
first two frames the density profiles of the normal and labeled native DNA samples
are shown. These were then mixed so that the concentration of each was 5 μg/ml. in
1·9 × SSC, and the heating and annealing cycle was carried out. At this concentration
only partial renaturation is expected. So low a concentration was chosen in order to

minimize the association of renatured DNA molecules since this would lead to an unnecessary smearing of the pattern. Consequently, at this concentration, one would expect to find 5 species of different density: heavy denatured DNA, heavy renatured DNA, hybrid DNA, light denatured DNA and light renatured DNA. The result, seen in the third frame of Fig. 5, is of precisely this character.

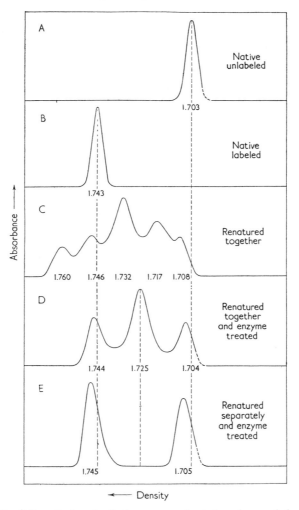

FIG. 5. The effect of *E. coli* phosphodiesterase on a heated and annealed mixture of heavy-labeled and normal *B. subtilis* DNA.

(A) and (B) Native samples.

(C) Mixed so that the final concentration of each was 5 μg/ml. and then heated and annealed.

(D) The sample shown in (C) was dialysed against 0·067 M-glycine buffer and incubated with the *E. coli* phosphodiesterase.

(E) Heated and annealed separately, treated with the phosphodiesterase and then mixed.

The application of the phosphodiesterase previously described would be expected to remove the first and fourth bands as well as shift the density of the remaining species back toward the native values. The band profile after this treatment is shown in the fourth frame and it is seen that the expected changes had been brought about.

As a control, the two native DNA samples were carried through the identical operations described above and then mixed and banded. The result is shown in the last frame where it is seen that the densities match within 0.001 g/cm^3 that of the outer bands of the frame above and that no hybrid is evident.

Similar experiments were carried out on *E. coli* DNA: the band profiles corresponding to frames C and D of Fig. 5 are shown in Fig. 6. The result is seen to be the same.

It is to be noted that the area under the hybrid band in these two experiments is about equal in area to that of the two outer bands.† Thus the conclusion can be drawn that the earlier and much more primitive experiment along these lines (Doty *et al.*, 1960) has been substantiated and the formation of hybrid DNA molecules by random pairing of complementary strands can be taken as proved.

It is useful to note in passing that the phosphodiesterase treatment had one other effect not evident in the band profiles. Without enzyme treatment part of the hybrid band would begin to form much earlier than the other, indicating high molecular weight material suggestive of some aggregation despite our efforts to avoid it. However, after enzyme treatment all three bands appear at the same rate. Thus, the enzymatic treatment not only improves the resolution by removing denatured material but appears to break up aggregated renatured DNA molecules as well.

(c) *Further investigation of 5-bromodeoxyuridine-substituted DNA*

With the successful conclusion of the above work, we have returned again to the 5-bromodeoxyuridine-labeled bacteriophage DNA. With the guide that the above work provides it has been possible to overcome some of the problems that appear to be associated with the greater inherent instability of this material. As a consequence results nearly as satisfactory as those shown in Fig. 5 and 6 are being obtained at present. This system has been used to study DNA homologies among the T-even bacteriophages (Schildkraut, Marmur, Wierzchowski, Green & Doty, 1961).

(d) *Kinetics of hybrid formation*

As a further check on the above interpretation and in order to show further details of the process of hybrid formation, samples were withdrawn from the annealing mixtures, held at 68°C, during renaturation. These were quickly cooled, in order to "freeze" the distribution present and band profiles were obtained in the ultracentrifuge. A number of profiles are shown in Fig. 7. From these, the progress of the renaturation can be followed and it is seen that it reaches about 50% completion in the first hour.

6. Hybrid DNA Formation among Various Strains of *E. coli*

It is to be expected from their close taxonomic, physiological and genetic relationships that all strains identified as *E. coli* should yield DNA which will form the 5-band pattern upon the renaturation discussed in the previous section. Hybrid formation in a number of pairs has been examined. The results are summarized in Plate III where the ultraviolet absorption photographs are reproduced. It is seen that the 5-band pattern is produced in all but the one case where sequence homology did not exist.

† The linearity of response of our optical system is not yet such as to justify a precise measurement of the area under these curves.

This DNA isolated from *E. coli* 11-IV-4, an alkali-producing form, had a GC content of 67%, which is far from that characteristic of *E. coli* DNA. Belozersky (1957) has reported that these cells were produced by an alteration of the properties of *E. coli* CM caused by growing it together with Breslau bacteria No. 70 killed by heat.

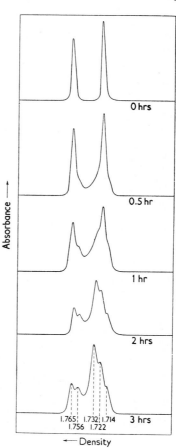

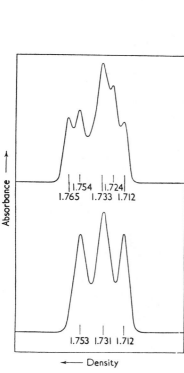

FIG. 6. The effect of *E. coli* phosphodiesterase on a heated and annealed mixture of heavy-labeled DNA from *E. coli* B and normal DNA from *E. coli* K12. The concentration of each sample was 5 μg/ml. each during the heating and annealing cycle.

FIG. 7. Kinetics of hybrid formation. Heavy-labeled and unlabeled DNA from *E. coli* B were heated together at 100°C for 10 min under the usual conditions of 5 μg/ml. each and in 1·9 × SSC. The mixture was placed at 68°C and quickly cooled portions were centrifuged after the addition of CsCl. The tracings are shown above. These samples have, of course, not been treated with the phosphodiesterase.

The failure to obtain a hybrid band confirms the gross difference from *E. coli* DNA, but of course does not help to eliminate the argument that these special cells may have been a contaminant rather than a variant. As seen in the photograph the location of the peaks and their intensities is variable. For this reason three of the samples were treated with phosphodiesterase. When this was done the patterns

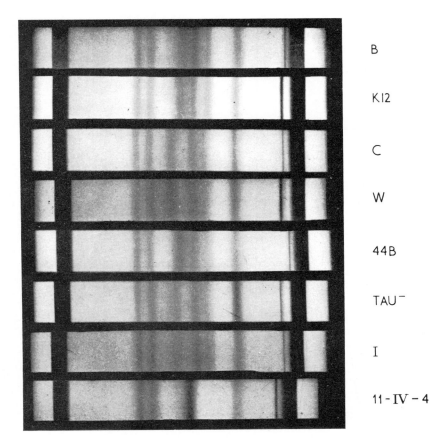

B

K12

C

W

44B

TAU⁻

I

11-IV-4

PLATE III. Hybrid formation between *E. coli* B and other *E. coli* strains. *E. coli* B DNA, labeled with ^{15}N and deuterium, was mixed with DNA from each of the strains listed above and heated and annealed in separate experiments. The concentrations were 5 μg/ml. each, and the other conditions were as described in the section on methods. Each of 8 different ultracentrifuge runs is represented above by a typical ultraviolet absorption photograph. Six DNA bands appear in all but the last example. The photographs have been lined up according to the position of the standard band at the far right, which is DNA from *Cl. perfringens*.

gave three bands and showed greatly reduced differentiation. We are, therefore, now in a position to decide if there are any small reproducible differences in the renaturation of various *E. coli* pairs that may indicate small differences in sequence, but the effect is at most quite small. We can conclude that the homology is essentially complete.

7. The Aggregation of Renatured DNA

In exploring further the extent to which hybrids can form between somewhat different DNA strands we encountered more examples of the way in which renatured DNA molecules could aggregate so as to indicate hybrid formation even though it was indeed an artifact. While these difficulties were always removed by phosphodiesterase treatment it is of interest to record a few examples in order to indicate the nature of this process and to emphasize the need for phosphodiesterase treatment.

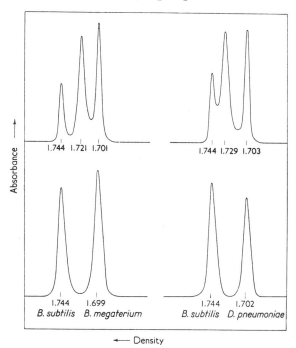

Fig. 8. Non-specific aggregation. ^{15}N-deuterated DNA from *B. subtilis* was heated with either DNA from *B. megaterium* or *D. pneumoniae* at a concentration of 50 μg/ml. each in 1·9 × SSC. The mixtures were annealed under the usual conditions. Portions were centrifuged in CsCl before (upper tracings) and after (lower tracings) treatment with the *E. coli* phosphodiesterase.

A number of experiments were carried out using the heavy-labeled *B. subtilis* DNA and various normal DNA samples at concentrations of about 50 μg/ml. where no denatured strands remain after renaturation. (This is in contrast to the standard procedure which involves tenfold lower concentrations.) Results are shown in Fig. 8 for renatured mixtures of *B. subtilis* DNA with either *B. megaterium* or *D. pneumoniae* DNA. The band profiles before enzyme treatment are shown at the top. Here we see three well-resolved band patterns highly suggestive of hybrid formation. However, the enzyme treatment completely eliminates the central band and increases the amount of material in the outer bands. Thus the center band must have been the result of homogeneous renatured DNA molecules being held together by the association of unpaired, protruding chains. The rather remarkable feature is that the aggregation must consist mostly of two such renatured molecules, one of each type, since otherwise a whole range of densities would be displayed.

It was also possible to form aggregates between *B. subtilis* and *B. brevis* DNA by heating and annealing at a concentration of 40 μg/ml. each, and between *B. subtilis* and *B. macerans* DNA by heating and annealing at 50 μg/ml. each. The GC content of *B. subtilis* DNA is similar to that of *B. brevis* while that of *B. macerans* is somewhat higher. In this case the density of the aggregates was considerably higher than the average of the two renatured "parent" samples. This was true for practically every case of aggregation except two, one shown in Fig. 8 and the other in Fig. 13. When *B. subtilis* and calf thymus DNA were heated together and annealed at a total concentration of 100 μg/ml., the heavy labeled *B. subtilis* DNA reformed completely while the calf thymus DNA maintained the denatured density. No intermediate band was visible.

As a final example we report the rather puzzling case of aggregation occurring upon mixing renatured solutions at room temperature. Heavy and light *B. subtilis* DNA at a concentration of 10 μg/ml. were heated and annealed separately; the band profiles are shown in the first 4 frames of Fig. 9. The two renatured solutions (C and D) were then mixed at room temperature and examined in the density gradient. The result is shown in Fig. 9 (E); it is seen that a central band has formed, apparently at the expense of the renatured molecules. Thus it would appear that unpaired chain ends which cannot be satisfied in the separated solutions can pair in the mixture. However, this artifact too was removed by phosphodiesterase treatment, as shown in the final frame of Fig. 9. It is important to note that this aggregation of renatured molecules at room temperature occurred between homologous molecules, not molecules from two different sources.

Thus the formation of bands of intermediate density that have the appearance of true hybrid DNA molecules has been clarified, and the importance of eliminating such false bands with phosphodiesterase treatment has been emphasized.

8. Interspecies Hybridization of DNA

We are now in a position to examine the possibility of forming DNA hybrids between strands of DNA coming from different bacterial species. By heating and annealing each of several DNA samples with the heavy-labeled *B. subtilis* DNA, it became evident, even without recourse to the enzyme treatment, that hybrid DNA molecules did not form. Such results are shown in Fig. 10. These and other results could be summarized by stating that hybrid DNA formation was not observed, even at double the usual DNA concentration, if the two DNA samples had significantly different base-compositions.

With the search for hybrid formation narrowed to DNA samples having essentially the same composition, it was natural to look first at DNA from two organisms of the same species which were known to be genetically related by virtue of their ability to undergo transformation with each other's DNA. This is true for *B. subtilis* and *B. natto* (Marmur, Seaman & Levine, 1961). The results for heating, annealing and enzyme treating this pair of DNA samples are shown in Fig. 11. It is seen that substantial

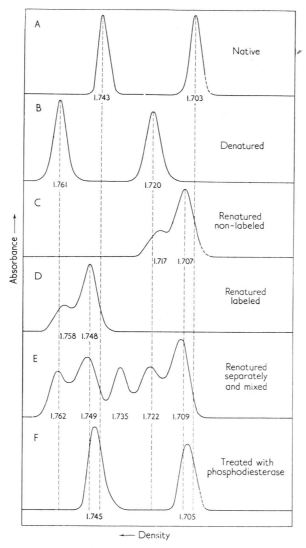

FIG. 9. Control experiment for hybrid formation with *B. subtilis* DNA. A stock solution in $1·9 \times$ SSC of normal *B. subtilis* DNA was prepared so that the concentration was 10 μg/ml. The same was done for an ^{15}N-deuterated sample. After portions of these solutions received the following treatments, they were made up to the proper density with CsCl and centrifuged. The tracings are shown above.

(A) Mixed while still native.

(B) Heated separately for 10 min at 100°C, quickly cooled and mixed.

(C) and (D) Heated separately, annealed under identical conditions and banded separately.

(E) Mixed immediately after annealing while still in the $1·9 \times$ SSC solvent. If the mixture was dialysed against a lower salt concentration (0·067 M-glycine buffer) before addition of CsCl or if each annealed sample was dialysed against 0·067 M-glycine buffer before mixing, 5 bands resulted.

(F) Mixed, dialysed against 0·067 M-glycine buffer and treated with the *E. coli* phosphodiesterase.

The same picture also resulted when portions of the annealed samples shown in B and C were separately treated with the enzyme and then mixed.

hybrid formation did occur although it may not have been as much as expected from random pairing of complementary strands since the three bands appear to have about equal areas.

For the next experiment we chose three members of the family Enterobacteriaceae which have similar base compositions (Lee, Wahl & Barbu, 1956) and which are genetically related. *E. coli* K12 shows a high degree of genetic exchange by conjugation and transduction with *E. coli* B and *Sh. dysenteriae* (Lennox, 1955; Luria & Burrous, 1957; Luria, Adams & Ting, 1960) whereas *Salm. typhimurium* mates well with K12 but is transduced only to a very limited extent, if at all (Zinder, 1960).

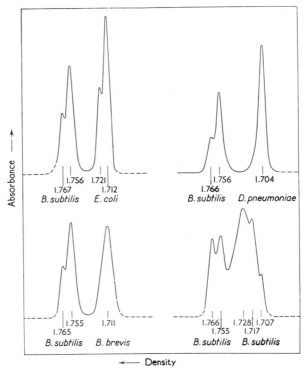

Fig. 10. The specific requirement for hybrid formation using heavy-labeled *B. subtilis* DNA. The three non-homologous samples were heated and annealed at a concentration of 10 μg/ml. each, under the same conditions already described. In the case where both samples were *B. subtilis* DNA, the concentrations were 5 μg/ml. each. Neither the labeled nor the unlabeled *B. subtilis* preparations were the same ones used to obtain the results shown in Fig. 5.

In this context, attempts were begun to prepare hybrid DNA molecules between ^{15}N-deuterated *E. coli* DNA and normal *Shigella* or *Salmonella* DNA. Unlabeled DNA isolated from *Sh. dysenteriae* was substituted for the unlabeled *E. coli* DNA in the procedure discussed in Section 5, and the heated and annealed mixture was treated with *E. coli* phosphodiesterase. Three bands were observed as shown in Fig. 12. The hybrid band does not contain twice as much DNA as either of the uniformly labeled renatured bands, as was observed in most previous examples. This indicates that not every *Sh. dysenteriae* DNA molecule is homologous to a corresponding *E. coli* DNA molecule. Genetic evidence also supports this partial degree of homology (Luria & Burrous, 1957).

Similar studies were carried out with DNA isolated from different strains and species of *Salmonella*. Heating and annealing was carried out with DNA isolated from each *Salmonella* organism listed in Table 1. The results for all samples were similar to those shown in Fig. 13 for the *Salm. typhimurium* and the heavy-labeled *E. coli* B DNA. No hybrid appears at 10 μg/ml. (top tracing) but an intermediate band does appear at 20 μg/ml. (middle tracing). It is, however, removed by enzyme treatment. The experiments were also repeated using heavy-labeled DNA from *E. coli* K12 since this strain has been used most successfully in the studies of genetic recombination. So far,

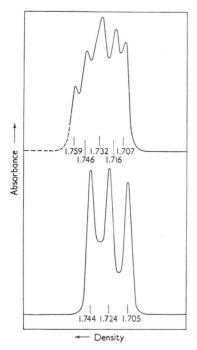

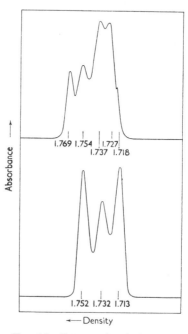

FIG. 11. Interspecies hybrid formation with *B. natto* DNA. Heavy-labeled *B. subtilis* DNA and unlabeled DNA from *B. natto* were heated and annealed at 5 μg/ml. each in 1·9 × SSC. The lower tracing shows the three bands produced when the mixture is treated with the *E. coli* phosphodiesterase.

FIG. 12. Interspecies hybrid formation with *Sh. dysenteriae* DNA. Heavy-labeled *E. coli* DNA and unlabeled DNA from *Sh. dysenteriae* were heated and annealed at 5 μg/ml. each in 1·9 × SSC. When the mixture is treated with the phosphodiesterase and centrifuged in CsCl, only three bands appear.

the DNA of only one of the strains of *Salmonella* (*Salm. typhimurium*) has been tried but no hybrid was formed in this case either. We may then conclude that, in general, there seems to be no indication of sequence complementarity between the DNA of *Salmonella* and *E. coli* as measured by hybrid formation.

Thus in this case, where some homology was to be expected, none was found. It is likely that there is homology in some regions of the DNA molecules, but its failure to be displayed suggests that the homologous regions are dispersed or exist in only a few of the several hundred different molecules. Low concentrations of hybrid molecules would not be observed in the analytical ultracentrifuge, but could be isolated by using larger amounts of interacting DNA and working with the preparative swinging-bucket rotor. In this way we plan to search very carefully for small amounts of hybrid that may be formed between ^{15}N-deuterated DNA from *E. coli* K12 and DNA from any one of the *Salmonella* strains.

Hybrid formation among other members of the family Enterobacteriaceae is now being investigated. Preliminary results with heavy *E. coli* B DNA and normal DNA from a strain of *E. freundii* (5610–52) or of *Erwinia carotovora* (ATCC 8061) show no

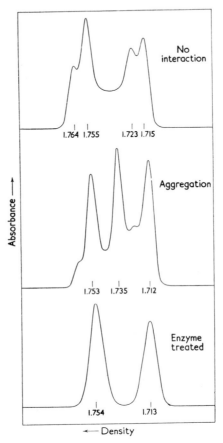

FIG. 13. Attempts at hybrid formation between *Salm. typhimurium* and *E. coli* DNA. The top tracing shows the results of heating and annealing DNA from *Salm. typhimurium* with ^{15}N-deuterated DNA from *E. coli*. Even at 10 μg/ml., which is double the usual concentration, no hybrids have been formed. In the central tracing an intermediate band appears as a result of heating at 20 μg/ml. each. This must be due to aggregates rather than double-stranded molecules since the band disappears after treatment with the phosphodiesterase (bottom tracing).

tendency toward hybrid formation. Unlabeled DNA isolated from another strain (17) of *E. freundii* did form hybrids with labeled *E. coli* B DNA. Since the classification of *E. freundii* is difficult, further studies with DNA of strains of this organism are necessary. Heavily-labeled DNA from *E. coli* K12, *Erwinia carotovora*, and *Salm. typhimurium* have also been prepared, allowing the study of other obvious possible relationships.

From this survey it can be concluded that equality of base composition and some genetic relation are necessary but not sufficient requirements for the formation of hybrid DNA molecules composed of strands from different organisms. The present work is only indicative in nature but it does suggest that, with the further development of this technique, the degree of sequential homology between two different DNA molecules can be quantitatively explored.

9. Discussion

Until now we have been talking strictly in terms of the concept of strand separation and recombination, in accordance with the evidence presented by Doty *et al.* (1960). It has been seen that some important new facts can now be added to this evidence. By using heavy-isotope-labeled and normal DNA for heating and annealing experiments, it has been demonstrated that the units that unite in renaturation could not be the same as those that were united in the native DNA. It has also been demonstrated that the hybrids formed are not simply physically entangled aggregates. The strong species specificity of hybrid formation indicates that the links that hold the subunits of the hybrid molecules together are determined by the base sequence in each subunit. These links break whenever the hydrogen bonded structure of the molecule is broken. In fact, the rate of separation into subunits corresponds very closely to what has been predicted for the rate of unwinding of the strands of a double helix. It is possible that certain unknown links are broken whenever the Watson-Crick structure is disrupted and that the base sequence determines the specific type of linkage. The most natural explanation, however, seems to be that the strands of the Watson-Crick double helix do separate completely and come back together again through the formation of hydrogen bonds between base-pairs of complementary strands.

The hypothesis that genetic information resides in the sequence of bases in DNA has created a demand for techniques that will allow the study of the linear order of nucleotides along the DNA strand. An approach to this problem has been made in the nearest neighbor studies of Josse, Kaiser & Kornberg (1961) which offers a statistical evaluation of sequences displayed in DNA of various sources. By inference it can be assumed that genetic compatibility is also a measure of similarity in sequences between the DNA of the two parental strains. The present report has outlined a method by which it is possible to measure the extent of renaturation between homologous and heterologous DNA strands and has attempted to add a new dimension to the study of sequences of DNA of microbial origin.

The methodology of molecular hybrid formation *in vitro* has been outlined and its use in the study of similarities of DNA of various groups of micro-organisms has been illustrated. The close correlation between genetic compatibility, taxonomy, and hybrid formation has been mentioned. These applications are discussed in much greater detail elsewhere (Marmur & Schildkraut, 1961b; Marmur, Schildkraut & Doty, 1961). Organisms whose taxonomic classification is in doubt might readily be classified by first determining their base composition and studying their interaction by heating and annealing. Such experiments are now being extended to other members of the family Enterobacteriaceae (*Aerobacter, Klebsiella*, and *Serratia*) as well as the Pseudomonadaceae (*Pseudomonas Xanthomonas* and *Acetobacter*). Moreover, it might also be possible, by collecting homogeneous fractions of DNA of animal or plant origin, to investigate relationships in a similar manner. Aside from its taxonomic importance the technique offers a rational approach to the study of genetic compatibility where genetic exchanges have not yet been demonstrated.

It is a pleasure to express our gratitude to Dr. L. Grossman for his generous gift of the *E. coli* phosphodiesterase and for his valuable advice. We are especially indebted to Mr. William Torrey for his expert technical assistance with many aspects of this work, and to Mrs. F. Seelig for aid in preparation of the manuscript. We would also like to thank Drs. L. Wierzchowski and Donald M. Green, and Mr. Robert Rownd for their aid. The authors are very grateful to Professor C. E. Hall of the Massachusetts Institute of Technology for the electron micrograph. This investigation was supported by a grant (C-2170) from the National Cancer Institute, United States Public Health Service.

Note added in proof

It can be argued that the DNA species of intermediate buoyant density observed in the CsCl density gradient could be formed by non-specific, end-to-end aggregation of heavy isotope labeled and unlabeled DNA. It is evident, however, that biological hybrids could not be used to form DNA species possessing buoyant densities characteristic of either renatured fully labeled or fully unlabeled DNA unless strand separation and subsequent recombination of similarly labeled strands occurs during the heating and annealing procedure. Preliminary experiments by R. Rownd and D. Green using biological hybrids labeled with both ^{15}N and deuterium in only one strand and isolated from either *Escherichia coli* or *Bacillus subtilis* have shown that the heating and annealing procedure does result in renatured labeled and fully unlabeled, as well as hybrid, DNA molecules. The proportions are the same as observed when a mixture of labeled and unlabeled DNA is used as the starting material.

REFERENCES

Belozersky, A. N. (1957). In *The Origin of Life on the Earth*. Reports on the International Symposium, ed. by A. Oparin, p. 194. U.S.S.R.: Academy of Sciences.
Belozersky, A. N. & Spirin, A. S. (1960). In *The Nucleic Acids*, ed. by E Chargaff & J. N. Davidson, Vol. III, p. 147. New York: Academic Press.
Cox, R., Marmur, J. & Doty, P. (1961). In preparation.
Davern, C. I. & Meselson, M. (1960). *J. Mol. Biol.* **2**, 153.
Doty, P., Marmur, J., Eigner, J. & Schildkraut, C. (1960). *Proc. Nat. Acad. Sci., Wash.* **46,** 461.
Dunn, D. B. & Smith, J. D. (1954). *Nature,* **174**, 305.
Eigner, J., Boedtker, H. & Michaels, G. (1961). *Biochim. biophys. Acta,* in the press.
Eigner, J. & Doty, P. (1961). In preparation.
Ephrati-Elizur, E. & Zamenhof, S. (1959). *Nature,* **184**, 472.
Ifft, J. B., Voet, D. H. & Vinograd, J. (1961). *J. Phys. Chem.,* in the press.
Josse, J., Kaiser, A. D. & Kornberg, A. (1961). *J. Biol. Chem.* **236**, 864.
Kuhn, W. (1957). *Experientia,* **13**, 301.
Lee, K. Y., Wahl, R. & Barbu, E. (1956). *Ann. Inst. Pasteur,* **91**, 212.
Lehman, I. R. (1960). *J. Biol. Chem.* **235**, 1479.
Lennox, E. S. (1955). *Virology,* **1**, 190.
Longuet-Higgins, H. C. & Zimm, B. H. (1960). *J. Mol. Biol.* **2**, 1.
Luria, S. E. & Burrous, J. W. (1957). *J. Bact.* **74**, 461.
Luria, S. E., Adams, J. N. & Ting, R. C. (1960). *Virology,* **12**, 348.
Marmur, J. (1961). *J. Mol. Biol.* **3**, 208.
Marmur, J. & Doty, P. (1959). *Nature,* **183**, 1427.
Marmur, J. & Doty, P. (1961). *J. Mol. Biol.* **3**, 585.
Marmur, J. & Lane, D. (1960). *Proc. Nat. Acad. Sci., Wash.* **46**, 453.
Marmur, J., Lane, D. & Doty, P. (1961). In preparation.
Marmur, J. & Schildkraut, C. L. (1961a). *Nature,* **189**, 636.

Marmur, J. & Schildkraut, C. L. (1961b). In *The Proceedings of the Vth Int. Congr. Biochem.*, London: Pergamon Press.

Marmur, J., Schildkraut, C. L. & Doty, P. (1961). In *The Molecular Basis of Neoplasia*. The Fifteenth Annual Symposium on Fundamental Cancer Research, ed. by S. Kit. Houston: Texas University Press.

Marmur, J., Seaman, E. & Levine, J. (1961). In preparation.

Marmur, J. & Ts'o, P. (1961). *Biochim. biophys. Acta*, in the press.

Meselson, M. & Stahl, F. W. (1958). *Proc. Nat. Acad. Sci., Wash.* 44, 671.

Meselson, M., Stahl, F. W. & Vinograd, J. (1957). *Proc. Nat. Acad. Sci., Wash.* 43, 581.

Rice, S. A. & Doty, P. (1957). *J. Amer. Chem. Soc.* 79, 3937.

Rolfe, R. & Meselson, M. (1959). *Proc. Nat. Acad. Sci., Wash.* 45, 1039.

Schildkraut, C. L., Marmur, J., Wierzchowski, L., Green, D. M. & Doty, P. (1961). In preparation.

Schildkraut, C. L., Wierzchowski, L. & Doty, P. (1961). In preparation.

Sinsheimer, R. L. (1959). *J. Mol. Biol.* 1, 43.

Sueoka, N., Marmur, J. & Doty, P. (1959). *Nature*, 183, 1429.

Szybalski, W., Opara-Kubinska, Z., Lorkiewicz, Z., Ephrati-Elizur, E. & Zamenhof, S. (1960). *Nature*, 188, 743.

Watson, J. D. & Crick, F. H. C. (1953). *Nature*, 171, 737.

Zamenhof, S. & Griboff, G. (1954). *Nature*, 174, 306.

Zinder, N. D. (1960). *Science*, 131, 813.

J. Mol. Biol. (1961) **3**, 756–761

An Estimate of the Length of the DNA Molecule of T2 Bacteriophage by Autoradiography

John Cairns†

Department of Genetics, Carnegie Institution of Washington,
Cold Spring Harbor, N.Y., U.S.A.

(*Received 29 June 1961*)

T2 bacteriophage, labelled with [³H]thymidine or [³H]thymine, is subject to suicide on storage. The efficiency of suicide from ³H-decay is apparently the same as that from ³²P-decay.

Autoradiography of T2 DNA, labelled with [³H]thymine and extracted in the presence of 1000-fold excess of cold T2, shows that the molecule can assume the form of an unbranched rod about 52 μ long. If the molecule is throughout its length a double helix in the B configuration, this indicates a molecular weight of 110×10^6.

1. Introduction

The decay of tritium gives rise to electrons whose mean range in autoradiographic emulsion is less than one micron (Fitzgerald, Eidinoff, Knoll & Simmel, 1951). It should therefore be possible to obtain a high resolution image of individual molecules of ³H-labelled DNA by autoradiography, using the very highly labelled [³H]thymine that is now available; thus DNA containing [³H]thymine of specific activity 10 c/m-mole will have roughly one disintegration per micron of double helix per week and should produce a near-continuous line of grains along its length after a few weeks' exposure.

Bacteriophage T2 seemed in most respects the best material with which to launch such a procedure. Most of the precursors for T-even thymine synthesis come from the medium after infection (Weed & Cohen, 1951; Kozloff, 1953) so there should be extensive incorporation of labelled thymine given at the time of infection; T2 DNA can be extracted in a pure and homogeneous state with phenol (Mandell & Hershey, 1960), each particle providing a single molecule with a molecular weight of over 100×10^6 (Rubinstein, Thomas & Hershey, 1961); lastly a molecule of such great size is ideal for testing a method of measuring molecular length the accuracy of which is, in theory at least, independent of length.

2. Materials and Methods

Phage. T2, strain T2H (Hershey), was used throughout.

Bacteria. Escherichia coli strain S was used for the production of stocks of unlabelled phage and for phage assays. Labelled phage was prepared in the thymineless strain B3 (Brenner).

† Present address: The Australian National University, Canberra, Australia.

Media. Stocks of unlabelled phage were prepared in M9 (Adams, 1959) supplemented with 0·5 g/l. NaCl. All experiments on the production of phage in the presence of limited thymine or thymidine were carried out using the glucose-ammonium medium described by Hershey (1955).

E. coli strain B3 was grown in the presence of 5 μg/ml. of thymine or thymidine. Phage assays were performed using the standard methods (Adams, 1959). Dilution of phage stocks was made in 10^{-3} M-MgCl$_2$, 0·05% NaCl, 0·001% gelatin, buffered with 0·01 M-tris (2-amino-2-hydroxymethylpropane-1:3-diol) pH 7·4.

[³H]Thymine and [³H]thymidine. These were obtained from the New England Nuclear Corp. and from Schwartz Inc. In the case of the former source these materials are prepared by reducing 5-hydroxymethyl-uracil with tritium so that the label is confined to one hydrogen atom in the methyl group; for this reason they can be specifically designated as 5-[³H]methyl-uracil and 5-[³H]methyldeoxyuridine.

Preparation of labelled phage. In order to ensure extensive incorporation of [³H]thymine (or thymidine) into phage it was necessary to engineer the situation so that phage would only be made if thymine was present and would then be made in an amount which was proportional to the amount of thymine present. The thymineless B3 strain of *E. coli* was used as host since this strain readily incorporates thymidine even at low concentrations, whereas the prototroph does not (Crawford, 1958). Since T2 infection causes the formation of thymidylate synthetase even in thymineless bacteria (Barner & Cohen 1959), it was necessary to block this enzyme by the addition of 5-fluorodeoxyuridine (FUDR) (Cohen, Flaks, Barner, Loeb & Lichtenstein, 1958). At the same time, uridine (UR) was added to ensure that FUDR derivatives were not incorporated into RNA. Thus the final procedure was as follows:

E. coli B3 was grown to 2×10^8 cells/ml. in glucose-ammonium medium with 5 μg/ml. thymidine, and then centrifuged and resuspended in one third volume of fresh medium without thymidine. FUDR and UR were added to give final concentrations of 10^{-5} and 10^{-4} M. Five minutes later the bacteria were infected with T2 at a multiplicity of 4. After 4 min, 0·02 ml. of these infected bacteria was added to 0·02 ml. of double strength medium (with 2×10^{-5} M-FUDR and 2×10^{-4} M-UR) and 0·02 ml. of [³H]thymidine in water. The final concentrations in this growth tube were 2×10^8 B3/ml. 10^{-5} M-FUDR, 10^{-4} M-UR, 2 to 8 μg/ml. thymidine (40 to 160 μC/ml.). (When thymine was the label, the growth tube was supplemented with 10^{-3} M-deoxyadenosine.) After aeration for a further 60 min, the bacteria were lysed with chloroform and the contents of the growth tube were made up to 10 ml. with diluting fluid.

The yield of phage from such a system was 1×10^{10} phage/μg thymidine and 6 to 8×10^9 phage/μg thymine; it was slightly less with [³H]thymidine and [³H]thymine. These yields are 2 to 4 times less than would be expected on the basis of the known thymine content of T2 (Hershey, Dixon & Chase, 1953), but they were not lowered further by raising the concentration of FUDR. In the absence of thymine or thymidine the yield was about 5 phage per bacterium. Purified phage prepared from cold thymidine in this way had the normal optical density per infective particle at 260 mμ. Phage prepared from hot or cold thymine or thymidine showed no rise in the frequency of r mutants.

Extraction of phage DNA. Phage was extracted with phenol according to the method of Mandell & Hershey (1960). When labelled phage was extracted, enough cold carrier phage was added to bring the concentration up to the requisite 3×10^{12} phage/ml.; the mixture was then packed in the centrifuge, resuspended and extracted with phenol.

Autoradiography. Once the DNA had been extracted it was diluted to a suitable concentration in various salt solutions and spread in various ways upon glass microscope slides which had previously been coated with various materials. These slides were, on occasion, then coated with chrome-gelatin (0·5% gelatin, 0·05% chrome alum). They were overlaid in the usual manner with Kodak autoradiographic stripping film, AR 10, and stored at 4°C over silica gel in an atmosphere of CO$_2$ to prevent latent image fading (Herz, 1959). After exposure the film was developed with Kodak D19b for 20 min at 16°C.

3. Results

(a) *Suicide of ³H-labelled phage*

Two lots of [³H]thymidine and one of [³H]thymine were used at various times for making labelled phage. In each case the resulting phage was diluted 10^2 to 10^4-fold, stored at 4°C and repeatedly assayed for surviving phage. Excess cold phage mixed with the hot phage and stored under the same conditions proved to be stable, as did phage prepared by the same procedure but with cold thymidine. Thus the observed

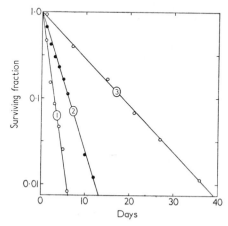

FIG. 1. The suicide of T2 labelled with (1) [³H]thymine 11·2 c/m-mole. (2) [³H]thymidine 5·2 c/m-mole and (3) [³H]thymidine 1·8 c/m-mole.

inactivation of labelled phage was due neither to indirect effects of radiation within the storage tube nor to any innate instability of phage prepared, for example, in the presence of FUDR. Since tritium has a half-life of 12 years, no correction has to be made for decline in radioactivity during the period of storage.

TABLE 1

The suicide of T2 phage labelled with [³H]thymine and [³H]thymidine.

Label	sp. A (c/m-mole)	k (lethals/phage. day)	$\dfrac{k.\text{m-mole}}{\text{c}}$	αN† ("lethal" methyl H atoms/phage)
(1) 5-(³H-methyl)-uracil	11·2	0·78	0·069	$3·9 \times 10^4$
(2) 5-(³H-methyl)-deoxyuridine	5·2	0·37	0·071	$4·1 \times 10^4$
(3) ³H-thymidine	1·8	0·13	0·072	$4·1 \times 10^4$

$$† \ \alpha N = \frac{\text{lethals}}{\text{disintegration}} \times \frac{\text{methyl H atoms}}{\text{phage}}$$

$$= \frac{\text{lethals}}{\text{phage . day}} \times \frac{\text{m-mole}}{\text{c}} \times \frac{\text{molecules}}{\text{m-mole}} \times \frac{\text{methyl H atoms}}{\text{molecule}} \times \frac{\text{Curie. day}}{\text{disintegration}}$$

$$= k \times \frac{\text{m-mole}}{\text{c}} \times 6·02 \times 10^{20} \times 3 \times \frac{1}{3·20 \times 10^{15}}$$

$$= 5·64 \times 10^5 \times \frac{k.\text{m-mole}}{\text{c}}$$

The results are shown in Fig. 1 and Table 1. The suicide of phage is seen to be a first order process with a rate constant (k) which is directly proportional to the specific activity (c/m-mole) of the thymine or thymidine. From this the number, αN, of "lethal" thymine methyl H atoms per phage may be calculated to be 4×10^4. Since the burst size in these experiments was around 120, the contribution of cold thymine from the pool of bacterial DNA (about 24 phage equivalents of thymine) (Hershey et al., 1953; Hershey & Melechen, 1957) will have lowered the specific activity of the incorporated thymine by 20%. Correction for this raises αN to 5×10^4.

This value is not significantly different from the number of "lethal" P atoms per phage, determined from the rate of ^{32}P-suicide (Hershey, Kamen, Kennedy & Gest, 1950; Stent & Fuerst, 1955). Since one third of the bases in T2 DNA are thymine, the total number of thymine methyl H atoms equals the total number of P atoms. It follows therefore that the efficiency of inactivation by decay of ^{32}P and ^3H are the same. This is an unexpected result. First, ^{32}P and ^3H differ greatly in the energy of the electrons they emit (max. energy 1700 and 17 kev respectively). Second, the sites of their incorporation into DNA seemingly could scarcely differ more; the decay of ^{32}P, in the sugar-phosphate chain, and its conversion to sulfur must necessarily break that chain; the decay of ^3H, in the methyl group of thymine, and its conversion to helium need not necessarily cause chain breakage nor perhaps any lasting local alteration in the DNA at all.

Practically, these results indicate that at least 99% of the phage is fairly uniformly labelled.

(b) Autoradiography of ^3H-labelled T2 DNA

Although the production and extraction of highly-labelled DNA presented no problem, there was little prior information on how best to fix this DNA in a sufficiently extended state so that its contours could be followed by autoradiography. Electron microscopy has shown that DNA can be adsorbed from phosphate-buffered solutions of pH 5 to 6 to a variety of surfaces as straight rods many microns long (Hall & Litt, 1958; Beer, 1961) and various methods have been used to ensure that at the time of adsorption the molecules are subject to sufficient shear to align them.

In the course of several months many combinations of DNA concentrations, suspending fluids, varieties of shearing force and adsorbent surfaces were tested. Interestingly, fibres drawn from DNA at high concentration show very poor extension of the minority of molecules that are labelled. The most satisfactory method was found by accident. On testing the appearance of labelled DNA adsorbed to slides partly coated with a co-polymer of polyvinylpyridine and styrene (generously supplied by Dr. Michael Beer), numerous straight molecules were seen adsorbed to the glass on either side of the area coated with polymer; this glass had been cleaned with chromic acid, coated with DNA by drawing the slide across the surface of a solution of DNA in M/15 phosphate buffer pH 5·6, drained and rinsed with distilled water, and then coated with chrome-gelatin (to ensure that the autoradiographic film remained stuck to the slide on drying). Even here, however, there were only localized regions where the DNA was suitably extended. In most regions, the individual molecules were apparently folded back on themselves several times to form a short "rod" of densely packed grains. It seems therefore that the best method for displaying DNA molecules —at least those as long as T2 DNA—has not yet been found.

The appearance of the extended DNA is shown in Plate I. Of the 13 labelled molecules (or fragments of molecules) shown, 7 have a length between 49 and 53 μ; of the remaining 6, one (immediately over the center of the scale) seems from its grain density and length to be folded about its center. Other samples of DNA prepared on different occasions likewise showed that the maximum length, when adsorbed on to glass, was slightly more than 50 μ.

As an estimate of length and hence of molecular weight this is subject to certain errors and variables. For example:

(i) There are several configurations which the molecules might assume, giving values of 2·55 to 3·46 Å per base pair (Langridge, Wilson, Hooper, Wilkins & Hamilton, 1960). Of these the most likely, particularly in the case of T2 DNA (Hamilton et al., 1959), is the B configuration with 3·4 Å per base pair.

(ii) Since it is the autoradiographic image and not the molecule itself which is seen, any stretching of the film between exposure and measuring will produce an apparent lengthening. This, however, seems to be a rare occurrence with stripping film.

(iii) Since the molecule is indicated as a series of grains which one may assume to be randomly placed along its length, it is simple to show for this case, where the mean number of grains (M) per molecule of length L is more than about 10, that the mode, mean and variance of observed lengths (that is, between the centres of the outermost grains) will be approximately $L(1 - 1/M)$, $L(1 - 2/M)$ and $L^2(2/M^2)$ respectively. One would therefore expect for molecules such as these, marked with 50 to 100 grains, that the length of the average molecule would be underestimated by 2 to 4%.

(iv) The resolution of the technique can best be judged by the fact that the grains appear to deviate little to either side of the apparent line of each labelled molecule. It is therefore unlikely that the length of any molecule is overestimated by more than 1 μ for the reason of poor resolution.

Thus if any single length has to be selected as the most likely for T2 DNA, that length is probably 52 μ. Taking a value of 3·4 Å per base pair and 357 as the average molecular weight of a base in the sodium salt of T2 DNA, this indicates a molecular weight of 110×10^6 and a phosphorus content of $3·0 \times 10^5$. These are slightly below the accepted values though probably not by enough to warrant, at this stage, postulating anything other than an uncomplicated double helix as the form of the T2 DNA molecule.

I am greatly indebted to Dr. A. D. Hershey for his advice and encouragement and for the hospitality of his laboratory, to Dr. Michael Beer for information and advice on spreading DNA, and to the National Institutes of Health (U.S.A.) for a post-doctoral fellowship during which this work was done.

REFERENCES

Adams, M. H. (1959). *The Bacteriophages*. New York: Interscience Publishers.

Barner, H. D. & Cohen, S. S. (1959). *J. Biol. Chem.* **234**, 2987.

Beer, M. (1961). *J. Mol. Biol.* **3**, 263.

Crawford, L. V. (1958). *Biochim. biophys. Acta*, **30**, 428.

Cohen, S. S., Flaks, J. G., Barner, H. D., Loeb, M. R. & Lichtenstein. J. (1958). *Proc. Nat. Acad. Sci.*, *Wash.* **44**, 1004.

Fitzgerald, P. J., Eidinoff, M. L., Knoll, J. E. & Simmel, E. B. (1951). *Science*, **114**, 494.

Hall, C. E. & Litt, M. (1958). *J. Biophys. Biochem. Cytol.* **4**, 1.

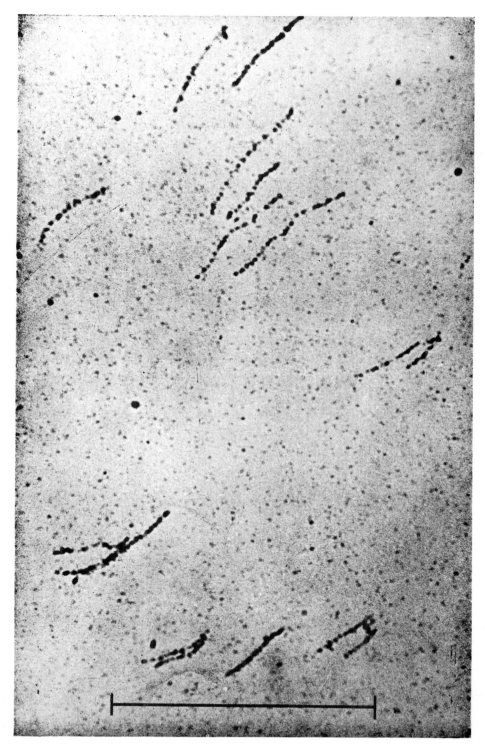

PLATE I. T2 DNA, labelled with [³H]thymine (11·2 c/m-mole), extracted with phenol in the presence of 1000-fold excess of cold T2, and adsorbed to glass at a total DNA concentration of 25 μg/ml. in M/15 phosphate buffer pH 5·6. The autoradiographic exposure was 63 days. The scale shows 100 μ.

152   J. CAIRNS

Hamilton, L. D., Barclay, R. K., Wilkins, M. H. F., Brown, G. L., Wilson, H. R., Marvin, D. A., Ephrussi-Taylor, H. & Simmons, N. S. (1959). *J. Biophys. Biochem. Cytol.* **5,** 397.

Hershey, A. D. (1955). *Virology,* **1,** 108.

Hershey, A. D., Dixon, J. & Chase, M. (1953). *J. Gen. Physiol.* **36,** 777.

Hershey, A. D., Kamen, M. D., Kennedy, J. W. & Gest, H. (1950). *J. Gen. Physiol.* **34,** 305.

Hershey, A. D. & Melechen, N. E. (1957). *Virology,* **3,** 207.

Herz, R. H. (1959). *Lab. Investigation,* **8,** 71.

Kozloff, L. M. (1953). *Cold Spr. Harb. Symp. Quant. Biol.* **18,** 209.

Langridge, R., Wilson, H. R., Hooper, C. W., Wilkins, M. H. F. & Hamilton, L. D. (1960). *J. Mol. Biol.* **2,** 19.

Mandell, J. D. & Hershey, A. D. (1960). *Analyt. Biochem.* **1,** 66.

Rubinstein, I., Thomas, C. A. & Hershey, A. D. (1961). *Proc. Nat. Acad. Sci., Wash.* **47,** 1113.

Stent, G. S. & Fuerst, C. R. (1955). *J. Gen. Physiol.* **38,** 441.

Weed, L. L. & Cohen, S. S. (1951). *J. Biol. Chem.* **192,** 693.

J. Mol. Biol. (1962) **4**, 288–292

An Active Cistron Fragment

Sewell P. Champe and Seymour Benzer

Department of Biological Sciences, Purdue University, Lafayette, Indiana, U.S.A.

(*Received 18 December 1961*)

An unusual *r*II mutant of phage T4 is described, in which a terminal portion of a cistron has been deleted without impairing the function of the remaining fragment. This suggests that the deleted section is non-essential. Nevertheless, mutations within this segment which do impair the function can be observed.

1. Introduction

The entire amino acid sequence of an enzyme is not always essential to its function; a fragment of the molecule may retain all or part of the catalytic activity as shown, for example, by the work of Hill & Smith (1956) on papain, and Kalnitsky & Rogers (1956) on ribonuclease. Assuming that the amino acid sequence of an enzyme is dictated by the nucleotide sequence in a genetic structure, one might expect that it would be possible, in some instances, to delete a part of the genetic structure without seriously impairing the activity of its product. The present paper describes a mutant of bacteriophage T4 which appears to represent such a case. In this mutant, although a sizable fraction of the B cistron of the *r*II region has been deleted, the remaining fragment retains the B cistron function.

2. Results

(a) *The two cistrons in the* r*II* *region of phage T4*

*r*II mutations in phage T4 prevent the phage from multiplying in strain K of *E. coli*. If a cell is also infected with the standard type phage, the function or functions needed by the mutant are supplied, and both mutant and standard type progeny are produced. Thus *r*II mutations are recessive.

*r*II mutants can be classified into either of two functional groups by means of the *cis-trans* test (Benzer, 1955, 1957), these groups being consistent with the location of the mutations either to the left or to the right of a unique divide in the genetic map constructed by recombination experiments.

Of some hundreds of extended ("deletion") mutations that have been mapped in the *r*II region, four (*r*1605, *r*NB7006, *r*1231 and *r*1589) span the divide between the A and B cistrons, leaving fragments of a cistron on either side. A genetic map showing these mutations (and others pertinent to the following discussion) is given in Fig. 1. *r*1605, *r*NB7006 and *r*1231 behave in complementation tests in the expected way, that is, they fail to complement any A or B cistron mutant. However, the behavior of *r*1589 is exceptional.

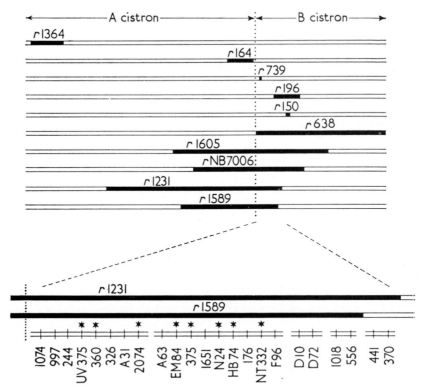

FIG. 1. Recombination map of the *r*II region showing the extent of mutations referred to in the text. A portion of the B cistron has been enlarged to show the individual sites. Asterisks indicate mutations that are strongly inducible to revert by 2-aminopurine to a form which shows the standard phenotype when plated on *E. coli* B. While the order of the five broken segments in the bottom line is known, the arrangement of sites within each segment is arbitrary. The method used for genetic mapping has been described (Benzer, 1961).

(b) *Properties of the exceptional mutant* r1589

The lesion in *r*1589 encompasses a large part of the A cistron and extends over 21 known sites (out of 129) of the B cistron. That is to say, when *r*1589 is crossed with any mutant located within or overlapping this span, no standard type recombinants have ever been detected in as many as 10^5 progeny. By the method of shrinkage of distance between outside markers (Nomura & Benzer, 1961) the lesion in *r*1589 has been shown (Bode, 1962) to be a true deletion.

Now *r*1589, in spite of the fact that a sizable portion of the B cistron is missing, complements strongly every B cistron mutant with which it has been tested, including ones with overlapping extended mutations, such as *r*638. The phage yields per infected cell due to complementation are comparable to the standard type phage. *r*1589 does not, however, complement any A cistron mutant so far tested. Data for various combinations are given in Fig. 2.

*r*1589 functions as if its B cistron (but not its A cistron) were intact. Thus, the portion of the B cistron deleted in *r*1589 appears to be non-essential for the activity of the remaining fragment. In mixed infection of K together with any B cistron mutant, *r*1589 supplies this active B fragment and the A activity is supplied by the other mutant, so that both necessary functions are present. Were it not for the fact that $r1589$ also has part of its lesion affecting the A cistron, it might not have appeared to be a mutant at all.

This confronts us with a paradox. Mutations have been detected in the non-essential portion of the B cistron at at least twenty-one different sites, and these mutants lack B cistron activity. This suggests that mutations occurring in a non-essential portion may nevertheless prevent the expression of the remainder of the cistron. For instance,

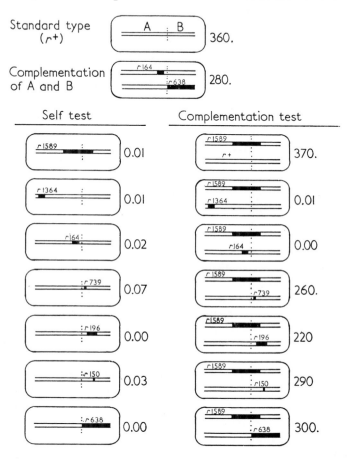

FIG. 2. Physiological activity of rII mutants in various combinations. Each oval represents *E. coli* strain K infected with the mutant or pair of mutants indicated. The burst size is given to the right of each figure.

Procedure: The bacteria were grown in broth (1% Difco bacto-tryptone plus 0·5% NaCl) to a cell concentration of 1·0 to 1·5 × 10^8/ml., and growth was arrested by adding NaCN to 5 × 10^{-3} M. 2 ml. of this culture was added to 1 ml. of broth containing 1·0 × 10^9/ml. particles of each phage mutant being tested, making the total input multiplicity 6 to 10 phage particles per cell. After 10 min of adsorption at 37°C two drops of T4 antiserum ($k = 25$/min) were added to eliminate the few percent of unadsorbed particles. Seven minutes later, the cells were diluted 10^4-fold, incubated in broth at 37°C for 60 min, then treated with chloroform. The phage yield was assayed by plating on *E. coli* B. The burst size was calculated as the ratio of phage yield to infective centers assayed by plating the cells (before burst) on strain B. In the case of complementing pairs, about half of the infected strain K cells survived to give infective centers. For non-complementing pairs, since the infective centers could not be assayed on B, the burst size was calculated on the basis of the average number of infective centers measured for complementing pairs.

an alteration of the nucleotide sequence such that the information becomes "nonsense" could interfere with the completion of a polypeptide chain.

The special properties of r1589 have been exploited by Crick, Barnett, Brenner & Watts-Tobin (1961) in a brilliant experiment. They have shown that the B cistron function of r1589 can be eliminated by inserting certain deficiencies in its A cistron fragment, while other deficiencies have no effect. Their interpretation, reinforced by an impressive series of other experiments with suppressor mutations, is that deficiencies corresponding to an integral number of amino acid-determining groups (nucleotide triplets) allow continued formation of a polypeptide chain while non-integral deficiencies do not. It would appear that the lesion of r1589 represents an instance where the joining of the remaining A and B cistron fragments is such that the continuity of the information is not interrupted.

It is to be noted that the mutations observable within the B-cistron span of r1589 include ones which are inducible to revert by the base analog 2-aminopurine, and therefore are probably due to substitutions of one base for another (Freese, 1959). Of the twenty-one known sites covered by r1589, eight are of this type, designated by asterisks in Fig. 1. This would seem to indicate that the genetic code is such that base pair substitutions are capable of converting sense to nonsense. In other words, the genetic code is not completely degenerate; not *every* triplet necessarily corresponds to an amino acid. Another possibility, however, is that certain amino acid substitutions, even in the non-essential region, could have drastic effects upon the folding of the rest of the polypeptide chain.

3. Discussion

In the standard type phage, the A and B cistrons function as independent units. Presumably, there is some information element between the two cistrons that acts as a continuity break in the transcription of the genetic information. In r1589 this element must be missing and fortuitous joining of the A and B cistron fragments enables continuity to be preserved.

The deletion r1231, which has also been shown by Bode (1962) to be a true deletion, extends only slightly further (two known sites) into the B cistron than does r1589. However, r1231 lacks B cistron activity, as do rNB7006 and r1605 which extend still further. The loss of the B cistron activity of r1231 could be due to essentiality of the additional tiny segment involved, or to a non-fortuitous joining of the A and B fragments. In the latter case one would anticipate that insertion of suitable deletions to the left of r1231 according to Crick *et al.*, might revive its B cistron activity.

r1589 is an illustration of how two cistrons can become joined together. One can easily imagine how the converse might also occur, i.e. the separation of one cistron into two parts, by a mutation which generates a continuity break at a suitable point. An example of an enzyme for which such cistron splitting could occur without deleterious effect is ribonuclease, since, as Richards (1958) has shown, the polypeptide chain can be cleaved into two fragments which, when both are present, still form an active enzyme. Genetically, this would then resemble the case of tryptophan synthetase in *E. coli* (Crawford & Yanofsky, 1958; Yanofsky & Crawford, 1959) where two dissociable chains of the enzyme are indeed controlled by two contiguous cistrons.

This work was supported by grants from the National Science Foundation and the National Institutes of Health.

REFERENCES

Benzer, S. (1955). *Proc. Nat. Acad. Sci., Wash.* **41**, 344.

Benzer, S. (1957). In *The Chemical Basis of Heredity*, ed. by W. D. McElroy & B. Glass. Baltimore: Johns Hopkins Press.

Benzer, S. (1961). *Proc. Nat. Acad. Sci., Wash.* **47**, 403.

Bode, W. (1962). Doctoral dissertation, University of Cologne, Germany, to be published.

Crawford, I. P. & Yanofsky, C. (1958). *Proc. Nat. Acad. Sci., Wash.* **44**, 1161.

Crick, F. H. C., Barnett, L., Brenner, S. & Watts-Tobin, R. J. (1961). *Nature*, **192**, 1227.

Freese, E. (1959). *Proc. Nat. Acad. Sci., Wash.* **45**, 622.

Hill, R. L. & Smith, E. L. (1956). *Biochim. biophys. Acta*, **19**, 376.

Kalnitsky, G. & Rogers, W. I. (1956). *Biochim. biophys. Acta*, **20**, 378.

Nomura, M. & Benzer, S. (1961). *J. Mol. Biol.* **3**, 684.

Richards, F. M. (1958). *Proc. Nat. Acad. Sci., Wash.* **44**, 162.

Yanofsky, C. & Crawford, I. P. (1959). *Proc. Nat. Acad. Sci., Wash.* **45**, 1016.

J. Mol. Biol. (1962) **5**, 18–36

Host Specificity of DNA Produced by *Escherichia Coli*

I. Host controlled modification of bacteriophage λ

Werner Arber and Daisy Dussoix

Biophysics Laboratory, University of Geneva, Switzerland

(*Received 23 January 1962*)

Lambda bacteriophage particles carry a "host specificity" determined by the bacterial strains on which they were produced. Upon infection of a different bacterial host (1) the phage DNA may be either accepted or rejected on the basis of this specificity, (2) if accepted, the phage multiplies and progeny phage are produced. Those progeny to which the parental phage DNA molecule is transferred, in either conserved or semi-conserved form, also receive the parental phage host specificity. All progeny containing only newly synthesized DNA receive only the specificity of the new bacterial host. It is concluded that host specificity is carried on the bacteriophage DNA.

Phage P1, present in a bacterial cell as either prophage or vegetative phage, imparts to λ DNA multiplying in the same cell a host specificity over and above that determined by the host itself. Such P1-induced specificity can be impressed equally well onto replicating and non-replicating λ DNA.

Introduction

Non-mutational changes of host range properties upon growth on new bacterial host strains have been found for many bacteriophages (see Luria, 1953). For bacteriophage λ such a modification occurs if *Escherichia coli* C is used as host instead of the usual *E. coli* K12 (Bertani & Weigle, 1953; Weigle & Bertani, 1953): phage λ, adapted to *E. coli* C, grows on *E. coli* K12 with a probability of only about 10^{-4}. Other systems producing host controlled modification in phage λ will be described in the present paper.

The reproduction of phage λ in a new sensitive host strain will be shown to be submitted to two successive host control mechanisms (Arber & Dussoix, 1961): (1) the infecting phage DNA is either recognized as incompatible with the host and degraded, or is accepted; (2) if fully accepted, it multiplies and its DNA replicas receive "host specificity", i.e. the particular non-heritable stamp given by the host. The present study will be concerned with the establishment of this host specificity.

Host controlled modification of phage DNA can be governed by the genetic material of the bacterial host cell. In some instances the controlling loci have been mapped on the bacterial chromosome (Zinder, 1960), while in other cases they are known to be contained in the genome of an unrelated prophage. Prophage P1, for example, induces modifications of phages T1, T3, T7 and P2 (Lederberg, 1957) and, as will be shown here, of phage λ. This modifying action will be shown to occur not only when λ phage multiplies in cells lysogenic for P1, but also following P1 superinfection of non-lysogenic cells in which λ is multiplying. The same holds true for phages T1 and P2 as found independently by Christensen (1961).

Materials and Methods

Notation. The "host specificity" of a phage will be represented as follows: the symbol of the phage and of its genotype (if relevant) will be followed by the name of the bacterial host strain from which the phage acquired its host specificity, phage and bacterial symbols being separated by a point. For example, $\lambda c \cdot K$ means: phage lambda, clear plaque type mutant, having the host specificity imparted by *E. coli* K12. Several host-specific characters may be found in one and the same phage particle, but we will give only the ones considered in the particular experiment. Classification of a phage stock with respect to host specificity is possible both by consideration of its history and by determination of its efficiency of plating on various hosts.

Bacterial strains. The following strains of *E. coli* and some derivatives of them have been used.

(a) *E. coli* K12. The K12 strains referred to in this paper are W3110, W3350 (see Arber, 1960a) and C600 (Appleyard, 1954). All of these strains were used as hosts for growth of phage λ; only C600 was used as a plating indicator strain.

(b) *E. coli* B. Bc (Cohen, 1959) and its derivative no. 251, a hybrid derived by transduction and carrying the mal^+-λ^s region from K12 (Arber & Lataste-Dorolle, 1961).

(c) *E. coli* C (Bertani & Weigle, 1953).

Bacteriophages

(a) Phage λ^{++}, wild type (Kaiser, 1957) and its mutants c (clear plaque, Jacob & Wollman, 1954), mi (minute, Kaiser, 1955) and b_2 (buoyant density mutant, Kellenberger, Zichichi & Weigle, 1960).

(b) Phage P1 adapted to K12 (see Arber, 1960b).

Media

(a) Tryptone broth: 1% Difco Bactotryptone, 0·5% NaCl, pH 6·9 to 7·1; for solid medium completed with 1·5% agar.

(b) Solid LB broth (Bertani, 1951), used for growth of phage P1.

(c) Synthetic medium M9: 0·7% Na_2HPO_4, 0·3% KH_2PO_4, 0·1% NH_4Cl, 0·05% NaCl, 10^{-4} M-$CaCl_2$, 10^{-3} M-$MgSO_4$, 0·4% glucose, 2×10^{-6} M-Fe^{3+} citrate, pH 7·0, completed if desired with 1% Bactocasamino acids (= M9a).

(d) H medium, a glycerol-lactate medium (Stent & Fuerst, 1955) with phosphorus supplied only by 0·05% casamino acids, pH 7·2.

(e) Adsorption medium for phage λ: 0·01 M-$MgSO_4$.

(f) Dilution media: tryptone broth or phosphate buffer: 0·7% Na_2HPO_4, 0·4% NaCl, 0·3% KH_2PO_4, 0·02% $MgSO_4$, pH 6·9.

(g) Tris buffer: 0·05 M-tris, 0·5% NaCl, 0·1% NH_4Cl, 10^{-3} M-$MgSO_4$, completed with a 10^{-3} dilution of phosphate buffer.

(h) 3XD medium: 0·45% KH_2PO_4, 1·05% Na_2HPO_4, 0·3% NH_4Cl, 0·03% $MgSO_4,7H_2O$, 1·5% casamino acids, 3% glycerol, 0·003% gelatin, 3×10^{-4} M-$CaCl_2$.

Preparation of ^{32}P-containing H medium: ^{32}P of the desired activity and casamino acids (half of which had been rendered P-free) were evaporated to dryness and then resuspended in the casamino acid-free H medium. Preparation of deuterated medium: M9 salt mixture (except $CaCl_2$) and casamino acids were evaporated to dryness, resuspended in 99·7% D_2O and completed with $CaCl_2$ and glucose.

The general *phage techniques* are described by Adams (1950) and those particular to λ by Arber (1958, 1960a). The CsCl density gradient centrifugation technique is described by Meselson, Stahl & Vinograd (1957) and by Weigle, Meselson & Paigen (1959). For all density gradients the swinging bucket rotor SW 39 of the Spinco preparative centrifuge was run for 16 hr at 22,000 rev./min. A hole was punched in the bottom of each tube and each drop was collected in a separate tube containing 1 ml. of tryptone broth.

As the lysogenic condition is not stable in many K12(P1) strains only young cultures grown from recently re-isolated single colonies were used in our experiments. These conditions assured a proportion of non-lysogenic segregants of less than 10^{-3}.

Results

Definition of systems giving host controlled modification of λ

The efficiencies of plating (e.o.p.) of some variants of phage λ are given in Table 1 for different host strains. The e.o.p. on the host strains on which the phage stock was produced is put equal to 1. For λ·K this procedure is justified, since the plaque-forming titer on K12 and the number of λ particles, as determined by electron microscopy, are approximately equal (Kellenberger & Arber, 1957). The similarity of the titers found in lysates of λ·K(P1), λ·B and λ·C and those in λ·K lysates suggests that the former lysates also do not contain many more λ particles than plaque-formers on their competent host.

TABLE 1

Efficiency of plating of phage λ variants on different host strains

Phage variant	Efficiency of plating on host strains			
	K12	K12(P1)	Bc 251	C
λ·K	1	2×10^{-5}	10^{-4}	1
λ·K(P1)	1	1	10^{-4}	1
λ·B	4×10^{-4}	7×10^{-7}	1	1
λ·C	4×10^{-4}	4×10^{-7}	2×10^{-4}	1

Indicator bacteria were grown in aerated tryptone broth to about 4×10^8 cells/ml. and then starved in 0·01 M-MgSO$_4$. Pre-adsorption was for 15 min at 37°C. For K12, either C600 or W3110 was used.

The physiological condition of the recipient bacteria does not greatly influence the e.o.p., although small fluctuations are found if the age of the bacteria or the medium are varied. Treatment of the infecting phage stock with DNase does not change the e.o.p.

Hosts containing several factors known to restrict the acceptance of a given phage result in a lower e.o.p. than hosts in which only one restricting factor is present, e.g. K12(P1) accepts phages λ·C and λ·B at a frequency of only 10^{-6} to 10^{-7}, while K12 non-lysogenic for P1 accepts the same phages at a frequency of 10^{-3} to 10^{-4}.

Modification of phage λ upon vegetative multiplication in a new host

Phage λ·K(P1) is accepted without restriction in the non-lysogenic host K12. The injected phage genomes multiply vegetatively and after the usual λ latent period the bacteria liberate active progeny phages with about the same burst size as if λ·K were used for infection. The great majority of the progeny phages show host controlled modification, i.e. are λ·K. A low proportion of λ·K(P1) is found, however, since plating of the lysate with the indicator strain K12(P1) gives a number of plaques distinctly higher than attributable to the usual e.o.p. of λ·K. This reappearance of apparently non-modified λ·K(P1) was studied in a number of one-cycle growth experiments in which non-lysogenic K12 was infected with λ·K(P1) and the resulting infective centres and subsequent lysates were assayed for total plaque-formers by plating with K12 bacteria and for λ·K(P1) by plating with K12(P1). In Table 2 it may be seen that bacteria infected with single λ·K(P1) particles in such an experiment

TABLE 2

One-cycle growth of $\lambda \cdot K(P1)$ on K12

Exp.	Growth medium	M.o.i.	Free $\lambda \cdot K(P1)$/ml. after washing	Infective centers/ml. (plaque titer before lysis)			Progeny phage/ml.		Burst size	
				on K12	on K12(P1)	$\dfrac{\lambda \cdot K(P1)}{\text{total } \lambda}$	on K12	on K12(P1)	Total	$\lambda \cdot K(P1)$
19/11	H	1·0	$2 \cdot 0 \times 10^2$	$4 \cdot 2 \times 10^7$	$2 \cdot 9 \times 10^7$	0·69	$3 \cdot 8 \times 10^9$	$1 \cdot 7 \times 10^7$	90	0·6
19/16	H	1·3	$2 \cdot 5 \times 10^2$	$3 \cdot 4 \times 10^7$	$2 \cdot 9 \times 10^7$	0·85	$8 \cdot 8 \times 10^9$	$3 \cdot 8 \times 10^7$	260	1·3
19/15	H	0·7	10^3	$5 \cdot 0 \times 10^7$	$3 \cdot 3 \times 10^7$	0·66	$1 \cdot 1 \times 10^{10}$	$4 \cdot 9 \times 10^7$	220	1·5
19/8	H	1·0	$2 \cdot 0 \times 10^2$	$4 \cdot 0 \times 10^7$	$2 \cdot 8 \times 10^7$	0·70	$4 \cdot 0 \times 10^9$	$2 \cdot 0 \times 10^7$	100	0·7
19/7	H	0·75	$2 \cdot 3 \times 10^4$	$4 \cdot 8 \times 10^7$	$3 \cdot 2 \times 10^7$	0·67	$7 \cdot 4 \times 10^9$	$2 \cdot 6 \times 10^7$	154	0·8
8/6	trypt.	0·2	—	$3 \cdot 0 \times 10^5$	$1 \cdot 0 \times 10^5$	0·33	$3 \cdot 6 \times 10^7$	$1 \cdot 2 \times 10^5$	120	1·2
8/6	trypt.	0·12	—	$8 \cdot 0 \times 10^3$	$3 \cdot 1 \times 10^3$	0·39	$8 \cdot 0 \times 10^5$	$2 \cdot 3 \times 10^3$	100	0·7

Phage $\lambda \cdot K(P1)$ (genetically cb_2), freed of contaminating P1 phages by treatment with anti-P1 serum, was adsorbed for 10 min to starved K12 bacteria (strain W3110). The mixture was diluted in aerated growth medium to assure a complete injection of the phage genomes, then treated with anti-λ serum and finally washed three times by low speed centrifugation. After resuspension in fresh growth medium, free phage was assayed by the chloroform technique and infective centers were determined on non-lysogenic and P1-lysogenic indicators. The same indicators were used to plate the lysates. Burst size is calculated as the number of liberated phages per infective center on the respective indicator. M.o.i. = multiplicity of infection.

have a very high probability (about 0·3 to almost 1, varying with different experimental conditions) of liberating at least one infective $\lambda \cdot K(P1)$, i.e. of forming a plaque on K12(P1). The average $\lambda \cdot K(P1)$ burst size per productive bacterium is only about one, whereas the burst size of $\lambda \cdot K$ is, as normally, between 100 and 250. This finding suggests that the $\lambda \cdot K(P1)$ progeny phage found after one cycle of growth on K12 are composed of material transferred from the parent $\lambda \cdot K(P1)$ phages.

Joint transfer of host specificity and DNA

Since it is known that λ parental phage DNA may be transferred to the progeny in one cycle of growth (Kellenberger, Zichichi & Weigle, 1961; Meselson & Weigle, 1961) the question arises as to whether transferred DNA and transferred host specificity are found in the same or in different phage progeny particles. Two types of experiments were designed to answer this question.

(a) Inactivation of parental and transferred $\lambda \cdot K(P1)$ by decay of incorporated ^{32}P

Heavily ^{32}P-labeled $\lambda \cdot K(P1)$ were prepared, either by infection of K12(P1) with $\lambda \cdot K(P1)$ or by u.v.-induction of K12(P1) $(\lambda)/\lambda$, in H medium containing ^{32}P at a specific activity between 200 and 800 mc/mg. Phages thus obtained and stored at 4°C in 3XD medium show an exponential inactivation due to the decay of the ^{32}P incorporated in the DNA (Hershey, Kamen, Kennedy & Gest, 1951). Aliquots of the ^{32}P-labeled $\lambda \cdot K(P1)$ stocks were allowed to grow for one cycle in unlabeled non-lysogenic K12 in unlabeled medium. The lysates thus obtained were also stored at 4°C in 3XD medium and assayed from day to day. Two typical experiments are represented in Fig. 1. The stock used for the experiment of Fig. 1(a) apparently contained a low proportion of the original parental phage which had not adsorbed and therefore remained unlabeled and stable to ^{32}P decay, and possibly also contained some progeny phages with a significantly lower than average labeling because of transferred parental ^{31}P atoms. The stock used for the experiment of Fig. 1(b) was obtained by induction of lysogenic cells and contained no detectable amount of non-labeled transferred prophage material, although the bacteria had been put in ^{32}P medium only after induction.

The results of these experiments, done at low multiplicity of infection (m.o.i.) (Table 3), are as follows:

(1) The total phage progeny after one growth cycle of ^{32}P-labeled $\lambda \cdot K(P1)$ phage on K12 non-labeled cells shows no detectable inactivation as a function of storage time, as is expected, since the bulk of the phage DNA is newly synthesized material.

(2) All progeny $\lambda \cdot K(P1)$ phages (plaque-formers on K12(P1)) are sensitive to ^{32}P decay. Hence the phages with transferred host specificity all contain transferred parental phosphorus. We conclude that the substances providing host specificity to the phage are physically linked to the DNA molecule.

(3) For the transferred $\lambda \cdot K(P1)$ progeny the rate of ^{32}P decay inactivation is about half that of the parental $\lambda \cdot K(P1)$: 45 and 46% for the experiments plotted in Figs. 1(a) and 1(b) respectively. Other experiments, not described here in detail, gave ^{32}P decay sensitivities of the transferred $\lambda \cdot K(P1)$ ranging from 40 to 50% of that of the parents. The most likely explanation of this finding is that the half-sensitive, and thus half-labeled, "hybrid" DNA results from replication according to the Watson-Crick model (Watson & Crick, 1953). One ^{32}P-labeled parental DNA strand would thus be

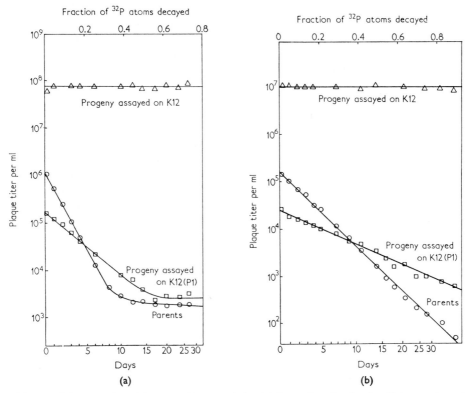

FIG. 1. *Joint transfer of host specificity and sensitivity to inactivation by disintegration of* ^{32}P *incorporated in the DNA.* Heavily ^{32}P-labeled stocks of $\lambda \cdot$K(P1) were prepared in H medium either by lytic multiplication of $\lambda cb_2 \cdot$K(P1) in strain W3350(P1) (experiment a) or by u.v.-induction of strain W3350(P1) (λ^{++})/λ (experiment b). The chloroformed lysates were diluted immediately in 3XD medium. Aliquots were stored at 4°C and assayed from day to day. Both indicators K12 and K12(P1) gave the same plaque titers ($\bigcirc$). Other aliquots were treated with anti-P1 serum and then allowed to grow for one lytic cycle in non ^{32}P-labeled, non-lysogenic K12 bacteria (strain W3110) in non-labeled tryptone medium (supplemented with 4×10^{-3}% gelatin and 0·2% glycerol), as described in Tables 2 and 3. Dilutions of the lysates in 3XD were also stored at 4°C and assayed from day to day on K12 ($\triangle$) and K12(P1) ($\square$). Titers obtained on K12(P1) were corrected for the e.o.p. of $\lambda \cdot$K. Titers of ^{32}P-labeled parent phage ($\bigcirc$) were adjusted for the plot, so that the initial titer coincides with the number of bacteria which were productive in the one-cycle growth.

TABLE 3

One-cycle growth of ^{32}P-labeled $\lambda \cdot K(P1)$ in non-labeled K12

(experiments plotted in Fig. 1)

Exp.	Phage	M.o.i.	Plaque titer before lysis		Plaque titer immediately after lysis		Burst size	
			on K12	on K12(P1)	on K12	on K12(P1)	Total	$\lambda \cdot$K(P1)
Fig. 1(a)	$\lambda cb_2 \cdot$K(P1)	0·008	$1 \cdot 1 \times 10^6$	$4 \cdot 7 \times 10^5$	$6 \cdot 4 \times 10^7$	$1 \cdot 6 \times 10^5$	58	0·34
Fig. 1(b)	$\lambda^{++} \cdot$K(P1)	0·006	$1 \cdot 5 \times 10^5$	$5 \cdot 3 \times 10^4$	$1 \cdot 2 \times 10^7$	$2 \cdot 8 \times 10^4$	80	0·53

transferred, associated with a newly synthesized unlabeled daughter strand, to each
$\lambda \cdot$K(P1) progeny phage. Although the parental DNA strand serves as template one
or more times in the course of vegetative phage multiplication the parental host
specificity does not separate from it.

That the transferred $\lambda \cdot$K(P1) is inactivated at slightly less than half the inactivation rate of the parental phage may be due to breakage and reunion recombination
between labeled parental DNA strands and unlabeled progeny strands. As will be seen
later some such recombinants can indeed be found.

(b) *Density labeling of parental* $\lambda \cdot K(P1)$ *with deuterium*

Preliminary experiments showed that differently modified variants of phage λ
(tested $\lambda \cdot$K, $\lambda \cdot$K(P1) and $\lambda \cdot$B) have the same buoyant density in CsCl, within the
precision of the determination by the drop-collecting method, whereas many genetic
mutants are known which differ in their density from the reference type (Kellenberger
et al., 1960).

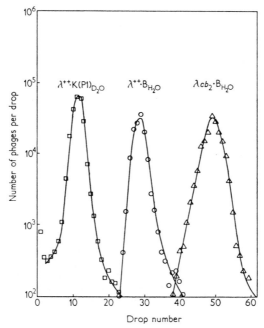

FIG. 2. *Density distribution of deuterated* $\lambda^{++} \cdot K(P1)$, *after centrifugation in CsCl density gradient.*
Deuterated $\lambda^{++} \cdot$K(P1) was obtained by u.v.-induction of strain W3350(λ) (P1)/λ, which had been
grown for many generations in M9a medium prepared with 99·7% D_2O. H_2O-grown $\lambda^{++} \cdot$B and
$\lambda cb_2 \cdot$B were added as density references and they were assayed on B (strain Bc 251).

In the experiments to be described here phage $\lambda \cdot$K(P1) was grown in a synthetic
M9a medium prepared with 99·7% D_2O instead of normal water. Such "heavy"
deuterium-labeled phages formed, upon centrifugation in a CsCl density gradient, a
band as sharp as normal (H_2O-grown) phages (Fig. 2). We conclude that our deuterated phage stocks were composed of uniformly dense phage particles. The exact
density was not measured but was compared with the densities (ρ) of phage λb_2 and
λb_2^+ grown with H_2O and having $\rho = 1\cdot491$ g/cm³ and $1\cdot508$ g/cm³ respectively
(Kellenberger *et al.*, 1960). In the centrifugation of heavy λcb_2, obtained by several

lytic growth steps on K12(P1), the normal λb_2 and λb_2^+ control bands were separated by 20 drops and the deuterated λb_2 formed a band 3 drops on the heavier side of the λb_2^+ control. In the centrifugation of heavy λb_2^+ (Fig. 2), obtained by u.v. induction of strain W3350(P1) $(\lambda)/\lambda$, the distance between the H_2O grown λb_2 and λb_2^+ controls was again 20 drops and the deuterated λb_2^+ banded at a distance of 17 drops on the heavier side of the λb_2^+ control.†

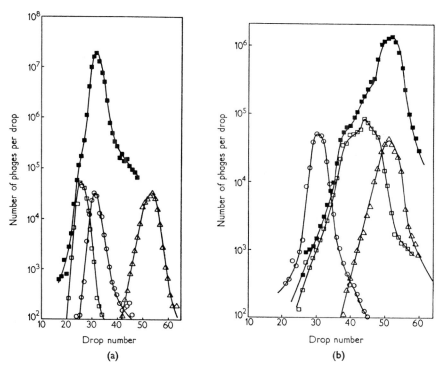

FIG. 3. *Density distribution of lysates obtained by one cycle growth of deuterated* $\lambda \cdot K(P1)$ *on normal, non-lysogenic K12 (strain W3110).*

(a) Lysate of experiment 23/5a (see Table 4): m.o.i. = 0·019. Aliquot centrifuged: 0·9 ml. of lysate, 82 drops collected.

□, $\lambda^{++} \cdot K(P1)$, assayed on K12(P1). ■, total λ^{++}, assayed on K12. ○, density reference $\lambda^{++} \cdot B$. △, density reference $\lambda c b_2 \cdot B$.

(b) Lysate of experiment 23/3d (see Table 4): m.o.i. = 28. Aliquot centrifuged: 0·05 ml. of lysate, 78 drops collected.

□, $\lambda c b_2 \cdot K(P1)$, assayed on K12(P1). ■, total $\lambda c b_2$, assayed on K12. ○, density reference $\lambda^{++} \cdot B$. △, density reference $\lambda c b_2 \cdot B$.

The deuterated $\lambda \cdot K(P1)$ phage stocks were allowed to grow for one cycle of multiplication on non-lysogenic K12 bacteria in medium prepared with H_2O (Table 4). The rate of transfer of the host specificity was as found for normal, non-deuterated $\lambda \cdot K(P1)$. The one-step lysates were centrifuged in a CsCl density gradient and the collected fractions were assayed for total phage and phage with K(P1) host specificity. Two typical distributions are reproduced in Fig. 3.

† The u.v.-induced stock of deuterated λ^{++} contained 1·5% plaque-type mutants, the stock of deuterated $\lambda c b_2$ obtained by infection contained less than 0·1% plaque-type mutants.

TABLE 4

One-cycle growth of deuterated $\lambda \cdot K(P1)$ on K12 in H_2O medium

Exp.	Phage $\lambda \cdot K(P1)$	Growth medium	M.o.i.	Free $\lambda \cdot K(P1)$/ml. after washing	Infective centers/ml. (plaque titer before lysis)			Progeny phage/ml.		Burst size	
					on K12	on K12(P1)	$\dfrac{\lambda \cdot K(P1)}{\text{total } \lambda}$	on K12	on K12(P1)	Total	$\lambda \cdot K(P1)$
23/3a	λ^{++}	M9a	0·014	$4·0 \times 10^2$	$1·4 \times 10^6$	$4·4 \times 10^5$	0·31	$5·6 \times 10^7$	$1·9 \times 10^5$	40	0·43
b	λcb_2	M9a	0·022	$4·0 \times 10^2$	$1·5 \times 10^6$	$6·0 \times 10^5$	0·40	$2·3 \times 10^8$	$5·0 \times 10^5$	153	0·83
c	λcb_2	M9a	7·1	$1·4 \times 10^3$	$1·8 \times 10^6$	$1·2 \times 10^6$	0·66	$2·8 \times 10^8$	$5·0 \times 10^6$	156	4·2
d	λcb_2	M9a	28	$6·0 \times 10^2$	$1·5 \times 10^6$	$1·7 \times 10^6$	1	$1·8 \times 10^8$	$1·2 \times 10^7$	120	7·1
23/5a	λ^{++}	H	0·019	$5·0 \times 10^2$	$1·8 \times 10^6$	$6·4 \times 10^5$	0·36	$7·0 \times 10^7$	$2·7 \times 10^5$	39	0·42
b	λcb_2	H	0·015	$8·0 \times 10^2$	$2·0 \times 10^6$	$8·2 \times 10^5$	0·41	$1·6 \times 10^8$	$5·0 \times 10^5$	80	0·61
c	λcb_2	H	2·8	$2·7 \times 10^3$	$6·4 \times 10^7$	$4·2 \times 10^7$	0·66	$7·3 \times 10^9$	$8·9 \times 10^7$	114	2·1

Deuterated $\lambda \cdot K(P1)$ phage stocks, treated with anti-P1 serum, were adsorbed to starved K12 bacteria (strain W3110). The complexes were then washed three times and, after resuspension in fresh growth medium, assayed for free $\lambda \cdot K(P1)$ and infective centers. The lysates were chloroformed 55 to 75 min after suspension in growth medium.

If the one cycle growth was initiated by an infection at low multiplicity, (1) all the transferred $\lambda \cdot K(P1)$ phages banded sharply at a density a few drops heavier than the reference carrying the same genotype, with a density corresponding to a transfer of roughly one-quarter of the initial "extra" density of the deuterated parent; (2) the majority of the $\lambda \cdot K$ banded sharply at the reference density. After infection at multiplicities higher than one, the band of the transferred $\lambda \cdot K(P1)$ appeared enlarged on its dense side and, with increasing multiplicity of infection, a second peak became more and more important (Fig. 3(b)). The maxima of the enlarged double band of $\lambda \cdot K(P1)$ were situated at about 25% and 50% of the initial distance separating the deuterated $\lambda \cdot K(P1)$ parent from the corresponding reference type. Qualitatively the same behavior was found for both $\lambda b_2^+ \cdot K(P1)$ and $\lambda b_2 \cdot K(P1)$ phage.

From these experiments we conclude that

(1) none of the phages having the transferred parental host specificity is composed uniquely of newly synthesized material,

(2) after low multiplicity of infection the transferred $\lambda \cdot K(P1)$ contain about half of the parental DNA, corresponding to about one-quarter of the parental "extra" density, since about half of this density is attributable to the protein coats. The parental DNA would thus be semi-conserved, probably having been replicated according to the Watson–Crick model.

(3) With high multiplicity of infection the DNA of some of the transferred $\lambda \cdot K(P1)$ is fully conserved, i.e. transferred intact, only the protein coat being new, non-deuterated. These findings are in agreement with similar observations made by Meselson & Weigle (1961) in transfer experiments using λ phage labeled with $^{13}C^{15}N$.

The question arises whether it is necessary for conservation of the parental host specificity that at least one complete parental DNA strand is transferred, or if a *partial transfer* of certain genome regions is sufficient. In order to answer this question, deuterated† $\lambda b_2 cmi^+ \cdot K(P1)$ was crossed with normal $\lambda b_2 c^+ mi \cdot K$ in a non-lysogenic K12 host in normal tryptone medium (Table 5). The lysate was then assayed on K12(P1) for the $\lambda \cdot K(P1)$ phages and on K12 for the total phage titer. As seen in Table 5, both parental types participated in vegetative multiplication and gave rise to recombinants at usual frequencies. But among the progeny phage particles, essentially only $\lambda \cdot K(P1)$ parental cmi^+ and some cmi recombinants were able to grow on K12(P1). The frequencies of phages showing transferred host specificity were 1·3% for λcmi^+ and 0·55% for λcmi genotypes. Hence the host specificity governed by P1 can be transferred by somewhat less than the whole λ genome. In our experiment a small proportion of $\lambda c^+ mi^+ \cdot K(P1)$ was found and most of such plaques were mottled, typical for c/c^+ heterozygotes. Some of them were further tested and their heterozygous nature was confirmed; none of them was mi, nor were they mi/mi^+ heterozygotes. The few phages of the $\lambda \cdot K$ parental genotype $(c^+ mi)$ which are found by assay on K12(P1) can be attributed to the rate of $\lambda \cdot K$ plaque formers on K(P1).

A CsCl density gradient centrifugation of the lysate of the cross described in Table 5 revealed that all the $\lambda \cdot K(P1)$, including the cmi recombinants and the few $c/c^+ mi^+$ heterozygotes, had a density corresponding approximately either to semi-conserved DNA transfer or, less frequently, to conserved DNA transfer, i.e. only

† A deuterated phage stock was used in this experiment in order to permit an analysis of the lysate by density gradient centrifugation. The results of the cross are, however, not influenced by the deuterium present in one of the parents, as has been shown by the same cross done with non-deuterated phages.

phages with deuterated parental material were able to grow on K12(P1). On the other hand, deuterium label was found among the phages restricted on K12(P1), eliminating the possibility that parental $\lambda \cdot K(P1)$ could not exchange material with $\lambda \cdot K$ genomes because of the difference in host specificity.

TABLE 5

Cross of deuterated $\lambda b_2 cmi^+ \cdot K(P1)$ *with normal* $\lambda b_2 c^+ mi \cdot K$ *on non-lysogenic K12 (strain W3110)*

	Genotypes			
	Parental		Recombinants	
	cmi^+	c^+mi	cmi	c^+mi^+
Multiplicity of infection	4·3	3·6		
Free phage/ml. after adsorption and washing	$8·0 \times 10^4$	$3·0 \times 10^5$		
Infective centers/ml. before lysis, assayed on { K12	$3·3 \times 10^7$	(35% mottled)		
{K12(P1)	$2·4 \times 10^7$	(0·6% mottled)		
Fraction of infective centers yielding $\lambda \cdot K(P1)$	0·73			
Progeny phage/ml., plated on K12	$8·5 \times 10^8$	$3·4 \times 10^8$	$3·3 \times 10^7$	$1·5 \times 10^7$
Fraction of recombinants per parental $\lambda cmi^+ \cdot K$			3·9%	1·8%
Progeny phage/ml., plated on K12(P1)	$1·1 \times 10^7$	$(10^4)†$	$1·8 \times 10^5$	$7·5 \times 10^3‡$
Fraction of recombinants per parental $\lambda cmi^+ \cdot K(P1)$			1·6%	0·07%

† Corresponds to the spontaneous modification rate of the $3·4 \times 10^8$ phages λc^+mi K/ml.

‡ Mostly heterozygotes c/c^+.

Production of $\lambda \cdot K(P1)$ *variants induced by superinfection with P1 of a K12-λ complex*

In the preceding sections, we studied the modification of host specificity in the progeny of $\lambda \cdot K(P1)$ which had infected a non-lysogenic K12 strain. The reverse experiment, to study the modification of $\lambda \cdot K$ multiplying in K12(P1), cannot be done in the same way since by virtue of the host specificity the infecting $\lambda \cdot K$ DNA is recognized and subsequently degraded (Dussoix & Arber, 1962). This restricted acceptance can, however, be surmounted if, instead of an already established P1-lysogenic acceptor K12(P1), a K12 recently infected with P1 is used or if phage P1 is introduced only after adsorption of λ on K12.

These superinfection experiments have been done by infecting K12 bacteria of strain W3350, grown in tryptone broth to 10^9 cells/ml. and subsequently starved in adsorption medium, with phage $\lambda c \cdot K$. After 10 minutes of adsorption at 37°C, fresh tryptone broth was added and the complexes were aerated at a concentration of 4×10^8 cells/ml. At various times aliquots were superinfected with phage P1 (grown on W3350). Good adsorption was insured by adding 1/500 M-CaCl$_2$ to the bacteria 10 minutes before the infection with P1. For P1-superinfections simultaneously, before, or shortly after the infection with λ, this general plan had to be slightly modified. After adsorption of both λ and P1 the non-adsorbed phages were measured and the infected bacteria were diluted in tryptone broth to a concentration of about 10^6 cells/ml. Except in the cases of very late superinfection with P1, infective centers were then measured on both K12 and K12(P1) indicators. At 50 to 60 minutes after the infection with λ all the

λ-producing bacteria had lysed, as revealed by periodically measuring intra- and extracellular λ phage. The lysates were chloroformed at 60 to 75 minutes after λ infection and assayed on K12 and K12(P1) indicators for the variant types of λ progeny. Non-lysogenic bacteria infected with both λ and P1 phages produce principally λ phage particles with a latent period characteristic for the K12–λ complexes. In our experimental conditions very few P1 progeny phages are found, 10^{-3} to 10^{-4} as many as λ.

TABLE 6

Induced modification of host specificity by superinfection with phage P1 of K12 bacteria infected with λ phage

Time of addition of superinfecting P1 (min)	Proportion $\dfrac{\lambda \cdot \text{K(P1)}}{\text{total } \lambda}$		
	Infective centers before lysis	Lysates chloroformed	
		at 60 min	at 75 min
0	0·97	0·77	0·74
10	0·50	0·17	0·15
15	0·80	0·33	0·45
20	0·91	0·24	0·39
25	0·52	0·22	0·30
30	0·13	0·083	0·18
35	0·025	0·005	0·11
40	0·019	$2·6 \times 10^{-5}$	0·02
45	0·022	$4·1 \times 10^{-5}$	0·007
Without superinfection	2×10^{-3}	3×10^{-5}	

Infection of K12 (strain W3350) with phage $\lambda cb_2 \cdot$ K at $t = 0$ min (m.o.i. $= 0·6$). Superinfection with P1 at various times (m.o.i. $= 5·4$). For procedure see text. First appearance of intracellular λ phage particles, liberated by chloroforming the culture, at 32 min. First appearance of extracellular λ phage particles (latent period) at 40 min. Massive liberation of first λ progeny completed at 50 min.

The results of a typical experiment are represented in Table 6. *Simultaneous infection* with λ and P1 usually induces the P1-specific modifications in almost all λ-producing complexes. In the lysates the fraction of phages showing the host specificity typical for K12(P1) is generally between 50% and almost 100% of the total phage output. As the superinfection with P1 is *delayed* the fraction of modified phages decreases. For the first 25 minutes irregular variations between 5 and 100% modifications were found, depending more, perhaps, on the metabolic conditions than on the effective time of superinfection. As the first intracellular λ phage particles appear, the possibility of modification drops rapidly and disappears almost completely with the beginning of lysis. It should be pointed out that the actual time of P1 penetration into the K12–λ complex is a few minutes after the measured P1 addition time (Table 6), since adsorption and injection are not immediate. In the experiments where the one-step lysates were chloroformed 20 or more minutes after the occurrence of massive lysis, very late P1 superinfections induced some detectable modifications. This might be due to a minor fraction of bacteria retarded in the production of λ or to newly infected bacteria producing a second generation of λ.

Higher frequencies of modification are found if a high multiplicity of P1 is used for superinfection. After simultaneous infection with $\lambda \cdot K$ and P1, maximum modification may already be attained with a multiplicity of 5 P1 phages per bacterium. At very low P1 multiplicity the $\lambda \cdot K(P1)$ producing centers increase linearly with multiplicity, and it is possible to calculate the probability of modification occurring in a cell infected by a single P1. In one particular experiment this probability was 34%, but it may depend on the experimental conditions. The multiplicity of $\lambda \cdot K$ does not seem to influence the frequency of modification.

These experiments show that the conversion of a $\lambda \cdot K$-producing complex into a $\lambda \cdot K(P1)$-producing complex occurs soon after the infection with phage P1. This conversion certainly does not depend on the presence of λ at the time of P1 infection, but may occur also if P1 is introduced into the K12 cells previous to the infection with λ. Experiments designed to explore this situation indicated that, indeed, late superinfection with $\lambda \cdot K$ of K12–P1 complexes is possible for a relatively long period, and that a high proportion of the λ progeny phages are thereby modified into $\lambda \cdot K(P1)$. These results suggest that the control of host specificity by the presence of P1 can be dissociated into (1) the control over the *acceptance* of non-adapted DNA and (2) the control over the *modification* of non-adapted DNA. The first control mechanism seems to be established only slowly, perhaps in parallel to the establishment of P1 lysogeny. The control of modification, however, is established rapidly, as already shown, and is also induced by the virulent mutants of P1 which do not yield stable lysogenics. More experiments are necessary to explore the connections between the establishment of these two control mechanisms on the one hand, and the steps leading to either vegetative multiplication or establishment of P1 lysogeny on the other hand.

The results presented in Table 6 suggest that λ-producing cells which are converted by superinfection with P1 may liberate mixed bursts of both non-modified $\lambda \cdot K$ and modified $\lambda \cdot K(P1)$ variant types, since the relative proportions of plaque formers on K12(P1) are smaller in the lysates than if the productive bacteria are plated before their lysis. The mixed nature of host specificity types liberated by individual bacteria has, indeed, been confirmed by single burst experiments.

It is of interest to know if the $\lambda \cdot K(P1)$ phage particles contain only genomes which were newly synthesized after the conversion by superinfection with P1, or if the new host specificity may be given even to finished, non-replicating phage genomes. We first tried to answer this question with genetic experiments of the following type: K12 was infected with λ and, after various times of incubation, superinfected with P1 and—simultaneously or after a short delay—with a second, genetically distinct type of λ. After lysis had occurred, the liberated phages were classified for their genotypes and their host-specificity types. The proportions of modification diminished as superinfection was retarded, but for all superinfection times the relative amount of modified $\lambda \cdot K(P1)$ among the genotypes of the first infecting parent and the superinfecting parent was roughly equal. The possibility that P1 host specificity may be given to a non-replicating DNA is thus suggested, although the experiments are not conclusive. Indeed, only very late superinfection at a moment where finished DNA is already withdrawn from the vegetative pool could give sure results, but by that time superinfection with a second λ is no longer successful, i.e. the superinfecting λ does not participate in the vegetative growth (Séchaud, 1960).

In order to surmount these difficulties, we designed another experiment, based on the fact that, after infection at high multiplicity, conserved phage DNA can be transferred to the progeny, apparently without active participation in the vegetative pool. Non-lysogenic K12 bacteria (strain W3110) were simultaneously infected with deuterated $\lambda cb_2 \cdot K$ (m.o.i. = 20) and normal P1 (m.o.i. = 6·7). After 15 minutes of adsorption, the complexes were washed and then incubated in tryptone broth. After 2 minutes, free phage titer and titer of productive centers were measured (Table 7).

TABLE 7

Modification of $\lambda \cdot K$ into $\lambda \cdot K(P1)$ by superinfection with phage P1

K12 bacteria (strain W3110), grown to a concentration of 10^9 cells/ml. and starved in 0·01 M-MgSO$_4$, were simultaneously infected with deuterated $\lambda cb_2 \cdot K$ and normal P1.

M.o.i. of deuterated $\lambda cb_2 \cdot K$	20
M.o.i. of P1	6·7
Free $\lambda \cdot K$ after adsorption and washing	$1·8 \times 10^5$/ml.
Infective centers/ml. (plaque titer before lysis) assayed on { K12	$9·3 \times 10^6$
K12(P1)	$8·7 \times 10^6 = 94\%$
Progeny phage/ml. assayed on { K12	$1·1 \times 10^9$
K12(P1)	$9·4 \times 10^8 = 85\%$
Total burst size	118

This lysate was centrifuged in CsCl density gradient, see Fig. 4.

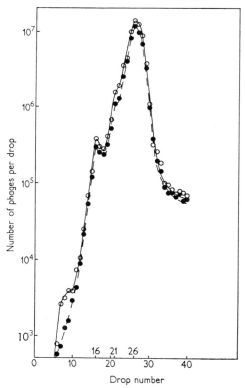

FIG. 4. Density distribution of the lysate obtained by one-step growth of deuterated $\lambda cb_2 \cdot K$ (m.o.i. = 20) on non-lysogenic K12 (strain W3110) simultaneously infected with P1 (m.o.i. = 6·7); see Table 7. Aliquot centrifuged: 0·05 ml. of lysate, i.e. $5·5 \times 10^7$ λ particles; 78 drops collected. Lysate assayed on K12 (○) and K12(P1) (●). Phages with conserved (drop 16), semi-conserved (drop 21) and newly synthesized (drop 26) DNA are modified in equal proportions. A few parental $\lambda \cdot K$ phages, which had not adsorbed and were not removed by washing (Table 7) are found at drops 7 to 9.

The lysate, chloroformed after 60 minutes of incubation, was assayed on K12 and K12(P1) and an aliquot was analysed by density gradient centrifugation (Fig. 4). It may be seen that not only phages containing newly synthesized DNA (either entirely new or half new–half conserved) showed the modified host specificity character, but also phages with fully conserved, transferred DNA were modified in about equal proportion as the total lysate. Similar results were obtained in two other experiments using $\lambda cb_2^+ \cdot K$ at m.o.i. of 25 and 12·5 respectively. These findings allow the conclusion that replication is not necessary for modifying $\lambda \cdot K$ into $\lambda \cdot K(P1)$ after conversion of the productive host cell by superinfection with phage P1.

Discussion

Many bacterial viruses show a pronounced specificity in respect to their host range. Several independent factors may restrict the number of hosts which can be infected successfully by a phage.

(1) The first condition for phage reproduction is its specific adsorption to the host. Tests for resistance to phage adsorption are easily performed and indicate the absence of appropriate phage receptors on the bacterial cell wall. It is possible to surmount this barrier either by isolation of bacterial mutants which can synthesize such phage receptors, or by isolation of phage mutants able to adsorb to the existing sites.

(2) After adsorption of a phage particle, its DNA must penetrate into the host (injection). Very little is known to date about aberrations which might occur at this step of infection, and whether such aberrations might be governed by bacterial or phage properties or both.

(3) As the phage DNA enters the bacterial host cell its "host specificity" is recognized by some at present unknown means. If the specificity is incompatible with the host, the DNA may be at least partially degraded. Such control over acceptance of DNA of foreign host specificity seems to be a widespread phenomenon and will be discussed in more detail in another article (Dussoix & Arber, 1962). This type of specificity is here shown to have its site on the DNA molecule, despite its non-mutational nature (Bertani & Weigle, 1953).

(4) Many processes are known or could be imagined which inhibit the replication of fully injected and non-degraded phage DNA, or which do not allow one of the other phage specific syntheses necessary for the production and liberation of new, infective phage particles. Such inhibitions may be governed either by bacterial or phage genes or by both, rendering the phage–bacterium complex incompetent. Examples are provided by the λ sus mutants (Campbell, 1959, 1961) and the T4 amber mutants (Epstein, personal communication) which grow on certain bacterial strains, but not on others. Many defective phages (Appleyard, 1954) are known, for which no competent host has been found and which can be reproduced only with the help of an active, superinfecting phage (Arber & Kellenberger, 1958).

Foreign DNA introduced into a new host is not necessarily degraded, and can in fact be fully accepted even though it differs in host specificity from phage DNA that is produced by the host cells in question. Full acceptance is found either (a) in a few cells of a bacterial population otherwise showing restriction, or (b) in the total population of certain bacterial hosts, as in the case described here in which $E.\ coli$ C accepts phages $\lambda \cdot K$ and $\lambda \cdot B$ as well as the adapted $\lambda \cdot C$. Such extended host range systems, which accept phage DNA of a foreign host specificity, enable us to study the process by which the new specificity is imparted during the growth of progeny particles.

The fate of the host specificity of $\lambda \cdot K(P1)$ was investigated in one-cycle growth experiments on *E. coli* K12, which fully accepts $\lambda \cdot K(P1)$ DNA. Roughly as many progeny phage particles were found to carry the parental host specificity as had been used for infection, and all other phages were $\lambda \cdot K$, carrying the host specificity of the new host K12. Host specificity of $\lambda \cdot K(P1)$ DNA is thus transferred to the progeny but is not replicated. Material linkage of transferred host specificity with transferred parental DNA could be demonstrated in experiments using two different techniques of labeling: (1) incorporation of enough ^{32}P to destroy the activity of the phage in the course of radioactive disintegration, (2) incorporation of deuterium to give the labeled phage a higher than normal density. These labeling experiments gave direct evidence that the transferred $\lambda \cdot K(P1)$ host specificity does not separate from the parental DNA molecule with which it was originally associated even in the course of DNA replication during vegetative phage growth. Indeed, all progeny phages, which were able successfully to infect K12(P1), were, in the case of ^{32}P labeling, still sensitive to disintegration of the ^{32}P contained in the backbone of the transferred DNA strand or were, in the case of deuterium labeling, at least partially heavy due to the label retained. The DNA of phage λ is believed to be two-stranded. When such DNA infects a host cell at low multiplicity we assume that the two parental strands separate, as replication and synthesis of new DNA occurs. This hypothesis predicts the formation of hybrid DNA molecules, carrying one parental and one new strand. Indeed, the phage particles obtained (in low m.o.i. experiments) which show joint transfer of parental DNA material and host specificity are found to be such hybrids. In the case of ^{32}P labeling, they are inactivated by decay of incorporated ^{32}P at about half the rate of the parents; in the case of deuterium label, they show a density intermediate between that corresponding to conserved DNA transfer and that of non-deuterated phage particles.

It is interesting to note that another sort of information concerning the sensitivity of DNA to decay of incorporated ^{32}P comes from the labeling experiments. This sensitivity seems directly proportional to the number of ^{32}P atoms present in the DNA, whether they are distributed over both strands or contained in only one of the strands of a hybrid DNA molecule. The inactivation by ^{32}P decay thus cannot be attributed to disintegrations occurring relatively close to each other, one on one strand and the other on the second strand, but must instead be a one-hit event.

The occurrence of phages with hybrid DNA molecules, which behave in respect to their host specificity like the parental $\lambda \cdot K(P1)$ phages, suggests that the specific characters given by the K12(P1) host are "dominant", i.e. phage DNA carrying the host specificity on one strand only is accepted and phage replication occurs. Genetic recombinants between the transferred $\lambda \cdot K(P1)$ parent and a superinfecting $\lambda \cdot K$ genome can be found for certain markers (e.g. the *mi* marker) without loss of the $\lambda \cdot K(P1)$ host specificity. When recombination involves the loss of longer sections of the parental genome, however, the $\lambda \cdot K(P1)$ host specificity is also lost.

After infection of K12 with a high multiplicity of $\lambda \cdot K(P1)$ phages some of the parental DNA molecules are transferred into the progeny without active replication. This conserved DNA is, however, found in new protein coats. After superinfection with genetically marked $\lambda \cdot K$, such complexes formed by infection of K12 with a high multiplicity of $\lambda \cdot K(P1)$ also liberate some recombinants showing almost fully conserved DNA. This confirms the similar findings of Meselson & Weigle (1961), who interpreted such recombinants as having been formed by a breakage and reunion mechanism. These recombinants, as well as those which are about semi-conserved, lose the original host specificity if by recombination they lose more than a small part of the parental genome.

It would be interesting to know if one parental DNA molecule may give rise to two or only one hybrid progeny molecule carrying the parental host specificity. In the present investigation, no experiment was designed to answer this question. Its resolution may be quite difficult, since transfer of ^{32}P label into the progeny DNA is known to be incomplete and so probably also is the transfer of parental host specificity. Part of the phage DNA, parental and replica, is indeed not yet assembled into finished phage particles at the moment of lysis, and is thus lost. Furthermore one does not know if parental $\lambda \cdot K(P1)$ DNA strands, conserved or semi-conserved, appearing in new infective phage particles, obligatorily contain their parental host specificity, or if they can lose it even without recombination with genomes showing the new host specificity. We could not find a definite answer to this question either, since many genetic recombinants in unmarked regions of the genome behave genotypically like the parental type and may have nevertheless lost their host specificity by recombination.

We have discussed the modification of $\lambda \cdot K(P1)$ into $\lambda \cdot K$ during growth on K12. In the opposite modification, from $\lambda \cdot K$ to $\lambda \cdot K(P1)$, which is easily induced by superinfection of the K12–$\lambda \cdot K$ complex with phage P1, the appearance of phages with conserved DNA but nevertheless modified $\lambda \cdot K(P1)$ host specificity is clearly shown by the data summarized in Fig. 4, which show that the proportion of modified progeny phages stays constant for the density gradient fractions representing λ with conserved, semi-conserved and completely new DNA. The presence of phage P1 can thus induce the modification of λ DNA even when the DNA is not replicating and when it is transferred as a conserved parental molecule.

The possibility that such modified, conserved genomes could be in reality "almost conserved" breakage and reunion recombinants formed between the heavy parent and a vegetative copy having acquired the new host specificity is excluded (1) by the high frequency of conserved particles showing modified host specificity and (2) by the fact that a large part of the $\lambda \cdot K(P1)$ DNA is needed to confer the K(P1) host specificity to a λ genome.

The rapid onset of modification of $\lambda \cdot K$ to $\lambda \cdot K(P1)$ after superinfection of the K12-λ complex with phage P1 is not paralleled by a rapid establishment of the control mechanism over acceptance of $\lambda \cdot K$ DNA. Indeed K12–P1 complexes may still be successfully infected with $\lambda \cdot K$ for a relatively long period after P1 infection. This asynchronous appearance of the two effects induced by the presence of phage P1 may be interpreted in two ways: (1) control over modification of $\lambda \cdot K$ into $\lambda \cdot K(P1)$ and control over acceptance of non-modified $\lambda \cdot K$ DNA are established by at least partly independent steps; or (2) establishment of the "modification" mechanism occurs at a low level of some P1-induced substance, whereas a much higher level of this substance is needed to establish the "acceptance" mechanism.

Nothing is known about the chemical nature of host specificity, about the way it is produced by the host, nor about the way it is recognized upon infection of a new host. This specificity could be due to the presence or absence, induction or repression of particular substances. The host specificity imparted to λ DNA in the presence of phage P1, for example, could be a substance which is produced under the direction of P1 and gets attached to the λ DNA. Upon multiplication of such DNA in K12 cells no

P1-specific substance would be produced because of the absence of a P1 genome, and all new DNA replicas would thus lack the parental P1 host specificity.

Tentative models for the explanation of the phenomena in relation to host specificity should be able to account for the following points established by our experimental data. (1) Every host cell produces DNA of its own host specificity: e.g. λ DNA produced by *E. coli* K12 carries the host specificity characteristic for K12; λ DNA produced by *E. coli* K12(P1) carries the host specificity characteristic for K12 and that characteristic for the presence of phage P1; λ DNA produced by *E. coli* B(P1) carries the host specificity characteristic for B and for the presence of phage P1, and so on. Only further studies will reveal if all these different host specificities are related in their nature, and to what extent they are additive rather than substitutive. (2) Host specific characters are closely associated with DNA and can be transferred with it in growth involving active participation in replication (semi-conserved transfer). New replicas, however, do not carry the parental host specificity unless the new host has the same specificity. (3) Genetic recombinants formed between λ genomes of different host specificities may or may not preserve the parental host specificity, probably depending on both the length and the location of the gene region involved. (4) Conversion of the host cell in respect to its specificity during λ replication, as by superinfection of a K12–λ complex with P1, can induce modification of both replicating and non-replicating λ DNA.

Host specificity may play a very important role in viral infections as suggested by the fact that host controlled modification is a widely spread phenomenon. It should be remembered here that our system is only a representative case and that for any given system some aspects of host specificity might be different. Luria & Human (1952) described, for example, a case where a modified phage seems to be no longer accepted by its own host. Further studies will have to be made in order to know if all these cases may be explained by any single model.

On the experimental level, host specificity gives us a useful biological label to follow the parental phage DNA molecule in phage infections.

We wish to thank Drs. E. Kellenberger and J. Weigle and Mrs. G. Kellenberger and M. Zichichi for their active interest in the work and Drs. D. Pratt and R. Epstein and Mr. E. Boy de la Tour for their help in the preparation of the manuscript.

The work was supported by a grant from the Swiss National Foundation for Scientific Research.

REFERENCES

Adams, M. H. (1950). In *Methods in Medical Research*, ed. by J. H. Comroe, vol. 2, p. 1. Chicago: The Year Book Publishers.
Appleyard, R. K. (1954). *Genetics*, **39**, 440.
Arber, W. (1958). *Arch. Sci., Geneva*, **11**, 259.
Arber, W. (1960a). *Virology*, **11**, 250.
Arber, W. (1960b). *Virology*, **11**, 273.
Arber, W. & Dussoix, D. (1961). Abstracts of the Int. Biophys. Cong., Stockholm, p. 291.
Arber, W. & Kellenberger, G. (1958). *Virology*, **5**, 458.
Arber, W. & Lataste-Dorolle, C. (1961). *Path. Microbiol.* **24**, 1012.
Bertani, G. (1951). *J. Bact.* **62**, 293.
Bertani, G. & Weigle, J. J. (1953). *J. Bact.* **65**, 113.
Campbell, A. (1959). *Virology*, **9**, 293.
Campbell, A. (1961). *Virology*, **14**, 22.
Christensen, J. R. (1961). *Virology*, **13**, 40.

Cohen, D. (1959). *Virology*, **7**, 112.

Dussoix, D. & Arber, W. (1962). *J. Mol. Biol.* **5**, 37.

Hershey, A. D., Kamen, M. D., Kennedy, J. W. & Gest, H. (1951). *J. Gen. Physiol.* **34**, 305.

Jacob, F. & Wollman, E. L. (1954). *Ann. Inst. Pasteur*, **87**, 653.

Kaiser, A. D. (1955). *Virology*, **1**, 424.

Kaiser, A. D. (1957). *Virology*, **3**, 42.

Kellenberger, E. & Arber, W. (1957). *Virology*, **3**, 245.

Kellenberger, G., Zichichi, M. L. & Weigle, J. (1960). *Nature*, **187**, 161.

Kellenberger, G., Zichichi, M. L. & Weigle, J. J. (1961). *Proc. Nat. Acad. Sci., Wash.* **47**, 869.

Lederberg, S. (1957). *Virology*, **3**, 496.

Luria, S. E. (1953). *Cold Spr. Harb. Symp. Quant. Biol.* **18**, 237.

Luria, S. E. & Human, M. L. (1952). *J. Bact.* **64**, 557.

Meselson, M., Stahl, F. W. & Vinograd, J. (1957). *Proc. Nat. Acad. Sci., Wash.* **43**, 581.

Meselson, M. & Weigle, J. J. (1961). *Proc. Nat. Acad. Sci., Wash.* **47**, 857.

Séchaud, J. (1960). *Arch. Sci., Geneva*, **13**, 427.

Stent, G. S. & Fuerst, C. R. (1955). *J. Gen. Physiol.* **38**, 441.

Watson, J. D. & Crick, F. H. C. (1953). *Nature*, **171**, 964.

Weigle, J. J. & Bertani, G. (1953). *Ann. Inst. Pasteur*, **84**, 175.

Weigle, J., Meselson, M. & Paigen, K. (1959). *J. Mol. Biol.* **1**, 379.

Zinder, N. D. (1960). *Science*, **131**, 813.

J. Mol. Biol. (1963) **6**, 208–213

The Bacterial Chromosome and its Manner of Replication as seen by Autoradiography

John Cairns

*Department of Microbiology, Australian National University,
Canberra, A.C.T., Australia*

(*Received 19 November 1962*)

In order to determine the form of replicating DNA, *E. coli* B3 and K12 Hfr were labelled for various periods with [^{3}H]thymidine. Their DNA was then extracted gently and observed by autoradiography. The results and conclusions can be summarized as follows.

(1) The chromosome of *E. coli* consists of a single piece of two-stranded DNA, 700 to 900 μ long.

(2) This DNA duplicates by forming a fork. The new (daughter) limbs of the fork each contain one strand of new material and one strand of old material.

(3) Each chromosome length of DNA is probably duplicated by one fork. Thus, when the bacterial generation time is 30 min, 20 to 30 μ of DNA is duplicated each minute.

(4) Totally unexpected was the finding that the distal ends of the two daughter molecules appear to be joined during the period of replication. The reason for this is obscure. Conceivably the mechanism that, *in vivo*, winds the daughter molecules lies at the point of their union rather than, as commonly supposed, in the fork itself.

(5) The chromosomes of both B3 (F$^-$) and K12 (Hfr) appear to exist as a circle which usually breaks during extraction.

1. Introduction

The semiconservative nature of DNA replication, predicted on structural grounds by Watson & Crick (1953), was demonstrated experimentally first for bacterial DNA (Meselson & Stahl, 1958), but how the two strands of the double helix come to separate during replication is not known. There must be formidable complexities to any process which, however feasible energetically (Levinthal & Crane, 1956), at once unwinds one molecule and winds two others. So a study was undertaken of the shape and form of replicating DNA.

Bacteria, despite their complexity, promised to be the most accessible source of replicating DNA. Each bacterium in an exponentially growing culture of *E. coli* makes DNA for more than 80% of the generation time (McFall & Stent, 1959; Schaechter, Bentzon & Maaløe, 1959); if, as seems likely, the bacterial chromosome is a single molecule of DNA, this molecule must be engaged in replication most of the time. A method had already been devised for extracting this DNA with little degradation (Cairns, 1962*a*) and it seemed probable that, with more care, the chromosome could be isolated intact and, caught in the act of replication, its DNA be displayed by autoradiography.

2. Materials and Methods

Bacteria. Since the chromosomes of F⁻ and Hfr bacteria differ in the type of their genetic linkage (Jacob & Wollman, 1958) and in the manner of their duplication (Nagata, 1962), two strains of *E. coli* were used, B3 (F⁻) (Brenner) and K12 3000 *thy⁻ B₁⁻* (Hfr). Both strains require thymine or thymidine.

Medium. The A medium of Meselson & Weigle (1961) was used. To this was added 3 mg/ml. casein hydrolysate, which had first been largely freed of thymine by steaming with charcoal. In this medium, supplemented with 2 μg/ml. TDR,† both strains have a generation time of 30 min.

Preparation of labelled bacteria for autoradiography. The bacteria were grown with aeration to 10^8/ml., centrifuged and resuspended in an equal volume of medium containing 2 μg/ml. [³H]TDR (9 c/m-mole). In pulse-labelling experiments, incorporation of label was stopped by diluting the bacteria either 50-fold into medium containing 20 μg/ml. TDR or 250-fold into cold 0·15 M-NaCl containing 0·01 M-KCN and 0·002% bovine serum albumin. In long-term experiments, the bacteria were labelled for two generations (1 hr) so that roughly half of the DNA would be fully labelled and half would be a hybrid of labelled and unlabelled strands.

Lysis of bacteria. Only in a few minor respects has the procedure been altered from that already published (Cairns, 1962a). Labelled bacteria are lysed after dilution to a final concentration of about 10^4/ml. Since it was important in certain experiments to be sure that DNA synthesis did not continue beyond the time the bacteria were sampled, 0·01 M-KCN was added to the lysis medium (1·5 M-sucrose, 0·05 M-NaCl, 0·01 M-EDTA). Various types of cold carrier DNA (4·7 μg/ml. calf thymus, *E. coli* or T2 DNA) were used at various times without apparently influencing the results. As before, lysis was obtained by dialysis against 1% Duponol C (Dupont, Wilmington, Delaware, U.S.A.) in lysis medium for 2 hr at 37°C. The Duponol was then removed by dialysis for 18 to 24 hr against repeated changes of 0·05 M-NaCl, 0·005 M-EDTA. As before, the DNA was collected on the dialysis membrane (VM Millipore filter, Millipore Filter Corporation, Bedford, Mass., U.S.A.).

Autoradiography. As before, Kodak AR10 stripping film was used and the exposure was about 2 months.

Thymidine incorporation experiments. To determine whether incorporation was delayed following transfer to a medium containing [³H]TDR, bacteria were grown in cold medium to 5×10^8/ml., centrifuged and resuspended in medium containing 2 μg/ml. [³H]TDR (1 c/m-mole). Samples were then removed into cold 5% TCA and washed on Oxoid membrane filters (average pore diameter 0·5 to 1·0 μ, Oxo Ltd, London, England) with cold TCA and finally with 1% acetic acid. The filters were dried, placed in scintillator fluid (0·4% 2,5-diphenyloxazole, 0·01% 1,4-bis-2[5-phenyloxazolyl]-benzene in toluene) and counted in a scintillation counter.

3. Results

In interpreting autoradiographs of extracted DNA certain assumptions are necessary. These can be stated at the outset.

(1) It is not clear why some molecules of bacterial DNA choose to untangle whereas others do not. However, the few that do are assumed to be a fairly representative sample; specifically we assume that they do not belong to some special class that is being duplicated in some special way.

(2) The ratio of mass to length for this untangled DNA is taken to be at least that of DNA in the B configuration, namely 2×10^6 daltons/μ (Langridge, Wilson, Hooper, Wilkins & Hamilton, 1960). We assume that single-stranded DNA will not be found in an extended state and so, even if present, will not contribute to the tally of untangled DNA.

† Abbreviation used: TDR stands for thymidine.

(3) The density of grains along these labelled molecules is assumed to be proportional to the amount of incorporated label. Specifically we assume that if one piece of DNA has twice the grain density of another this shows that it is labelled in twice as many strands.

(a) Pulse labelling experiments

Simple pulse-labelling experiments could tell much about the process of DNA replication. As pointed out already, DNA synthesis in *E. coli* is virtually continuous. If, therefore, the bacterial chromosome is truly a single piece of two-stranded DNA and if, at any moment, duplication is occurring at only a single point on this molecule, then the length of DNA labelled by a short pulse will be just that fraction of the total length of the chromosome that the duration of the pulse is a fraction of the generation time; if there are several points of simultaneous duplication on the single molecule, or several molecules which are duplicated in parallel, then the length of DNA labelled will be appropriately less. Further, from such pulse experiments it should be possible to determine whether one or two new strands are being made in each region of replication.

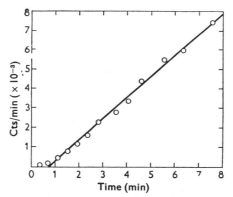

FIG. 1. Incorporation of ^{3}H into cold TCA-insoluble material, following transfer of *E. coli* B3 to medium containing [^{3}H]TDR.

However, first it was necessary to find out how quickly [^{3}H]TDR gets into DNA when the bacteria are transferred from a medium containing cold TDR. The result of such an experiment with *E. coli* B3 is shown in Fig. 1 (see also Materials and Methods). A similar result was obtained for *E. coli* K12. So it seems that, under these particular conditions, the error in timing a pulse will be less than 1 minute.

Autoradiographs of *E. coli* B3 DNA were prepared following (i) a 3 minute pulse of [^{3}H]TDR, (ii) a 3 minute pulse, followed by 15 minutes in cold TDR, and (iii) a 6 minute pulse, followed by 15 minutes in cold TDR. Examples of what was found are shown in Plate I. Similar results were obtained using *E. coli* K12.

From these experiments the following conclusions can be drawn.

(1) Immediately after a 3 minute pulse of label (Plate I(a), (b) and (c)) the labelled DNA consists of two pieces lying in fairly close association. Fifteen minutes later (Plate I(d), (e) and (f)) these pieces have moved apart and can be photographed separately. It seems therefore that *two* labelled molecules are being formed in the region of replication.

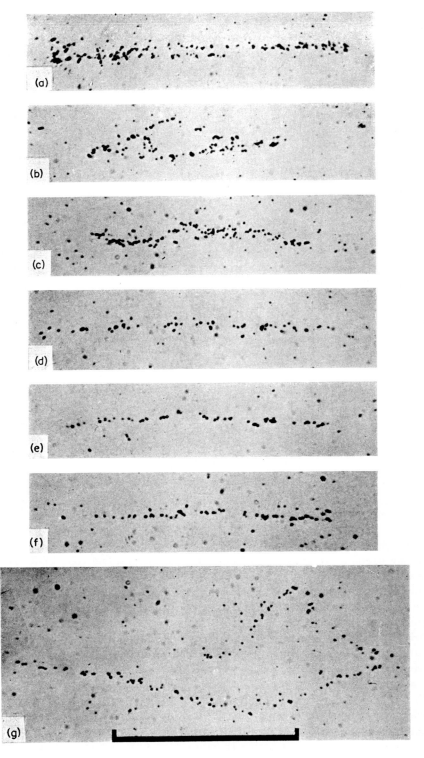

PLATE I. Autoradiographs of *E. coli* B3 DNA labelled by a pulse of [³H]TDR. Exposure time was 61 days. The scale shows 50μ.

(a), (b), (c): Immediately after a 3 min pulse; (d), (e), (f): 15 min after a 3 min pulse; (g): 15 min after a 6 min pulse.

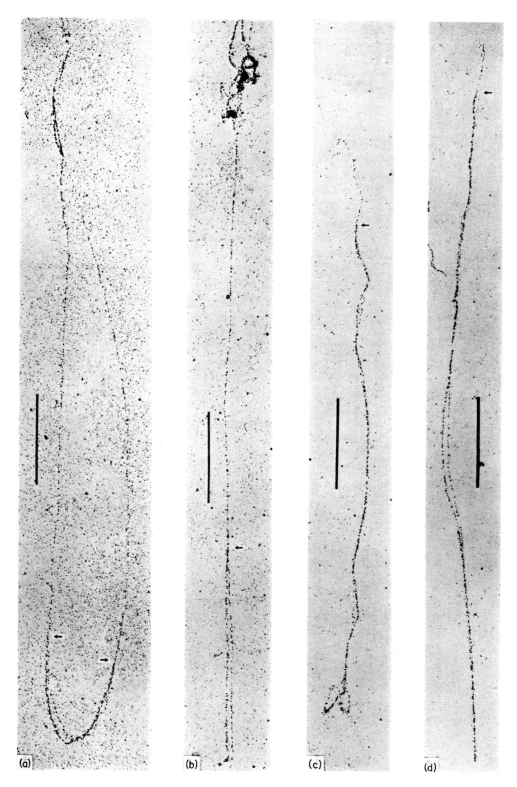

PLATE II. Autoradiographs of *E. coli* B3 (a), (b), (c) and *E. coli* K12 (d) DNA following incorporation of [³H]TDR for a period of 1 hr (two generations). The arrows show the point of replication. Exposure time was 61 days. The scales show 100 μ.

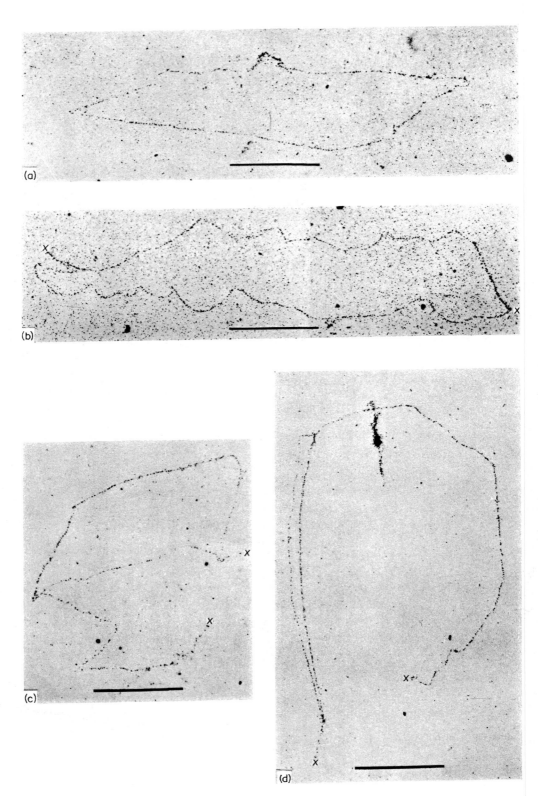

PLATE III. Autoradiographs of *E. coli* B3 (a), (b), (c) and *E. coli* K12 (d) DNA following incorporation of [³H]TDR for a period of 1 hr. In (b), (c), and (d), the postulated break is marked ×—×. Exposure time 61 days. The scales show 100 μ.

(2) There is approximately 1 grain per micron in these labelled regions. Since similarly labelled T2 and λ bacteriophage DNA, both of which are known to be largely two-stranded, show about twice this grain density (Cairns, 1962b), the two molecules being created at the point of duplication must each be labelled in one strand.

(3) A 3 minute pulse labels two pieces of DNA each 60 to 80 μ long (Plate I(a) to (f)); a 6 minute pulse labels about twice this length (Plate I(g)). In one generation time (30 minutes) the process responsible for this could cover 600 to 800 μ, or slightly more than this if the duration of the pulse is being over-estimated (see Fig. 1). There is unfortunately no precise estimate for the DNA content of the *E. coli* chromosome. When the generation time is 1 hour, each cell contains 4×10^9 daltons DNA (Hershey & Melechen, 1957). If each cell contained only one nucleus, this value would have to be divided by 1·44 (1/ln 2) to correct for continuous DNA synthesis. However, as such cells are usually multinucleate (see Schaechter, Maaløe & Kjeldgaard, 1958), this corrected value of $2·8 \times 10^9$ daltons (or 1400 μ DNA) must be too high. There is therefore no marked discrepancy between the total length of DNA that has to be duplicated ($< 1400 \mu$) and the distance traversed by the replication process in one generation (at least 600 to 800 μ). This suggests that one or at the most two regions of the chromosome are being duplicated at any moment.

These various conclusions are reinforced by the results reported in the next section.

(b) *Two-generation labelling experiments*

One conspicuous feature of the pulse experiments is not apparent from the photographs and was not listed. The proportion of the labelled DNA that had untangled, and could therefore be measured, was far lower immediately after the 3 minute pulse than 15 minutes later. Thus the replicating region of a DNA molecule is apparently less readily displayed than the rest. It was not surprising therefore that considerable search was necessary before untangled and replicating molecules were found in the two-generation experiments. Since the products of such searches are generally somewhat suspect, the guiding principles of this particular search will be given.

As pointed out earlier there are reasons for assigning the hypothetical DNA molecule of *E. coli* a length of less than 1400 μ. Therefore no molecules were accepted whose length, presumably through breakage, was much less than 700 μ. No replicating (forking) molecules were accepted unless both limbs of their fork were the same length. Lastly many extended molecules were excluded because of the complexity of their form; although perhaps interpretable in terms of a known scheme they could not be used to provide that scheme. Samples of what remained are shown in Plate II. They have been selected to illustrate what appear to be various stages of DNA replication.

First, the rough agreement between the observed length of *E. coli* DNA (up to 900 μ) and the estimated DNA content of its chromosome ($< 1400 \mu$) supports the conclusion, arising from the pulse experiments, that this chromosome contains a piece of two-stranded DNA.

Second, it seems clear that this DNA replicates by forking and that new material is formed along both limbs of the fork. This latter was shown by the pulse experiments and is confirmed here. In each of the forks shown, one limb plainly has about twice the grain density of the other and of the remainder of the molecule. The simplest hypothesis is that we are watching the conversion of a molecule of hybrid (hot–cold)

DNA into one hybrid and one fully-substituted (hot–hot) molecule. It is not surprising that, nominally after two generations of labelling, the process is seen at
various stages of completion. In Plate II(a), duplication has covered about a sixth of
the visible distance in what appear to be two sister molecules. In Plate II(b) and
(c), duplication has gone about a third and three-quarters of the distance, respectively. In Plate II(d), duplication is almost complete, about 800 μ of DNA having
been replicated. So these pictures support what seemed likely from the pulse
experiments—namely, that the act of replication proceeds from one end of the
molecule to the other.

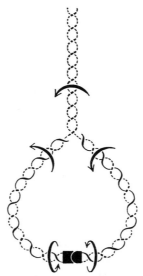

Fig. 2. The consequence of uniting the ends of the replicating fork. The arrows show the
direction of rotation, as the parent molecule unwinds and the two daughter molecules are formed
(modified from Delbrück & Stent, 1957).

The most conspicuous and totally unexpected feature of these pictures has been
left for discussion last. In the case of each replicating molecule, the ends of the fork
are joined. This complication seems to have been taken one stage further in Plate
II(a); here the two limbs of the fork may be joined to each other but they also appear
to be joined to their opposite numbers which are being formed from the sister molecule.
Conceivably such terminal union of the new double helices (which must be alike in
their base sequence) is the artificial consequence of a freedom to unite that only
comes with lysis; this union may not exist inside the bacterium. Alternatively,
terminal union may be the rule during the period of replication. If so, whatever
unites the two ends must have the freedom to rotate so that the new helices can
rotate as they are formed (Fig. 2). This uniting structure, or swivel, could in fact
be the site of the mechanism that, *in vivo*, spins the parent molecule and its two
daughters.

(c) *The bacterial chromosome*

The primary object of this work was to determine the form of DNA when it is
replicating, not the form of the entire bacterial chromosome. The pictures presented
so far give little indication of the latter as they show molecules that are either broken

(Plate II(c)) or partly tangled (Plate II(a) and (b)). It seemed possible, however' that out of all the material that had been collected some model for the shape of the whole chromosome might emerge.

In searching for such a model that could account for all the kinds of structure seen, the premise was adopted that since excess cold carrier DNA was invariably present these structures must be related to the model by breakage, if needs be, but not by end-to-end aggregation. Granted this premise, there seems to be only one model that

FIG. 3. Two stages in the duplication of a circular chromosome. (B) and (D) mark the positions of the breaks postulated to have produced the structures shown in Plate III(b) and (d), respectively.

could generate every structure merely by breakage. This model supposes that the chromosome exists as a circle. Duplication, as in Fig. 2, proceeds by elongation of a loop at the expense of the remainder of the molecule; since, however, the distal end of the molecule is also attached to the swivel, duplication creates a figure 8 each half of which ultimately constitutes a finished daughter molecule (Fig. 3). Depending on how this structure breaks at the time of extraction, it may form a rod with a terminal loop (Plate II), a rod with a subterminal loop (Plate III(c) and (d)), or a circle (Plate III(a) and (b)); in the case of a circle, the circumference may be up to twice the length of the chromosome. All structures seen, including circles of varying circumference, can be readily derived from this model whereas they do not conform to any other obvious scheme. It is, however, possible that the process of duplication may vary; that, for example, the structure shown in Plate II(a) was genuinely an exceptional case. In any event, here as elsewhere no significant difference was detected between the chromosomes of *E. coli* B3 (F⁻) and *E. coli* K12 (Hfr).

I am greatly indebted to Dr. A. D. Hershey and Professor Max Delbrück for helpful criticism and to Miss Rosemary Henry for able technical assistance.

REFERENCES

Cairns, J. (1962*a*). *J. Mol. Biol.* **4**, 407.
Cairns, J. (1962*b*). *Cold Spr. Harb. Symp. Quant. Biol.* **27**. In the press.
Delbrück, M. & Stent, G. S. (1957). In *The Chemical Basis of Heredity*, ed. by W. D. McElroy & B. Glass. Baltimore: Johns Hopkins Press.
Hershey, A. D. & Melechen, N. E. (1957). *Virology*, **3**, 207.
Jacob, F. & Wollman, E. L. (1958). *Symp. Soc. Exp. Biol.* **12**, 75.
Langridge, R., Wilson, H. R., Hooper, C. W., Wilkins, M.H.F. & Hamilton, L.D. (1960). *J. Mol. Biol.* **2**, 19.
Levinthal, C. & Crane, H. R. (1956). *Proc. Nat. Acad. Sci., Wash.* **42**, 436.
McFall, E. & Stent, G. S. (1959). *Biochim. biophys. Acta*, **34**, 580.
Meselson, M. & Stahl, F. W. (1958). *Proc. Nat. Acad. Sci., Wash.* **44**, 671.
Meselson, M. & Weigle, J. J. (1961). *Proc. Nat. Acad. Sci., Wash.* **47**, 857.
Nagata, T. (1962). *Biochem. Biophys. Res. Comm.* **8**, 348.
Schaechter, M., Bentzon, M. W. & Maaløe, O. (1959). *Nature*, **183**, 1207.
Schaechter, M., Maaløe, O. & Kjeldgaard, N. O. (1958). *J. Gen. Microbiol.* **19**, 592.
Watson, J. D. & Crick, F. H. C. (1953). *Cold Spr. Harb. Symp. Quant. Biol.* **18**, 123.

J. Mol. Biol. (1963) **6**, 306–329

Allosteric Proteins and Cellular Control Systems

Jacques Monod, Jean-Pierre Changeux and François Jacob

Services de Biochimie Cellulaire et de Génétique Microbienne,
Institut Pasteur, Paris, France

(Received 19 December 1962)

The biological activity of many proteins is controlled by specific metabolites which do not interact directly with the substrates or products of the reactions. The effect of these regulatory agents appears to result exclusively from a conformational alteration (allosteric transition) induced in the protein when it binds the agent. It is suggested that this mechanism plays an essential role in the regulation of metabolic activity and also possibly in the specific control of protein synthesis.

1. Introduction

Considerable progress has been made during the past few years in the study of regulation and control of cellular metabolism. It is now established that even in the simplest organisms, such as bacteria, complex circuits of regulation play an essential role, governing not only the rate of flow of metabolites through different pathways but also the synthesis of proteins and other macromolecules. Most of these control systems involve a sequence of reactions and interactions and their physiological diversity is extreme. However, in several instances the components of such systems have been resolved, allowing identification and study of the elementary controlling interaction. In virtually all of the systems which have been analysed in sufficient detail, this elementary interaction involves a protein endowed with a specific biological activity and an active agent, generally a low-molecular weight metabolite, in whose presence the specific process governed by this protein is either accelerated or inhibited.

It would appear, in other words, that certain proteins, acting at critical metabolic steps, are electively endowed with specific functions of regulation and coordination; through the agency of these proteins, a given biochemical reaction is eventually controlled by a metabolite acting apparently as a physiological "signal" rather than as a chemically necessary component of the reaction itself (Monod & Jacob, 1961; Jacob & Monod, 1962).

It is hardly necessary to point out the critical role, indeed the physiological necessity, of such metabolic interconnections. In this paper we will not be concerned with the physiological interpretation of individual systems but rather with the mechanism of the controlling interaction. Our aim will be to enquire whether, in spite of the extreme diversity of these systems, it may be possible to formulate certain generalizations concerning the functional structures responsible for the regulatory competence of the controlling proteins, allowing them to act as specific mediators of these essential interactions. At the outset we should like to make it clear that we will not be proposing a new theory, nor any original interpretation of individual facts, but only comparing various examples and attempting to see to what extent and in what way a general description of these systems might be valid and useful.

For the sake of clarifying the discussion and defining the terminology to be used, it is convenient to state *a priori* some of the conclusions at which we shall arrive. This may be done in the form of a general model schematizing the functional structures of controlling proteins. These proteins are assumed to possess two, or at least two, stereospecifically different, non-overlapping receptor sites. One of these, the *active site*, binds the substrate† and is responsible for the biological activity of the protein. The other, or *allosteric site*, is complementary to the structure of another metabolite, the *allosteric effector*, which it binds specifically and reversibly. The formation of the enzyme–allosteric effector complex does not activate a reaction involving the effector itself: it is assumed only to bring about a discrete reversible alteration of the molecular structure of the protein or *allosteric transition*, which modifies the properties of the active site, changing one or several of the kinetic parameters which characterize the biological activity of the protein.

An absolutely essential, albeit negative, assumption implicit in this description is that an allosteric effector, since it binds at a site altogether distinct from the active site and since it does not participate at any stage in the reaction activated by the protein, need not bear any particular chemical or metabolic relation of any sort with the substrate itself. The specificity of any allosteric effect and its actual manifestation is therefore considered to result exclusively from the specific construction of the protein molecule itself, allowing it to undergo a particular, discrete, reversible conformational alteration, triggered by the binding of the allosteric effector. The *absence* of any inherent obligatory chemical analogy or reactivity between substrate and allosteric effector appears to be a fact of extreme biological importance, and in a sense it is the main subject of the present paper. In addition, it is evidently essential to a proper definition of allosteric effects as distinct from the action of coenzymes, secondary substrates or substrate analogues, all of which react with the substrate or substitute for the substrate and therefore must bear some structural relation with or chemical reactivity towards it. This being said, one should certainly not exclude the possibility that the action of certain coenzymes or other enzyme effectors may involve allosteric effects in addition to their classical role as transient reactants or transporters. Nor should one forget the possibility, suggested by Koshland (1958), that the binding of substrate involves an induced alteration of the shape of the enzyme site. These possibilities will be discussed only briefly in this paper which will deal exclusively with typical cases of regulatory allosteric effects, limitatively defined as above.

Since our purpose is to enquire whether any generalizations can be made concerning the functional structures of regulatory proteins, we will have to compare different systems, controlling different reactions and endowed with different physiological functions, in widely different organisms. Unfortunately, the nature of the information concerning different systems is very heterogeneous, rarely allowing detailed parallel comparisons, and the generalized picture will have to be sorted out of this experimental puzzle. We will first discuss the properties of certain bacterial enzymes that act as regulators of biosynthetic pathways. The kinetic properties of some of these systems have been well studied and their regulatory role is perfectly clear, but little is known

---

† In the present context, we shall use the word "substrate" in a somewhat wider sense than is usual, to designate the specific compound upon which a protein exerts its biological activity, whether or not the protein in question is an enzyme *sensu stricto* or not.

about the molecular properties of the proteins. In subsequent sections we will consider certain mammalian enzymes, subject to different regulatory effects, the precise physiological role of which is not always clear but in which conformational alterations have been directly observed. In the last section we shall discuss the validity, qualifications and limitations of the "allosteric" model, and the biological significance of this mechanism.

2. Allosteric Proteins as Metabolic Regulators

(a) *Specificity and kinetics of allosteric effects*

The biosynthetic pathways of bacteria have afforded some of the clearest instances of metabolic regulation. We refer to the so-called "feed-back" or "end-product" inhibition effect discovered by Novick & Szilard (1954), whose early observations on the synthesis of tryptophan, followed by the enzymological work of Umbarger (1956), Yates & Pardee (1956) and others, have now been extended to most, if not all, pathways leading to the synthesis of essential metabolites. Actually it appears to be a rule in bacteria that the terminal metabolite synthesized in any given pathway is a powerful and specific inhibitor of its own synthesis. It is also a rule that only one enzyme (usually the first one in each specialized pathway) is responsible for this effect.

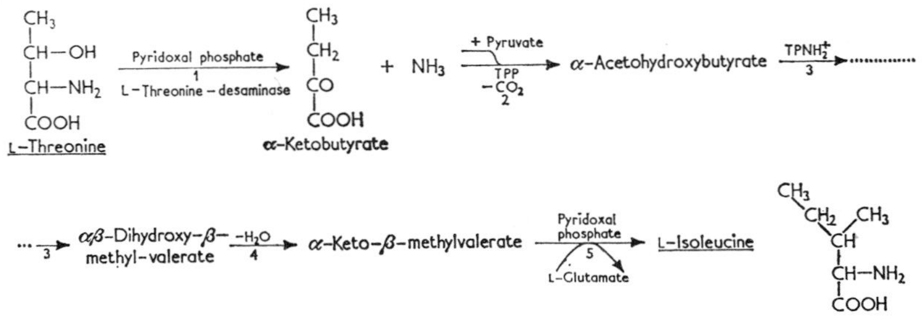

FIG. 1(a). (See p. 309.)

Several of these enzymes have now been studied in some detail and proved to possess certain remarkable and even, at first sight, paradoxical properties which as we shall see actually depend upon and reveal the allosteric construction of these proteins. Similar feed-back effects have been observed in various metabolic pathways of higher organisms.†

Six of these enzymes: threonine-deaminase (Umbarger & Brown, 1958; Changeux, 1961, 1962); aspartic-transcarbamylase (ATCase)‡ (Gerhart & Pardee, 1962); phosphoribosyl-ATP-pyrophosphorylase (PRPP-ATP-PPase) (Martin, 1962); aspartokinases I and II (Stadtman, Cohen, Le Bras & de Robichon-Szulmajster, 1961) and homoserine-dehydrogenase (Patte, Le Bras, Loviny & Cohen, 1962) will especially be mentioned here. The biosynthetic pathways in which they respectively operate are shown in Fig. 1.

---

† Actually, one of the very first clearly recognized instances of a regulatory feed-back mechanism of this type appears to have been the inhibition of glucose phosphorylation by phosphoglyceric acid in erythrocytes. This was described by Dische over 20 years ago (Dische, 1941) in a paper which came only very recently to our attention.

‡ The following abbreviations are used in this paper: ATCase = aspartic transcarbamylase; PRPP-ATP-PPase = phosphoribosyl-ATP-pyrophosphorylase; GDH = glutamic dehydrogenase; S-RNA = soluble RNA.

All these systems obey the rules stated above, namely:

(a) the regulatory enzymes (each of them acting immediately *after* a metabolic branching point) are all strongly and specifically inhibited by the terminal metabolite of the pathway in which each of them operates; intermediary metabolites in each pathway do not inhibit the regulatory enzyme;

(b) the enzymes which intervene *after* the regulatory one in each pathway are not significantly sensitive to inhibition by the terminal metabolite.

These facts alone suffice to demonstrate that the inhibitory effects must be due to highly specialized molecular structures present in the sensitive enzymes and cannot be accounted for by considerations of steric analogy between substrate and inhibitor. In the case of threonine-deaminase, for instance, it might be considered that substrate and inhibitor are steric analogues to the extent that both are α-amino acids; but certain α-amino acids are devoid of any inhibitory action, while others are activators, as we shall see later. Moreover, *E. coli* is known to synthesize two different threonine-deaminases, one of which, as shown by Umbarger, is a degradative enzyme and is completely insensitive to inhibition by isoleucine (Umbarger & Brown, 1957). Finally, the coexistence in *E. coli* of two different aspartokinases catalysing identical reactions, respectively inhibited by threonine and by lysine (Stadtman *et al.*, 1961), offers a striking illustration of the fact that the nature and structure of the inhibitor is, in a sense, irrelevant to the interpretation of the effect. Clearly, such an interpretation must be sought exclusively in the functional structure of the regulatory protein itself.

Since the inhibitory metabolites must act by forming a stereospecific complex with the enzyme, the first questions to consider concern the relationship of the inhibitor binding site to the substrate binding site. Clearly, the same system of binding groups† cannot be involved for both, since inhibitor and substrate are not *isosteric*, but rather *allosteric* with respect to each other. This conclusion, based exclusively upon structural considerations, has been directly confirmed by the discovery (Changeux, 1961; Gerhart & Pardee, 1961, 1962) that under certain conditions or after treatment by certain agents, or also by mutation, a regulatory enzyme may lose its sensitivity to the inhibiting metabolite while retaining its activity towards substrate. This observation has now been made with at least five different systems and therefore appears to be of great significance for the interpretation of allosteric interactions. We shall discuss it in more detail later. For the time being we use it only as proof that the binding of substrate and inhibitor do not involve the same groups.

This being established, it will be useful to distinguish *a priori* three possible types of interaction between substrate and inhibitor. These are shown schematically in Fig. 2.

In the first type, the binding sites actually overlap (although the binding *groups* are not the same). The binding of substrate and inhibitor are therefore exclusive of one another, because of *steric hindrance*. In the second type the two sites lie so close to one another that *direct interactions* (either attractive or repulsive) between substrate and inhibitor occur. In the third type no direct interactions are involved, the two sites are completely separate; the effect is therefore mediated entirely by the protein, presumably through a conformational alteration resulting from the binding of the

† We define "binding site" as the particular area covered by a substrate or effector on the protein surface; "binding groups" as the atoms or groups of the protein which form actual bonds with the substrate and/or effector.

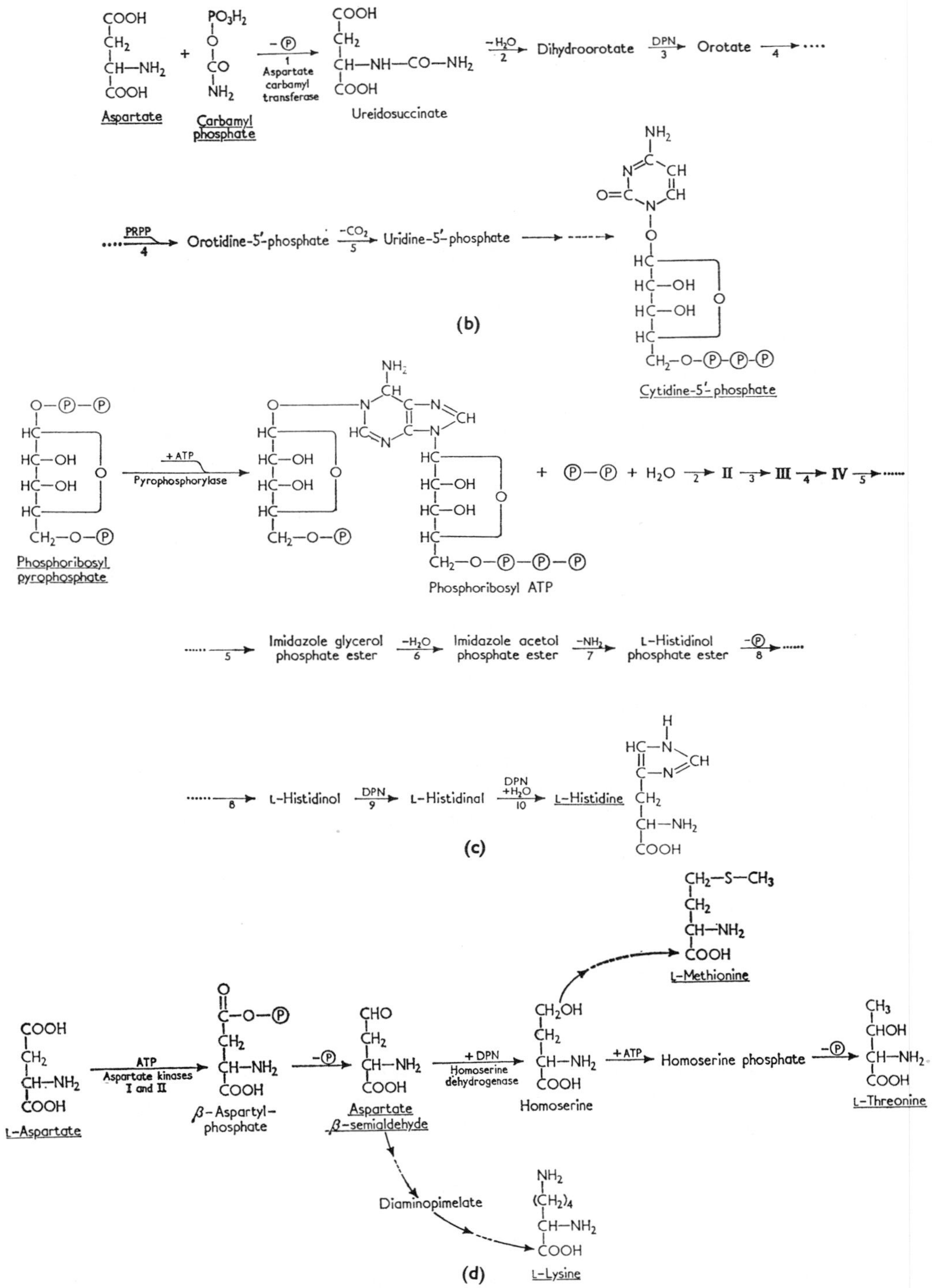

Aspartate + Carbamyl phosphate →[1 Aspartate carbamyl transferase][− $\circ$P] Ureidosuccinate →[2][$−H_2O$] Dihydroorotate →[3][DPN] Orotate →[4] ····

····→[4][PRPP] Orotidine-5′-phosphate →[5][$−CO_2$] Uridine-5′-phosphate → ----→ Cytidine-5′-phosphate

(b)

Phosphoribosyl pyrophosphate →[Pyrophosphorylase][+ ATP] Phosphoribosyl ATP + P−P + H_2O →[2] II →[3] III →[4] IV →[5] ······

······→[5] Imidazole glycerol phosphate ester →[6][$−H_2O$] Imidazole acetol phosphate ester →[7][$−NH_2$] L-Histidinol phosphate ester →[8][$−$P] ······

······→[8] L-Histidinol →[9][DPN] L-Histidinal →[10][DPN $+H_2O$] L-Histidine

(c)

L-Aspartate →[Aspartate kinases I and II][ATP] β-Aspartyl-phosphate →[$−$P] Aspartate β-semialdehyde →[Homoserine dehydrogenase][+ DPN] Homoserine →[+ ATP] Homoserine phosphate →[$−$P] L-Threonine

Aspartate β-semialdehyde ⟶ L-Methionine

Aspartate β-semialdehyde ⟶ Diaminopimelate ⟶ L-Lysine

(d)

FIG. 1. (a), (b), (c), (d). Examples of biosynthetic pathways subject to the "end-product inhibition" effect. The inhibitory metabolites and the substrates of the sensitive enzyme(s) in each pathway are underlined. In the pathway beginning with aspartate, the aspartate-kinase reaction is inhibited by both L-threonine and L-lysine; the homoserine-dehydrogenase reaction is inhibited by L-threonine.

inhibitor, i.e. through an *allosteric transition*. Let us briefly consider the predictions of each of these models as far as the kinetics of inhibition are concerned.

In the case of model I the interaction can *only* be of the "strict competitive" type; that is:

(a) the presence of inhibitor affects *exclusively* the apparent enzyme–substrate dissociation;

(b) the inhibition curve will be asymptotic to 100% at high inhibitor concentration. Therefore, any other result, such as non-competitive inhibition (apparent dissociation constant unaffected), mixed inhibition (apparent dissociation and maximal activity both altered) or incomplete competitive inhibition (inhibition curve asymptotic to a finite value), would eliminate model I.

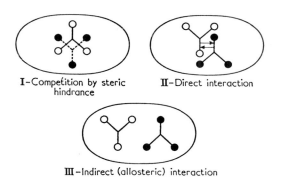

Fig. 2. Three models of interaction between a substrate and an inhibitor binding respectively with different groups on enzyme surface.

Assuming that model I did not apply, one would have to distinguish between models II and III, which is evidently far more difficult, both models being compatible with a variety of kinetics.† However, model II is more restrictive than model III since the former requires that any interaction be *reciprocal* while the latter does not. It follows from this that strictly non-competitive effects are not expected on the basis of model II, which implies that the affinities of the inhibitor for the free enzyme and for the enzyme–substrate complex should be different.

The available data concerning the kinetics of inhibition of bacterial enzymes do not in every case allow application of these criteria. However, as we shall see, these data when taken in conjunction with other lines of evidence appear to be incompatible with model I and difficult to reconcile with model II.

To begin with, we may note that in the cases of PRPP-ATP-PPase, aspartokinase II and homoserine-dehydrogenase inhibition is strictly non-competitive, eliminating model I and also contradicting model II. With aspartic-transcarbamylase and aspartokinase II, the inhibition is competitive (in the sense that only the apparent affinity is affected) but it is incomplete. This may be taken also to eliminate model I (but not model II). This evidence is somewhat questionable, however, because it has repeatedly been observed that even mild treatments (such as are involved in careful purification)

---

† The fact that an interaction is *kinetically* "strictly competitive" does not constitute proof that the competition is actually for the same site. In any instance where the interacting compounds are structurally unrelated such an interpretation should be considered with suspicion.

may result in partial desensitization of allosteric enzymes. It is therefore conceivable that a small spontaneously desensitized fraction may be responsible for the inhibition being incomplete.

With threonine-deaminase the inhibition is competitive and reaches 100%. None of the models could therefore be eliminated on this basis alone. If, however, on the strength of other evidence, model I proved inadequate also for threonine-deaminase, this enzyme would offer an example of an interesting limit-case, where the interaction, albeit not due to steric hindrance, is of such strength as to make the simultaneous binding of substrate and inhibitor (on the native enzyme) impossible.

We may now turn to another line of evidence. Since two systems of specific groups are involved in the enzyme–substrate and enzyme–(allosteric) effector complexes respectively, one expects to find two series of compounds able to complex with the protein, namely analogues of the substrate and analogues of the effector. The effector analogues, as well as the substrate analogues, should behave as strict competitive inhibitors according to model I. On the basis of models II and III, one may expect different effector analogues to behave in different ways:

(a) some analogues should behave like the natural allosteric inhibitor;

(b) others, able to displace the allosteric effector, while failing to interact with the substrate, should reactivate the inhibited enzyme while exerting no effect in the absence of inhibitor.

Both types of behaviour are observed with different analogues of isoleucine assayed for their effect upon the threonine-deaminase reaction (Changeux, 1962). For instance, norleucine strongly restores the activity when added in the presence of isoleucine. L-Leucine alone inhibits, and cooperates with isoleucine when added in its presence. These results are incompatible with model I, and prove that the inhibitory action of isoleucine on this enzyme, although "strictly competitive", cannot be due to binding at the active site.

The effects of valine are particularly interesting. When assayed alone at low concentrations of substrate, valine actually *activates* the reaction by increasing the affinity of the enzyme for threonine. Since valine is apparently an isoleucine analogue, one might believe that valine binds at the same site as isoleucine. However, when assayed in the presence of different concentrations of isoleucine, valine behaves as "partially competitive" towards the inhibitor, i.e. it reactivates the enzyme only to a finite value, which depends upon the isoleucine concentration (Fig. 3). These observations inevitably force the conclusion that threonine-deaminase bears not two, but at least three different sites; the active site, the isoleucine or "inhibitor" site and a valine or "activator" site. Binding of isoleucine at its site results in virtually abolishing the affinity of threonine for the active site, and the reverse must necessarily be true. Binding of valine at its site increases the affinity of the active site and simultaneously decreases the affinity of the isoleucine site.

Very similar observations have been made by Gerhart & Pardee (1962) with ATCase, where ATP acts both as an activator in its own right and as an antagonist of the inhibitor, GTP.

These complex "ternary" interactions would evidently be extremely difficult to account for by a direct interaction model, and we may conclude from this discussion of the kinetics of inhibition (and activation) of "controlling" enzymes of bacteria that the regulatory metabolites do not, in any case, act by steric hindrance (model I), and probably not by direct interaction (model II) between substrate, inhibitor and/or activator. By elimination of other possible mechanisms, these findings constitute evidence in favour of the conclusion that the regulatory metabolites act indirectly by triggering an allosteric transition of the protein molecule.

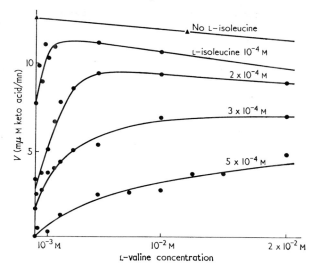

FIG. 3. Antagonistic effects of L-isoleucine and of L-valine on the activity of L-threonine-deaminase, in the presence of a constant concentration of L-threonine (2×10^{-2} M).

(b) *The "desensitization" effect*

We have already noted the fact that the sensitivity of regulatory enzymes to the inhibiting metabolite is, as a rule, an extremely labile property which may be lost as a result of various treatments, with little or no loss of activity. Complete "desensitization" has been obtained with threonine-deaminase (Changeux, 1961), ATCase (Gerhart & Pardee, 1962), homoserine-dehydrogenase (Patte *et al.*, 1962), PRPP-ATP-PPase (Martin, 1962), α-acetolactate synthetase (Martin & Cohen, unpublished results) in particular by treating with mercurials or urea and/or by gentle heating (cf. Fig. 4). Desensitization without loss of activity has also been observed as a result of mutations of the specific gene which controls the structure of threonine-deaminase (Changeux, unpublished results).

At first site, the simplest interpretation might appear to be that the desensitizing agents, or the mutations, destroy the inhibitor binding site itself. This interpretation is not satisfactory, however, because it does not account for the generality of the effect, nor for the exceptional lability of the sensitive state. In view of this, it seems far more likely that the action of the effector depends not only upon the integrity of its binding sites, but upon complete conservation of the native state. If this were the case, a slight disorganization of the protein as a whole (which might be brought about by a variety of attacks at different points on the molecule) would result in desensitization by uncoupling of the interaction without destroying the effector site or the active site. This interpretation has been validated by the important observation of Martin (1962) that desensitized PRPP-ATP-PPase still binds histidine, as tested by equilibrium dialysis. Similar tests have not yet been performed on any of the other systems discussed here, but certain observations of an entirely different kind made with threonine-deaminase and with ATCase lead to similar conclusions.

A remarkable feature common to both of these systems is that, at substrate concentrations below half saturation, the reaction velocity increases faster than the substrate concentration, while at low inhibitor concentrations the rate decreases faster than the inhibitor concentration; this means that the enzyme molecule can bind more than one substrate or one inhibitor molecule at a time and that in the *native* enzyme there is some sort of cooperative interaction between the homologous

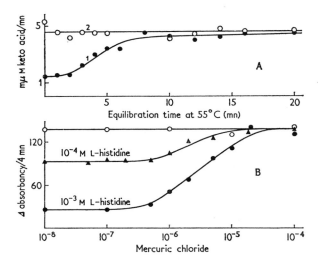

FIG. 4. (a) "Desensitization" of L-threonine-deaminase by heat treatment. *Curve* 1: activity measured with 2×10^{-2} M-L-threonine and 10^{-2} M-L-isoleucine. *Curve* 2: same L-threonine concentration, no L-isoleucine (after Changeux, 1961).

(b) "Desensitization" of PRPP-ATP-PPase by treatment with mercuric chloride (substrate added simultaneously with $HgCl_2$ at concentration indicated). *Upper curve*: enzyme assayed without addition of inhibitor. *Lower curves*: enzyme assayed at the two indicated concentrations of L-histidine (after Martin, 1962).

binding sites. These cooperative effects are closely comparable to the classical haem–haem interaction in haemoglobin to which we shall return later (Figs. 5 and 6). The striking fact which we wish to emphasize here is that the "desensitized" enzyme, in both cases, exhibits no trace of the *substrate cooperative effect*. As may be seen from Fig. 6 the kinetics of the reaction catalysed by desensitized ATCase are "normalized" and obey the Michaelis–Henri relation while, in the presence of native enzyme, the rate-concentration curve is sigmoid. Desensitization and "normalization" also exhibit a parallel dependence upon pH and ionic strength. Both the cooperative effect of substrate and the inhibitory effect of isoleucine on threonine-deaminase are maximal at pH values between 7 and 8, while both are abolished (reversibly) around pH 10. Finally, mutations which desensitize the enzyme also partially abolish or alter the cooperative effect of substrate.

These results show that both the cooperative interaction between substrate binding sites *and* the antagonistic interaction between inhibitor and substrate sites depend largely upon the same features of protein structure and are both similarly related to the integrity of the native state. The presence of several substrate and several inhibitor sites on each molecule of ATCase and threonine-deaminase actually suggests a more specific hypothesis as to which structures present in the native state may be

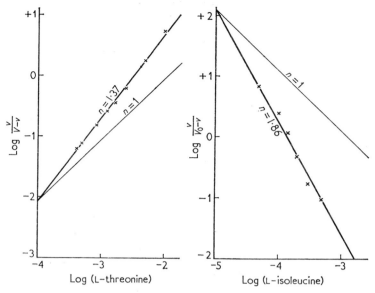

FIG. 5. L-threonine-deaminase activity as a function of (A) substrate and (B) allosteric inhibitor concentration. Both relations are seen to be conveniently represented by expressions of the form:

$$\log \frac{v}{V_{\max} - v} = n \log S - \log K \quad \text{(for substrate)}$$

$$\log \frac{v}{V_0 - v} = \log K' - n' \log I \quad \text{(for inhibitor)}$$

These equations are formally identical with Hill's empirical relation for the binding of oxygen to haemoglobin. It is seen that for both substrate and inhibitor $2 > n > 1$ indicating cooperative interactions between homologous sites and also showing that the reaction is *not* truly bimolecular.

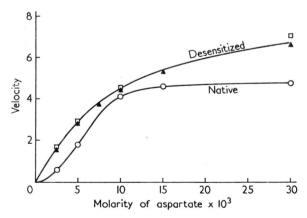

FIG. 6. Effects of desensitization of ATCase upon reaction kinetics. *Lower curve*: native enzyme, assayed in the absence of CTP. *Upper curve*: enzyme desensitized by heat treatment, assayed without inhibitor (squares) and with inhibitor (triangles) (CTP 2×10^{-4} M). (After Gerhart & Pardee, 1962.) Compare with Fig. 9.

primarily involved; namely that the native enzyme is made up of subunits and that the antagonistic as well as the cooperative interactions depend upon the relationships between these units. Desensitization might then result from alteration of these relations, i.e. from the rupture of bonds between the subunits; and this might of course apply also to systems which do not exhibit any substrate–substrate cooperation.

There is some fairly good, albeit incomplete, evidence in favour of this interpretation. Gerhart & Pardee (1962) have found that the sedimentation velocity of desensitized ATCase is decreased (from 11·6 to 5·9) by comparison with the native (sensitive) enzyme, and similar observations have been made by Patte et al. (1962) with homoserine-dehydrogenase. No detectable alterations of sedimentation velocity were observed, following desensitization, with PRPP-ATP-PPase nor with threonine-deaminase. The positive evidence of course carries more weight than the negative, especially since an incomplete separation of subunits need not entail a detectable alteration of the sedimentation velocity.

In any case we may conclude that the sum of the observations concerning the desensitization effect would be exceedingly difficult to reconcile with model II (not to mention model I) or more generally with any model which exclusively involves direct substrate–inhibitor interactions. On the other hand, the assumption that the interaction is due to an allosteric transition involving the protein molecule as a whole in its native state accounts very well for the characteristic lability of the sensitive state in these regulatory proteins and for the peculiar alteration of the kinetic parameters which attends desensitization of ATCase and threonine-deaminase.

3. Allosteric Effects as Conformational Alterations

Direct evidence of reversible conformational alterations provoked by the binding of a "regulatory" metabolite has been obtained with several proteins of higher organisms. We will discuss here only the best known and most significant systems, namely, beef liver glutamic-dehydrogenase, acetyl-CoA carboxylase from adipose tissue, muscle phosphorylase b and haemoglobin. Given the biochemical and physiological diversity of these systems, we will consider the properties of each of them in turn, reserving any general discussion for the next section.

As isolated in crystalline form from beef liver, the enzyme glutamic-dehydrogenase has a molecular weight of 10^6. The important discovery was made by Frieden (1959) some years ago that NADH provoked dissociation of the protein into subunits of molecular weight 250,000, while ADP antagonized the dissociation. Further work by Frieden (1961, 1962a,b,c), Yielding and Tomkins (1960, 1962), Tomkins, Yielding & Curran (1961), Tomkins & Yielding (1961) and Wolff (1962) has shown that this reversible dissociation is favoured or antagonized specifically by a somewhat bewildering variety of metabolites and also by non-specific agents, notably pH. The dissociative agents appear invariably to inhibit glutamic-dehydrogenase activity, while the associative agents exert the opposite effect. The great significance of this system as a model of physiological interactions at the molecular level was indicated in particular by the discovery that estrogens are among the most potent dissociative agents and that inhibition of glutamic-dehydrogenase is attended by concomitant *activation* of alanine-dehydrogenase. Tomkins et al. (1961) and Wolff (1962) later showed that thyroxine also is a potent inhibitor of glutamic-dehydrogenase, as well as a dissociative agent.

The specific inhibitors and the activators of glutamic-dehydrogenase include metabolites which are neither substrates nor coenzymes of GDH (estrogens, thyroxine, ADP, ATP, GDP, etc.) but the list also includes NADH, NAD^+, NADPH and $NADP^+$, which are coenzymes of the system, and the amino acids leucine, isoleucine and methionine, which may be considered as "secondary" substrates of the enzyme. This might seem to exclude the latter compounds from our initial, strictly limitative, definition of allosteric effectors as compounds which *do not* participate in the reaction. The definition, however, does not imply that one and the same compound cannot contribute both as an allosteric effector and as a participant of some kind in a reaction, but it does require that the two contributions should be distinct. The operational validity of this definition is in fact illustrated by the GDH system since, according to

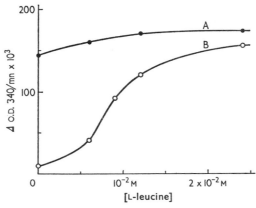

FIG. 7. Reversal by L-leucine of the inhibition of glutamic dehydrogenase by diethylstilbestrol. *Curve* A: no inhibitor. *Curve* B: enzyme assayed with $8·0 \times 10^{-6}$ M-diethylstilbestrol. (After Tomkins & Yielding, 1961.)

Tomkins & Yielding (1961), the allosteric effect of leucine, for example, is due to binding at a site *other* than the active substrate site; while according to Frieden (1961) the dissociative effects of the reduced pyridine nucleotides are due to binding at sites distinct from the active coenzyme site. Moreover, glutamate itself has no effect on the dissociation. It is interesting to note that both the activation by leucine and the inactivation by diethylstilbestrol show evidence of cooperative (multimolecular) effects (Fig. 7). The similarity with ATCase, threonine-deaminase and other allosteric enzymes of bacteria is obvious. It is further strengthened by the finding that treatment of the enzyme with SH reagents renders it insensitive to diethylstilbestrol, and to ADP as well, without modifying the activity (Tomkins & Yielding, 1961).

By contrast with the bacterial systems, whose functions are simple and obvious, the physiological interpretation of the multiple sensitivities and activities of GDH appears exceedingly difficult. But while one may wonder whether each and all of the metabolites which act upon it *in vitro* have any significant role *in vivo*, it cannot be doubted that the complex allosteric reactivity of GDH does reflect its central, multivalent role in cellular metabolism. In any case the observations of Frieden and of Tomkins and his colleagues leave no doubt that the effectors which activate or inhibit the two potential activities of GDH act primarily by inducing a conformational alteration, eventually expressed as a dissociation of the protein.

Another remarkable example where a typical allosteric effect has been directly demonstrated to involve a conformational alteration has been provided by the recent work of Martin & Vagelos (1962) and Vagelos, Alberts & Martin (1962a,b). The enzyme is acetyl-CoA-carboxylase, which catalyses the two-stage reaction

$$\text{biotin E} + CO_2 + \text{ATP} \longrightarrow CO_2\text{-biotin E} + \text{ADP} + P$$
$$CO_2\text{-biotin E} + \text{acetyl-CoA} \longrightarrow \text{malonyl-CoA} + \text{biotin E}$$

Sum: $CO_2 + \text{acetyl-CoA} + \text{ATP} \longrightarrow \text{malonyl-CoA} + \text{ADP} + P$

It had been known for a long time that the biosynthesis of fatty acids is activated by citrate and it was assumed that citrate acted as a metabolic source of NADPH. When it was found that citrate specifically activates the enzyme acetyl-CoA-carboxylase, the metabolite was rather naturally believed to participate at some stage of the reaction itself. The exhaustive experiments of Vagelos et al. (1962a,b) have proved

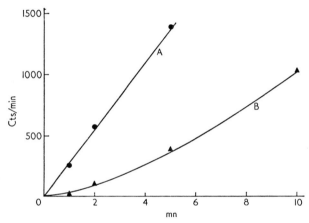

Fig. 8. Activation of acetyl-CoA-carboxylase by citrate. *Curve* A: citrate (5×10^{-3} M) added 30 min at 30°C before zero time. *Curve* B: citrate added at zero time (after Vagelos et al., 1962b).

this assumption to be incorrect. No evidence was found that citrate participates in any way in the reaction. Fluorocitrate, as a matter of fact, is just as active as citrate itself, while other Krebs-cycle intermediates, except fumarate, are relatively inactive.[†] Since citrate modifies only the velocity of the reaction and not the substrate affinities, no direct interaction with substrate is apparently involved. The kinetics of the activation show that it is not immediate. Under the conditions of the experiment illustrated by Fig. 8, full activity is reached only after 30 minutes of incubation with citrate. By contrast, dilution of the activator brings about rapid inactivation of the enzyme. Centrifugation of the enzyme, in sucrose gradients, with and without previous incubation with citrate, revealed that activation is attended by an increase of sedimentation coefficient from 18 s to about 43 s. Although actual molecular weights have not yet been determined, it seems highly probable (in view particularly of the kinetics of activation and deactivation) that this large alteration of sedimentation coefficient is due to the formation of an active polymer (probably a trimer) from inactive monomers rather than to the folding up of the protein from an extended into a more globular form.

† Using a preparation from another source, Waite & Wakil (1962) have also found other Krebs-cycle intermediates to be active.

Probably the first allosteric enzyme mechanism to have been discovered and analysed in detail is the effect of 5′-AMP on muscle phosphorylase b. It was shown by Cori and his school (Cori, Colowick & Cori, 1938; Cori & Green, 1943; further references in Krebs & Fischer, 1962), already many years ago, that this enzyme which is almost inactive in the absence of 5′-AMP is instantaneously and reversibly activated in its presence. ADP, ATP and other nucleotides (except IMP) are inactive. It was naturally supposed at first that 5′-AMP played the role of a coenzyme. But further thorough experiments showed that the nucleotide does not participate in any detectable way in the reaction. Moreover, since the effector does not alter the affinity but only the velocity constant of the reaction, direct interaction with substrate (cf. model II above) appears improbable.

Now, as is well known, phosphorylase b may also be converted to an active state by an entirely different process, also discovered and analysed by the Cori school, namely phosphorylation by ATP (in the presence of a specific kinase) automatically attended by dimerization to phosphorylase a. This mechanism, by contrast with the 5′-AMP effect, is irreversible. Phosphorylase a is stable and reconversion of a to b follows a different course, namely hydrolytic dephosphorylation by a specific phosphatase.

This system therefore provides the proof that a reversible, presumably non-covalent, interaction between an allosteric effector and a protein may mimic, in part at least, the effects of an irreversible stable modification of protein structure involving initially a covalent reaction, followed by a reassociation of subunits. It seems inevitable to conclude that the transition induced by 5′-AMP in phosphorylase b is, in some respects at least, "equivalent" to the phosphorylation reaction. This conclusion is greatly strengthened by the demonstration by Kent, Krebs & Fischer (1958) that phosphorylase b dimerizes in the presence of 5′-AMP and also under suitable conditions crystallizes as such with the nucleotide (Fischer & Krebs, 1958; Kent et al., 1958). It should be added that phosphorylase b is known to be made of two subunits, hence phosphorylase a of four; and it has been found that, by treatment with p-chloromercuribenzene sulfonate, phosphorylases a and b dissociate into inactive subunits (Madsen & Cori, 1956). There is little doubt therefore that the activity of phosphorylase is dependent upon its "quaternary" structure and that certain acid (thiol) groups play a critical role in maintaining this structure. Nor is there any doubt that the activation of phosphorylase b by AMP results from a conformational alteration. However, a very significant question remains to be solved, namely whether this alteration is induced *directly* by the binding of the nucleotide, the dimerization reaction being then a result of this primary effect, or whether the activating alteration results from the dimerization itself.

All the examples which we have considered until now relate to enzymic proteins. Haemoglobin is an example of a non-enzymic protein whose specific regulatory competence has long been recognized. As is well known, the dissociation curve of oxyhaemoglobin as a function of oxygen tension is sigmoid, demonstrating a cooperative effect of the binding sites. By contrast, in the case of myoglobin the dissociation function is a simple adsorption isotherm (i.e. identical to the Michaelis–Henri relation). When the two curves are plotted on the same graph (Fig. 9), the analogy between this situation and that of native (sensitive) and desensitized threonine-deaminase or ATCase is obvious. This functional difference between haemoglobin and myoglobin is of course known to depend upon the tetrameric structure of the former and the monomeric state of the latter.

As is also well known, haemoglobin is subject to another effect endowed with regulatory significance, namely the Bohr effect, which consists of an increase of the oxygen dissociation as the pH is lowered (i.e. *in vivo* as CO_2 tension increases). Wyman (1947) demonstrated several years ago that the Bohr effect is due to a discharge of protons provoked by the binding of oxygen. The recent work of Riggs (1959) seems to identify the acidic groups responsible for the Bohr effect with cysteinyl residues which are apparently also involved in the haem–haem interaction, but it is also possible that the actual "oxygen-linked" acid groups are imidazole residues, presumably closely associated with the thiol groups (Benesch & Benesch, 1961). In any case, the blocking of the latter groups (by mercurials or N-ethyl maleimide) alters both the Bohr effect and the haem–haem interaction, just as similar treatments have been found to abolish allosteric effects in bacterial and other regulatory enzymes.

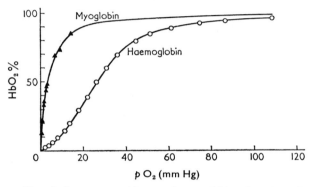

FIG. 9. Oxygen dissociation curves of human haemoglobin (data from Morgan & Chichester, 1935) and of horse heart myoglobin (data from Theorell, 1934). Compare with Fig. 6.

The interactions may also be altered as a result of mutations, as in the case of threonine-deaminase. In haemoglobin H, both the Bohr effect and the haem–haem interactions are abolished. This is particularly interesting, since this haemoglobin contains only β chains: it suggests that α–β chain inter-relations play an important role in both effects (Benesch, Ranney, Benesch & Smith, 1961).

It was believed for a long time that the cooperative binding of oxygen molecules on haemoglobin was due to direct interactions (of the kind shown by model II) between haem groups, presumed to lie very close to one another in the protein. The work of Perutz *et al.* (1960) has demonstrated that the four haem groups actually are wide apart, excluding any possibility of direct interaction and imposing the only alternative interpretation, namely that the interaction is indirect, therefore presumably due to a conformational alteration. As a matter of fact, it had been known for several years that oxyhaemoglobin and reduced haemoglobin occur in different crystal forms (Haurowitz, 1938). The recent crystallographic work of Muirhead & Perutz (personal communication) has indicated that the *distance* between certain SH residues in the molecule may be shifted by about 19% upon oxygenation, providing direct though still tentative evidence of a conformational alteration. Thus, in the case of haemoglobin, there is complete evidence that the regulatory effect, i.e. the cooperative binding of oxygen, is related to a reversible, discrete conformational alteration of the protein, i.e. in our nomenclature, to an allosteric transition. Actually, thanks to the considerable work which has been devoted to it, the haemoglobin system provides the most valuable model from which to start in the further analysis and interpretation of allosteric effects in general.

4. General Discussion and Conclusions

(a) *Validity, qualifications and limitations of the allosteric model*

We may now reconsider the general model proposed in the Introduction for the functional structures of "controlling" proteins. Our aim should be to inquire whether, in the light of the experimental evidence analysed in the preceding sections, the allosteric model appears to be valid and whether it could be further specified and qualified.

In its most general form the allosteric mechanism is defined by two statements, one negative, the other positive.

1. No direct interactions of any kind need occur between the substrate(s) of an allosteric protein and the regulatory metabolite which controls its activity.

2. The effect is *entirely due* to a reversible conformational alteration induced in the protein when it binds the specific effector.

These two statements are not independent. Besides its biological significance (which we shall discuss below), the first of these statements is essential because, if and where it can be proved correct, the second statement must also be correct. Conversely, in those cases where the first statement is inadequate or unproved the second one could hardly ever be proved, even if direct evidence of a conformational alteration were obtained.

As we have seen, direct evidence that the action of a regulatory metabolite involves a conformational alteration of protein structure is available only in four cases (GDH, acetyl-CoA-carboxylase, phosphorylase *a* and haemoglobin). In one at least of these instances (haemoglobin) "effector" and "substrate" are one and the same molecule. It should perhaps be emphasized again that this does not invalidate the allosteric model, provided the substrate function and the allosteric effector function can be operationally distinguished. This is certainly the case for haemoglobin, and also for the effects of leucine and of NADPH on glutamic-dehydrogenase.

Most of the evidence available at present concerns the structural specificity and the kinetics of action of regulatory metabolites upon certain enzymes. This evidence can be used to test the validity of the first statement, but since this statement is a strictly negative one the evidence also can only be negative. Putting it more precisely: the allosteric model is compatible with virtually any kinetics while models involving direct interactions at the active site are more restrictive. In spite of this logical difficulty, the bulk of the evidence (concerning in particular the bacterial systems) seems to be overwhelmingly in favour of the allosteric model, because neither the chemical properties, nor the structural specificity, nor the kinetics of action of the regulatory metabolites appear compatible with "direct interaction" models.

In addition, the generality of the desensitization effect, which is a positive prediction of the allosteric model, is evidently exceedingly difficult to account for by a direct interaction model. The occurrence of desensitized states of a regulatory enzyme (whether as a result of mutation, or of the action of denaturing agents) therefore constitutes in any given instance one of the most specific tests of the validity of the allosteric model and may be considered to prove it when (as is the case in several systems which we discussed) the other evidence independently points to the same conclusion.

This statement should, however, be qualified by carefully defining the operational meaning of the expression "allosteric transition". Throughout the preceding discussion, we have treated it as equivalent to "specifically inducible conformational alteration of protein structure". However, the only conclusion which can be drawn from kinetic data together with the occurrence of desensitization is that a given effect is indirect, due to the binding of substrate and effector at sites remote from each other and whose interaction must therefore be mediated through the protein. Such mediation would not necessarily involve a conformational alteration *sensu stricto*. It might conceivably be due, for instance, to a redistribution of charge within the molecule without *detectable* alteration of its spatial configuration. In a protein molecule, however, any redistribution of charge might be expected to involve or to facilitate a true conformational alteration. Actually, as we have seen, indirect evidence suggests in many cases and direct observations prove in a few instances that allosteric transitions involve the breaking, or formation, or substitution of bonds between subunits in the protein. Whether this may be considered a general rule is evidently a question of great importance, which might perhaps profitably be stated also in the following way: do allosteric transitions occur in monomeric proteins containing a single polypeptide chain? This problem is related to a more general one, which is the role of quaternary structures in the biological activity of proteins. Following the discovery by Cori of the phosphorylase conversions it has become increasingly evident during the past few years that many proteins, particularly enzymes, are homo- or heteropolymers, and that their activity is dependent upon correct association between their subunits (cf. Lwoff & Lwoff, 1962). In any such protein, disorientation or re-orientation, however slight, of the subunits with respect to each other would entail loss or gain of activity. Given these facts and the evidence which we have discussed here concerning regulatory proteins, one may feel that more complete observations, once available, might justify the conclusion that allosteric transitions frequently involve alterations of quaternary structure.

Even if this assumption were generally valid, the role of the effector itself would remain to be accounted for. In the best studied and also probably the simplest case, haemoglobin, the role of the effector-substrate, oxygen, in inducing the transition is far from being completely understood. It is certain, however, that the binding of oxygen to a haem induces within the molecule a redistribution of charge, expressed as a discharge of protons by an acidic group; hence *motu contrario* the pH effect on oxygen affinity. Similar pH effects have been observed, as we already noted, with several other allosteric systems. In the case of threonine-deaminase both the positive interaction between active sites and the negative interaction between active and allosteric sites are abolished at high pH, suggesting that the allosteric transition ultimately depends upon the ionization of certain critical acid groups (or their conjugate base). In addition, let us recall the fact that in most systems the allosteric effect is blocked by reagents known to attack certain acidic groups (thiol and imidazole), also suggesting that such groups play a critical role in the transition. It should be clear, however, that the effect could not be ascribed solely or primarily to the charge or polarity of the effector itself, but only to the specific type of bonding which it forms with the protein. Again consider the case of threonine-deaminase, where valine increases and isoleucine decreases the affinity of the active site, although both amino acids carry the same charge with the same absence of polarity in their side-chain.

No contradiction need be seen between the extreme specificity of these effects and the fact that similar or identical transitions of structure may *also* result in certain systems from the action of non-specific agents (cf. glutamic-dehydrogenase). It would evidently be very misleading to consider reversible discrete conformational alterations, attended by modifications of biological activity, as a privilege of regulatory proteins. There is of course ample experimental evidence showing that such reversible alterations of structure occur in many proteins under the action of non-specific conditions or chemical agents, including in particular pH, ionic strength, hydrogen-bonding or hydrophobic compounds, and mercurials. The essential properties of typical regulatory proteins (i.e. the capacity to undergo an allosteric transition triggered by the stereospecific binding of a particular metabolite) are to be understood as highly specialized manifestations of general properties, shared by all or most proteins. In other words, an allosteric protein represents the outcome of a process of selective development of a molecular species where the flexibility of protein structure assumes the specific functional role of mediating certain chemical signals.

The "induced-fit" theory of enzyme action, proposed by Koshland (1958, 1960), involved the following postulate:

(a) a precise orientation of catalytic groups is required for enzyme action;

(b) the substrate may cause an appreciable change in the three-dimensional relationship of the amino acids at the active site;

(c) the changes in protein structure caused by a substrate will bring the catalytic groups into proper orientation for reaction, whereas a non substrate will not.

While the purpose of this model is to account for certain anomalous features of enzyme specificity, its central postulate is similar to the basic assumption of the allosteric model, to the extent that it invokes a functional role for the flexibility of protein structure. The evidence concerning allosteric systems shows that the binding of substrate (or coenzyme) does provoke conformational alterations in certain proteins. Such observations must be interpreted with caution, however, since as we have seen in the case of glutamic-dehydrogenase the allosteric sites for leucine and NADPH are distinct from the active sites, and the substrate itself, glutamate, does not affect the dissociation. On the other hand, in the case of ATCase and threonine-deaminase (actually in *any* case where the allosteric effect results in a decrease of substrate affinity) the substrates must be considered to provoke a transition *opposed* to the transition corresponding to the binding of inhibitor. One is tempted to suggest that in those cases where the binding of substrate does provoke a detectable conformational alteration of an enzyme the effect may often turn out to be interpretable in terms of a regulatory allosteric transition. The possibility that the "induced-fit" model might be extended to involve regulatory effects has in fact been mentioned by Koshland himself in a recent theoretical paper (Koshland, 1962).

(b) *The biological significance of allosteric control systems*

Even granting that allosteric mechanisms exist and intervene at many stages of cellular metabolism it might be asked whether one would be justified in considering that this particular class of interactions plays a special, uniquely significant role in the control of living systems.

Other types of mechanisms contribute to cellular regulation. We need only mention mass action; while it inevitably intervenes, a living system is constantly fighting against, rather than relying upon, thermodynamic equilibration. The thermodynamic significance of specific cellular control systems precisely is that they successfully circumvent thermodynamic equilibration (until the organism dies, at least). An illustration of this statement is given by certain metabolic pathways which are both thermodynamically and physiologically reversible, such as the synthesis of glycogen from glucose-1-phosphate. It is now established that the cells do not use the same pathways for synthesis and degradation of glycogen, and that each of these pathways is submitted to different specific controls, involving hormones and other metabolites, none of which participate directly in the reactions themselves (cf. Krebs & Fischer, 1962; Rall & Sutherland, 1961; Leloir, 1961); all this evidently because the physiological requirements could not be satisfied otherwise, certainly not by simply obeying mass action.

Competition between enzymes for common substrates evidently plays a role in the balance of metabolism by distributing certain important metabolites, such as coenzymes, between different pathways. Such mechanisms, however, would by themselves be unable to govern and control, that is to say to *modify*, the distribution of building blocks or chemical potential according to the requirements of remote pathways, or to chemical alterations of the environment, or to the physiological meaning of chemical signals issued by other cells. For the chemical activities of a cell to be precisely adjusted to its own requirements, adapted to the environment and directed towards the performance of a particular function, the specific activity of those proteins which are responsible for critical metabolic steps must be altered electively in response to the presence of certain metabolites playing the role, not of substrates for the reaction in question, but of chemical signals.

The primary reason for considering allosteric proteins as essential and characteristic constituents of biochemical control systems is their capacity to respond immediately and reversibly to specific chemical signals, effectors, *which may be totally unrelated to their own substrates, coenzymes or products.*

We have discussed several examples which illustrated this point (and need not be recalled here), leading us to the paradoxical conclusion that the structure and *sui generis* reactivity of an allosteric effector is "irrelevant" to the interpretation of its effects. There remains no real chemical paradox, once it is recognized that an allosteric effect is indirect, being mediated entirely by the protein and due to a specific transition of its structure. Still, the arbitrariness, chemically speaking, of certain allosteric effects appears almost shocking at first sight, but it is this very arbitrariness which confers upon them a unique physiological significance, and the biological interpretation of the apparent paradox is obvious. The specific structure of any enzyme-protein is of course a pure product of selection, necessarily limited, however, by the structure and chemical properties of the actual reactants. No selective pressure, however strong, could build an enzyme able to activate a chemically impossible reaction. In the construction of an allosteric protein this limitation is abolished, since the effector does not react or interact directly with the substrates or products of the reaction but only with the protein itself. A regulatory allosteric protein therefore is to be considered as a specialized product of selective engineering, allowing an indirect interaction, positive or negative, to take place between metabolites which otherwise would not or even could not interact in any way, thus eventually bringing a particular reaction under the control of a chemically foreign or indifferent compound. In this way it is possible to understand how, by selection of adequate allosteric protein structures, any physiologically useful controlling connection between any pathways in a cell or any tissues in an organism may have become established. It is hardly necessary to point out that the integrated chemical functioning of a cell requires that such controlling systems should exist. The important point for our present discussion is that these circuits of control could not operate, i.e. could not have evolved, if their elementary mechanisms had been restricted to direct chemical interactions (including direct interactions *on* an enzyme site) between different pathways. By using certain proteins not only as catalysts or transporters but as molecular receivers and transducers of chemical signals, freedom is gained from otherwise insuperable chemical constraints, allowing selection to develop and interconnect the immensely complex circuitry of living organisms. It is in this sense that allosteric interactions are to be recognized as the most characteristic and essential components of cellular control systems.

This brings up the problem of hormones as allosteric effectors; we are now referring mostly to hormones of small molecular weight. The specificity of hormones; their capacity of simultaneously activating or inhibiting a variety of metabolic processes and of exerting different effects on different tissues; the surprisingly small number of reactions in which they have been proved to take part as reactants opposed to the large number of enzymes upon which they have been found to act; the lack of chemical reactivity of certain hormones, such as the steroids; all in fact of these physiologically essential and chemically bewildering properties could be accounted for by the assumption that hormones in general (but not necessarily in all of their manifestations) act as allosteric effectors, each of them able specifically to trigger allosteric transitions in a variety of different proteins, mostly enzymes, but possibly also genetic repressors. In fact it seems difficult to imagine any biochemical mechanism other than allosteric which could allow a single chemical signal to be understood and interpreted simultaneously in different ways by entirely different systems; although this appears to be the case for many hormones.

Unfortunately, glutamic-dehydrogenase is one of the very few enzyme systems where hormones (thyroxine and estrogens) have undoubtedly been proved to act as allosteric effectors and we shall resist the temptation to make sweeping generalizations. The most serious objection to the concept of allosteric control is that it *could* be used to "explain away" almost any mysterious physiological mechanism.

While the possession of this universal key raises serious latent dangers for the experimenter, it is of such value to living beings that natural selection must have used it to the limit. Structural mutations occurring in a "classical" (non-regulatory) enzyme may alter one or both of its kinetic constants with respect to its normal substrate. Mutations occurring in an allosteric protein may modify its functional properties in a much larger number of ways. Figure 10 illustrates the properties of a number of mutants of the gene which determines the structure of threonine-deaminase in *E. coli*. As it may be seen, some of the mutant proteins have partially lost their sensitivity to isoleucine, others have increased it; the shape of the inhibition curves are different, indicating that the degree of cooperation between allosteric sites has been altered, as well as their interactions with the active sites. How these exquisite possibilities may be used for adjusting allosteric systems precisely to their functions is illustrated by Fig. 11 taken from a paper by Riggs (1959) which shows that the Bohr effect exhibited by haemoglobins of different mammals is closely correlated to their size.

In the present paper, we have exclusively discussed allosteric systems which control the *activity* of enzymes or other proteins. It is of extreme interest to enquire whether allosteric effects may also be involved in the specific control of protein *synthesis*, i.e. in the mechanisms of "genetic repression".

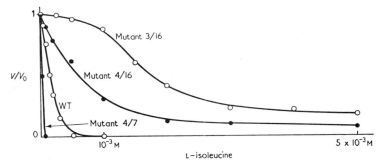

FIG. 10. Inhibition by L-isoleucine of various structural mutants of L-threonine-deaminase.

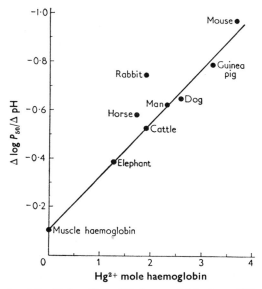

FIG. 11. Magnitude of the Bohr effect with haemoglobin from different mammals correlated with the number of Hg atoms taken up per molecule of protein (after Riggs, 1959).

Let us briefly recall the organization of this control system as it is understood today (see Fig. 12). The structural information written as a sequence of deoxyribonucleotides in a gene is first transcribed into a ribonucleotide sequence, the messenger. The messenger attaches to a ribosome, where the transcription into a polypeptide sequence takes place, the amino acids being transferred over from amino acyl S-RNA and positioned along the sequence by appropriate base-pairing between messenger and S-RNA.

This system is controlled at the level of messenger synthesis by specific agents, the repressors, able to recognize and bind electively certain genetic loci, called operators, which apparently function as exclusive initiation points for the first transcription.

The DNA segment whose transcription is thus "coordinated" by a given operator may involve one or several genes (or cistrons); it constitutes a unit of genetic expression called an operon. The synthesis of the corresponding protein(s) is therefore governed by the homologous repressor which, in turn, is synthesized under the control of a specialized "regulator" gene. In most, if not all cases, the activity of the repressor, i.e. presumably its ability to bind the corresponding operator, is controlled

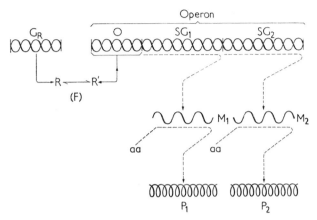

Fig. 12. General model of protein synthesis controlled by genetic repression (Jacob & Monod, 1961).

by specific small molecular compounds acting either as positive effectors (activating the repressor and thereby blocking messenger and protein synthesis) or as negative effectors (inhibiting the repressor and thereby inducing the synthesis of the messenger and of the protein(s)). The positive repression effectors are called "co-repressors". The negative repression effectors are called "inducers" (Jacob & Monod, 1961; Monod, Jacob & Gros, 1961).

We are interested here exclusively in the nature of the inducer–repressor–operator interaction. It was still considered likely not long ago that the repressor might be a polyribonucleotide (Jacob & Monod, 1961). This assumption, which did not by itself account for the repressor–inducer interaction (Jacob, Sussman & Monod, 1962), has met with further serious difficulties, while several lines of indirect experimental evidence suggest that the active product of a regulator gene is a protein, present in exceedingly minute amounts in cells. Since it has not proved possible so far to isolate a repressor and to observe its interactions *in vitro*, what knowledge we have comes from *in vivo* experiments, which can, however, be conducted under rigorous conditions, excluding many complications and ambiguities. On this basis, the following conclusions may be considered established.

1. The stereospecificity of the interaction is extreme (Monod, Cohen-Bazire & Cohn, 1951).

2. The interaction is virtually immediate and reversible, being completed both ways in less than 15 seconds according to the recent elegant work of Képès (1962).

3. The genetic–biochemical evidence shows that a single gene, therefore presumably a single specific macromolecular constituent, the repressor, is responsible for the specificity of the interaction (cf. Jacob & Monod, 1962).

4. Single mutations of this gene abolish the repressor–effector interaction while conserving the repressor–operator affinity (Willson, Perrin, Jacob & Monod, 1963).

It is evident that all these properties are immediately accounted for if the repressor is an allosteric protein possessing two sites, one of which binds the operator, the other the (positive or negative) effector. Almost any other model, by contrast, meets with extreme difficulties which need not be gone into here.

There are therefore strong reasons to assume as a working hypothesis that the specific effects of small molecules in activating or inhibiting, at the genetic level, the synthesis of messenger RNA and protein are mediated by an allosteric transition of the repressor.

For the time being only a few, perhaps a dozen, regulatory proteins have clearly been shown to exert their function by virtue of undergoing an allosteric transition. All of these proteins (except haemoglobin) are metabolic enzymes. One may rather confidently expect the number of metabolic control systems experimentally identified as allosteric mechanisms to increase considerably in the next few years. If, as one may hope, genetic repressors can eventually also be isolated and directly tested as to their indirectly inferred properties, it may be found that the fundamental elements or biological control systems, whether governing the activity or the synthesis of specific macromolecules, are allosteric proteins, those most elaborate products of molecular evolution.

Note added in proof.

With respect to the problem of the correlation between allosteric effects and association–dissociation reactions of proteins, two recent observations should be mentioned:

Frieden (manuscript in preparation) has found that the molecular weight of GDH at high dilutions (such as are used for assay of enzyme activity) is 250,000. Therefore no direct and immediate correlation appears to exist between the state of aggregation of this enzyme and its activity. Similarly, the effect of AMP on the state of aggregation of phosphorylase b has been studied at protein concentrations corresponding to those used for enzyme assay. Centrifugation in sucrose gradients showed that the sedimentation velocity was the same both in the presence and in the absence of AMP (Ullmann, Vagelos & Monod, unpublished). It would appear therefore that the activation of phosphorylase b by AMP does not directly depend upon dimerization of the protein. Thus while allosteric agents frequently appear to affect the state of aggregation of the sensitive proteins, the activating or inhibitory effects of the same agents do not seem necessarily to depend upon the association–dissociation reaction itself. The nature of the indirect correlation which appears, nevertheless, to exist between the two classes of effects remains to be explored and interpreted.

This work has been aided by grants from the Jane Coffin Childs Memorial Fund for Medical Research, the National Institutes of Health, the National Science Foundation, the "Commissariat à l'Energie Atomique", and the "Délégation Générale à la Recherche Scientifique".

REFERENCES

Benesch, R. & Benesch, R. E. (1961). *J. Biol. Chem.* **236**, 405.
Benesch, R. E., Ranney, H. M., Benesch, R. & Smith, G. M. (1961). *J. Biol. Chem.* **236**, 2926.
Changeux, J. P. (1961). *Cold Spr. Harb. Symp. Quant. Biol.* **26**, 313.
Changeux, J. P. (1962). *J. Mol. Biol.* **4**, 220.
Cori, G. T., Colowick, S. P. & Cori, C. F. (1938). *J. Biol. Chem.* **123**, 381.
Cori, G. T. & Green, A. A. (1943). *J. Biol. Chem.* **151**, 31.
Dische, Z. (1941). *Bull. Soc. Chim. Biol.* **23**, 1140.

Fischer, E. H. & Krebs, E. G. (1958). *J. Biol. Chem.* **231**, 65.
Frieden, C. (1959). *J. Biol. Chem.* **234**, 809.
Frieden, C. (1961). *Biochim. biophys. Acta*, **47**, 428.
Frieden, C. (1962a). *Biochim. biophys. Acta*, **62**, 421.
Frieden, C. (1962b). *Biochim. biophys. Acta*, **59**, 484.
Frieden, C. (1962c). *J. Biol. Chem.* **237**, 2396.
Gerhart, J. C. & Pardee, A. B. (1961). *Fed. Proc.* **20**, 224.
Gerhart, J. C. & Pardee, A. B. (1962). *J. Biol. Chem.* **237**, 891.
Haurowitz, F. (1938). *Z. Physiol. Chem.* **254**, 268.
Jacob, F. & Monod, J. (1961). *J. Mol. Biol.* **3**, 318.
Jacob, F. & Monod, J. (1962). In *Cytodifferentiation and Macromolecular Synthesis*, 21st Growth Symposium. New York: Academic Press Inc., in the press.
Jacob, F., Sussman, R. & Monod, J. (1962). *C.R. Acad. Sci., Paris*, **254**, 214.
Kent, A. B., Krebs, E. G. & Fischer, E. H. (1958). *J. Biol. Chem.* **232**, 549.
Képès, A. (1962). *Biochim. biophys. Acta*, in the press.
Koshland, D. E. (1958). *Proc. Nat. Acad. Sci., Wash.* **44**, 98.
Koshland, D. E. (1960). *Advanc. Enzymol.* **22**, 45.
Koshland, D. E. (1962). In *Horizons in Biochemistry*, p. 265. New York: Academic Press Inc.
Krebs, E. G. & Fischer, E. H. (1962). *Advanc. Enzymol.* **24**, 263.
Leloir, L. F. (1961). *Harvey Lect.* ser. **56**, 23.
Lwoff, A. & Lwoff, M. (1962). *J. Theoret. Biol.* **2**, 48.
Madsen, N. B. & Cori, C. F. (1956). *J. Biol. Chem.* **223**, 1055.
Martin, D. B. & Vagelos, P. R. (1962). *J. Biol. Chem.* **237**, 1787.
Martin, R. G. (1962). *J. Biol. Chem.* **237**, in the press.
Monod, J., Cohen-Bazire, G. & Cohn, M. (1951). *Biochim. biophys. Acta*, **7**, 585.
Monod, J. & Jacob, F. (1961). *Cold Spr. Harb. Symp. Quant. Biol.* **26**, 389.
Monod, J., Jacob, F. & Gros, F. (1961). *Biochem. Soc. Symp.* **21**, 104.
Morgan, V. E. & Chichester, D. F. (1935). *J. Biol. Chem.* **110**, 285.
Novick, A. & Szilard, L. (1954). In *Dynamics of Growth Processes*. Princeton: Univ. Press, 21.
Patte, J. C., Le Bras, G., Loviny, T. & Cohen, G. N. (1962). *Biochim. biophys. Acta*, in the press.
Perutz, M. F., Rossmann, M. G., Cullis, A. F., Muirhead, H., Will, G. & North, A. C. T. (1960). *Nature*, **185**, 416.
Rall, T. W. & Sutherland, E. W. (1961). *Cold Spr. Harb. Symp. Quant. Biol.* **26**, 347.
Riggs, A. (1959). *Nature*, **183**, 1037.
Stadtman, E. R., Cohen, G. N., Le Bras, G. & de Robichon-Szulmajster, H. (1961). *J. Biol. Chem.* **236**, 2033.
Theorell, H. (1934). *Biochem. Z.* **268**, 73.
Tomkins, G. M. & Yielding, K. L. (1961). *Cold Spr. Harb. Symp. Quant. Biol.* **26**, 331.
Tomkins, G. M., Yielding, K. L. & Curran, J. (1961). *Proc. Nat. Acad. Sci., Wash.* **47**, 270.
Umbarger, H. E. (1956). *Science*, **123**, 848.
Umbarger, H. E. & Brown, B. (1957). *J. Bact.* **73**, 105.
Umbarger, H. E. & Brown, B. (1958). *J. Biol. Chem.* **233**, 415.
Vagelos, P. R., Alberts, A. W. & Martin, D. B. (1962a). *Biochem. Biophys. Res. Comm.* **8**, 4.
Vagelos, P. R., Alberts, A. W. & Martin, D. B. (1962b). *J. Biol. Chem.* in the press.
Waite, M. & Wakil, S. J. (1962). *J. Biol. Chem.* **237**, 2750.
Willson, C., Perrin, D., Jacob, F. & Monod, J. (1963). *Biochem. Biophys. Res. Comm.* in the press.
Wolff, J. (1962). *J. Biol. Chem.* **237**, 230.
Wyman, J. (1947). *Advanc. Protein Chem.* **4**, 420.
Yates, R. A. & Pardee, A. B. (1956). *J. Biol. Chem.* **221**, 757.
Yielding, K. L. & Tomkins, G. M. (1960). *Proc. Nat. Acad. Sci., Wash.* **46**, 1483.
Yielding, K. L. & Tomkins, G. M. (1962). *Biochim. biophys. Acta*, **62**, 327.

J. Mol. Biol. (1963) **7**, 1–12

Hybrid Protein Formation of *E. coli* Alkaline Phosphatase Leading to *in vitro* Complementation

Milton J. Schlesinger and Cyrus Levinthal

Department of Biology
Massachusetts Institute of Technology,
Cambridge, Mass., U.S.A.

(Received 23 January 1963)

In vitro complementation has been demonstrated for alkaline phosphatase-negative mutants of *E. coli*. Enzymically active alkaline phosphatase can be formed by mixing monomer subunits derived from inactive purified proteins which are antigenically related to the wild-type enzyme of *E. coli*. Preparations of these proteins purified from four different phosphatase-negative mutants of *E. coli* could be reacted in pairs to yield partially active enzyme. Experimental evidence indicates that the active protein is a hybrid molecule composed of a monomer from each of the mutant proteins used in the reaction. The monomers, which can be prepared from the native protein by mild acid treatment or reduction with thioglycollate in urea, undergo a temperature and metal-dependent bimolecular reaction to yield a product distinct from normal enzyme but with partial enzymic activity.

The normal alkaline phosphatase protein is composed of two identical subunits whose structure is determined by a single functioning genetic unit. These results thus support the theory of hybrid protein formation which has been proposed to account for *intra-cistron* complementation.

1. Introduction

Mutations in a genetic region which determines the structure of a specific enzyme can either prevent the formation of any recognizable protein or can lead to the production of an altered protein with reduced enzymic activity. The altered or missing biological activity in such cases can be restored by genetic recombination, by a second suppressor mutation either within the same gene or elsewhere in the chromosome, by a reverse mutation to the wild-type configuration, or by genetic complementation (cf. review Levinthal & Davison, 1961). This last process can be demonstrated when two chromosomes, each derived from different cells and carrying different mutations in the same genetic region, are introduced into the same cell. The two mutants are said to complement each other if the biochemical function is restored in the heterozygous cell or, in other terms, if the amount of enzymic activity produced in this cell is greater than the sum of the two gene products acting separately. This phenomenon is distinguished from genetic recombination by the fact that this property is not inherited once the two chromosomes have segregated to different daughter cells.

Complementation is not a rare event. Many complementing systems have been described in recent years (Catcheside, 1960; Fincham & Pateman, 1957; Giles, Partridge & Nelson, 1957; Lacy & Bonner, 1961; Gross, 1962; Schwartz, 1962).

However, two fundamentally different types of complementation must be distinguished. If the two mutations are in functionally different regions of the genetic map (different cistrons) then *all* such mutants will complement (Benzer, 1957). This situation can be explained if it is assumed that each cistron determines the structure of a separate polypeptide chain and the cell can function if each polypeptide chain is made in its normal configuration by at least one of the two chromosomes. Complete complementation would then be expected if the different polypeptide chains were part of a single enzyme or determined different enzymes in the same biochemical pathway. In the second type of complementation, the two mutations are in the same cistron and affect the same polypeptide chain. In this case, which we will call *intra-cistron* complementation, recovery of enzymic activity only occurs between *some* pairs of mutants. The amount of active enzyme formed is dependent on the particular pair of mutants used and is generally much lower than in the case of *inter-cistron* complementation described above.

Several mechanisms have been proposed to account for *intra-cistron* complementation. The possibility has been considered that molecular rearrangement or crossing-over at the template RNA or protein level could be responsible for the production of active enzyme. However, the finding that the enzyme made by a complementing cell can have different physical–chemical properties from the wild-type enzyme (Fincham, 1962; Partridge, 1960; Giles *et al.*, 1957) seems to eliminate such mechanisms unless it is also postulated that unequal crossing-over is very common under these circumstances. Another hypothesis, which seems to account in a more satisfactory way for all the known facts, is that protein subunits derived from the same cistron mutated at different points can sometimes combine in a heterozygous cell to produce a hybrid form of the protein with partial enzymic activity (Brenner, 1959; Fincham, 1960). According to this model, *intra-cistron* complementation can only occur in those cases in which the enzymes are composed of two or more identical subunits. The mutant cells would produce altered monomers which could not combine to produce active enzyme, although they might combine to produce a protein antigenically related to the normal enzyme. The complementing cell would produce each of the altered monomers and these could, under some circumstances, combine to produce active enzyme. Only certain pairs of mutants might be expected to complement and the affected protein would be different from that made by the wild-type cell.

Recent experimental results from studies of the alkaline phosphatase of *E. coli* suggested that this system could serve to test the hybrid protein theory of *intra-cistron* complementation. Rothman & Byrne (1963) have shown that this enzyme is a dimer composed of identical subunits and *in vivo* complementation has been reported by Garen & Garen (1963). Levinthal, Signer & Fetherolf (1962) found that the enzyme could be reversibly denatured by treatment with thioglycollate in urea, and studies on the reactivation process indicated that the reduction reaction led to the formation of monomer subunits. Accordingly, proteins (CRM)† antigenically related to alkaline phosphatase were purified from several phosphatase-negative mutants of *E. coli*. Monomer subunits were then prepared and examined for their ability to complement. The results of these studies are presented in this paper. They support the hybrid protein hypothesis of complementation and, in addition, have led to information concerning the structure of the alkaline phosphatase enzyme and on the process by which active enzyme is formed from the monomer.

† Abbreviations used: CRM = cross-reacting material; TCA = trichloroacetic acid.

2. Materials and Methods

Strains. Four phosphatase-negative mutants previously isolated from *E. coli* K10 strain Hfr either by ultraviolet irradiation (U9, U13, U47) or by treatment with ethyl methane sulfonate (S33) were used in these studies. The mutant genes were transferred to a female (F$^-$) strain which contained a mutation in the *R2* control region leading to constitutive production of phosphatase activity (Echols, Garen, Garen & Torriani, 1961).

Cell growth and enzyme purification. Cells were grown overnight in batches of 50 l. under the conditions described previously (Levinthal *et al.*, 1962). After harvesting in a Sharples centrifuge, the cells were washed once in 6.7×10^{-2} M-tris buffer, pH 9.0, and resuspended in the same buffer at a concentration of 10^{10} cells/ml. All mutants had a low but detectable amount of enzymic activity and the extraction and purification procedures were identical to those reported for wild-type enzyme (Levinthal *et al.*, 1962). The purified proteins were tested for their antigenic properties by the use of the double diffusion technique of Ouchterlony (1948). In Table 1 the properties of the purified CRM prepared from four

TABLE 1

Properties of purified mutant CRMs

Strain	Specific activity†	% CRM activity	Relative electrophoretic mobility
U47	0·004	100	Fast
U13	0·01	50	Same as wild
U9	2·7	50	Fast
S33	0·35	100	Slow
Wild-type	1200	100‡	—

† Enzyme units/mg protein.   ‡ Wild-type is 100 by definition.

phosphatase-negative mutants are presented. These preparations were estimated to be at least 90% pure from the fact that no protein bands were observed on starch gel electrophoresis, except those usually associated with the alkaline phosphatase, and from the fact that the fractions—which were collected from gradient elution of a DEAE-cellulose column—contained protein and CRM in a constant ratio.

Preparation of monomers. Monomers were prepared by the reduction procedure recently described (Levinthal *et al.*, 1962). As indicated below, they could also be formed by incubating the protein in 5×10^{-2} M-sodium acetate buffer, pH 4.0, at 0°C for 15 min.

Electrophoresis. Starch gel electrophoresis was carried out vertically by a modification of the method of Smithies (1955; Signer, 1962). The techniques for detecting protein and enzymic activity were identical to those already described (Levinthal *et al.*, 1962).

Sedimentation analysis of the monomer and hybrid proteins were performed using the sucrose gradient technique described by Martin & Ames (1961). [^{14}C]Carboxymethyl alkaline phosphatase, prepared by alkylation of reduced alkaline phosphatase with [^{14}C]iodoacetate (M. Schlesinger, unpublished experiments), was added to the material layered on the gradient (5 to 20%) as a marker for the sedimentation behavior of the monomers. After centrifugation at 35,000 rev./min in the Spinco SW 39 rotor for 15 hr at 0°C, the centrifuge tube was punctured and fractions of about 0·125 ml. were collected for measurement of TCA-precipitable ^{14}C and measurement of either direct enzymic activity or of ability to complement and produce active enzyme.

Enzymic activity was measured by the rate of hydrolysis of *p*-nitrophenyl phosphate (Sigma 104, Sigma Chemical Co., St. Louis, Mo.), 0·02% in M-tris, pH 8.0. Samples of the reacting mixtures were added to 2·0 ml. of assay solution at 37°C and, after a short time, 0·5 ml. of K$_2$HPO$_4$ (13%) was added to stop the reaction. Absorbancy at 410 mμ was recorded. One unit of enzymic activity is that amount of enzyme which leads to a change of absorbancy of 1·0 units/min. Protein was determined by absorbancy at 280 mμ assuming an absorbancy of 0·770 for a protein concentration of 1 mg/ml. and a light path of 1 cm (Rothman & Byrne, 1963).

Determination of radioactivity was carried out by collecting TCA-precipitated protein on membrane filters (0·45 micron pore size, from Millipore, Inc., Bedford, Mass.). The filters were glued onto planchets with rubber cement, dried and counted in a low background gas-flow Geiger counter.

Antiserum against alkaline phosphatase was prepared by inoculation of rabbits with purified wild-type alkaline phosphatase.

Pronase-P (streptomyces griseus protease), B grade, was obtained from California Corporation for Biochemical Research. Trypsin was obtained from Worthington Biochemical Corporation.

3. Results

Initial attempts to detect enzymic activity by mixing monomers prepared by reduction in urea of CRM under the conditions described as optimal for reactivation of wild-type reduced enzyme (Levinthal *et al.*, 1962) were unsuccessful. It was found, however, that if the reduced CRMs were first acidified to pH 4·0 and the urea removed by dialysis, enzymic activity greater than that observable by the individual CRMs could be obtained after mixing at 37°C in a reactivation buffer. On further investigation it was noted that pre-treatment of the CRM with pH 4·0 sodium acetate buffer alone was sufficient to bring about formation of active enzyme after mixing and incubating the acidified proteins at 37°C. Table 2 compares the enzymic activity

TABLE 2

Complementation of U9–S33 monomers

Protein	Concentration (μg/ml.)	Enzymic activity units/ml.	
		zero time	50 min
U9 mon (R)	320	0·003	0·02
S33 mon (R)	320	0·003	0·003
U9 mon (R) + S33 mon (R)	320	0·003	0·48
U9 mon (A)	320	0·004	0·03
S33 mon (A)	320	0·005	0·3
U9 (A) + S33(A)	320	0·007	5·0
U9 native	200	0·2	0·2
S33 native	200	0·04	0·02
U9 native + S33 native	400	0·2	0·2

The reaction mixtures contained 0·7 ml. of reactivation buffer (10^{-2} M-MgC$_2$H$_3$O$_2$, 6×10^{-2} M-KCl and 6×10^{-3} M-mercaptoethanol), 0·1 ml. of M-tris, pH 7·8, and protein, in a total volume of 1·0 ml. Samples were taken at indicated times after incubation at 37°C for determination of enzymic activity. (R) and (A) refer to preparations of monomers by reduction or acidification, respectively.

observed with the different combination of CRMs and their monomers for the pair of mutants U9 and S33. This pair was chosen as Garen & Garen (1963) had found that it gave relatively high *in vivo* complementation. Only the mixture of monomers led to formation of enzymically active protein and the greatest activity was obtained from those monomers prepared by acid treatment alone.

A detailed study of the conditions for producing enzymically active protein from the U9–S33 inactive CRMs revealed that the reaction proceeded rapidly at 37°C at pH 7·8 and required the presence of a metal ion. As indicated in Table 3, zinc at a

TABLE 3

Effect of metals on U9–S33 complementation

Metal	Concentration	Enzyme units per ml.	
		0 min	60 min
Zn^{2+}	10^{-3}	0·23	5·4
	10^{-4}	0·24	7·2
	10^{-5}	0·21	0·23
Co^{2+}	10^{-3}	0·24	5·4
	10^{-4}	0·22	0·47
Mn^{2+}	10^{-3}	0·31	0·90
	10^{-4}	0·22	0·47
Mg^{2+}	10^{-3}	0·23	0·21
	10^{-4}	0·22	0·22
	10^{-5}	0·20	0·25
K^+	10^{-3}	0·18	0·20

Incubation mixtures containing acid-prepared monomers from U9 (80 μg) and S33 (80 μg) and 50 μmoles tris, pH 7·8, in 0·5 ml. were reacted at 37°C.

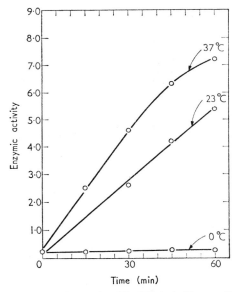

FIG. 1. Formation of enzymically active U9–S33 hybrid as a function of temperature. The reaction mixtures contained 80 μg each of U9 and S33 acid-prepared monomers in 10^{-1} M-tris, pH 7·8, and 10^{-4} M-ZnCl$_2$.

molar ratio of metal ion/protein of about 25:1 was the most active. Of the other cations tested only cobalt showed activity. The pH of the reactivation buffer was not critical between 6·0 and 9·0 but the rate of reaction was temperature dependent and varied with concentration of the monomer proteins. The rate of reactivation *v.* temperature is indicated in Fig. 1. An activation energy of 6·3 kcal. for complement enzyme formation can be calculated using the rate constants determined at 23 and 37°C. A plot of the initial rate of active enzyme formation *v.* concentration of protein over a sixteenfold range is noted in Fig. 2. This rate is found to be proportional to the square of the protein concentration as would be expected for a bimolecular reaction.

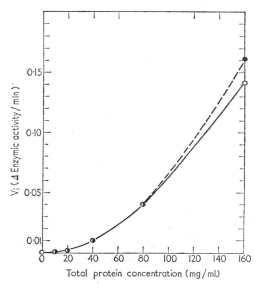

FIG. 2. Initial rate of enzymically active U9–S33 hybrid formation as a function of total protein concentration. The reaction mixtures contained the acid-prepared monomers in 10^{-1} M-tris, pH 7·8, 10^{-4} M-$ZnCl_2$ in a total vol. of 1·0 ml. Assays for enzymic activity were carried out after 0, 5, 10, 20 and 40 min incubation at 37°C. ● – – – ● theoretical calculations for bimolecular reaction; ○——○ actual values.

As a further test of the nature of the complementation phenomenon it was important to examine the active protein to determine whether indeed it was a hybrid molecule. Three sets of criteria indicate that this protein is a dimer containing one monomer of each CRM.

(1) *Analysis by starch electrophoresis.* Native alkaline phosphatase prepared from *E. coli* behaves as several electrophoretic species when subjected to starch gel electrophoresis. The reason for this banding is not yet understood but available evidence indicates that it is not due to a technical problem involving extraction or analysis on starch gel (Signer, 1962). Furthermore, the purified enzyme behaves as a single species in analytical ultracentrifugation and a single mutation which changes the charge or enzymic activity of the protein affects all the bands identically (Bach, Signer, Levinthal & Sizer, 1961). When the U9–S33 reaction mixture after formation of active enzyme is subjected to electrophoresis in starch gel, the enzymically active material is also found to have bands and these move to a position intermediate between those of the two parental CRMs (Plate I, no. 4 *v.* 2 and 3), as revealed by histochemical stain for enzymic activity. Although the most active bands are detectable in this intermediate position, all the bands overlap some of those of the U9 and S33 protein. From these results, the active enzyme appears to be an electrophoretic hybrid of these CRMs.

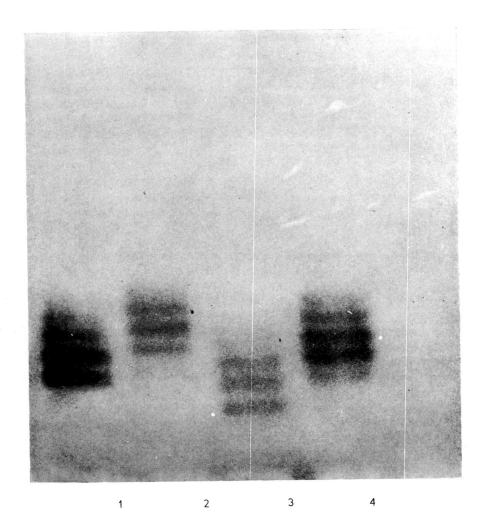

1 2 3 4

PLATE I. Starch gel electrophoresis. (1) Wild-type enzyme; (2) U9; (3) S33; (4) U9–S33.
All samples, previously acidified to pH 4·0, were incubated 3 hr at 37°C in the presence of 10^{-1} M-tris, pH 7·8, 10^{-4} M-ZnCl$_2$. They were dialysed 16 hr at 4°C against 5×10^{-3} M-NH$_4$HCO$_3$ and lyophilized. The samples were dissolved in 4×10^{-2} M-tris, pH 8·0, and the amounts added to the gel were as follows: wild-type; 12·5 μg, specific activity 732; U9: 370 μg, specific activity 0·8; S33: 320 μg, specific activity 0·43; U9 – S33: 180 μg, specific activity 23.

(2) *Sedimentation studies.* When the acidified S33 CRM is subjected to zone sedimentation in a sucrose gradient, the material which is able to carry out the complementation reaction with U9 sediments at a velocity identical to that of the alkylated monomer. In this experiment (Fig. 3(c)), portions of each fraction collected

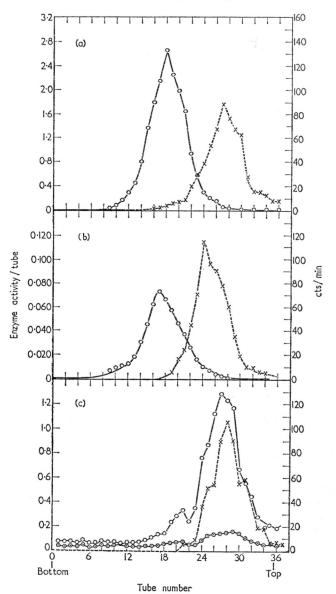

FIG. 3. Sucrose gradient analysis. All samples contained about 3000 cts/min of alkylated wild-type protein ($\times$ – – – $\times$).

(a) Wild-type enzyme ($\bigcirc$——$\bigcirc$) in 10^{-2} M-tris, pH 7·4, 10^{-3} M-MgCl$_2$; (b) U9–S33 hybrid ($\bigcirc$——$\bigcirc$) in 10^{-2} M-tris, pH 7·4, 10^{-3} M-MgCl$_2$; (c) S33 acid-prepared monomer in 0·05 M-sodium acetate, pH 4·0 ($\circ$——$\circ$ assayed at 0 time, $\bigcirc$——$\bigcirc$ assayed after 60 min in the presence of U9 monomers).

Details in text.

from the gradient were incubated with 40 μg of U9 monomer in 10^{-1} M-tris, pH 7·8, and 10^{-4} M-ZnCl$_2$ in a total volume of 0·5 ml. Enzymic activity was determined immediately and after 60 minutes' incubation at 37°C. A sedimentation analysis of the enzyme product prepared in the normal complementation reaction mixture of U9–S33 is depicted in Fig. 3(b). By a direct enzyme assay of each fraction it is seen that this material has a sedimentation behavior identical to that of wild-type alkaline phosphatase whose pattern is plotted in Fig. 3(a).

These analyses indicate that the material which participates in the complementation reaction is a monomer having the same sedimentation coefficient as the alkylated material. The molecular weight of alkylated normal enzyme has been shown to be 40,000, half that of the native protein (Rothman & Byrne, 1963).

(3) *Quantitative hybrid formation.* The extent of hybridization as determined enzymically is a function of the relative concentrations of the two monomers. When the ratio of concentration of S33 and U9 is varied from 1:9 to 9:1, the amount of active enzyme produced varies (Table 4). A value of the specific activity of each

TABLE 4

Extent of complementation as a function of relative proportions of monomers

Ratio S33:U9	Enzymic activity units/ml.	Theoretical amount of hybrid protein (μg)	Specific[†] activity units/mg
1:9	2·0	36	55
1:4	2·7	64	42
1:3	4·1	75	55
7:13	3·5	91	39
2:3	5·5	96	57
1:1	5·4	100	54
3:2	5·7	96	59
13:7	4·3	91	47
3:1	3·5	75	47
4:1	3·0	64	47
9:1	1·6	36	45

The total concentration of protein in each reaction tube was 200 μg/ml. Incubations were at 37°C in 10^{-1} M-tris, 10^{-4} M-ZnCl$_2$. See text for details of calculations.

† Enzymic activity/theoretical amount of hybrid protein.

mixture cannot be determined directly since there is no way of measuring the amount of hybrid protein formed. One can, however, determine the theoretical amount of hybrid enzyme in these various mixtures by assuming that the hybrid protein has the same probability of being formed as the separate mutant proteins and that the relative concentration of the various species will depend only on the probability of the different types of collisions. In this case the concentration of the hybrid can be calculated from the binomial expansion formula. When one uses this theoretical amount of hybrid enzyme formed as a basis for expressing specific activity, the values obtained are relatively constant (column 3, Table 4) in spite of the extreme variation in proportions of monomer CRMs used. For the complement enzyme of this pair, the specific activity is 5% that of the wild-type enzyme. These data are thus consistent with the hypothesis that the monomers of these two mutant proteins produce dimers by random collisions.

Complementation of these mutants has provided a convenient assay for the dimerization process, and preliminary information concerning the nature of the alkaline phosphatase monomer has been obtained. In Table 5, data are presented

TABLE 5

Effects of various reagents on complementation

Ratio reagent : protein	Reagent	Enzyme activity units/ml.	% inhibited
	None	3·45	0
2000	0·01 M-IAA†	2·4	30
2000	0·01 M-CMB†	0·78	80
400	0·002 M-CMB†	2·8	20
1/20	Pronase	0·0	100
1/10	Trypsin	0·0	100

The reaction mixtures contained acid-prepared monomers of U9 (50μg) and S33 (50 μg) and reagent in 10^{-1} M-tris, pH 7·8, 10^{-4} M-ZnCl$_2$ in a total vol. of 0·5 ml. Enzymic activity was measured after 80 min at 37°C.

† IAA refers to iodoacetate, CMB refers to p-chloromercuribenzoate.

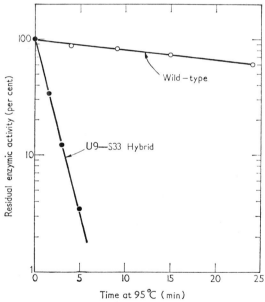

Fig. 4. Heat stability of U9–S33 hybrid enzyme. The reaction mixtures contained 100 μg protein in 10^{-1} M-tris, pH 8·0, 10^{-4} M-ZnCl$_2$ in a total vol. of 0·5 ml.

which show the effect of the proteolytic enzymes pronase and trypsin and of sulfhydryl binding agents on the ability of the monomers to dimerize. The monomers are sensitive to low concentrations of trypsin and pronase. On the other hand, dimerization is relatively insensitive to high concentrations of iodoacetate and p-chloromercuribenzoate.

Studies on the nature of the hybrid dimer show that it is distinct from wild-type alkaline phosphatase. The specific activity appears to be greatly reduced and it is considerably more heat labile. Three minutes at 95°C leads to an 88% loss of activity of the hybrid enzyme in contrast to the wild-type enzyme which shows a loss of 12% after 10 minutes at the same temperature (Fig. 4).

Monomers prepared from purified CRMs of two other mutants have been tested for their ability to complement. The results of all possible pairs for the four mutants are presented in Table 6. All reactions were carried out at the same protein concentrations, and the values therefore indicate the relative specific activities of the hybrid molecules, assuming that the probability of hybrid formation is the same between all pairs. All the CRM pairs except S33–U13 show complementation. From preliminary genetic mapping which indicates that U13 and U47 are extremely closely linked (A. Torriani, unpublished experiments) there appears to be no simple correlation between the values noted here and the genetic map distances.

TABLE 6

In vitro *complementation between monomers derived from proteins of p-strains of* Escherichia coli

Strain	U47	U13	S33	U9
U47	0·06	8·4	4·2	5·7
U13		0·01	0·01	2·5
S33			0·05	4·4
U9				0·4

Complementation was examined in reaction mixtures containing 200 μg of total protein, 100 μmoles tris, pH 7·8, and 0·1 μmole $ZnCl_2$ in a total vol. of 1·0 ml. Values are enzyme units/ml. observed after 60 min incubation at 37°C.

4. Discussion

The complementation assay has made it possible to carry out initial studies on the dimerization process for the formation of active enzyme and also to determine certain properties of the monomer subunits. Dimerization of CRM subunits has been found to be a zinc-activated, temperature-dependent bimolecular reaction that proceeds within a broad pH range (6 to 9). Monomers can be formed by treatment of the protein at low pH as well as by reduction in urea. Preliminary results (Fetherolf & Schlesinger, unpublished experiments) indicate that these same properties for monomer formation and dimerization are applicable to the wild-type alkaline phosphatase. The dimer therefore appears to be held together by ionic bonds stabilized by the presence of zinc which is a normal component of the enzyme (Plocke, Levinthal & Vallee, 1962). The disulfide bridges in the enzyme are *intra*-chain since reduction is not necessary to effect monomer formation nor do sulfhydryl binding agents block dimerization. However, the monomer subunits are considerably more labile than the intact protein as they are readily attacked by proteolytic enzymes which do not affect the dimer.

Although only one pair of CRM-forming mutants has been studied in detail, formation of enzymically active hybrid enzymes has been examined among pairs of monomer subunits from four purified CRMs which are synthesized by mutants having different mutation sites within the gene. Of the six possible hybrids, five are partially active enzymes. From these results and preliminary studies of the relative position of these mutants on the gene map, there does not appear to be a simple colinear relation. Two mutants which are very close on the genetic map gave more active hybrids than another pair which map farther apart. *In vitro* complementation can be carried out with crude extracts of mutants and experiments are in progress to determine whether or not any correlation can be found between gene-map position and the ability to complement.

There have been previous reports of *in vitro* complementation for several enzymes (Woodward, 1959; Loper, 1961; Glassman, 1962). In these cases, the mixing of extracts or of chilled cells prior to the extraction of proteins gave partially active enzyme. Results of these experiments suggested cytoplasmic interaction but no definitive conclusion could be made concerning the molecular basis of the phenomenon. The enzyme product could be shown to be different from the normal wild-type enzyme in one of these systems (Partridge, 1960). However, insufficient information was available as to the structure of the enzymes to establish the hybrid protein nature of the product.

Experimental evidence now exists for the alkaline phosphatase of *E. coli*, both with respect to the *in vivo* complementation and the molecular basis of *in vitro* complementation. The structural gene of this enzyme is a single cistron (Garen & Garen, 1963) and the protein, whose structure is determined by this genetic region, is composed of two identical subunits. The experiments reported here indicate that enzymically active proteins can be formed from monomer subunits obtained from inactive alkaline phosphatase proteins (CRM). With the use of highly purified CRMs it has been possible to demonstrate that this active material is a hybrid molecule composed of one monomer from each CRM. Furthermore, the active protein product is distinct from the normal enzyme; it appears to have a much lower specific activity and is considerably more heat labile. The *in vivo* complementation as reported by Garen & Garen (1963) also leads to partially active and heat-sensitive enzyme. These properties are not unexpected since the hybrid protein contains two "lesions" in its structure.

The ability of any particular pair of mutants to hybridize and yield active enzyme would be difficult to predict without detailed knowledge of the tertiary and quaternary structure of the protein and of the requirements for the active site of the enzyme. Furthermore, without such knowledge it is not possible to determine whether the active sites are confined to a single monomer subunit or whether they span across the two monomers. The fact that the dimer formation is necessary for activity could imply either that the active site does bridge the dimer or that the formation of the dimer modifies the tertiary structure of the monomer in such a way as to stabilize the active site in its required three-dimensional configuration. Nevertheless, it is apparent from these studies that the active-site of a "mutated" inactive enzyme composed of multiple subunits can sometimes be restored by the interaction of different mutated subunits which occurs through molecular hybridization.

These experimental results thus support the hypothesis that *intra-cistron* complementation can be accounted for by hybrid protein formation.

This work was supported by grant E-2028 (C5) from the U.S. Public Health Service, and grant G-8816 from the National Science Foundation.

REFERENCES

Bach, M. L., Signer, E. R., Levinthal, C. & Sizer, I. W. (1961). *Fed. Proc.* **20**, 255.

Benzer, S. (1957). In *Symp. on Chemical Basis of Heredity*, ed. by W. D. McElroy & B. Glass, p. 70. Baltimore: The Johns Hopkins Press.

Brenner, S. (1959). In *Symp. on Biochemistry of Human Genetics*, CIBA Found. and Internatl. Union of Biol. Sciences, ed. by G. E. W. Wolstenholme & C. M. O'Connor, p. 304. London: Churchill.

Catcheside, D. G. (1960). In *Microbial Genetics*, Tenth Symp. Soc. for Gen. Microbiol., p. 181. Cambridge: The University Press.

Echols, H., Garen, A., Garen, S. & Torriani, A. (1961). *J. Mol. Biol.* **3**, 425.

Fincham, J. R. S. (1960). In *Advances in Enzymology*, **22**, ed. by F. F. Nord, p. 1. New York: Interscience.

Fincham, J. R. S. (1962), *J. Mol. Biol.* **4**, 257.

Fincham, J. R. S. & Pateman, J. A. (1957). *Nature*, **179**, 741.

Garen, A. & Garen, S. (1963). *J. Mol. Biol.* **7**, 13.

Giles, N. H., Partridge, C. W. H. & Nelson, M. J. (1957). *Proc. Nat. Acad. Sci., Wash.* **43**, 305.

Glassman, E. (1962). *Proc. Nat. Acad. Sci., Wash.* **48**, 1491.

Gross, S. R. (1962). *Proc. Nat. Acad. Sci., Wash.* **48**, 922.

Lacy, A. M. & Bonner, D. M. (1961). *Proc. Nat. Acad. Sci., Wash.* **47**, 72.

Levinthal, C. & Davison, P. F. (1961). *Ann. Rev. Biochem.* **30**, 641.

Levinthal, C., Signer, E. R. & Fetherolf, K. (1962). *Proc. Nat. Acad. Sci., Wash.* **48**, 1230.

Loper, J. C. (1961). *Proc. Nat. Acad. Sci., Wash.* **47**, 1440.

Martin, R. G. & Ames, B. N. (1961). *J. Biol. Chem.* **236**, 1372.

Ouchterlony, C. (1948). *Acta path. microbiol. scand.* **25**, 186.

Partridge, C. W. H. (1960). *Biochem. Biophys. Res. Comm.* **3**, 613.

Plocke, D. J., Levinthal, C. & Vallee, B. L. (1962). *Biochemistry*, **1**, 373.

Rothman, F. & Byrne, R. (1963). *J. Mol. Biol.* **6**, 330.

Schwartz, D. (1962). *Proc. Nat. Acad. Sci., Wash.* **48**, 750.

Signer, E. (1962). Ph.D. dissertation. M.I.T., Cambridge, Mass.

Smithies, O. (1955). *Biochem. J.* **61**, 629.

Woodward, D. O. (1959). *Proc. Nat. Acad. Sci., Wash.* **45**, 846.

J. Mol. Biol. (1963) **7**, 281–308

Electron Microscope Studies on the Structure of Natural and Synthetic Protein Filaments from Striated Muscle

H. E. HUXLEY

*Medical Research Council Laboratory of Molecular Biology,
Hills Road, Cambridge, England*

(*Received 22 April 1963*)

A technique has been developed for fragmenting striated muscle into its component thick and thin filaments by homogenization in a "relaxing medium". Such preparations are very suitable for examination by the negative-staining technique. The thick filaments so obtained correspond closely in their structure to those seen in the A-bands of sectioned muscles. The thin filaments, often still attached to residual Z-line structures, also resemble closely those seen in sectioned tissue; they show the same characteristic internal structure, first seen and analysed in filaments from smooth muscles by Hanson and Lowy, as filaments of purified actin. Preparations of purified myosin, precipitated at low ionic strength, are found to contain spindle-shaped aggregates similar in appearance and dimensions to the thick filaments, and having a remarkable differentiated appearance along their length. Along part of the length of the filaments numerous projections are visible, probably corresponding to the cross-bridges seen in sectioned material. However, in a central zone, always about 0·15 to 0·2 μ in length, the projections are absent. The same is the case in the naturally occurring thick filaments. This appearance can be accounted for if the myosin molecule has a projection at one end, and if the myosin molecules in either half of the filaments are arrayed with opposite polarities. In confirmation of some recent results of others it is found that isolated myosin molecules, when examined by the shadow-casting technique, do indeed show such a structure. Observations on heavy and light meromyosin confirm this picture of the structure of the thick filaments.

The thin filaments and filaments of purified actin are both found to form the same very characteristic complex structure when allowed to combine with myosin, or with heavy meromyosin. The results obtained show that the filaments are structurally polarized, and in muscle are arranged so that all of them attached on one side of a given Z-line point in one direction, whilst those on the other are oppositely oriented.

A close similarity is found between the lattice structure seen in crystals of tropomyosin B and that formed by the interconnecting system of filaments at the Z-lines.

The functional implications of these results, particularly those concerned with the polarity of the filaments, is discussed.

1. Introduction

Previous studies on muscle structure in the electron microscope (Huxley, H. E., 1957), using the thin sectioning technique, have provided a picture of the general arrangement of the filaments which is in accord with the sliding filament model of muscular contraction (Huxley, H. E. & Hanson, 1954; Huxley, A. F. & Niedergerke, 1954). However, this technique, in its usual form at least, shows very little detail of the internal structure of the filaments, and attempts to improve the technique in this direction by using alternative fixing, staining or embedding methods, have not been particularly successful (Huxley, H. E., unpublished observations). Moreover, the thin sectioning technique is not particularly well adapted to the study of purified muscle proteins in the electron microscope. It has therefore been difficult to extend the structural picture of muscle so as to shed more light on the *detailed* mechanism of contraction, i.e. the details of the process by which a relative force is developed between the cross-bridged filaments of actin and myosin.

In recent years, the negative-staining technique (Hall, 1955; Huxley, 1956; Brenner & Horne, 1959) has proved to be of great value in showing up fine structural detail in biological specimens in the electron microscope. In this technique the specimen is dried down in a thin film of a solution of some dense salt such as sodium phosphotungstate, which comes out of solution as an almost completely amorphous layer in which the specimen is embedded. The relatively low density of the biological specimen (or at least those parts of it which exclude the negative stain) then enables it to be seen by negative contrast. This technique has proved outstandingly successful in the case of viruses, where a great deal of structural detail, which does not show up at all by normal staining techniques, can be resolved (e.g. Horne, Brenner, Waterson & Wildy, 1959).

However, the degree of detail that can be resolved by this technique is limited by the thickness of the specimen. In the case of virus particles a few hundred Ångstroms in diameter, partially embedded in a very thin layer of the negative stain, this limitation is not serious. But clearly the technique is not applicable to intact pieces of tissue, and even muscle fibrils 1μ in diameter are far too thick to be usefully examined. The muscle must first be broken down in some way.

When glycerinated rabbit psoas muscle is mechanically disrupted in a high-speed blendor, it fragments into separated myofibrils, but normally these are very resistant to further breakdown. Presumably the cross-links between the filaments hold them together in a very robust lattice. According to the sliding filament theory, these cross-links represent sites of actin–myosin interaction. A number of systems have been discovered in which the ATP-induced contraction of glycerinated muscle is inhibited, and in which the fibres become much more readily extensible. These are usually known as relaxing systems, and they are believed to operate by inhibiting the enzymic breakdown of ATP and causing the ATP to dissociate actomyosin. In terms of the sliding filament model this would correspond to detaching all the cross-bridges. Thus we might expect the structure of the fibrils to be greatly weakened by such treatment, and to be much more susceptible to mechanical disruption. This indeed proves to be the case, and fibrils treated with the ATP–EDTA relaxing system are readily broken down into their constituent filaments. These filaments are found to be very satisfactory subjects for examination by the negative-staining technique and some preliminary notes on the technique and some early results have already been published (Huxley, 1961, 1962). Hanson & Lowy (1963) have reported a very elegant study of the structure of actin filaments by the same general technique, and they have also studied the various kinds of filaments in a variety of muscles from invertebrate animals (Lowy & Hanson, 1962; Hanson & Lowy, 1962). In the present paper a detailed study of the filaments from striated muscles of rabbit and chicken will be described, and the structures seen compared with those found in preparations of purified muscle proteins. A study of the interaction between the filaments and specific antibodies against the various muscle proteins is described elsewhere (Pepe & Huxley, 1963).

2. Methods

Preparation of separated filaments

First of all, glycerinated muscle was prepared according to the procedure of Szent-György (1951) as modified slightly by Huxley & Hanson (1957). Strips of rabbit psoas muscle, or of chicken breast or thigh muscle, approximately 4 mm × 2 mm in section and 5 to 10 cm in length, were dissected out from the freshly killed animals, tied with wool onto Perspex strips at the required length, and placed in a medium containing 50 vol. glycerol, 40 vol. water, 10 vol. 0·067 M-phosphate buffer, pH 7·0, at 4°C. The glycerol–water mixture had previously been passed over a column of Amberlite IR120H resin (to remove heavy-metal contaminants), as had the water used to make up the phosphate buffer; the pH of the whole mixture was adjusted before use. After 24 hr fresh medium at +4°C was substituted, and after a further 24 hr the preparation was transferred to the deep freeze at −20°C and stored there for at least three weeks, but usually much longer, until required for use.

A portion of one strip, weighing approximately 0·3 to 0·4 g (wet), was then transferred to a solution containing 15 vol. glycerol, 85 vol. standard salt solution for ½ hr at 0°C and then shredded into very thin bundles of fibres with a needle. (The standard salt solution contained 0·1 M-KCl, 0·001 M-MgCl₂, 0·0067 M-phosphate buffer, pH 7·0.) These fibre bundles were then transferred to standard salt solution for a further period of ½ to 1 hr to wash out the rest of the glycerol.

The fibre bundles were then cut up into short lengths (about 2 mm) with scissors (by picking up and cutting a lot of them together) and blended in an MSE homogenizer (cooled in ice and running at top speed) in standard salt solution containing in addition 10^{-3} M-EDTA and 10^{-2} M-MgCl₂. Blending for three periods of 20 sec, interspersed with periods of 15 sec. low-speed stirring to allow cooling to occur, was usually sufficient to break up nearly all the filaments into separated myofibrils. This was checked in the phase contrast light microscope.

The fibril preparation was then spun down at about 650 g in a bench angle centrifuge for 3 min and the supernatant solution, containing unwanted soluble proteins, was discarded. The fibrils were resuspended in the same volume of ordinary standard salt solution and spun down as before, the supernatant again being discarded. They were then resuspended in a very small volume (< 1 ml.) of the EDTA medium described above, and carefully cooled to 0°C in melting ice. The relaxing medium, also cooled to 0°C, was then added to make up the normal volume for blending (7 ml.). The relaxing medium consisted of standard salt solution containing in addition 10^{-3} M-EDTA, 10^{-2} M-MgCl₂, 3 to 5 × 10^{-3} M-ATP, pH adjusted (if necessary) to 7·0. The preparation was immediately homogenized again according to the same schedule as before. Any intact fibrils or other debris were spun out as before; the supernatant solution contained the separated filaments.

Such preparations were stable for about 24 hr at 0°C. After longer periods of time, the filaments tended to aggregate again, but could often be satisfactorily dispersed if 1 mg ATP/ml. was added and the preparation shaken violently by hand for a few seconds.

Satisfactory filament preparations were also made in which the standard salt solution contained 0·15 or 0·20 M-KCl.

Preparations of filaments from fresh insect indirect flight muscle, and from fresh mytilus adductor muscle were made by blending up the tissue directly in the relaxing medium. Such preparations were not as clean as the one described above, but were adequate for the particular purposes for which they were required.

Preparation of muscle proteins

Myosin

Myosin was prepared from rabbit muscle essentially by the method of Szent-György (1951). Extraction of the tissue, previously passed once through a meat mincer with 5 mm holes, was carried out for 10 min at 0°C in a medium containing 0·3 M-KCl, 0·15 M-phosphate buffer, pH 6·5. All subsequent operations were also carried out at 0°C. Coarse debris was strained off through muslin, finer debris was removed by centrifugation at 1000 g for $\frac{1}{2}$ hr, and lipid by filtration through glass wool.† Myosin was precipitated by dilution to $\mu = 0\cdot05$, separated by centrifugation (finally being packed down into a small vol. by $\frac{1}{2}$ hr spin at 40,000 g), and redissolved in a minimum vol. of 1·9 M-KCl, 0·1 M-phosphate buffer, pH 7·0. Contaminating actomyosin (normally present in only very small amounts) was removed by dilution to $\mu = 0\cdot28$ and centrifugation at 20,000 g for $\frac{1}{2}$ hr. The cycle of precipitation and re-solution was then repeated again. Residual lipid contaminants were filtered out with glass wool again. Such preparations of myosin were checked in the ultracentrifuge and found to give a single hypersharp boundary.

Actomyosin

"Natural" actomyosin was prepared according to the method of Szent-Györgyi (1951).

Heavy and light meromyosins (HMM and LMM)

These were prepared by the method of Szent-Györgyi (1953) except that the digestion medium was buffered with 0·1 M-phosphate at pH 7·0 (instead of 0·01 M-borate at pH 8·8), and the meromyosins separated by dialysing the preparation against 50 vol. of 0·05 M-KCl, 0·003 M-phosphate buffer, pH 7·0, and centrifuging out the precipitated LMM (modifications similar to those used by Lowey & Holtzer (1959)). The LMM was then redissolved in 0·5 M-KCl, 0·1 M-phosphate buffer, pH 7·0. LMM Fraction I was prepared according to the method of Szent-Györgyi, Cohen & Philpott (1960).

Tropomyosin

This was prepared according to the method of Bailey (1948) except that acetone-dried, butanol-treated muscle powder, prepared according to the method of Tsao & Bailey (1953), was the material which was treated with 1 M-KCl in the initial extraction.

Actin

Actin was extracted according to the method of Tsao & Bailey (1953) from butanol–acetone dried powder except that water (instead of 30% acetone), containing 10^{-4} M neutralized ATP but no ascorbic acid, was used as the extracting medium. After the initial isoelectric precipitation, redispersion and dialysis against water containing 10^{-4} M-ATP, the actin was further purified by ultracentrifugation (2 hr at $\sim 100,000$ g), first in the absence of salt, when the supernatant fraction was preserved, then in the presence of 0·1 M-KCl, 10^{-3} M-MgCl$_2$, 0·0067 M-phosphate buffer, pH 7·0, when the resultant pellet was redispersed in the same medium.

Negative staining

The basic method used was that in which a drop of the suspension under examination is applied to a carbon-filmed specimen grid, the grid then rinsed with pure solvent so that only particles adhering to the grid remain (still immersed in solvent), and the solvent replaced (without drying) by the negative-staining solution, which is then allowed to dry. A number of variations of this technique were used; some of them have already been mentioned (Huxley & Zubay, 1960). The more important ones were as follows.

(1) The substance most usually used as a negative stain was uranyl acetate. The advantageous properties of this as a negative stain were noticed during the work just referred to, and became even more apparent when working on muscle. Muscle filaments are largely destroyed if attempts are made to stain them negatively in sodium phosphotungstate without prior fixation, whereas they are well preserved in uranyl acetate. Moreover, better contrast was obtained using uranyl acetate, particularly over holes (see below).

The media in which the preparations were suspended often contained phosphate buffer, and, if allowed to mix with the uranyl acetate solution, this tended to produce a precipitate which adversely affected the negative staining. Such preparations were therefore rinsed with 0·1 M-KCl alone before applying the negative stain, which was generally used in unbuffered 1% aqueous solution, pH around 4·25 to 4·5. There was no sign in general that the preparations were adversely affected by this pH.

† Caution: glass wool should be washed free of possible heavy-metal contaminants.

(2) When maximum resolutions and contrast were required, together with absence of background "noise" due to surface structure in the supporting film, preparations were made on perforated carbon film, as described by Huxley & Zubay (1960), so that the specimens were embedded in thin films of the negative stain extending over the holes.

(3) Preparations destined for negative staining with sodium phosphotungstate were always fixed beforehand, either in formalin as below, or by floating the grid on which they had been placed face downwards on 0·1% osmium tetroxide solution buffered in the usual way with M/15 veronal acetate, pH 7·0, or by brief (20 sec) exposure of the grid, still wet with a thin film of solvent, to the confined vapour of a 2% aqueous solution of osmium tetroxide.

(4) Even when uranyl acetate was used as a negative stain, some details of the structure were best preserved by prior fixation. This included, in particular, the cross-bridges on the natural thick filaments. Fixation was effected by mixing the preparation with an equal volume of 10% formalin (4% formaldehyde) in M/15 veronal acetate buffer, pH 7·0, leaving it 1 hr at 0°C and then proceeding in the normal way.

(5) In some cases, the whole process was carried out in a cold room at 0°C and the drying down of the negative stain took place over a water bath so that it was very slow. Although it was not always a decisive effect, one often had the impression that preparations made in this way were superior to those made at room temperature, particularly in the case of the LMM lattices. It appeared, however, that it was the stage when the drop of suspension was applied to the grid, rather than the final drying, which was more sensitive.

Shadow casting

The procedure originated by Hall (1956) and applied to myosin preparations by Rice (1961a,b) was followed closely, except that an extremely low shadowing angle (giving a shadow length/height ratio of about 15 : 1) was used.

Electron microscopy

Preparations were examined in a Siemens Elmiskop I using an accelerating voltage of 80 kv, beam current 13 μA, double condenser illumination (15 clicks of K1), a 200 μ aperture in condenser II, and a 50 μ molybdenum objective aperture.

3. Results

(i) *General characteristics of the muscle preparations*

After the first stage of homogenization described above, the glycerinated muscle is broken down into isolated myofibrils, which can be examined in the phase-contrast light microscope. These are still visible, slightly swollen, immediately after they have been resuspended in the relaxing medium containing ATP and EDTA. At first they are uncontracted but, after standing for a few minutes on a slide under a coverslip, they begin to shorten and eventually contract very strongly into tightly packed and unrecognizable masses. The same process takes several hours in the bulk suspension and probably occurs as a result of the gradual lowering of the ATP concentration by slow hydrolysis to a level which is insufficient to cause relaxation.

After the second stage of homogenization, i.e. in the relaxing medium, a considerable proportion (which varies somewhat from preparation to preparation) of the fibrils are broken down into fragments invisible in the light microscope, and examination in the electron microscope shows that they have dispersed into separated filaments. A typical preparation of these is shown in Plate I.

A number of different components is present. These comprise thick filaments, thin filaments, I-segments and vesicles.

(a) *Thick filaments*

These are present in great profusion and show up very clearly in negatively stained preparations. The filaments are 100 to 120 Å in diameter and are predominantly about 1·5 to 1·6 μ in length. The ends of the filaments are tapered, and often fragments of thin filaments are still attached to them, so that it is not easy to give a precise estimate of length; but filaments are not found, for example, with lengths of 1·8 μ or 2·0 μ or longer. The filaments have large numbers of irregular-looking projections on their surfaces (see Plate II). These are present right out to the tips of the filaments, but they appear to be absent from a central zone about 0·15 to 0·2 μ in length. In the middle of this bare central zone, a slight thickening of the filament can often be seen. Some thick filaments are found having lengths less than 1·5 μ, e.g. as short as 0·6 to 0·8 μ, but these show signs of having been formed by mechanical fracture of the longer filaments; thus they have the same diameter as the longer filaments, but the bare zone is no longer centrally disposed, and the tapering is either uneven or absent.

Thus the appearance of these separated thick filaments corresponds in all respects with the appearance of the thick A-band filaments seen in muscles which have been fixed, positively stained and sectioned for examination in the electron microscope; some examples of the latter preparations showing the tapering, the projections or cross-bridges present on the filaments right out to their tips, and the bare central zone with the central thickening are reproduced in Plate III. It is clear that the homogenizing procedure has released these A-band filaments virtually intact.

(b) *Thin filaments and I-segments*

A second characteristic type of filament is present, having a diameter of 60 to 70 Å. These occur predominantly as separated filaments of somewhat variable length. They are often 0·5 to 1·0 μ long, but substantially longer ones can also be found. We will see later that these are probably formed by the polymerization of the shorter fragments. However, a certain proportion of them, which varies from one preparation to another, are present as bundles of filaments, of the kinds shown in Plates IV and V. In these, the filaments seem to be attached together at a central plate and to extend about 1 μ on either side of it. Both the attached and the separated thin filaments have the same characteristic beaded appearance, first noticed by Hanson & Lowy (1962, 1963) in preparations of filaments from smooth muscles.

It is clear that the bundles of thin filaments with this central zone of adhesion correspond to the arrays of thin filaments seen in sectioned material attached to a Z-line and extending on either side of it to terminate at the edges of the H-zones. The lengths of these separated bundles, or "I-segments", correspond to the value we should expect from the model of muscle proposed earlier and their general appearance is unmistakable. Thus we can conclude that the separated thin filaments are to be identified with the thin filaments seen in thin sections of muscle. The diameter of the filaments measured by the latter technique (about 50 Å) is somewhat smaller than that seen here, but it is not clear at present whether the difference is a significant one; certainly the treatment of the material has been very different.

(c) *Vesicles*

In addition to the filamentous components, membrane-bound vesicles of a wide variety of sizes (300 to 5000 Å) are also present. Many of these no doubt derive from the sarcoplasmic reticulum, but the method of preparation—including the use of glycerinated material—is ill-suited to preserve its original structure. We will not be concerned any further with this component here.

(ii) *Other properties of the separated filaments*

(1) If the ionic strength of the medium in which the filaments are suspended is increased (by adding KCl) to $\mu = 0.6$, the thick filaments dissolve, but the thin ones remain. If they are suspended in a medium containing 0.6 M-KI, however, both types of filament dissolve. Thus the separated filaments have solubility properties which are the same as those deduced earlier (Hanson & Huxley, 1953, 1955) from observations in the phase-contrast light microscope of the behaviour of intact muscle fibrils when treated with salt solutions of various kinds. These observations led to the conclusion that the protein myosin was located in the thick filaments, and actin in the thin ones.

(2) If the original fibril preparation is treated with solutions known to extract myosin selectively (e.g. Hasselbach & Schneider's solution (1951)), and the residue then resuspended in the standard salt medium and a filament preparation made either in the normal way, or simply by homogenization in the absence of a relaxing medium, then large numbers of the thin filaments, showing their characteristic structure, are released, but no thick filaments are found.

(3) The length of the thick filaments, and of the I-segments, is the same whether they are prepared from muscles at rest length, or stretched by up to 25%; and their length is still constant when the preparation has stood for a length of time sufficient to lower the ATP concentration, so that fibrils to which this preparation is added will undergo strong contraction. The filaments and I-segments also remain unchanged when the grids bearing them are, before negative staining, rinsed with solutions containing 10^{-3} M-ATP, 10^{-1} M-KCl, 10^{-3} M-MgCl$_2$ (which cause fibrils to contract violently), even when the filaments are lying in holes in the carbon film and are therefore unrestrained.

(4) In many instances thick filaments are to be seen with thin filaments lying alongside them (Plate IV(b) and (c)). The thick and thin filaments do not lie with their surfaces in close contact, however; they are usually 100 to 200 Å apart and the projections on the thick filaments extend across this gap and touch the thin filaments. Again, this corresponds closely to the structure of muscle structure derived earlier by other techniques.

To proceed further with our observations and deductions, we must now turn to the examination of purified muscle proteins, the structures which can be formed from them *in vitro*, and the comparison of these with the naturally occurring thick filaments.

4. Examination of Preparations of Purified Myosin

(i) *Myosin aggregates*

If a solution of purified myosin in 0.6 M-KCl is examined by the normal negative-staining technique, no particles are usually visible. The reason for this remains obscure, and we shall not be concerned with it further at this point; the molecules can be rendered visible by use of the shadow-casting technique, as has been demonstrated by Rice (1961a), and we shall illustrate and discuss their appearance later.

However, if the ionic strength is lowered to 0·2 or 0·1 μ, then rod-shaped particles are readily visible in negatively stained preparations. These particles may be up to 2 μ in length and 150 Å in diameter, and clearly represent aggregates of myosin molecules, for a single myosin molecule has dimensions of the order of 1500 Å × 20 to 40 Å. More commonly the aggregates are 0·5 to 1·0 μ in length and 60 to 100 Å in diameter, and typical preparations are shown in Plates VI and VII. Short aggregates are formed by rapid dilution of the stock myosin solution in 0·6 M-KCl with standard salt solution, longer aggregates by dialysis of the stock myosin solution against standard salt solution. The preparation shown in Plate VI gave a rapidly spreading peak in the ultracentrifuge with a sedimentation coefficient of the general order of magnitude of 50 s. The aggregates are characteristically somewhat spindle-shaped and, like the naturally occurring thick filaments, have very rough surfaces due to the large number of projections on them (Plate VIII). Similar aggregates of myosin to these were observed by Jakus & Hall (1947) using the shadow-casting technique. If 10^{-3} or 10^{-4} M-ATP is added to such a suspension of synthetic filaments in 0·1 or 0·2 M-KCl, no change is observed in their appearance as seen by negative staining. If the myosin is precipitated at lower ionic strength, say at 0·05 M—as is done during the usual myosin preparation—the denser aggregates then formed consist of large clumps of the same kind of needle-shaped filaments as are found at higher ionic strength, rather than of larger, thicker filaments.

Closer examination of these synthetic filaments of myosin reveals some rather remarkable and significant features. The shortest filaments seen are about 0·25 to 0·3 μ in length, and in many cases consist of a relatively bare central shaft, about 0·15 to 0·2 μ in length, with irregular projections at either end, as shown in Plate IX (top). Larger filaments, say 0·5 μ in length, again frequently have a bare central shaft of about the same length as before, but now there is a longer section on either side of the centre where the irregular projections are visible all the way out to either end of the particle (see Plates IX and X). The character of these projections is somewhat variable. Often there are many fewer of them, but they are much larger, than on the naturally occurring thick filaments; it may be that they arise from clumps of myosin molecules nearly in register rather than a succession of exactly spaced ones as is believed to be the case in vivo. The same characteristic appearance may be found in even larger filaments still, even in ones as long as 1·5 μ (see Plate X). Throughout the whole range of lengths the bare shaft has always about the same length, i.e. 0·15 to 0·2 μ, and is situated near to the centre of the filament, never at one end.

That a structure differentiated in this way along its length can form spontaneously by the aggregation of an assumedly pure protein seems at first remarkable. But it can be accounted for in a very simple way, which it will be convenient to describe here rather than in the Discussion; the significance of some of the other results will then be easier to see.

The myosin molecule is generally supposed to be of the order of 1500 Å in length. It can be cleaved by brief tryptic digestion into two well-defined types of fragment, light meromyosin (LMM) and heavy meromyosin (HMM); and the actin-combining ability and ATPase activity reside in the HMM moiety (Szent-Györgyi, 1953). It appears probable that each myosin molecule contains only one subunit having the properties of HMM, at least if we accept the straightforward view that the purified HMM molecules do not consist of dimers or higher polymers of units which were originally separated in the intact myosin. In the intact muscle, cross-bridges can be seen in the electron microscope extending from the thick (myosin) filaments and attaching to the thin (actin) filaments. It was suggested (Huxley, H. E., 1957, 1960) that these corresponded to the HMM part of myosin (or a part of it) and that the LMM part was contained in the backbone of the thick filaments.

If the projections seen on the synthetic myosin filaments are formed of HMM too, then the appearance of the shortest aggregates seen can be explained in two possible ways. Either:

(1) the myosin molecules could have an HMM unit at both ends (to form the projections), could be somewhat longer than current estimates, and could be laid down side by side with no restrictions on their polarity; or

(2) the myosin molecules could have an HMM at one end only, could be 1500 Å long, and could form a short filament with the molecules at one end all pointing in one direction (with the HMM end outwards) with those at the other end pointing in the reverse direction, and with the anti-parallel molecules partially overlapped in the centre. In this way the projections due to the HMM could be separated by a bare central shaft whose length could be up to twice the length of the straight part (as opposed to the "HMM projection") of the myosin molecule. All the LMM would be in the core of the filaments. This arrangement is illustrated in Fig. 1(a).

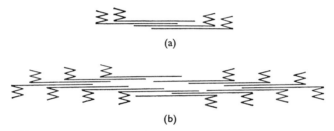

(a)

(b)

Fig. 1 (a). Possible arrangement of myosin molecules, with globular region at one end only, to produce short filaments of type observed, with globular region at either end and straight shaft in centre. The polarity of the myosin molecules is simply reversed on either side of the centre. (b). Possible arrangement of same myosin molecules to produce longer filaments in which the straight shaft in the centre is still present but in which a longer region on either side now has globular projections on it. The polarity of the myosin molecules is reversed on either side of the centre, but all the molecules on the same side have the same polarity.

The first of these possibilities conflicts with the model of myosin having only one HMM subunit per molecule and seems to be ruled out completely by the appearance of the somewhat longer filaments (i.e. those with lengths 0·5 to 0·7 μ), which still show the bare central shaft. If there were projections at either end of the myosin molecule then the only way in which a central bare shaft could be preserved as more and more myosin molecules were added to the original group of molecules which formed the shortest filament would be to displace these additional molecules along the filament, relative to the original group, by a distance only slightly less than their length, i.e. to place them so that the projections at their ends fell just at the edges of the original bare shaft and so did not fill it up. However, if this were done, then filaments whose over-all length was around 0·6 to 0·75 μ (i.e. three times the length of the shortest filaments) should show a second bare shaft on either side of the central bare shaft, separated from it by a cluster of projections, with another cluster of projections at the ends of the filament. These other bare shafts could not be filled up with projections until the length of the filament reached about 1 μ. Such effects are not observed. Outside the central zone, projections are observed all the way along to the ends of the filaments, whatever the length of the filament may be. Thus this type of electron microscope observation provides independent evidence that a myosin molecule does not have an HMM subunit projecting out at either end of it.

Let us now consider the alternative way of accounting for the visible structure of the filaments, i.e. to suppose that the myosin molecules, with a projection at one end only, are assembled together, each one staggered longitudinally somewhat relative to its neighbours, but all pointing in the same direction on one side of the central zone of the filament, and having their polarities all reversed on the other side. The polarity is such that the projections are at the ends of the molecules nearer the ends of the filament, and the straight part of the molecules point towards the centre of the filament. The molecules in the central zone which overlap in antiparallel are placed so that the projections at their ends are 0·15 to 0·20 μ apart (i.e. if the straight part of the myosin molecule is, say, 1200 Å long, then the overlap would be 400 to 900 Å). Filaments of any desired length can then be constructed simply by adding more and more myosin molecules by the same rule. This model, illustrated in Fig. 1(b), seems entirely satisfactory.

It will also be noticed that it accounts for the tapered appearance of both the synthetic and natural filaments, for there must always be progressively fewer and fewer molecules in parallel over the last length l_{ms} of the filament, where l_{ms} is the length of the non-projecting part of the myosin molecule. Furthermore, as the cross-bridges can be seen in both cases (but especially clearly in the case of the naturally occurring thick filaments (see Plate III)) to occur right out to the tips of the filaments, it requires that the projection on the myosin molecule be very near to the end of the molecule.

There are of course important functional implications of this particular model. These will be dealt with in the Discussion.

It became clear during this work that it was important to try to confirm the apparently rather compelling conclusion that a myosin molecule has a projection at one end of it only, in case some artifact, or some impurity, was being overlooked. Attempts to do this by examination of single myosin molecules by the negative-staining technique were largely unsuccessful. In some experiments, apparently because of unknown and unreproducible combinations of factors in the preparation of the carbon-coated supporting grid and of the myosin itself, filamentous structures having the appropriate dimensions to be myosin molecules could be seen in certain areas; but the evidence of degradation was so obvious as to make such occasional observations of limited value. A better technique is required.

(ii) *Examination of individual myosin molecules by shadow casting*

We have already mentioned that myosin molecules—or, more accurately, structures which may be single, intact myosin molecules—can be seen using the shadow-casting technique. Although this technique gives substantially lower resolution than negative staining when used to examine the *internal* structure of particles and filaments and is a good deal more difficult and complicated to use, it does have the advantage that particles of approximately the expected dimensions can be seen, quite consistently, in preparations of isolated myosin molecules, whereas none can be seen by normal negative-staining methods; and I am indebted to Dr. Rice for kindly showing me details of how it is carried out. Rice (1961a,b; 1963) has observed a number of characteristic structures in such preparations, consisting of rod-shaped molecules of somewhat variable length with one or more globular regions attached, and has lately favoured the view that the myosin molecule, about 1000 to 1500 Å long, has only one globular region on it. He has supported this view with the observations that HMM molecules, visualized by the same technique, often appear as globules of the same size as those on the intact myosin, with short tails (up to 800 Å) attached, whereas LMM molecules appear principally as simple rods up to 900 Å in length. It therefore seemed worth while to find out how consistently such structures could be found by a different observer and, if practicable, to photograph a sufficient number of them to do a statistical study at least of the lengths of the different particles seen.

The patterns seen are at first disappointing, and it is clear that most of the material is not present as distinct, separate and identical molecules. Large aggregates of material are deposited at the centre of the droplets as they dry up. However, there is a region around many of the droplets in which a considerable number of particles of similar appearance can be seen. Presumably these represent protein salted out from the droplet as the salt concentration rises during drying (they do not appear to be deposited during the initial impact of the droplets, for when such a heavy spraying is used that the droplets merge and then separate into a different droplet pattern as the preparation dries, the particles are found deposited mainly around the final droplets). Many of these particles, as others have described, consist of a rod-shaped structure with a globular region at one end (Plates XI and XII). In many other cases, however, particles lacking the globular region are present in considerable numbers. Under such conditions, a bias on the part of the investigator as to which structures most closely resemble the normal ones can influence the collection of data. In the present study fields were photographed in which at least a few of the particles with a globular end were present (it was very easy to find such areas) but when the photographs were analysed, *all* the particles in the field photographed (perhaps 10 times as many as were used to select the field) were included in the measurements. Thus there was some bias, but a limited one.

Altogether, 43 fields containing about 1700 particles were analysed. The predominating species of particle was rod-shaped with a globular region at one end. For every 100 such particles, there were about 20 particles with no globular region, and about 20 either with more than one globular region, or with a globular region not at the end. This seems to be quite strong confirmation that a rod with a single globular region at one end is the most likely approximation to the intact myosin molecule. Only about one in five of the rod-shaped particles without globules was longer than the *average* length of the rods with globules, so it seems unlikely that the latter particles are formed from the former by the folding up of one end; it is much more probable that the simple rods are formed by breakage of the longer compound particles during preparation for electron microscopy.

The lengths of the 1200 or so "typical" particles are plotted in the histogram shown in Fig. 2. The lengths of the particles showed a wide scatter. The average length was 1520 Å and 50% of the particles had lengths within $\pm$ 260 Å of this value (i.e. within the range 1260 to 1780 Å). Particles in the better-looking fields (i.e. those with least debris and aggregation) often tended to be slightly longer than the over-all average.

Thus the average length of the particles shown in Plate XII, which were obtained from the best fields, is 1680 Å. No doubt many of the shorter particles were produced by breakage during the drying of the sprayed droplets, and the longer particles may equally well represent aggregates of broken particles and intact ones. Alternatively, they may represent particles which have been stretched by surface-tension forces during drying. They may also represent intact particles at their true length, for these longer particles are not anomalous in any other respects and all the shorter particles,

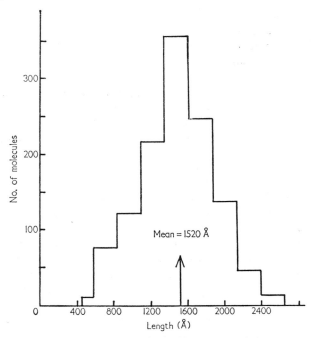

FIG. 2. Histogram showing the distribution in lengths of about 1200 myosin molecules. Only molecules consisting of a straight rod with a globular region at one end only are included in the distribution.

including those near the average length, may be breakdown products. In these circumstances, it is not possible to say what is the relation between the initial length of the particles—assuming their initial lengths were all identical—and the average length measured here. But as there appears to be an excess of rods without globules left over at the end of these events, the average length of the observed rods with globules must represent an underestimate of the initial length of these structures; thus it is probably safe to say that the length of the molecule is very likely to lie between 1400 and 2000 Å.

The diameter of the rod-shaped region was about 15 to 20 Å, that of the globular region about 40 Å. These are of course the dimensions of the *dried* particle. The length of the globular region was about 150 to 250 Å.

The general conclusion from these observations agrees with that described by Rice (1961*a*,*b*; 1963) and lately by Zobel & Carlson (1963), namely that myosin, as seen in the electron microscope, is a rod-shaped molecule of the general order of 1500 Å long, with a globular region at one end. This is in good accord with hydrodynamic

LEGENDS TO PLATES

PLATE I. Typical preparation of separated filaments from glycerinated rabbit psoas muscle, negatively stained with uranyl acetate without previous fixation. A number of intact thick filaments, about 1·5 to 1·6 μ in length, may be seen, together with some broken ones. The surfaces of the filaments seem to be covered with small projections and the ends of the filaments are tapered. Also visible are separated thin filaments, some thin filaments still attached together in large (bottom) or small (upper left) bundles, and a few vesicles. Magnification: 45,000 ×.

PLATE II (a) to (e). Thick filaments from muscle, showing projections all the way along their length except for a short central region about 0·2 μ long, which is comparatively bare but has a slight thickening in the middle. These specimens were fixed in formalin before negative staining, since it was found that the projections were better preserved under these conditions. (f). Synthetic "thick filament" made from purified myosin. Note close similarity to appearance of natural thick filaments, including the bare central shaft. Magnification: 105,000 ×.

PLATE III (a) to (c). Arrays of thick filaments as seen in the A-bands of fixed, embedded and sectioned tissue. The tapered appearance of the filaments can be seen (e.g. in (a)), together with cross-bridges all the way along the filaments, except for a central region, which can be seen well in (c), which is relatively bare but which has a zone of greater thickness half-way along it. The filaments are foreshortened by compression during sectioning. Magnification: 150,000 ×.

PLATE IV (a). Isolated I-segments, from muscle filament preparation, consisting of bundle of thin filaments held together at the Z-line and extending a distance of about 1 μ on either side of it. Magnification: 50,000 ×. (b), (c). Thick and thin filaments in homogenized muscle preparation still attached together by cross-bridges. Magnification: 150,000 ×.

PLATE V. Group of thin filaments, still held together in register at Z-line showing characteristic beaded structure along their length. Magnification: 150,000 ×.

PLATE VI. Preparation of purified myosin, at an ionic strength $\mu = 0.15$, showing spindle-shaped aggregates of the protein, predominantly around 0·5 μ in length. These were formed by rapid lowering of the ionic strength from $\mu = 0.6$. Magnification: 50,000 ×.

PLATE VII. Another preparation of purified myosin at ionic strength $\mu = 0.15$, showing the longer aggregates formed when the ionic strength is reduced slowly from $\mu = 0.6$ by dialysis. Filaments up to 1·5 μ and longer may be found in such preparations. Magnification: 50,000 ×.

PLATE VIII. General view of typical field showing a number of filaments prepared by lowering the ionic strength of purified myosin solution to $\mu = 0.15$. The filaments show the characteristic bare central zone, and the irregular projections at either end. Magnification: 145,000 ×.

PLATE IX. Synthetic myosin filaments of different lengths ranging from 2500 to 4500 Å, showing characteristic bare central shaft of approximately constant length (1500 to 2000 Å), together with irregular projections along the rest of the filament. Magnification: 145,000 ×.

PLATE X. Synthetic myosin filaments with lengths ranging from 5000 to 9000 Å, together with one of length greater than 14,000 Å (1·4 μ). All of them show the same characteristic pattern of a bare central shaft and projections all the way along the rest of the length of the filament. The appearance of the projections is rather variable, perhaps due to clumping together in some cases. Magnification: 145,000 ×.

PLATE XI. A number of fields showing appearance of individual myosin molecules examined by the shadow-casting technique. A high proportion of the rod-shaped particles seen have a characteristic thickened region at one end. Particles are varied in length, often around 1500 to 1700 Å, and sometimes stick together, but the separated ones have a very characteristic appearance, being rod-shaped with a thickened region at one end. The diameter of the long shaft of the molecule measures 10 to 20 Å, that of the head about 40 Å. TMV rods are for location and calibration purposes. Magnification: 38,600 ×.

PLATE XII. A number of selected myosin molecules from shadowed preparations, demonstrating the existence of the "head and tail" structure in a large number of instances. The average length of the particles in this particular group is 1670 Å. Magnification: 100,000 ×.

PLATE XIII. Preparations of light meromyosin (LMM) as seen by the shadow-casting technique. The molecules generally appear as simple rods, without globular regions, often 600 to 700 Å in length. Magnification: 30,880 ×.

PLATE XIV. Preparations of heavy meromyosin (HMM) as seen by shadowing technique. The particles very frequently resemble those visible in myosin preparations, in being rod-shaped with a globular region at one end, but are much shorter. The actual lengths are very variable, ranging from 100 to 200 Å (when only the thickened region, or the "head", is seen) to about 800 Å. Possibly the HMM is particularly sensitive to breakage during the drying of these sprayed preparations. Magnification: 38,600 × .

PLATE XV (a). Light meromyosin "crystal", negatively stained (with uranyl acetate). Wide banded structures like this are readily formed when the ionic strength of LMM solutions is reduced slowly (by dialysis) from $\mu = 0.6$ to $\mu = 0.15$. There are indications of longitudinal filaments within the fibre; the axial periodicity is about 430 Å. Magnification: 155,000 × . (b). Light meromyosin filament formed when ionic strength is reduced rapidly to $\mu = 0.15$, by dilution. The filament appears to be composed of well-oriented, rod-shaped molecules, and the absence of an axial periodicity suggests that in this case they are not aligned in transverse register in any way. The very sharp outline of the filament, which is free of projections, indicates that the width of the molecules is very uniform, and that they are straight. Magnification: 145,000 × .

PLATE XVI. Normal appearance of LMM preparation when ionic strength reduced quickly to $\mu = 0.15$. The protein aggregates into needle-shaped structures, frequently up to 500 Å more in width, which may be many microns in length. The surfaces of these filaments are completely smooth. Occasionally, a lattice structure is observed (bottom left) (see also Plates XXVII and XXVIII). Magnification: 38,600 × .

PLATE XVII. Purified F-actin, prepared from acetone-dried muscle powder. The actin forms filaments of indefinite length but constant width 60 to 70 Å, with the characteristic beaded structure. Magnification: 155,000 × .

PLATE XVIII. Purified F-actin preparation, prepared by homogenizing muscle from which all myosin had been previously extracted. The field shown here is one in which the F-actin is embedded in a film of uranyl acetate over a hole in the carbon supporting film, a procedure which gives very good contrast and detail. In many of the filaments, the double-helical structure can be seen (e.g. bottom right indicated by arrows). Magnification: 155,000 × .

PLATE XIX. Filaments of F-actin, treated with solution of heavy meromyosin. A very characteristic compound structure is formed, showing a strong axial periodicity of approximately the same value (350 to 370 Å) as that seen in F-actin. The filaments also have a well-marked structural polarity ("arrowheads") which has the same sense all the way along any given filament. Magnification: 155,000 × .

PLATE XX (a), (b). Thin filaments from muscle, treated with solution of heavy meromyosin. The filaments show the characteristic compound structure (see Plate XXI), and in the upper picture (a) two filaments pointing in opposite directions may be seen. Magnification: 155,000 × . (c). Preparation of "natural" actomyosin, deposited from solution at an ionic strength of $\mu = 0.6$. The same type of "arrowhead" complex filament structures are seen as in the (actin + HMM) filaments. This preparation was made using a normal supporting film, whereas (a) and (b) are photographed over holes, and the greater degree of flattening, etc., in the present case produces a somewhat different appearance. All preparations negatively stained with uranyl acetate. Magnification: × 150,000 Å.

PLATE XXI. Another preparation of "natural" actomyosin at $\mu = 0.6$, fixed in formalin and negatively stained with sodium phosphotungstate, again showing characteristically polarized compound structure. Many of the compound filaments are about 1 μ long. They frequently line up in small parallel arrays in which all the filaments have the same polarity and in which the pattern seen in adjacent filaments is displaced by about half a period, so that an approximate lattice is formed. Magnification: 145,000 × .

PLATE XXII. Untreated I-segment, showing bundle of thin filaments extending out on either side of their Z-line attachment. This micrograph is a control for Plates XXV and XXVI. Magnification: 108,000 × .

PLATE XXIII. I-segment treated with solution of heavy meromyosin. The thin filaments now show the characteristic polarized structure, and all filaments on the same side of a Z-line point in the same direction. Those on the other side of the Z-line point in the opposite direction. Magnification: 108,000 × .

PLATE XXIV. Another small I-segment, treated with heavy meromyosin solution, again showing all filaments with same polarity on given side of Z-line region, and reverse polarity on other side. All I-segments examined show identical arrangement. Magnification: 108,000 × .

PLATE XXV. Lattice structure sometimes formed by light meromyosin in 0·15 M-KCl, negatively stained with uranyl acetate. Lattice spacing measures 550 to 600 Å. Magnification: 150,000×.

PLATE XXVI (a). Small light meromyosin lattice, with a filament of LMM growing out of it along one of the principal axes. (b), (c). Light meromyosin just beginning to aggregate, showing isolated "stars" (with diameters up to about 800 Å) and some of the more normal filamentous aggregates, together with indications of individual LMM molecules in background. Magnification: 150,000×.

PLATE XXVII. Lattice structure, obtained by homogenization of tropomyosin B crystals, negatively stained with uranyl acetate. Lattice spacing ∼ 200 Å. The "woven" appearance arises because alternate lattice points along either of the two principal lattice directions are displaced alternately slightly to either side of the mean position. Possibly the lattice points do not all originally lie in one plane, but in two, and the displacement seen is a consequence of the drying down of this structure. The lines joining the lattice points frequently appear double, consisting of two fine filaments measuring about 10 to 15 Å in diameter and lying about 25 Å apart. Magnification: 150,000×.

PLATE XXVIII (a). Tropomyosin B preparation, showing some of the typical lattices, with fibrous aggregates growing out of them along the principal lattice directions. Magnification: 150,000×. (b). Another example of tropomyosin B lattice. The lattice is always rhombic, but the angle by which it differs from an accurately square lattice is somewhat variable in the range 5° to 15°. Magnification: 150,000×. (c). Cross-section through a number of fibrils of rabbit psoas muscle in I-band and Z-line region; the denser fibril in the lower part of the picture has been sectioned through the Z-line, and the approximately square lattice in that region can be seen. On the left-hand side of this fibril one is seeing the lattice on one side of the Z-line, but at the top right of the fibril one is seeing the lattice on both sides simultaneously, and a lattice with $1/\sqrt{2}$ the spacing of the first one and rotated 45° to it is seen (see text). Magnification: 50,000×.

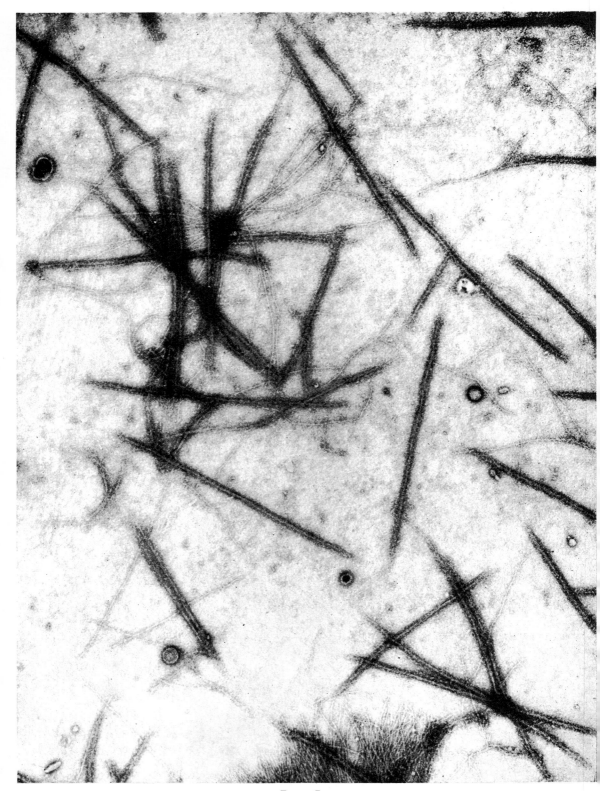

PLATE I

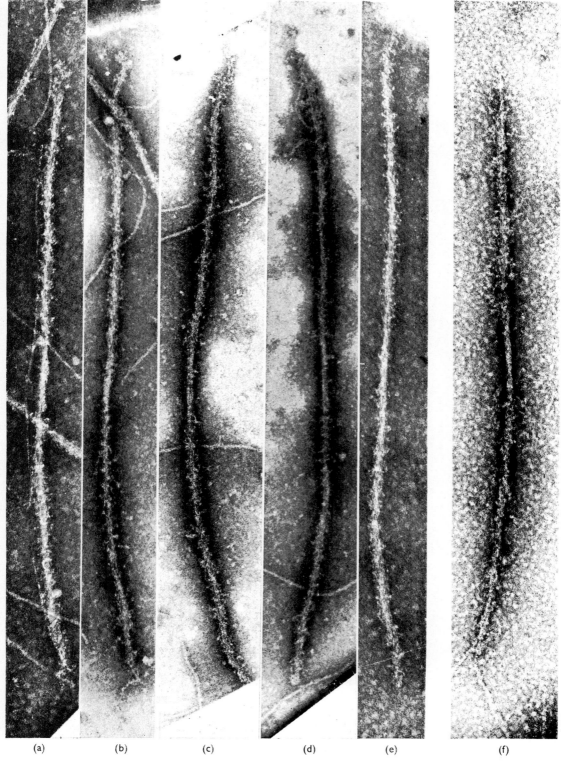

(a)   (b)   (c)   (d)   (e)   (f)

PLATE II

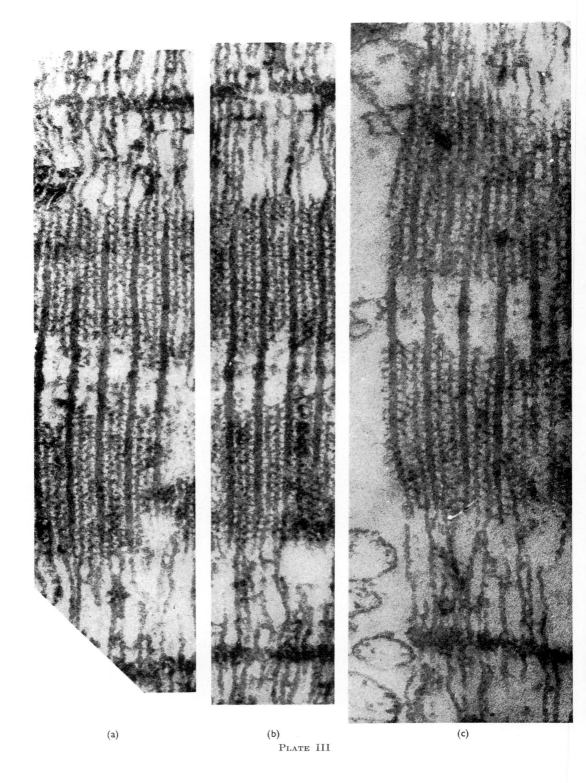

(a)   (b)   (c)

PLATE III

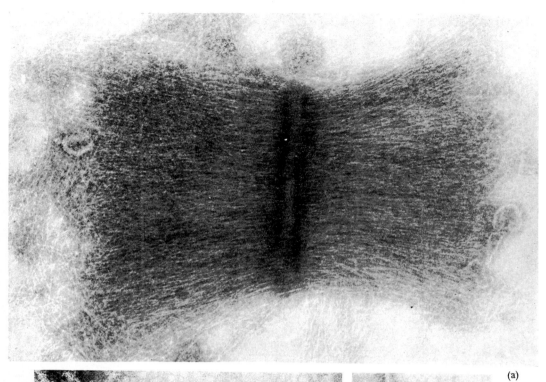

(a)

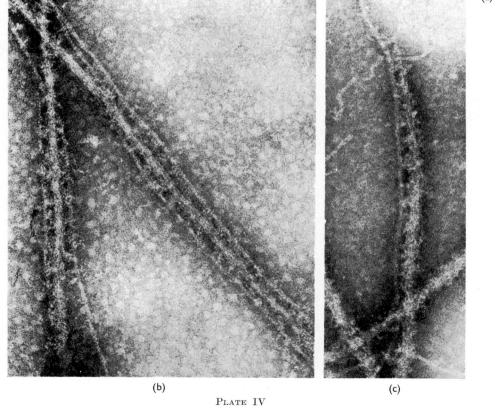

(b)

(c)

PLATE IV

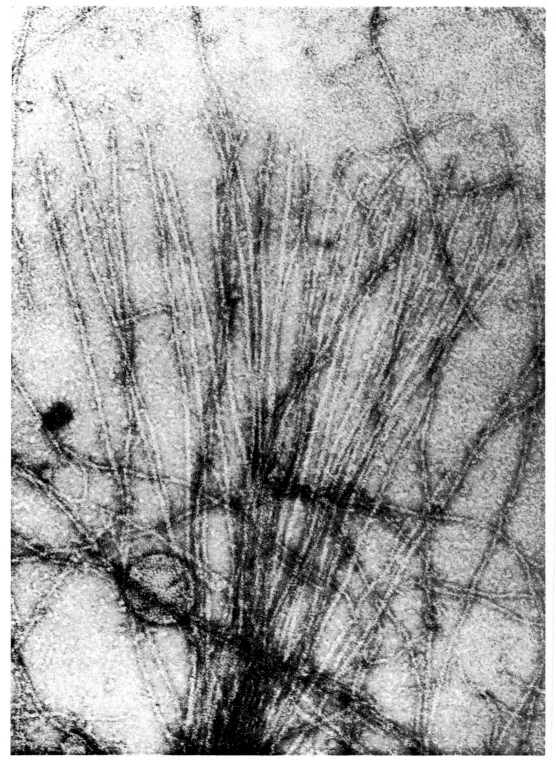

PLATE V

PLATE VI

PLATE VII

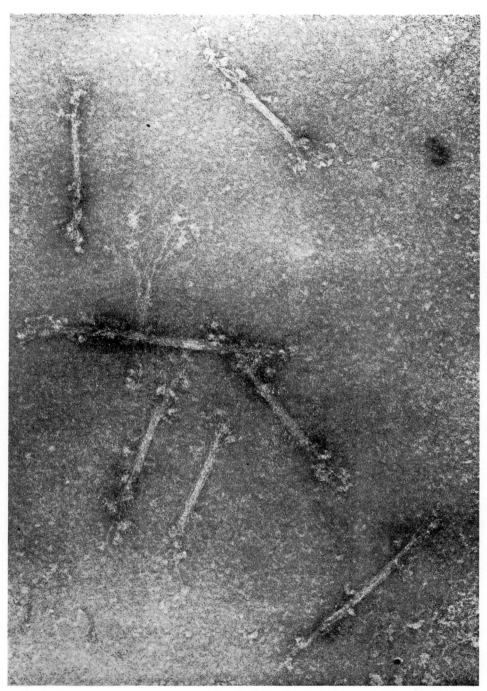

PLATE VIII

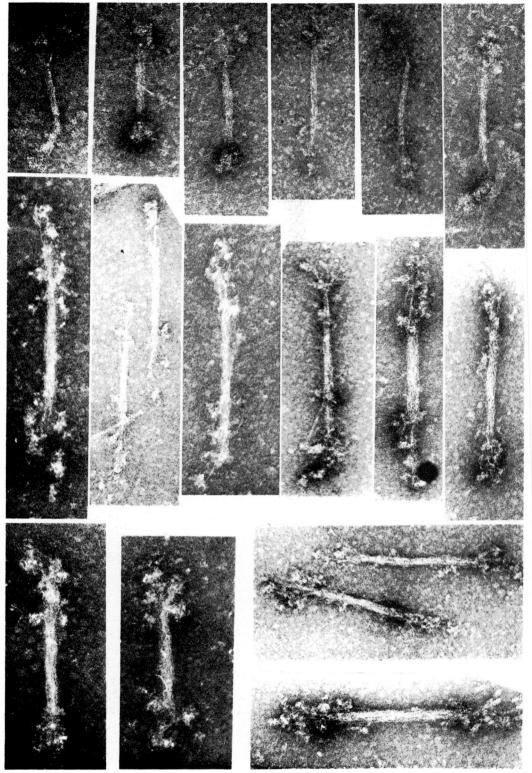

PLATE IX

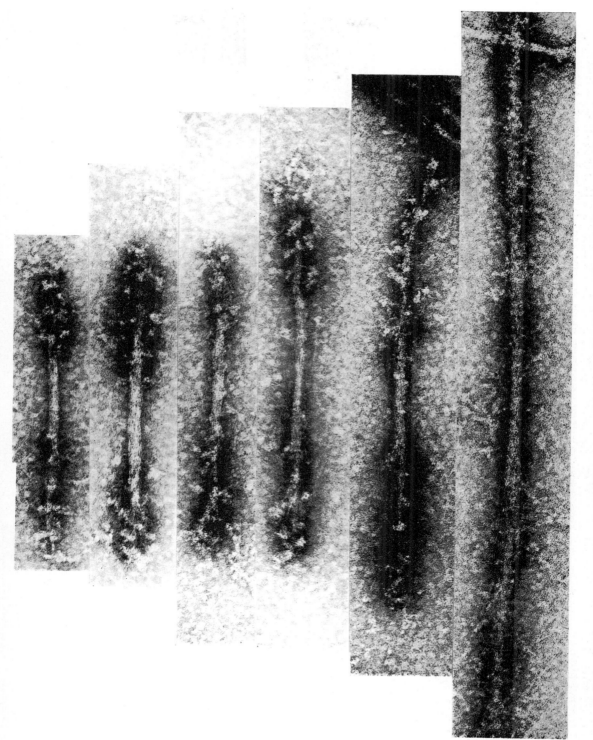

PLATE X

PLATE XI

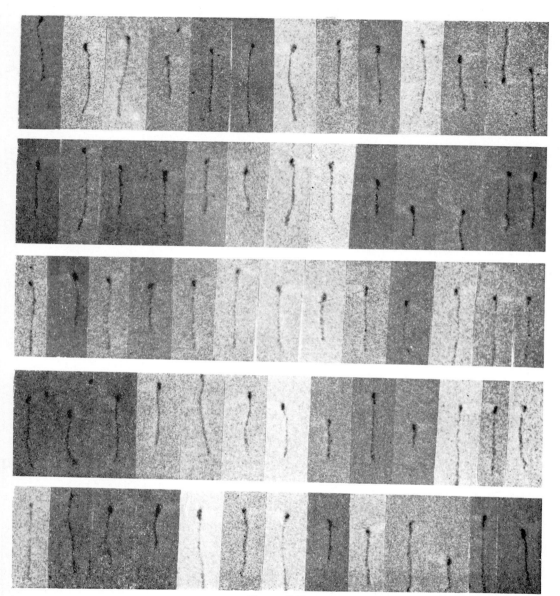

PLATE XII

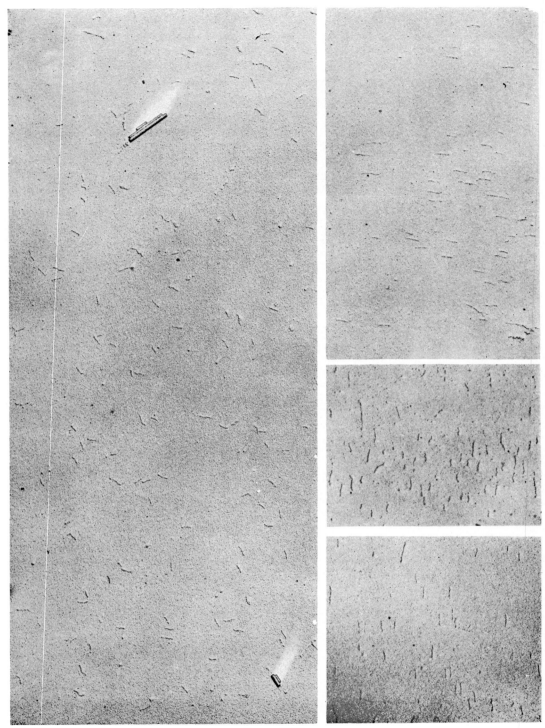

PLATE XIII

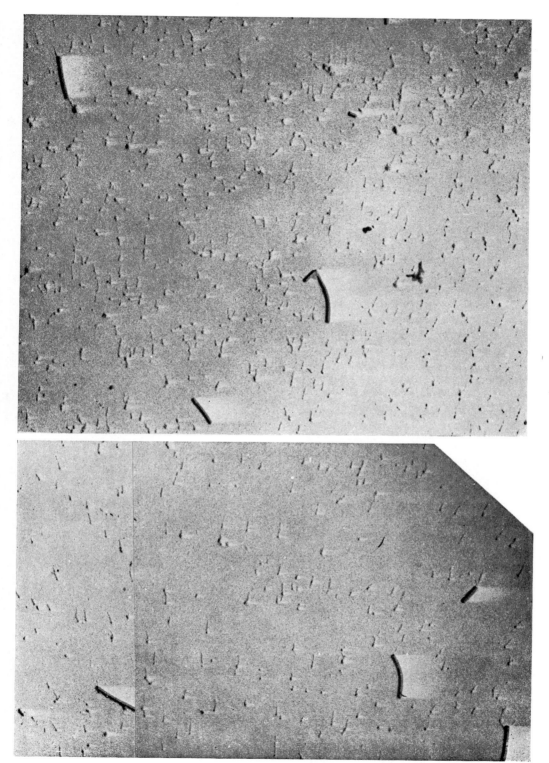

PLATE XIV

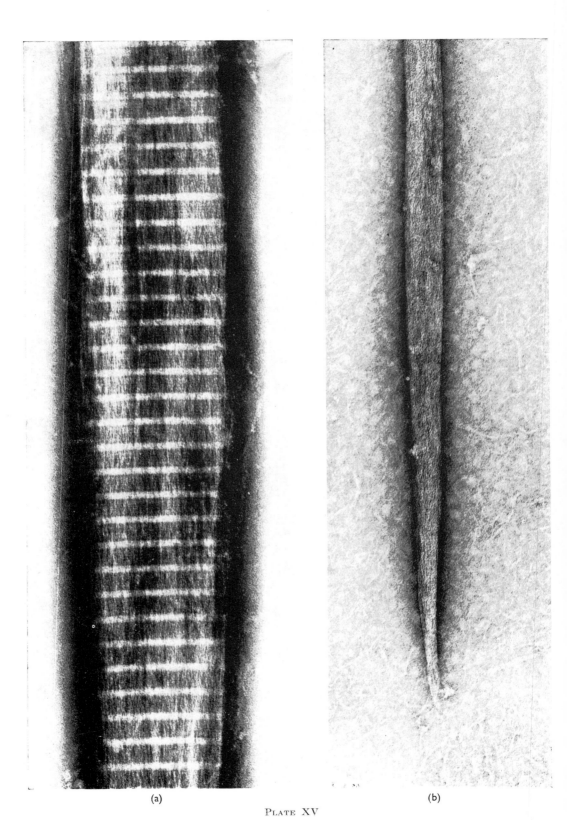

(a)   (b)

PLATE XV

PLATE XVI

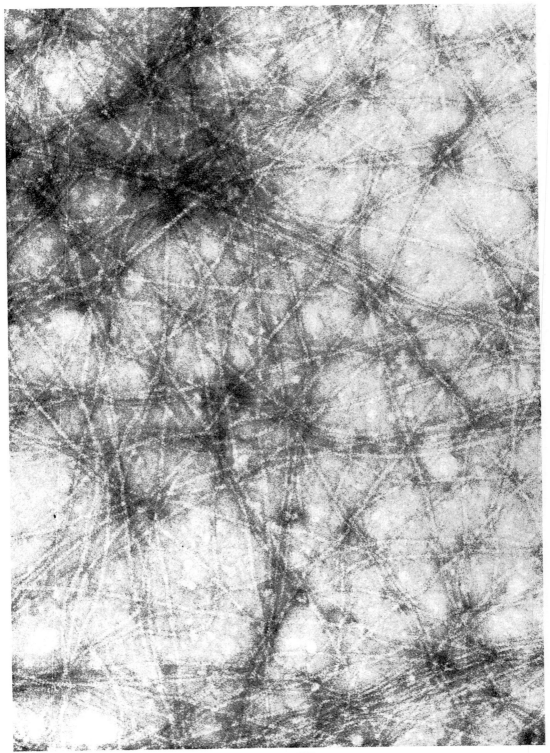

PLATE XVII

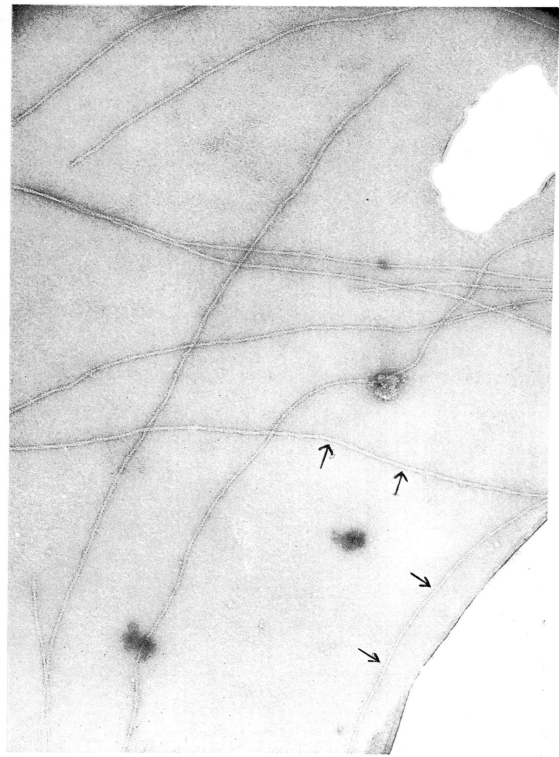

PLATE XVIII

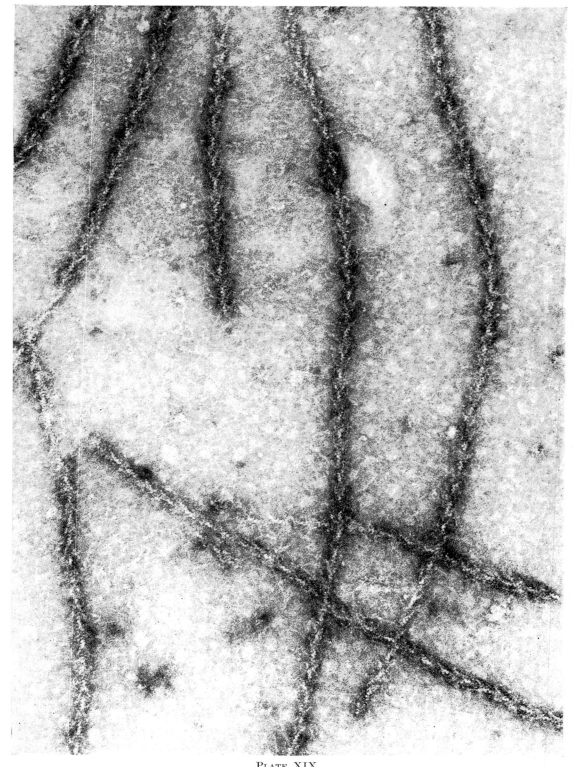

PLATE XIX

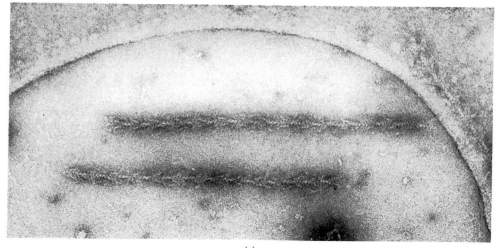

(a)

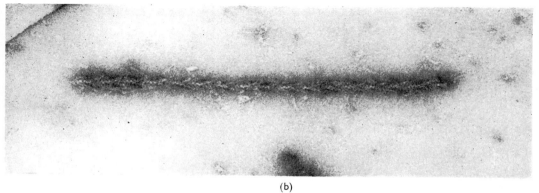

(b)

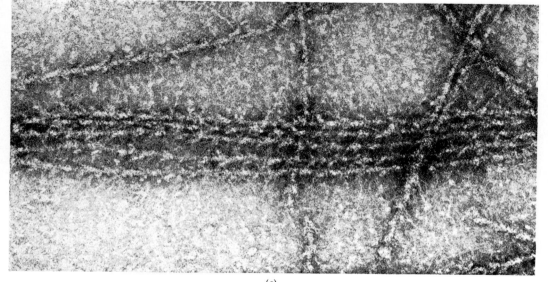

(c)

PLATE XX

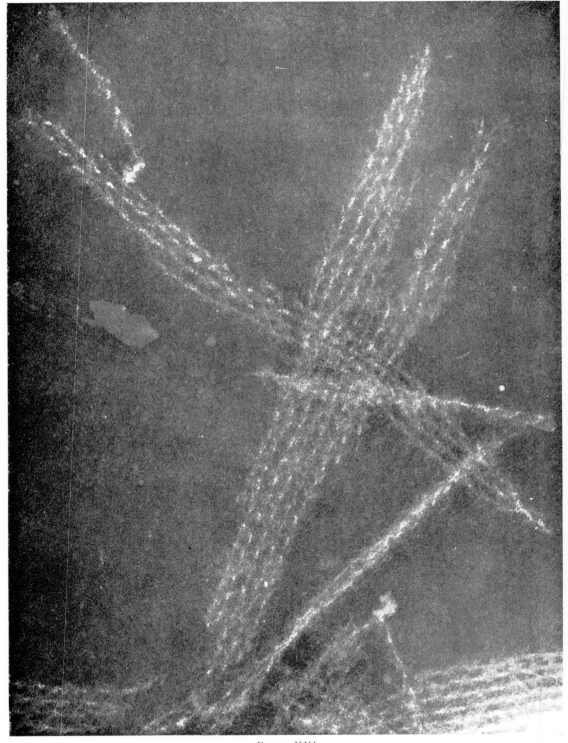

PLATE XXI

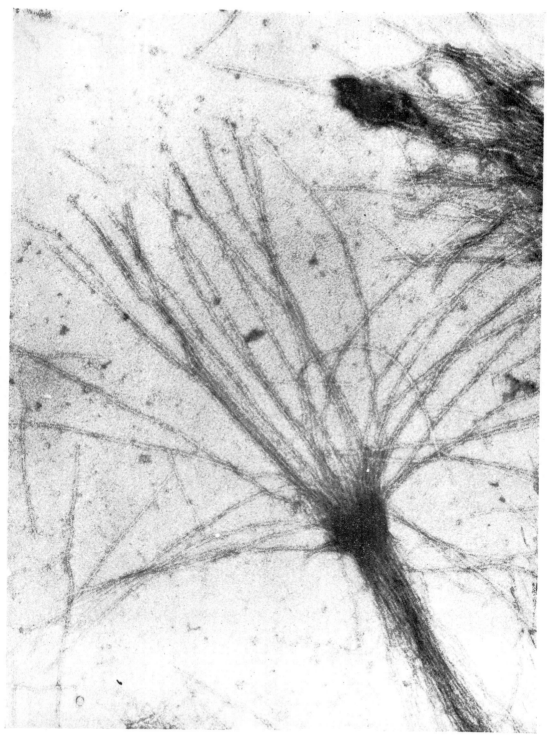

PLATE XXII

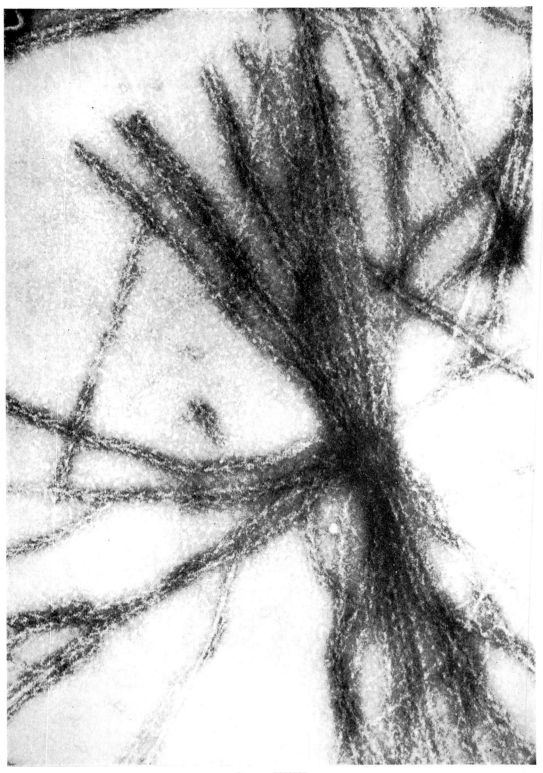

PLATE XXIII

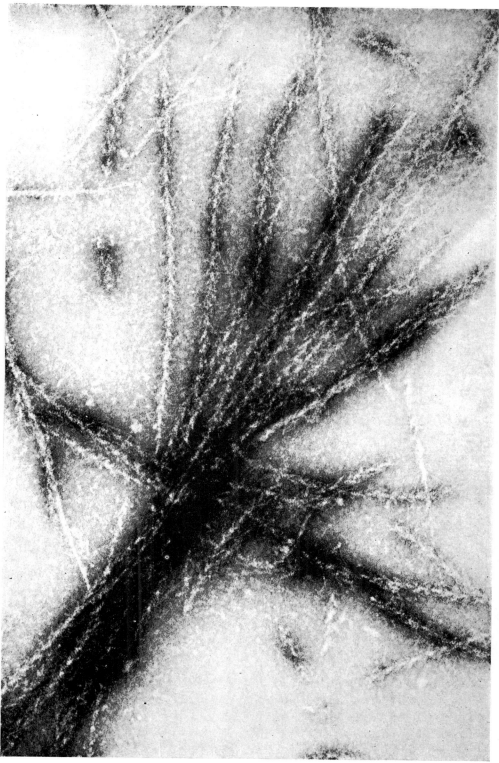

PLATE XXIV

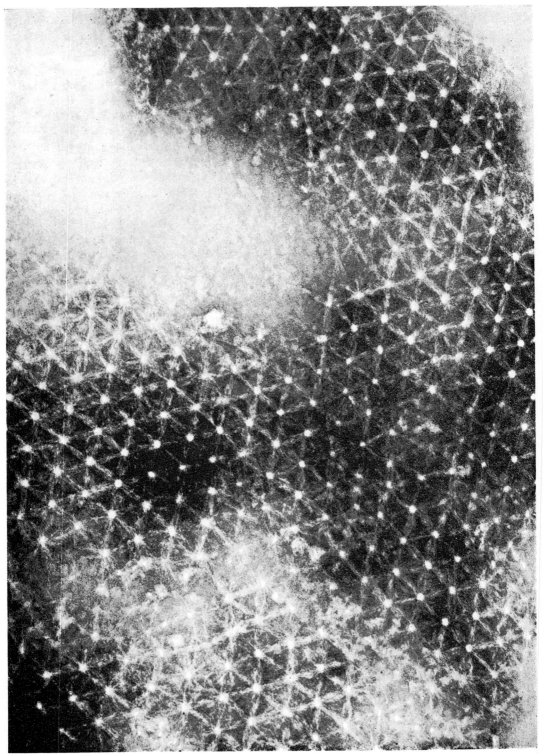

PLATE XXV

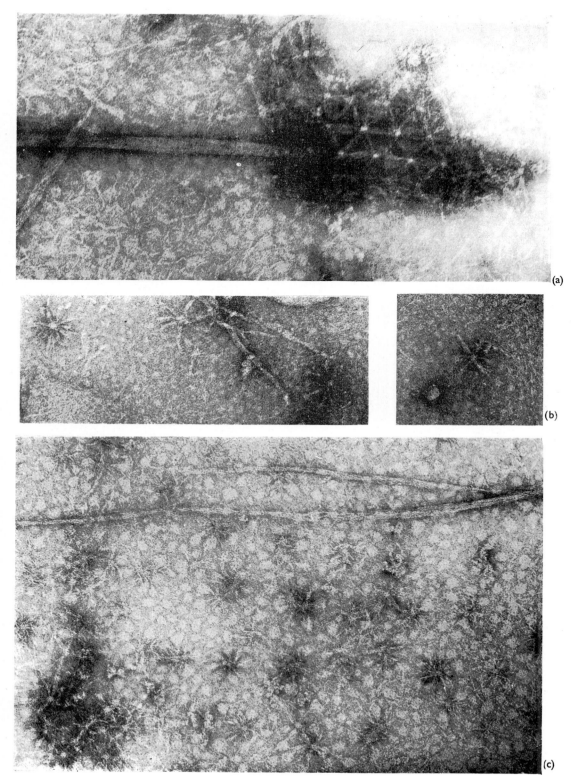

PLATE XXVI

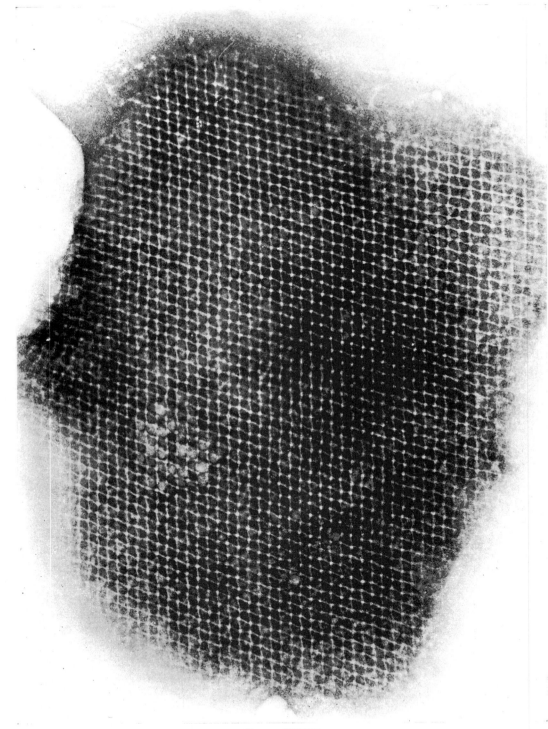

PLATE XXVII

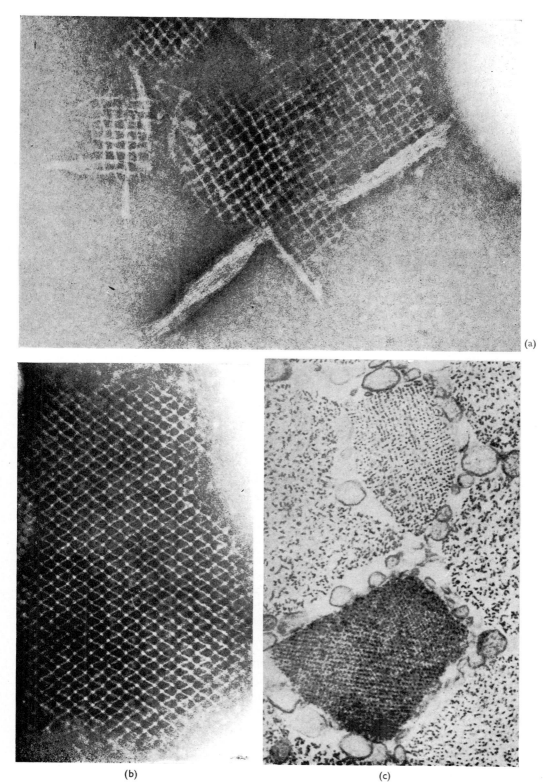

Plate XXVIII

studies (e.g. Lowey & Cohen, 1962), which we will mention later, and of course it fits in very well with the conclusions we have already reached on the basis of the negative-staining studies of myosin aggregates, if we equate the projections seen there with the globular regions seen here. This picture is further confirmed by observation on HMM and LMM.

Appearance of LMM and HMM when studied by the shadow-casting technique

When LMM in 0·6 M-ammonium acetate is sprayed onto mica and examined in the usual way, rod-shaped particles 15 to 20 Å in diameter are seen, as well as larger aggregates (see Plate XIII). These particles always lack the thicker globular region seen in the intact myosin molecules. Their length shows a very wide scatter indeed (possibly the structure has been weakened by the trypsin) and one is even more uncertain than in the case of myosin what significance to attach to the average value (for 1112 particles), which was found to be 610 Å ± 100 Å. We shall see later that other electron microscope evidence suggests a molecular length of about 530 Å, 960 Å, or 1390 Å, etc. (i.e. 430n Å + 100 Å).

The significant observation is the absence of the globular region seen in the intact myosin and the fact that the molecule is very much shorter. The fact that the diameter of the molecules is about the same as the measured diameter of the intact myosin *may* be significant, but it is possible in principle that the intact myosin could contain two LMM side by side so that the height of the particle, and hence the shadow-length—which is what we are measuring here—was the same.

When HMM is examined in the same way, globular particles, often with a short rod-shaped tail attached to them (Plate XIV), are seen. The globular regions are identical in size with those seen in the intact myosin. The over-all length of the particles is again extremely variable, too much so to allow average values to have very much significance, but the length often lies in the general range 600 to 900 Å. The globular region most commonly occurs at the end of the rod, not near the middle of it. These observations on LMM and HMM, made on a large number of preparations, are in general agreement with those of Rice (1963).

The observations naturally suggest, as Rice (1963) has pointed out, a very simple model for the myosin molecule, consisting of HMM and LMM joined end to end, LMM being a linear molecule, and HMM being linear for part of its length but with a globular region at one end, which also lies at the end of the intact myosin molecule. This model is in excellent accord with the model which was deduced from the observation on aggregates of myosin, and it also fits in reasonably well with the hydrodynamic data. However, the amount of degradation and aggregation that has clearly occurred in the preparations is disturbing, and had the results turned out to be in conflict with other types of observations, one would have hesitated to place complete reliance on them.

Observations on light meromyosin (LMM) aggregates by negative staining

As with myosin, LMM, when dissolved in 0·6 M-KCl and so present as single molecules, gives very disappointing results when attempts are made to visualize it by the negative-staining technique. Sometimes tangled and degraded-looking filaments of 10 to 20 Å in diameter can be seen, but no useful conclusions can be drawn. However, when the ionic strength is reduced to 0·1 or 0·2 μ, then spindle-shaped aggregates of LMM are found, and these can grow to microscopic dimensions, as was shown by Szent-Györgyi (1953) and by Philpott & Szent-Györgyi (1954), who drew attention to the fine axial periodicity of 425 Å often seen in them. These aggregates show up very well when negatively stained, and a typical preparation is shown in Plate XV(a). They seem to be able to grow indefinitely in length and diameter, in contrast to the myosin filaments which rarely exceed 100 to 200 Å in diameter and 2 μ in length. Their surfaces are completely smooth—as can be seen in Plate XV(b)— and one forms the impression that the aggregates are formed from straight and uniformly rod-shaped molecules lacking in any projections which would interfere with their orderly stacking into arrays of any size. Very fine filaments 10 to 20 Å in diameter and 20 to 30 Å apart can often be discerned within the spindle-shaped aggregates. According to the concepts developed earlier, the rod-shaped part of the myosin molecule, which includes the LMM moiety, packs together neatly to form the bare central shaft of both the synthetic and the natural thick filaments, and also form the backbone of the rest of the thick filaments. The present findings are clearly in very good agreement with this picture.

When the aggregates are formed by rapid lowering of the ionic strength (by dilution), they rarely show any sign of an axial periodicity, and a typical such preparation is shown in Plate XVI. However, when the ionic strength is reduced more slowly (by dialysis), many of the filaments show one very pronounced axial period (of about 430 Å) by the negative-staining technique, as already shown in Plate XV(a). The main feature of this repeat is the light band, about 100 Å in width. The band shows up most strongly in filaments which are deeply immersed in negative stain, indicating that it does not arise from positive staining of the rest of the repeating pattern. This could correspond to a region of greater protein density than the average. It may reflect a difference in mass per unit length along the length of each molecule. Alternatively, it could arise from an assembly of molecules, which were relatively uniform along this length, but which were arrayed in register with successive sets of molecules overlapping by 100 Å. This model would imply a molecular length of $430n$ Å $+ 100$ Å where $n = 1, 2, 3 \ldots$ depending on whether there is approximately a whole, half, one-third ... molecular length displacement between successive sets of molecules.

Again, the staining pattern may arise because a certain region of the molecule has a specific charge distribution which tends to exclude the negative stain. In that case, it would be more difficult to draw conclusions about molecular length from the band pattern.

Light meromyosin can also aggregate in a very different manner, into two- or three-dimensional open lattices, which we will describe later; they are not immediately relevant here.

Attempts to examine HMM molecules by negative staining have been unsuccessful so far, and it appears that their structure is either destroyed or completely infiltrated by the negative stain.

Observations on natural and synthetic actin filaments·

When preparations of purified F-actin are examined by the negative-staining technique, they are found to consist of filaments all having the same diameter (60 to 70 Å) and of indefinite length (see Plate XVII). These filaments show the same characteristic beaded structure as the thin filaments in the homogenized muscle preparations and have the same diameter. This structure has been analysed in detail by Hanson & Lowy (1963) and seems to consist of two chains of units (probably G-actin monomers) wound around each other in a double helix. The pitch of the helical path pursued by each chain is approximately 700 Å and there are 13 monomers per turn, with a spacing of about 55 Å. The two helices are displaced relative to each other by half a turn, and the monomers in the two helices are displaced relative to each by half the separation of the monomer along either helix. Thus the resultant structure repeats after 700/2 Å = 350 Å. The clarity with which this double helix can be seen is somewhat variable. Some filaments which show it rather well are shown in Plate XVIII. The fact that the thin filaments from striated muscles show exactly the same structure confirms conclusively that they contain actin, in the form of a uniform filament of F-actin, about 1 micron or so in length. The appearance in the muscle homogenate of some thin filaments longer than 1 μ is probably due to the ease with which F-actin filaments can join end to end to form extremely long polymers. Some experiments were done to find out whether the filaments changed in length at all during the negative-staining procedure. First of all, the lengths of a number of I segments from rabbit psoas muscle, which were sufficiently thin for the ends of the filaments in them to be accurately defined, were measured. The average value for 35 such measurements was 1·06 μ ± 0·02 μ, for the distance from the Z-line to the end of the filaments, giving a length of 2·12 μ for the total length of the I-segment.

A rather accurate value for the length of the I-segments in intact, wet fibrils was found in the following way. When the sarcomere length of a fibril is such that a gap exists between the ends of the thin filaments in the centre of the A-band, then a narrow lighter band, the H-zone, is seen there. When the sarcomere length is such that the ends of the thin filaments overlap slightly in the centre of the A-band, then a darker line appears there. Experience showed (1) that in many fibrils the sarcomere length was apparently very uniform over 10 or 20 sarcomeres, which all exhibited identical band patterns and (2) that the point at which the light H-zone disappeared, and was replaced by a dark line, could be judged quite accurately with a little practice. Thus it was found that in fibrils where the average sarcomere length was 2·27 μ or more, the H-zone could be seen in all the sarcomeres. In ones with a sarcomere length between 2·26 and 2·23 μ, neither an H-zone nor a dark line could be seen. In fibrils with an average sarcomere length of 2·22 μ or less, a dark line could be seen in all sarcomeres. This then gives a length of about 2·24 μ for the length of the intact hydrated I-segment of rabbit psoas muscle, to be compared with the value 2·12 μ of segments after negative staining. The amount of shrinkage that can have taken place is thus about 5%. It may in fact be less, for the I-filaments measured in the electron microscope were never perfectly straight and it was not practicable to try to correct for slight kinks in them. Thus if the periodicity of the I-filaments is 350 Å measured in negatively stained preparations, that in the intact muscle, assuming that all that can have happened is that the filaments may have changed in length slightly, can at a maximum be only 5% higher, i.e. about 372 Å.

Actin filaments treated with HMM and with myosin

It was shown by Szent-Györgyi (1953) that HMM retains the ability of the parent myosin to form a complex with actin, and it was observed subsequently in the phase-contrast microscope (Hanson & Huxley, 1955) that the I-segments of muscle fibrils from which the myosin had been extracted would take up considerable amounts of myosin and of HMM from solution, undergoing a large increase in density all along their length in the process. This experiment provided another line of evidence that actin was located in the I-filaments and was uniformly distributed along their length. It therefore seemed worth while to examine these·actin–myosin or actin–HMM complexes by the negative-staining technique. Composite filaments of actin and HMM are readily formed by placing a drop of a solution of HMM in 0·1 M-KCl onto a carbon-filmed specimen grid previously treated with either an F-actin solution or a muscle-filament preparation, and rinsed but not dried, so that many filaments remain attached to the carbon film and still evidently able to combine with HMM. This method was less successful in the case of myosin, but composite filaments are readily obtained here by mixing the two proteins in solution, or from preparations of natural actomyosin.

The appearance of the actin–HMM filaments as seen by negative staining is shown in Plates XIX and XX. This remarkable composite structure, whose diameter (200 to 300 Å) is considerably greater than that of the actin filaments, has a very strongly defined long axial period. Measurements of a total of 870 such periods in 118 different filaments negatively stained with uranyl acetate gave an average value of 366 Å $\pm$ 15 Å, very close to the period of the double helix in the F-actin structure. Measurements in the electron microscope of the lengths of intact I-segments treated with HMM show that they are the same (within 2%) as those of untreated I-segments. Measurements in the light microscope, of the kind already described, of the lengths of the I-segments in extracted muscle fibrils treated with HMM show that here too the addition of the HMM alters the length of the I-filaments by less than 5%. The composite filaments also show a well-marked shorter axial period of 50 to 60 Å, similar to the periodicity of the monomers along the F-actin helix. The filaments have a very definite structural polarity, perhaps best seen in terms of the tendency of their rather complicated internal structure to give an appearance of arrowheads which all point in the same direction along the whole length of any given filament.

Exactly the same appearance is seen whether the complex is formed with purified actin or with the naturally occurring thin filaments from striated muscle or from insect indirect flight muscle, or from mytilus adductor muscle (a smooth muscle), filaments found by Hanson & Lowy (1963) to show the actin structure. No significant difference in period is observed in any instance, including experiments in which the complex was fixed in either formalin or osmium tetroxide before negative staining, or in which the filaments were suspended over holes in the carbon film so that they were not in contact with a carbon substrate.

If unfixed preparations of the actin–HMM complexes are treated with standard salt solution containing 10^{-3} M-neutralized ATP and then examined by the negative-staining technique, the normal actin structure is seen again; the complex has apparently been dissociated by the ATP, as we should expect.

The appearance of actin–myosin complexes when deposited onto the specimen support from 0·6 M-KCl (see Plates XX(c) and XXI) is rather similar to that of actin–HMM. The composite filaments show the same polarized "arrowhead" structure, and have approximately the same periodicities. Unlike the actin–HMM filaments, however, they tend to associate together in rafts, in which the side-to-side distance is about 220 Å, the filaments all point in the same direction, and the long axial periodicities of adjacent filaments are displaced relative to each other by about half a period. When such actomyosin complexes are derived from natural actomyosin, a considerable proportion of the filaments are around 1 μ in length, as though they were individual I-filaments to which the myosin had become attached. These filaments probably correspond to those seen by Hall, Jakus & Schmitt (1946) in positively stained preparations of (acto)myosin.

In the presence of 10^{-3} M-ATP (in 0·6 M-KCl), only ordinary actin filaments could be obtained from actin–myosin solutions.

At lower ionic strengths (e.g. in 0·15 M-KCl), the same type of actin–myosin complex was observed as in 0·6 M-KCl, provided, of course, that the complex was initially formed at high ionic strengths, where myosin is in the form of single molecules. If the myosin was in the form of aggregates at the lower ionic strength, before actin was added, then the compound filaments were not formed, but merely associations between the myosin and actin filaments. The existence of these two quite distinct forms of actomyosin should be borne in mind. When 10^{-3} M-ATP was added to the compound filament type of preparations (in very dilute solution), the actin–myosin complexes all apparently dissociated, and only actin filaments and some myosin aggregates could be found by the negative-staining technique.

The natural interpretation of these observations is that an HMM (or myosin) molecule has combined with each of the G-actin units in the F-actin structure, giving rise to a double helix of HMM molecules wound around the outside of the original actin filament. The looser packing of the molecules at the greater radius would enable the periodicities in the structure to be seen more easily.

The exact details of the structure of the molecules of HMM and the manner in which they are attached are not evident from the electron micrographs obtained so far. Isolated molecules of HMM are not visible by the negative-staining technique, which perhaps degrades them in some way. Thus the structure seen in the composite filaments may contain only the degraded remnants of the HMM molecules, remnants large enough, however, still to be visible. Filaments of actomyosin (deposited on the grid from 0·6 M-KCl) have a very similar appearance to those of actin–HMM, and, except in a few places, show little sign of 1500 Å long myosin molecules attached to the actin monomer; possibly a good deal of the myosin molecule has been degraded too.

There are two conclusions, however, which do emerge unambiguously from the general appearance of these composite filaments. One is a relatively trivial one, that one HMM molecule attaches per G-actin subunit; if this were not the case if, for instance, only one HMM molecule was present per two G-actin units and was shared between two G-actins on opposite chains, or on the same chain, then the long period would have twice the observed value. The second is more important, and follows from the structural polarity of the complex filaments. This polarity is not a consequence of a unidirectional growth of the complex. Actin filaments treated very briefly with low concentration of HMM will show the composite structures along short stretches of their length, separated by long stretches of normal actin. These stretches of composite structure are all polarized in the same direction. Thus the structural polarity must be imposed by the underlying actin structure, and so the actin filament itself must have a structural polarity. This means that the monomeric units in the F-actin structure in both chains must all be oriented in the same direction (of course, if every n^{th} monomer, where $n > 2$, along either or both chains was reversed, then a polarized structure would still be produced, but such a structure should then show other periodicities in electron micrographs, and these are in fact not observed).

If the actin filaments have a polarized structure, the question immediately arises as to how this is arranged *in situ* in the muscle. It is possible to use the same technique to answer this question. As already mentioned, preparations of separated filaments often contain bundles of thin filaments still attached to a Z-line. When such a preparation is treated with HMM, the filaments take up the HMM and form the complex structure as before, and it can be seen (Plates XXIII and XXIV) that all the filaments on one side of the Z-line are structurally polarized in the same direction, and all the ones on the other side in the reverse direction. The sense of these polarizations (the "arrowheads" pointing away from the Z-line) is the same for all I-segments examined (several hundred). As in the case of the myosin filaments, the functional implications of this will be considered in the Discussion. There are some further structural implications, however, which it will be convenient to mention before describing observations concerning the nature of the Z-line.

It is quite clear that the actin filaments do not pass uninterrupted and unchanged through the Z-line. This is apparent from many electron micrographs and the nature of the interconnections at the Z-line has recently been admirably described by Knappeis & Carlson (1962). The basic observation is that actin filaments on either side of the Z-line do not lie in the same straight line. They are all displaced relative to each other and interconnected by very thin diagonally oriented filaments. What the present observations show is that these connections cannot be, in whole or in part, simply a continuation of the actin filaments in, say, an untwisted form. Such a continuity is ruled out by the reversal of polarity on either side of the Z-line. The actin structure has to be reversed in some way within the width of the Z-line, and one might think that the best way of achieving this was to connect the actin filaments to some other structure which possessed a twofold rotation axis perpendicular to the long axis of the muscle.

We shall describe in the next sections some observations which bear on the possible nature of this substance. The observations made on tropomyosin B were stimulated by the appearance of certain structures in LMM preparations, and so we shall describe these first, though they may not have any direct relevance to the structure of muscle.

Lattice structures in LMM aggregates

The most usual type of aggregate formed when the ionic strength of a solution of LMM was reduced to below $0 \cdot 2 \, \mu$ is the spindle-shaped filament already described. However, occasional preparations—and the exact conditions have not been successfully defined so far—contained, in addition to the usual filaments, remarkable open lattice structures, as shown in Plate XXV. These hexagonal lattices seem to be formed by three groups of molecules or filaments, meeting or crossing each other at 120° angles at each lattice point. Within each group, a small amount of scatter in direction and position seems to be present, after negative staining anyway. Sometimes isolated groups or "stars" are seen (Plate XXVI) and in these the scatter is considerable. The separation of the lattice points is slightly variable between 550 and 600 Å, and the width of the structures which cross at the lattice points is 100 to 150 Å. This is of course much greater than the width of a single LMM molecule. The width of the isolated stars is 800 to 900 Å. Sometimes normal-type LMM filaments grow out of the lattices along one of the principal axes (see Plate XXVIII). Preparations of LMM which show both types of structure in their aggregates give only a single peak in the ultracentrifuge. Preparations of LMM Fraction 1 can also give both types of aggregate. It is not apparent at present whether there is any sort of a lattice structure present in the third dimension, or whether one is simply looking at a simple two-dimensional lattice or a number of such lattices in register and in contact.

These observations demonstrate a rather novel way in which a fibrous protein can aggregate, but they probably have no direct relevance to the way the LMM units of the myosin molecules are arranged in muscle. However, they did serve to remind the author of the open framework lattices seen by Hodge (1959) in positively stained and unstained crystals of tropomyosin B, and to suggest that these might with advantage be examined for them by the negative-staining technique.

Observation on aggregates of tropomyosin B

Tropomyosin B was precipitated by dialysis against ammonium sulphate as described by Bailey (1948) so as to give crystals of the protein. These were dispersed in a mechanical blendor into smaller fragments suitable for negative staining. When examined in this way in the electron microscope, the preparations were found to contain lattices of the type shown in Plates XXVII and XXVIII. The lattices are approximately square and the lattice spacing in all cases examined was about 200 Å. Two filaments cross at each lattice point, as opposed to three in LMM. The structure is clearly the same as that seen by Hodge in sectioned crystals, but some additional features can now be discerned. The structures which cross at the lattice points are very narrow and measure 20 to 30 Å across. In many places they can be seen to consist of two fine filaments about 25 Å apart, and measuring about 10 to 15 Å in diameter. It is tempting to suggest that these filaments represent single α-helices, or coiled-coils of two α-helices. In some instances, filamentous aggregates may be seen growing out of the lattices (Plate XXVIII(a)).

On close examination, some curious features of the lattice itself become apparent. First of all, no cases have been observed where the lattice is accurately square; the relevant angle is always 5 to 15° different from 90°. Secondly, the structure often presents a somewhat "woven" appearance. This occurs because successive lattice points along each of the two principal directions do not lie precisely on a straight line. They lie, or have a tendency to lie, alternately on either side of such a straight line. Thus the lattice points fall into two groups and define two new lattices. These "superlattices" are again of course approximately square, have a spacing $\sqrt{2}$ times that of the original lattice, are rotated relative to it by $\sim 45°$, and are displaced relative to each other as shown in Fig. 3.

The simplest way of accounting for this effect is to suppose that the two sets of lattice points lie in two different planes—or lay in two different planes before the preparation was dried in the negative stain—and that this has given rise to a fairly systematic displacement of the lattice points during the preparation, depending on which of the two planes the lattice point belongs to.

This structure seen in the fragmented tropomyosin crystals therefore bears a startling resemblance to the arrangement at the Z-line (Plate XXVIII(c)). The general organization of filaments there into an approximately square lattice was noted by the author some time ago, and the exact structure has recently been worked out in detail by Knappeis & Carlson (1962) and by Reedy (1962 in preparation). We will go into the matter in a little more detail in the Discussion, but mention at this point that the structure found by them is defined by two sets of lattice points in two planes, with the same spacings and arrangement of connecting filaments between them as seen in the present negatively stained preparations of tropomyosin. It seems too much of a coincidence for this not to suggest that some at least of the tropomyosin B is located in the Z-line filaments.

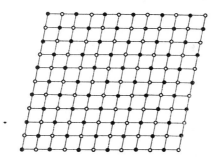

FIG. 3. Lattice structure seen in tropomyosin crystals and resembling closely that seen in cross-sections through the Z-line. The two types of lattice point in the crystal may lie in two separate planes, in each of which the lattice spacing is $\sqrt{2}$ times the spacing in the lattice formed by their superposition. The two lattices would correspond to the attachment points of the two sets of actin filaments on either side of the Z-line.

5. Discussion

(i) *General structure of muscle*

The finding that discrete filaments of well-defined length, equal to that of the A-bands, can be recovered from homogenized preparations of striated myofibrils provides a new kind of evidence in support of the view that an array of such filaments is present in the A-bands of striated muscle and is responsible for the characteristic high density and birefringence of that zone (Hanson & Huxley, 1953). Admittedly such discrete filaments have already been seen clearly in osmium tetroxide-fixed PTA-stained (positive-staining) sections of muscle (Huxley, H. E., 1957), but no harm is done by confirming such observations by a very different technique. The further finding that such homogenized preparations also contain structures, again of a clearly defined length, which are obviously sets of I-filaments still connected to a Z-line and ending, in register, about 1 μ on either side of it, confirms the other part of the picture of muscle structure put forward earlier.

The results described above are also in full agreement with the earlier conclusions concerning the location of actin and myosin in muscle, the general nature of the filaments they form, and the invariance in length of those filaments under different physiological conditions. These various issues have already been fairly well covered during the description of the results, and the points made will not be laboured further here. The results provide further evidence that the cross-bridges are to be associated with the HMM component of myosin, with the LMM component present in the backbone of the thick filaments, and they indicate that it may be worth while exploring further the possibility that part of the tropomyosin B in muscle is located in the Z-lines.

On the question of whether additional protein components besides actin and myosin are present in the thin and thick filaments, the evidence is still incomplete. Thus the remarks in the following two sections are necessarily of a somewhat speculative and indecisive nature. However, it may be worth while to set down such arguments as there are, before passing on to the more substantial issues concerning structural polarity. Let us consider first the thin filaments.

(ii) *Other possible components in thin filaments*

Hanson & Huxley (1955) showed that when muscle fibrils, from which myosin had already been extracted (so that they consisted simply of a succession of I-segments), were further extracted with 0·6 M-KI (known to depolymerize and extract actin), much of the material of the thin filaments was removed. However, a very tenuous backbone structure still remained when the extraction was observed, in phase contrast, taking place in fibrils attached to the coverslip or slide. This indicated that some other filamentous substance as well as actin might be present in the thin filaments, but the evidence could not be said to be completely conclusive because of the possible presence of denatured actin filaments in these glycerinated fibrils.

Later work (Huxley & Hanson, 1957) showed that the I-substance in such fibrils accounted for some 24% of the total muscle protein (36% of fibril protein) and the "Z-substance" (i.e. the material responsible for the additional density of the Z-line) about 4% (6% of fibril protein). Estimates by Hasselbach & Schneider (1951) had shown that the quantity of actin which could be extracted from whole muscle amounted to some 14% of the total protein. The difference between these two values could be ascribed to the presence of some material besides actin in the thin filaments; but it might also be at least partly due to incomplete extraction of actin.

It has been shown that actin and tropomyosin tend to be extracted together from muscle (Corsi & Perry, 1958) and also that they tend to remain associated together in some way once extracted (Martonosi, 1962). Again this suggests the possibility of actin and tropomyosin being located together in the thin filaments, but it makes quantitative arguments based on earlier estimates of actin content (which may have included an appreciable amount of tropomyosin as well) less precise than ever. Another type of argument is that advanced earlier by the present author (1960) where it is calculated that if the thin filaments were built out of the helical arrangement of nodes described by Selby & Bear (1956) and if a node represented one actin monomer, then the quantity of actin present should be either 16·5% or 20·25% of the total muscle protein, depending on whether a molecular weight of 57,000 or 70,000 was taken for actin. This figure represented the *maximum* quantity of actin which could be present, if the entire cross-section of the muscle were occupied by fibrils. There is thus a distinct discrepancy between this value and the value of 24% for the amount of I-substance deduced from interference microscope measurements and, as suggested earlier, this difference might be accounted for by the presence of tropomyosin.

In that case, tropomyosin would have to be present side by side with actin all the way along the thin filaments, for the interference microscope observations include measurements of density within the I-heads but outside the region of the Z-line (that is, they were not all measurements of density integrated over the whole I-region), and the density throughout the I-region is too high. The possibility that tropomyosin is also present in the form of a lattice in the Z-line has already been mentioned, and there is a certain attraction in the idea that just as lattices and filaments of both tropomyosin and of L-meromyosin may be observed growing out of each other *in vitro*, so filaments of tropomyosin extend out from either side of a double layer of tropomyosin lattice at a Z-line. It is perfectly possible for such a lattice to have the appropriate symmetry requirements mentioned earlier (2-fold rotation axis in plane of lattice) such that the projecting tropomyosin filaments will have opposite polarity. Along these filaments actin is laid down again with opposite polarity on either side of the Z-line. The dimensions required for such a lattice may be easily calculated from the known spacing of the hexagonal array in the A-bands of muscle. Taking a value of 440 Å for this, and remembering that there are two thin filaments for every thick filament in the array, then, if the Z-line lattice is 10° out from being a true square, the I-filament separation in the region on either side of the Z-line will be about 292 Å and the separation of the lattice points in the Z-line, seen face on, so that the two layers of the lattice are superimposed, will be $\sqrt{2} \times 292 \cong 206$ Å. This value agrees very well with the observed lattice spacing in the tropomyosin crystals.

It is shown elsewhere (Page & Huxley, in press) that the periodicity visible in the thin filaments in sectioned muscle, when corrected for shrinkage effects, lies very close to 406 Å, a value decisively different from the helical period found by Hanson & Lowy (1963) in filaments of actin (~ 350 Å), and that found here in the actin–HMM filaments (~ 366 Å). As we have already seen, this difference is not a consequence of any length change in the thin filaments when they are negatively stained, or when they combine with HMM and are negatively stained. It is very interesting and a little disappointing that no obvious changes in the actin structure take place when it combines with myosin or HMM; but such changes may lie below the resolution of the present techniques. It is of course by no means established that the pitch of the actin helix in intact muscle is of the order of 406 Å rather than 350 Å (although recent very accurate X-ray studies by Worthington (1959) do support a value of 411 Å), only that *some* structure in the I-bands, probably superimposed on the actin structure, has this period. This possibility is discussed by Hanson & Lowy (1963). All these considerations point to the desirability of studying the interactions of tropomyosin B and actin, particularly by the negative-staining technique. Of course, it may simply turn out that the difference in period arises merely from a change in helical parameters unaccompanied by significant length changes of the filaments during negative staining.

If the four very fine filaments seen by Knappeis & Carlson (1962) to continue on from the ends of the thin filaments near the Z-line represent continuations of tropo-myosin filaments extending throughout the I-filaments, we can place limits on the amount of tropomyosin present. If tropomyosin consists of two α-helical poly-peptide chains side by side (Tsao, Bailey & Adair, 1951) then a filament containing four such molecules in parallel, and also containing 13 actin monomers of molecular weight 62,000 per 350 Å, would have an actin/tropomyosin ratio of approximately 3·6 : 1. Perry (1960) has estimated that tropomyosin B may represent 6 to 10% of the total muscle protein (cf. about 14% of actin) so there is certainly enough tropo-myosin present to build such a structure; on the basis of these figures, there would be almost enough to have at least 8 molecules in parallel in the thin filaments, and still have enough left over to build a lattice at the Z-lines.†

Another question that presents itself, and which we can do little to answer at present, is how in a muscle all the thin filaments come to be laid down with such a

† It should be appreciated that only a part of the Z-line density is accounted for by the very fine filaments passing through it; a good deal of other material appears to be present between the fine filaments.

precisely determined length. In rabbit, 26 full periods (of ~406 Å) can be counted on either side of the Z-line structure; other species give a number one or two different from this. F-actin filaments *in vitro*, on the other hand, seem to grow indefinitely. It is difficult to see how a single protein molecule such as G-actin, about 50 Å in diameter, can have the information built into its structure to form helical structures of determined length around 10,000 Å. Conceivably, two species of molecules which could form structures having slightly different periods (say about 4% different) might tend to aggregate together into structures which were particularly stable when their length corresponded to the lowest common multiple of the two individual periodicities.

(iii) *Other possible components in the thick filaments*

In the case of the A-band filaments in muscle, the question of whether myosin is the sole protein in them can again not be decided satisfactorily at present.

There is no evidence from the electron micrographs that the naturally occurring thick filaments and those reconstituted from purified myosin differ in their appearance in such a way as to suggest that the former contained a substantial amount of some other protein, although such differences would not necessarily be apparent. There is one small difference in that the thick filament seen in sectioned material, and often the isolated ones, have a slight thickening half-way along their length (in the middle of the central bare zone) but the amount of material involved would at most be 2 to 3% of their total mass. There is no sign of any residual core when the thick filaments are dissolved at higher salt concentrations. There are nevertheless some rather stubborn discrepancies, not large but perhaps significant just the same, which arise when one tries to describe the situation on a strict quantitative basis.

If one takes values of 110 Å for the diameter of the thick filaments, 40 Å for the diameter of the cross-bridges and 180 Å for the length of the cross-bridges, then, knowing the number of filaments per ml. of muscle (about 2.4×10^{14}), the contribution made to the total wet weight of the muscle by the thick filaments may be calculated to be 6.3 mg/ml. Taking a value for the myosin content of 34% of the total protein, and a protein content of 20% of the wet weight, then the myosin content may be calculated to be 6.8 mg/g or 7.1 mg/ml. On this basis then the agreement is reasonably good.

On the other hand, using the same dimensions for the filaments, one may calculate that the "molecular weight" of each of them is about 160×10^6. Now the number of cross-bridges associated with one thick filament appears to be about 200 to 220 (Huxley, H. E., 1957). If there is one cross-bridge per myosin molecule, one can calculate the contribution made by myosin to the molecular weight of the filament, if one knows the molecular weight of myosin. Current values for myosin lie around either 470,000 (Lowey & Cohen, 1962) or 620,000 (Kielley & Harrington, 1960). This leads to contributions of either about 100×10^6 or 130×10^6, leaving a substantial contribution to be made by some other substance to bring the value up to 160×10^6. And the difficulty cannot be avoided satisfactorily merely by accepting that one has assessed the dimensions of the filaments and cross-bridges incorrectly, for then the agreement with the total "myosin" content will no-longer hold. If a contribution from some other substance is admitted, then it also has to be accepted that this substance is erroneously estimated with the myosin. Alternatively, one might suppose that there were two myosin molecules per cross-bridge, that the size of the filaments was being underestimated, and their number overestimated.

Perhaps it is pushing the data too far at present to expect perfect numerical agreement, but it does appear that at least one of the values being used must be wrong.

As in the case of the thin filaments, there exists the question of what determines the lengths of these filaments, in muscle, within such close limits; again, it is difficult to see how this could reliably be achieved by the controlled polymerization of myosin on its own, although preparations of synthetic myosin filaments can be made in which most of the filaments have lengths of the right order of magnitude. Again, one wonders about the possibility of a copolymer with some other structure of a slightly different period taking place so that stable structures were formed only at lengths equal to the lowest common multiple of the two periods. There exists the following curious numerical coincidence. The periodicity of the myosin filaments seems to be 435 Å, whilst the other axial periodicity in muscle which can be determined very accurately by X-ray diffraction is 411 Å (Worthington, 1959). (This second period is probably equivalent to the approximately 406 Å period in the I-filaments mentioned earlier.) The lowest common multiple (to a good approximation) of these two lengths is 7400 Å ($435 \times 17 = 7395$, $411 \times 18 = 7398$) which lies close to one-half the length of the A-band.

(iv) *The structural polarity of the actin and myosin filaments*

When a structure consists of two helical chains of molecules wound round each other, then it by no means necessarily follows that the resultant structure has to be directionally polarized; it would not be, for instance, if alternate molecules in each of the two chains were reversed, or if all the molecules in each chain pointed in the same direction but the two chains pointed in opposite directions. It is all the more interesting then to find that when the G-actin molecules polymerize to give F-actin, it is a polarized structure that they form, in which all the molecules in both chains are oriented in the same equivalent direction.

Furthermore, the actin filaments at the Z-line do not just simply run straight through it, but are interrupted and connected to a fairly elaborate structure there in such a way that all of them on one side point in one direction and all on the opposite side of the Z-line in the opposite direction. This all suggests very strongly that there is a functional requirement that all the actin *monomers* on one side of a Z-line should point in one direction, and all those on the other in the reverse direction.

According to the sliding filament model, the relative force that is developed between the actin and myosin filaments on either side of a given Z-line has to pull all the myosin filaments towards that Z-line. If the force is generated as a consequence of short-range interactions between the actin monomers and the cross-bridges, and if spatial activation effects are not invoked to provide directionality—and the fact that myofibrils seem to contract quite normally when merely irrigated with solutions of ATP strongly indicates that such effects are not involved—then the direction of the force developed at any given site must be locally determined. It could be determined by the orientation of the actin monomer, or by the orientation of the myosin molecule of which the cross-bridge is a part, or by the orientation of both. Furthermore, the presumably very specific interaction of two protein molecules, involving, amongst other things, changes in enzymic activity, is likely to occur only in one critical relative orientation; so that if one molecule is reversed, the other must be too.

Thus the findings concerning the strict polarity of the actin monomers are not merely consistent with the sliding filament model, but seem to be an essential requirement of it.

In the same way, the arrangement of myosin molecules in the thick filaments, with their polarity reversed in either half of the A-band, seems again to be an essential requirement of the sliding filament model. It is particularly interesting that myosin molecules aggregate to form this type of structure spontaneously *in vitro*, and it suggests that the built-in ability of myosin molecules to form filaments with the appropriate pattern of polarity may be a requirement of the processes by which muscle structure is built up *in vivo*. It also suggests the beginnings of possible ways out of the conceptual difficulties which arise when one tries to apply the sliding filament theory to the contraction of a synthetic actomyosin fibre. It has generally been assumed that such a fibre contains none of the precisely ordered structures present in a striated muscle, and so it has been a little difficult to see how the sliding mechanism could still work; and yet it would be surprising if a different mechanism was also available. However, it is now becoming apparent that such synthetic fibres will contain myosin molecules not only organized into ordered filaments but even into filaments whose polarity is reversed at either end, and so even if the difficulties are not fully overcome, they are at least reduced.

Thus while the negative-staining results have so far not added a great deal to our knowledge of the detailed molecular architecture of the cross-bridges, they have revealed a number of new items of information about the structure of the thick and thin filaments, all of which either confirm or are in agreement with various aspects of the sliding filament theory, and which may help us to approach the problem of contraction a little more closely.

(v) *Speculations on contraction mechanisms*

Whilst questions of structural polarity have always been an inherent feature of sliding filament models which depend on short-range interaction between protein molecules in the two types of filament, the present results, demonstrating that such polarities actually exist and can be seen in the electron microscope, are perhaps helpful in that they decrease the size of the speculative element in discussions about the possible mechanisms involved in contraction. The problem can now be stated as follows.

Two types of filaments, built up of different proteins, lie side by side in an aqueous medium of ionic strength 0·1 to 0·2 μ. The filaments have diameters of about 65 Å and 110 Å and their surfaces are about 170 Å apart. The thick filaments contain myosin, the thin ones actin. There are projections, 30 to 40 Å in diameter, on the thick filaments which extend out sideways and touch the thin filaments. These projections probably contain the enzymic site of the myosin molecule, and also the site responsible for combination with actin. These sites are not necessarily the same, but have at least to interact with each other. The sites on all the projections are oriented in the same sense. It is not known whether the projections are rigid or flexible, nor whether they move during activity. There is a projection between a given pair of myosin and actin filaments at regular intervals, and the interval is probably 435 Å. The actin filaments contain two helical chains of actin monomers, and the pitch of the helix is probably 410 Å, though there is a possibility it may be only about 350 Å. The actin monomers in each chain are about 55 Å apart; they are all oriented in the same direction and each of them can interact with the cross-bridges from the myosin filaments. Each actin filament has three myosin filaments around it, with which it interacts, and probably the cross-bridges from these filaments touch the actin filament at intervals of $435°/3 = 145$ Å (rather than all in register at 435 Å intervals). The cross-bridges provide the only apparent mechanical connection between the two types of filament, and so they are likely to be responsible in some way for the relative force which can be developed between the filaments in an axial direction.

A relative force is developed between the actin and myosin filaments when ATP is split by the system; the filaments can slide past each other by distances which are very large compared to the separation of the active sites on the actin and on the myosin and continue to generate tension. The problem is to try to define the different types of mechanism which could exist to produce the relative force, and to think of means of differentiating between them. This is not the place to embark on such a general discussion, but there are a few specific points which could be made now that we can think of the system in concrete terms.

The difference in period between the actin and myosin filaments is probably about 25 Å, and there are 3 bridges directed at the actin filament in each 435 Å period. Thus, on average, successive bridges will be progressively further out of register with the active sites on the actin monomers by 8 Å per bridge. Thus if the interaction at one bridge can result in the actin filament moving along by as little as 8 Å, the next site can start to interact. Provided that there are at least seven bridges in the initial distance of overlap ($\geqslant 1000$ Å), the actin can therefore be pulled along by a distance equal to the separation of monomers in each chain; another cycle of events can then start at the first cross-bridge involved and a continuous sliding could occur. The system could probably work with even a shorter average initial overlap than this, as a result of longitudinal Brownian motion of entire thick filaments, which would not necessarily always stay exactly in register with each other. All this is not to say that the amount of movement produced by one cross-bridge *is* only 8 Å, but merely that it could in principle be as little as this.

There are a number of ways in which the interaction of actin and myosin at the cross-bridges could produce movement and these have been dealt with in various degrees of detail by several authors (Hanson & Huxley, 1955; Huxley, A. F., 1957; Huxley, H. E., 1960; Spencer & Worthington, 1960). Clearly, the relatively small amount of movement required, and the precise orientation of the active sites, make it perfectly possible to think of the two molecules having a certain area of mutual contact and moving relative to each other during a reaction in which ATP is split. Alternatively, the actual observation of two different periodicities in muscle, apparently arising from the actin and myosin filaments (Worthington, 1959), provides a continuing stimulus to think of models of the type (amongst others) discussed by Hanson & Huxley (1955) which specifically depend on such a difference in period to stretch the actin filaments and interrupt the actin helix when actin–myosin combination takes place. This type of mechanism has also recently been favoured by Oosawa, Asakura & Ooi (1961) on the basis of their work on actin.

A final point may be made on the subject of the interaction between actin and myosin filaments. If an actin filament can only interact with myosin filaments which are appropriately oriented, and if the result of this interaction is a relative force directed along the actin filament, then such a filament would tend to be maintained in motion in a constant direction when placed in a suspension of myosin filaments, in the presence of ATP. This might have interesting and measurable consequences on the diffusion of actin under these conditions. Furthermore, an oriented gel of actin in which a preponderance of filaments was polarized in one direction might tend to propel itself along under similar circumstances. Possibly some such mechanism is involved in the streaming of cytoplasm.

I am indebted to Dr. M. F. Perutz and Dr. J. C. Kendrew, and to the Medical Research Council, for their encouragement and support, and to Miss Pamela Dyer for her expert and unfailing technical assistance.

REFERENCES

Bailey, K. (1948). *Biochem. J.* **43**, 271.
Brenner, S. & Horne, R. W. (1959). *Biochim. biophys. Acta*, **34**, 103.
Corsi, A. & Perry, S. V. (1958). *Biochem. J.* **68**, 12.
Hall, C. E. (1955). *J. Biophys. Biochem. Cytol.* **1**, 1.
Hall, C. E. (1956). *J. Biophys. Biochem. Cytol.* **2**, 625.
Hall, C. E., Jakus, M. A. & Schmitt, F. O. (1946). *Biol. Bull.* **90**, 32.
Hanson, J. & Huxley, H. E. (1953). *Nature*, **172**, 530.
Hanson, J. & Huxley, H. E. (1955). *Symp. Lond. Soc. Exp. Biol.* **9**, 228.
Hanson, J. & Lowy, J. (1962). *Proc. 5th Int. Cong. Electron Microscopy*, Abstract No. 09. New York: Academic Press.
Hanson, J. & Lowy, J. (1963). *J. Mol. Biol.* **6**, 46.
Hasselbach, W. & Schneider, G. (1951). *Biochem. Z.* **321**, 462.
Hodge, A. J. (1959). *Rev. Mod. Phys.* **31**, 409.
Horne, R. W., Brenner, S., Waterson, A. P. & Wildy, P. (1959). *J. Mol. Biol.* **1**, 84.
Huxley, A. F. (1957). *Progr. Biophys. Biophys. Chem.* **7**, 255.
Huxley, A. F. & Niedergerke, R. (1954). *Nature*, **173**, 971.
Huxley, H. E. (1953). *Biochim. biophys. Acta*, **12**, 387.
Huxley, H. E. (1956). In *Proc. Stockholm Congr. Electron Microscopy*, p. 260. Stockholm: Almqvist and Wiksell.
Huxley, H. E. (1957). *J. Biophys. Biochem. Cytol.* **3**, 631.
Huxley, H. E. (1960). In *The Cell*, ed. by J. Brachet and A. E. Mirsky, vol. IV, p. 365. New York: Academic Press.
Huxley, H. E. (1961). *Circulation*, **24**, 328.
Huxley, H. E. (1962). *Proc. 5th Int. Cong. Electron Microscopy*, Abstract No. 01. New York: Academic Press.
Huxley, H. E. & Hanson, J. (1954). *Nature*, **173**, 973.
Huxley, H. E. & Hanson, J. (1957). *Biochim. biophys. Acta*, **23**, 229.
Huxley, H. E. & Zubay, G. (1960). *J. Mol. Biol.* **2**, 10.
Jakus, M. A. & Hall, C. E. (1947). *J. Biol. Chem.* **167**, 705.
Kielley, W. W. & Harrington, W. F. (1960). *Biochim. biophys. Acta*, **41**, 401.
Knappeis, G. G. & Carlson, F. (1962). *J. Cell. Biol.* **13**, 323.
Lowey, S. & Cohen, C. (1962). *J. Mol. Biol.* **4**, 293.
Lowey, S. & Holtzer, A. (1959). *Biochim. biophys. Acta*, **34**, 470.
Lowy, J. & Hanson, J. (1962). *Physiol. Rev.* **42**, Suppl. 5, 34.
Martonosi, A. (1962). *J. Biol. Chem.* **237**, 2795.
Oosawa, F., Asakura, S. & Ooi, T. (1961). *Progr. Theoret. Phys.* (*Kyoto*), Suppl. no. 17, 14.
Page, S. & Huxley, H. E. (1963). *J. Cell. Biol.* (in press).
Pepe, F. A. & Huxley, H. E. (1963). *Proc. Arden House Conf. Muscle* (in press).
Perry, S. V. (1960). *Ann. Rep. Chem. Soc.* **56**, 343.

Philpott, D. E. & Szent-Györgyi, A. G. (1954). *Biochim. biophys. Acta*, **15**, 165.

Reedy, M. K. (1962). Medical Thesis. University of Washington School of Medicine.

Rice, R. V. (1961*a*). *Biochim. biophys. Acta*, **52**, 602.

Rice, R. V. (1961*b*). *Biochim. biophys. Acta*, **53**, 29.

Rice, R. V. (1963). *Proc. Arden House Conf. Muscle* (in press).

Selby, C. C. & Bear, R. S. (1956). *J. Biophys. Biochem. Cytol.* **2**, 71.

Spencer, M. & Worthington, C. R. (1960). *Nature*, **187**, 388.

Szent-Györgyi, A. (1951). *Chemistry of Muscular Contraction*, 2nd Ed. New York: Academic Press.

Szent-Györgyi, A. G. (1953). *Arch. Biochem. Biophys.* **42**, 305.

Szent-Györgyi, A. G., Cohen, C. & Philpott, D. E. (1960). *J. Mol. Biol.* **2**, 133.

Tsao, T-C. & Bailey, K. (1953). *Biochim. biophys. Acta*, **11**, 102.

Tsao, T-C., Bailey, K. & Adair, G. S. (1951). *Biochem. J.* **49**, 27.

Worthington, C. R. (1959). *J. Mol. Biol.* **1**, 398.

Zobel, C. R. & Carlson, F. D. (1963). *J. Mol. Biol.* **7**, 78.

J. Mol. Biol. (1964) **8**, 835–840

N-Formyl-methionyl-S-RNA

K. Marcker and F. Sanger

Medical Research Council Laboratory of Molecular Biology
Hills Road, Cambridge, England

(*Received 17 February 1964*)

The reaction between methionine and S-RNA of *Escherichia coli* has been investigated. It has been demonstrated that, after the initial attachment to its specific S-RNA, the free α-amino group of the attached methionine may become formylated.

1. Introduction

The general role of S-RNA in protein biosynthesis is now well established. The amino acid is first activated by interaction with an enzyme and ATP to form a bound amino acyl-AMP. It is then transferred from this compound to form an ester linkage on the 2'- or 3'-hydroxyl group of the ribosyl moiety of the terminal adenosyl unit of S-RNA. This reaction is catalysed by the same enzyme which activates the amino acid. The amino acyl-S-RNA then reacts with the ribosomes, where the amino acid is incorporated into a nascent protein, starting from the amino end.

Zachau, Acs & Lipmann (1958) showed that the adenosyl ester of leucine was liberated when leucyl-S-RNA was digested with ribonuclease, thus demonstrating that ribonuclease acted upon the amino acyl-S-RNA in the following way:

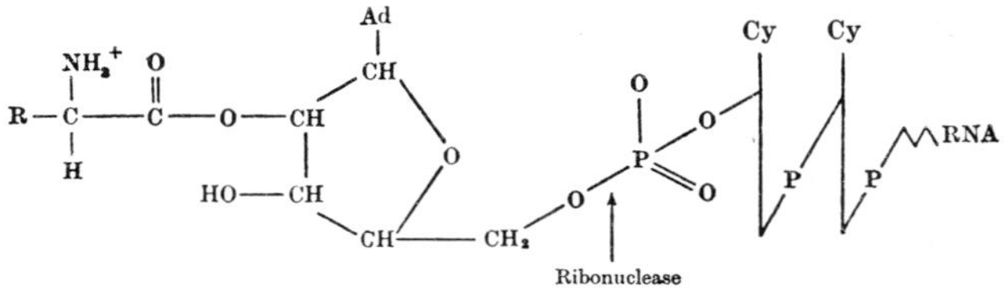

The authors have carried out similar experiments with [^{35}S]methionine in order to characterize the complex formed between methionine and its specific S-RNA. The adenosyl esters liberated by ribonuclease digestion were separated from the other nucleotides by high-voltage ionophoresis at pH 3·5. Besides the normal methionine ester, a second radioactive substance was present which was identified as the adenosyl ester of *N*-formyl-methionine. This indicates that the reaction between methionine and its specific S-RNA leads to the formation of the normal methionyl-S-RNA ester and, surprisingly, to *N*-formyl-methionyl-S-RNA ester. It is also shown that the formylation of the free α-amino group of methionine only takes place after methionine has become esterified to the S-RNA.

2. Materials and Methods

S-RNA was prepared from frozen cells of *E. coli* strain B according to the method of Zubay (1962). The supernatant solution containing the amino acid activating enzymes was prepared by grinding frozen cells with twice their weight of alumina and extracting with 2·5 vol. of 0·001 M-tris buffer pH 7·8 containing 0·01 M-magnesium acetate, 0·006 M-mercaptoethanol and 1 to 2 μg/ml. DNase. All manipulations were carried out at

4°C. The crude extract was clarified by two centrifugations (10,000 rev./min for 20 min). The ribosome-free supernatant solution was obtained from the clarified extract by 2 hr centrifugation at 105,000 g. The top two-thirds of the supernatant solution were withdrawn and stored in small tubes at $-20°C$ without prior dialysis.

Methionine labelled with ^{35}S was made according to the method of Sanger, Bretscher & Hocquard (1963). The ATP, GTP and PEP† were all used as their sodium salts. Pyruvate kinase (*Krystallsuspension*) was purchased from Boehringer and Soehne. The RNase used was a lyophilized Worthington preparation. The S-RNA labelled with ^{14}C in the terminal adenosine was a gift from Mr. M. Bretscher and had a specific activity of 9·3 mc/ m-mole (Bretscher, 1963). Formyl-DL-methionine (m.p. 100°C) was prepared by the method of Windus & Marvel (1931). Two sets of conditions were used for charging S-RNA with methionine:

(1) 0·2 ml. M-tris (pH 7·4) 0·15 M-MgCl$_2$ buffer, 0·1 ml. 0·1 M-ATP, 40 μl. 2 M-NH$_4$OOC.CH$_3$, approximately 5 μc [^{35}S]methionine, 4 to 5 mg S-RNA, 40 to 50 μl. *E. coli* supernatant solution, water to 2 ml. The mixture was then incubated for 5 min at 37°C. When using these conditions, it was found that mainly [^{35}S]methionyl-S-RNA was formed.

(2) 0·2 ml. M-tris (pH 7·4) 0·15 M-MgCl$_2$ buffer, 0·1 ml. 0·1 M-ATP, 40 μl. 2 M-NH$_4$OOC.CH$_3$, approximately 5 μl. [^{35}S]methionine, 4 to 5 mg S-RNA, 200 to 400 μl. *E. coli* supernatant solution, water to 2 ml.; incubation for 10 min at 37°C. Under these conditions it was found that a high proportion of the methionyl-S-RNA was transformed into formyl[^{35}S]methionyl-S-RNA, although there were considerable differences between different batches of *E. coli* supernatant solution. After the incubation, the mixture was chilled in ice and 0·1 vol. of 20% CH$_3$COOK (pH 5) was added. It was then shaken in the cold with 2 ml. water-saturated phenol and the RNA in the aqueous layer was precipitated by addition of 3 vol. ethanol. The precipitate was washed once with ethanol. If the charged S-RNA was to be used for characterization purposes, it was then dissolved in a small amount of water and incubated for 5 min at room temperature with 5 to 10 μg RNase. The mixture was put on Whatman no. 3 MM paper and subjected to ionophoresis for 1 hr at pH 3·5 and 3000 v. Finally, a radioautograph was prepared, usually by leaving the paper in contact with X-ray film for one night.

If the charged S-RNA was to be used for incorporation studies into proteins, it was re-dissolved in 1 ml. of water, re-precipitated and washed with ethanol as described above.

Artificially formylated [^{35}S]methionyl-RNA was prepared as follows: 2 mg of S-RNA were incubated as described above using (1). After isolation, the RNA was washed twice with ethanol and dried. It was then dissolved in 50 μl. water and slowly put into 1 ml. of dimethylformamide containing a trace of pyridine. A 200-fold excess of *p*-nitrophenyl-formate (prepared by coupling *p*-nitrophenol with formic acid in the presence of dicyclohexylcarbodiimide) in 1 ml. of dimethylformamide was then added. The mixture was shaken for 1 hr at room temperature, and the S-RNA was isolated by addition of 0·1 vol. 20% potassium acetate and 3 vol. ethanol.

To study the transfer of methionine from its S-RNA derivatives into protein, the following incubation mixture was used: 50 μl. M-tris buffer (pH 7·8), 40 μl. 0·2 M-Mg(CH$_3$COO)$_2$, 50 μl. 2 M-NH$_4$OOC.CH$_3$, 10 μl. M-β-mercaptoethanol, 10 μl. 0·1 M-ATP, 20 μl. 0·01 M-GTP, 80 μl. 0·05 M-PEP, 20 μl. pyruvate kinase suspension (2 mg/ml.). To each tube was added 100 μl. of an undialysed S-30 fraction prepared from freshly grown *E. coli* cells according to the method of Nirenberg & Matthaei (1961). Finally, the respective RNA's were added and the final volume made up to 0·5 ml. by the addition of water. After incubation for 15 min at 37°C, an equal volume of 10% TCA was added. The precipitate was then heated for 15 min at 90 to 95°C and filtered through a Millipore filter. It was washed several times with 5% TCA and twice with 1 : 1 ethanol–ether and finally plated and counted.

3. Results

Plate I shows a typical radioautograph of a ribonuclease digest of S-RNA labelled with [^{35}S]methionine. There are two distinct bands (A and B) travelling towards the cathode, in addition to free methionine. That the presence of these two bands

† Abbreviations used: PEP, phosphoenolpyruvate; TCA, trichloroacetic acid.

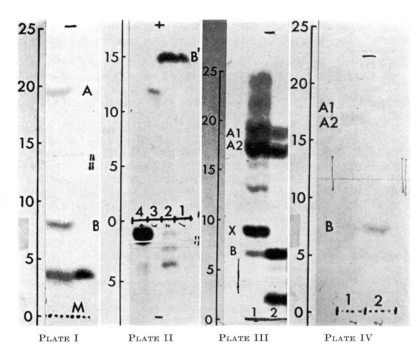

PLATE I.   PLATE II.   PLATE III.   PLATE IV.

PLATE I. Ribonuclease digest of methionyl-S-RNA. 2 mg *E. coli* S-RNA were incubated for 10 min at 37°C in 1 ml. of the reaction mixture described in the Materials and Methods section. The amount of supernatant solution was 100 μl. The S-RNA was then isolated and digested with RNase as described. The digest was subjected to ionophoresis for 1 hr pH 3·5, 3000 v. After drying, the paper was radioautographed overnight. M refers to free methionine used as marker.

PLATE II. Radioautograph of ionogram (pH 3·5, 3000 v, 1 hr) of: (1) band B (Plate I) eluted and incubated 1 hr in dilute NH_3 (pH 10·5) at 37°C; (2) formyl[^{35}S]methionine made according to Sheehan & Young (1958); (3) N-acetyl[^{35}S]methionine made by acetylation of [^{35}S]methionine with acetyl chloride; (4) [^{35}S]methionine.

PLATE III. Radioautograph of ionogram (pH 3·5, 3000 v, 1 hr) of: (1) ribonuclease digest of methionyl-S-RNA carrying a ^{14}C label in the terminal adenosine position: [^{14}C]S-RNA (0·25 μc) was incubated for 10 min at 37°C in a mixture containing tris, $MgCl_2$, ATP and $NH_4OOC.CH_3$ in the concentrations described under Materials and Methods, 5 μl. 0·1 M-methionine and 75 μl. *E. coli* supernatant solution. The RNA was isolated and digested with ribonuclease as described; (2) ribonuclease digest of [^{35}S]methionyl-S-RNA, prepared as above but using [^{35}S]methionine (0·5 μc) and unlabelled S-RNA. Band X has been eluted and shown to be free adenosine.

PLATE IV. Radioautograph of ionogram (pH 3·5, 3000 v, 1 hr) of: (1) ribonuclease digest of [^{35}S]methionyl-S-RNA, prepared as described under (1) of Materials and Methods; (2) ribonuclease digest of [^{35}S]methionyl-S-RNA from the same preparation as above after formylation with *p*-nitrophenyl formate.

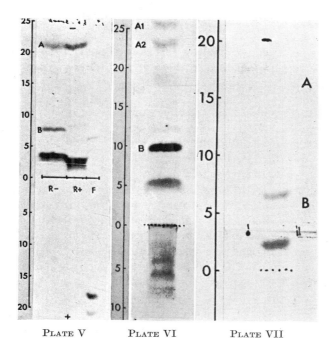

PLATE V   PLATE VI   PLATE VII

PLATE V. Effect of ribonuclease on the formation of formyl-methionyl-S-RNA.

(R+), medium preincubated with ribonuclease. A mixture containing 10 μl. M-tris buffer (pH 7·4), 0·15 M-MgCl$_2$, 5 μl. 0·1 M-ATP, 2 μl. 2 M-NH$_4$OOC.CH$_3$, [^{35}S]methionine (0·5 μC), 0·5 mg S-RNA, water to 100 μl. was incubated for 5 min at 37°C with 10 μg RNase. 20 μl. *E. coli* supernatant solution containing 100 μg RNase/ml. was then added. The mixture was incubated for an additional 10 min at 37°C and applied directly to the paper.

(R–), incubation as above except that ribonuclease was omitted in every step, but after the final incubation 10 μg RNase was added and the mixture applied to the paper.

(F), a synthetic mixture obtained by formylation of [^{32}S]methionine, containing formylmethionine and formyl-methionine sulphoxide. Ionophoresis at pH 3·5, 3000 v, 1 hr.

Band A in both R– and R+ is believed to be largely *S*-adenosyl methionine. Its mobility does not alter on treatment at pH 10·5. Band B in R– was eluted and shown to give *N*-formyl-methionine after treatment at pH 10·5.

PLATE VI. Ribonuclease digest of [^{35}S]methionyl-S-RNA isolated from growing *E. coli*. *E. coli* strain B cells were grown overnight in the synthetic medium described by McQuillen (1955). Next morning, growth was restarted by addition of 10% glucose to a concentration of 0·8%. When the cells were in the log phase, 100 μC ^{35}SO$_4^=$ were added. After 3 min, chloromycetin was added to a final concentration of 50 μg/ml. (This was found to lead to an accumulation of counts in the S-RNA fraction.) After an additional 3 min, the cells were chilled in ice, centrifuged and washed once with 0·001 M-tris (pH 7·4), 0·01 M-Mg(CH$_3$COO)$_2$ buffer. The cells were then resuspended in approximately 2 ml. of the buffer, and 1 mg of solid lysozyme was added. The suspension was left for 5 to 10 min at 4°C, then frozen and thawed 5 to 6 times, 4 ml. of water-saturated phenol were added and the mixture shaken for 15 min at 4°C. The S-RNA was isolated from the aqueous phase, digested with RNase and subjected to ionophoresis at pH 3·5 as described in the Materials and Methods section. Band B has been eluted and found to contain *N*-formyl-methionine.

PLATE VII. 5 mg S-RNA were incubated as described in method (2) of Materials and Methods. The amount of supernatant solution was 400 μl. After 30 min incubation at 37°C, a sample of 0·2 ml. was withdrawn and precipitated without phenol-shaking by 0·1 vol. 20% CH$_3$COOK (pH 5) and 3 vol. ethanol. The isolated S-RNA was then digested with RNase and run at pH 3·5, 3000 v for 1 hr.

was not due to incomplete splitting of the methionyl-S-RNA by ribonuclease has been shown by using ^{32}P-labelled S-RNA charged with unlabelled methionine. Ionophoresis of a ribonuclease digest of this S-RNA under the same conditions did not give any bands corresponding to those obtained from [^{35}S]methionyl-S-RNA. Thus neither A nor B contained phosphorus, whereas any incomplete digestion product would contain phosphorus.

The two bands were eluted from the paper and incubated with dilute ammonia at pH 10·5 for one hour at 37°C. Band A was converted to methionine. This was consistent with the known alkali-lability of the amino acyl-S-RNA ester linkage and suggested that it was the methionyl adenosine ester. As shown below (Plate III) this band contains the terminal adenosine residue of the RNA.

In many experiments (e.g. Plates III and VI) there are two bands close together near the position of band A, the relative amounts varying considerably from one experiment to another. Both behave similarly on treatment at pH 10·5. The reason why two compounds are obtained is not entirely clear. In the case of leucyl-adenosine ester, only one such compound was demonstrated. We have found that ribonuclease digests of other amino acid S-RNA's such as the proline and phenylalanine RNA's give two similar bands. It seems probable that the effect is due to a migration of the amino acid between the 2'- and 3'-hydroxyl groups of the ribosyl moiety of the terminal adenosine. Rammler & Khorana (1963) have recently shown that such isomers do in fact become rearranged in the above way, and they also found that it was possible to separate them on silicic acid columns.

The substance from band B, on treatment at pH 10·5, did not give methionine but a compound B' which by ionophoresis was found to be rather acidic at pH 3·5 (Plate II). This showed that it had gained an additional negative charge by the alkali treatment, since before it was slightly basic. It was found, however, that by mild acidic hydrolysis (15 min 100°C, N-HCl) B' was converted quantitatively into methionine. This suggested that it was a derivative of methionine in which the α-amino group was blocked. It was therefore subjected to ionophoresis at various pH values in parallel with synthetic N-acetyl methionine. At pH 2·1 it was neutral, but at pH 3·5, 6·5 or 8·9 it travelled ahead of N-acetyl methionine, and the distance between the two bands was approximately constant at these pH values. Attempts to acetylate it did not alter its ionophoretic mobility, thus showing that it did not have a free α-amino group. Since it moved faster than N-acetyl methionine on ionophoresis at pH 3·5, it seemed most likely that it was N-formyl methionine. This was especially supported by the finding that the derivative was converted into methionine by mild acidic hydrolysis. The acidic lability of the N-formyl group has long been known (Fischer & Warburg, 1905). In fact, this group has been used in peptide synthesis as an amino-blocking group because of its easy removal by mild acidic hydrolysis. Compound B' was therefore submitted to ionophoresis at pH 3·5 together with synthetic N-formyl[^{35}S]methionine. The result is shown in Plate II. As can be seen from the radioautograph, the synthetic N-formyl-methionine and the unknown compound B' do travel together, confirming that the latter is indeed N-formyl-methionine. As further confirmation, samples of band B' were mixed with synthetic non-radioactive formyl-DL-methionine and subjected to ionophoresis at pH 3·5 and 6·5 and to chromatography in water-saturated benzyl alcohol. After preparing radio-autographs, the papers were developed with the platinic iodide reagent (Toennies & Kolb, 1951) to locate the synthetic compound. In each case it was found to be co-extensive with the radioactive material. In some experiments a band moving slightly faster than band B' at pH 3·5 was present. This appeared to be formyl-methionine sulphoxide.

That *N*-formyl-methionine is attached to S-RNA is shown by the following facts. *N*-formyl-methionine is released by a very mild alkali treatment from compound B (Plate II), thus showing that it is bound through its carboxyl group by a very alkali-labile bond. When undigested *N*-formyl-methionyl-RNA was subjected to iono-phoresis at pH 3·5, most of the radioactivity stayed at the origin, with streaking extending towards the cathode. No radioactivity could be found corresponding to the bands shown in Plate I. Finally, when S-RNA labelled with ^{14}C in the terminal adenosine unit was charged with unlabelled methionine, digested with ribonuclease and subjected to ionophoresis together with a ribonuclease digest of $[^{35}S]$methionyl-S-RNA, band B, as well as the two forms of band A, was found to contain ^{14}C-labelled adenosine, as shown in Plate III. When $[^{35}S]$methionyl-S-RNA was formylated in dimethyl-formamide by *p*-nitrophenyl-formate ester and subsequently digested with ribonuclease, it was found that the two A band substances were converted into the band B substance, as shown in Plate IV.

The above results indicate that *N*-formyl-methionyl-S-RNA may be formed in cell-free systems from *E. coli*. There are two possible ways in which this compound could be made: (1) by formylation of methionine followed by combination of the *N*-formyl-methionine with S-RNA; (2) by formylation of methionyl-S-RNA. In order to distinguish between these two possibilities, the effect of pretreating the components of the incubation mixture with ribonuclease was studied. After the incubation, the whole mixture was submitted to ionophoresis at pH 3·5 as usual, together with a control in which ribonuclease was added after the 10 minutes incubation. As shown in Plate V, no compound corresponding to *N*-formyl-methionyl adenosine ester (band B) was present. Furthermore, no free *N*-formyl-methionine could be found. The same result was obtained when the S-RNA was omitted from the reaction mixture. Attempts to label the S-RNA with synthetic formyl$[^{35}S]$methionine did not lead to any significant incorporation. It therefore appears that reaction (1) can be ruled out and that formylation takes place on the methionyl-S-RNA.

It was of interest to see whether or not the same compound could be isolated from growing cells. Accordingly *E. coli* cells brought into the log phase were grown for a short time with $[^{35}S]$methionine or $^{35}SO_4^=$ in the growth medium. The S-RNA was then isolated, digested with ribonuclease and subjected to ionophoresis at pH 3·5. The result of such an experiment is shown in Plate VI. As can be seen, the same three bands appear as were found in the *in vitro* system. The radioactivity in band B has definitely been shown to be due to *N*-formyl-methionine; it can therefore be con-cluded that the same compound is formed in growing cells.

Attempts have been made to see whether or not other amino acids are formylated on their respective S-RNA's. So far, all these attempts have failed both *in vivo* and *in vitro*, with the possible exception of cysteine. A ribonuclease digest of S-RNA isolated from a $^{35}SO_4^=$ grown methionine-requiring *E. coli* mutant, which should not contain radioactive methionine, did give a band with the same mobility as band B; but the band has not yet been positively shown to contain *N*-formyl-cysteine. The fact that the formyl derivative of other amino acyl-S-RNA's could not be detected cannot of course be regarded as evidence that the reaction does not occur in these cases, since the derivatives may be too rapidly destroyed.

Experiments have also been undertaken to see if organisms other than *E. coli* make *N*-formyl-methionyl-S-RNA. This has definitely been shown to be the case for yeast. S-RNA isolated from yeast grown on $^{35}SO_4^=$ was treated with ribonuclease and shown to give rise to bands A and B. S-RNA from rat liver or hen oviduct that had been incubated with [^{35}S]methionine, gave only band A (methionyl adenosine ester) and no detectable amount of the formyl derivative.

Since it is possible that the formation of *N*-formyl-methionyl-S-RNA is connected in some way with protein biosynthesis, experiments were undertaken in order to see whether or not methionine in the form of *N*-formyl-methionyl-S-RNA was incorporated into protein. Before doing so, it was thought necessary to develop a method for obtaining *N*-formyl-methionyl-S-RNA essentially free from methionyl-S-RNA. This could best be done by using more *E. coli* supernatant solution and incubating for a longer time, or by carrying out the incubation as described under (2) and then isolating the S-RNA and re-incubating once more without the addition of free methionine. Plate VII shows a ribonuclease digest of formyl-methionyl-S-RNA prepared by one of the methods described above. It is seen that it is possible to convert methionyl-S-RNA almost completely into *N*-formyl-methionyl-S-RNA.

Incorporation studies were carried out using an *E. coli* cell-free system (Nirenberg & Matthaei, 1961). Such systems would incorporate about 2 to 8% of methionine when present as the S-RNA derivative. Samples of S-RNA containing predominantly methionine or formyl-methionine were prepared as above. Incorporation into hot TCA-insoluble material was obtained in both cases, and in general methionine was somewhat more efficiently incorporated when it was in the form of formyl-methionyl-S-RNA than when in the form of methionyl-S-RNA. For example, in a typical experiment in which 10,000 counts per minute of the S-RNA's were used, 379 counts were incorporated from the methionine derivative and 621 from the formyl methionine derivative. This incorporation is dependent on the energy system, since no significant incorporation occurred when this was omitted, and is inhibited by addition of puromycin at a concentration (5×10^{-4} M) which is known to inhibit protein biosynthesis. It was also shown that the transfer from formyl-methionyl-S-RNA was not due to stimulation of the incorporation of the small amount of contaminating methionyl-S-RNA by some unknown factor present in the preparation, since no stimulation from [^{35}S]methionyl-S-RNA was obtained by adding a non-radioactive formyl-methionyl-S-RNA preparation.

4. Discussion

The above results indicate that *N*-formyl-methionyl-S-RNA is formed by an enzymic reaction from methionyl-S-RNA. It is also found in growing cells of *E. coli* and yeast, the two derivatives being present in about the same concentration, as measured by the yields of the adenosyl derivatives. This may, however, be deceptive, as the *N*-formyl derivative is more stable than the methionyl-S-RNA and therefore its relative yield may appear higher due to greater losses of the methionine derivative during isolation.

It is highly probable that the unknown compound attached to S-RNA isolated from $^{35}SO_4$-grown *E. coli* cells by Aronson *et al.* (1959) is identical with *N*-formyl-methionine. By incubation of the isolated S-RNA with alkali, they found that in addition to [^{35}S]methionine another radioactive compound was liberated, which on chromatography in butanol–formic acid travelled with the front. Although no further attempt at identification was undertaken, its chromatographic behaviour is consistent with the idea that the unknown compound is in fact *N*-formyl-methionine, which under those conditions would be expected to be non-polar.

Whether or not all the methionyl-S-RNA can be formylated is not certain. Preparations could be made which contained almost exclusively the formyl derivative (as shown in Plate VII); however, relatively long incubation periods were used and it may be that the more labile methionyl-S-RNA was broken down, leaving only the formyl-methionyl-S-RNA. The nature of the formyl donor has not been investigated in detail. It has been found that addition of [^{14}C]sodium formate to the incubation mixture does not lead to the incorporation of any radioactive formyl groups into formyl-methionyl-S-RNA. It is, therefore, unlikely that the formyl donor is readily generated from free formate under the conditions used. However, further work is clearly necessary on this point.

Since the reaction involves S-RNA, which is generally considered to be an essential intermediate in the synthesis of protein, the question arises as to whether the formylation has any role in protein biosynthesis. Studies with the cell-free system suggested that the formyl-methionyl-S-RNA was incorporated as well as, if not better than, the methionyl-S-RNA into the hot TCA precipitate. Experiments are in progress to decide whether this incorporation does in fact constitute the true formation of peptide linkages. If this is so, it is difficult to avoid the conclusion that the formyl-methionyl-S-RNA is in fact involved in the normal pathway of incorporation of methionine into protein. It is felt, however, that this would be a premature conclusion and that further work is necessary to evaluate the biological significance of the formylation reaction.

We are very grateful to various members of this laboratory for advice and help during the course of this work, and especially to Mr. M. S. Bretscher and Dr. R. Monroe. One of the authors (K. M.) wishes to thank the Carlsberg–Wellcome Foundation for a scholarship.

REFERENCES

Aronson, A. I., Bolton, E. T., Britten, R. S., Cowie, D. B., Duerksen, J. D., McCarthy, B. J., McQuillen, K. & Roberts, R. B. (1959). *Yearb. Carneg. Instn*, **59**, 270.
Bretscher, M. S. (1963). *J. Mol. Biol.* **7**, 446.
Fischer, E. & Warburg O. (1905). *Ber.* **38**, 3997.
McQuillen, K. (1955). *Biochim. biophys. Acta*, **17**, 382.
Nirenberg, M. W. & Matthaei, H. J. (1961). *Proc. Nat. Acad. Sci., Wash.* **47**, 1588.
Rammler D. H. & Khorana H. G. (1963). *J. Amer. Chem. Soc.* **85**, 1997.
Sanger, F., Bretscher, M. S. & Hocquard, E. J. (1963). *J. Mol. Biol.* **8**, 38.
Sheehan, J. C. & Young, D. D. H. (1958). *J. Amer. Chem. Soc.* **80**, 1154.
Toennies, G. & Kolb, J. J. (1951). *Analyt. Chem.* **23**, 823.
Windus, W. & Marvel, C. S. (1931). *J. Amer. Chem. Soc.* **53**, 3490.
Zachau, H. G., Acs, G. & Lipmann, F. (1958). *Proc. Nat. Acad. Sci., Wash.* **44**, 885.
Zubay, G. (1962). *J. Mol. Biol.* **4**, 347.

J. Mol. Biol. (1964) **9**, 734–745

On the Mechanism of Genetic Recombination between DNA Molecules

MATTHEW MESELSON

Biological Laboratories, Harvard University
Cambridge, Massachusetts, U.S.A.

(*Received 24 April 1964*)

A two-factor cross was performed between bacteriophages labeled with heavy isotopes. Recombinants were found with chromosomes formed entirely or almost entirely of parental DNA. This and other features of the distribution of parental DNA among recombinant phages and among their descendants show that genetic recombination occurs by breakage and joining of double-stranded DNA molecules. Also, there is some indication that a small amount of DNA is removed and resynthesized in the formation of recombinant molecules.

1. Introduction

Genetic recombination between bacteriophages was discovered by Delbrück & Bailey (1946) in the course of experiments designed to determine whether more than one infecting phage can multiply in a single bacterium. They examined phages produced by cells jointly infected with two types of phage differing from one another in two genetic characters and found not only the two parental phage types but also two types with combinations of the parental characters. For a time, it was suspected that the new types resulted from a novel process more akin to mutation than to recombination. However, Hershey & Rotman (1948) showed that the phenomenon was indeed genetic recombination, for its operation closely followed rules already worked out for recombination in higher organisms.

Although the details of the mechanism are not known, there is strong evidence in the case of bacteriophage λ that recombination results from the breakage and joining of DNA molecules. The initial evidence for breakage and joining came from measurements of the amount of labeled DNA in selected recombinants yielded by cells infected jointly with isotopically labeled and unlabeled phages (Meselson & Weigle, 1961; Kellenberger, Zichichi & Weigle, 1961). Recombinant phages were found to contain labeled DNA in discrete proportions, corresponding to the proportion of the λ genetic map lying beyond the selected site of recombination in the direction of the allele contributed by the labeled parent phage. Since the chromosome of λ is a single DNA molecule along which hereditary determinants are spaced at least approximately as they are along the λ genetic map (Kaiser, 1962 and personal communication), this simple result may be explained as the result of breakage and joining of chromosomes at the selected site of recombination. Alternatively, it could reflect the operation of a rather less plausible mechanism whereby a fragment of a parental chromosome is rebuilt by synthesis of its missing length on a template provided by the homologous portion of a chromosome of different parentage. The latter mechanism has been called breakage and copying.

Breakage and joining may be distinguished from breakage and copying by an examination of the amount of parental DNA in recombinant phages from a cross between parents *both* of which contain labeled DNA. Only the former of the two mechanisms can produce a recombinant containing label from both parents. Using host-induced modification as a label, Ihler & Meselson (1963) obtained evidence that both parents do contribute lengths of DNA which become joined together to form recombinant chromosomes.

The mechanism of recombination is specified more conclusively and in greater detail by the present results, a preliminary report of which has appeared elsewhere (Meselson, 1962). The occurrence of recombination by breakage and joining of λ chromosomes is confirmed, and the process is shown to be independent of chromosome replication in the sense that recombinants may be formed entirely, or almost entirely, of unreplicated chromosome fragments. However, there is some indication that a small amount of DNA is removed and resynthesized in the formation of recombinant chromosomes.

2. Materials and Methods

(a) *Bacteriophages*

Phage λ, the wild type of Kaiser (1957), and λhc were used as parents for all crosses. Phage λhc is a recombinant isolated from a cross of the λh of Kaiser (1962) by λc_{26}, a nitrous acid-induced c_1 mutant of λ.

All of these phages have buoyant density 1.508 g cm^{-3} in pure CsCl solution at 20°C. To insure homology of the two parental phages, the h allele was passed through five consecutive crosses alternately to λc_{26} and λ, selecting in the first case for λhc and in the second for λh. Although many thousands of plaques were examined, there was no indication of segregation of characters other than h and c.

Stocks of λ were prepared by the induction of lysogenic bacteria with ultraviolet light. Stocks of λhc were produced by lytic multiplication. Phages uniformly labeled with ^{13}C and ^{15}N were prepared and purified according to Meselson & Weigle (1961). The uniformity of labeling of each stock was verified by density-gradient centrifugation. The frequency of λh was found to be less than 0.01% in each stock.

(b) *Bacteria*

Escherichia coli K12 strain 3110 (E. M. Lederberg) was used for the preparation of all parental phage stocks. Strain C600·5, a mutant of *E. coli* K12 C600 (Appleyard, 1954), was used as host for all crosses. The host modification and restriction properties of this mutant are identical to those of *E. coli* strain C (Bertani & Weigle, 1953). Strain C 600·5 was used in preference to C, because λ is rapidly inactivated in lysates of our strain C. Strains C600, C600·5, and *E. coli* K12/λ (Kaiser, 1962) were used as indicators.

(c) *Methods*

Crosses and low-multiplicity growth cycles were performed on strain C600·5 following Ihler & Meselson (1963). In crosses, the multiplicity of infection was approximately three of each parental type. Recombinants λh were scored as turbid plaques on a mixed indicator containing equal concentrations of C600·5 and ultraviolet-irradiated K/λ. On this indicator λhc produces clear plaques whereas λ and λc give rise to barely discernable ghost plaques. Irradiation removes the barrier which otherwise would restrict phages produced on strain C600·5 from infecting K/λ. For the assay of unrestricted phages, a mixture of C600 and non-irradiated K/λ was used and is referred to as restricting indicator. The validity of both scoring procedures was established by appropriate reconstruction experiments. Only one parental type and one recombinant type, λhc and λh, were scored. Density-gradient centrifugation was performed according to Meselson & Weigle (1961).

3. Outline of the Experiments

When several λ phages simultaneously infect a cell, some of the chromosomes replicate semi-conservatively while others remain unreplicated (Meselson & Weigle, 1961). Accordingly, a cell infected with several isotopically labeled phages will contain chromosomes with both polynucleotide strands labeled, others with only one labeled strand, and still others with no label. If genetic recombination occurs by simple breakage and joining, then recombination at a given site followed by phage maturation could yield altogether nine discrete phage types with respect to the proportion of labeled DNA they contain. If the site of recombination is chosen at the center of the chromosome, the number of possible classes is reduced to five: fully labeled, three-quarter, one-half, one-quarter and unlabeled. The fully-labeled and three-quarter-labeled classes deserve special attention, because they are expected to result from breakage and joining but not from breakage and copying. An important result of the experiments described below is the discovery of these two classes among the recombinant phages from a suitably designed cross. However, this result does not prove that segments from different chromosomes have become permanently joined to one another. Instead, they could be only temporarily associated. That this is not the case is shown by the finding of half-labeled recombinant phages among the progeny yielded by cells singly infected with fully-labeled, three-quarter-labeled or half-labeled recombinant phages. This result shows that recombination results in an association between polynucleotide chains from different parent chromosomes which is not disrupted by the events of chromosome injection, replication or maturation.

The genetic markers h (host range) and c (clear plaque) were chosen for these experiments, because of their symmetrical location about the center of the linkage map of λ. That they also straddle the center of the λ chromosome is shown both by the present results and by the behavior of infectious molecules of λ DNA when they are broken in half by hydrodynamic shear (Kaiser, 1962). The frequency of recombination between h and c is approximately 7%, and the hc interval comprises about one third of the λ genetic map.

In order to improve the resolution of certain classes of labeled phages in the presence of a great excess of unlabeled phages, several gradients were assayed on restricting indicator as described under Materials and Methods. On this indicator, phages from strain C600·5 plate with very low efficiency unless they contain at least one strand composed of parental DNA (i.e., DNA synthesized on strain 3110). This is an example of the phenomena of host-induced modification and restriction elucidated by Arber & Dussoix (1962). It should be emphasized that we have relied on host-induced modification only as a means of suppressing the appearance of unlabeled phages, and not as a label for parental DNA.

4. Results

(a) *A cross with isotopically unlabeled phages*

Figure 1 shows the density distribution of phages from a cross of unlabeled λ with unlabeled λhc. It is seen that both λhc and λh form a single sharp band in the gradient. Bands of exactly the same shape and location were found from assays performed on restricting indicator bacteria. The complete absence of phages with atypically high density provides assurance that variations in the amount of isotopic label are the only cause of the density variations observed in the remainder of these experiments. The light tail seen on each band results from mixing caused by the method of collecting fractions.

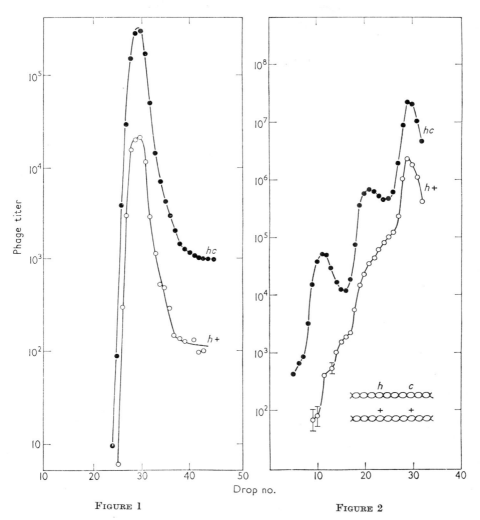

FIGURE 1   FIGURE 2

FIG. 1. Density distributions of λhc and λh from the cross λhc × λ.

FIG. 2. Density distributions of λhc and λh from the cross [$^{13}C^{15}N$]λhc × [$^{13}C^{15}N$]λ. Fractions 12 and 13 were pooled for the assay of λh. Error tags indicate 90% confidence limits. The three modes in the distribution of λhc are formed by phages with conserved, semi-conserved, and newly synthesized chromosomes, in order of decreasing density. The location of mutations h and c on the chromosome is indicated schematically at the lower right of the Figure.

(b) Crosses with isotopically labeled phages

The density distributions of λhc and λh from a cross of $^{13}C^{15}N$-labeled λhc by $^{13}C^{15}N$-labeled λ are shown in Fig. 2. The parental type λhc is seen to occur primarily in three discrete modes, corresponding to phages with conserved, semi-conserved and completely new chromosomes (Meselson & Weigle, 1961). The distribution of the recombinant type λh shows inflections at the positions expected for phages with fully-labeled, three-quarter-labeled, and one-half-labeled chromosomes; but lack of resolution prevents the positive identification of these categories.

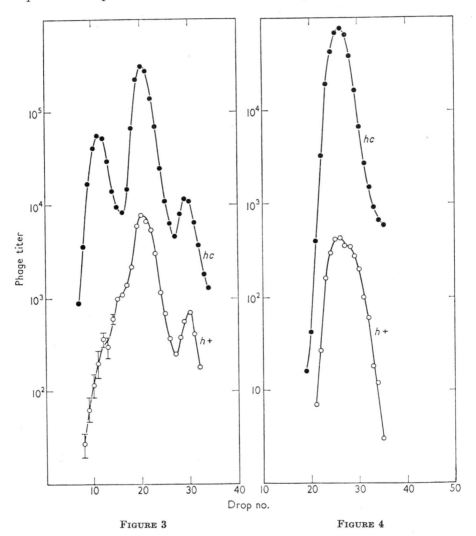

FIGURE 3   FIGURE 4

FIG. 3. Density distributions of λhc and λh in the gradient depicted in Fig. 2 found by assays on restricting indicator bacteria. Error tags indicate 90% confidence limits.

FIG. 4. Density distributions of λhc and λh from the fully-labeled region (fractions 10 to 13) of the gradient shown in Fig. 2.

An attempt was made to obtain increased resolution by assaying the distribution on restricting indicator bacteria, so as to suppress the appearance of phages lacking a completely labeled polynucleotide chain. Figure 3 shows that this procedure serves to reveal a discrete mode of half-labeled recombinants but that fully-labeled and three-quarter-labeled λh remain unresolved. The slight peaks at fraction 30 are formed by unlabeled phages plating at the low efficiency with which λ from strain C600·5 infects most K12 strains.

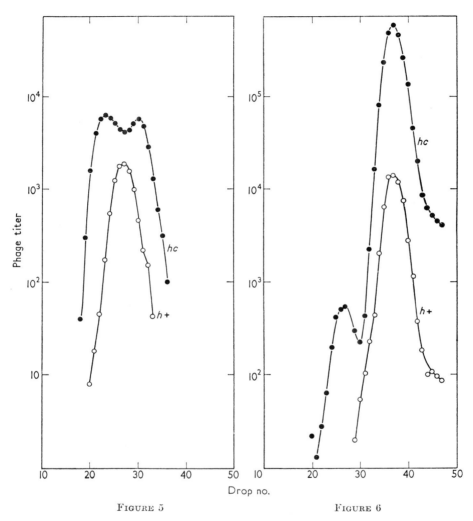

FIGURE 5   FIGURE 6

FIG. 5. Density distributions of λhc and λh from the three-quarter-labeled region (fractions 14 to 17) of the gradient shown in Fig. 2. Conserved and semi-conserved λhc are seen to be barely resolved.

FIG. 6. Density distributions of λhc and λh from the one-half-labeled region (fractions 19 to 21) of the gradient shown in Fig. 2.

The difficulty caused by insufficient resolution was overcome by subjecting fractions from each region of interest in the original gradient to further density-gradient analysis. The distributions of phages in pooled fractions taken from the fully-, three-quarter-, and one-half-labeled regions of the original gradient are shown in Figs 4, 5 and 6 respectively. Discrete bands of recombinant phages are found in all three cases. The distributions of Fig. 5 are of special interest. The contrast between the distinct uni-modal band of three-quarter-labeled recombinants and the broad bimodal distribution of the accompanying non-recombinant phages shows most strikingly that the labeled recombinants are indeed a discrete density species formed by the breakage and joining of double-stranded chromosomes. (It may be noted that the band of three-quarter-labeled recombinants appears broadened by about two drops, probably as a result of the considerable length of the *hc* interval, anywhere within which recombination presumably may occur. If this explanation is correct, then the sharpness of the band

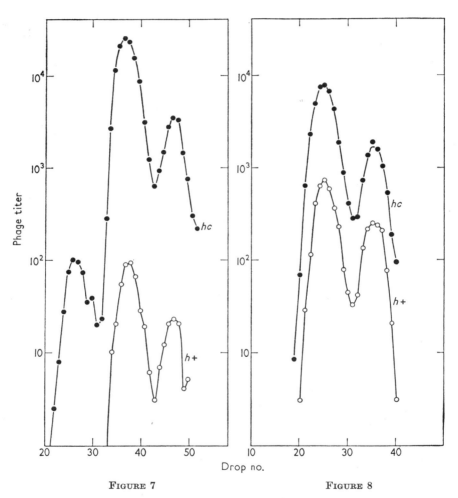

FIGURE 7   FIGURE 8

FIG. 7. Distributions of λ*hc* and λ*h* yielded by cells singly infected with fully-labeled phages from the gradient of Fig. 2. Assays were performed on restricting indicator bacteria. The most dense band of λ*hc* contains unadsorbed phages with fully-labeled chromosomes. The bands at fractions 37 and 47 contain phages with one-half-labeled and unlabeled chromosomes respectively.

FIG. 8. Distributions of λ*hc* and λ*h* yielded by cells singly infected with three-quarter-labeled phages from the gradient of Fig. 2. Assays were performed on restricting indicator bacteria. The bands at fractions 25 and 35 are formed by phages with half-labeled and unlabeled chromosomes respectively. One-quarter-labeled recombinant phages are not expected to plate on the restricting indicator and, accordingly, no indication of their presence is seen.

of half-labeled recombinants would suggest that most of them arose when the composition of the vegetative phage pool favored recombination between two conserved or two hybrid chromosomes more than recombination between a conserved and a completely new chromosome. In this regard, the relative amounts of the various labeled classes of parental and recombinant phages found in these experiments are compatible with the generally accepted view that recombination occurs throughout the latent period.)

Certain details of the distributions of Figs 4 to 6 should be noted. A light shoulder appears on the band of fully-labeled recombinants and possibly also on the band of three-quarter-labeled recombinants, whereas the half-labeled recombinants form a

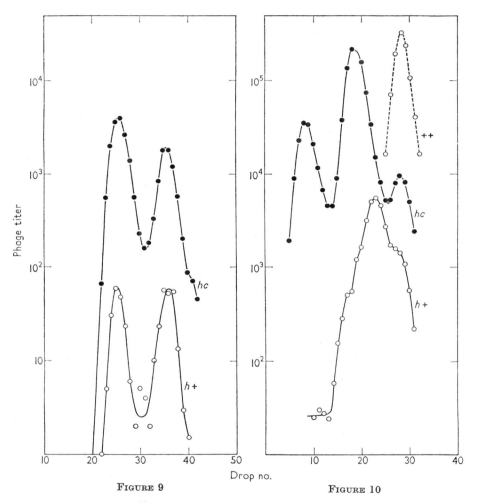

FIGURE 9   FIGURE 10

FIG. 9. Distributions of λhc and λh yielded by cells singly infected with one-half-labeled phages from the gradient of Fig. 2. Assays were performed on restricting indicator bacteria. The bands at fractions 26 and 36 are formed by phages with half-labeled and unlabeled chromosomes respectively.

FIG. 10. Distributions of λ, λhc and λh from the cross [^{13}C^{15}N]λhc × λ. Assays were performed on restricting indicator bacteria.

band as sharp as do non-recombinant phages. Upon repetition of the entire experiment, these features of the various distributions were found to recur.

(c) One-step growth of the cross progeny

Phages in reserve portions of the fully-, three-quarter-, and one-half-labeled fractions described above were allowed to multiply (one cycle) on strain C600·5. The total multiplicity of infection was 0·01 phage per bacterium. Each of the three lysates was analysed by density-gradient centrifugation with results shown in Figs 7, 8 and 9. Assays were performed on restricting indicator bacteria in order to suppress the appearance of unlabeled phages, which otherwise would obscure the discrete bands of half-labeled λhc and λh seen in each gradient. Comparison of Figs 7 to 9 with Figs 4 to 6 shows that, in each case, the recombinant phages give rise to half-labeled progeny with approximately the same efficiency as do the parent type phages.

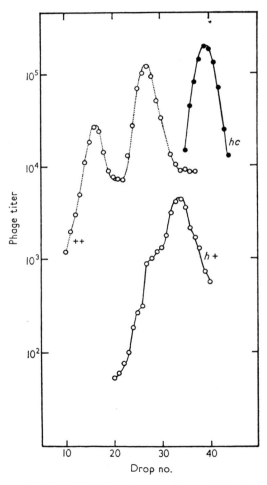

FIG. 11. Distributions of λ, λhc, and λh from the cross $\lambda hc \times [^{13}C^{15}N]\lambda$. Assays were performed on restricting indicator bacteria. The shoulder of λh at fractions 27 to 30 probably results from multiple recombination between hybrid and light chromosomes.

(d) Crosses of labeled and unlabeled phages

Two crosses were performed in which only one of the parental types was labeled with heavy isotopes. The results provide additional assurance that the fully-, three-quarter- and one-half-labeled recombinants examined above do in fact derive their label jointly from chromosomes of both parental types, as expected for recombination by simple breaking and joining. Figures 10 and 11 show the density distributions of progeny from the crosses $[^{13}C^{15}N]\lambda hc \times \lambda$ and $\lambda hc \times [^{13}C^{15}N]\lambda$ respectively. Assays were performed on restricting indicator bacteria in order to suppress the appearance of unlabeled phages. The prominent band of unrestricted λh found in both gradients occurs near the position expected for phages with one-quarter-labeled chromosomes. The exact location of the band corresponds to slightly less than one-quarter label when λ is the labeled parent, and slightly more when λhc is labeled. If the probability of recombination is relatively uniform in the region between h and c, then we may conclude that these markers are equidistant from a point about 55% of the way in from the "left" end of the chromosome.

Apart from the prominent band of one-quarter-labeled λh, there is an inflection in the distributions of Figs 10 and 11 suggesting the presence of half-labeled λh. No three-quarter- or fully-labeled recombinants are expected, nor is there any indication of their presence.

5. Discussion

Parental DNA is found in recombinant phages in discrete amounts which cannot be accounted for by breakage and copying, copy choice, or any process which would alter the genotype of a chromosome without substantial replacement of chromosomic material. Instead, the observed distribution of parental DNA among the progeny of crosses and among their descendants shows that genetic recombination in λ occurs by the breakage and joining of double-stranded phage chromosomes. However, a reservation must be placed on the interpretation of the evidence for the joining of chromosomes. The discovery of phages with half-labeled chromosomes among the descendants of labeled recombinants shows only that the two parental contributions to at least one strand of the recombinant chromosome are able to remain associated throughout the processes of injection, replication and phage maturation. Whether both strands retain their integrity in this fashion and whether the association results from the formation of a co-valent bond can not be decided.

It may be inquired whether breakage and joining are confined to a site or sites between h and c, while other mechanisms of recombination operate elsewhere on the λ chromosome. This possibility is rendered very unlikely by a number of observations, including the finding that chromosome breakage is associated with recombination in regions not overlapping the hc interval (Meselson & Weigle, 1961; Jordan & Meselson, unpublished experiments).

Considerable evidence for recombination by breakage and joining of DNA molecules has been obtained from experiments with T-even bacteriophages and with several bacteria. In the former case, it is well established that fragments of parental chromosomes appear in progeny phages (see Delbrück & Stent, 1957; Hershey & Burgi, 1956; Levinthal & Thomas, 1957; Roller, 1961; Kosinski, 1961; Kahn, 1964). Although other explanations can be devised, the implication is that recombination by breakage and joining is occurring in these cases. This interpretation is strengthened by the recent findings of Tomizawa & Anraku (1964), who have extracted from T4-infected cells structures containing DNA fragments from different parental chromosomes. In recombinants produced by bacterial conjugation, Siddiqi (1963) found an association between a paternal genetic character and radioisotope which had been used to label the female cells before mating. In bacterial transformation, structures which are either true recombinants or zygotes may be recovered very soon after the uptake of transforming DNA and under conditions which considerably inhibit DNA synthesis (Fox, 1960; Voll & Goodgal, 1961). It appears that these structures are inactivated at the same initial rate as in donor DNA if the latter is extensively labeled with ^{32}P (Fox, 1962). These various observations have been interpreted as evidence that double-stranded donor DNA is integrated directly into the recipient chromosome. This would be in accord with the finding of recombination by breakage and joining between unreplicated double-stranded λ DNA molecules. More recently, however, Lacks (1962), Bodmer & Ganesan (personal communication) and Fox (personal communication) have been led by studies of the fate of donor DNA after uptake by recipient cells to favor the possibility that only one strand of donor DNA is integrated during transformation. These apparently conflicting views might be reconciled with each other, and with our knowledge of the mechanism of recombination in λ, if joining were to take place through the agency of relatively extensive hybrid regions as described below.

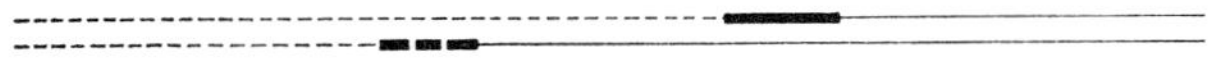

FIG. 12. Possible structure of an unreplicated recombinant DNA molecule. At least in mature phages, the possibility that the region of joining contains more than two DNA strands is made unlikely by the absence of heavy shoulders on all bands of recombinant phages and by the observation that λ heterozygotes are not more dense than other λ particles (Kellenberger, Zichichi & Epstein, 1962; Meselson, unpublished experiments).
Light solid and dashed lines represent polynucleotide strands contributed by two different parent molecules. Heavy lines represent either regions from which DNA has been removed and resynthesized along the opposing strand, or else regions in which missing DNA is not yet replaced. In the latter case, replacement is thought of as occurring during replication. The removal and repair of DNA may take place on only one strand rather than on both, as depicted in the Figure.

There is some suggestion in the present experiments that joining is accompanied by a small amount of DNA synthesis. A light shoulder appears with the band of fully-labeled recombinants; the band of three-quarter-labeled recombinants exhibits a slight and possibly significant light shoulder; whereas the one-half-labeled and un-labeled recombinants form bands as sharp in appearance as those of non-recombinant phages. These various features may be explained by assuming that up to five or ten per cent of the DNA of the λ chromosome is removed and resynthesized in the course of genetic recombination. The replacement of labeled DNA with unlabeled DNA would explain why a light shoulder appears with bands of fully-labeled recombinants, but not with unlabeled recombinants. The bands of three-quarter- and one-half-labeled re-combinants would possess light shoulders with intermediate displacements, which in the case of the latter may be so slight as to have escaped detection. An alternative explanation of the light shoulders is that they contain phages from which up to one or two per cent of the DNA has been removed without replacement. If most of the one-half-labeled and unlabeled recombinants replicate at least once before maturation, the missing DNA might usually be restored, or less likely, the deficient strands might be lost, explaining the absence of substantial light shoulders in these cases. Although neither explanation of the light shoulders can be substantiated without additional information, both entail the removal and probably the resynthesis of a small amount of DNA in recombinant chromosomes.

If we consider (1) the present demonstration that recombination occurs by breakage and joining of double-stranded DNA molecules, (2) the indication that some DNA is removed and resynthesized in the course of recombination, and (3) the extensive evidence that bacteriophage recombinants arise through the formation of partially heterozygous structures (see Luria, 1962), we are led to imagine that the unreplicated recombinant chromosome resembles the structure depicted in Fig. 12. The possibility that such structures are responsible for recombination in higher organisms as well as in bacteria and viruses has been discussed by Whitehouse (1963) and by the author (Meselson, 1963).

I am grateful to Mrs Miriam Wright for superb technical assistance in performing these experiments. This work was supported by grants from the U.S. National Science Foundation.

REFERENCES

Appleyard, R. K. (1954). *Genetics*, **39**, 440.

Arbor, W. & Dussoix, D. (1962). *J. Mol. Biol.* **5**, 18.

Bertani, G. & Weigle, J. (1953). *J. Bact.* **65**, 113.

Delbrück, M. & Bailey, W. T. (1946). *Cold Spr. Harb. Symp. Quant. Biol.* **11**, 33.

Delbrück, M. & Stent, G. S. (1957). In *The Chemical Basis of Heredity*, ed. by W. D. McElroy & B. Glass, p. 699. Baltimore: Johns Hopkins Press.

Fox, M. S. (1960). *Nature*, **187**, 1004.

Fox, M. S. (1962). *Proc. Nat. Acad. Sci., Wash.* **48**, 1043.

Hershey, A. D. & Burgi, E. (1956). *Cold Spr. Harb. Symp. Quant. Biol.* **21**, 91.

Hershey, A. D. & Rotman, R. (1948). *Proc. Nat. Acad. Sci., Wash.* **34**, 89.

Ihler, G. & Meselson, M. (1963). *Virology*, **21**, 7.

Kahn, P. L. (1964). *J. Mol. Biol.* **8**, 392.

Kaiser, A. D. (1957). *Virology*, **3**, 42.

Kaiser, A. D. (1962). *J. Mol. Biol.* **4**, 275.

Kellenberger, G., Zichichi, M. L. & Epstein, H. T. (1962). *Virology*, **17**, 44.

Kellenberger, G., Zichichi, M. L. & Weigle, J. J. (1961). *Proc. Nat. Acad. Sci., Wash.* **47**, 869.

Kosinski, A. W. (1961). *Virology*, **13**, 124.

Lacks, S. (1962). *J. Mol. Biol.* **5**, 119.

Leventhal, C. & Thomas, C. A. Jr. (1957). *Biochim. biophys. Acta*, **23**, 453.

Luria, S. E. (1962). *Ann. Rev. Microbiol.* **16**, 205.

Meselson, M. (1962). *Pontificiae Acad. Sci. Scripta Varia*, **22**, 173.

Meselson, M. (1963). *Symposium of 16th International Congress of Zoology.* New York: Doubleday & Co.

Meselson, M. & Weigle, J. J. (1961). *Proc. Nat. Acad. Sci., Wash.* **47**, 857.

Roller, A. (1961). Ph.D. Thesis, California Institute of Technology.

Siddiqi, O. H. (1963). *Proc. Nat. Acad. Sci., Wash.* **49**, 589.

Tomizawa, J. & Anraku, N. (1964). *J. Mol. Biol.* **8**, 516.

Voll, M. J. & Goodgal, S. H. (1961). *Proc. Nat. Acad. Sci., Wash.* **47**, 505.

Whitehouse, H. L. K. (1963). *Nature*, **199**, 1034.

J. Mol. Biol. (1964) **10**, 565–569

An Optical Method for the Analysis of Periodicities in Electron Micrographs, and Some Observations on the Mechanism of Negative Staining

Recently Markham has described a new method for the enhancement of image detail in electron micrographs of objects with regularly repeating structure. Two types of apparatus have been developed. In the first (Markham, Frey & Hills, 1963), the rotational periodicity of an object may be investigated by repeatedly rotating its electron micrograph through a given angle and making a photographic superposition of the set so obtained. If the angle has been chosen correctly, the rotational structure of the object will be much enhanced. A more sophisticated stroboscopic technique has also been developed (Markham, 1963), but in principle the method is still one of superposition of the image in various angular positions. Linear objects are treated analogously (Markham, Hitchborn, Hills & Frey, 1964); in essence, the image is translated successively through an appropriate given distance and a photographic superposition made. In each case the correct angle or translation must be determined by trial and error.

Markham's method, as well as our own problems in the electron microscopy of viruses and other biological particles, have stimulated us to consider other methods which might reveal periodicities present in the image of any object under investigation. Clearly what is required is a decomposition of the image into its harmonic components, i.e. a complete Fourier analysis. The Fraunhofer diffraction pattern of the image would automatically provide just such an analysis. In practice, one might use as the subject of diffraction either the image as it appears on the plate or a transparency of it made in reverse contrast. A monochromatic plane wavefront of visible light incident upon the subject will be broken up into a number of diffracted plane waves. Each such beam corresponds to one of the Fourier components into which the transmitting power of the subject can be resolved. The diffracted rays can be brought to a focus in the back focal plane of a lens situated behind the subject and the entire diffraction pattern can thus be observed or photographed. It may be helpful to point out a parallel between Markham's method and the one described here. Each spot on the diffraction pattern corresponds to a direction in which a regular periodicity occurs in the subject, and the distance of the spot from the centre corresponds to the inverse of the spacing. The optical method thus may be thought of as testing, simultaneously, all possible translations. Moreover, the testing is carried out effectively over the entire length of the subject and a good estimate is obtained of the strength of all possible harmonic components. Fortunately, the development of such an "optical diffractometer" has already been carried out by Lipson and Taylor and their collaborators in Manchester, to assist in the solution of X-ray diffraction problems (Taylor, Hinde & Lipson, 1951; Hughes & Taylor, 1953; Taylor & Thompson, 1957).

Our initial idea was to use the image on the electron-microscope plate itself as the subject and we have carried out some experiments in Manchester to show that good diffraction patterns can be obtained in this way. However, the scale of the periodicities being investigated is something of the order of 20 to 100 Å at a magnification of, say,

40,000. This gives rise to spacings of the order of a fraction of a millimetre; this is rather small for use with the Manchester optical diffractometer, which incorporates lenses of approximately 150 cm focal length. With this instrument we have therefore found it more convenient to work, not with the original plates, but with a photographic enlargement of the image. Given the existing apparatus, the diffraction pattern is relatively easy to form and to photograph. The crucial step in the procedure is to mount the photographic film or plate in immersion oil between optical flats; this is the technique introduced by Bragg & Stokes (1945) to iron out the optical path differences which arise because of the surface irregularities in photographic film.

In the present communication we report the application of this optical method to the analysis of the electron-microscope images of three linear biological objects of widely different origin and appearance; tobacco mosaic virus, actin filaments from muscle and catalase crystals. In the following paper (Finch, Klug & Stretton, 1964), we describe the application of the method to the analysis of a hitherto unknown structure, that of the "polyheads" of T4 bacteriophage. We have also examined images of various spherical viruses viewed along their symmetry axes and "end-on" views of discs of reaggregated TMV protein. We hope to discuss these elsewhere, but it should be stated at once that the patterns are more complex and difficult to interpret.

All the images we have examined were produced by the negative-staining technique and our experiments have been able to throw some light on the mechanism of that process. It appears that contrast is generally developed on both sides of the particle so that the stain must, in fact, envelop the entire particle. We are able to distinguish between the contributions to the prevailing contrast arising from the "near" surface of an actual particle and those arising from the "far" surface, and to estimate their relative strengths and the levels at which contrast is effectively developed.

We begin with a very simple example, the two-dimensional lattice observed in catalase crystals under negative staining. The micrographs were taken and supplied to us by Dr W. Longley, who is engaged in a crystallographic study of this material. Plate I(a) is a positive replica of the transparency used as subject in the diffractometer, and Plate I(b) is an enlargement of the diffraction pattern obtained. The positions of the spots give immediately the major periodicities present in the original image; the variation in intensity of the spots reflects more subtle periodicities present. In this example, the structure stands out more or less perfectly in the subject itself ; and the diffraction method does not add very much to the analysis, although it does provide an objective and accurate method of measurement.

Some general features of this diffraction pattern are worth commenting upon. The intense central region is contributed to not only by the zero-order diffracted beam, but also by the halation ring arising from reflection of this beam at the back surface of the photographic film. This phenomenon is commonly met with in the photography of strong sources of light. The two, approximately perpendicular, spikes of high intensity which pass through the centre of the pattern represent the aperture function of the subject, i.e. the diffraction pattern of the mask used to frame the region illuminated. Because the short end of the mask has not always been set strictly perpendicular to the long edge, these spikes are not necessarily mutually orthogonal. The very fine periodicities within the spikes are functions of the mask dimensions and are analogous to effects that arise in electron diffraction from very small crystals.

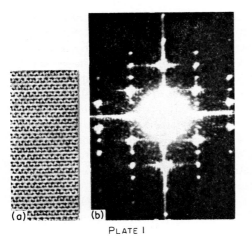

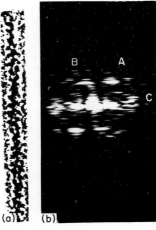

PLATE I

PLATE II

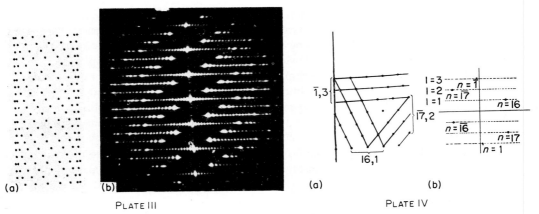

PLATE III

PLATE IV

PLATE I. (a) Catalase crystal, phosphotungstate stain, × 240,000. Full-scale positive replica of the image used in the diffractometer. Diffraction subjects shown in other Plates are likewise reproduced to actual scale.

(b) Optical diffraction pattern (enlarged) of (a). 1 mm corresponds to 0·075mm^{-1} in this and all other cases.

PLATE II. (a) Part of actin filament, uranyl acetate stain, × 333,000.
(b) Optical diffraction pattern of (a).

PLATE III. (a) Contact print of mask of holes representing one side of a projected "right-handed" helical structure having the parameters of TMV.
(b) Optical diffraction pattern of (a).

PLATE IV. (a) and (b) Diagrammatic representations of Plates III(a) and (b).

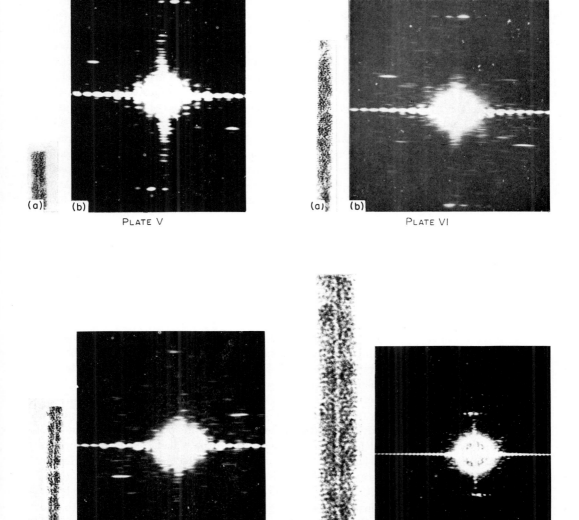

(a)   (b)   PLATE V

(a)   (b)   PLATE VI

(a)   (b)   PLATE VII

(a)   (b)   PLATE VIII

PLATES V TO VII. (a) Electron-micrographs of TMV negatively stained with uranyl formate, × 230,000.
(b) Corresponding optical diffraction patterns.

PLATE VIII. (a) TMV, uranyl formate stain, × 540,000.
(b) Optical diffraction pattern of (a).

Plate II(a) is part of the image of an actin molecule taken from an electron-micrograph supplied by Dr H. E. Huxley (1963). The aperture used appears as the black rectangular region bordering the molecular image. The diffraction pattern is shown in Plate II(b) and, well above the noise level, the characteristic pitch of the actin filament is immediately evident as a layer line of spacing 57 Å; intense peaks A and B occur on *both* sides of the meridian and correspond to the first maxima of the Bessel function J_1. The maxima marked A may be thought of as arising from one side of the original particle (which, on other evidence, we believe to be the surface closest to the electron-microscope grid) and correspond to a radius of 21 Å. They are stronger than the maxima labelled B which apparently arise from the other side, and which correspond to a radius of about 59 Å. This is a very strong indication that, in this case, the contrast of negative staining is produced on *both* sides of the particle but to a different extent on each. Moreover, the stronger contrast seems to be associated with the deeper penetration of the stain. With a particle as thin as this, one would obviously not have a sharp demarcation between two "sides" of such different effective radii. It is much more likely that the radius contrasted by the stain varies continuously around the particle. The two radii actually determined are thus averages over the two "sides". Diffraction in the regions marked C arises from the organization of the actin molecule into twin cables of steep pitch, twining around the long axis of the molecule. This can be clearly distinguished from the equatorial spots of the aperture function.

The next application we describe is to tobacco mosaic virus (TMV), on which most of our preliminary efforts have been concentrated. We have used electron micrographs provided by Dr J. T. Finch (1964) and the results are presented in Plates V to VIII. The marked axial periodicity of 23 Å corresponding to the pitch of the TMV helix is immediately apparent in the diffraction patterns as a near-meridional maximum and, indeed, these patterns do provide a very precise measurement of the average value of this spacing over the whole length of the particle image illuminated. Much more information can be extracted from this diffraction pattern by use of helical diffraction theory (see Plate III) or even of the rough physical version of it illustrated in Plate IV, and, for a fuller appreciation, we must discuss this.

Plate III(a) is a contact print of a mask of holes representing one side of the surface lattice of TMV drawn for a radius of 100 Å and projected on to a plane. Plate III(b) shows the diffraction pattern obtained. It is not possible to describe this pattern in simple terms although certain features can be elucidated by reference to the diagram in Plate IV, which represents the surface lattice unrolled from a cylinder into a plane. In this diagram the helical lines which correspond to the largest three spacings in the TMV lattice are represented by three families of straight lines. Considered as a two-dimensional diffracting object, this plane net gives rise to the diffraction spots shown in Plate IV(b). However, as one proceeds across the *projected* lattice of the actual structure, the spacing of these lines is not constant (Plate III(a)) and it is this feature which gives rise to diffraction patterns of greater complexity; the single spots now become sets of peaks, viz. Bessel functions. Formally, in the terminology of Klug, Crick & Wyckoff (1958), these three families of maxima correspond to the helices ($\bar{1}$,3), (16,$\bar{1}$) and ($\bar{17}$,2) where the first number represents the order, n, of the Bessel function and the second number the layer line, l, on which it occurs. The first figure also gives the number of co-axial parallel helices comprising a family. Thus, (1,3) corresponds to the basic single helix of 23 Å pitch. In what follows we will simply refer to the diffraction maxima by their n index, which is invariant under small deformations of the helix (Franklin & Klug, 1955). The convention is such that $(\bar{1},3)$, for instance, corresponds to a right-handed helix and $(1,3)$ to a left-handed helix. In the diffraction pattern of the projection of a *complete* structure, contributions of both types would seem to appear. But if only one "side" of the structure is present (or, for our present purposes, imaged), then only one of these directions contributes. The pattern arising from the other "side" would, of course, be the mirror-image reflected in the meridional line.

Plate III(a) represents the ideal case which would arise if the particle remains undistorted under the conditions of negative staining and the exact helical lattice of one "side" is preserved and imaged. If the particle is considerably flattened, then the image would approximate more to that of a plane lattice such as that shown in Plate IV(a). The details in Plate III(b) and Plate IV(b) therefore represent the two extremes of optical diffraction patterns to be expected in our experiments on helical structures.

To obtain the optical diffraction pattern of tobacco mosaic virus, a region of a particle image was selected and isolated by a rectangular mask. The long side of the mask was kept as close and as parallel as possible to the sides of the particle. The long edges therefore indicate the direction of the particle axis and the sharp horizontal spike arising from them in the diffraction pattern therefore represents a base line from which angles in the pattern can be measured. In particular, the meridian is taken as the line perpendicular to this horizontal spike. The diffraction patterns show strong contributions of $n = 1$ and $n = 16$ on the third and first layer lines respectively. The maximum $n = 17$ on the second layer line is only weakly developed in most images. This is what we should expect by comparison with the low-angle X-ray diffraction pattern of oriented gels of TMV which show that the contribution to $n = 17$ by radii near the outside of the particle is much weaker than their contributions to $n = 16$ and $n = 1$. This means that the grooves (17,2) on the surface of the particle are not so marked as are the grooves (1,3) and (16,1). It appears that this feature of the structure is preserved under negative staining.

All the diffraction patterns appear to show a contribution from *both* the "near" and "far" sides of the particle, since each layer line has spots on both sides of the meridian. In the case of most of the particle images we have examined, these spots (with indices (n, l) and $(\bar{n}, l)$) are unequal in intensity. This indicates that the contrast has been *unequally* developed on the two "sides". Moreover, the stronger spot of the pair is generally further from the meridian than the weaker, just as we have seen in the actin pattern. It corresponds, in most cases, to a radius of about 60 Å and this presumably gives an indication of the depth of penetration of the negative stain into the particle. The weaker spot corresponds to a radius of about 90 to 100 Å, a figure which is close to the maximum diameter of the particle. This confirms our conclusions from the actin pattern that the stronger contrast is associated with deeper penetration of the stain.

Another striking fact which has emerged is that the extremely disordered appearance of some of the particles is only an apparent phenomenon due to the superposition of contrast developed on the two sides of the particle. Thus, for example, the image in Plate VIII(a) appears to the eye to have the basic helix right-handed at the bottom but more or less disordered at the top with a tendency towards left-handedness. The diffraction pattern at right shows, as compared with the other diffraction patterns reproduced, a greater proportion of contrast from the "other side" of the particle, and it is this effect which makes the electron-microscope image appear so disordered. There is no appreciable disorder whatever in the particle itself, since there is no diffraction on the meridian in the 23 Å region. This would be expected if the structure were actually changing from a right-handed to a left-handed one through a region of true disorder.

Of the other particle images reproduced here, two are predominantly "left-handed" and one is predominantly "right-handed"; but here again the diffraction pattern reveals that parts of the "other side" are showing through on the image. It is thus clear that the determination of the absolute hand of the particle will involve more than a simple determination of the dominant screw sense in any given particle or series of particles. We hope, however, from a more thorough study of the negative staining process, to resolve this question.

While the applications discussed here have all been to images produced by the negative staining technique, it should be stressed that the method described is applicable to the analysis of any image produced by any other technique. The analytical procedures which have been developed for a wide range of X-ray crystallographic problems may as well be employed in interpretation of these optical diffraction patterns. The difference is that we are here dealing with the diffraction pattern of an individual particle or part of a particle. In principle, this sort of information could be obtained directly by electron diffraction combined with electron-microscopy, provided it were possible to produce electron beams small enough to illuminate a single biological macromolecule, but at present this is not feasible.

We wish to express our gratitude to Professor H. Lipson and Dr C. A. Taylor of the Department of Physics, Manchester College of Science and Technology, for making freely available to us their optical diffractometer and the general facilities of their laboratory. One of us (J. E. B.) has benefited greatly from having worked previously at Manchester under Dr Taylor's supervision.

We also thank our colleagues, Drs H. E. Huxley, J. T. Finch and W. Longley for providing us with the electron micrographs on which these studies were carried out.

This investigation was supported in part by a Public Health Service fellowship no. 7-F3-GM-19813-02 from the Institute of General Medical Sciences, U.S. Public Health Service to one of the authors (J. E. B.).

Medical Research Council Laboratory of Molecular Biology   A. KLUG
Hills Road   J. E. BERGER
Cambridge, England

Received 22 October 1964

REFERENCES

Bragg, W. L. & Stokes, A. R. (1945). *Nature*, **156**, 332.
Finch, J. T. (1964). *J. Mol. Biol.* **8**, 872.
Finch, J. T., Klug, A. & Stretton, A. O. W. (1964). *J. Mol. Biol.* **10**, 570.
Franklin, R. E. & Klug, A. (1955). *Acta Cryst.* **8**, 777
Hughes, W. & Taylor, C. A. (1953). *J. Sci. Instrum.* **30**, 105.
Huxley, H. E. (1963). *J. Mol. Biol.* **7**, 281, Plate XVIII.
Klug, A., Crick, F. H. C. & Wyckoff, H. W. (1958). *Acta Cryst.* **11**, 199.
Markham, R. (1963). In *Viruses, Nucleic Acids and Cancer*, p. 180. Baltimore: The
    Williams & Wilkins Company.
Markham, R., Frey, S. & Hills, G. J. (1963). *Virology*, **20**, 88.
Markham, R., Hitchborn, J. H., Hills, G. J. & Frey, S. (1964). *Virology*, **22**, 342.
Taylor, C. A., Hinde, R. M. & Lipson, H. (1951). *Acta Cryst.* **4**, 261.
Taylor, C. A. & Thompson, B. J. (1957). *J. Sci. Instrum.* **34**, 439.

J. Mol. Biol. (1965) **11**, 90–96

The Detachment and Maturation of Conserved Lambda Prophage DNA

Mark Ptashne

Biological Laboratories, Harvard University
Cambridge, Massachusetts, U.S.A.

(*Received 15 September 1964*)

Isotopically labeled *Escherichia coli* cells lysogenic for phage λ were infected with phage 434hy. Among the resulting phage progeny there was found a class of λ phages containing DNA made predominantly before infection. Experiments are presented which indicate that this DNA is attached to the bacterial chromosome before infection with 434hy. It is argued that the lysogenized bacterial chromosome contains a complete λ chromosome which may be detached and matured intact.

1. Introduction

Genetic experiments have shown that *Escherichia coli* bacteria lysogenic for the temperate phage λ maintain the genetic determinants of the phage (the prophage) at a specific location on the host chromosome (Jacob & Wollman, 1961). ^{32}P suicide experiments suggest that the prophage is DNA (Stent, Feurst & Jacob, 1957), but the nature of the attachment of the prophage to the host DNA is not known. Campbell (1962) has suggested that the prophage is λ DNA which is inserted into the host genome by a single reciprocal cross-over between circular λ and *E. coli* DNA molecules. If this recombinational event occurs by a break-and-join mechanism, the original lysogen would contain both strands of the phage DNA. Detachment of the prophage, on the Campbell model, proceeds by a reversal of this process. If this picture is correct, we might expect that the prophage DNA may appear in viable phage without replication or recombination with other DNA molecules. This expectation is confirmed by the experiments reported in this paper.

2. Plan of the Experiment

The experiments described here exploit the observation that superinfection of a cell lysogenic for λ by the closely related phage 434hy (see Materials and Methods section) induces the production of an average of 0·01 to 1 phage of the λ immunity type in addition to the normal burst of 434hy phages. Single-burst experiments show that each induced cell yields about 1 phage with λ immunity (Thomas & Bertani, 1964; Ptashne, unpublished experiments). The small yield of λ phages suggests that under these conditions extensive replication of the λ genomes does not occur and that some of these phages may contain prophage DNA. To test this possibility, lysogenic cells were grown in heavy isotopes medium and then transferred to light medium and immediately infected with phage 434hy. Density-gradient centrifugation of the progeny reveals three distinct classes of λ phages containing, respectively, heavy, hybrid and light DNA. An experiment is presented which indicates that the DNA of the heavy phages is attached to the bacterial chromosome as prophage before infection with 434hy. Also, the superinfecting phage is genetically marked so that the density distribution of a certain class of recombinants formed between the λ and 434hy phages may be analyzed.

3. Materials and Methods

Phage and bacterial strains

Phage λ is the "wild type" of Kaiser (1957). Phage 434hy is described by Kaiser & Jacob (1957). It contains the immunity region of phage 434 crossed into a predominantly λ genetic background. It is approximately 0.005 g/cm³ less dense in CsCl solution than λ. Phage $hi^{434}c$ was prepared by successively crossing 434hy with λh of Kaiser (1962), and with λc67 (Kaiser, 1957). i^{434} denotes immunity region of phage 434. E. coli K12 strain W3110 (from E. M. Lederberg) and E. coli K12 strain C600 (Appleyard, 1954) were lysogenized with $λind^-$ (Jacob & Campbell, 1959) and with λ. W3110 ($λind^-$) was made /λ by selection for resistance to λvir. This ensures that few if any of the $λind^-$ phages produced by spontaneous induction during growth of the lysogenic host will readsorb and inject their DNA. W3110 ($λind^-$)/λ plates $hi^{434}c$, but not $i^{434}c$. Phages λ and 434hy were distinguished by plating on the indicators C600(434hy) and C600(λ). Recombinants λh were detected by plating on K(hi^{434})/λ (Kaiser, 1962).

Methods

Strains W3110($λind^-$)/λ and C600(λ) were grown at 37°C to saturation in heavy isotopes medium containing $^{13}C^{15}N$ (Meselson & Weigle, 1961). The cells were centrifuged, resuspended in 0.01 M-MgSO₄ at about 10^8 cells/ml. and infected with 10^9 $hi^{434}c$/ml. After a 10-min adsorption period at 37°C, an equal volume of double-strength Tryptone broth was added and the cells were reincubated at 37°C for another 5 min. The cells were then chilled, washed with 0.01 M-MgSO₄ on a Millipore filter to remove unadsorbed phage, and resuspended in Tryptone broth at about 5×10^6 cells/ml. After 55 min growth at 37°C with aeration, the cultures were lysed with chloroform, cleared of debris by centrifugation, and banded in CsCl according to Meselson & Weigle (1961).

For the experiment shown in Fig. 4, W3110($λind^-$)/λ was grown to about 1.6×10^8 in heavy isotopes medium. A portion was removed and treated as above. The remainder was centrifuged and the cells resuspended in broth at 8×10^7 cells/ml. and aerated at 37°C. After 0.7 and 2.2 generations of growth as determined by cell counts, equal fractions were removed and treated as above. A distinct lag period was observed following transfer to the light medium, and so these generation times may be only approximately equal to the number of DNA doublings (Schaechter, 1961).

4. Results

Superinfection of heavy cells lysogenic for λ

Figure 1 shows the density distribution of progeny phage when heavy C600(λ) is superinfected with light $hi^{434}c$ in light medium at a multiplicity of infection of about 10. Three peaks of phages with λ immunity are observed at the densities expected for λ phages with light protein coats containing the normal complement of heavy, hybrid and light DNA, respectively (Meselson & Weigle, 1961). The $hi^{434}c$ phages are distributed in a single band at the density characteristic of light 434hy. The absence of a substantial shoulder or band of denser $hi^{434}c$ indicates that DNA replication occurring after superinfection uses mainly the light isotopes. The heavy phages contain DNA made either entirely or predominantly before infection.

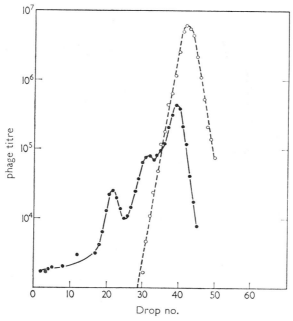

FIG. 1 Density-gradient analysis showing the progeny phage when C600(λ) labeled with $^{13}C^{15}N$ is infected in light medium with light $hi^{434}c$ at a multiplicity of infection of about 10. $hi^{434}c$ burst size: 15; λ burst size: 0·25. —●—●—: λ. —○—○—: $hi^{434}c$.

Superinfection of heavy cells lysogenic for λind⁻

Figure 2 shows the results of superinfecting cells lysogenic for λ carrying the mutation ind^-. This phage, described by Jacob & Campbell (1959), cannot be induced by u.v. light, apparently because it produces an unusually effective or abundant repressor. The positions of the bands here are identical to those in Fig. 1, but in this case the heavy peak contains nearly one-third of all the λ's released, whereas in the first experiment only about one-twentieth of the λ's are found in the conserved peak. The heights of the peaks of Fig. 2 indicate that one-third of the $λind^-$ prophages replicate an average of two to three times after infection with $hi^{434}c$. The differences observed between Figs 1 and 2 are reproducible and are determined by the repressor genotype. When W3110(λ) is infected with $hi^{434}c$, the density distribution of the induced λ's resembles that seen in Fig. 1.

Distribution of the λh recombinants

The distribution of the λ's seen in Figs 1 and 2 strongly suggests that induction by $hi^{434}c$ is not merely the rescue of $i^λ$ markers by genetic recombination, because this process would not be expected to produce three distinct density types of $i^λ$ phages. Further confirmation of this is seen in the density distribution of those λ phages which have acquired the h marker by recombination with the infecting $hi^{434}c$ (Fig. 3). No λh recombinants are found in the heavy peak. In fact, the distribution of the recombinants is very similar to that observed for crosses performed with isotopically labeled and unlabeled phages carrying markers similar to the ones used here (Meselson, 1964).

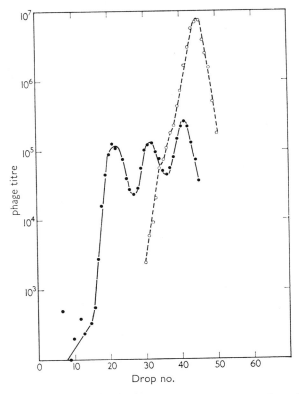

FIG. 2. Density-gradient analysis showing the progeny phage when isotopically labeled W3110(λind^-)/λ is infected in light medium with light $hi^{434}c$, multiplicity of infection about 10. $hi^{434}c$ burst size: 30; λ burst size: 1. —●—●—: λ. —○—○—: $hi^{434}c$.

Origin of the conserved phage

The experiments described above show that unreplicated λ DNA is found in mature phages following infection of a λ lysogen with $hi^{434}c$. However, the question arises whether the heavy DNA was attached to the bacterial chromosome before infection with $hi^{434}c$, or whether it was instead free λ DNA present in the cytoplasm. Wolf & Meselson (1963) have shown that repressed cytoplasmic λ DNA, introduced by superinfecting a λ lysogen with λ phage, does not replicate. In contrast, prophages which are known to be attached to the host chromosome replicate once for every doubling of the bacterial DNA. The criterion of replication during cellular division may be used to indicate the origin of the DNA found in the heavy phages. This is accomplished by growing heavy lysogenic cells in light medium for various times before superinfection with $hi^{434}c$. If the heavy phages carry DNA which was replicating in synchrony with the host DNA, they should be completely replaced by hybrid phages when the bacterial DNA has replicated just once. Unintegrated repressed λ DNA is simply diluted by cell division, however, and any contribution to the heavy peak from this class would only be halved after one duplication in light medium before superinfection.

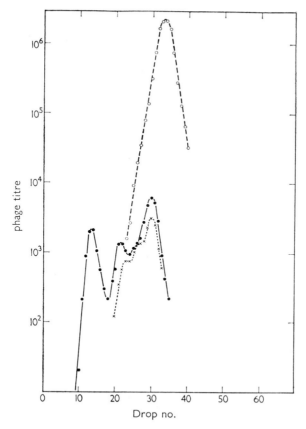

FIG. 3. Density distribution of λh recombinants produced by infection of W3110(λind⁻)/λ with $hi^{434}c$, multiplicity of infection 10. $hi^{434}c$ burst size: 15; λ burst size: 0·03. —●—●—: λ; —○—○—: $hi^{434}c$; —×—×—: λh.

The results of such an experiment are presented in Fig. 4. It is seen that if the heavy cells are infected immediately upon transfer to light medium, the usual pattern of three nearly equal peaks of λ phages is observed. However, if superinfection is preceded by about one generation of growth, the class of heavy phages drops by a factor of at least 10³. After another 1·5 generations of growth, the ratios of light to hybrid phages increase by the amount expected if the prophages are dividing with the bacterial host and if the extent of replication of the induced prophages after induction is about that indicated by the heights of the peaks in Fig. 2.

5. Discussion

The experiments described in this paper show that infection of a cell lysogenic for λ with $hi^{434}c$ produces phages containing DNA made either entirely or predominantly before infection. The fact that this DNA has all replicated when the bacterial DNA has replicated about once suggests that the heavy DNA is derived directly

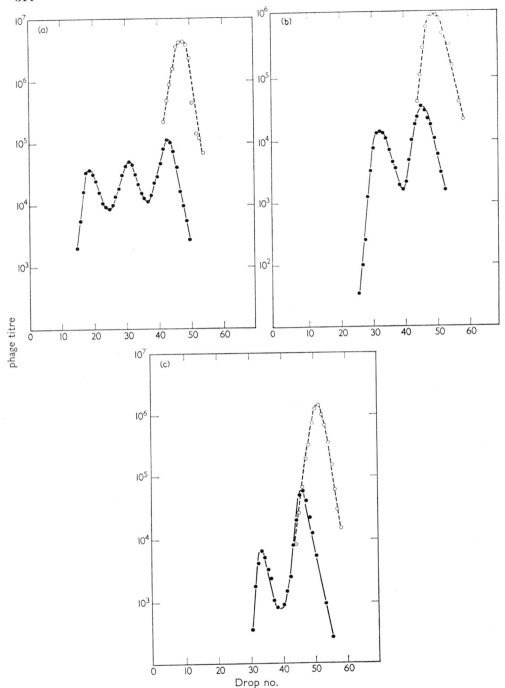

Fig. 4. Density-gradient analysis showing the replacement of the heavy phages seen in Figs 1 and 2 after growth of W3110(λind^-)/λ in light medium. The isotopically labeled cells were infected with $hi^{434}c$, multiplicity of infection about 10, immediately following transfer (a), and after about 1 and 2·5 generations of growth (b) and (c). In these experiments, $hi^{434}c$ burst size: 15 to 20; λ burst size: about 0·3. —●—●—: λ. —○—○—: $hi^{434}c$.

from the prophage which is attached to the bacterial chromosome. We cannot eliminate the possibility that the prophage DNA might spend some fraction of each cell generation in an unattached state, and that it is only in this state that it may be matured by infection with $hi^{434}c$. With this reservation, we may conclude that the lysogenized bacterial chromosome contains a complete λ chromosome which may be detached and matured without replication to form a viable phage. These results do not, of course, tell us anything about the nature of the attachment of the λ DNA molecule to the chromosome of the lysogenic bacterium.

It is not known whether the mechanism of induction by 434hy is similar to that observed with other inducing agents such as u.v. light. The maturation of conserved DNA might plausibly be explained by assuming that the prophage is detached late in the latent period, and, in the presence of a large number of 434hy genomes, under-goes only limited replication before maturation. The higher percentage of conserved phages observed when a λind^- lysogen is induced by 434hy might be explained by assuming the ind^- repressor either delays detachment or inhibits replication after detachment.

The author gratefully acknowledges the cogent criticisms of Professor Matthew Meselson. This investigation was carried out while the author held an NSF co-operative pre-doctoral fellowship, and was supported in part by a grant from the National Science Foundation.

REFERENCES

Appleyard, R. K. (1954). *Genetics*, **39**, 440.
Campbell, A. M. (1962). *Advanc. Genetics*, **11**, 101.
Jacob, F. & Campbell, A. M. (1959). *C.R. Acad. Sci. Paris*, **248**, 3219.
Jacob, F. & Wollman, E. L. (1961). *Sexuality and the Genetics of Bacteria*. New York: Academic Press.
Kaiser, A. D. (1957). *Virology*, **3**, 42.
Kaiser, A. D. (1962). *J. Mol. Biol.* **4**, 275.
Kaiser, A. D. & Jacob, F. (1957). *Virology*, **4**, 509.
Meselson, M. (1964). *J. Mol. Biol.* **10**, 734.
Meselson, M. & Weigle, J. J. (1961). *Proc. Nat. Acad. Sci., Wash.* **47**, 857.
Schaechter, M. (1961). *Cold Spr. Harb. Symp. Quant. Biol.* **26**, 53.
Stent, G. S., Fuerst, C. R. & Jacob, F. (1957). *C.R. Acad. Sci. Paris*, **244**, 1840.
Thomas, R. & Bertani, G. (1964). *Virology*, in the press.
Wolf, B. & Meselson, M. (1963). *J. Mol. Biol.* **7**, 636.

J. Mol. Biol. (1965) **12**, 456–465

Molecular Consequences of the Amber Mutation and its Suppression

A. O. W. Stretton and S. Brenner

Medical Research Council Laboratory of Molecular Biology, Cambridge, England

(Received 15 March 1965)

Each amber mutant of the head protein of bacteriophage T4D produces a characteristic fragment of the polypeptide chain when grown on su^- strains of *Escherichia coli*. On su^+ strains chain propagation occurs, but chain termination is not completely prevented. The structures of the relevant regions of the head protein have been determined in wild-type T4D and in the amber mutant H36 grown on su^- and su^+ bacteria. The results show that the polypeptide formed by H36 is an N-terminal fragment produced by termination of the chain at the site of mutation. Suppression of the mutant by one suppressor, su_1^+, leads to the insertion of serine at this site, replacing a glutamine present in the wild type.

1. Introduction

Amber mutants of the head protein of bacteriophage T4D produce fragments of the polypeptide chain when grown in non-permissive bacteria. Each mutant produces a fragment of distinctive size corresponding to its position on the genetic map, thus showing that the gene is colinear with the polypeptide chain (Sarabhai, Stretton, Brenner & Bolle, 1964). Strains of *Escherichia coli* are known which permit the growth of amber mutants; these bacteria contain one of a number of different suppressors which allow propagation of the polypeptide chain with different efficiencies (Kaplan, Stretton & Brenner, 1965, manuscript in preparation). It is likely that the effects of the amber mutation are expressed at the level of protein synthesis, since amber mutants vanish when the phase of reading of the genetic message is altered (Brenner & Stretton, manuscript in preparation). In order to understand the molecular consequence of this type of mutant and the mechanism of its suppression, we have made a detailed study of the relevant region of the head protein in the amber mutant H36. In this paper we show that the polypeptide produced by this mutant in the su^- host is an N-terminal fragment of the head protein, which allows us to conclude that termination of the synthesis of the polypeptide chain occurs at the site of the mutation. We also show that suppression of this mutant by a strain carrying the suppressor su_1^+ leads to the introduction of a serine residue in the propagated chain and that this replaces a glutamine residue present in the wild type. This result is in excellent agreement with the results of Weigert & Garen (1965) and of Notani, Engelhardt, Konigsberg & Zinder (1965).

2. Materials and Methods

Phages. T4D and the amber mutant H36 (Sarabhai *et al.*, 1964).

Bacteria. *E. coli* B was used as the su^- strain, and *E. coli* CR63 as the su^+ strain. The latter contains a suppressor which we call su_1^+ and which appears to be indistinguishable from the *Su–1* of Weigert & Garen (1965).

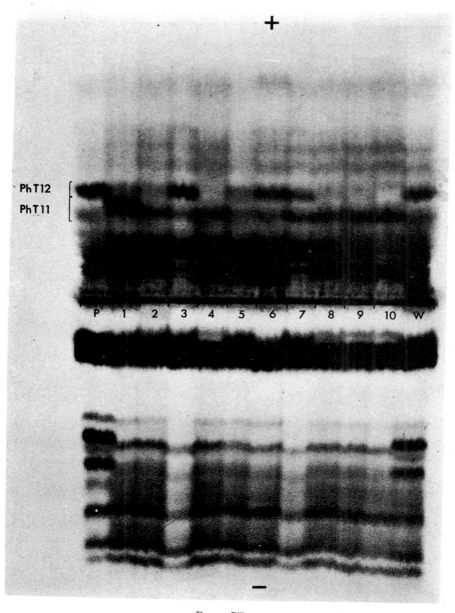

PLATE I(i).

PLATE I. Autoradiographs of tryptic peptides of [^{14}C]phenylalanine-labelled protein. P is wild-type phage, W is wild-type infected cells, and 1 to 10 are amber mutant-infected cells as follows: 1 = H36, 2 = H32, 3 = C208, 4 = C140, 5 = C137, 6 = A489, 7 = B278, 8 = B272, 9 = B17, and 10 = H11.

(i) Ionophoresis at pH 6·4 of total tryptic digest. (ii) Ionophoresis at pH 3·6 of the zone containing peptide PhT11 as marked in (i). (iii) Ionophoresis at pH 3·6 of the zone containing peptide PhT12 as marked in (i). Peptide PhT12 typically shows severe trailing at pH 3·6.

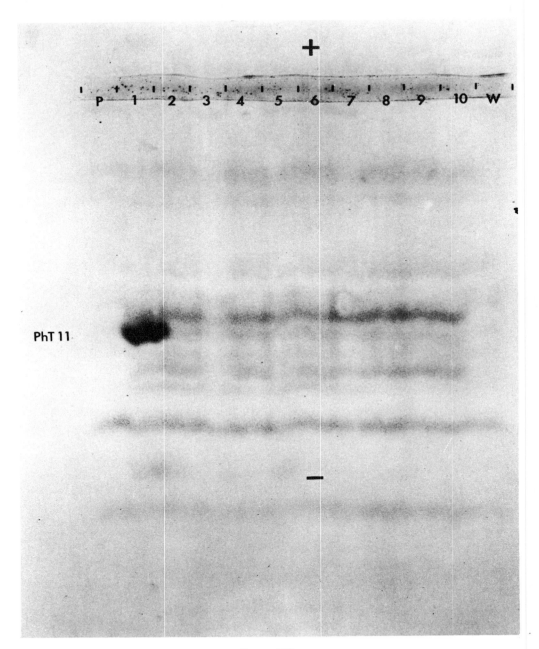

+

P   1   2   3   4   5   6   7   8   9   10   W

PhT 11

−

PLATE I(ii).

PLATE I(iii).

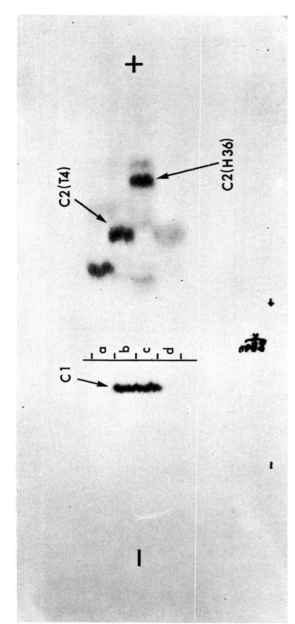

(i)

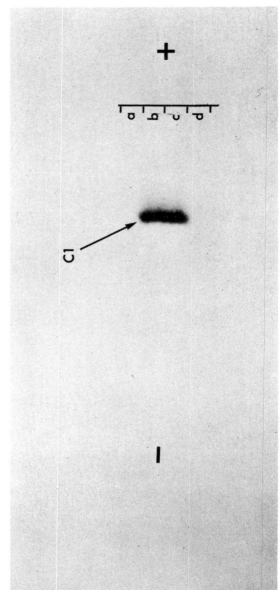

(ii)

PLATE II.

Autoradiographs of labelled peptides as follows. (a) Untreated [¹⁴C]phenylalanine-labelled peptide PhT12 from T4D. (b) Chymotryptic digest of [¹⁴C]phenylalanine-labelled peptide PhT12 from T4D. (c) Chymotryptic digest of [¹⁴C]phenylalanine-labelled peptide PhT11 from H36. (d) Chymotryptic digest of [¹⁴C]arginine-labelled peptide PhT12 from T4D.

(i) Ionophoresis at pH 6·4.
(ii) Ionophoresis at pH 3·6 of zone C1 from (i).

Media. M9 (Adams, 1959) and 3XD (Fraser & Jerrell, 1953). M9 buffer is M9 without NH$_4$Cl and glucose.

Buffers. Electrophoresis buffer pH 6·4 contains 100 ml. pyridine, 4 ml. glacial acetic acid made up to 1 litre with water; the pH 3·6 buffer contains 10 ml. pyridine, 100 ml. glacial acetic acid/l.

[^{14}C]Amino acids were obtained from the Radiochemical Centre, Amersham.

Enzymes. Trypsin (TCA purified), chymotrypsin, pepsin, DNase, RNase, and lysozyme were obtained from the Worthington Biochemical Company, Freehold, N.J. Pronase was a product of CalBiochem. Leucine aminopeptidase was obtained from Seravac Pty., Ltd.

General methods

(a) *Preparation and digestion of phage protein labelled with [^{14}C]amino acids*

Ten ml. cultures of *E. coli* B or CR63 grown in M9 to 5 × 10^8 cells/ml. at 37°C were infected with phage at a multiplicity of 5. After 5 min of aeration, the cultures were superinfected with the same phage at the same multiplicity to produce lysis inhibition. Five min later 5 μc of the [^{14}C]amino acid was added. [^{14}C]Phenylalanine, proline, arginine and isoleucine are not appreciably converted into other amino acids in *E. coli*, but in the case of valine, 50 μg/ml. [^{12}C]leucine was added to quench conversion into leucine, and for glycine, 50 μg/ml. uracil and hypoxanthine were added to prevent incorporation into nucleic acids. The cultures were then handled in one of two ways. To prepare *protein from infected bacteria*, aeration was continued for 60 to 90 min and the cells removed by centrifugation. These were resuspended in 2 ml. of distilled water and the suspensions heated at 100°C for 20 min. After cooling, DNase, RNase and lysozyme were added to final concentrations of 10, 10 and 50 μg/ml., respectively, and digestion allowed to proceed at 37°C for 2 hr. The suspensions of precipitated protein were dialysed overnight against 1000 vol. distilled water, and ammonium bicarbonate added to give a final concentration of 1%.

To prepare *protein from phage particles*, the infected cultures were aerated for 4 hr and then lysed by the addition of chloroform. The phages were purified by 2 cycles of low- and high-speed centrifugation and finally resuspended in 2 ml. of M9 buffer. This phage suspension was heated at 100°C for 20 min, and treated with DNase (10 μg/ml.) for 2 hr at room temperature. The precipitated protein was collected by centrifugation, washed twice with distilled water and resuspended in 2 ml. of 1% ammonium bicarbonate.

Trypsin was added to the suspensions to a final concentration of 5 μg/ml. and digestion allowed to proceed for 7 hr at 37°C. Additional enzyme (5 μg/ml.) was added at 1 and 2 hr. The digestion was terminated by the addition of one drop of glacial acetic acid, and the digests taken to dryness.

(b) *Preparation and digestion of large amounts of phage protein*

A New Brunswick Fermenter was used to prepare 36-l. batches of phage lysate. Three vessels, each containing 12 l. of 3XD were inoculated with *E. coli* B or CR63, and aerated vigorously at 37°C. When the cultures had reached a concentration of 1 to 2 × 10^9 cells/ml. they were infected with about 10^7 phage/ml. Artificial lysis with chloroform at 4 hr after infection yielded extremely viscous lysates which were allowed to stand overnight at room temperature. The lysates were decanted from the chloroform, and debris removed by continuous flow centrifugation in a Sharples centrifuge. Two procedures were used to purify the phage protein. In the first, the phage was purified by two cycles of high- and low-speed centrifugation and finally resuspended in about 500 ml. of M9 buffer. In the second, the phage was precipitated from the clarified lysate by the addition of ammonium sulphate to a final concentration of 50% saturation (340 g/l. of lysate). After standing for 2 hr, the precipitate was removed by continuous flow centrifugation. It was then resuspended in 1 litre of M9 buffer which results in osmotic shocking of some of the phage particles. The extremely viscous suspension was then treated with DNase (about 5 μg/ml.), and then centrifuged at low speed. High-speed centrifugation of the supernatant fluid yielded a pellet of phage particles and phage ghosts, which were resuspended in 500 ml. of M9 buffer and clarified by low-speed centrifugation. Each large-scale batch yielded between 0·5 and 2 g of phage protein.

The suspensions of phage were pelleted by high-speed centrifugation and lyophilized and ground into a powder; 100 ml. of 10^{-3} M-MgSO$_4$ was added, followed by DNase (10 μg/ml.), and the suspensions digested with stirring for 4 to 6 hr at room temperature. The phage ghosts were removed by centrifugation, washed with acetone and ether and dried; 500-mg batches of phage protein were suspended in 30 ml. of water and heated at 100°C for 20 min. After cooling, ammonium bicarbonate was added to a final concentration of 1% followed by 2·5 mg of trypsin and 2 drops of toluene. Digestion was allowed to proceed for 18 hr at 37°C with stirring. The digest was lyophilized and resuspended in 6 ml. pyridine acetate buffer, pH 6·4, and the insoluble material removed by centrifugation.

It is unnecessary to purify the head protein which constitutes about 90% of the phage ghost. The other major protein component, the sheath, is not denatured by these procedures and remains resistant to proteolytic digestion.

(c) Fractionation of ^{14}C-labelled digests

This was achieved by paper ionophoresis. The dried digest was suspended in water and an amount containing about 1 μc of ^{14}C applied as a 1-in. band to Whatman 3MM paper. Ionophoresis was carried out at pH 6·4 for 1·25 hr with a voltage gradient of 70 v/cm. After drying, the peptides were detected by autoradiography using Ilford X-ray film. For further fractionation, peptide zones were cut out and stitched into a second piece of Whatman 3MM paper, which was subjected to ionophoresis at pH 3·6 for 1·5 hr with a voltage gradient of 70 v/cm, and the peptides located by autoradiography.

Methods of peptide analysis

(a) Fractionation of peptides on a large scale

This was achieved by gel filtration on a column of Sephadex G50 (2·5 cm × 120 cm) in pyridine–acetate buffer, pH 6·4, followed by ion exchange chromatography on a column of Dowex 1 X2 (2·5 cm × 45 cm) using an exponential gradient from 500 ml. pyridine–acetate buffer, pH 6·4, to 1 M-acetic acid for elution. The jacketed column was held at 40°C.

(b) Amino acid analysis

Peptides were hydrolysed in twice-distilled constant boiling HCl in sealed tubes, usually for 16 hr at 105°C. After removal of the HCl in vacuo, the amino acids present were analysed by means of a Beckman–Spinco automatic amino acid analyser fitted with a high-sensitivity photometer, recording at 570 mμ (Evans Electroselenium, Ltd., Halstead, Essex), which increased the sensitivity about 10 times above that of the unmodified machine.

When small quantites of peptide were analysed, proline was determined by measuring the peak height from the chart recording and extrapolating the band width from that of glutamic acid (or another amino acid) and scaling from standard runs. At the 5 mμmole level, an accuracy of $\pm$ 20% could be achieved.

All peptides analysed were purified by ionophoresis on Whatman no. 3 paper and eluted from a 1 cm × 10 cm strip with pH 6·4 buffer. The following amounts of amino acids were eluted from a 1 cm × 10 cm strip of blank paper, and were subtracted from the experimental values obtained for peptides: aspartic acid 0·5 mμmole; serine 1·2 mμmoles; glutamic acid 0·9 mμmole; glycine 1·6 mμmoles; alanine 0·7 mμmole.

(c) Amino acid sequence determination

(i) Edman degradation was carried out according to the method of Gray & Hartley (1963) except that cleavage of the phenylthiocarbamyl peptide was brought about by treatment with trifluoroacetic acid (Konigsberg & Hill, 1962) for 1 hr at room temperature. The stepwise progress of the degradation was followed by the identification of the N-terminal residue after each cycle.

(ii) End group determination. The fluorescent amino end-group reagent 1-dimethyl amino-napthalene-5-sulphonyl chloride (dansyl chloride) was used according to the method of Gray & Hartley (1963).

(iii) *Enzyme digestions.* Peptides were dissolved in 0·3 ml. of 1% ammonium bicarbonate for chymotrypsin and pronase digestions. Both enzymes were used at 50 μg/ml. and digestion allowed to proceed for 8 hr at 37°C. For digestion with pepsin, the peptide was dissolved in 0·3 ml. of pH 1·9 buffer (87 ml. glacial acetic acid, 25 ml. formic acid/l. of distilled water), enzyme added at 80 μg/ml. and incubated at 37°C for 18 hr. The peptides were digested with leucine amino peptidase (30 μg/ml.) in 0·3 ml. of tris–magnesium buffer (0·005 M-MgCl$_2$ in 0·005 M-tris–HCl, pH 8·0) for 24 hr.

(iv) *Partial acid hydrolysis.* Peptides were incubated at 37°C for 48 hr in constant boiling HCl.

(v) *Hydrolysis with acetic acid.* Peptides were dissolved in 0·3 ml. of 0·25 M-acetic acid and heated in a sealed tube at 110°C for 14 hr (Partridge, 1948).

(vi) *Hydrazinolysis.* This was carried out using a technique devised by R. Offord (personal communication). The peptides were dissolved in 50 μl. of anhydrous hydrazine and heated in an evacuated tube for 18 hr at 70°C.

(d) *Fractionation of peptides following enzyme digestion*

The digests were fractionated by paper ionophoresis at pH 6·4 on Whatman no. 3 paper, followed, if necessary, by ionophoresis at pH 3·6.

3. Results

Tryptic digests of the [^{14}C]phenylalanine-labelled protein synthesized by *am* H36 in the *su*$^-$ strain contain a peptide (PhT11) which is not found in the wild type nor in any other amber mutant. We surmised that this unique peptide was a fragment of the phenylalanine-containing tryptic peptide PhT12, which is absent in H36 and is present only in those amber mutants which map to the right of H36 (Plate I).

[^{14}C]Phenylalanine-labelled PhT11 and PhT12 were digested with chymotrypsin. Both gave rise to a radioactive peptide (C1) which was neutral at pH 6·4 (Plate II). At pH 3·6, peptide C1 from both PhT11 and PhT12 had the same electrophoretic mobility (Plate II), and on subjection to partial acid hydrolysis, both gave rise to an identical pattern of peptide fragments containing [^{14}C]phenylalanine. Taken together, this evidence suggests very strongly that the neutral peptides (C1) have the same amino acid sequence. In addition to the common peptide C1, both chymotryptic digests gave phenylalanine-containing peptides which moved towards the anode at pH 6·4, but that derived from PhT11 had a higher mobility than that from PhT12 (Plate II).

Similar experiments carried out with [^{14}C]arginine-labelled material showed that PhT11 contained no arginine, whereas PhT12 did. Chymotryptic digestion of PhT12 gave an arginine-containing peptide having the same mobility as the acidic phenylalanine-containing chymotryptic peptide of PhT12 (Plate II). Since trypsin cleaves peptide bonds involving the carboxyl groups of arginine (or lysine), these experiments suggest that PhT11 comprises the N-terminal portion of PhT12.

Structure of PhT12 and PhT11

In order to prepare peptide PhT12 in large enough quantities to carry out amino acid sequence determination, purified [^{14}C]phenylalanine-labelled peptide PhT12 was added to tryptic digests of 500 mg phage protein. Purification was achieved by gel filtration on G50 Sephadex followed by ion exchange chromatography on Dowex 1 (X2), the fractionation being monitored by measuring the radioactivity in fractions collected from columns (Fig. 1). Finally, after paper electrophoresis at pH 6·4, the radioactive peptide present corresponded to the major peptide located with ninhydrin. Specific staining reactions carried out on the paper confirmed that it contained arginine.

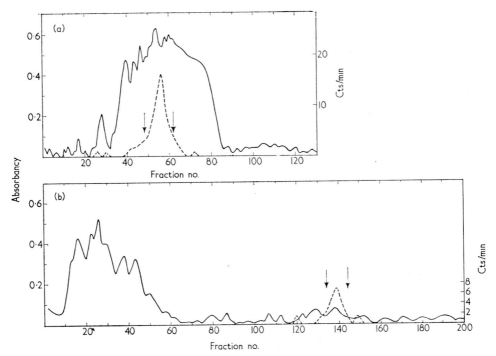

Fig. 1. Fractionation of peptide PhT12. The solid line shows absorbancy at 570 mμ following ninhydrin analysis after alkaline hydrolysis (cf. Hirs, Moore & Stein, 1956). The dashed line shows radioactivity.

(a) Gel filtration on G50 Sephadex; 8-ml. fractions were collected; 0·2-ml. samples were taken for ninhydrin analysis and 0·3-ml. samples for ¹⁴C analysis.

(b) Fractions between the arrows in (a) were pooled and further separated on Dowex 1 X2. 17-ml. fractions were collected: 0·2-ml. portions were taken for ninhydrin analysis and 0·5-ml. samples for ¹⁴C analysis.

The amino acid analysis of PhT12 shows it is a tridecapeptide containing aspartic acid, glutamic acid, proline, glycine, alanine, valine, isoleucine, phenylalanine, and arginine (Table 1).

Stepwise degradation from the N-terminal end using the dansyl Edman technique gave alanine as the first amino acid, and glycine as the second. Leucine amino-peptidase removed one residue each of aspartic acid, glutamine, glycine, alanine and valine, and two residues of phenylalanine (Table 2). The aminopeptidase digest contained an acidic arginine-containing peptide (LAP2) which was (Asp$_2$ Pro, Ile$_2$ Arg). Stepwise degradation of this peptide showed that aspartic acid was N-terminal, followed by proline in the second position.

The distribution of amide groups now becomes clear: both aspartic acid residues in peptide LAP2 must be in the free acid form since this peptide contains arginine and is negatively charged at pH 6·4. The release by LAP of one residue of glutamine, together with one residue of aspartic acid enables us to conclude that the net charge of PhT12 at pH 6·4 must be $-3 + 1 = -2$.

On hydrolysis of peptide PhT12 with dilute acetic acid, three dipeptides were found, (Phe, Glu), (Pro, Ile) and (Ile, Arg) in addition to the tetrapeptide, (Ala, Gly, Val, Phe) and, in low yield, the pentapeptide (Ala, Gly, Val, Phe, Asp). These peptides

TABLE 1

Amino acid analysis of peptide PhT12 from the wild-type phage T4, and from H36 grown on CR63

	T4D grown on *E. coli* B PhT12 24 hr hydrolysis	T4D grown on *E. coli* B PhT12 72 hr hydrolysis	T4D grown on *E. coli* CR63 PhT12 24 hr hydrolysis	H36 grown on *E. coli* CR63 PhT12 24 hr hydrolysis
Lys	N	0·08	0·08	0·08
His	N	<0·02	<0·02	<0·02
Arg	N	1·13	0·74	1·08
Asp	2·94	3·09	2·93	3·16
Thr	0·06	0·07	0·06	0·07
Ser	0·05	0·09	0·07	1·10
Glu	1·10	1·04	1·11	0·21
Pro	0·99	1·08	1·02	1·16
Gly	1·01	1·06	1·06	1·15
Ala	0·99	0·98	1·04	1·04
Val	0·91	0·92	1·03	0·76
Met	0·01	<0·02	<0·02	<0·02
Ile	1·94	1·96	1·92	1·85
Leu	0·05	0·05	0·07	0·05
Tyr	<0·01	<0·02	<0·01	<0·02
Phe	1·92	1·92	1·79	1·62

N, not estimated.

The results are expressed as molar ratios.

are called HA4, HA2–1, HA1, HA2–2 and HA3, respectively, in Table 2. The (Pro Ile) peptide had N-terminal proline.

Summarizing the sequence data thus far, we may write: Ala . Gly (Val, Phe) Asp. (Gln, Phe) Asp . Pro . Ile . Asp . Ile . Arg. Digestion of PhT12 with chymotrypsin gave a neutral tetrapeptide C1 (Ala, Gly, Val, Phe) and a nonapeptide C2 which after stepwise degradation was shown to be Asp . Phe (Gln, Asp, Pro, Ile, Asp, Ile, Arg) (Table 2).

Finally, the tetrapeptide Pr2 (Gln, Asp, Pro, Ile) was isolated from a pronase digest of PhT12. The amino acid sequence of peptide PhT12 is therefore:

Ala . Gly . (Val, Phe) Asp . Phe . Gln . Asp . Pro . Ile . Asp . Ile . Arg.

In confirmation of this sequence, peptic digestion gave the peptides P1, P3, P4, and P2, which on analysis were shown to be (Ala, Gly, Val, Phe), (Ala, Gly, Val, Phe, Asp), (Asp, Phe, Gln, Asp, Pro, Ile, Asp, Ile, Arg) and (Phe, Gln, Asp, Pro, Ile, Asp, Ile, Arg), respectively.

We suppose that the (Val, Phe) sequence is as shown with the brackets removed since chymotrypsin gives rise to the peptide Ala . Gly (Val, Phe), and from the known specificity of the enzyme it is likely that phenylalanine is C-terminal.

It is interesting to note that chymotryptic cleavage at the second phenylalanine has not been observed despite the use of a range of enzyme concentrations and times of digestion (50 to 500 μg/ml., up to 24 hours digestion).

TABLE 2†

Amino acid acid analyses of peptides derived from wild type peptide PhT12

	Amino acids released by LAP	LAP2	HA1	HA2-1	HA2-2	HA3	HA4	C1	C2	Pr2	Pr3	P1	P2	P3	P4
Lys	N	N	<0.05	N	N	<0.1	<0.1	N	<0.02	N	N	N	N	N	N
His	N	N	<0.05	N	N	<0.1	<0.1	N	<0.02	N	N	N	N	N	N
Arg	N−	N+	0.9	N−	N−	<0.1	<0.1	N	0.85	N−	N−	N−	N+	N−	N+
Asp	1.00	2.02	0.05	0.08	0.10	0.9	<0.1	<0.05	2.98	1.11	0.99	<0.1	2.05	1.14	2.84
Thr	<0.02	<0.02	<0.05	<0.02	<0.05	<0.1	<0.1	<0.05	<0.02	<0.05	0.04	<0.1	0.07	<0.05	0.04
Ser	See Gln	<0.02	0.05	0.05	0.05	<0.1	<0.1	<0.05	<0.02	<0.02	0.04	<0.1	<0.02	<0.05	<0.01
Glu	0.05	0.08	<0.05	0.05	0.10	<0.1	0.8	<0.05	1.02	0.82	0.25	<0.1	0.91	<0.05	1.01
Pro	—	1.2	—	0.9	—	—	—	—	1.0	1.0	—	—	1.0	—	0.82
Gly	0.85	<0.02	0.05	0.04	1.13	1.0	<0.1	0.85	0.07	0.16	0.07	0.93	0.12	1.07	0.03
Ala	1.12	<0.02	<0.05	0.03	1.00	1.0	<0.1	1.0	0.06	0.07	0.02	1.21	0.16	1.03	0.08
Val	1.16	0.11	<0.05	<0.02	1.02	1.0	<0.1	1.1	0.07	0.13	0.07	0.86	0.16	0.79	0.04
Met	<0.02	<0.02	<0.05	<0.02	<0.05	<0.1	<0.1	<0.05	<0.02	<0.02	<0.02	<0.1	<0.02	<0.05	<0.01
Ile	0.05	1.96	1.1	1.00	0.15	<0.1	<0.1	<0.05	2.04	0.97	1.01	<0.1	2.21	0.13	1.66
Leu	<0.02	<0.02	<0.05	<0.02	<0.05	<0.1	<0.1	<0.05	<0.01	0.06	0.06	<0.1	<0.02	0.10	0.02
Tyr	<0.02	<0.02	<0.05	<0.02	<0.05	<0.1	<0.1	<0.05	<0.02	0.11	0.06	<0.1	<0.02	<0.05	<0.01
Phe	1.77	0.11	<0.05	<0.02	0.86	0.9	1.2	1.1	0.93	0.13	0.05	0.93	0.81	0.82	0.94
Gln‡	1.12	—	—	—	—	—	—	—	—	—	—	—	—	—	—

N, not estimated; +, present; −, absent; as judged by specific staining reactions on paper (Block, Durrum & Zweig, 1958).

† The peptides in Tables 2 and 3 are derived as follows: LAP from leucine aminopeptidase digestion, and peptides C, Pr, and P from digestion with chymotrypsin, pronase and pepsin, respectively. Peptides HA are produced by hydrolysis with dilute acetic acid.

‡ Serine and glutamine are not resolved by the amino acid analyser; the presence of glutamine has been confirmed by paper ionophoresis of the neutral amino acids released by LAP.

PLATE III. Identification of the C-terminal residue of peptide PhT11 after hydrazinolysis and separation of the free amino acids by ionophoresis at pH 1·9. Amino acid hydrazides were removed by prior ionophoresis at pH 6·4. Phe is a [14C]phenylalanine marker; H36 is the hydrazinolysate of [14C]phenylalanine-labelled peptide PhT11.

We can now compare the structure of PhT11 with the amino acid sequence of PhT12. First we recall that chymotryptic digestion of [^{14}C]phenylalanine-labelled PhT11 gave a neutral peptide C1 which has the same structure as the C1 peptide released from PhT12, the sequence of which we now know to be Ala . Gly (Val, Phe). PhT11 yields another phenylalanine-containing peptide and must therefore include the sequence of PhT12 at least up to the next phenylalanine residue. To confirm this, T4D and H36-infected cells were labelled in separate experiments with the following ^{14}C-labelled amino acids: proline, glycine, isoleucine, valine and phenylalanine. Fractionation of the peptides showed that PhT11 contained glycine, valine and phenylalanine, as expected, while proline and isoleucine were absent. This means that PhT11 does not contain the Pro . Ile . Asp . Ile . Arg sequence of PhT12. The mobility of PhT11 at pH 6·4 is slightly less than that of PhT12 (which has a net charge of − 2). Since PhT11 is a smaller peptide and lacks arginine, it cannot have two acidic groups. This suggests that the aspartic acid residue adjacent to the proline residue is also absent. It follows then that the C-terminal residue must either be phenylalanine or glutamine. Hydrazinolysis of [^{14}C]phenylalanine-labelled PhT11 shows quite clearly that phenylalanine is C-terminal (Plate III). The acidic peptide released by chymotrypsin is therefore Asp . Phe. This is confirmed by partial acid hydrolysis of this peptide which yielded unchanged material and free phenylalanine. We can therefore deduce that the amino acid sequence of PhT11 is

Ala . Gly . Val . Phe . Asp . Phe.

Suppression of am H36

When am H36 is grown on E. coli CR63, peptides PhT11 and PhT12 are both present in digests of the protein from the infected cell, and the two peptides are present in about equal amounts.

Peptide PhT12 from am H36 grown on E. coli CR63 was prepared on a large scale using the same fractionation procedures as were used for the wild-type peptide. Analysis of the acid-hydrolysed peptide showed that the glutamine had been replaced by a serine residue (Table 1). This difference is specific to am H36 since the composition of PhT12 is unaltered in wild-type phage grown on E. coli CR63 (Table 1).

Methods similar to those already described for the wild-type peptide were used to determine the amino acid sequence of this peptide. Stepwise degradation showed that the N-terminal sequence was Ala . Gly . Val. Leucine aminopeptidase removed one residue each of aspartic acid, serine, glycine, alanine and valine, and two residues of phenylalanine, leaving an arginine-containing acidic hexapeptide (LAP2) which had the N-terminal sequence Asp . Pro. The amino acid analysis of this and other peptides is given in Table 3.

Hydrolysis with dilute acetic acid gave two dipeptides (Ile, Arg) and (Pro, Ile) and a tetrapeptide (Ala, Gly, Val, Phe); proline was N-terminal in the (Pro, Ile) dipeptide. Chymotrypsin yielded the tetrapeptide (Ala, Gly, Val, Phe) and an acidic arginine-containing nonapeptide (C2) in which the N-terminal sequence was found to be Asp . Phe. Peptic digestion of peptide PhT12 gave the three peptides (Ala, Gly, Val, Phe), (Ala, Gly, Val, Phe, Asp) and (Asp, Phe, Ser, Asp, Pro, Ile, Asp, Ile, Arg).

The sequence of peptide PhT12 from suppressed am H36 may therefore be written:

Ala . Gly . Val . Phe . Asp . Phe . Ser . Asp . Pro . Ile . Asp . Ile . Arg

TABLE 3 †

Amino acid analyses of peptides derived from peptide PhT12 of suppressed am *H36*

Amino acids	released by LAP (LAP2)	LAP2	HA1	HA2-1	HA2-2	C1	C2	P2	P3	P4
Lys	N	0·03	<0·01	N	N	N	<0·02	N	N	N
His	N	<0·02	<0·01	N	N	N	<0·02	N	N	N
Arg	N	0·90	1·01	N−	N−	N−	1·01	N+	N−	N+
Asp	0·98	2·12	0·03	0·07	0·16	0·03	3·24	2·22	1·19	3·05
Thr	<0·10	<0·01	<0·01	0·04	0·04	0·02	0·03	0·07	0·10	0·18
Ser	0·87	<0·01	<0·01	<0·01	0·19	<0·01	0·74	0·91	0·13	1·05
Glu	0·17	<0·01	<0·01	0·01	0·09	<0·01	0·08	0·09	0·10	0·22
Pro	—	1·00	—	1·06	—	<0·05	1·00	0·96	<0·10	0·90
Gly	1·08	<0·01	<0·01	0·04	1·10	1·01	0·11	0·13	0·93	0·31
Ala	1·09	<0·01	<0·01	0·03	0·81	0·90	0·05	0·06	0·92	0·24
Val	1·07	<0·02	<0·01	<0·01	1·05	1·06	0·05	0·09	0·93	0·17
Met	<0·05	<0·01	<0·01	<0·01	<0·01	<0·01	<0·02	<0·01	<0·01	<0·05
Ile	0·27	2·00	0·99	0·94	0·06	0·01	1·98	2·02	0·10	1·73
Leu	0·20	0·02	<0·01	<0·01	0·04	<0·01	<0·02	0·03	0·09	0·08
Tyr	<0·05	<0·01	<0·01	<0·01	<0·02	<0·01	<0·02	<0·02	<0·02	<0·05
Phe	1·91	<0·01	<0·01	<0·01	1·03	0·94	0·95	0·89	0·88	1·19

† See footnote to Table 2.

4. Discussion

Our results show very clearly that peptide PhT11, which is uniquely found in *am* H36, is a fragment of the wild-type peptide PhT12. Furthermore, since we have demonstrated that PhT11 is the N-terminal portion of PhT12, the protein made by H36 must be an N-terminal fragment of the whole head protein. Since proteins are synthesized from their N-terminal ends (Bishop, Leahy & Schweet, 1960; Dintzis, 1961; Goldstein & Brown, 1961), the amber mutation must result in *termination* of polypeptide chain synthesis.

We also show that glutamine is the amino acid which is altered in the amber mutant since, on the suppressing host, su_1^+, the glutamine of peptide PhT12 is replaced by serine (Fig. 2). Since the C-terminal amino acid of the H36 fragment is

T4D	Ala . Gly . (Val, Phe) Asp . Phe . Gln . Asp . Pro . Ile . Asp . Ile . Arg
H36	Ala . Gly . Val . Phe . Asp . Phe . Ser . Asp . Pro . Ile . Asp . Ile . Arg
H36 fragment	Ala . Gly . Val . Phe . Asp . Phe

FIG. 2. Amino acid sequences of peptide PhT12 from wild-type phage and from *am* H36 grown on *E. coli* CR63, and of the fragmented peptide found when *am* H36 is grown on *E. coli* B.

the amino acid immediately preceding the glutamine/serine position, chain termination must occur *at the site of mutation*. We can discount any mechanism which involves, for example, a change in the phase of reading of the genetic message (Crick, Barnett, Brenner & Watts-Tobin, 1961) at the site of mutation followed by chain termination further down the chain due to the generation of a nonsense signal in the shifted phase.

Our finding that an amber mutant arises from a glutamine triplet agrees well with the results of Weigert & Garen (1965), and of Notani *et al.* (1965). In at least two other amber mutants of the head gene (B278 and E161), it has been shown that the amino acid which is altered is tryptophan (Kaplan, Stretton & Brenner, unpublished results) which again agrees with the findings of Weigert & Garen (1965).

The expression of the amber mutation is prevented when the phase of reading of the genetic message is altered (Brenner & Stretton, manuscript in preparation). This suggests that the effects of this mutation are most likely to occur at the level of protein synthesis, since it is hard to imagine a process other than the reading of messenger RNA by S-RNA, which would be phase sensitive. We believe that the amber configuration is, in fact, a triplet, and we suggest that chain termination is achieved by the recognition of this codon by a specific S-RNA.

When amber mutants are suppressed, chain termination is not completely prevented. Different suppressors have different efficiencies with which they prevent termination and allow chain propagation to occur (Kaplan, Stretton & Brenner, manuscript in preparation). The suppressors therefore provide a mechanism for competing with chain termination. It is possible that the suppressor arises by a change of an activating enzyme which allows it to attach an amino acid to the postulated chain-terminating S-RNA. We discuss this suggestion in detail in another paper (Brenner, Stretton & Kaplan, 1965).

We should like to thank Dr A. S. Sarabhai, who participated in the early stages of this work. We also thank Dr A. B. Edmundson and Mr G. Lee for performing some of the amino acid analyses and Mr R. Offord for advice on his method of hydrazinolysis.

The excellent technical assistance of Mrs R. Fishpool and Mrs E. Southgate is gratefully acknowledged.

REFERENCES

Adams, M. H. (1959). *Bacteriophage.* New York: Interscience.
Bishop, J., Leahy, J. & Schweet, R. (1960). *Proc. Nat. Acad. Sci., Wash.* **46**, 1030.
Block, R. J., Durrum, E. L. & Zweig, G. (1958). In *Paper Chromatography and Electrophoresis.* New York: Academic Press.
Brenner, S., Stretton, A. O. W. & Kaplan, S. (1965). *Nature*, in the press.
Crick, F. H. C., Barnett, L., Brenner, S. & Watts-Tobin, R. J. (1961). *Nature*, **192**, 1227.
Dintzis, H. M. (1961). *Proc. Nat. Acad. Sci., Wash.* **47**, 247.
Fraser, D. & Jerrell, E. A. (1953). *J. Biol. Chem.* **205**, 291.
Goldstein, A. & Brown, B. J. (1961). *Biochim. biophys. Acta,* **53**, 438.
Gray, W. R. & Hartley, B. S. (1963). *Biochem. J.* **89**, 379.
Hirs, C. H. W., Moore, S. & Stein, W. H. (1956). *J. Biol. Chem.* **219**, 623.
Konigsberg, W. & Hill, R. J. (1962). *J. Biol. Chem.* **237**, 2547.
Notani, G. W., Engelhardt, D. L., Konigsberg, W. & Zinder, N. (1965). *J. Mol. Biol.* **12**, 439.
Partridge, S. M. (1948). *Biochem. J.* **42**, 238.
Sarabhai, A. S., Stretton, A. O. W., Brenner, S. & Bolle, A. (1964). *Nature*, **201**, 13.
Weigert, M. C. & Garen, A. (1965). *J. Mol. Biol.* **12**, 448.

J. Mol. Biol. (1965) **13**, 373–398

A Two-dimensional Fractionation Procedure for Radioactive Nucleotides

F. SANGER, G. G. BROWNLEE AND B. G. BARRELL

Medical Research Council Laboratory of Molecular Biology
Cambridge, England

(*Received 21 April 1965*)

A method is described for the two-dimensional fractionation of ribonuclease digests of ^{32}P-labelled RNA. High-voltage ionophoresis is used in both dimensions. The first is on cellulose acetate at pH 3·5, the second on DEAE-paper at an acid pH. The method is capable of resolving the di- and tri- and most of the tetra-nucleotides in digests prepared by the action of ribonuclease T_1 or pancreatic ribonuclease. It has been applied to the 16 s and 23 s components of ribosomal RNA which show significant quantitative differences, and to sRNA from *Escherichia coli* and from yeast. A method is described for the determination of the sequence of a nucleotide by partial digestion with spleen phosphodiesterase.

1. Introduction

Progress in the determination of amino acid sequences in proteins has been largely dependent on the development of methods for the fractionation of the peptides produced by partial hydrolysis. Similarly, the study of nucleotide sequences in nucleic acids is likely to be limited by the techniques available for the fractionation of oligonucleotides. The methods most used at present are the two-dimensional procedure of Rushizky & Knight (1960), which uses ionophoresis and chromatography on paper (Rushizky & Sober, 1962; Armstrong, Hagopian, Ingram, Sjöquist & Sjöquist, 1964), and ion-exchange column chromatography, particularly on DEAE-cellulose or DEAE-Sephadex (Khorana & Vizsolyi, 1961; Staehelin, 1961,1964*a*; Rushizky, Bartos & Sober, 1964). Although the latter method is probably the most efficient separation procedure and is the most widely employed, paper techniques are often preferable from the point of view of speed and ease of operation. The advantages of both techniques may be combined by the use of ion-exchange papers. Tyndall, Jacobson & Teeter (1964) have described a two-dimensional technique using chromatography on DEAE-paper at two different pH values. We have found that good separations may be obtained by high-voltage ionophoresis on DEAE-paper at an acid pH, and this system is used as the second dimension for the two-dimensional technique. It appears that fractionation is due both to ion-exchange and to electrophoretic effects. The positive charges on the paper cause a rapid electro-endosmotic flow of buffer from the cathode to the anode which carries the nucleotides through the paper, thus subjecting them to ion-exchange chromatography. Superimposed on this is the electrophoretic fractionation which is probably an important factor in the fractionation, since considerably better separations are obtained by this technique than by simple chromatography on the DEAE-paper. The ionophoresis essentially opposes the ion-exchange effect, the more acidic components moving faster by ionophoresis but slower by ion-exchange.

331

In conjunction with this method, we have used ionophoresis on cellulose acetate as the basis for a two-dimensional technique. Ionophoresis of nucleotides on paper is somewhat limited by a tendency of the larger ones to streak. Previous experiments with peptides had indicated that considerably less streaking occurred if cellulose acetate was used instead of paper, and the same was found to be true for nucleotides. A disadvantage of the use of cellulose acetate for many purposes is that it is only suitable for small amounts of materials. Nucleotides are normally detected by their absorption of ultraviolet light, and this limits the scale on which one can work. In order to lower this scale, we have studied nucleotides which have been labelled with ^{32}P and have detected and estimated them by radioautography and counting techniques. Many nucleic acids can be prepared in a radioactive form by biological labelling and some, such as messenger RNA, can be studied only in this form.

The following paper describes the two-dimensional fractionation procedure and its application to ribonuclease digests of the two main components of ribosomal RNA from *Escherichia coli*. This material, which may readily be prepared, has been used to study the potentialities of the method and to determine the position of the different nucleotides on the two-dimensional fingerprint. We have studied digests prepared by the action of ribonuclease T_1 (Sato & Egami, 1957) because of its high specificity, splitting only at G† residues, but have also done preliminary work with pancreatic ribonuclease. Digests of mixed sRNA have also been studied. These are complicated by the presence of minor components and have not been investigated so completely. During this work a new minor component was detected and identified as the alkali degradation product of dihydrouridylic acid. In the meantime, the presence of dihydrouridylic acid as a component of sRNA has been reported by Madison & Holley (1965).

The structure of the nucleotides has been determined by conventional methods and by a new technique involving partial digestion with spleen phosphodiesterase. A somewhat similar method using venom phosphodiesterase was recently described by Holley, Madison & Zamir (1964). Spleen phosphodiesterase splits off 3'-mononucleotides sequentially from the 5' terminal end of an oligonucleotide (Razzell & Khorana, 1961). If we consider an oligonucleotide obtained from a ribonuclease T_1 digest and suppose its structure is ABCG, where A, B and C are unknown residues, on partial digestion with spleen phosphodiesterase the following degradation products will be found: BCG, CG and mononucleotides. It has been found that on the DEAE-paper–pH 1·9 system used for the two-dimensional procedure, any nucleotide will move faster than a corresponding nucleotide having the same structure but with one extra residue added to its 5' terminal end; i.e., BCG will move faster than ABCG, etc. Consequently, the various degradation products of an oligonucleotide will be arranged in order of their size on fractionation in this system, the larger ones moving slower than the smaller ones. The distance between any two products which differ by only one residue will depend on the nature of that residue. Thus from these distances one may deduce the nature of the various residues, and hence it is frequently possible to determine a complete sequence from a single degradation and ionophoretic fractionation.

† Abbreviations used: The letters A, C, G and U will be used for 3' nucleotide residues. Thus G refers to guanosine-3'-phosphate and ACG refers to the trinucleotide ApCpGp (Shuster, Khorana & Heppel, 1959). The residue bearing the free 5'-hydroxyl group will be referred to as the 5' terminal residue and that with the 3'-monoesterified phosphate as the 3' terminal residue.

2. Materials and Methods

(a) *Materials*

Most of the samples of cellulose acetate electrophoresis strips were obtained from British Celanese Ltd., Plastics Division, through the kindness of Mr Prescott and Mr Hughes. Different batches varied considerably in their resolving power, and this appeared to be related to the method of production. Batch 1625 was a particularly suitable one, and material similar to this may now be obtained in sheets 95 cm × 25 cm from Oxo Ltd., Oxoid Division, 20 Southwark Bridge Road, London, S.E.1.

The DEAE-paper was Whatman Chromedia DE81 (originally known as DE20).

Ribonuclease T_1 was prepared by Mr R. E. Offord of this laboratory by the method of Rushizky & Sober (1962). Pancreatic ribonuclease was obtained from the Nutritional Biochemicals Corporation. Bacterial alkaline phosphomonoesterase, snake venom phosphodiesterase, micrococcal nuclease and spleen phosphodiesterase were obtained from the Worthington Biochemical Corporation.

In order to follow the course of an ionophoresis, it was convenient to apply a mixture of coloured dyes to be run side by side with the unknowns to act as markers. The dye mixture used in this work contained equal volumes of 1% Xylene Cyanol F.F (blue), 2% Orange G (yellow) and 1% Acid Fuchsin (pink) (all from George T. Gurr, Ltd., London, S.W.6).

(b) *Ionophoresis*

High-voltage ionophoresis was normally carried out by the method of Michl (1951). This, however, could not be used with the DEAE-paper, which is very fragile and difficult to handle when wet. Satisfactory results were obtained in an apparatus similar to that used by Naughton & Hagopian (1962) in which the paper is supported on a rack that can be lifted in and out of the tank. The centre partition was 15 cm high and the rods on the rack were arranged to accommodate papers 57 or 85 cm long.

The following buffer systems were used: pH 3·5, 0·5% pyridine–5% acetic acid (v/v); pH 1·9, 2·5% formic acid–8·7% acetic acid (v/v); pH 6·5, 10% pyridine–0·3% acetic acid (v/v). For the pH 6·5 system the coolant was toluene; for the other systems it was white spirit 100 (Esso Petroleum Company).

For ionophoresis at intermediate values of pH, a water-cooled flat plate apparatus, as described by Gross (1961), was used. The buffer systems were 5% acetic acid adjusted with ammonia to pH 3·5; 1% acetic acid adjusted with ammonia to pH 4·1; 0·5% acetic acid adjusted with ammonia to pH 4·4; 0·25% acetic acid adjusted with ammonia to pH 4·9; and 1·7% *N*-ethyl morpholine adjusted with acetic acid to pH 7·9. The separations took about 1 hr at 80 v/cm.

Due to the small amounts of nucleotides being studied, it was necessary to take all possible precautions against contamination with nucleases (Apgar, Holley & Merrill, 1962). Rubber gloves were worn when there was danger of contamination from the fingers. Small test tubes and centrifuge tubes were treated with silicone and washed with glass-distilled water. Non-radioactive "carrier" nucleotides were usually added to isolated samples to avoid too low concentrations.

(c) *Preparation of ^{32}P-ribosomal RNA*

The 16 s and 23 s components of ribosomal RNA were prepared from *E. coli* (strain MRE600) that had been grown on 100 ml. of the medium of Garen & Levinthal (1960), which contained 10^{-4} M-β-glycerophosphate and about 10 mc carrier-free [^{32}P]phosphate. This organism, which was kindly supplied by Dr H. E. Wade, of M.R.E. Experimental Station, Porton, Wilts., is reported to be free of ribonuclease.

A large inoculum was used, and after about 5 hr growth the bacteria were collected in a thick-walled centrifuge tube and frozen. They were then thawed and ground up for 5 min at 0°C with alumina, using a glass rod with a diameter about half that of the inside of the centrifuge tube. The broken cells were then extracted with 2 ml. of buffer (0·01 M-magnesium acetate, 0·01 M-tris–acetate (pH 7·4), 0·006 M-mercaptoethanol). After sedimenting the debris at 10,000 **g** for 10 min, the supernatant solution was centrifuged for 90 min at 39,000 rev./min in the Spinco model L ultracentrifuge. The ribosomal pellet, which was just visible, was dissolved in about 0·1 ml. 0·1 M-LiCl, 0·01 M-tris buffer (pH 7·4) containing 0·5% sodium dodecyl sulphate. It was then subjected to density-gradient centrifugation at 10°C in 5 ml. of the above medium using a gradient from 5 to 20% sucrose. It was centrifuged for 4 hr at 39,000 rev./min in the SW39 rotor. Fractions were collected and samples counted. The peak fractions for the 23 s and 16 s components were combined and the RNA precipitated with 2·5 vol. ethanol. After standing in the deep-freeze overnight, the precipitates were collected. In some experiments the peaks were re-run on the same density-gradient system to ensure purification. The ethanol precipitates were then treated with 1 ml. 2% sodium acetate (adjusted to pH 5·4) and 1 ml. water-saturated phenol. After shaking for 1 hr at room temperature, the aqueous layer was collected and the RNA precipitated with 2·5 vol. ethanol. It was washed with 75% ethanol and with absolute ethanol and dried in a desiccator. The residue was then dissolved in 0·1 ml. water and kept frozen until used. The yields varied considerably, but on the average the final solutions contained about 1 μc ^{32}P and 1 μg RNA per μl.

(d) Preparation of ^{32}P-sRNA

(i) From E. coli

E. coli was grown on 5 to 20 mc carrier-free ^{32}P on the medium described above. After centrifugation, the cells were transferred in a volume of 1 ml. to a small centrifuge tube and 1 ml. water-saturated phenol and 0·5 mg carrier sRNA were added. The tube was tightly stoppered and shaken vigorously at 5°C for 1 hr. After centrifugation, the aqueous layer was separated and treated with 0·1 vol. 2 M-sodium acetate (pH 5) and 2 vol. 95% ethanol. The precipitate was collected after standing in the cold, washed with 95% ethanol and ether and dried. It was then dissolved in 0·1 ml. 0·1 M-tris–HCl (pH 8·0), 0·1 M-LiCl, 0·5% sodium dodecyl sulphate and subjected to density-gradient centrifugation in 2·5 ml. of the above medium at 10°C using a gradient from 5 to 20% sucrose for 10 hr at 39,000 rev./min. Ten-drop fractions were collected and the position of the 4 s peak found by counting 0·1-μl. samples. The main peak sedimented with a maximum near the middle of the gradient. The peak tubes were pooled, and the RNA was precipitated with 2 vol. ethanol and collected after standing overnight at 4°C. The yield of [^{32}P]-sRNA was between 1 and 2% (counts) of the radioactive material added to the medium. The method is modified from that described by Gilbert (1963).

(ii) From yeast

Yeast was grown in a medium which was modified from that of Williams & Dawson (1952). (NH$_4$)$_2$HPO$_4$ and KH$_2$PO$_4$ in their medium were replaced by triammonium citrate (3·25 g/l.). The medium was adjusted to pH 5·2 with citric acid. 10 mc of [^{32}P]phosphate and 1 to 5 mg bakers' yeast were added to 50 ml. medium which was incubated for about 6 hr at 30°C with efficient aeration. The uptake of radioactivity from the solution into the yeast was 95% complete. The cells were collected by centrifugation, 1 mg yeast sRNA was added as carrier and the [^{32}P]sRNA was isolated by the method of Holley et al. (1961). The yield was about 1% (counts) of the radioactive material added to the medium.

(e) Preparation of [^{14}C]methyl-labelled sRNA

E. coli strain K12, RC^{rel}, met$^-$, leu$^-$, thr$^-$ was used (CB3 of Dr S. Brenner). This is a multiple auxotrophic derivative of K12 W6, an RNA relaxed strain. Borek, Ryan & Rockenbach (1955) described the properties of K12 W6; Stent & Brenner (1961) prepared the multiple auxotrophic derivative of it. CB3 was grown overnight on 100 ml. of M9 (which contained/l. 5·8 g Na$_2$HPO$_4$, 3·0 g KH$_2$PO$_4$, 0·5 g NaCl, 1·0 g NH$_4$Cl, 0·02% glucose and 10^{-3} M-MgSO$_4$), supplemented with 2 μg/ml. L-methionine, 2 μg/ml. L-threo-nine and 2 μg/ml. L-leucine. Under these conditions glucose is limiting. In the morning, 2 ml. of 20% glucose was added, and growth started exponentially almost immediately. After 2 hr the cells reached an optical density at 550 mμ of 0·5 and were harvested by centrifugation, washed and resuspended in M9 with no amino acids added. The cells were bubbled for 15 min, after which 10 μc of [^{14}C]methyl-labelled methionine (30 μc/μmole) was added. Growth was allowed to go on for another 3 hr, when the turbidity had reached

0·95 and the uptake (counts) of radioactive material was 40%. [^{14}C]methyl-labelled sRNA was prepared by the method of Holley *et al.* (1961) and the yield was about 1·5% (counts) of the radioactive material added to the medium.

(f) *Enzymic digestion*

In preliminary experiments with ribonuclease T$_1$, difficulty was experienced in obtaining reproducible fingerprints, and it was necessary to define the hydrolysis conditions rather closely. If too mild conditions were used, cyclic nucleotides were produced which led to a duplication of each spot. Under too violent conditions, there was some splitting of bonds not involving G residues. This may have been partly due to contaminants in the ribonuclease T$_1$ preparation, but was probably more often due to contaminants in the RNA preparations. In order to minimize this latter effect, it was considered advisable to digest with relatively high concentrations of enzyme for a short time. The following conditions were found to give satisfactory results with ribosomal RNA and sRNA: enzyme/substrate ratio 1:20, 0·02 M-tris buffer (pH 7·4) containing 0·002 M-EDTA (neutralized). For amounts of RNA from 5 to 20 μg the digestion was carried out in a volume of 5 to 10 μl. in a small capillary tube. The digest could then be applied directly to the cellulose acetate for fractionation. Incubation was for 30 min at 37°C. The same conditions were also used for digestion with pancreatic ribonuclease.

(g) *Fractionation procedure*

The digest, which should not contain more than 0·1 mg nucleotides in a volume of less than 10 μl., was first fractionated by high-voltage ionophoresis at pH 3·5 according to the method of Michl (1951) on strips of cellulose acetate. The strip should be about 3 cm wide and the length will depend on the fractionation required. To obtain a fingerprint with all the smaller nucleotides, as in Plates I and II, the strip should be about 60 cm long. For better fractionation of larger nucleotides, as in Plates III and IV, strips of 95 cm can be used in conjunction with the "moving up" technique described below. The cellulose acetate was first wetted with the buffer. To avoid the inclusion of air bubbles, this should be done from one side by floating it on the buffer contained in a Petri dish. The point of application, which should be about 10 cm from the cathode end of the strip, was blotted and the digest applied as a spot and allowed to soak in. Samples of the coloured marker were applied each side of the digest. Care had to be taken to avoid the strip drying out while the sample was being applied. This was done by having the ends dipping in the buffer or covered with wet Kleenex tissue-paper. Excess buffer was then removed from the strip by blotting and the strip was rapidly put into the ionophoresis tank with the point of application near the negative electrode vessel, which was at the top of the tank. In the tanks normally used, the distance between the two electrode vessels was about 50 cm. With the longer strips the lower end was allowed to lie in the buffer at the bottom of the tank (anode compartment). Ionophoresis was carried out at 3000 v. For the shorter strips it was continued until the yellow marker had almost reached the anode buffer (about 2 hr). It was found that the majority of the nucleotides migrate slower.than the pink and faster than the blue marker. In order to achieve better resolution in the area between these two markers, the following "moving up" technique was used with the longer strips. Ionophoresis was continued until the pink marker was near the anode buffer. The strip was then taken out and excess buffer blotted off the end that was lying in the anode buffer. It was then replaced in the tank so that the blue marker was just outside the buffer in the cathode compartment. The part of the strip behind the blue marker, which contained the origin, could be cut off. Ionophoresis was then continued until the pink marker had again approached the buffer. The blue marker had by then moved away from the cathode end and the moving up procedure could be repeated. The process was continued until the pink marker was near the end of the strip and there was a distance of about 40 cm between the two markers.

Fractionation in the second dimension was carried out on sheets of DEAE-paper. Normally sheets 57 cm × 46 cm were used, but it may be obtained in rolls, and longer sheets (e.g. 85 cm) have also been used with satisfactory results. This material is very fragile and difficult to handle when wet. The cellulose acetate, on the other hand, is easy to handle when wet but brittle when dry. Attempts to sew the cellulose acetate strip on to the DEAE-paper were thus unsuccessful. A good transfer of the material from the cellulose acetate could be obtained, however, by the following blotting procedure. A sheet of the DEAE-paper was laid on a glass plate. The strip of cellulose acetate on which the nucleotides had been fractionated was removed from the tank and hung up so that the excess white spirit dripped off. Before the buffer dried, the wet strip was laid on the DEAE-cellulose sheet about 10 cm from one of the short sides. A pad of four strips (46 cm × 2 cm) of Whatman 3 MM paper that had been soaked in water was then put on top of the cellulose acetate strip and a glass plate placed on top to press the strips together evenly. Water from the paper pad passes through the cellulose acetate and into the DEAE-paper, carrying the nucleotides with it. Being acidic, they are held on to the DEAE-paper by ion-exchange and remain in the position in which they are first washed on. One thus obtains a good transfer of the fractionated nucleotides without any smearing of the spots as usually occurs with blotting techniques. By this means a strip of the DEAE-paper about 6 cm wide was wetted. To ensure a more complete transfer, more water could be added from a pipette to the paper pad while still in position. In a control experiment, it was shown that 90 to 95% of a ribonuclease T_1 digest of RNA could be transferred from cellulose acetate to DEAE-paper by this technique. After removal of the pad and cellulose acetate strip, the position of the coloured spots was marked on the DEAE-paper, which was hung up to dry in a current of warm air.

After drying, the DEAE-paper was wetted with the pH 1·9 buffer as follows. Wetting was started on either side of the line on which the nucleotides were applied, and the fronts of the solution were allowed to meet along this line. After wetting the rest of the cathode end of the paper (i.e. one-third of the total length of paper), it was put on the rack for running in the special ionophoresis apparatus. The rest of the paper was then wetted and the rack lowered slowly into the tank with the applied nucleotides near the cathode compartment. The system has a high conductivity and there is considerable heating if very high voltages are used. To obtain a fingerprint as in Plate I, about 1500 v was used until the blue marker was almost at the top of the rack (about 4 hr). For better separations, as in Plate III, the ionophoresis was carried out overnight at 1500 v, by which time the blue marker had usually run off the end of the paper. The DEAE-paper was finally dried in air while still on the rack. To prepare radioautographs, the paper was marked with radioactive ink (^{35}S) and put in a folder with one or two sheets of Ilford industrial G X-ray film. The folders were covered on one side with a thin sheet (0·5 mm) of lead so that they could be stacked together in a light-proof cabinet. Where more than 1 µc of [^{32}P]RNA had been used, the film could be developed after 1 day and there would be sufficient material to study the structure of the nucleotides. If it was only desired to obtain a fingerprint, less material could be used (about 0·01 µc for a short fractionation as in Plate I) and development was after 1 to 2 weeks.

The fractionation of a T_1 ribonuclease digest of [^{14}C]methyl-labelled sRNA was done exactly as for [^{32}P]sRNA, which was always run in parallel with it.

The fractionation of an alkaline hydrolysate of [^{14}C]methyl-labelled sRNA was done on a two-dimensional system which differed from that described above. The digest was loaded as a spot on to Whatman 3 MM paper and run at pH 3·5 for 2 hr at 60 v/cm, after which the strip of paper (bounded on each side by the blue marker) was sewn on to a sheet of Whatman no. 1 paper for chromatography in the second dimension using propan-2-ol–HCl–water as the developer. After running for 24 hr, the paper was dried and a radioautograph made. A hydrolysate of [^{32}P]sRNA was run in parallel.

(h) *Preparation of* ^{32}*P-β-ureidopropionic acid* N-*ribotide*

The method used was adapted from that published by Cohn & Doherty (1956) in that the intermediate, dihydrouridylic acid, was not isolated. A solution containing 1 mg of rhodium catalyst and 1 µc of [^{32}P]uridylic acid in 0·5 ml. of 0·01 M-sodium acetate (pH 5·0) was bubbled with hydrogen for 2 hr. The catalyst was removed by centrifugation and the supernatant fraction transferred on to a piece of polythene and dried down in a vacuum desiccator. It was taken up in about 0·1 ml. of 0·2 N-NaOH and incubated at 37°C for 1 hr. The solution was then loaded on to the origin of the Whatman 3 MM paper for ionophoresis at pH 4·4.

(i) Structure of the nucleotides

The positions of the nucleotides on the DEAE-paper was determined from the radio-autograph, using the marks from the radioactive ink to line up the film with the paper. The spots were cut out and eluted with alkaline triethylamine carbonate prepared as follows: CO_2 was passed into a mixture of 70 ml. water and 30 ml. redistilled triethylamine until saturated. More triethylamine was then added to about pH 10. In order to minimize losses due to the very small amounts of nucleotides present, non-radioactive nucleotides were added as carrier to the eluting medium. These were prepared as follows: 1·0 g yeast RNA was treated with 25 ml. 1·0 N-KOH at 37°C for 20 min. The KOH was then neutralized with perchloric acid and the precipitate removed by centrifugation. The supernatant solution was made up to 50 ml. and thus contained 20 mg nucleotides/ml. 2 ml. of this was added to each 100 ml. of the eluting mixture.

Elution was carried out as described by Sanger & Tuppy (1951), except that thin-walled capillary tubes were used. These were prepared by drawing out melting point tubes (Gray, 1964). Elution was very rapid by this method and was complete within 20 min. The eluted material was usually put on to polythene sheets and taken to dryness in a desiccator in a partial vacuum. If a good vacuum is put on the desiccator, the triethylamine bubbles badly. In order to ensure complete removal of the triethylamine carbonate, water was added to the spots and they were taken to dryness several times.

(i) Composition

In order to identify the mononucleotides present in an oligonucleotide, it was digested with about 10 μl. 0·2 N-NaOH at 37°C for 16 hr. This was done in drawn-out melting point tubes. The solution was drawn into the tube using a piece of polythene tubing and the ends were sealed off. After incubation, the digest was applied to Whatman no. 52 paper for ionophoresis at pH 3·5 and 60 v/cm for 1 to 1½ hr. Each digest was applied over 2 cm of the paper with 1-cm gaps between them. The four mononucleotides were well separated on this system and the composition of each digest could be determined from the radio-autograph. With nucleotides from ribonuclease T_1 digests, G was present as a single spot, but with those from pancreatic ribonuclease digests it usually formed two spots, which were presumably the 2′- and 3′-phosphates. With simple nucleotides it was usually possible to deduce how many residues of each mononucleotide were present by estimating the intensity of the bands on the radioautograph; however, this was usually checked by direct estimation on the scintillation counter. Areas of equal size for the various bands were cut out from the paper, using the radioautograph as a guide. The area was usually 1·8 cm × 2·6 cm and was chosen to take in the whole of each spot. If the spots contained less than about 100 cts/min the results were not very reliable, but above this about 10% accuracy was achieved.

sRNA contains a number of minor components. Many of these separated from the four major components on the pH 3·5 ionophoresis. They were investigated further by paper chromatography using the following systems on Whatman no. 1 paper: propan-2-ol (680 ml.), concn HCl (176 ml.) and water to 1 l. (Wyatt, 1951); and propan-2-ol (70 ml.), water (30 ml.), concn NH_3 (1 ml.) (Markham & Smith, 1952).

(ii) Digestion of oligonucleotides with enzymes

To determine the sequence of residues in oligonucleotides, they were further degraded with nucleases and the products identified. The most useful enzyme for the investigation of nucleotides from ribonuclease T_1 digests was pancreatic ribonuclease and, conversely, ribonuclease T_1 was used to study nucleotides from pancreatic ribonuclease digests. In both cases the products that can be obtained are the same and consist of the mono-nucleotides G, C and U and oligonucleotides with various numbers of A residues terminated at the 3′ end by G, C or U. Most of these could be satisfactorily separated by ionophoresis at pH 3·5 (Fig. 1).

Pancreatic ribonuclease. Samples of the nucleotides that had been dried down on poly-thene were treated with 5 to 10 μl. of 0·001 M-EDTA–0·01 M-tris buffer (pH 7·4) containing 0·2 mg pancreatic ribonuclease/ml. and transferred to capillary tubes which were sealed

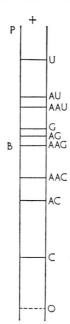

FIG. 1. Diagram showing the position of products from pancreatic digests of nucleotides from ribonuclease T_1 digests on ionophoresis at pH 3·5. B and P show the positions of the blue and pink markers, respectively.

at the drawn out end and incubated at 37°C for 30 min. The samples were then treated with 0·1 N-HCl to break down any cyclic phosphates as follows. 2 µl. 0·5 N-HCl was put on a polythene strip. The tip of the capillary was broken and the digest squirted out, mixed with the HCl and sucked back into the capillary, which was resealed and incubated at 37°C for 1 hr. The digest was then transferred to Whatman no. 52 paper and subjected to ionophoresis at pH 3·5 at 3 kv until the pink marker was near the end of the paper. The products could usually be identified by their position (Fig. 1), but if there was any doubt the nucleotide was eluted with water and subjected to alkaline hydrolysis for analysis. This was most frequently necessary with AG, AAG and AAAG, which are only just resolved on the ionophoresis.

Ribonuclease T_1. Products from pancreatic ribonuclease digests were digested with ribonuclease T_1 for 30 min at 37°C, using the following mixture: 0·001 M-EDTA–0·01 M-tris buffer (pH 7·4), 0·1 mg ribonuclease T_1/ml.

Micrococcal nuclease. Digestion was carried out in 0·05 M-borate buffer (pH 8·8) containing 0·01 M-CaCl$_2$, with an enzyme concentration of 0·1 mg/ml. Incubation was for 2 hr at 37°C and the products were fractionated at pH 3·5. Since the specificity of this enzyme is not clearly defined (Cunningham, Catlin & Privat de Garilhe, 1956; Reddi, 1959), it was not possible to identify the products from their positions, and it was necessary to analyse them and in certain cases to determine the terminal residues as described below with phosphomonoesterase and phosphodiesterase.

(iii) *Determination of terminal residues*

The nucleotides were first treated with bacterial alkaline phosphomonoesterase to remove the terminal 3′-phosphate group and the product purified by ionophoresis at pH 6·5. The 3′-terminal residue was usually known, due to the specificity of the ribonucleases; but if it was desired to determine it, the dephosphorylated nucleotide was digested with alkali. The 3′-terminal residue would be present as a nucleoside and would not be seen on the radioautograph. It was thus identified as the mononucleotide that was present in an alkaline digest of the oligonucleotide, but not in that of the dephosphorylated product.

To identify the 5′ terminal residue, the dephosphorylated product was treated with snake venom phosphodiesterase. This breaks the 3′ phosphoester bonds (Razzell & Khorana, 1959) so that the 5′ terminal residue is now present as a nucleoside and is identified as the residue that is missing from the digest.

The digestion with phosphomonoesterase was carried out for 30 min at 37°C in capillary tubes with 5 to 10 μl. of 0·005 M-MgCl$_2$–0·1 M-NH$_4$HCO$_3$ containing 0·1 mg enzyme/ml. It was then taken to dryness to remove the ammonium bicarbonate and subjected to ionophoresis at pH 6·5 (50 v/cm, 1·5 hr) on Whatman no. 52 paper. The bands were located by radiography, cut out and eluted with water containing carrier nucleotides. If more than one main band was present, they were also cut out and studied.

Treatment of the dephosphorylated product or a sample of it with venom phosphodiesterase was carried out with 5 to 10 μl. of the following solution: 0·01 M-magnesium acetate–0·025 M-tris buffer (pH 8·5)–0·1 mg enzyme/ml. After incubation for 2 hr at 37°C, the 5′-mononucleotides were separated by ionophoresis at pH 3·5.

(iv) *Partial digestion with spleen phosphodiesterase*

In order to know the position of the unchanged material on the final DEAE-paper ionophoresis, each nucleotide was divided into two samples: a large sample which was digested, and a smaller one containing about half as much material which was used as a control.

Conditions of digestion were varied somewhat for different nucleotides and sometimes samples were studied under different conditions. The following conditions were found to be the most generally useful and were used in most preliminary experiments with products from ribonuclease T$_1$ digests. The nucleotide was treated with 5 to 10 μl. of the following mixture in a capillary tube for 30 min at 37°C: 0·1 M-ammonium acetate (pH 5·7), 0·002 M-EDTA, 0·05% Tween 80, 0·2 mg enzyme/ml. For nucleotides containing more than one C residue, digestion was usually carried out for 60 min. Nucleotides from pancreatic ribonuclease digests were digested more rapidly, and for satisfactory results an enzyme concentration of 0·1 mg/ml. was used.

After digestion, the material was applied directly to DEAE-paper. The untreated sample was applied next to it, with no space between. A sample of an unfractionated digest of ribosomal RNA was also applied to each paper as well as the coloured markers. After wetting with the pH 1·9 mixture, ionophoresis was carried out as described above until the blue marker had reached the top of the rack (about 4 hr at 1·5 kv).

3. Results

(a) *Ribonuclease T$_1$ digests of ribosomal RNA*

Plates I to IV show two-dimensional fractionations of ribonuclease T$_1$ digests of the 16 s and 23 s ribosomal components. The fractionations shown in Plates I and II have been run for shorter times to include all the small nucleotides, except G, whereas those in Plates III and IV have been run for longer times to obtain better fractionation of the larger nucleotides. The system of numbering the spots is shown in Fig. 2 and the results of the analyses of the different nucleotides are summarized in Table 1. These combined results were obtained in several different experiments. It was found that the distribution of the spots was reproducible if identical conditions of digestion were used. All the nucleotides recorded in Table 1 have G as the 3′ terminal residue, indicating good specificity of ribonuclease T$_1$. Occasionally nucleotides not containing a G residue were obtained. This was probably due to contamination with other nucleases. Thus, in the fingerprint obtained with the 23 s material there were a number of faint spots which were not present in the 16 s material. Most of these gave only A

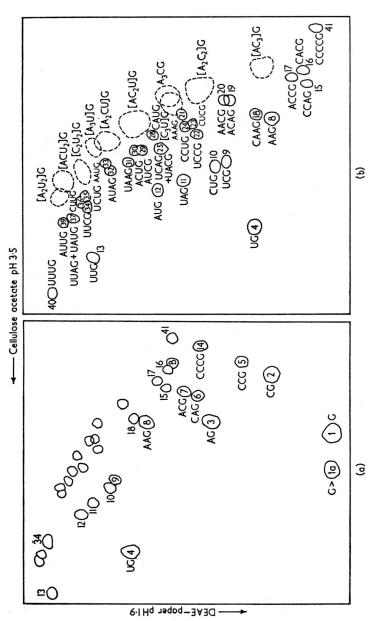

FIG. 2. Diagram showing the position of nucleotides from a ribonuclease T_1 digest of RNA on the two-dimensional system. (a) Illustrates a fractionation that has been run for a short time on the DEAE-paper; (b) one that has been run for a longer time. B marks the position of the blue marker.

TABLE 1

Nucleotides from ribonuclease T_1 digests

Spot no. (Fig. 2)	Composition† C	A	U	5′ terminal‡ residue	Products obtained with pancreatic ribonuclease	Products obtained with micrococcal nuclease	Structure deduced
1	—	—	—				G
1a	—	—	—				G>
2	x	—	—	C			CG
3	—	0·9	—	A			AG
4	—	—	x	U			UG
5	2·1	—	—				CCG
6	0·9	0·7	—		C, AG		CAG
7	1·0	1·0	—		G, AC	G, AC	ACG
8	—	1·9	—	A	unchanged		AAG
9	0·95	—	0·9	U			UCG
10	1·1	—	0·9	C			CUG
11	—	0·9	0·85	U	U, AG		UAG
12	—	1·2	1·0	A	G, AU		AUG
13	—	—	1·9				UUG
14	2·9	—	—				CCCG
15	xx	x	—	C	C, AG		CCAG
16	xx	x	—	C	C, G, AC		CACG
17	1·9	1·1	—	A	C, G, AC		ACCG
18	1·0	2·1	—		C, AAG	AG, C, A	CAAG
19	1·0	2·0	—		G, AC, AG, AAC		ACAG+AACG
21	—	3·0	—		unchanged		AAAG
22	xx	—	x	U		C, G, UC, CC	UCCG
23 } 24 }	1·9	—	1·0 {	C C		C, G, CG, CU CC, UG	CUCG CCUG
25	1·2	1·1	1·0	U	C, U, AG, AC		UCAG+UACG
28	1·15	1·0	1·0	C	C, G, AU		CAUG
29	1·0	1·0	1·1	A	C, U, G, AC, AU		ACUG+AUCG
31	—	2·2	1·2		U, AAG		UAAG
32 } 33 }	—	2·1	1·0 {		AU, AG G, AAU		AUAG AAUG
34 } 35 } 36 }	1·1	—	2·0 {	U U C		C, G, UC, UU, CG C, U, UC, UG C, U, G, CU, UG	UUCG UCUG CUUG
37	—	1·0	1·9		U, G, AU, AG		UUAG+UAUG
39	—	x	xx	A	U, G, AU		AUUG
40	—	—	3·1				UUUG
41	4·0	—	—				CCCCG

† Compositions are expressed as the yield of a given mononucleotide relative to one mole of G and were estimated in the scintillation counter as described in the text. Where the results are given in x's instead of figures, the composition was determined only by inspection of the radio-autograph. All nucleotides contained G, which is not recorded.

‡ 5′ terminal residues (column 5) were determined by degradation with phosphomonoesterase and phosphodiesterase as described in the text.

1a. On alkaline digestion gave a double spot corresponding to the 2′- and 3′-guanylic acids.

25 may contain a small amount of C terminal residue.

34, 35, 36. There was considerable overlapping between these components. The sequences of the (CU) dinucleotides from the micrococcal nuclease digest were determined using phosphomono-esterase and phosphodiesterase.

and U on hydrolysis. This suggested that the 23 s RNA was slightly contaminated with a nuclease which was not present in the 16 s RNA.

Using an enzyme to substrate ratio of 1:20, the only 2',3'-cyclic phosphate present was that of guanylic acid (spot 1a, Fig. 2), which appears to be more stable than the cyclic phosphates of the oligonucleotides. When less enzyme was used, cyclic forms of other nucleotides were present, which led to a considerably more complex pattern. The cyclic forms moved faster than the non-cyclic phosphates on the cellulose acetate ionophoresis dimension, but at approximately the same rate in the DEAE-paper dimension.

The results obtained by pancreatic ribonuclease digestion were very clear and gave considerable information about structure. Some difficulty was experienced in determining the 5' terminal residues using phosphomonoesterase and phosphodiesterase, since the various preparations of phosphomonoesterase which were used appeared to contain a nuclease, so that on purification of the dephosphorylated product several spots were frequently present and it was necessary to study them all to decide which was the required one.

The probable structures of the components of the different spots are shown in the last column of Table 1, and these results are summarized in Fig. 2. The di- and tri-nucleotides are all readily separated by this technique. Of the 27 possible tetranucleotides, 10 are present as single pure spots. Very slight overlapping usually occurs between CCUG(23) and CUCG(24) and between AAUG(32) and AUAG(33). The following pairs occur in single spots, where there is slight evidence of separation; AUCG(29) and ACUG(30), ACAG(19) and AACG(20). No separation has been observed between the pairs UCAG(25) and UACG(26), and UUAG(37) and UAUG(38). The three isomers having the composition $(CU_2)G$ (34–36) are in a large spot which shows partial resolution of all three. We have not yet detected the nucleotide CUAG in ribosomal RNA, which would be expected to occur in or near spot 28. In both the 16 s and the 23 s material, spot 28 appears to contain only CAUG, giving AU and no AG when treated with pancreatic ribonuclease. It is possible that CAUG may run with spot 25.

The composition of the pentanucleotide spots has been determined in many cases, but the sequences have not been studied in detail and it seems probable that most spots are mixtures of isomers. There are 81 possible pentanucleotides, and for some compositions, e.g. $(C_2AU)G$, there are 12 possible isomers. There is a considerable spread of isomers, as is indicated by the difference in the patterns in the pentanucleotide area obtained from the 16 s and 23 s components. The areas occupied by the different isomers are indicated in Fig. 2 with dashed lines.

The position of a nucleotide on the fingerprint is determined largely by its composition. Figure 3 shows the relationship between the composition of a nucleotide and its position. It was prepared as follows. The "centre of gravity" of each set of isomers was first marked with a point. Lines were then drawn to link up the points representing nucleotides that differed only in the number of C residues they contained. Thus, for instance, a line was drawn joining UG, (CU)G, $(C_2U)G$ and $(C_3U)G$, etc. Similarly, lines were drawn connecting nucleotides that differed only in A residues. The mononucleotide that has the greatest effect on mobility in the DEAE-paper dimension is U, so that the fingerprint may be regarded as being composed of three sections representing nucleotides with two, one or no U residues, respectively. The lines joining the spots form three graticules corresponding to the different sections, and one axis on each

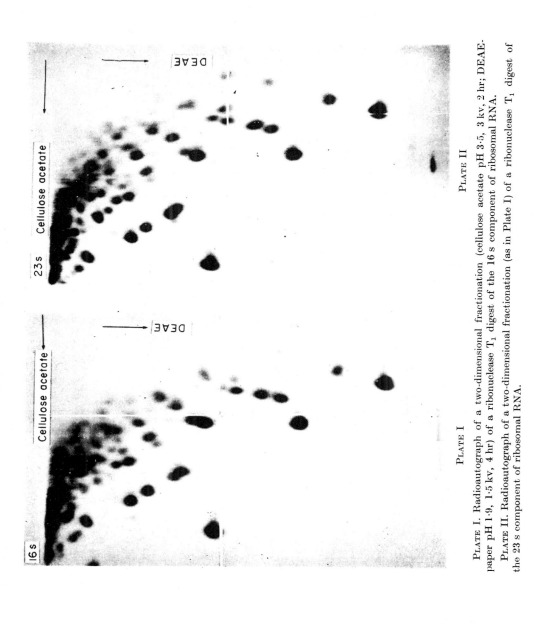

PLATE I

PLATE I. Radioautograph of a two-dimensional fractionation (cellulose acetate pH 3·5, 3 kv, 2 hr; DEAE-paper pH 1·9, 1·5 kv, 4 hr) of a ribonuclease T₁ digest of the 16 s component of ribosomal RNA.

PLATE II. Radioautograph of a two-dimensional fractionation (as in Plate I) of a ribonuclease T₁ digest of the 23 s component of ribosomal RNA.

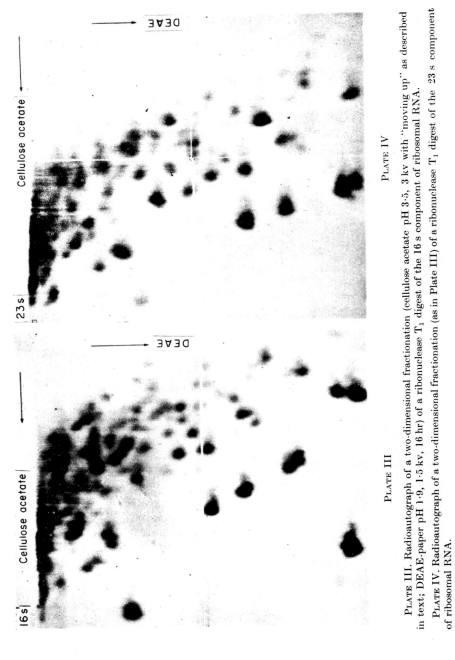

PLATE III

PLATE IV

PLATE III. Radioautograph of a two-dimensional fractionation (cellulose acetate pH 3·5, 3 kv with "moving up" as described in text; DEAE-paper pH 1·9, 1·5 kv, 16 hr) of a ribonuclease T_1 digest of the 16 s component of ribosomal RNA.

PLATE IV. Radioautograph of a two-dimensional fractionation (as in Plate III) of a ribonuclease T_1 digest of the 23 s component of ribosomal RNA.

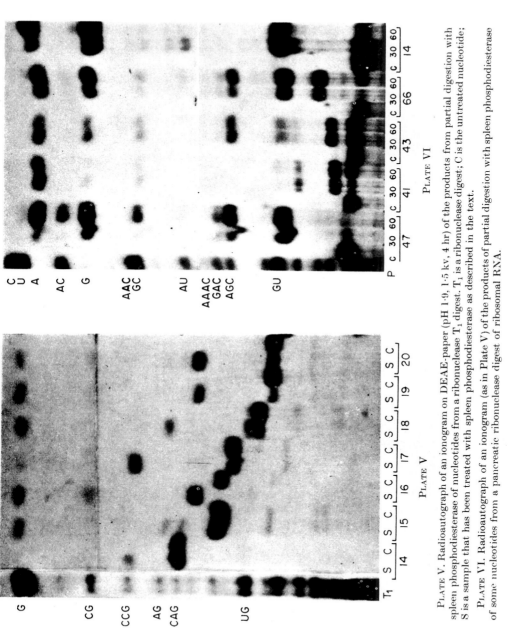

PLATE V. Radioautograph of an ionogram on DEAE-paper (pH 1·9, 1·5 kv, 4 hr) of the products from partial digestion with spleen phosphodiesterase of nucleotides from a ribonuclease T_1 digest. T_1 is a ribonuclease digest; C is the untreated nucleotide; S is a sample that has been treated with spleen phosphodiesterase as described in the text.

PLATE VI. Radioautograph of an ionogram (as in Plate V) of the products of partial digestion with spleen phosphodiesterase of some nucleotides from a pancreatic ribonuclease digest of ribosomal RNA.

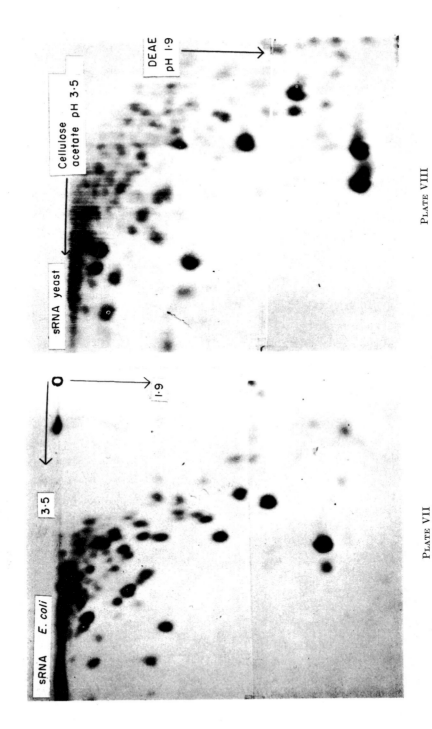

PLATE VII

PLATE VII. Radioautograph of a two-dimensional fractionation (as in Plate I) of a ribonuclease T₁ digest of sRNA from *E. coli.*

PLATE VIII. Radioautograph of a two-dimensional fractionation (as in Plate I) of a ribonuclease T₁ digest of sRNA from bakers' yeast.

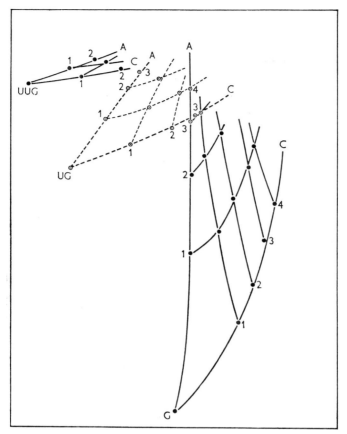

FIG. 3. Diagram illustrating the relationship between the composition of a nucleotide and its position on the two-dimensional system using ionophoresis at pH 1·9 for the DEAE-paper dimension. For explanation see text.

graticule represents the number of A residues, whereas the other gives the number of C residues. In this way the probable composition of a nucleotide may be determined from its position on the map. In the pentanucleotide area there is considerable overlapping between the different graticules. Thus, for instance $(A_3C)G$ is in the same area as $(C_3U)G$. This overlapping may be almost completely avoided by carrying out the fractionation on DEAE-paper using between 7 and 10% formic acid instead of the pH 1·9 mixture (2·5% formic acid, 8·7% acetic acid). This increases the separation of the sections containing different numbers of U residues, and a distribution as shown in Fig. 4 is obtained. The resolution of isomers of the faster moving spots is probably less efficient on this system, although it is probably better for the slow-moving ones. The system has not yet been investigated in sufficient detail to say precisely to what extent the composition of a nucleotide can be determined from its position on this map, and it seems probable that some overlapping may occur between the different isomers. Thus the tetranucleotide CAAG comes very close to the trinucleotide AAG, and it seems that larger nucleotides that bear this type of relationship to one another may also overlap.

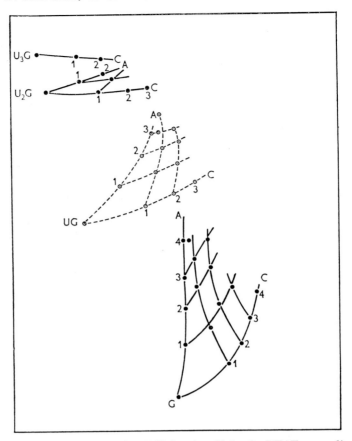

Fig. 4. Diagram as in Fig. 3 but using 10% formic acid for the DEAE-paper dimension.

The fact that there is a separation of isomers indicates that the sequence affects the behaviour on the two-dimensional system. One generalization that seems possible is that isomers which have A as the 5′ terminal residue move slower in the DEAE dimension than those having other 5′ terminal residues.

(b) *Partial digestion with spleen phosphodiesterase*

The results listed in Table 1 were obtained before the introduction of the method using spleen phosphodiesterase. This method has been applied to most of the spots the sequence of which had been determined by the normal methods, and the results have been used as confirmation and to demonstrate the general applicability of the method. Some of the results are illustrated in Plate V and Fig. 5.

If we consider the nucleotide ABCG, where A, B and C are unknown residues, its breakdown products will be BCG and CG. Suppose the distance moved from the origin by ABCG on the DEAE-paper is y and the distance between it and its first degradation product (BCG) is x, we can then define a value M for ABCG as x/y. This value depends on the nature of the 5′ terminal residue A. For all the nucleotides studied from the ribonuclease T_1 digests, it has been found that the values of M lie within the limits

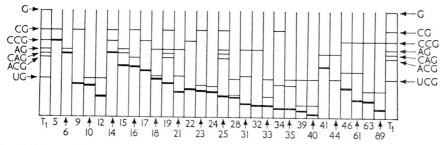

Fig. 5. Diagram illustrating the fractionation on DEAE-cellulose (pH 1·9) of the products of partial digestion with spleen phosphodiesterase of oligonucleotides from ribonuclease T_1 digests of ribosomal RNA. The thick line represents the position of the unchanged nucleotide which was run as a control. Mononucleotides are not included.

shown in Table 2 and there is no overlapping between the values for the different residues; thus the various residues may be identified directly from the values of M. The dinucleotides and many of the trinucleotides can be identified from their position on the DEAE-paper by comparison with the ribonuclease T_1 digest that is run as a marker.

TABLE 2

Values of M *on DEAE-paper ionophoresis (pH 1·9) for nucleotides from ribonuclease* T_1 *digests*

5′ terminal residue	Range of M-values
C	0·05–0·3
A	0·5 –1·0
U	1·5 –2·5

As examples, we may consider the results with the three isomers $(C_2A)G$, numbers 15, 16 and 17 (see Plate V and Fig. 5). The value of M for the first degradation step for 16 is 0·18. This falls within the limits for C (Table 2), which is therefore the 5′ terminal residue. The value of M for the second step is 0·58, establishing it as an A residue. The product produced at this stage is rapidly moving and can only be the dinucleotide CG, thus confirming the structure CACG. For no. 15, the values of M for the first two steps are 0·22 and 0·07, respectively, and the dinucleotide moves at the rate of AG, thus establishing the sequence CCAG. Nucleotide no. 17 gives only one visible degradation product under the conditions used. This is related to the unchanged material by a value of 0·64 and occupies a unique position which identifies it as the trinucleotide CCG, thus confirming the structure ACCG.

For most of the nucleotides studied it was possible to deduce the complete sequence by this method. Difficulties might be experienced if the nucleotides studied were mixtures. Thus the results with no. 19 could be interpreted in various ways; from the values for M of the different bands, it could for instance fit a sequence ACCACG. This, however, is clearly impossible from its position on the fingerprint and is precluded by a knowledge of the positions of the different nucleotides on the DEAE-paper system. Thus the presence of two dinucleotides CG and AG indicates that it is a mixture. The first degradation step corresponds to a value of 0·64 for M, establishing A as the only terminal residue. The two bands close together resulting from this first degradation can be identified from their position as CAG and ACG, thus showing it to be a mixture of ACAG and AACG. Similarly, spot 25 is a mixture of UACG and UCAG. Occasionally extra bands were present in the untreated nucleotide control, suggesting that some degradation or side reaction had occurred during elution or storage of the nucleotide. Such bands were ignored in interpreting the results.

One limitation of the method lies in the different rates at which the different residues are released by the enzyme, U residues being split off more rapidly than A, and A more rapidly than C. Thus it may be that not all the possible degradation products of a nucleotide are present in a particular digest; this may lead to a misinterpretation in the case of larger nucleotides. For this reason it is advisable not to rely completely on this method, but to have some other data such as a composition on all nucleotides. Thus in Fig. 5, CG was not observed in nucleotides 17 and 22, UG was not observed in 24 and in 32 the intermediate UAG was missing. Of the pentanucleotides studied, no. 41 loses a 5′ terminal residue giving a band which corresponds to band 14 (CCCG) and no further breakdown products under the conditions used; it is therefore CCCCG. Similarly 44 is ACCCG, 46 is AACCG, the final product being CCG, which is uniquely identified from its position, 63 is AUCCG and 89 UUCCG.

(c) *Pancreatic ribonuclease digests of ribosomal RNA*

Digests prepared by the action of pancreatic ribonuclease have not been studied in such detail as those obtained with ribonuclease T_1. However, the simplicity in the composition of the nucleotides and the smaller number of larger ones made a study of these digests somewhat simpler. Figure 6 shows the pattern of spots obtained, and the sequence determinations are summarized in Table 3. The results were obtained from various digests of ribosomal RNA components. Of the 16 tetranucleotides, 10 have been identified as single pure spots. The following pairs of isomers show partial separation but there is usually some overlapping: AAGU(18) and AGAU(20), GAGC(25) and GGAC(27), GAGU(26) and GGAU(28).

The determination of the 5′ terminal residue by digestion with phosphomono-esterase and venom phosphodiesterase was more generally satisfactory than with the nucleotides from ribonuclease T_1 digests, since only one product was usually obtained after phosphomonoesterase treatment. This suggested that the contaminant in the preparations of phosphomonoesterase causing degradation of the nucleotides in ribo-nuclease T_1 digests had a specificity similar to pancreatic ribonuclease. In fact it was always possible to identify the 5′ terminal residues from the positions of the nucleotides on the fingerprint, since isomers with A end groups move slower on the DEAE-paper than those with G end groups.

The degradation method with spleen phosphodiesterase also gave rather clear results. In this case one is only concerned with A and G residues, and these are both split off at a similar speed, so that a complete spectrum of degradation products is usually produced. For the pancreatic ribonuclease digestion products, the values of M associated with an A 5′ terminal residue varied from 0·4 to 1·1 and for a G terminal residue from 1·2 to 3·1. There is a wide spread and the limits are rather close; thus for values between 0·9 to 1·5 it is advisable to have confirmation from other data. The

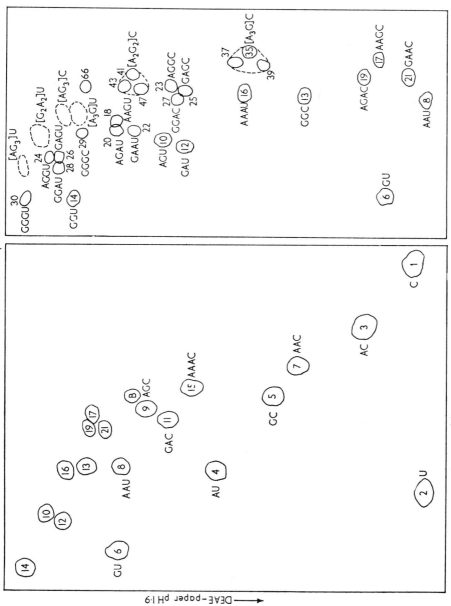

— Cellulose acetate pH 3·5

(a)

(b)

DEAE-paper pH 1·9

Fig. 6. Diagram showing the position of nucleotides from a pancreatic ribonuclease digest on the two-dimensional system. (a) Illustrates a fractionation that has been run for a short time on the DEAE-paper dimension; (b) one that has been run for a long time. B marks the position of the blue marker.

TABLE 3

Nucleotides from pancreatic ribonuclease digests

Spot no. (Fig. 6)	Composition†				5′ terminal residue	Products obtained with ribonuclease T$_1$	Structure deduced
	C	U	A	G			
1	x	—	—	—			C
2	—	x	—	—			U
3	x	—	x	—			AC
4	—	x	x	—			AU
5	x	—	—	x			GC
6	—	x	—	x			GU
7	x	—	1·9	—			AAC
8	—	x	1·7	—			AAU
9	x	—	x	x		C, AG	AGC
10	—	x	1·0	1·0	A	U, AG	AGU
11	x	—	x	x		G, AC	GAC
12	—	x	x	x	G	G, AU	GAU
13	x	—	—	2·0			GGC
14	—	x	—	1·7			GGU
15	x	—	2·6	—			AAAC
16	—	x	2·8	—			AAAU
17	x	—	1·7	1·0	A	C, AAG	AAGC
18‡	—	x	2·1	1·0		U, AAG	AAGU
19	x	—	xx	x	A	AG, AC	AGAC
20‡	—	x	2·1	1·0		AG, AU	AGAU
21	x	—	1·9	0·9	G	G, AAC	GAAC
22	—	x	2·0	1·2		G, AAU	GAAU
23	x	—	1·0	1·9	A	C, G, AG	AGGC
24	—	x	x	xx	A	U, G, AG	AGGU
25‡	x	—	1·1	2·0	G	C, G, AG	GAGC
26‡	—	x	1·0	1·9	G		GAGU
27‡	x	—	1·1	2·0	G	G, AC	GGAC
28‡	—	x	1·0	1·9	G	G, AU	GGAU
29	x	—	—	xxx			GGGC
30	—	x	—	2·7			GGGU
33	x	—	2·9	0·9		C, AC, AAG, AAAG	AAAGC+AAGAC
37	x	—	xxx	x		AG, AAC	AGAAC
39	x	—	2·6	1·0		G, AAAC	GAAAC
41	x	—	2·0	1·9		G, C, AAG	AAGGC§
43	x	—	xx	xx		C, AG	AGAGC
47	x	—	2·1	1·9		G, AC, AG, C, AAG	GAGAC+GAAGC§

† Compositions are expressed relative to the amount of C or U present.

‡ The following pairs show slight overlapping: 18 and 20, 25 and 27, 26 and 28. In each case the composition was determined on a mixture of both components.

§ In deducing these structures, the 5′ terminal residues have been determined from the position on the DEAE-paper, assuming that isomers with an A terminal residue move slower than those with a G.

extreme values are found among the smaller nucleotides, which in any event are best identified from their position on the DEAE-paper.

Some results with some larger nucleotides are shown in Plate VI. The values of M for the various steps in no. 41 are 0·46, 0·68 and 1·8. This latter step gives rise to the dinucleotide GC, demonstrating the sequence AAGGC. Two dinucleotides, AC and GC, are present in the digest of no. 47, indicating it is a mixture. The first two degradation steps are clear, giving GA (M being 1·7, 0·52). Two bands are then produced close together which correspond to the positions of GAC and AGC on the marker. It thus appears to be a mixture of GAGAC and GAAGC. Number 43 could be interpreted as AAAAGC; however, this would not fit with its position and it is probable that the third band from the origin is due to contamination with spot 21 and that the actual sequence is AGAGC. Number 66 appears to analyse clearly as the hexanucleotide GAAAGC.

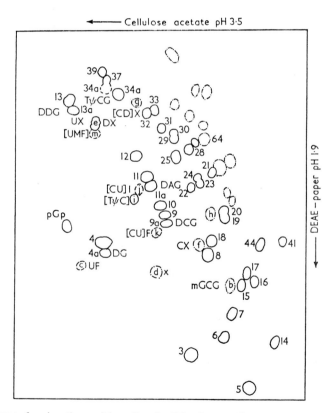

FIG. 7. Diagram showing the position of nucleotides from a ribonuclease T_1 digest of sRNA on the two-dimensional system. The sequence or composition is shown for those nucleotides, symbolized by letters a to m, containing minor bases. The numbered spots refer to nucleotides observed in digests of ribosomal RNA, the sequence of which is shown in Fig. 2. Those not mentioned there are referred to in the text.

(————) Indicates that the nucleotide was detected in both yeast and *E. coli*, (– – –) that it was detected only in yeast, and (– · – · –) that it was seen only in *E. coli*. See Table 4 and text for further explanation.

TABLE 4

Nucleotides containing minor bases from ribonuclease T_1 digests of sRNA

Spot no. (Fig. 7)	Composition†					Structure deduced	Source (yeast (y) or E. coli (c))
	C	A	G	U	Minor nucleotides‡		
4a	—	—	x	—	(Y) x	DG	
9a	x	—	x	—	(Y) x	DCG	
11a	—	x	x	—	(Y) x	DAG	y+c
13a	—	—	x	—	(Y) xx	DDG	
34a	x	—	x	—	(T) x (ψ) x	TψCG	
34a>	x	—	x	—	(T) x (ψ) x	TψCG >	
b	x	—	x	—	(mG) x	mGCG	c
c	—	—	—	x	(F) x	UF	y
d	—	—	—	—	(X)	X	c
e	—	—	—	x	(X) xx	UX	y+c
e	—	—	—	—	(Y) x, (X) xx	DX	
f	1·0	—	—	—	(X) 1·8	CX	c
g	1·0	—	—	—	(X) 1·7, (Y) 1·0	(CD)X	c
h	x	—	x	x	(mG) x	—	c
i	x	—	—	—	(T) x, (ψ) x	TψC	y+c
j	x	—	—	x	(I) x	(CU)I	y
k	x	—	—	x	(F) x	(CU)F	y
m	—	—	—	x	(M) x, (F) x	(UMF)	y
z	—	0·9	1·0	—	(Y) 2·1	(D₂A)G	c

† Compositions were determined by inspection of the intensity of blackening of the radio-autograph and are symbolized by x's. Sometimes the yields of mononucleotides were estimated in a scintillation counter as described in the text and the result expressed per one mole of G, or per one mole of C if G was absent.

‡ The most probable structure of the nucleotides symbolized by the letters is recorded in Table 5 and in the text.

4a gives D (see Table 5) and G with pancreatic ribonuclease.

9a, 11a and 13a give D as 5'-terminal mononucleotides as found by degradation with phospho-monoesterase and phosphodiesterase as described in the text.

34a. This sequence was found by partial alkaline hydrolysis.

b. This sequence was found by partial degradation with a spleen phosphodiesterase as described in the text.

e. UX and DX are clearly separated from one another when 10% formic acid is used for ionophoresis in the second dimension of the two-dimensional fractionation system.

g and z were isolated from a 10% formic acid run as in e above. z is not shown in Fig. 7.

h is probably a mixture.

i. Sequence was not determined; it was presumed to be a breakdown product of 34a.

j and k. I and F are presumed to be 3'-terminal from published observations on the specificity of T_1 ribonuclease for minor nucleotides (Holley et al., 1963; Staehelin, 1964b).

(d) sRNA

Plates VII and VIII show two-dimensional fractionations of ribonuclease T_1 digests of mixed sRNA from *E. coli* and from yeast. They show considerable differences from ribosomal RNA. There are those spots, identified by letters of the alphabet in Fig. 7 and Table 4, which are not present in ribosomal RNA and are oligonucleotides containing minor bases, and also there are certain of the larger oligonucleotides that are particularly strong, suggesting that there are some sequences of bases common to many of the sRNA molecules.

(i) *Minor components*

Those minor mononucleotides which separated from G and U on ionophoresis at pH 3·5 were further analysed by paper chromatography. The properties of these various components are listed in Table 5. Identification is based on their mobility on ionophoresis at pH 3·5, on their chromatographic behaviour and on the published composition of sRNA. T and ψ are ribothymidylic acid and pseudo-uridylic acid. The most probable structure for X is 2′-O-methylguanylyl guanylic acid.

TABLE 5

Ionophoretic and chromatographic properties of minor nucleotides and guanylic acid

| Symbol | R_U† on ionophoresis at pH 3·5 | R_U† on chromatograms | | Probable structure (see text) |
		Propan-2-ol–HCl	Propan-2-ol–NH$_3$	
Y	1·07	0·93, 1·1	0·63	β-Ureidopropionyl N-ribotide
D	1·0	0·90	—	4,5-Dihydrouridylic acid
T	0·98	1·1	1·7	Thymidylic acid
ψ	0·98	0·79	0·63	Pseudouridylic acid
I	0·90	0·54	—	Inosinic acid
X	0·89	0·22	0·20	2′-O-Methylguanylyl guanylic acid
mG	0·82	0·64	0·57	—
F	0·76	0·64	—	N^2-Dimethylguanylic acid
G	0·73	0·55	0·50	Guanylic acid
M	0·70	0·72	—	N^1-, or N^2-Methylguanylic acid
A$_2$	0·44	0·83	2·0	6-Methyladenylic acid
A$_1$	0·25	0·68	1·5	2-Methyladenylic acid

† R_U is the distance moved relative to uridylic acid.

Dihydrouridylic acid. One of the minor components (Y) appeared to be present in relatively large amounts and did not have the properties of any of the then known minor components. This was identified as a breakdown product of dihydrouridylic acid, the presence of which in sRNA has recently been reported by Madison & Holley (1965). Our identification is briefly reported here since the methods used are somewhat different from those of Madison & Holley and confirm their work.

Figure 8 shows the pH–mobility curve for Y plotted relative to U. The best fit to the experimental values, indicated by ×'s, was the theoretical titration curve of an acidic group of pK 4·1. These results suggested that Y had a carboxyl group of pK about 4·1.

Compound Y was resolved into two spots on the propan-2-ol–HCl system. The faster and stronger spot was identified as a mixture of ribose phosphate and inorganic

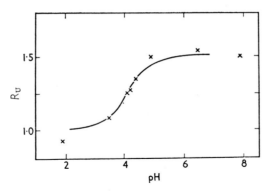

F<small>IG</small>. 8. pH–mobility curve for Y. The mobility is expressed as an R_U value which is the distance of Y from the origin divided by the distance of U. The curve is a theoretical titration curve of an acidic group of pK 4·1.

phosphate by ionophoresis at pH 3·5. The slower and weaker spot was shown to re-run in the same position on chromatography and to run in a position between uridylic acid and compound Y on ionophoresis at pH 3·5. The results suggested Y was acid-labile, and gave ribose phosphate as the major product and an unidentified substance as the minor product as soon as the acid developer touched Y on the paper. The purine nucleotides, adenylic acid and guanylic acid, do not cleave under these conditions. Compound Y also degraded slightly to ribose phosphate on ionophoresis at pH 1·9. Again this was interpreted as a partial acid hydrolysis occurring because of the acidity of the buffer used for the ionophoresis.

It was of interest to establish whether Y had a methyl group. An alkaline digest of mixed [^{14}C]methyl-labelled sRNA from *E. coli* was fractionated on the two-dimensional system described in section 2(g). No radioactive spot was seen in the known position of Y. The results suggested that Y was not methylated.

The presence of a carboxyl group in compound Y and its acid lability suggested that Y might be a nucleotide with the purine or pyrimidine ring opened by alkali.

Further evidence that Y was a degradation product was obtained from a study of the digestion products obtained from nucleotide 11a. With alkali, 11a gave A, G and Y, but on digestion with pancreatic ribonuclease it gave AG and a band moving at the same rate as U on ionophoresis at pH 3·5 and slower than U on paper chromatography in propan-2-ol–HCl. This band was referred to as D. It was clear that D was very similar to U, since trinucleotides containing it were always found on the two-dimensional system very close to corresponding trinucleotides in which the D was replaced by U. Also D was not methylated since the D-containing spots from a ribonuclease T$_1$ digest of [^{14}C]methyl-labelled sRNA were not radioactive.

These results suggested that D was dihydrouridylic acid, which decomposes with alkali to give β-ureidopropionic acid *N*-ribosyl phosphate (Y) (Cohn & Doherty, 1956). To confirm this, [^{32}P]β-ureidopropionic acid *N*-ribosyl phosphate was prepared from [^{32}P]U by the method of Cohn & Doherty (1956) and shown to have the same iono-phoretic mobility at pH 4·4 and 3·5, and the same chromatographic behaviour on propan-2-ol–HCl–water or propan-2-ol–NH$_3$–water as compound Y. Cohn & Doherty also reported that it is labile to acid, which fits with the properties of Y.

Compound X. Compound X is a dinucleotide because equimolar amounts (that is equal counts) of inorganic phosphate and a new band in the expected position on ionophoresis of 2'-O-methylguanylyl guanosine are formed after digestion with bacterial alkaline phosphomonoesterase. This new band gave guanosine 5'-phosphate after hydrolysis with snake venom phosphodiesterase. X is methylated, because a compound with its properties was isolated in good yield from an alkaline digest of [^{14}C]methyl-labelled sRNA from *E. coli* (strain CB3). Also free X occurred on the two-dimensional separation system in a position compatible with this structure. When X occurred in oligonucleotides isolated from a ribonuclease T_1 digest of *E. coli* sRNA (see Table 4), it had twice as much radioactive material in it as the other mononucleotides. This suggested there were two phosphate groups in X.

Smith & Dunn (1959) first reported the occurrence of 2'(3)-O-methylribose in alka-line-resistant dinucleotides isolated from the RNA of rat liver and plants. Hall (1963) reported that all four of the 2'-O-methylribosides were present in soluble RNA from *E. coli*, but that 2'-O-methylguanosine was the commonest of the four and isolated it in a yield of 0·1%. We would add to these reports that we have not detected any other alkali-resistant dinucleotides in the products of a T_1 ribonuclease digest of sRNA from *E. coli* or yeast. Presumably they must be present in very much smaller amounts than 2'-O-methylguanylyl guanylic acid.

Other minor components. We are less sure of the identity of mG. It is acid-labile and methylated, and may possibly be 4-amino-5(N-methyl)-formamido-isocytosine ribotide 1, which is the alkaline degradation product of N^1-methyl guanylic acid.

Other minor mononucleotides have been detected in oligonucleotides isolated from a T_1 digest of [^{14}C]methyl-labelled sRNA and in an alkaline hydrolysis of [^{14}C]methyl-labelled sRNA. These are listed in Table 5 as A_1, which is probably 2-methyladenylic acid, and A_2, which is probably 6-methyladenylic acid (Dunn, Smith & Spahr, 1960). These components are indistinguishable from C and A in their R_U values (the distance moved relative to uridylic acid) on ionophoresis at pH 3·5 and are therefore missed in digests of [^{32}P]sRNA.

F, I and M have been detected in sRNA of yeast and are assigned from their chromatographic properties (Holley *et al.*, 1963), as N^2-dimethylguanylic acid, inosinic acid, and N^1- or N^2-monomethylguanylic acid.

Table 4 and Fig. 7 show only those oligonucleotides which have been separated in a pure state. There are many others that are present as mixtures. For instance, a T_1 ribonuclease digest of [^{14}C]methyl-labelled sRNA from *E. coli* shows that there are methylated nucleotides covering all those areas of the fingerprint where phosphate-labelled nucleotides are found.

sRNA from yeast and *E. coli* show many similarities to one another. Both contain the same dihydrouridylic acid-containing nucleotides and the sequence TψCG. Dihydrouridylic acid occurs in relatively high proportions in sRNA from both sources. It is probably about as common as pseudouridylic acid in *E. coli*. In addition to those T_1 sequences recorded in Table 4, it has been detected in the sequences GGD and AGD from pancreatic ribonuclease digests. TψCG is remarkable in that it gives the strongest radioactive spot excepting guanylic acid in a T_1 digest of sRNA. The sequence GTψCG must be common to many or all of the different amino acid-accepting species of sRNA in both yeast and *E. coli* (Holley, Everett, Madison, Marquisee & Zamir, 1964, *Abstr. 6th Int. Congr. Biochem.*, section 1, p. 9). Similarly, the commonest sequence containing dihydrouridylic acid is DAG. It is stronger than UAG and about one-third as strong as TψCG and would therefore appear to be present in many different sRNA's.

(ii) *Other nucleotides*

A comparison of a T_1 digest of sRNA (Plates VII and VIII) with ribosomal RNA (Plates I and II) shows that some of the larger oligonucleotides are particularly strong, and others, present in ribosomal RNA, are absent from sRNA. The following are some examples seen in sRNA from *E. coli*: spot 64 (Fig. 7) is mainly CUCAG, and the other 11 isomers of the composition $(C_2UA)G$ are absent or in much smaller amounts. Spot 44 is CACCG and is the only nucleotide of the composition $(C_3A)G$ present. Of the (CAU)G isomers, UCAG(25) is the commonest and UACG(26) and CUAG have not been detected. The three isomers $(AC_2)G$ (15, 16, 17) form a triangle in ribosomal RNA, but ACCG(17) is weak in *E. coli* sRNA. Even among the trinucleotides, the amounts of the two isomers are unequal in certain cases and there are nearly twice as much CUG as UCG and twice as much CAG as ACG. We suspect from these results that there are some sequences which are common even in the mixed sRNA. GCUCAG, for example, must be present in more than half the different sRNA's in *E. coli*. It is not common, however, in sRNA from yeast, and this is one of the most obvious distinctions between them.

4. Discussion

The fractionation procedure described in this paper is capable of distinguishing the di-, tri- and most of the tetranucleotides in digests prepared by the action of ribonuclease T_1 or pancreatic ribonuclease. The resolution of larger nucleotides depends on the complexity of the mixture. It should thus prove a useful tool in the study of sequences in smaller RNA's. Most of the information required about a nucleotide mixture may be obtained by running it on the two-dimensional system using 10% formic acid for the DEAE-paper dimension (Fig. 4). The approximate composition of a spot is determined from its position, and the sequence may be determined by partial degradation with spleen phosphodiesterase.

The method may also be used as a fingerprinting technique to characterize RNA's and perhaps to detect small differences as between species and mutants. A comparison of the patterns obtained for the ribonuclease T_1 digests of the two ribosomal RNA components (Plates I to IV) shows definite differences among the tetranucleotides. Thus UAAG(31) is almost completely absent in the 16 s component but present in the 23 s, whereas AUUG(39) is present in the 16 s but absent from the 23 s. The three isomers $(AC_2)G$ (15, 16, 17) form a triangle with each one in about equal strength in the 23 s component; but in the 16 s component CCAG is much stronger than the other two. The general pattern of spots is very different in the pentanucleotide areas, indicating that there must be many different sequences. That this is the type of behaviour that would be expected is illustrated by the following very approximate calculations.

The yield of G from a ribonuclease T_1 digest will depend on the frequency of the sequence GG. If, as a rough approximation, one assumes an RNA to be a random arrangement of the 4 mononucleotides present in equal amounts, the sequence G will occur once every 4 residues and GG once every 16 residues. A sequence such as GAG, which gives rise to the dinucleotide AG, will occur once every 64 residues. Similarly, a trinucleotide will be obtained once from a run of 250 residues, a tetranucleotide once from 1000 residues, and a pentanucleotide once from every 4000 residues. The 16 s and 23 s components contain about 1600 and 3200 residues, respectively; thus the average yield of a tetranucleotide would be 2 to 3 moles and of a pentanucleotide would be less than one. It is thus not surprising that certain tetranucleotides and many of the pentanucleotides are absent from the digests. This illustrates that the method can be an effective fingerprinting method for RNA's of this size, since characteristic patterns are obtained.

The conditions for the DEAE-paper ionophoresis are rather acidic and some depurination of the nucleotides might be expected. That this cannot be very extensive is apparent from the fact that all expected nucleotides are found in the digests. It would seem that the effective pH on an ion-exchange material is not necessarily that of the buffer with which it is washed. Occasionally after a long run at a high voltage gradient the tank may become warm and streaking may be observed behind some of the spots; this may be due to depurination.

We wish to thank Mr J. R. Prescott and Mr R. P. Hughes of British Celanese Ltd. for generous supplies of cellulose acetate, Dr F. Egami for a gift of ribonuclease T_1, Mr R. E. Offord for the preparation of ribonuclease T_1 and Drs D. B. Dunn and J. D. Smith for discussions and helpful advice.

REFERENCES

Apgar, J., Holley, R. W. & Merrill, S. H. (1962). *J. Biol. Chem.* **237**, 796.
Armstrong, A., Hagopian, H., Ingram, V. M., Sjöquist, I. & Sjöquist, J. (1964). *Biochemistry*, **3**, 1194.
Borek, E., Ryan, A. & Rockenbach, J. (1955). *J. Bacteriol*, **69**, 460.
Cohn, W. E. & Doherty, D. G. (1956). *J. Amer. Chem. Soc.* **78**, 2863.
Cunningham, L., Catlin, B. W. & Privat de Garilhe, M. (1956). *J. Amer. Chem. Soc.* **78**, 4642.
Dunn, D. B., Smith, J. D. & Spahr, P. F. (1960). *J. Mol. Biol.* **2**, 113.
Garen, A. & Levinthal, C. (1960). *Biochim. biophys. Acta*, **38**, 470.
Gilbert, W. (1963). *J. Mol. Biol.* **6**, 389.
Gray, W. R. (1964). Ph.D. Dissertation, Cambridge.
Gross, D. (1961). *J. Chromatography*, **5**, 194.
Hall, R. H. (1963). *Biochem. Biophys. Res. Comm.* **12**, 429.
Holley, R. W., Apgar, J., Doctor, B. P., Farrow, J., Marini, M. A. & Merrill, S. H. (1961). *J. Biol. Chem.* **236**, 200.
Holley, R. W., Apgar, J., Everett, G. A., Madison, J. T., Merrill, S. H. & Zamir, A. (1963). *Cold Spr. Harb. Symp. Quant. Biol.* **28**, 117.
Holley, R. W., Madison, J. T. & Zamir, A. (1964). *Biochem. Biophys. Res. Comm.* **17**, 389.
Khorana, H. G. & Vizsolyi, J. P. (1961). *J. Amer. Chem. Soc.* **83**, 675.
Madison, J. T. & Holley, R. W. (1965). *Biochem. Biophys. Res. Comm.* **18**, 153.
Markham, R. & Smith, J. D. (1952). *Biochem. J.* **52**, 552, 558.
Michl, H. (1951). *Monatsh. Chem.* **82**, 489.
Naughton, M. A. & Hagopian, H. (1962). *Analyt. Biochem.* **3**, 276.
Razzell, W. E. & Khorana, H. G. (1959). *J. Biol. Chem.* **234**, 2114.
Razzell, W. E. & Khorana, H. G. (1961). *J. Biol. Chem.* **236**, 1144.
Reddi, K. K. (1959). *Biochim. biophys. Acta*, **36**, 132.
Rushizky, G. W., Bartos, E. M. & Sober, H. A. (1964). *Biochemistry*, **3**, 626.
Rushizky, G. W. & Knight, C. A. (1960). *Virology*, **11**, 236.
Rushizky, G. W. & Sober, H. A. (1962). *J. Biol. Chem.* **237**, 834.
Sanger, F. & Tuppy, H. (1951). *Biochem. J.* **49**, 463.
Sato, K. & Egami, F. (1957). *J. Biochem.*, *Tokyo*, **44**, 753.
Shuster, L., Khorana, H. G. & Heppel, L. A. (1959). *Biochim. biophys. Acta*, **33**, 452.
Smith, J. D. & Dunn, D. B. (1959). *Biochim. biophys. Acta*, **31**, 573.

Staehelin, M. (1961). *Biochim. biophys. Acta*, **49**, 11.

Staehelin, M. (1964a). *J. Mol. Biol.* **8**, 470.

Staehelin, M. (1964b). *Biochim. biophys. Acta*, **87**, 493.

Stent, G. S. & Brenner, S. (1961). *Proc. Nat. Acad. Sci., Wash.* **47**, 2005.

Tyndall, R. L., Jacobson, K. B. & Teeter, E. (1964). *Biochim. biophys. Acta*, **87**, 335.

Williams, R. B. & Dawson, R. M. C. (1952). *Biochem. J.* **52**, 314.

Wyatt, G. R. (1951). *Biochem. J.* **48**, 584.

J. Mol. Biol. (1965) **13**, 629–637

Ochre Mutants, a New Class of Suppressible Nonsense Mutants

S. Brenner and J. R. Beckwith†

Medical Research Council Laboratory of Molecular Biology, Cambridge, England

(*Received 28 April 1965, and in revised form 21 June 1965*)

Suppressors of a strongly polar mutant of the *lac* gene in *Escherichia coli* also suppress mutants of the r_{II} genes of bacteriophage T4. Included amongst this set are some *amber* mutants, but there are also many other mutants which are not suppressed by *amber* suppressors. These mutants are defined as *ochre* mutants, and the suppressors as *ochre* suppressors. *Ochre* mutants are nonsense mutants because, like the *amber* mutants, they abolish the B activity of the deletion *r*1589. This effect is not suppressed by *amber* suppressors, which shows that the *ochre* mutants are different from the *amber* mutants.

1. Introduction

Suppressors which affect mutations in many different genes are of great interest, since it is likely that they act at one of the steps in the translation of the nucleotide sequence of DNA into protein (Yanofsky & St. Lawrence, 1960; Benzer & Champe, 1961). Recently, progress has been made in analysing one type of suppressible mutation, the *amber* mutation. These mutations, which have been found in a large number of different bacterial and bacteriophage genes, can be defined by pairs of isogenic strains differing only at the *su* locus. The ambivalent subset I mutants of Benzer & Champe (1961) are *amber* mutants of the *r*II region. Earlier experiments suggested that they were nonsense mutants, interrupting the reading of the genetic message (Benzer & Champe, 1962), and we now know that *amber* mutations of the head protein gene of T4 produce NH_2-terminal fragments of the protein, the polypeptide chain terminating at the site of the mutation (Sarabhai, Stretton, Brenner & Bolle, 1964; Stretton & Brenner, 1965). In permissive strains, the chain is propagated with an efficiency which depends on the su^+ locus in the strain (Kaplan, Stretton & Brenner, unpublished results).

In a preliminary report, Beckwith (1964*a*) described two new suppressors which restore activity to certain mutants of the *lac* region of *Escherichia coli*. These mutants had been termed operator negative or O^0 mutants by Jacob & Monod (1961). However, it was shown by Beckwith (1964*b*) that the mutations do not lie in the *lac* operator and are probably only extreme examples of *polar* mutations which map in the z gene and reduce or abolish galactoside permease activity. Since it had been proposed that these mutations produced their effects at the level of messenger RNA transcription (Jacob & Monod, 1961), it became important to study their suppressors. We discovered that these suppressors permitted growth of some T4*r*II mutants, amongst which were included *amber* mutants (Beckwith, 1964*a*). Similar results have been reported by Orias & Gartner (1964). This suggested that the suppression of the strongly polar mutants was in some way related to the suppression of *amber* mutants, and that the polar mutants might exert their effects at the level of protein synthesis. Brenner & Stretton (1965) have presented strong evidence that suppressible mutations do act at

† Present address: Department of Bacteriology, Harvard Medical School, Boston, Mass., U.S.A.

this level. In this paper we show that the suppressors of strongly polar mutants are suppressors of nonsense mutants, and that these suppressors allow us to define a new set of nonsense mutants, which are different from the *amber* mutants and which we propose calling *ochre* mutants.

2. Materials and Methods

(a) Bacterial strains

The strain used by Beckwith is an F^- *E. coli* K12 derivative, 2320, which carries the lac_2 mutation (formerly called $O^0{}_2$) and which was isolated by Dr F. Jacob. From this strain (XA1 in our culture collection) Beckwith isolated XA2 and XA3, which carry $su^+{}_C$ and $su^+{}_B$, respectively (Beckwith, 1964a). Since they grew T4 very poorly, the suppressors were transferred to an Hfr strain by transduction with P1kc, and all suppressor loci are in this strain. The following are the derivatives of Hfr H *B1*⁻ which have been used.

CA154: i^- lac_2 (λ), from CA150: i^- lac_2, obtained from Dr F. Jacob.

CA165: i^- $lac_2su^+{}_B$ (λ), by transduction of CA154 with P1kc from XA3.

CA167: i^- $lac_2su^+{}_C$ (λ), by transduction of CA154 with P1kc from XA2.

CA248: i^- $lac_2su^+{}_D$ (λ) and CA254: i^- $lac_2su^+{}_E$ (λ) are spontaneous revertants of CA154.

The amounts of β-galactosidase activity restored by the suppressors of lac_2 are, relative to CA159 (i^- lac^+): 100. CA165 ($su^+{}_B$): 3·85. CA167 ($su^+{}_C$): 4·45. CA248 ($su^+{}_D$): 6·1. CA254 ($si^+{}_D$): 1·95.

CA180: i^+ $lac_{Y14}su^+{}_{II}$ (λ) derived from CA161: i^+ $lac_{Y14}si^+{}_{II}$. This strain contains an *amber* mutation of the *z* gene and the *amber* suppressor of *E. coli* C600. It was constructed by transducing CA85 (lac_{Y14}, obtained from Dr F. Jacob) with P1kc grown on C600.

CA265: i^+ $lac_{L125}su^+{}_{III}$ (λ) and CA267: i^+ $lac_{L125}su^+{}_I$ (λ) are spontaneous revertants of CA244: i^+ lac_{L125} (λ). The *amber* mutant lac_{L125} was isolated by Dr D. Zipser. The suppressors in these strains were characterized by suppression pattern.

E. coli strains B, B (Berkley), K12(λ) and CR63(λh). The latter strain contains the *amber* suppressor $su^+{}_I$.

(b) Bacteriophages

T4Br^+ and *rII* mutants from the Cambridge collection and from Dr S. Benzer. T4D and *amber* mutants from the Cambridge collection and from Drs R. H. Epstein and R. Edgar.

(c) Media

B broth is 1% Difco bacto-tryptone plus 0·5% NaCl; BG broth is B broth supplemented with 0·4% glucose; bottom agar is B broth solidified with 1·2% agar, top agar is B broth containing 0·7% agar.

(d) Methods

(i) Spot testing of mutants

Lysates (4×10^9/ml.) in B broth were streaked with paper strips on the surface of double-layer plates containing the bacterial strain in the top agar.

(ii) Burst sizes

Bacteria grown to 3×10^8/ml. in B broth were infected with about 10^7 phages/ml. After 8 min at 37°C, the bacteria were diluted 10^{-3} into B broth and lysed with chloroform after 60 min at 37°C.

(iii) Double mutants

These were constructed by crossing the mutant to be studied with *r1589* on strain B. *r* progeny were picked and backcrossed to both parents. The doubles were purified and checked for their content of *r1589* by crossing with point mutants. The genetic methods used are those described by Benzer (1961).

(iv) Complementation tests with r1589 and its derivatives

These were carried out by a modification of the method of Benzer & Champe (1962). The bacteria were grown in BG broth to 2×10^8/ml. and growth stopped by addition of KCN to a final concentration of M/500. 1 ml. was added to 1 ml. of broth containing 10^9 particles of r1299 and 10^9 particles of r1589 or its derivatives. After adsorption for 10 min at 37°C, the mixedly infected cells were diluted 10^{-3} into broth and lysed with chloroform after 60 min at 37°C. The titre of r^+ recombinants was estimated by plating on *E. coli* K12(λ). r1299 is a B deletion covering segments B5 and B6. It yields approximately 0·3% recombinants when crossed with r1589 on *E. coli* B. In K strains, recombinants only emerge from those cells in which r1299 and the r1589 derivative can complement each other. This procedure dispenses with the need to inactivate the unadsorbed phages with antiserum and gives results which are comparable to those obtained by measurements of burst size.

3. Results

(a) *Definition of* amber *and* ochre *mutants*

A large number of rII mutants were screened by spot tests on CA154, the su^- strain, and on CA165 and CA167, the strains carrying su^+_B and su^+_C, respectively. We found a number of mutants which could grow on one or both of the su^+ strains but which were still unable to make plaques on the su^- strain. While this set included some, but not all, of the rII amber mutants, the majority of the mutants were not suppressed by strains carrying *amber* suppressors. This suggested that these constituted a different set of suppressible mutants. We propose calling these mutants *ochre* mutants, and their suppressors, *ochre* suppressors.

Table 1 contains a summary of the qualitative suppression patterns of the rII

TABLE 1

Qualitative suppression patterns of r*II* amber *and* ochre *mutants*

A. *Amber* mutants

Segment	Site	Amber suppressors			Ochre suppressors			
		su^+_I	su^+_{II}	su^+_{III}	su^+_B	su^+_C	su^+_D	su^+_E
A2	HD120	+	+	+	+	0	+	0
A3	N97	+	+	+	Poor	Poor	+	0
	S116	0	0	+	0	+	+	0
	HF103	+	+	+	Poor	0	+	0
A4	N11	+	+	+	Poor	+	+	0
A5	S172	+	+	+	Poor	+	+	0
	S24	+	+	+	0	0	+	0
	HB129	+	+	+	+	0	+	0
A6	S99	+	+	+	+	0	+	0
	N19	+	Poor	+	Poor	0	+	0
	N34	+	+	+	+	+	+	0
B1	HE122	+	+	+	Poor	0	+	0
	2074	+	+	+	+	0	+	0
	EM84	+	+	+	+	+	+	0
	HB74	+	+	+	+	+	+	+
	NT332	+	+	+	+	0	+	0
B4	X237	+	Poor	+	0	+	+	0
	AP164	+	+	+	+	+	+	+
B7	HB232	+	+	+	+	0	+	+
	X417	+	+	+	0	0	+	0
	HD231	+	+	+	+	+	+	+

TABLE 1 (*continued*)

B. *Ochre* mutants

Segment	Site	Amber suppressors			Ochre suppressors			
		su^+_I	su^+_{II}	su^+_{III}	su^+_B	su^+_C	su^+_D	su^+_E
A1	HD147	0	0	0	+	+	+	+
	N55	0	0	0	+	+	+	0
A2	X20	0	0	0	+	0	+	0
	X220	0	0	0	+	+	+	+
A3	X372	0	0	0	+	+	+	0
A4	X352	0	0	0	+	+	+	0
	N31	0	0	0	+	+	+	0
A5	X25	0	0	0	+	+	+	0
	X170	0	0	0	+	0	+	0
	X319	0	0	0	+	+	+	0
	X337	0	0	0	+	+	+	+
	X358	0	0	0	+	+	+	0
A6	X164	0	0	0	+	0	+	0
	X558	0	0	0	+	+	+	0
	N21	0	0	0	+	0	0	+
B1	UV375	0	0	0	+	+	+	+
	360	0	0	0	+	0	+	+
	X511	0	0	0	+	+	+	+
	X27	0	0	0	+	+	+	+
	X375	0	0	0	+	+	+	0
	N24	0	0	0	+	+	+	+
B4	N17	0	0	0	+	+	+	+
	X528	0	0	0	+	+	+	+
B6	N7	0	0	0	+	+	+	+
	X321	0	0	0	+	0	+	0
B7	X234	0	0	0	+	+	+	+
	X191	0	0	0	+	+	+	+
	N12	0	0	0	+	Poor	+	+
	UV191	0	0	0	+	+	+	+
	UV199	0	0	0	+	+	+	+
	SD160	0	0	0	+	+	+	+
B8	N29	0	0	0	0	0	+	0
B9	AP53	0	0	0	+	+	+	+

amber and *ochre* mutants. Four different *amber* suppressor strains are known (Brenner, unpublished results), which suppress different but overlapping sets of *r*II *amber* mutants, but do *not* suppress any *ochre* mutant. This is the feature which operationally distinguishes the two kinds of mutants. The *ochre* mutants are suppressed by one or more of the *ochre* suppressors. Two of these, su^+_B and su^+_C, were isolated by Beckwith (1964a); the other two, su^+_D and su^+_E, have been isolated as new suppressors of lac_2 (see Materials and Methods). The different *ochre* suppressors show different suppression patterns not only of *ochre* but also of *amber* mutants. Thus far, we have been unable to isolate suppressors which are specific for *ochre* mutants.

The *ochre* suppressors also restore activity to *amber* mutants in other genes of bacteriophage T4. We have screened a large number of *amber* mutants on CA165 and CA167 and have found suppression in many cases. For example, of the *amber* mutants of cistron 42 which controls the deoxycytidylic acid hydroxymethylase (Epstein *et al.*, 1963), all but one (N122) are suppressed by su^+_B or su^+_C or by both. On the other hand, none of the 14 different *amber* mutants of the head protein (cistron 23) nor any of the mutants in cistrons 34 and 37 is visibly suppressed by any of the four *ochre* suppressors. Since these genes control structural components of the phage, this suggests that the *ochre* suppressors are not as efficient as some of the *amber* suppressors. We have also isolated *ochre* mutants of other genes in the phage, confirming again that the action of the suppressors is general and does not depend on any specific function a gene may have.

(b) Ochre *mutants: a new class of nonsense mutants*

The operational distinction between *ochre* and *amber* mutants depends on suppression. The suppression patterns shown by the different suppressors can be explained by quantitative and qualitative differences in the suppressed product. It could be argued that our classification is incorrect and that there is no real difference between the two classes; the *ochre* mutants just happen to be *amber* mutants in which suppression by an *amber* suppressor leads to an inactive protein product. In fact, if we look at the properties of mutant S116 (Table 1) we see that, had we not possessed the strain su^+_{III}, this mutant would have been classified as an *ochre* mutant. We believe this to be the exception rather than the rule, as the number of *ochre* sites in the two genes is impressively large.

In order to characterize the mutants more fully, we have performed experiments with *ochre* mutants analogous to those done by Benzer & Champe (1962) with *amber* mutants. A deletion, $r1589$, which joins part of the A cistron to part of the B cistron, still has B function although its A activity is abolished. Benzer & Champe (1962) combined A cistron *amber* mutants with this deletion and showed that the double mutants had no activity on the su^- strain. The su^+ strain restored activity with varying efficiency to these mutants. This suggested that the *amber* mutations produced nonsense codons which could not be translated in the su^- strain. They also found other mutations which did not have this effect of interrupting the reading of the genetic message, and concluded that these were *missense* mutations.

We have investigated 8 *ochre* mutants and 3 *amber* mutants in the A cistron. The burst sizes of these mutants on different su^+ strains is shown in Table 2. Table 3 shows the results of complementation experiments. Both su^- and su^+ cells, infected simultaneously with the B deletion mutant $r1299$ and with $r1589$, produce a burst of recombinants, confirming that $r1589$ can provide the B function in the complex. The yield is reduced to less than 0.1% of this value in su^- cells infected with the double mutants in which $r1589$ is combined with either an *amber* or an *ochre* mutant. This suggests that, like the *amber* mutants, *ochre* mutants also produce nonsense. The strain carrying the *amber* suppressor su^+_I restores varying degrees of activity to the three *amber* mutants. We confirm the finding of Benzer & Champe (1962) that restoration is efficient for N11 but poor for N97, which we know to be identical to the mutant HB122 used by them. Activity is also restored to S116 by su^+_I, although there is no suppression by this strain of the mutant by itself. None of the *ochre* mutants shows any increase in activity on su^+_I, which is evidence supporting their recognition as a distinctive class of nonsense mutant. The restoration of activity by the ochre suppressor, su^+_B, is, if it exists at all, very poor; this may be attributed to the known weakness of these suppressors.

Further evidence that *ochre* mutants are a class of mutants distinct from *amber* mutants comes from studies on the suppression of the lac_2 mutant by the *amber*

TABLE 2

Burst sizes of mutants on su⁻ *and* su⁺ *strains*

			Amber suppressors		*Ochre* suppressors	
		su^-	$su^+{}_I$	$su^+{}_{II}$	$su^+{}_B$	$su^+{}_C$
		CA154	QA1	CA180	CA165	CA167
Wild type	r^+	110	160	130	100	105
Deletion	r1589	0·1	0·08	0·2	0·3	0·15
Non-suppressible	S108	0·3	0·2	0·2	0·3	0·3
	S88	0·0	1·2	1·3	1·8	1·8
	N14	0·3	0·3	0·7	0·6	0·3
Amber	N97	0·1	175	10	2	3
	S116	0·3	0·2	0·8	0·4	36
	N11	0·8	115	24 ´	3·3	8
	S172	0·2	116	12	3	6
	X237	0·1	155	0·6	0·5	2
	X417	0·2	133	6	0·5	0·8
Ochre	N55.	0·1	0·08	0·2	100	68
	X20	0·1	0·1	0·3	7	1
	X220	0·3	0·2	0·2	90	25
	X352	0·2	0·1	0·1	17	20
	X372	0·1	0·2	0·1	15	6
	N31	0·3	0·1	0·4	27	10
	X25	0·04	0·05	0·04	80	27
	X170	0·03	0·05	0·15	8	1
	X319	0·1	0·08	0·2	13	4
	X528	0·2	0·14	0·3	73	40
	N24	0·1	0·1	0·1	108	50
	N7	0·3	0·16	0·3	29	3
	X321	0·2	0·2	0·2	2·7	0·8
	N12	0·25	0·13	0·08	5·9	1

TABLE 3

Complementation on su⁻ *and* su⁺ *strains*

		Recombinants per 10^3 infected cells		
		su^-	$su^+{}_I$	$su^+{}_B$
		CA154	CA267	CA165
Control	1589	215	320	232
Ochre mutants	N55+1589	0·0	0·0	0·0
	X20+1589	0·0	0·0	0·0
	X220+1589	0·0	0·0	0·0
	X25+1589	0·0	0·0	0·0
	X352+1589	0·0	0·0	0·0
	X372+1589	0·0	0·0	0·2
	N31+1589	0·0	0·0	0·2
	X170+1589	0·0	0·0	0·2
Amber mutants	N97+1589	0·0	2·0	0·0
	S116+1589	0·0	10·0	0·0
	N11+1589	0·0	110·0	0·0

suppressors. Strains were constructed with the genotypes $lac_2\ su^+{}_I$ and $lac_2\ su^+{}_{II}$. Neither suppressor restored β-galactosidase activity to this mutant; but, more important, in neither strain was permease activity, as determined by growth on melibiose, restored. It cannot be argued that the suppressors fail to suppress polarity, since the polarity of lac_{L125}, a strongly polar *amber* mutant, is suppressed by both $su^+{}_I$ and $su^+{}_{II}$. Thus, in the case of *ochre* mutants such as $lac^-{}_2$ there is no action of *amber* suppressors on the mutated site.

4. Discussion

The abolition of the B activity of $r1589$ by *ochre* mutants strongly suggests that these are nonsense mutants, leading to an interruption of the reading of the genetic message. They have this property in common with the *amber* mutants and also with mutants which alter the phase of reading of the genetic message (Crick, Barnett, Brenner & Watts-Tobin, 1961). All the *ochre* mutants studied can be induced to revert to standard type by 2-aminopurine; this shows that they are more likely to be due to nucleotide substitutions than to additions or deletions. In this respect, they resemble the *amber* mutants and are different from the acridine-type mutants. We conclude that, like the *amber* mutants, they act by producing a nonsense signal in the message. This conclusion depends on the assumption that *all* amino acid substitutions in the A region will be acceptable for the B activity of $r1589$, that is, no missense mutant will be mistakenly diagnosed as nonsense. Benzer & Champe (1962) have, in fact, found mutants which do not abolish B function when combined with the deletion; we know that two of these, N74 and AP129, are not suppressed by either *amber* or *ochre* suppressors. We have also tested five bromouracil-induced A cistron mutants in our collection, S30, N14, S88, S108 and S9. All are non-suppressible and all are missense when combined with $r1589$. On the other hand, we direct attention to the fact that $su^+{}_I$ restores activity with varying efficiency to $r1589$ combined with different *amber* mutants; the effect of this suppressor on N97 is particularly weak. This does suggest that, in these experiments, it is not merely the amount of polypeptide chain made which is important, but also its quality. Some missense mutations might therefore generate a configuration which is completely unacceptable. However, it is highly significant that *amber* and *ochre* mutants, defined in the first place by suppression, behave uniformly as nonsense in the genetic test.

Neither the *ochre* suppressor, $su^+{}_B$, nor the *amber* suppressor, $su^+{}_I$, restores B activity to *ochre* mutants combined with $r1589$. We attribute the first result to the weakness of the ochre suppressor, the efficiency of which is in the range of 4 to 8% (Kaplan, Stretton & Brenner, unpublished results). It should be noted, however, that this is sufficient to restore the full burst size of some r_{II} *ochre* mutants, for example, N55. There are other results which indicate that the r_{II} product is in excess and that the burst size is not proportional to the amount of product present (see Champe & Benzer, 1962). The amount of B activity provided by the deletion $r1589$ could be very much less than that of the intact B cistron, and yet give a full burst size; inefficient suppression of mutants combined with this deletion may then very sharply reduce the activity.

The failure of $su^+{}_I$ to restore activity to the *ochre* mutants combined with the deletion cannot be explained in this way, because this suppressor is very efficient (Kaplan, Stretton & Brenner, unpublished results). This result indicates that the *ochre* mutants constitute a class of nonsense mutants different from the *amber* mutants. More precisely, we suggest that the two types of mutants contain two different nonsense triplets and that the *ochre* suppressors recognize both triplets, whereas *amber* suppressors recognize only the *amber* triplet. Recently, this distinction has been directly demonstrated. *Ochre* mutants can be converted into *amber* mutants by one-step mutations, which shows that the triplets are not only different, but also connected (Brenner, Stretton & Kaplan, 1965). The classification of the mutants by suppression is therefore not trivial.

Amber mutations of the head protein result in termination of the polypeptide chain at the site of the mutation (Sarabhai *et al.*, 1964; Stretton & Brenner, 1965). The relative weakness of the *ochre* suppressors makes the isolation of *ochre* mutants of the head protein impossible. We have therefore been unable to investigate what such mutants do; but it is reasonable to surmise that they, too, result in chain termination. Moreover, it is likely that the effects of both *amber* and *ochre* mutants are exerted at the level of protein synthesis, since both kinds of mutants vanish when the frame of reading of the message is altered (Brenner & Stretton, 1965). It is reasonable to suppose that the molecular consequences of the mutations will not depend on the gene in which they occur, and that the classification of the mutants by suppression is unitary. It follows then that the lac_2 which was used to isolate the suppressors is a nonsense mutant of the *ochre* class, and that its effects are also exerted at the level of protein synthesis. This mutation, which maps in the *z* gene, not only produces no β-galactosidase, but it has the additional property of being strongly *polar* and abolishes the synthesis of the β-galactoside permease as well. Strong polarity is not unique to *ochre* mutants, since *amber* mutants of the *z* gene are known with the same property. But, more important, strong polarity is not a necessary consequence of nonsense mutants, since there are other *ochre* and *amber* mutants of the *z* gene which are not strongly polar (Newton, Zipser, Beckwith & Brenner, unpublished results). To a first approximation this appears to depend on the map position.

We are unable to give a unique explanation of the polarity effect. Ames & Hartman (1963) have suggested that polarity might be explained if a polycistronic messenger RNA possessed only one obligatory starting point for ribosome attachment. All nonsense mutants would then be expected to be strongly polar; but in the *lac* operon this is patently not the case. We suspect that the polarity shown by some nonsense mutants will require a special explanation, perhaps involving the destruction of the messenger RNA, but there is not much point in pursuing this question further at the present time. It is clear, however, that the present results, taken in conjunction with those of Beckwith (1964b), forbid the definition of such mutants as operator-negative, and suggest that these mutants do not act directly at the level of messenger RNA transcription.

One of us (J. R. B.) was a Fellow of the Jane Coffin Childs Memorial Fund for Medical Research. We thank Miss M. I. Wigby for technical assistance.

REFERENCES

Ames, B. N. & Hartman, P. E. (1963). *Cold Spr. Harb. Symp. Quant. Biol.* **28**, 349.
Beckwith, J. R. (1964a). *Struktur und Funktion des genetischen Materials, Erwin-Bauer Gedächtnisvorlesungen* III, p. 119. Berlin: Akademie-Verlag.
Beckwith, J. R. (1964b). *J. Mol. Biol.* **8**, 427.
Benzer, S. (1961). *Proc. Nat. Acad. Sci., Wash.* **47**, 403.
Benzer, S. & Champe, S. P. (1961). *Proc. Nat. Acad. Sci., Wash.* **47**, 1025.
Benzer, S. & Champe, S. P. (1962). *Proc. Nat. Acad. Sci., Wash.* **48**, 1114.

Brenner, S. & Stretton, A. O. W. (1965). *J. Mol. Biol.* **13**, 944.

Brenner, S., Stretton, A. O. W. & Kaplan, S. (1965). *Nature*, **206**, 994.

Champe, S. P. & Benzer, S. (1962). *Proc. Nat. Acad. Sci., Wash.* **48**, 532.

Crick, F. H. C., Barnett, L., Brenner, S. & Watts-Tobin, R. J. (1961). *Nature*, **192**, 1277.

Epstein, R. H., Bolle, A., Steinberg, C. M., Kellenberger, E., Boy de la Tour, E. & Chevalley, R., Edgar, R. S., Susman, M., Denhardt, G. & Lielausis, A. (1963). *Cold Spr. Harb. Symp. Quant. Biol.* **28**, 375.

Jacob, R. & Monod, J. (1961). *Cold Spr. Harb. Symp. Quant. Biol.* **26**, 193.

Orias, E. & Gartner, R. K. (1964). *Proc. Nat. Acad. Sci., Wash.* **52**, 859.

Sarabhai, A. S., Stretton, A. O. W., Brenner, S. & Bolle, A. (1964). *Nature*, **201**, 13.

Stretton, A. O. W. & Brenner, S. (1965). *J. Mol. Biol.* **12**, 456.

Yanofsky, C. & St. Lawrence, P. (1960). *Ann. Rev. Microbiol.* **14**, 311.

J. Mol. Biol. (1966) **19**, 548–555

Codon—Anticodon Pairing:

The Wobble Hypothesis

F. H. C. Crick

Medical Research Council, Laboratory of Molecular Biology

Hills Road, Cambridge, England

(*Received 14 February 1966*)

It is suggested that while the standard base pairs may be used rather strictly in the first two positions of the triplet, there may be some wobble in the pairing of the third base. This hypothesis is explored systematically, and it is shown that such a wobble could explain the general nature of the degeneracy of the genetic code.

Now that most of the genetic code is known and the base-sequences of sRNA molecules are coming out, it seems a proper time to consider the possible base-pairing between codons on mRNA and the presumed anticodons on the sRNA.

The obvious assumption to adopt is that sRNA molecules will have certain common features, and that the ribosome will ensure that all sRNA molecules are presented to the mRNA in the same way. In short, that the pairing between one codon–anticodon matching pair will to a first approximation be "equivalent" to that between any other matching pair.

As far as I know, if this condition has to be obeyed, and if all four bases must be distinguished in any one position in the codon, then the pairing in this position is *highly likely* to be the standard one; that is:[†]

$$G ==== C$$
$$\text{and} \quad A ==== U$$

or some equivalent ones such as, for example,

$$I ==== C$$
$$\text{and} \quad A ==== T$$

since this is the only type of pairing which allows all four bases to be distinguished in a strictly equivalent way.

We now know enough of the genetic code to say that in the *first two* positions of the codon the four bases are clearly distinguished; certainly in many cases, and probably in all of them. I thus deduce that the pairings in the first two positions are likely to be the standard ones.

[†] Throughout this paper the sign $====$ is used to mean "pairs with". If two bases are equivalent in their coding properties, this is written $\genfrac{}{}{0pt}{}{U}{C}$ or $\left.\genfrac{}{}{0pt}{}{U}{C}\right\}$

However, what we know about the code has already suggested two generalizations about the third place of the codon. These are:

(1) U⎫† this already appears true in about a dozen cases out of the possible 16,
     C⎭ and there are no data to suggest any exceptions.

(2) A⎫ probably true in about half of the possible 16 cases, but the evidence
     G⎭ suggests it may perhaps be incorrect in several other cases.

The detailed experimental evidence is rather complicated and will not be discussed here. (For details of the code see, for example, Nirenberg *et al.*, 1965; and Söll *et al.*, 1965.) It suffices that these rules *may* be true, as suggested by Eck (1963) a little time ago. Alternatively, only the first one may be true.

This naturally raises the question: Does *one* sRNA molecule recognize more than one codon, e.g. both UUU *and* UUC. Some evidence for this was first presented by Bernfield & Nirenberg (1965). They showed that *all* the sRNA for phenylalanine can be bound by poly U, although this sRNA also recognizes the triplet UUC, at least in part. More recent evidence along these lines is presented in Söll *et al.* (1966) and Kellogg *et al.* (1966). Again I do not wish to discuss here the evidence in detail, but simply to ask: If one sRNA codes both XYU and XYC, how is this done?

Now if we do not know anything about the geometry of the situation, it might be thought that almost any base pairs might be used, since it is well known that the bases can be paired (i.e. form at least two hydrogen bonds) in many different ways. However, it occurred to me that if the first two bases in the codon paired in the standard way, the pairing in the third position might be *close* to the standard ones.

We therefore ask: How many base pairs are there in which the glycosidic bonds occur in a position close to the standard one? Possible pairs are:

$$G ==== A \tag{1}$$

In my opinion this will not occur, because the NH_2 group of guanine cannot make one of its hydrogen bonds, even to water (see Fig. 1).

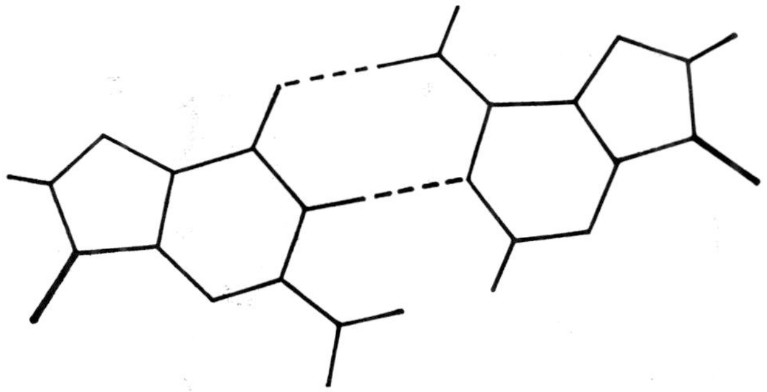

Fig. 1. The unlikely pair guanine–adenine.

$$U ==== C \tag{2}$$

This brings the two keto groups rather close together and also the two glycosidic bonds, but it may be possible (see Fig. 2).

† This symbol implies that both U and C code the same amino acid.

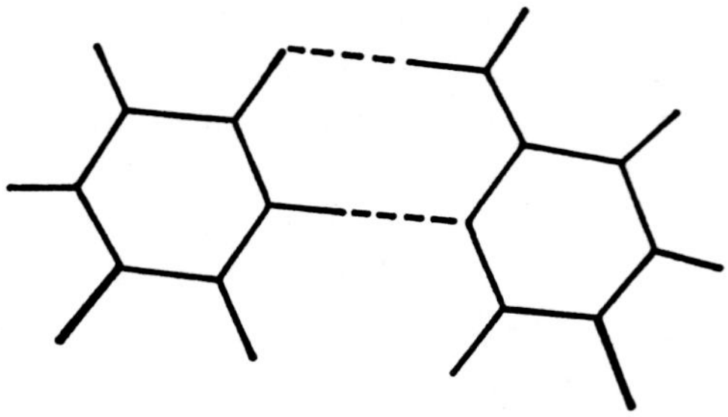

FIG. 2. The close pair uracil–cytosine.

$$U ==== U \tag{3}$$

Again rather close together (see Fig. 3).

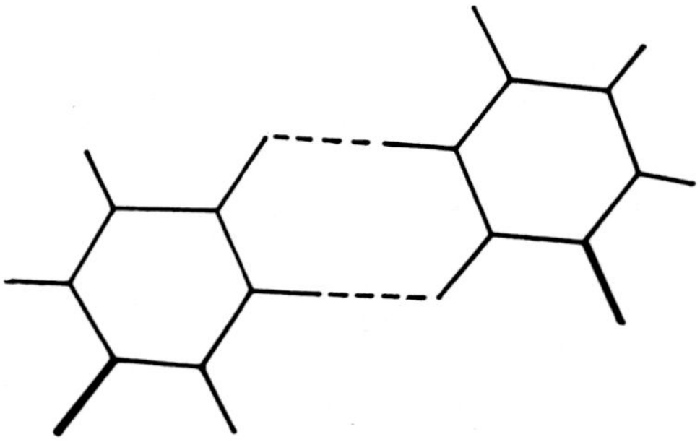

FIG. 3. The close pair uracil–uracil.

$$G ==== U \tag{4}$$
$$\text{or } I ==== U$$

These only require the bond to move about 2·5 Å from the standard position (see Fig. 4).

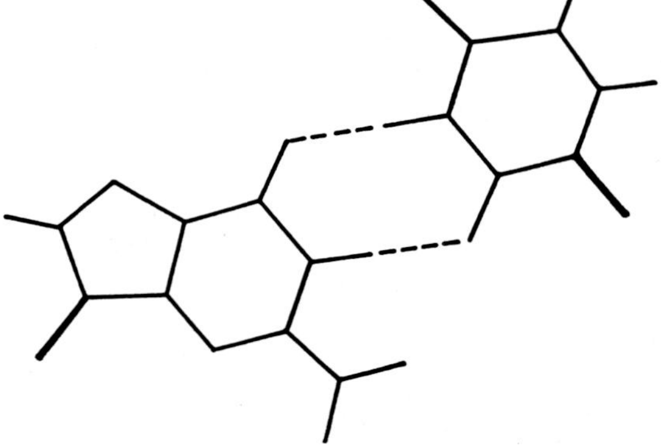

FIG. 4. The pair guanine–uracil (the pair inosine–uracil is similar).

$$I ==== A \qquad (5)$$

This is perfectly possible. Poly I and poly A will form a double helix. The distance between the glycosidic bonds is increased (see Fig. 5).

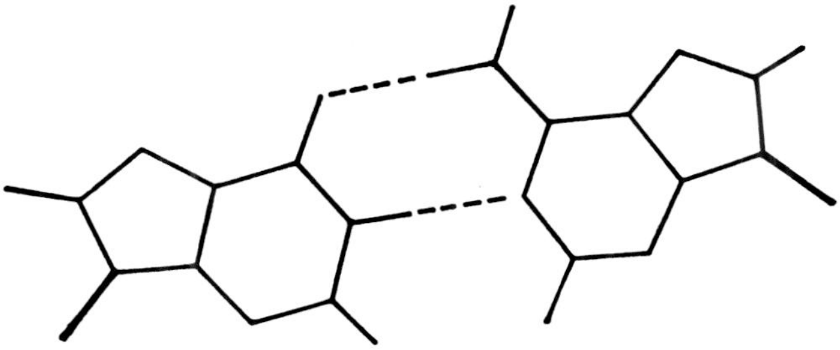

Fig. 5. The pair inosine–adenine.

As far as I know, these are all the possible solutions if it is assumed that the bases are in their usual tautomeric forms.

I now postulate that in the base-pairing of the third base of the codon there is a certain amount of play, or wobble, such that more than one position of pairing is possible.

As can be seen from Fig. 6, there are seven possible positions which might be reached by wobbling. However, it by no means follows that all seven are accessible, since the molecular structure is very likely to impose limits to the wobble. We should therefore strictly consider all possible *combinations of allowed positions*. There are 127 of these, but most of them are trivial. If we adopt the rule that *all four bases* on the codon (in the third position) must be recognized (that is, paired with) we are left with 51 different combinations. This is too many for easy consideration, but fortunately we can eliminate most of them by only accepting combinations which do not violate the broad features of the code. If we assume:

(a) that all four bases must be recognizable;

(b) that the code must *in some cases* distinguish between

$\left.\begin{matrix} U \\ C \end{matrix}\right\}$ and $\left.\begin{matrix} A \\ G \end{matrix}\right\}$ as it appears to do for the pairs

Phe	Tyr	His	Asn	Asp
Leu	C.T.†	Gln	Lys	Glu

(not all of which are likely to be wrong)

then by strictly logical argument it can be shown both that the standard position must be used, and that the three positions on the left of Fig. 6 cannot be used.

This leaves us with only four possible sites to consider one of which—the standard one—must be included. There are therefore only seven possible combinations. I have examined all these, but I shall restrict myself here to the case in which all four positions are used, as this is structurally the most likely and also seems to give the code (called code 4 in the note privately circulated) which best fits the experimental data.

† C.T., Chain termination.

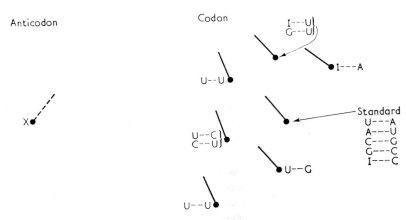

FIG. 6. The point X represents the position of the C_1' atom of the glycosidic bond (shown dotted) in the anticodon. The other points show where the C_1' atom and the glycosidic bond fall for the various base pairs. (Pairs with inosine in the codon have been omitted for simplicity.) The wobble code suggested uses the four positions to the right of the diagram, but not the three close positions.

The rules for pairing between the third base on the codon and the corresponding base on the anticodon are set out in Table 1. It can be seen that these rules make several strong predictions:

(1) it is not possible to code for either C alone, or for A alone.

For example, at the moment the codon UGA has not been decisively allocated. Wobble theory states that UGA might either:

    (a) code for cysteine, which has UGU and UGC; or
    (b) code for trypotophan, which has UGG; or
    (c) not be recognized.

TABLE 1

Pairing at the third position of the codon

Base on the anticodon	Bases recognized on the codon
U	A G
C	G
A†	U
G	U C
I	U C A

† It seems likely that inosine will be formed enzymically from an adenine in the nascent sRNA. This may mean that A in this position will be rare or absent, depending upon the exact specificity of the enzyme(s) involved.

However it does *not* permit UGA to code for any amino acid other than cysteine or tryptophan. This rule could also explain why no suppressor has yet been found which suppresses only *ochre* mutants (UAA), although suppressors exist which suppress both *ochre* and *amber* mutants (UA$_G^A$).

(2) If an sRNA has inosine in the place at the relevant position on the anticodon (i.e. enabling it to pair with the third base of the codon), then it must recognize U, C and A in the third place of the codon. Conversely, those amino acids coded only by XY$_C^U$ (such as Phe, Tyr, His, etc.) cannot have inosine in that place on their sRNA.

(3) Wobble theory does not state exactly how many different types of sRNA will actually be found for any amino acid. However if an amino acid is coded for by all four bases in the third position (as are Pro, Thr, Val, etc.), then wobble theory predicts that there will be at least two sRNA's. These can have the recognition pattern:

$$\left.\begin{matrix}U\\C\end{matrix}\right\} \text{ plus } \left.\begin{matrix}A\\G\end{matrix}\right\}$$

$$\text{or} \quad \left.\begin{matrix}U\\C\\A\end{matrix}\right\} \text{ plus } G$$

Note that the sets actually used for any amino acid may well vary from species to species.

The Anticodons

At this point it is useful to examine the experimental evidence for the anticodon. In the sRNA for alanine from yeast, Holley *et al.* (1965) have the following sequences:

$$--- \text{pUpUpIp Gp CpMeIp}\Psi\text{p} ---$$

$$\text{position} --- 36 \ 37 \ 38 ---$$

Zachau and his colleagues (Dütting, Karan, Melchers & Zachau, 1965) have for one of the serine sRNA's from yeast:

$$--- \text{p}\Psi\text{pUpIpGpApA}^+\text{p}\Psi\text{p} ---$$

(A$^+$ stands for a modified A)

For the valine sRNA from yeast, Ingram & Sjöquist (1963) have shown that the only inosine occurs in the sequence:

$$--- \text{pIpApCp} ---$$

Holley *et al.* (1965) have already pointed out that IGC is a possible anticodon for alanine, and the additional evidence makes it almost certain to my mind that this is correct, and that the anticodons are as given in the Table below†:

† *Note added 26 April 1966.* Drs J. T. Madison, G. A. Everett and H. Kung (personal communication) have completed the sequence of the tyrosine sRNA from yeast. The sequence strongly suggests that the anticodon in this case is GΨA, corresponding to the known codons UA^U. Since Ψ can form the same base pairs as U, this is in excellent agreement with the previous data.

Yeast sRNA		
	Anticodon	Codon
Ala	I G C	G C ?
Ser	I G A	U C ?
Val	I A C	G U ?

remembering that the pairing proposed between codon and anticodon is *anti*-parallel. Thus I confidently predict: the anticodon is a triplet at (or very near) positions 36–37–38 on every sRNA, and that the *first two bases* in the codon pair with this (in an anti-parallel manner) *using the standard base pairs.*

However, inosine does not occur in every sRNA. In particular Holley et al. (1963) (and personal communication) have reported that the tyrosine sRNA has two peaks, neither of which contains inosine. Moreover, Sanger (personal communication) tells me that there is rather little inosine in the total sRNA from *E. coli.*

Testing the Theory

Two obvious tests present themselves:

(1) To find which triplets are bound by any one type of sRNA. This is being done by Khorana and his colleagues (Söll *et al.*, 1966), and also by Nirenberg's group (Kellogg, Doctor, Loebel & Nirenberg, 1966). The difficulty here is to be sure that the sRNA used is pure, and not a mixture.

(2) To discover unambiguously the position of the anticodon on sRNA, and to find further anticodons. This will certainly happen as our knowledge of the base sequence of sRNA molecules develops. The absence of inosine from any anticodon is obviously of special interest.

In conclusion it seems to me that the preliminary evidence seems rather favourable to the theory. I shall not be surprised if it proves correct.

I thank my colleagues for many useful discussions and the following for sending me material in advance of publication: Dr M. W. Nirenberg, Dr H. G. Khorana, Dr G. Streisinger, Dr W. Holley, Dr J. Fresco, Dr H. G. Zachau, Dr C. Yanofsky, Dr H. G. Wittmann, Dr H. Lehmann and Dr J. D. Watson.

REFERENCES

Bernfield, M. R. & Nirenberg, M. W. (1965). *Science,* **147,** 479.
Dütting, D., Karan, W., Melchers, F. & Zachau, H. G. (1965). *Biochim. biophys. Acta,* **108,** 194.
Eck, R. V. (1963). *Science,* **140,** 477.
Holley, R. W., Apgar, J., Everett, G. A., Madison, J. T., Marquisee, M., Merrill, S. H., Penswick, J. R. & Zamir, A. (1965). *Science,* **147,** 1462.
Holley, R. W., Apgar, J. Everett, G. A., Madison, J. T., Merrill, S. H. & Zamir, A. (1963). *Cold Spr. Harb. Symp. Quant. Biol.* **28,** 117.
Ingram, V. M. & Sjöquist, J. A. (1963). *Cold. Spr. Harb. Symp. Quant. Biol.* **28,** 133.

Kellogg, D. A., Doctor, B. P., Loebel, J. E. & Nirenberg, M. W. (1966). *Proc. Nat. Acad. Sci., Wash.* **55**, 912.

Nirenberg, M., Leder, P., Bernfield, M., Brimacombe, R., Trupin, J., Rottman, F. & O'Neal, C. (1965). *Proc. Nat. Acad. Sci., Wash.* **53**, 1161.

Söll, D., Jones, D. S., Ohtsuka, E., Faulkner, R. D., Lohrmann, R., Hayatsu, H., Khorana, H. G., Cherayil, J. D., Hampel, A. & Bock, R. M. (1966). *J. Mol. Biol.* **19**, 556.

Söll, D., Ohtsuka, E., Jones, D. S., Lohrmann, R., Hayatsu, H., Nishimura, S. & Khorana, H. G. (1965). *Proc. Nat. Acad. Sci., Wash.* **54**, 1378.

J. Mol. Biol. (1966) **19**, 556–573

Specificity of sRNA for Recognition of Codons as Studied by the Ribosomal Binding Technique†

D. Söll, D. S. Jones, E. Ohtsuka, R. D. Faulkner, R. Lohrmann,
H. Hayatsu, H. G. Khorana

Institute for Enzyme Research, University of Wisconsin

J. D. Cherayil, A. Hampel and Robert M. Bock

Department of Biochemistry, University of Wisconsin, U.S.A.

(*Received 14 February 1966*)

Multiple sRNA species which are specific for individual amino acids have been purified from yeast and *Escherichia coli* sRNA using techniques of column chromatography or countercurrent distribution. The purified sRNA fractions, after charging with radioactive amino acid, have been tested for binding to ribosomes in the presence of the trinucleotides assigned as codons for the respective amino acids. From the stimulation of binding, it is concluded that (1) an sRNA species shows strict specificity for the recognition of the first letter of a codon; and (2) an sRNA species often can recognize multiple codons differing in the third letter. The total results, which are summarized in Table 4, appear to provide support for the postulates of the wobble hypothesis (Crick, 1966).

1. Introduction

Fractionation of sRNA from different organisms frequently yields multiple peaks corresponding to one amino acid, and, as was first shown by Weisblum, Benzer & Holley (1962) for leucine-sRNA, the different sRNA peaks, presumably, are specific for certain codons of that amino acid (Bennett, Goldstein & Lipmann, 1965; Weisblum, Gonano, von Ehrenstein & Benzer, 1965). Recent progress in the assignment of codons to different amino acids (see e.g. Nirenberg *et al.*, 1965; Söll *et al.*, 1965) indicates that most of the possible sixty-four triplets are meaningful codons. This conclusion has focused attention on the question: Is there invariably a discrete sRNA species for the recognition of each codon, or can one sRNA species, under certain conditions, recognize more than one codon? That the latter may be the case is suggested, for example, by the work on alanine-sRNA, where purification and structural analysis have shown only one sRNA species in yeast for this amino acid.

In the present work, the specificity of sRNA for recognition of codons has been studied by the trinucleotide-stimulated binding (Nirenberg & Leder, 1964) to ribosomes of aminoacyl-sRNA's prepared from purified sRNA fractions. The results show that while there is strict specificity of sRNA for the first two letters of a codon, one sRNA can recognize multiple codons differing in the third letters only. The patterns for multiple codon recognition which emerge are consistent with the postulates of the wobble hypothesis (Crick, 1966).

---

† This is paper LVII in the series "Studies on Polynucleotides". Paper LVI is by D. S. Jones, S. Nishimura & H. G. Khorana (1966), *J. Mol. Biol.* **16**, 454.

2. Materials and Methods

(a) Oligo- and polynucleotides and amino acids

All of the ribotrinucleotides used were prepared by chemical methods which have been described elsewhere (Lohrmann, Söll, Hayatsu, Ohtsuka & Khorana, 1966). Poly AAG was prepared as described previously (Nishimura et al., 1965b). Poly AG was prepared as described by Nishimura, Jones & Khorana (1965a) (see also Jones, Nishimura & Khorana, 1966). Radioactive amino acids were purchased commercially. Their specific activities (μc/μmole) were as follows: [^{14}C]Ala, 123; [^{14}C]Arg, 222; [^{3}H]Arg, 350; [^{14}C]Gly, 74; [^{3}H]Gly, 1136; [^{14}C]Ile, 246; [^{14}C]Leu, 222; [^{14}C]Phe, 333; [^{14}C]Ser, 123; [^{3}H]Ser, 860.

(b) sRNA and radioactive aminoacyl-sRNA

(i) Escherichia coli B sRNA

E. coli B sRNA was prepared by Zubay's method (Zubay, 1962), except that chromatography on a DEAE-cellulose column was included as the final step.

Radioactively labeled aminoacyl-sRNA's were prepared from the purified fractions of E. coli B sRNA by the method of Nishimura et al. (1965b). The incubation mixture contained, per ml.: 1 mg of sRNA, 100 μmoles of Tris-chloride (pH 7·3), 10 μmoles of magnesium acetate, 10 μmoles of potassium chloride, 2 μmoles of ATP, 300 μg of aminoacyl-sRNA synthetase and radioactive amino acids (4 to 22 mμmoles, of specific activities described above).

(ii) Yeast sRNA and radioactive aminoacyl-yeast sRNA

Yeast sRNA was prepared by phenol extraction of 9·8 kg of washed, glucose-grown, log-phase Saccharomyces lactis Y14 by the method of Holley (1963). The aqueous phase was made 2% with sodium dodecyl sulfate and again extracted with phenol. The aqueous phase was precipitated with 2 vol. of cold 95% ethanol, washed with 75% ethanol containing 0·05% potassium acetate and chromatographed on DEAE-cellulose in batches (Holley, 1963). The appropriate fractions were collected over phenol, and after shaking and removing the phenol phase, the aqueous phase was precipitated from ethanol. The precipitate, after 2 washings with ethanol, was vacuum dried over phosphorus pentoxide and stored at $-$ 20°C until used for fractionation.

Radioactive aminoacyl-yeast sRNA was prepared using the following incubation mixture: 0·1 vol. buffer (0·4 M-potassium maleate (pH 7·0), 0·05 M-magnesium chloride, 0·005 M-EDTA), 0·1 vol. of 0·01 M-ATP (pH 7), 0·05 vol. of 0·005 M-CTP, 0·025 vol. of 0·05 M-dithiothreitol, 0·025 vol. of 1 M-potassium chloride, 0·1 vol. of dialyzed yeast 100,000 g supernatant fraction, radioactive amino acid (5 μc of ^{14}C or 20 μc of ^{3}H/ml. of final mixture) and sRNA (10 o.d.$_{260}$ units†/ml. of fractions with purity 60% or above, and 30 o.d. units/ml. for fractions less than 30% pure with respect to acceptor activity for the amino acid under study). Water was added to bring the total volume to one. After 25 min incubation at room temperature, 0·5 vol. of phenol equilibrated with the above buffer was added, the mixture was shaken and then centrifuged. After repetition of the phenol deproteinization step, the aminoacyl-sRNA was precipitated with 1·2 vol. of cold 95% ethanol containing 0·02 M-magnesium chloride and 0·1% unlabeled amino acid. The precipitate was collected by centrifugation in the cold and washed twice with ethanol–water (2 : 1).

The precipitate was dissolved in a small volume of 0·005 M-magnesium chloride + 0·01 M-sodium acetate (pH 4·2) and the solution frozen in small plastic vials until used.

(c) Purification of sRNA's specific for individual amino acids

(i) Isoleucine-, leucine- and serine-specific sRNA's from E. coli B

E. coli B sRNA was fractionated by the technique of countercurrent distribution, using the method of Holley et al. (1963). sRNA (2·0 g) was added to the first 5 tubes of a 200-transfer apparatus (40 ml. of each phase/tube). The final fractions were combined

---

† o.d.$_{260}$ unit refers to absorbance measurements in 1 ml. 1-cm light path at 260 mμ and at neutral pH values.

into groups of 2 tubes each and sRNA was recovered by the procedure of Apgar, Holley & Merrill (1962). The total ultraviolet absorbance and the total acceptor activity (for iso-leucine, leucine and serine) present in the different fractions is given in Fig. 1. Isoleucine acceptor activity gave 3 peaks and, for binding studies, fractions 64, 88 and 118 were used, the material in these fractions being designated peaks I, II and III, respectively. Leucine acceptor activity gave 3 peaks and, in the present study, material in fraction 114 was used, it being designated peak I. The material of fraction 134 is peak II. Fractions 174 to 190 (peak III) were combined for redistribution by the method of Apgar *et al.* (1962). In the present work, peak III is designated as that material which was insoluble in the redistribution solvent system. Serine acceptor activity gave two peaks; the material in fraction 76 (peak I) and that in fraction 118 (peak II) were used.

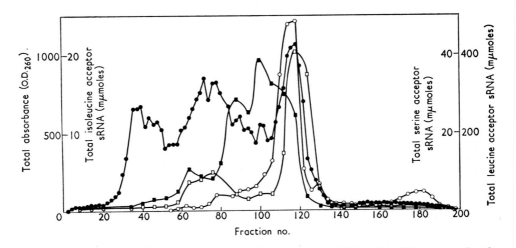

FIG. 1. Distribution of leucine, serine and isoleucine *E. coli* sRNA after 200 transfers by the countercurrent distribution technique. For details see Materials and Methods.
—○—○—, Leucine; —□—□—, serine; —■—■—, isoleucine; —●—●—, optical density at 260 mμ.

(ii) *Alanine-sRNA from yeast*

This was purified by the countercurrent distribution method of Apgar *et al.* (1962). The material finally obtained after 3 stages of countercurrent distribution was at least 85% pure.

(iii) *Arginine-sRNA from yeast*

Yeast sRNA was fractionated for arginine acceptor activity by repeated chromatography on DEAE-cellulose columns as described by Cherayil & Bock (1965). Elution of arginine acceptor activity from DEAE-cellulose at pH 7·5 and rechromatography of the arginine-acceptor peak at pH 4·5 on DEAE-cellulose with a urea gradient gave two peaks which are designated as arginine-sRNA-I and arginine-sRNA-II (Fig. 2). Peak I was approximately 20% pure and peak II was further purified on DEAE-Sephadex with a salt gradient at pH 4·5, until more than 80% pure. When arginine-sRNA-I was charged with [14C]arginine and arginine-sRNA-II with [3H]arginine and the mixture chromatographed on a DEAE-Sephadex column, they were clearly resolved (Fig. 3).

(iv) *Phenylalanine-sRNA from yeast*

This was prepared by countercurrent distribution according to the procedure of Hoskin-son & Khorana (1965). The material used in this study was at least 85% pure; it was generously donated by Drs U. L. RajBhandary and A. Stuart.

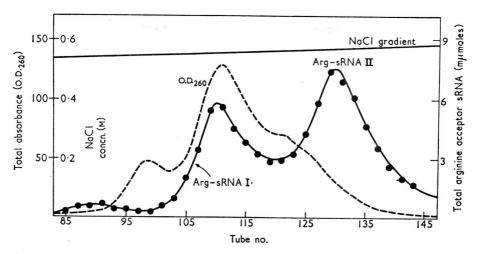

Fig. 2. Chromatography of 125 mg of partially purified yeast sRNA at pH 4·5. The DEAE-Sephadex A50 column (1·5 cm × 100 cm) was equilibrated with 0·42 M-NaCl in 7 M-urea and 0·02 M-acetate buffer (pH 4·5). Elution was carried out with a linear gradient of NaCl, 0·48 to 0·64 M, in a total volume of 1200 ml., in the presence of 7 M-urea and buffer. Fractions of 6 ml. were collected every 30 min.

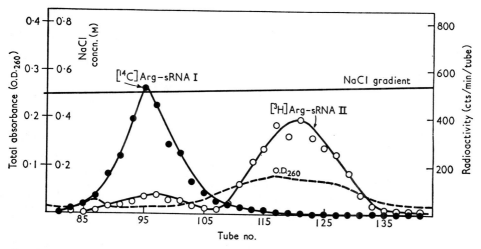

Fig. 3. Chromatography of partially purified, radioactive arginine-sRNA of yeast at pH 4·5. The DEAE-Sephadex A50 column (0·7 cm × 48 cm) was equilibrated with 0·42 M-NaCl in 7 M-urea and 0·02 M-acetate buffer (pH 4·5). A sample of [14C]Arg-sRNA I (see Materials and Methods) (10,000 cts/min) and [3H]Arg-sRNA II (14,000 cts/min) was applied to the column and eluted with a linear gradient of NaCl, 0·45 to 0·64 M, in a total volume of 250 ml., in the presence of 7 M-urea and buffer. The flow rate was 3·2 ml./hr and fractions were collected every 15 min. A measured portion from each fraction was pipetted onto a 2 cm × 10 cm strip of filter paper, dried in a stream of air, washed in 70% ethanol which is 0·01 M in magnesium chloride. The dried papers were counted in a liquid-scintillation counter.

(v) *Glycine-sRNA from yeast*

Chromatography of yeast sRNA on DEAE-Sephadex has previously been shown to give multiple peaks for glycine-acceptor activity (Cherayil & Bock, 1965). Chromatography under the conditions used for purification of arginine-sRNA gave two widely separated glycine-sRNA fractions (Fig. 4). The first fraction, designated glycine-sRNA "A", was better than 70% pure. The later fraction (glycine-sRNA "B") had a purity of about 25%. A portion of fraction A was charged with [³H]glycine, and fraction B was charged with [¹⁴C]glycine. Portions of charged fractions were combined and rechromatographed. The pattern of distribution of radioactivity (Fig. 5) showed that the above fractions were again well resolved. The bulk of [¹⁴C]glycyl-sRNA fraction B was chromatographed and gave 3 peaks which are designated peaks II to IV. The material in peak A appeared as a single peak which is designated as peak I. Appropriate fractions were pooled and after dilution (4 times) with water, they were absorbed on a DEAE-cellulose bed of 0·2 ml. total volume. Elution with 0·5 ml. of 0·8 M-sodium chloride in 7 M-urea was followed by precipitation with 1·2 vol. of ethanol containing 0·02 M-magnesium chloride. The precipitates were collected by centrifugation and washed repeatedly with ice-cold 75% ethanol and dissolved in the minimum volume of 0·005 M-magnesium acetate, previously adjusted to pH 5.

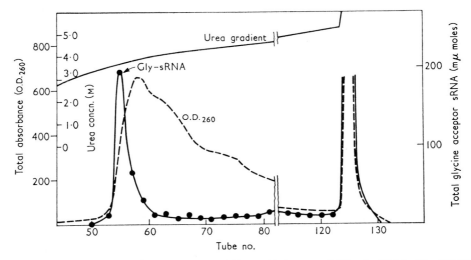

Fɪɢ. 4. Chromatography of 990 mg of partially purified yeast-sRNA at pH 4·5. The DEAE-cellulose column (4·5 cm × 108 cm) was equilibrated with 0·35 M-NaCl in 1·5 M-urea and 0·03 M-acetate buffer (pH 4·5). Elution was carried out by an exponential gradient of urea; 6 M-urea in 0·4 M-NaCl and buffer was added to a constant-volume mixer containing 3000 ml. of the mixture used for equilibration. Fractions of 60 ml. were collected every 30 min. Tubes 54 to 55 and 124 to 128 were pooled to give fractions A and B (Fig. 5).

(d) *Preparation of ribosomes and the supernatant fractions*

(i) *From E. coli*

These were prepared as described previously (Nishimura *et al.*, 1965*b*), except that the ribosomes were washed three times with 2 M-potassium chloride buffer by gentle suspension and sedimentation by centrifugation.

(ii) *From yeast Y42*

The supernatant fraction was prepared as described previously by Bretthauer, Marcus, Chaloupka, Halvorson & Bock (1963), with the modification that the buffer used was: 0·03 M-Tris-acetate (pH 7·2), 0·005 M-magnesium acetate, 0·10 M-potassium chloride, and

0·01 M-mercaptoethanol. Yeast ribosomes were prepared by layering the 20,000 **g** × 20 min supernatant fraction of a cell extract over a 2 M-sucrose solution, centrifugation for 4 hr at 40,000 rev./min and collecting the lower phase (Bruening, 1965). All solutions were 0·5 M in ammonium chloride, 0·1 M in magnesium chloride and 0·03 M in pH 7·2 Tris-chloride. Sucrose was removed by passing the ribosome suspension through a Sephadex G25 column. The ribosomes were stored in the above-mentioned buffer in liquid nitrogen and resedimented before use.

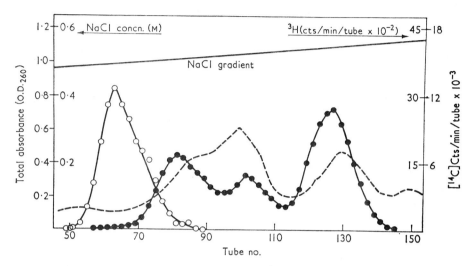

FIG. 5. Chromatography of partially purified radioactive glycine-sRNA of yeast on a DEAE-Sephadex A50 (0·7 cm × 40 cm) column. The conditions used were the same as described in Fig. 3, except that fractions were collected every 20 min. 60,000 cts/min of [³H]Gly-sRNA A and 200,000 cts/min of [¹⁴C]Gly-sRNA B were applied. —○—○—, [³H]Gly-sRNA; —●—●—, [¹⁴C]Gly-sRNA; ————, optical density at 260 mμ.

(e) Assay of aminoacyl-sRNA binding to ribosomes

The general procedure of Nirenberg & Leder (1964) was used. The conditions of assay used were as follows.

(i) Method A (for E. coli ribosomes)

The incubation mixture (0·05 ml.) contained 0·1 M-Tris-acetate (pH 7·2), 0·05 M-potassium chloride, 0·02 M-magnesium acetate, 1·5 to 2·0 O.D.$_{260}$ units of ribosomes and the amount of RNA-template and labeled aminoacyl-sRNA's ($\mu\mu$moles) as shown in the Tables and Figures. Incubation was carried out at 25°C for 20 min.

(ii) Method B (for E. coli ribosomes)

The incubation mixture (0·05 ml.) contained 0·1 M-Tris-chloride (pH 7·5), 0·05 M-potassium chloride, 0·01 M-magnesium acetate, 1·5 to 2·0 O.D.$_{260}$ units of ribosomes, and the amount of RNA-template and labeled aminoacyl-sRNA's ($\mu\mu$moles) as shown in the Tables and Figures. Incubation was carried out at 37°C for 7·5 min.

(iii) Method C (for yeast ribosomes)

The incubation mixture (0·05 ml.) contained 0·1 M-Tris-acetate (pH 7·2), 0·05 M-potassium chloride, 0·03 M-magnesium acetate, 1·5 to 2·0 O.D.$_{260}$ units of ribosomes, and the amount of RNA-template and labeled aminoacyl-sRNA's ($\mu\mu$moles) as shown in Table 1. Incubation was carried out at 25°C for 30 min.

TABLE 1

Specificity of aminoacyl-sRNA's in trinucleotide-stimulated binding to yeast ribosomes

Trinucleotide	Aminoacyl-sRNA bound to ribosomes ($\mu\mu$moles)			
	− trinucleotide	+ trinucleotide	− trinucleotide	+ trinucleotide
	[^{14}C]Arginyl-sRNA-I (6·2 $\mu\mu$moles)		[^{14}C]Arginyl-sRNA-II (6·2 $\mu\mu$moles)	
CpGpU	0·16	0·40	0·35	0·33
CpGpC	0·16	0·19	0·35	0·36
CpGpA	0·16	0·40	0·35	0·34
CpGpG	0·16	0·18	0·35	0·32
ApGpA	0·16	0·18	0·35	0·46
ApGpG	0·16	0·17	0·35	0·34
	[^{14}C]Alanyl-sRNA (5·4 $\mu\mu$moles)			
GpCpU	0·23	0·41		
GpCpC	0·23	0·31		
GpCpA	0·23	0·33		
GpCpG	0·23	0·23		
	[^{14}C]Phenylalanyl-sRNA (11·3 $\mu\mu$moles)			
Poly U	0·58	4·33		
UpUpU	0·58	0·68		
UpUpC	0·58	0·73		

The reaction mixtures were set up according to Method C for binding experiments. The trinucleotides were used in a concentration of 6 mμmoles, except for the experiment with [^{14}C]-phenylalanyl-sRNA in which 20 mμmoles in base residues were used. The amounts of the labeled aminoacyl-sRNA's used are shown in brackets after each sRNA.

(f) *In vitro amino acid-incorporating system*

(i) E. coli

The system is similar to that of Nirenberg & Matthaei (1961) and has been described before (Nishimura *et al.*, 1965b) except for the following modification. Instead of adding uncharged sRNA, the following amounts of precharged (with yeast aminoacyl-sRNA synthetase) arginine-sRNA were added per 1 ml. reaction mixture: unfractionated [^{14}C]arginyl-sRNA, 164 $\mu\mu$moles (40,000 cts/min), [^{14}C]arginyl-sRNA-I, 184 $\mu\mu$moles (45,000 cts/min) and [^{14}C]arginyl-sRNA-II, 184 $\mu\mu$moles (45,000 cts/min). In the control reaction without sRNA, 4100 $\mu\mu$moles of [14C]arginine (1,000,000 cts/min) were added. After incubation at 37°C for 15 min, arginine incorporation was measured by the tungstate–trichloroacetic acid procedure as described previously (Nishimura *et al.*, 1965b).

(ii) *Yeast Y42*

The reaction mixture contained 50 μmoles ammonium maleate (pH 6·5), 1 μmole spermidine, 12·5 μmoles magnesium acetate, 5 μmoles potassium chloride, 0·25 μmole GTP, 1 μmole ATP, 2·5 μmoles phosphoenolpyruvate, 100 μg phosphoenolpyruvate kinase, 6 μmoles mercaptoethanol, 1 μc [^{14}C]arginine, 500 μg of 100,000 **g** supernatant fraction, 15 o.d. units of ribosomes, 200 mμmoles of poly AAG and the amount of sRNA specified below, in a total volume of 1·0 ml. Of the sRNA fractions, the following amounts were added: crude sRNA 30 o.d. units; Arg-sRNA-I 6·4 o.d. units, Arg-sRNA-II 1·6 o.d. units. (The amounts added all had the same amount of arginyl-sRNA acceptor activity.) The incubation was carried out at 20°C for 45 min. Portions (0·025 to 0·05 ml.) of the reaction mixture (usually 0·1 ml. total volume) were taken out at different intervals and assayed for arginine incorporation as above.

3. Results

(a) *Specificity of sRNA for codons differing in first letter*

(i) *Leucine sRNA from* E. coli

Degeneracy in the code at the sRNA level has been demonstrated previously by studying leucine incorporation in a cell-free amino acid-incorporating system using randomly linked poly UG and poly UC (Weisblum *et al.*, 1962). One leucine-sRNA fraction obtained by countercurrent distribution stimulated specifically the incorporation of leucine under the direction of poly UC, whereas a second leucine-sRNA fraction caused leucine incorporation in the presence of poly UG. The three leucine-specific sRNA peaks (peaks I to III) obtained in Fig. 1 were charged with [14C]leucine and tested for stimulation of their binding to ribosomes in the presence of the six trinucleotides, UpUpA, UpUpG, CpUpU, CpUpC, CpUpA and CpUpG, proposed as codons for leucine. None of these trinucleotides stimulated the binding of [14C]leucyl-sRNA prepared from leucine-sRNA-II, and only UpUpG and CpUpG gave significant binding with peaks I and III. The failure of many trinucleotides, which are authentic codons, to stimulate the binding of the appropriate aminoacyl-sRNA has been previously observed and discussed (Bernfield & Nirenberg, 1965; Söll *et al.*, 1965; Brimacombe *et al.*, 1965). With UpUpG and CpUpG, striking specificity was observed when leucine sRNA peaks I and III were tested in the binding reaction. As seen in Fig. 6, the binding of leucyl-sRNA of peak I was specifically stimulated by CpUpG, whereas the binding of leucyl-sRNA of peak III was specifically stimulated by the trinucleotide UpUpG. It may further be noted that, since no stimulation was given by UpUpA and CpUpA, the two leucine-sRNA peaks (I and III) are, apparently, both specific for G in the third letter of the codons.

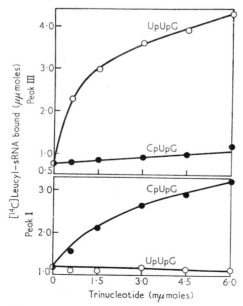

FIG. 6. Stimulation of the binding of *E. coli* [14C]leucyl-sRNA species I and III to ribosomes by the trinucleotides UpUpG and CpUpG. Method A (see Materials and Methods) was used. Each reaction mixture contained 14·7 μμmoles of [14C]Leu-sRNA.

(ii) *Arginine-sRNA from yeast. Specificity in binding and in arginine incorporation*

The two purified peaks of arginine-sRNA (Figs 2 and 3) were charged with [^{14}C]arginine and tested in binding experiments using the trinucleotides, ApGpA, ApGpG, CpGpA, CpGpU, CpGpC and CpGpG, all of which have been concluded to be codons for arginine (Nishimura *et al.*, 1965*b*; Brimacombe *et al.*, 1965; Söll *et al.*, 1965). The results of the binding experiments (Fig. 7) showed that the trinucleotides ApGpA

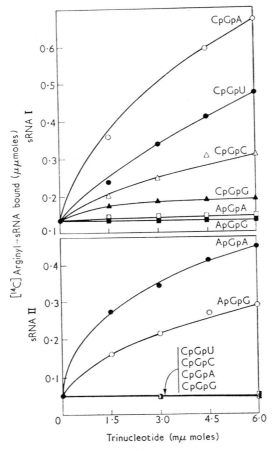

Fig. 7. Stimulation of the binding of yeast [^{14}C]arginyl-sRNA's I and II to ribosomes by the trinucleotides CpGpU, CpGpC, CpGpA, CpGpG, ApGpA and ApGpG. Method B (see Materials and Methods) was used. 2·0 μμmoles [^{14}C]Arg-sRNA I or 1·7 μμmoles [^{14}C]Arg-sRNA II was added to each reaction mixture.

and ApGpG specifically stimulated the binding of [^{14}C]arginyl-sRNA prepared from peak II, while the preparation from peak I was bound by the trinucleotides with the first letter C (CpGpA, CpGpU, CpGpC). While the experiments of Fig. 7 were carried out using *E. coli* ribosomes, corresponding experiments using yeast ribosomes showed similar specificity (Table 1). It is thus clear that the specificity pattern was not altered when the source of ribosomes was varied.

The results of further binding experiments using poly AAG and poly AG are given in Table 2. These were also in agreement with the results with the trinucleotides, in that poly AAG and poly AG specifically stimulated the binding of [^{14}C]arginyl-sRNA prepared from peak II (Fig. 2).

TABLE 2

Specificity of aminoacyl-sRNA's in trinucleotide-stimulated binding to E. coli *ribosomes*

Template (concentration)	Aminoacyl-sRNA bound to ribosomes ($\mu\mu$moles)			
	− template	+ template	− template	+ template
	[^{14}C]Arginyl-sRNA-I (6·2 $\mu\mu$moles)†		[^{14}C]Arginyl-sRNA-II (6·2 $\mu\mu$moles)†	
Poly AAG (5 mμmoles)	0·16	0·21	0·05	1·37
Poly AG (5 mμmoles)	0·16	0·19	0·05	0·80
ApGpA (6 mμmoles)	0·16	0·17	0·05	0·45
CpGpA (6 mμmoles)	0·16	0·51	0·05	0·06
	[^{14}C]Isoleucyl-sRNA-II (4·1 $\mu\mu$moles)‡		[^{14}C]Isoleucyl-sRNA-III (4·1 $\mu\mu$moles)‡	
ApUpU (20 mμmoles)§	0·11	0·40	0·10	0·50
ApUpC (20 mμmoles)§	0·11	0·24	0·10	0·24
	[^{14}C]Phenylalanyl-sRNA (24 $\mu\mu$moles)†			
Poly U (20 mμmoles)§	0·41	9·2		
UpUpU (20 mμmoles)§	0·41	0·79		
UpUpC (20 mμmoles)§	0·41	1·01		

† Incubation conditions were those of "Method B" for binding assay.
‡ Incubation conditions were those of "Method A" for binding assay.
§ In base residues.

The availability of the polynucleotide, poly AAG, offered an opportunity to check results of the binding experiments by testing amino acid incorporation. The results on the incorporation of arginine into polypeptidic material in response to yeast arginine-sRNA fractions are shown in Table 3, experiments having been carried out using both the yeast and *E. coli* amino acid-incorporating systems. In the yeast system, although the backgrounds were high and significant incorporation was observed even in the absence of poly AAG, marked stimulation of arginine incorporation was obtained specifically when arginine-sRNA-II was used. Confirmatory results with lower backgrounds were obtained using the more stable *E. coli* incorporating system (Table 3). In the latter experiment, aminoacylated sRNA was used, so that the aminoacyl-sRNA synthetases of *E. coli* would not influence the sRNA specificity. The results of Table 3 also demonstrate the dependence of the system on addition of sRNA, since when [^{14}C]arginyl-sRNA was replaced by the free radioactive amino acid, the incorporations were close to the background (see especially last line of Table 3).

TABLE 3

[^{14}C]*Arginine incorporation directed by poly AAG in response to purified yeast arginine sRNA fractions*

Incorporation system	[^{14}C]Arginine incorporated ($\mu\mu$moles/ml.)					
Yeast	Crude sRNA		Arginine-sRNA-I		Arginine-sRNA-II	
	0 min	45 min	0 min	45 min	0 min	45 min
Complete	11	66	13	53	13	*223*
— poly AAG	10	43	11	37	9	43
— sRNA	14	34	14	34	14	34
E. coli	Crude [^{14}C]arginyl-sRNA		[^{14}C]Arginyl-sRNA-I		[^{14}C]Arginyl-sRNA-II	
	0 min	15 min	0 min	15 min	0 min	15 min
Complete	2	*74*	2	6	3	*145*
— poly AAG	3	5	2	3	2	3
— [^{14}C]arginyl-sRNA						
+ [^{14}C]arginine	10	15	10	15	10	15

Incubation conditions used are as described under Materials and Methods. In the experiment with the yeast system, uncharged sRNA (crude or fractions) was used and the system was supplemented with [^{14}C]arginine. In the *E. coli* system, [^{14}C]arginyl-sRNA was used. No [^{14}C]arginine was added except in the reaction mixture of the last line.

The numbers which show clear stimulation of arginine incorporation are in italics.

(b) *Specificity of sRNA for codons differing in third letter*

(i) *Arginine sRNA from yeast*

As seen above, the results of Fig. 7 also show that arginine sRNA of peak II can be bound to ribosomes by both of the codons ApGpA and ApGpG, and similarly peak I shows response to the set of codons CpGpX in which X may be U, C or A. It is not certain whether the effect given by CpGpG (Fig. 7) can be ignored.

(ii) *Alanine-sRNA from yeast*

The results of the binding experiments with alanine sRNA using GpCpU, GpCpC, GpCpA and GpCpG are shown in Fig. 8. It is clear that at least three of these trinucleotides gave a marked stimulation in the binding reaction.

(iii) *Glycine sRNA from yeast*

[^{14}C]Glycyl-sRNA preparations were tested for binding of *E. coli* ribosomes in the presence of the trinucleotides, GpGpX (X = U, C, A or G), which are the known codons for glycine. The results are shown in Fig. 9. Thus, glycine-sRNA-I was essentially specific in being bound in the presence of GpGpA and GpGpG, while glycine-sRNA-II was bound by GpGpU and GpGpC preferentially. Glycine-sRNA-III was bound very strongly in the presence of GpGpG and not of GpGpA. Whether the additional stimulation observed in this experiment in the presence of GpGpU and GpGpC is due to contamination by glycine-sRNA-II, or is an intrinsic property of glycine-sRNA-III, remains to be determined. Glycine-sRNA-IV was bound well with

GpGpU and GpGpC, but not in the presence of GpGpA and GpGpG (Fig. 9). Glycine-sRNA-peak IV thus had binding properties similar to those of peak II, and it is possible that peak IV represents an aggregate of peak II (Cherayil & Bock, 1965). Because of this possibility, peak IV will not be considered further, even though it appears to be of higher purity than peak II. The chromatographic and codon binding behavior of glycine-sRNA peaks II and III are interpreted as indicating contamination of each of these peaks with the peak that preceded it.

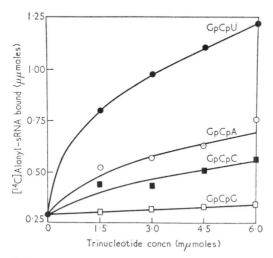

Fig. 8. Stimulation of the binding of yeast [^{14}C]alanyl-sRNA to ribosomes by the trinucleotides GpCpU, GpCpC, GpCpA and GpCpG. Method B (see Materials and Methods) was used. Each reaction mixture contained 5·3 $\mu\mu$moles of [^{14}C]alanyl-sRNA.

37°C for 7·5 min.

(iv) *Phenylalanine sRNA from yeast*

When this sRNA species was tested for binding to *E. coli* and yeast ribosomes in the presence of UpUpU and UpUpC, comparable stimulation was observed (Tables 1 and 2). The stimulation of binding was very marked when poly U was used as the template (Table 2).

(v) *Serine sRNA from* E. coli

[^{14}C]Seryl-sRNA preparations obtained by charging of serine-sRNA peaks I and II (Fig. 1) were tested for binding to *E. coli* ribosomes in the presence of the trinucleotides, UpCpX (X = U, C, A or G) and ApGpU and ApGpC which have been assigned as serine codons. As Fig. 10 shows, serine-sRNA-I was specific for binding of UpCpU and UpCpC, whereas serine-sRNA-II bound almost exclusively in the presence of UpCpA and UpCpG. (The very good binding behavior of serine-sRNA peak II to UpCpG suggests that there may possibly be two serine-sRNA species in that fraction, one being specific for UpCpA and UpCpG and the other one only for UpCpG.)

ApGpU and ApGpC failed to stimulate the binding of either of the serine-sRNA peaks to ribosomes. The lack of binding to the purified species as well as the reported low binding of unfractionated serine-sRNA (Nirenberg *et al.*, 1965; Söll *et al.*, 1965) may be due to inactivation (Muench & Berg, 1966) of the aminoacyl-sRNA synthetase specific for the corresponding sRNA species.

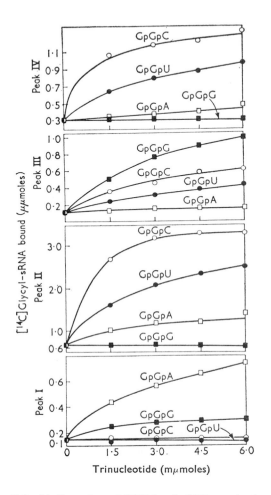

Fig. 9. Stimulation of the binding of yeast [14C]glycyl-sRNA species to ribosomes by the trinucleotides GpGpU, GpGpC, GpGpA and GpGpG. Method B (see Materials and Methods) was used. 21·0 μμmoles of [14C]Gly-sRNA I or II or IV, or 11·4 μμmoles of [14C]Gly-sRNA III was added to each reaction mixture.

(vi) *Isoleucine sRNA from* E. coli

Experiments using isoleucine-sRNA peaks I to III of Fig. 1 showed that the binding of isoleucine-sRNA-I was not stimulated by either ApUpU or ApUpC. The response given with these trinucleotides by isoleucine-sRNA-II and III is shown in Table 2. Both sRNA's were found to possess the same specificity and each recognized both trinucleotides equally well. Possibly peak I-Ile-sRNA can recognize ApUpA (Söll *et al.*, 1965).

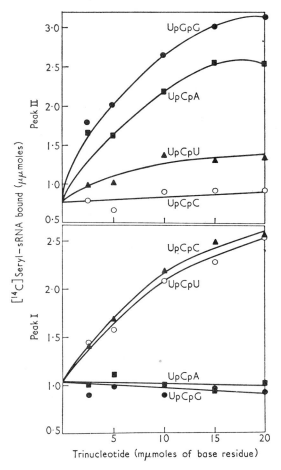

Fig. 10. Stimulation of the binding of *E. coli* [14C]seryl-sRNA peaks I and II to ribosomes by the trinucleotides UpCpU, UpCpC, UpCpA and UpCpG. Method A was used. Each reaction mixture contained 7·0 $\mu\mu$moles of [14C]Ser-sRNA.

4. Discussion

(a) *Stimulation of the binding of aminoacyl-sRNA's to* E. coli *and yeast ribosomes*

It was desirable to support the conclusions about the specificity of aminoacyl-sRNA recognition of codons drawn from using yeast aminoacyl-sRNA's and *E. coli* ribosomes by binding experiments using the homologous yeast system. However, while the trinucleotide-directed binding of aminoacyl-sRNA to *E. coli* ribosomes has been studied extensively (see e.g. Nirenberg *et al.*, 1965; Söll *et al.*, 1965), corresponding experiments with yeast ribosomes have not been reported previously. Unpublished experiments showed that, under the usual conditions, long ribopolynucleotides but not trinucleotides brought about significant specific aminoacyl-sRNA binding to yeast ribosomes. By increasing the magnesium ion concentration and lowering the incubation temperature, even trinucleotides could be induced to stimulate specific binding of aminoacyl-sRNA's to yeast ribosomes. The effects observed (Table 1) were small compared to those usually observed with *E. coli* ribosomes but they were reproducible. When taken together with the results obtained with *E. coli* ribosomes, they appear to be significant; and from the trends in stimulation, it is concluded that the heterologous *E. coli* ribosomes did not change the specificity of codon recognition.

(b) *sRNA specificity for first codon letter*

While most of the degeneracy in the code involves the third letter in the codons (Nirenberg *et al.*, 1965; Söll *et al.*, 1965), there are three amino acids in which degeneracy involves an additional change of first letter (leucine and arginine) or of the first and second letter (serine). In the present work, when fractionated leucine- and arginine-sRNA's were tested with the triplet sets assigned as codons, strict specificity for the first letter in the codon set was found. Thus UpUpG specifically stimulated the binding of leucyl-sRNA-I, whereas CpUpG specifically stimulated the binding of leucyl-sRNA-III. Thus U and C in the first position were not interchangeable. Similarly, the binding of arginyl-sRNA from one peak was specifically stimulated by the trinucleotides, ApGpA and ApGpG, while the binding of arginyl-sRNA from the other peak was specifically stimulated by the trinucleotide set CpGpX (X = U, C, A or G).

In the case of arginyl-sRNA, the binding results were further supported by incorporation experiments. Thus the incorporation of arginine into polypeptides in the presence of poly AAG was specifically mediated by the arginyl-sRNA from peak II but not that from peak I.

From the above results we conclude that wherever a change in the first letter of the codon occurs, a new sRNA species would be required to recognize it. In the case of the two sets of codons for serine (UpCpX and ApGpU, ApGpC), where a change in the middle letter is also involved, one would certainly predict that separate sRNA species would be required.

(c) *Recognition by one sRNA species of multiple codons differing in the third letter*

Before discussing the results bearing on this question, it is well to point out that the sRNA fractions used may not be homogeneous. Because the recognition process could be altered by a minimum of a single base change, there may be no satisfactory proof for homogeneity short of deducing the complete base sequence. Furthermore, caution is also necessary in drawing very firm conclusions from the binding results as obtained using messengers as short as trinucleotides. Detailed comments on the possible artifacts and ambiguities in the binding results have been made previously (Söll *et al.*, 1965). Nevertheless, it is likely that the conclusions reached below on the patterns of multiple codon recognition will prove to be correct, because the evidence came from not one but several cases and, furthermore, clear-cut specificities often were observed. The total results which are presented in Table 4 may be summarized as follows.

(i) *An sRNA species recognizes both U and C in the third position*

Several examples of this type of recognition were observed with both purified and partially purified *E. coli* and yeast sRNA. Thus purified phenylalanine-sRNA recognizes both trinucleotides, UpUpU and UpUpC. One species of glycine-sRNA recognizes both GpGpU and GpGpC and, similarly, one species of serine-sRNA binds

TABLE 4

Summary of specificity data on sRNA recognition of codons

sRNA	Codons recognized
Specificity for first letter of codon	
Arginine-sRNA-I	CpGpX (X = U, C, A)
Arginine-sRNA-II	ApGpA, ApGpG
Leucine-sRNA-I	CpUpG
Leucine-sRNA-III	UpUpG
Multiple recognition of third letter of codons	
Recognition of U and C	
Glycine-sRNA-II	GpGpU and GpGpC
Isoleucine-sRNA-II and III	ApUpU and ApUpC
Phenylalanine-sRNA	UpUpU and UpUpC
Serine-sRNA-I	UpCpU and UpCpC
Recognition of A and G	
Arginine-sRNA-II	ApGpA and ApGpG
Glycine-sRNA-I	GpGpA and GpGpG
Serine-sRNA-II	UpCpA and UpCpG
Recognition of U, C and A	
Alanine-sRNA	GpCpX (X = U, C and A)
Arginine-sRNA-I	CpGpX (X = U, C and A)
Specific recognition of G	
Glycine-sRNA-III	GpGpG
Leucine-sRNA-I	CpUpG
Leucine-sRNA-III	UpUpG
Methionine-sRNA	ApUpG†
Tryptophan-sRNA	UpUpG†

† Previous findings (Söll *et al.*, 1965).

with UpCpU and UpCpC. Furthermore, two species of isoleucine-sRNA respond in the same way to the two codons ApUpU and ApUpC.

(ii) *An sRNA species recognizes both A and G in the third position*

Several cases of this type of recognition were found. Serine-sRNA-II recognizes both UpCpA and UpCpG. Arginine-sRNA-II gives very good binding to ribosomes in response to ApGpA and ApGpG, and similarly glycine-sRNA-I recognized both GpGpA and GpGpG.

(iii) *An sRNA species recognizes U, C and A (and possibly G) in the third position*

This type of specificity, which seems most interesting, was found twice. A pure alanine-sRNA species was bound to ribosomes in the presence of GpCpU, GpCpA and GpCpC; GpCpG gave very little or no stimulation. Similarly, arginine-sRNA-I was bound in the presence of CpGpU, CpGpC and CpGpA but only poorly with CpGpG. Both of these cases were found in yeast sRNA. Whether this codon recognition pattern exists in other organisms is not yet known.

(iv) *An sRNA species recognizes only G in the third position*

This type of specificity was frequently encountered. Thus, glycine-sRNA-III, leucine-sRNA-I and leucine-sRNA-III recognize only the codons which have G in the third place, GpGpG, CpUpG and UpUpG, respectively. From the previous work (Söll *et al.*, 1965), two more examples can be added. Thus methionine-sRNA recognizes only ApUpG and not ApUpA and, similarly, tryptophan-sRNA recognizes UpGpG and not UpGpA.

(d) *General comments*

A theory for codon–anticodon pairing has been put forward by Crick (1966) which not only can account for the recognition of more than one codon by an sRNA species, but which also predicts the types of multiple codon-recognition patterns that might occur. Thus, according to this "wobble" theory, the base U (the first base, assuming antiparallel base-pairing) on the anticodon can recognize both A and G in the third place on the codon; the base C on the anticodon can recognize only G in the third place; the base A can only recognize U; the base G can recognize both U and C, and the base hypoxanthine can recognize U, C and A in the third place. The results described above at least establish that an sRNA species can recognize multiple codons differing in the third letter, and the patterns of multiple recognition observed could, within the limitations of the binding technique, be regarded as consistent with the postulates of the wobble hypothesis.

It also follows from the present work that, if one assumes the validity of the wobble hypothesis (Crick, 1966) and the determination of the codon recognition pattern of an sRNA species by the binding and/or incorporation experiments, one can predict the nature of the anticodon in the sRNA species. Thus, for example, in the case of yeast alanine-sRNA, the sequence IpGpC is an excellent candidate for the anticodon (see also Crick, 1966; Leder, Nirenberg, Holley & Keller, in preparation).

A comparison of the binding data obtained using yeast and *E. coli* sRNA's leads to another interesting conclusion. Thus, the results of Fig. 8 show that purified yeast alanine-sRNA gave very poor binding with GpCpG, whereas the results reported previously with *E. coli* alanine-sRNA showed excellent binding with this trinucleotide. In the case of *E. coli* arginine-sRNA, neither the trinucleotide ApGpA nor poly AAG gave good binding, but the binding of yeast arginine-sRNA with ApGpA was very marked. These observations point to the conclusion that while the codon assignments do not change from organism to organism, the amount to which the individual codons in each set is actually used may differ markedly. This may have an important relationship to regulatory roles of sRNA or to the evolution of the code.

Finally, from the concept of multiple codon recognition by an sRNA species, it may be concluded that the minimum number of sRNA molecules required for recognition of all of the meaningful codons is relatively small, and this conclusion in turn raises the question of redundancy in the sRNA pool of a cell. Thus, in yeast two of the three purified glycine-sRNA species show the following specificity: glycine-sRNA-I recognizes GpGpG and GpGpA and glycine-sRNA-III responds to GpGpG only. So we have two different sRNA's both of which can recognize the same codon. This to our knowledge is the first demonstration of this type of redundancy. Redundancy at the sRNA level could have an important role in the elaboration of genetic suppressors and also be a safeguard against some mutations.

This work has been supported by grants from the National Institutes of Health (grants no. CA–05178 and GM12395), National Science Foundation (grant no. GB–976 and GB–387) and the Life Insurance Medical Research Fund (grants no. G–62–54). We wish to acknowledge the advice and assistance of LaJean Chaffin and Shih-eh Wang in preparation of yeast ribosomes.

REFERENCES

Apgar, J., Holley, R. W. & Merrill, S. H. (1962). *J. Biol. Chem.* **237**, 796.

Bennett, T. P., Goldstein, J. & Lipmann, F. (1965). *Proc. Nat. Acad. Sci., Wash.* **53**, 385.

Bernfield, M. & Nirenberg, M. (1965). *Science*, **147**, 479.

Bretthauer, R. K., Marcus, L., Chaloupka, J., Halvorson, H. O. & Bock, R. M. (1963). *Biochemistry*, **2**, 1079.

Bruening, G. (1965). Doctoral Dissertation, University of Wisconsin.

Brimacombe, R., Trupin, J., Nirenberg, M., Leder, P., Bernfield, M. & Jaouni, T. (1965). *Proc. Nat. Acad. Sci., Wash.* **54**, 954.

Cherayil, J. D. & Bock, R. M. (1965). *Biochemistry*, **4**, 1174.

Crick, F. H. C. (1966). *J. Mol. Biol.* **19**, 548.

Goldstein, J., Bennett, T. P. & Craig, L. C. (1964). *Proc. Nat. Acad. Sci., Wash.* **51**, 119.

Holley, R. W. (1963). *Biochem. Biophys. Res. Comm.* **10**, 186.

Holley, R. W., Apgar, J., Everett, G. A., Madison, J. T., Marquisee, M., Merrill, S. H., Penswick, J. R. & Zamir, A. (1965). *Science*, **147**, 1462.

Holley, R. W., Apgar, J., Everett, G. A., Madison, J. T., Merrill, S. H. & Zamir, A. (1963). *Cold Spr. Harb. Symp. Quant. Biol.* **28**, 117.

Hoskinson, R. M. & Khorana, H. G. (1965). *J. Biol. Chem.* **240**, 2129.

Jones, D. S., Nishimura, S. & Khorana, H. G. (1966). *J. Mol. Biol.* **16**, 454.

Lohrmann, R., Söll, D., Hayatsu, H., Ohtsuka, E. & Khorana, H. G. (1966). *J. Amer. Chem. Soc.* **88**, 819.

Muench, K. H. & Berg, P. (1966). *Procedures in Nucleic Acid Research*; ed. by G. L. Cantoni & D. R. Davies, p. 375. New York: Harper and Row.

Nirenberg, M. & Leder, P. (1964). *Science*, **145**, 1399.

Nirenberg, M., Leder, P., Bernfield, M., Brimacombe, R., Trupin, J., Rottman, F. & O'Neal, C. (1965). *Proc. Nat. Acad. Sci., Wash.* **53**, 1161.

Nirenberg, M. W. & Matthaei, J. H. (1961). *Proc. Nat. Acad. Sci., Wash.* **47**, 1588.

Nishimura, S., Jones, D. S. & Khorana, H. G. (1965*a*). *J. Mol. Biol.* **13**, 302.

Nishimura, S., Jones, D. S., Ohtsuka, E., Hayatsu, H., Jacob, T. M. & Khorana, H. G. (1965*b*). *J. Mol. Biol.* **13**, 283.

Söll, D., Ohtsuka, E., Jones, D. S., Lohrmann, R., Hayatsu, H. & Khorana, H. G. (1965). *Proc. Nat. Acad. Sci., Wash.* **54**, 1378.

Weisblum, B., Benzer, S. & Holley, R. W. (1962). *Proc. Nat. Acad. Sci., Wash.* **48**, 1449.

Weisblum, B., Gonano, F., von Ehrenstein, G. & Benzer, S. (1965). *Proc. Nat. Acad. Sci., Wash.* **53**, 328.

Zubay, G. (1962). *J. Mol. Biol.* **4**, 347.

J. Mol. Biol. (1968) **31**, 1–12

Structure of Viruses of the Papilloma–Polyoma Type

IV.† Analysis of Tilting Experiments in the Electron Microscope

A. Klug and J. T. Finch

Medical Research Council, Laboratory of Molecular Biology
Hills Road, Cambridge, England

(*Received 27 June 1967*)

Earlier papers in this series on the electron microscopy of negatively-stained human and rabbit papilloma viruses have shown both particles to be composed of 72 morphological units arranged on the $T = 7$ icosahedral surface lattice. This surface lattice is skew and can exist in either right- or left-handed forms: the hand of the human virus is *dextro* ($T = 7d$) and that of the rabbit virus is *laevo* ($T = 7l$). The absolute hand of the structure was determined in each case from images of particles which were stained dominantly on one side, but these "one-side" images form only a small proportion of the total.

This paper presents the results of experiments in which "two-side" electron microscope images of the same virus particles obtained before and after tilting the supporting grid are compared with simulated images, i.e. superposition patterns from an appropriate model tilted through the same angle. The electron microscope images used were taken from special fields of view—close-packed arrays of particles in a sheet of negative stain—in which the particle structure is well preserved. The superposition patterns were computed and photographed from a display screen, the parameters of the model being varied until the detail in certain distinctive patterns agreed best with that in the corresponding electron microscope images (Plate I). Once the best model was obtained, a gallery of superposition patterns was calculated as a function of the spherical co-ordinates of the axis of view (e.g. Plate IV). When two superposition patterns are selected from the gallery which individually give the best fit for the pair of electron microscope images obtained from the same virus particle before and after tilting the grid, the axis and the angle of rotation to get from one pattern to the other, is, within experimental error, exactly the same as the axis and angle of tilt used in the experiment (Plate V). Moreover, the observed sense of rotation is consistent only with a *dextro* structure for the human virus and the *laevo* structure for the rabbit virus.

There is very good correlation in detail between the electron microscope images and the simulated images computed for the model. This correspondence extends right to the periphery of the computed superposition patterns (Plate VI) and shows the faint outermost detail in the images to be meaningful, attesting to the faithfulness with which the negative stain preserves and outlines the structure.

In the Appendix, the symmetry relationships between the superposition patterns obtained along different axes of view are discussed with the aid of a stereogram.

---

† Paper III in this series is Finch & Klug, 1965.

Terminology

The terminology used in this paper is essentially that originally proposed on theoretical grounds by Caspar & Klug (1962, 1963), and it has been used in a number of practical applications (Klug & Finch, 1965; Finch & Klug, 1966).

The subdivision of a spherical surface into a large number of as nearly equivalent triangles as possible is one with icosahedral symmetry (Caspar & Klug, 1962). This subdivision of a closed surface results in a network of approximately equivalent spherical triangles, which are taken to be the *unit cells* of the surface lattice, and the vertices of the triangles are called the *lattice points* of the surface lattice. The vertices are not all strictly equivalent as would be required by strict crystallographic symmetry, but are of two types: 5-vertices and 6-vertices according to the number of triangles meeting there. These vertices are geometrically quasi-equivalent. There are always twelve 5-vertices, and for a surface lattice specified by the triangulation number T, there are $10(T-1)$ 6-vertices. The set of $12+10(T-1)=10T+2$ vertices is referred to as the *surface lattice points*.

The surface lattice possesses all the rotational symmetry axes pertaining to strict icosahedral symmetry, and has, in addition, extra axes of symmetry which have only a local character referring to the immediate neighbourhood and not to the whole array. The local or quasi-symmetry is always that of the plane group $p6$, which has 2-fold, 3-fold and 6-fold axes of symmetry.

From the point of view of symmetry, the term *structure unit* is analogous to the term *asymmetric unit* in the classical theory of symmetry groups, and is an element of the structure which, when operated on by both the strict and local axes of symmetry, generates the whole structure. In terms of construction of a shell, the structure units are the smallest functionally equivalent building units. In the case of a protein shell, the structure unit may usually be identified with a single protein molecule, but it might consist of an aggregate of identical or different chemical subunits. The shape and location of structure units are not determined in any way by the symmetry of the surface lattice and must be found by observation in any particular case.

The term *morphological unit* (Klug & Caspar, 1960) refers to an element of the whole structure which can be readily seen in the electron microscope. It may consist of a single structure unit, but more frequently consists of a small symmetrical cluster of structure units which pack so tightly as to form a distinct grouping easily resolved in the electron microscope from the similar neighbouring groupings. The resolution of the individual structure units within a cluster requires a higher order of resolution and may not be possible in the electron microscope. The number of structure units in a cluster depends on the order of the rotational symmetry axis about which they are grouped. In a structure with $T=1$ icosahedral symmetry the number can be 2, 3 or 5; when $T>1$, it can be 2, 3 or 5 and 6.

1. Introduction

In the first paper of this series (Klug & Finch, 1965), we described the results of an electron microscope study of negatively stained human wart virus. We interpreted most of the images as "two-sided", arising from the superposition of detail from the near and far sides of virus particles. On this basis, good correspondence was obtained between the most distinctive types of image and superposition patterns made from a model of the virus viewed at particular angles, the model being composed of 72 hollow or ring-shaped morphological units arranged according to the $T=7$ icosahedral surface lattice. Two-side images were also obtained in a similar study of turnip yellow mosaic virus (Finch & Klug, 1966) and these again corresponded well with the appropriate superposition patterns from models of that virus. In the latter case it was further shown that the relationship between the two-side images obtained from the same virus particle, before and after tilting the latter through a known angle, was exactly that between the two corresponding superposition patterns.

We present here the results of similar tilting experiments carried out with HWV†
which confirm our earlier interpretation of the structure, by showing that the relationship between two particular characteristic two-side images is that required by
the $T = 7d$ icosahedral surface lattice.

In the TYMV work we produced superposition patterns by making shadows of
skeletal models. Since that time we have developed a computational method for
producing simulated images of trial structures (Finch & Klug, 1967). The main
disadvantage of shadowgraphs is that they are made by using a physical model,
and this must be built for each variation of model tried. Moreover, the shadowgraphs
necessarily include the shadows of extraneous components used in holding the model
together, and there is no connection between the blackness of the shadow of an
opaque model and the thickness of material producing it. These disadvantages
are eliminated by our present method, which computes and displays the superposition patterns arising from a given model.

We must stress, however (Finch & Klug, 1967), that the patterns obtained are still not
analogue images such as those produced by Caspar (1966); they are equivalent to direct
radiographs of a model and not to negatives of radiographs of a negatively-stained model.
This deficiency shows itself particularly acutely at the periphery of the pattern which is
too intense relative to the centre. But the facility of being able to compute and display
superposition patterns quickly at any desired angle, from a model which can be easily
varied, outweighs this last consideration, and in any event it can be allowed for mentally.
The method has proved invaluable in this work. For a structure as complex as that of
the papilloma viruses, the two-side images, and *pari passu* the superposition patterns,
can change markedly when the angle of view is changed by as little as 1°.

For the tilting experiments, fields of view were generally chosen in which the virus
particles formed close-packed arrays within an extended sheet of stain, since in such
arrays the particles appeared to be well preserved; and, in fact, this was proved by
the tilting experiments as described below. At the same time, the particles in these
regions were completely embedded in stain and the images appeared uniformly
two-sided: this was confirmed by the comparison of images with superposition
patterns computed on the basis of equal contributions from the two sides of the
model (Plate IV). In addition, of course, the higher density of particles increases the
incidence of that small proportion of particles which lie with their axes in an orientation favourable for the analysis described below.

2. Experimental Procedure

The methods of preparation and electron microscopy of human wart virus were as
previously described (Klug & Finch, 1965), and the conventions of grid loading and photographic reproduction of the electron micrographs were also the same. Thus our final prints
reproduce the images of particles as the particles are seen from the electron source. The
specimen grids were tilted in the electron microscope using a Valdrè stereo-cartridge
(Valdrè, 1962).

Superposition patterns were computed on an Argus computer and photographed from a
display screen as described by Finch & Klug (1967). The setting of the icosahedron
relative to the spherical co-ordinates used is shown in Plate IV(a).

3. Calculation of Superposition Patterns

(a) *Choice of model*

The great advantage of the method of computing and displaying superposition
patterns is the ability to test trial structures rapidly. In Plate I are shown a selection

† Abbreviations used: HWV, human wart virus; RPV, rabbit papilloma virus; TYMV, turnip
yellow mosaic virus.

of the patterns obtained in the course of a search for the best size of morphological units to represent the surface structure of HWV: by "best" we mean the choice that produces patterns which are most like the electron microscope images, within the limitation of the method referred to above, namely, that the computation does not produce a true, analogue negative-stain image. The view of the model chosen for this comparison was in the direction of a 3-fold axis, because for this view the two-side images are the most distinctive. A selection of these images was shown in our earlier paper (Klug & Finch, 1965, Plate XI), and some are reproduced in Plate I(i). These images show three central, apparently normal, morphological units which arise by the almost exact superposition of the three central morphological units on one side of the virus particle upon the corresponding three morphological units on the opposite side of the particle. Surrounding these are three terraces of apparently smaller "units", which are, in fact, regions of overlap of the projections of the other morphological units on the two sides of the particle. The innermost terrace contains 18 of these "smaller units". Both this and the second terrace are approximately triangular in shape; they are clearly distinguished from each other and from the inner three large units. As discussed in more detail later, the outermost terrace appears much fainter; it is not everywhere clearly separated from the second terrace and is circular rather than triangular in shape, forming a boundary to the image. These are the features of the image that we sought to bring out in the superposition patterns.

The two main parameters that determine the over-all morphology of a model are the diameter of a morphological unit and its radial extent. A morphological unit is composed of structure units the centres of which lie on a circle—the diameter of this circle can be thought of as a mean diameter of the morphological unit. In our earlier paper (Klug & Finch, 1965), we concluded that this mean diameter was about two-thirds of the distance between lattice points. In Plate I(iii) are shown the one- and two-side computed patterns from models in which the mean diameter of the morphological units varies from two-thirds to a half of the distance between lattice points. In the models used to produce these patterns, the morphological units are built from sets of points with particular co-ordinates in the unit cell; but to the resolution with which we are concerned here, the shape of a morphological unit is only slightly dependent on the azimuthal orientation of these points around the lattice points. That is, we are not here concerned with the exact co-ordinates of the structure units, and the points used to build the model do not represent them except in their distance from the nearest surface lattice points. The investigation of the finer features would require a longer and more detailed analysis than that which we describe here.

In Plate I(iii), (a) is shown the two-side superposition pattern when the morphological units have a mean diameter which is equal to two-thirds of the lattice distances: there is no division into terraces and in the one-side pattern the morphological units are not evident. When the mean diameter is equal to half the lattice distance, the morphological units appear too small on the one-side pattern, and although there is now a clear division into terraces in the two-side pattern, the innermost terrace itself is not divided into 18 bright patches as is required by the electron micrographs.

The patterns in (b) are from a model midway between (a) and (d), and, although the innermost terrace now has 18 patches, the division into distinct terraces has been lost, and in any event the morphological units are too large. In (c), however, which is midway between (b) and (d), the patterns agree fairly well with the electron microscope images: the terraces are clearly visible in the superposition pattern and the innermost terrace is broken into 18 bright patches. We have therefore used the co-ordinates (c) in the following work.

The patterns in Plate I(iv), (e) and (f) show the effect of giving radial extent to the morphological units in the model (c), by including another set of points with the same spherical co-ordinates at a smaller radius. In (e) the difference in radii is 10% and in the two-side pattern the terraces are no longer clearly resolved from each other. In (f) the difference in radii is 5% and the terraces appear with the correct clarity, and this is essentially the model which has been used in the following work. In the case of some particles, however, the stain does not always appear to penetrate the central hole of the morphological unit completely; to represent this kind of image, we have also used model (g) which has some density in the centres of the morphological units.

Another common two-side image, namely that with a "hooped" appearance, arises from views of the particle in a direction midway between a 5- and a 6- co-ordinated morphological unit (see Plate X of Klug & Finch, 1965). Some examples are shown in Plate II(a), together with the best matches selected from a gallery of superposition patterns made at 2° intervals in the neighbourhood of the axis of view with the spherical co-ordinates $\theta = 90°$, $\phi = 135°$ (Plate II(b)). As can be seen from this gallery, the over-all hooped appearance is rather insensitive to small changes in the direction of the axis of view, but in a number of cases we have been able to identify views which are only about 2° off from the others. These are the images (v) and (vi) in the Plate. All the images except (i), which was obtained further from focus than the others, show ring-shaped units over the bottom half, where the centres of morphological units on the two sides of the particles superpose. This means that the morphological units possess a deep central hole which can be filled by stain.

The differences between the diameters of the morphological units in the models (b), (c) and (d) in Plate I, would be about 4 Å for the papilloma viruses. It may seem surprising that one can discriminate between the models when making the comparisons with the electron-microscope images in which the *effective point-to-point* resolution is nowhere near 4 Å. But it should be remembered that the detail in the structure is repetitive, albeit on the surface of a sphere, so that one has, as it were, repeated images of a single feature. This is quite analogous to the case of a truly periodic structure, e.g. a crystal or helix, where by making an average, for example by means of optical diffraction, one can deduce the dimensions of significant features accurately to say 1 or 2 Å even though the effective point-to-point resolution may be, as above, only about 20 Å.

This detection of small differences is further facilitated by working with superposition patterns of two-side images; these patterns are extremely sensitive to small changes in the structure. As a crude explanation one may think of the superposition pattern being dominated by the overlap regions where density occurs on both sides of the particle; the linear dimensions of these regions can vary much more than the change in dimensions of the individual regions of density on one side. Although the linear dimensions of the overlap region in a direction joining the centres of the two individual areas which overlap can change only by twice the change in dimensions of these areas, the absolute changes in directions other than this can be much greater.

(b) *A gallery of superposition patterns*

The 3-fold terraced image and the hooped image arise from views of the particle in special directions; but they are both members of a class of two-side images which occur for views in directions perpendicular to a 2-fold axis of the particle and as a consequence have mirror symmetry about a line parallel to this axis. If one sees an image with this symmetry, one knows that the direction of view lies in a **plane** perpendicular to a 2-fold axis of the particle. This fact is of great help in the identification of an image, since to our working resolution there are some 200 distinct superposition patterns; marked changes in the patterns occur for changes in particle orientation of 2° and as little as 1° for some directions of view. In Plate III is shown a gallery of computed two-side superposition patterns for directions of view lying in a plane perpendicular to a 2-fold axis; in terms of the spherical co-ordinate frame (see Plate IV(a)) the plane is specified by $\theta = 90°$, and the patterns are shown at 2° intervals from $\phi = -4°$ to $+94°$. For directions of view within this plane, the patterns from $\phi = 90°$ to 180° are identical to those for $\phi = 90°$ to 0°, taken in that order, but are rotated by 180° in the plane of the paper; the patterns from $\phi = 180°$ to 360° are identical of course to those from $\phi = 0$ to 180°. Thus the patterns at $\phi = -4°$ and 94° are respectively the same as those with $\phi = 4°$ and 86°, but rotated by 180° about the axis of view. The symmetry relationships are discussed in more detail in the Appendix.

It is hard to describe most of the superposition patterns in words, although a number are so distinctive that they can be ascribed names, as we have already done for the 3-fold terraced and hooped images. Once one commits these patterns to visual memory—helped by giving them descriptive names—it is easy to pick them out in a field of particle images. As further examples, we cite the following superposition patterns: (1) $\theta = 90°$, $\phi = 62°$ (local 2-fold axis)—"two eyes separated by two bands". Electron microscope images of this type are shown in Plates V(a),A and V(b),E. Note that one eye is closer to the periphery than the other eye, which has a further band on its peripheral side: this distinguishes it from the rather similar pattern at $\theta = 90°$, $\phi = 90°$, which is formed in the direction of a strict 2-fold axis and which has the eyes and bands symmetrically placed. (2) $\theta = 90°$, $\phi = 78°$—"the wedge and stripes" (stripes running vertically in the upper half and a wedge or lozenge of clearly resolved apparent morphological units in the lower half). An electron microscope image of this type is shown in Plate V(b),A. The features which make a pattern distinctive are usually relatively insensitive to small changes in angle of view so that the name can be retained over a small range of orientations. In fact, if one wishes, one may define a distinctive pattern as one on which the gross detail is stationary with respect to changes in aspect, although examination of the finer detail enables one to find the exact orientation exactly (cf. Plate II(a)).

The limited gallery shown in Plate III enables one to follow the change in appearance of a two-side image of a particle as it is tilted only if the particle has one of its 2-fold axes in the direction of the tilt axis. Within the angular resolution of 1° to which we are working, a departure of the 2-fold axis of the particle from the direction of the tilt axis of up to 3° would be permitted when the angle of tilt is 20°. However, only about 2% of the particles in an array with random orientation would satisfy this condition. This fraction will be increased if we allow a greater angle between the 2-fold axis of the particle and the tilt axis, but the direction of view will not then be confined to the plane $\theta = 90°$. We have therefore calculated companion galleries to Plate III, varying ϕ for each of the values $\theta = 86°, 88°, 92°$ and 94°. Sections of the over-all resulting gallery are shown in Plates II(b) and IV(c). It is clear from this gallery that for a given ϕ the superposition patterns for values of $\theta = 90°+\alpha$ and $90°-\alpha$ are related to each other by a mirror line parallel to the 2-fold axis $\theta = 0$ (see Appendix).

The galleries have been calculated for a $T = 7d$ arrangement. By relabelling the

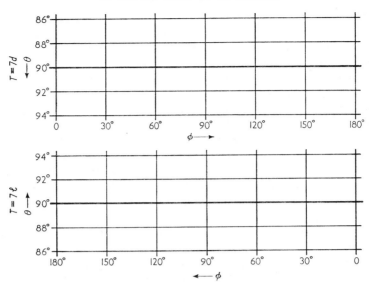

FIG. 1. The relabelling of co-ordinates which enables the gallery of superposition patterns computed for the *dextro* $T=7$ structure to be used for the *laevo* structure, without change in the lay-out.

axes as shown in Fig. 1, they can also be used for a $T = 7l$ arrangement, which gives, of course, identical superposition patterns but in a different order. Thus the superposition pattern corresponding to the spherical co-ordinates θ and ϕ for $T = 7l$ is the same as that for $T = 7d$ for the co-ordinates $\theta_d = 180° - \theta$, $\phi_d = 180° - \phi$.

4. Tilting Experiments

In Plate V(i) are shown electron micrographs of the same field of a close-packed array of negatively-stained HWV, before and after tilting the grid through an angle of about 20°. The direction of the tilt axis is shown by the arrow and the sense of the tilt from Plate V(a) to (b) is anti-clockwise, when one looks in the direction of the arrow. All the particle images are markedly changed after tilting. For the reasons described above, we have selected for analysis images which show a mirror line of symmetry, or an approximate line of symmetry, within one or two degrees of the direction of the tilt axis. Some of these images are marked in Plate V and enlarged in column (ii). We have selected from our gallery the computed superposition patterns which are closest in detail to these images, and these are shown alongside. In each case the change in image on tilting from (a) to (b) corresponds to an increase in ϕ for the superposition pattern of about 15°. The close similarity between the individual images and the superposition patterns and the fact that the relationship between the images of the same particle before and after tilting is the same as that for the $T = 7$ lattice confirms our earlier conclusions that this is the lattice on which the surface structure of HWV is based. Further, the sense of tilt to go from (a) to (b) in the electron micrographs is the same as the sense of rotation of the model to go from the patterns in (a) to those in (b), and thus the hand of the $T = 7$ lattice present in HWV is the same as that of the model, i.e. $T = 7d$.

Since the tilting stage used is not accurately calibrated, the actual angle through which the grid was tilted in this experiment was determined by measuring changes in distances between fixed points in the electron micrographs, and was found to be within the range $20° \pm 5°$. The magnitude of the rotation of the model to give the two superposition patterns which correlate best with the images obtained before and after tilting is $15° \pm 1°$. This value lies within the expected range and we conclude that there is no detectable flattening of the particle structure: if there were appreciable flattening, the difference between the values of ϕ of the patterns corresponding to the two images would be smaller. Thus, within close-packed arrays of particles, such as that shown in Plate V, the particle structure is well preserved.

Similar tilting experiments on isolated particles show the same angular relationship as that found for particles in arrays, provided that they are enclosed in fairly deep stain. In regions of thin stain, however, the images tend to be one-sided; it is these that were selected in our earlier tilting experiments (Finch & Klug, 1965) and, as shown by our measurements, appreciable flattening of the particle can occur in this case.

We have performed similar tilting experiments with rabbit papilloma virus and find in this case that, for the same sense of tilt as in Plate V(a) to (b), the matching superposition patterns selected from the gallery described above have values of θ which decrease instead of increase. Since the gallery was computed for a right-handed particle, this shows that the structure of RPV is based on the $T = 7l$ lattice,

5. Discussion

(a) Surface lattice of the papilloma viruses

The detailed correspondence observed between the simulated two-side images and the electron microscope images for the many different views identified is so good that there is no doubt in our minds that we are dealing with a unique structure based on the $T = 7$ icosahedral surface lattice. No other surface lattice could give rise to the peculiar types of superposition patterns we have noted in this and earlier papers. Even if, by chance, there were some degree of similarity between any one image and a superposition pattern computed for a model based on another icosahedral surface lattice, the tilting experiments would rule out the latter. When the gallery of superposition patterns computed for our $T = 7$ model is used to select the two which individually give the best fit for each of the pair of electron microscope images

---

PLATE I. (i) "Two-side" electron microscope images of negatively-stained particles of human wart virus seen in a direction close to a 3-fold axis.

(ii) Simulated image of trial structure [(iv)f], which gives best agreement with (i).

(iii) and (iv) Computed superposition pattern in the direction of a 3-fold axis ($\theta = 90°$, $\phi = 69°5'$) for models based on the lattice $T = 7d$ in which the size and radial extent of the morphological units are varied. The models themselves are shown as orthogonal projections of the upper side in the same direction.

In (iii) the points representing the morphological units are all at one radius, the mean diameters of the morphological units being 0·67, 0·58, 0·54 and 0·5 of the lattice edge, in (a) to (d), respectively. The model (c) best represents the detail in electron micrographs.

In (iv) the morphological units in (c) are given radial extent, equal to 10% of the external radius in (e), and 5% in (f). The latter corresponds best with the electron micrographs and is therefore shown for comparison in (ii). In (g), the central holes of the morphological units are filled in lightly to represent better those images in which the stain has not penetrated in these regions.

All superposition patterns shown are computed on the basis of equal contributions from both sides of the model.

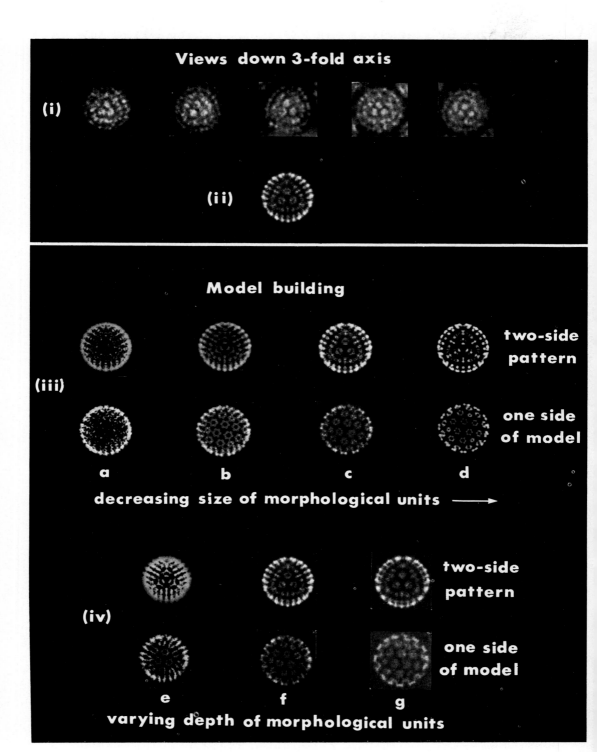

Views down 3-fold axis

(i)

(ii)

Model building

(iii)

two-side pattern

one side of model

a   b   c   d

decreasing size of morphological units ⟶

(iv)

two-side pattern

one side of model

e   f   g

varying depth of morphological units

PLATE I

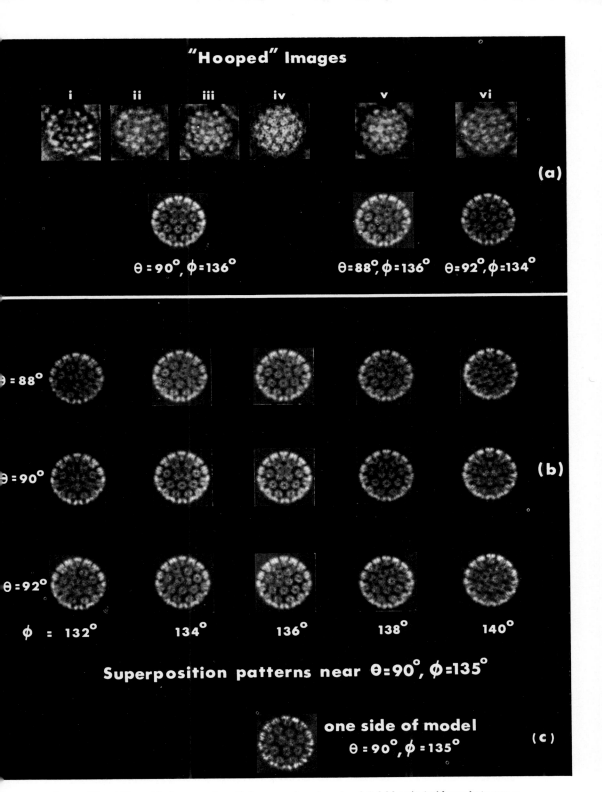

PLATE II. (a) Two-side images of particles seen close to a local 2-fold axis (midway between a 5- and a 6- co-ordinated morphological unit), compared with the closest corresponding simulated images selected from (b).

(b) Part of the array of superposition patterns for directions of view in the neighbourhood of $\theta = 90°$, $\phi = 135°$, computed for model (f) in Plate I.

(c) Model (f) in Plate I, viewed as a projection of one side along the direction $\theta = 90°$, $\phi = 135°$.

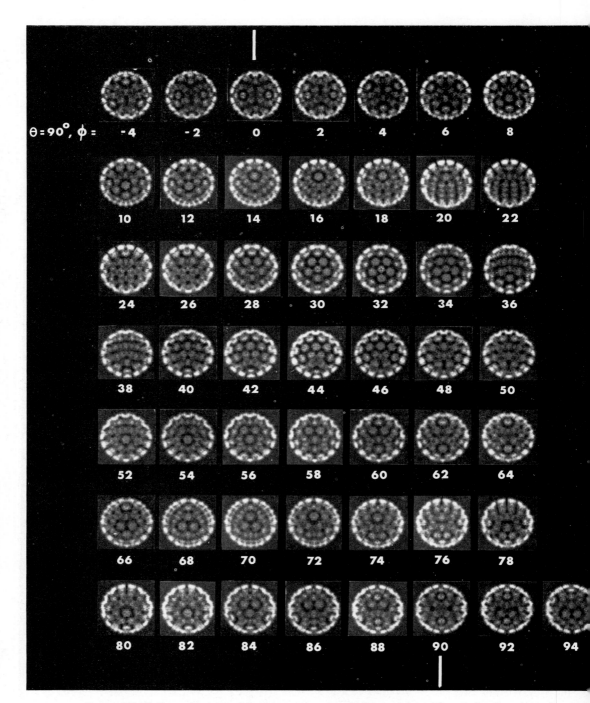

PLATE III. Gallery of superposition patterns computed for model (g) Plate I, for directions of view perpendicular to a 2-fold axis of the model (i.e. within the plane $\theta = 90°$).

The patterns are calculated at $2°$ intervals in ϕ. The patterns at $\phi = 0°$ and $90°$ correspond to views exactly along 2-fold axes of the model; the patterns at $32°$ and $70°$ correspond to views almost along the 5-fold and 3-fold axes, respectively. The "hooped" patterns occur in the region of $\phi = 45°$, which is close to the direction of a local 2-fold axis. Another local 2-fold axis occurs near $\phi = 62°$.

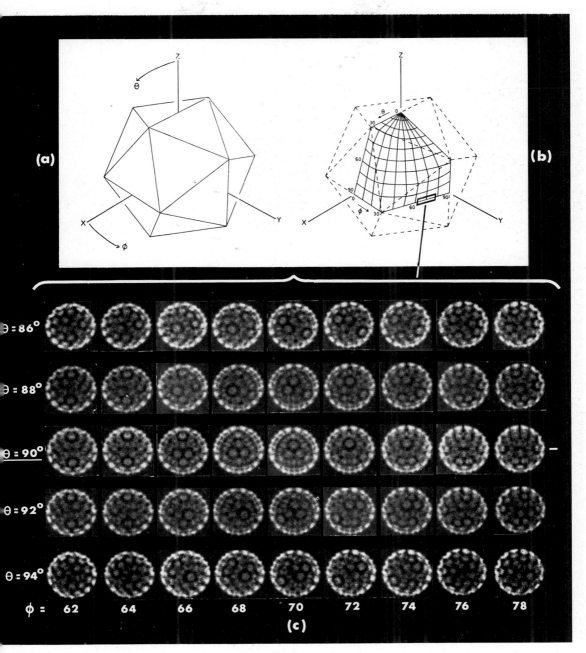

PLATE IV. (a) The relationship between the spherical co-ordinate frame and the setting of the icosahedron.

(b) An octant of the icosahedron divided by lines of equal θ and ϕ.

(c) Part of the array of computed superposition patterns from model (g) in Plate I, for directions of view lying in the region shown boxed in (b). This region encompasses the 3-fold axis at $\theta = 90°$, $\phi = 69°5'$.

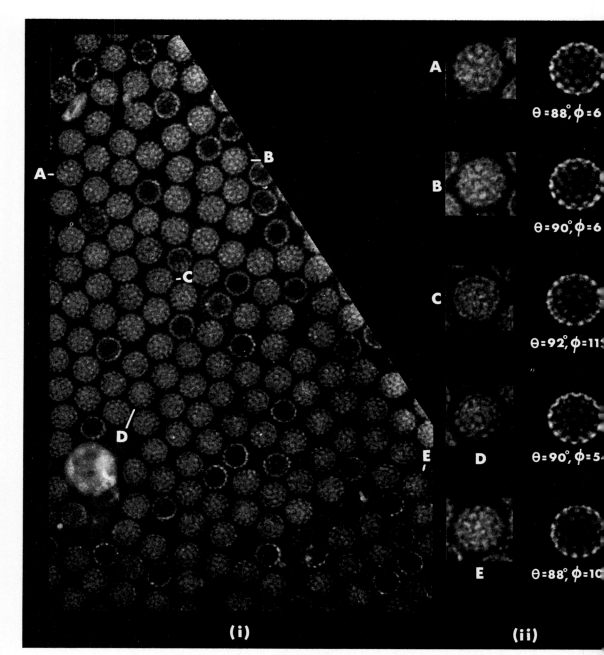

θ=88°, φ=6

θ=90°, φ=6

θ=92°, φ=11

θ=90°, φ=5

θ=88°, φ=1C

(i)   (ii)

PLATE V(a)

PLATE V. (i) Electron micrographs of the same field of close-packed, negatively-stained HWV particles (a) before and (b) after tilting the specimen grid through an angle not far from 20° about an axis parallel to the arrow. The rotation is in an anticlockwise sense, looking in the direction of the arrow.

If one wishes to consider these pictures as a stereo-pair, then the stereoscopic baseline is at right angles to the arrow, and the left eye should look at (a) and the right eye at (b).

(ii) Enlargements of the lettered images in (i) preserving the orientation, together with the computed superposition patterns which correspond to them most closely; most of these patterns appear in Plate IV(c).

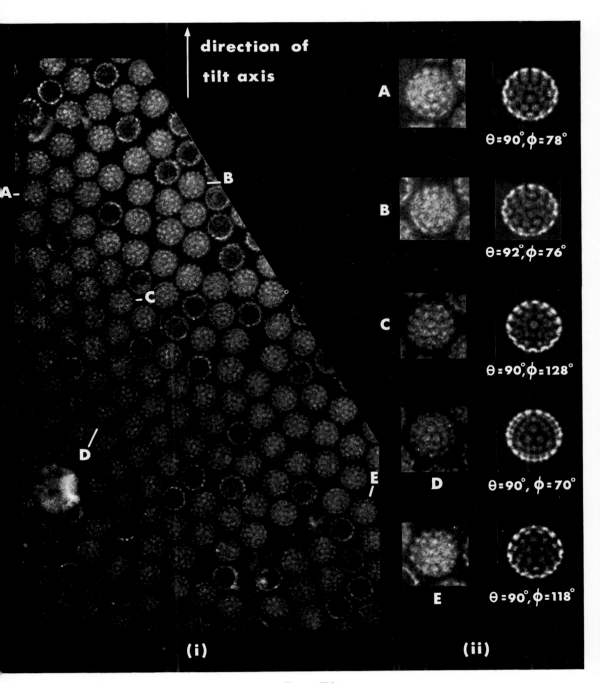

direction of tilt axis

A

B

C

D

E

(i)

A   θ=90°, φ=78°

B   θ=92°, φ=76°

C   θ=90°, φ=128°

D   θ=90°, φ=70°

E   θ=90°, φ=118°

(ii)

PLATE V(b)

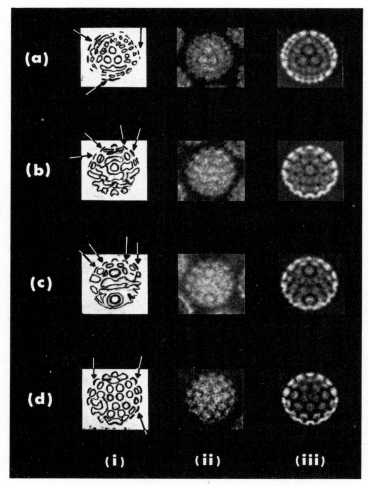

PLATE VI. (ii) Some distinctive electron microscope images of views of negatively-stained HWV particles.

(iii) Computed superposition patterns corresponding to the views in (ii).

(i) Drawings of main details of the images in (ii). The arrows indicate the faint details on the outermost boundary of the image which correspond closely to the peripheral parts of the superposition patterns.

The top three particles come from the array reproduced in Plate V. The peripheral features of some of their neighbours can also be discerned, and it can be deduced that neighbouring particles are touching.

obtained from the same virus particle before and after tilting the grid, the axis and the angle of rotation to get from one pattern to the other is, within experimental error, exactly the same as the axis and angle of tilt used in the experiment.

The results of the tilting experiments on HWV and RPV not only show that both viruses have the $T = 7$ lattice but also that the two viruses have different hands, *dextro* for the human virus and *laevo* for the rabbit virus. This confirms our earlier conclusions both about the surface lattice and the different sense of hand (Klug & Finch, 1965; Finch & Klug, 1965). Moreover, since the absolute sense of the hand agrees with that deduced from our earlier tilting experiments in which use was made of particles stained dominantly on one side (Finch & Klug, 1965), the present work also necessarily confirms the related conclusion on the staining of the papilloma viruses, namely, that when one side is dominantly stained, it is the side nearest the carbon substrate.

(b) *An assessment of the results of negative staining*

In the above work we have made use of the detailed similarity between the appearances of a number of two-side images with the corresponding computed superposition patterns. For all views identified, the strong correlation between the central features of an image and the matched superposition pattern strikes one's eye immediately. In fact, on closer inspection, this detailed correlation between the two can be followed right to the periphery, if allowance is made for the over-brightness of the computed pattern in this region. Four examples illustrating this are shown in Plate VI; in each case the images are accompanied by drawings showing their main recognizable features. In (a), the view is in the direction of the 3-fold axis of the particle; apart from the three central large apparent morphological units and the two innermost terraces of smaller units, there is a faint circular boundary to the image which corresponds to the third, outermost terrace of the superposition pattern. Although its appearance in the electron microscope image is softened by the dark stain, this third terrace can be clearly discerned as three symmetrically disposed regions, where, because of its circularity, it lies clear of the second, more triangularly shaped, terrace (these regions are arrowed in the drawing).

Likewise, the extreme periphery of the particle can be seen in almost all other types of image, when they come from particles which lie in regions where the preservation is good and which are not too heavily obscured by stain. As in the case of the 3-fold terraced image, the variation in density of the periphery can be recognized most easily in views which give rise to a type of image in which the peripheral features do not merge continuously into the body of the image. In (b), (c) and (d), some of the apparent units close to the periphery of each image are accompanied by fainter bright patches still farther out, while others appear doubled. Again these correspond to the extreme peripheral details on the computed superposition patterns, It is thus evident that the faint and often "fuzzy" boundary to the image is meaningful.

We therefore conclude that the particle structure can be extremely well preserved by the negative stain, and that the fine detail is faithfully recorded in the electron microscope images. We find these cases of good preservation more particularly in arrays where the particles are touching, or in regions of deep stain on the carbon substrate from which these examples were selected. But, as we have shown, the great drawback is that it is precisely in these cases that both sides of a particle are being recorded at once.

Other evidence on the faithfulness of negative staining comes from optical diffraction patterns on electron micrographs of negatively-stained helical particles (Klug & Berger, 1964; Finch, Klug & Stretton, 1964; Klug & DeRosier, 1966) which show that the periodic detail is generally preserved to a resolution of about 20 Å in intact, well-stained particles, and on occasions even to 15 Å. The detail corresponding to even lower spacings is lost due to the background "noise" produced by the irregularity of the staining and by lack of preservation of the particles. The ultimate resolution would in any event be limited by the discrete molecular structure of the stain: one would expect this resolution to be dictated by the diameter of the staining molecules, i.e. roughly about 10 Å for the phospho-tungstate ion and about 6 Å for the uranyl acetate, but the actual values would depend on the state of aggregation of, or interactions between, the ions.

The tilting experiments in this paper are similar to those we carried out on TYMV (Finch & Klug, 1966). There is a difference, however, in that the specimens of TYMV used were negatively stained with uranyl acetate and the fields were chosen over holes in the carbon substrate where the virus particles were enveloped in a film of stain. In that investigation we were concerned with finer detail than we have been studying here, and it was essential to minimize the background noise by omitting the carbon substrate. But unfortunately particles over holes are subject to a fair amount of distortion due to the stretching or contraction of the film of stain, though in some fields the particle shape was sufficiently well preserved to make it possible to follow the changes in the images upon tilting the grid. In the case of the papilloma virus particles, one can tolerate less distortion in the surface lattice because, even though the morphological units are about the same size as those of TYMV, there are 72 of them on the sphere instead of 32, making the superposition patterns much more sensitive to distortion.

(c) *The morphology of the virus particle*

The close correspondence between the electron microscope images and the superposition patterns attests not only to the faithfulness of the negative staining but also of course to the validity of the structural detail embodied in the model. We have already reported the lateral and radial dimensions of the morphological units as deduced from direct measurements on the electron micrographs (Klug & Finch, 1965). The distance apart of the morphological units at the radius of mean contrast is about 100 Å. The radius of mean contrast is about 220 Å compared with a value of 280 Å for the maximum radial extent of undeformed particles. The difference of 60 Å between these two radii corresponds to the mean depth of the grooves between morphological units. The morphological units have a central hole or depression of about 40 Å diameter and of a depth possibly equal to that of the grooves between morphological units. These details have been confirmed by the work of Caspar (1966) in which a model of the virus particle, constructed of 72 morphological units with these specifications was coated in plaster of Paris and radiographed. There is a very close similarity between the analogue negative-stain images thus obtained and the electron micrographs.

APPENDIX

Symmetry Relations between Superposition Patterns along Different Axes of View

Because of the considerable amount of computing required to produce an array of superposition patterns, it is important to know what range of angles θ, ϕ needs to be explored in order to obtain all independent superposition patterns. If the object under study is considered to be held fixed, i.e. at rest in a spherical frame of co-ordinates θ, ϕ, a superposition pattern arises by constructing the orthogonal projection in the direction θ, ϕ of the axis of view.

If we consider a spherical surface of large radius drawn about the object, then we can think of each superposition pattern as residing on this *reference sphere* at the associated point θ, ϕ. The number of superposition patterns on the sphere will depend on the angular difference between adjacent axes of view, and the fineness of the subdivision must be decided according to the complexity of the original object and the resolution with which we wish to study it.

When the object has some symmetry, the same superposition pattern will be obtained along more than one axis of view. The array of superposition patterns on the reference sphere will express the point group symmetry of the original object if, when associating a pattern with a point θ, ϕ on the sphere, we keep its orientation fixed in the fixed frame of reference of the object. That is, any one pattern at the point θ, ϕ will be operated on by the symmetry elements of the point group to produce identical patterns, in the appropriate orientations, at the points related by symmetry to the point θ, ϕ. The range of θ and ϕ required for investigation, therefore, need bound only the asymmetric unit of the point group.

In fact, only half of this asymmetric unit needs to be explored, because of the fact that two patterns at diametrically opposite points on the reference sphere are mirror images of each other. This is because they correspond to orthogonal projections in opposite directions along the same line. In order to express this relationship between the patterns, we must introduce for our array of patterns on the sphere, additional operations not normally considered among the relationships of point group symmetry. These are the so-called operations of *antisymmetry* (Shubnikov, 1964), in which a change of position produced by a normal symmetry element is accompanied by a change of quality; e.g. "black" is changed into "white" or "spin up" into "spin down". In the present case, the additional symmetry element is an "anticentre", i.e. a centre of symmetry, operating on both the point θ, ϕ and its associated pattern, plus—and here we have the local change of quality appropriate for our problem— the operation of turning the pattern around by 180° about its own centre.

This is illustrated for the icosahedral point group in Fig. 2. A stereographic representation of the point group 532 is shown in Fig. 2(a), including the great circles perpendicular to 2-fold axes. These great circles divide the stereogram into two sets of 60 equivalent spherical triangles. In the point group 53m (or, in full, $\bar{5}\,\bar{3}\,\frac{2}{m}$) which has mirror planes perpendicular to 2-fold axes, these 120 triangles would all be equivalent; however, in the point group 532 the asymmetric unit consists of a pair of these triangles, one from each set. If with one point θ, ϕ in an elementary triangle, we associate a pattern symbolically represented by the letter F, the icosahedral symmetry elements reproduce F in the appropriate orientation at the corresponding points in the other 59 triangles of the set (Fig. 2(b)). The introduction of the anticentre produces the mirror related image ꟻ at the diametrically opposite points which lie in the 60 triangles of the other set. The combination of the anticentre with the 2-fold axes of the point group produces mirror planes of antisymmetry (m') perpendicular to these axes: these again involve the operation of a normal mirror plane plus the change of quality, i.e. a local rotation of 180°. The symmetry of the whole array is thus 53m' (or in full $\bar{5}'\,\bar{3}'\,\frac{2}{m}$,) and the asymmetric unit is one of the small 120 triangles. A mirror plane of antisymmetry can be seen at work along the lines $\theta = 90°$ in Plates II(b) and IV(c), where a pattern at a given ϕ and say $\theta = 90° + \alpha$ is related to that at $\theta = 90° - \alpha$ by reflection through the line $\theta = 90°$ followed by a rotation of the pattern through 180°.

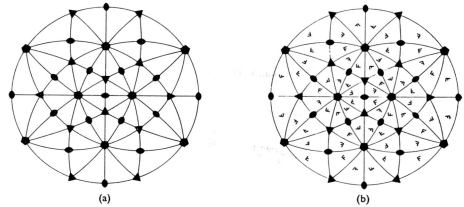

(a)   (b)

FIG. 2. (a) Stereographic representation of the point group 532. The 15 great circles shown are perpendicular to the 2-fold axes.

(b) Stereogram showing how a general superposition pattern, symbolised by the letter **F**, is repeated for 60 symmetrically related directions of view. The pattern obtained for views along the same axes, but looking in the opposite direction, is the mirror image ꟻ, which also occurs 60 times, in the appropriate orientations, as shown. The two systems of patterns, F and ꟻ, are related by planes of mirror "anti-symmetry", the traces of which are the great circles shown in the stereogram.

A stereographic representation of the symmetry relations between patterns over the whole sphere like that of Fig. 2(b) is very useful in following changes of pattern when the axis of tilt is a general one. When the axis of tilt is close to a 2-fold axis of the particle, a Mercator projection of the array, part of which is shown in Plates II(b) and IV(c), is the most convenient two-dimensional representation.

The relationships discussed above are for the point group 532, but from the treatment given it is obvious how to construct the appropriate antisymmetric point group relating the superposition patterns of particles whose symmetry is that of other point groups—one simply adds an anticentre, as defined above, to the point group. When the point group has a centre of symmetry—a case sometimes appropriate for a biological particle viewed at low resolution—the superposition patterns are also centrosymmetric and so the effect of the change of quality is nil. The array of superposition patterns will in this case have the same symmetry as the original point group.

We thank Miss Angela Campbell and Miss Marion Holder for much technical assistance in this work.

REFERENCES

Caspar, D. L. D. (1966). *J. Mol. Biol.* **15**, 365.
Caspar, D. L. D. & Klug, A. (1962). *Cold Spr. Harb. Symp. Quant. Biol.* **27**, 1.
Caspar, D. L. D. & Klug, A. (1963). *Viruses, Nucleic Acids and Cancer* (17th M.D. Anderson Symposium), p. 27. Baltimore: Williams & Wilkins Co.
Finch, J. T. & Klug, A. (1965). *J. Mol. Biol.* **13**, 1.
Finch, J. T. & Klug, A. (1966). *J. Mol. Biol.* **15**, 344.
Finch, J. T. & Klug, A. (1967). *J. Mol. Biol.* **24**, 289.
Finch, J. T., Klug, A. & Stretton, A. O. W. (1964). *J. Mol. Biol.* **10**, 570.
Klug, A. & Berger, J. E. (1964). *J. Mol. Biol.* **10**, 565.
Klug, A. & Caspar, D. L. D. (1960) *Advanc. Virus Res.* **7**, 225.
Klug, A. & DeRosier, D. J. (1966). *Nature*, **212**, 29.
Klug, A. & Finch, J. T. (1965). *J. Mol. Biol.* **11**, 403.
Shubnikov, A. V. (1964). In *Colored Symmetry*, ed. by W. T. Holser. Oxford: Pergamon Press.
Valdrè, U. (1962). *J. Sci. Inst.* **39**, 278.

J. Mol. Biol. (1968) **35**, 143–164

Structure of Crystalline α-Chymotrypsin

II.† A Preliminary Report Including a Hypothesis for the Activation Mechanism

P. B. Sigler‡, D. M. Blow, B. W. Matthews§ and R. Henderson

*Medical Research Council Laboratory of Molecular Biology
Hills Road, Cambridge, England*

(*Received 2 February 1968*)

An electron density map of tosyl-α-chymotrypsin at 2 Å resolution is presented, which shows the conformation of the polypeptide chain. An electron density map of the differences between the tosylated and the native enzyme has also been calculated. These maps lead to the following conclusions.

Histidine 57 and serine 195 are known to be part of the active site. In the native enzyme, these are in an environment open to the solvent and in a conformation consistent with the existence of a hydrogen bond between them. In the tosylated enzyme, small movements of histidine 57, serine 195 and methionine 192 bring histidine 57 into a position where it can interact with the sulphonyl group. No other conformational changes are observed.

The enzyme contains an ion pair between the α-amino group of isoleucine 16 and the β-carboxyl group of aspartate 194, located in an otherwise non-polar cavity. In conjunction with kinetic and spectroscopic studies (especially Oppenheimer, Labouesse & Hess, 1966) and X-ray analysis of chymotrypsinogen and δ-chymotrypsin at low resolution (Kraut, Sieker, High & Freer, 1962; Kraut, Wright, Kellerman & Freer, 1967), our results lead to a hypothesis for the stereochemistry of the activation process. It is proposed that (i) activation of the zymogen involves no gross reorganization of the main chain nor a significant helix–coil transition; (ii) activation involves a structural change of the enzyme, caused by the formation of an ion-pair between isoleucine 16 and aspartate 194. This structural change would be reversed when the positively charged α-amino group of isoleucine 16 is deprotonated at high pH. The reversal seems to be sterically blocked when the enzyme is inhibited by a bulky group on serine 195.

In the light of our model, the location of the disulphide bridges in the chemical sequence of trypsin suggests that trypsin and chymotrypsin have nearly identical tertiary structures, and that their disulphide bridges serve to stabilize rather than to determine the structure.

The interactions between the molecules related by the non-crystallographic dyad axes in the crystal are described. One of these involves the active site, but does not inhibit enzymic activity. Of the heavy atoms used for phase determination by the method of isomorphous replacement, platinum(II)- and mercury-containing substituents are bound to sites containing cystine and methionine residues.

† Paper I in this series is Matthews, Sigler, Henderson & Blow (1967).
‡ On leave from the Department of Biophysics, University of Chicago, Chicago, Ill., U.S.A.
§ Present address: National Institute of Arthritis and Metabolic Diseases, Bethesda, Md., U.S.A.

1. Introduction

Alpha-chymotrypsin is a protease of molecular weight 25,000 which crystallizes from 2 M-ammonium sulphate at pH 4 in the monoclinic space group $P2_1$ (Bernal, Fankuchen & Perutz, 1938). Tosyl-α-chymotrypsin is an irreversibly inhibited derivative in which the active serine is sulphonylated (Fahrney & Gold, 1963; Strumeyer, White & Koshland, 1963; Kallos & Rizok, 1964). Crystals of tosyl-α-chymotrypsin are isomorphous with the native enzyme, having unit cell dimensions $a = 49.3$ Å, $b = 67.3$ Å, $c = 65.9$ Å, $\beta = 101.8°$ (Sigler, Jeffery, Matthews & Blow, 1966).

We recently published a preliminary account of an electron density map of tosyl-α-chymotrypsin at 2 Å resolution (Matthews, Sigler, Henderson & Blow, 1967) (paper I). The electron density map was obtained by the isomorphous replacement method. It provided an image of the molecular structure in which the course of the polypeptide chains could be traced, and many side chains, especially those of cystine and the aromatic amino acids, could be clearly recognized. Some parts of the map were very much clearer than others. The detailed interpretation of the map relied heavily on the determination of the amino-acid sequence by Hartley (1964).

The crystallographic asymmetric unit of α-chymotrypsin contains two molecules which are related by a system of local twofold axes parallel to the crystallographic a^* direction (Blow, Rossmann & Jeffery, 1964). An electron density map (Fig. 1) obtained by averaging the crystallographically independent electron densities at pairs of points related by such a local twofold axis was found to be considerably easier to interpret than the map which showed the independent molecules separately. This indicates that many of the difficulties of interpretation are still due to inaccuracies of the crystallographic analysis, and before a final interpretation is made we wish to make further measurements. In the meanwhile, we wish to make available more complete information about our present results.

FIG. 1. Electron density map obtained by averaging the electron density from a Fourier synthesis for crystalline α-chymotrypsin about an appropriate local twofold axis (paper I, Matthews *et al.*, 1967).

The contours represent increments of approximately 0.3 el/Å³ and the lowest contour is drawn at approximately 1.0 el/Å³. The round black dots represent the positions of the local twofold axes, which pass perpendicular to the sections on the lines $z = 0.156x + 0.475$, $y = 0$ and $z = 0.156x + 0.975$, $y = 0$. Dyad A is at the top.

Since the local twofold axes are perpendicular to the y- and z-axes, and parallel to a^*, these directions form a convenient orthogonal system which has been used for measurements on the "averaged" electron density map. The vertical dashed line is drawn at $y = 0$. The horizontal dashed lines serve to delineate a rectangular block in the electron-density map which contains almost the whole of one molecule, and which was used in model building. The number on each section represents the x co-ordinate in 1/64 of the unit cell; the perpendicular spacing between the sections is 0.75 Å. The positive direction of x is into the paper.

To assist in converting the co-ordinate system back to conventional crystallographic co-ordinates, small crosses (+) are drawn on each section at the intersections of $y = 0$, ± 0.20 with $z = 0.45$, 1.05. These are the crystallographic co-ordinates in one of the two independent molecules (designated "Molecule 2") corresponding to the indicated position in the averaged map. The indicated directions for positive y and z refer to this molecule.

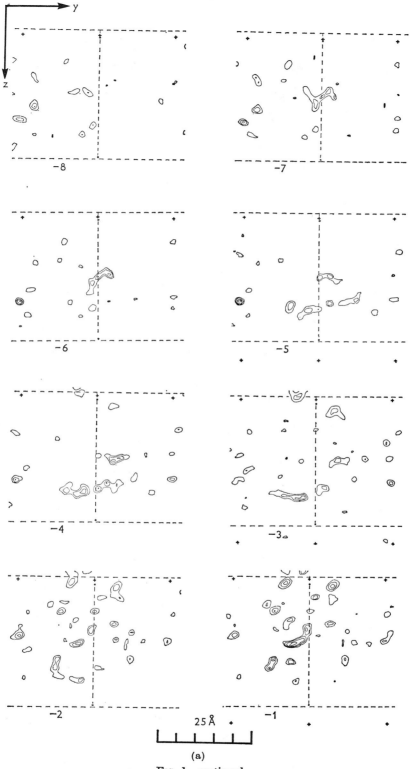

(a)

Fɪɢ. 1—*continued*

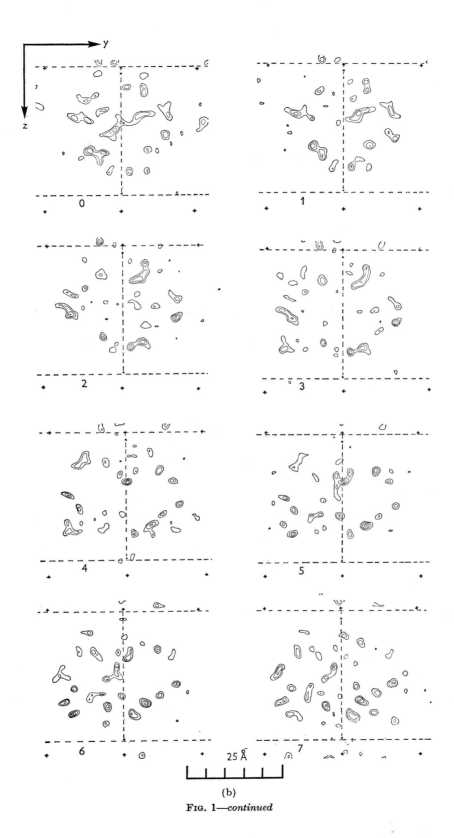

(b)

FIG. 1—*continued*

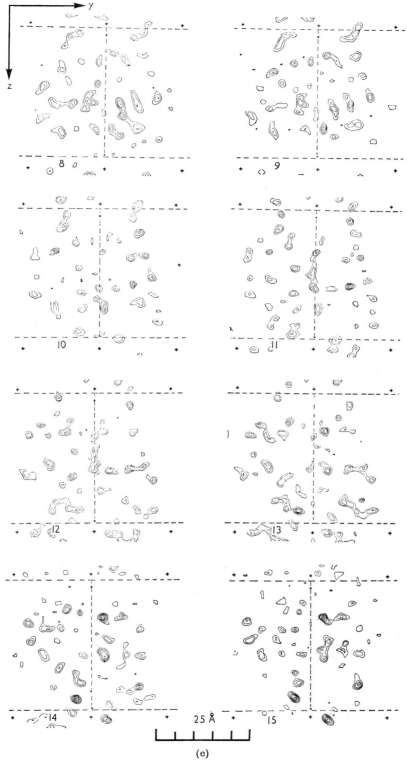

(c)

FIG. 1—*continued*

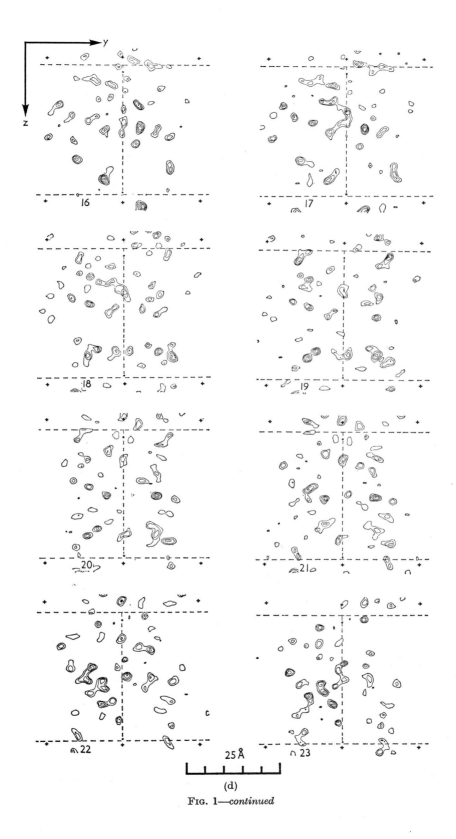

(d)

FIG. 1—*continued*

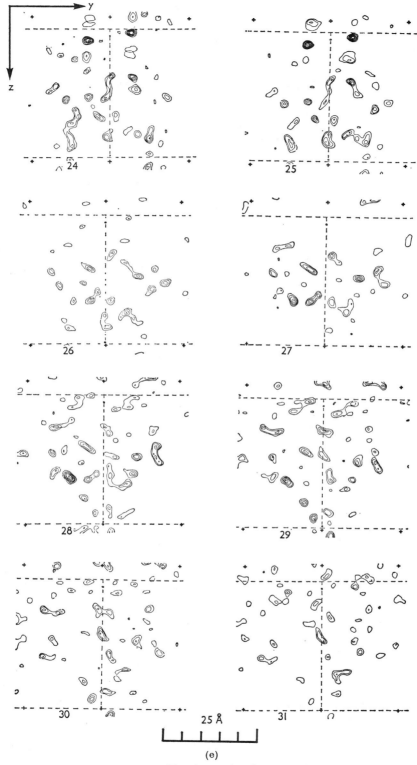

(e)

FIG. 1—*continued*

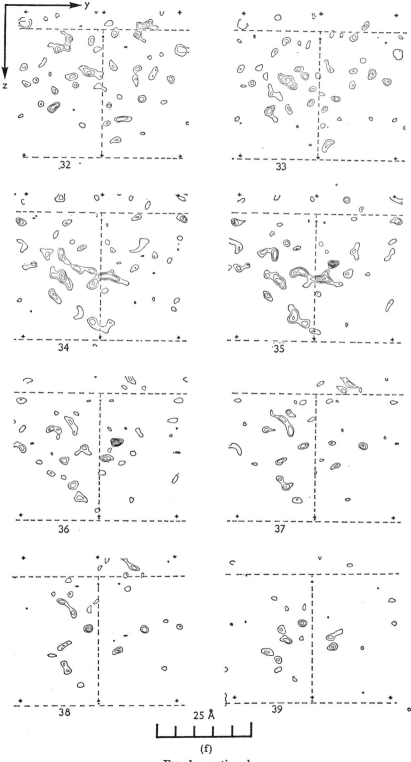

25 Å

(f)

Fɪɢ. 1—continued

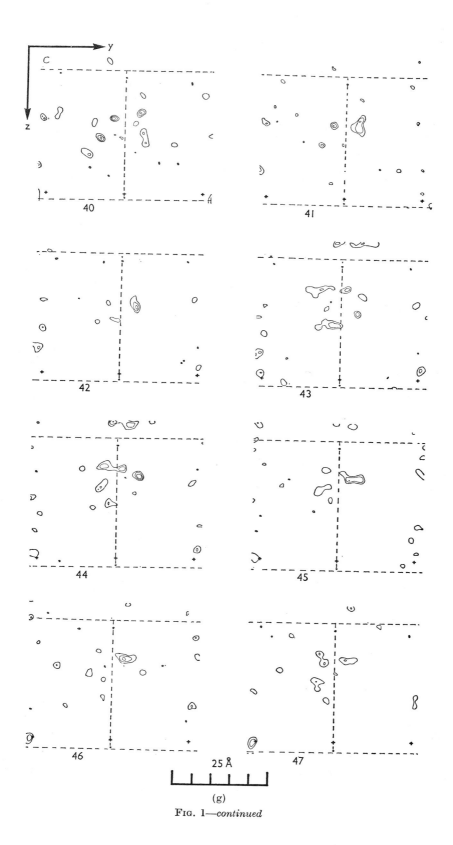

25 Å

(g)

FIG. 1—continued

2. Electron Density Map

Figure 2 indicates the course of the polypeptide chains through the molecule. This Figure summarizes our current interpretation of the map of the average electron density of the two independent molecules. We can see no significant difference in the folding of the main chain of the two crystallographically independent molecules, even in regions where the local environments are different. The electron density variations in the map become very much weaker at the extreme "top" (low values of x) and "bottom" (high values of x) of the molecule as viewed in Figure 6. This leads to some uncertainty about the exact course of the main chain in these regions, indicated by a dashed line in Figure 2.

Also indicated on Figure 2 are the *approximate* positions of most of the α-carbon atoms of the molecule. It must be emphasized that in addition to uncertainties of interpretation, the model-building technique will have introduced inaccuracies, and some of these positions may be wrong by as much as 1·5 Å. The positions of α-carbon atoms 30 to 40, 70 to 91 and 172 to 179 (numbering system as for chymotrypsinogen A, see Hartley, 1964; Hartley & Kauffman, 1966) are uncertain, and are omitted from the Figure.

3. Heavy-atom Binding Sites

Two of the isomorphous derivatives used in the crystallographic analysis depend upon binding of ions to specific points in the crystal structure. In all cases but one, these co-ordination complexes can be interpreted in terms of the interaction of the ion with specific functional groups of the protein.

(a) *The chloroplatinite ion* $(PtCl_4^{2-})$†

There are three distinct binding sites for Pt(II) on each enzyme molecule. Because of the local symmetry, these appear as pairs of binding sites in the crystal structure. In one pair of sites (A and B in the notation of Sigler *et al.*, 1966), Pt(II) appears to interact with the terminal amino group of cystine 1–127 and at least one of the two sulphur atoms of the disulphide bridge. In another pair of sites (D and E), the Pt(II) is situated close to the sulphur atoms of the methionine 192 side chains of two adjacent molecules. These sites are only 4 Å apart, straddling the local twofold axis. The occupancy in sites D and E is approximately 0·5, suggesting that the binding of Pt(II) to one site excludes it from the other. The binding of Pt(II) at site C is now known to consist of two distinct binding sites centred less than 1 Å apart. The occupancy at these two sites is similar to that observed at sites D and E. The interaction of Pt(II) with the protein at site C is not understood at present.

(b) *Phenylmercuric acetate*

These sites are close to sites A and B and evidently also involve interactions with the terminal α-amino group and the disulphide bridge 1–127.

All these sites, therefore, involve the interaction of a heavy-metal ion with a sulphur-containing group. The binding of Pt(II) to methionine and cystine in chymotrypsin is consistent with the co-ordination behaviour observed between Pt(II) and sulpho-organic compounds (McAuliffe, 1967; Haake & Turley, 1967). Binding of Pt(II) to methionine has also been noted in ribonuclease-S (Wyckoff *et al.*, 1967).

† The formula $PtCl_4^{2-}$ refers to the chemical composition of the ion when added to the crystal system and is not meant to imply that this is necessarily the Pt(II)-complex bound in the crystal structure.

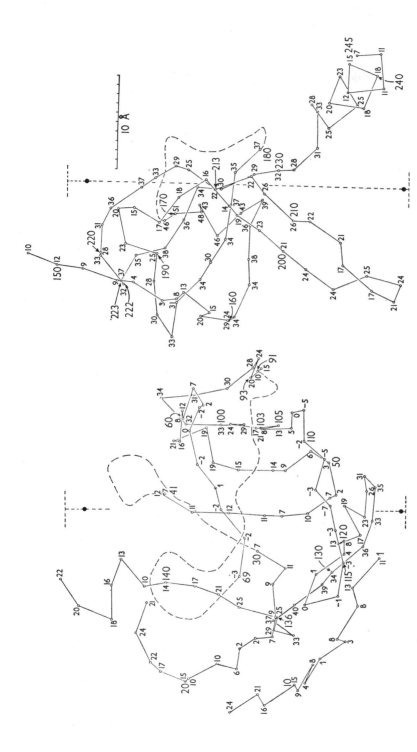

FIG. 2. Each circle represents the *approximate* position of an α-carbon atom, and the figure beside it represents the *x* co-ordinate in units of 1/64 of the unit cell (0·75 Å). The positions of the local twofold axes are indicated, and the larger figures indicate positions in the polypeptide chain (Hartley, 1964; Hartley & Kauffman, 1966).

Although this Figure represents the general course of the polypeptide chain with reasonable accuracy, the positions of the α-carbon atoms do not carry any meaningful implications of the orientations of the peptide bonds or of the amino-acid side chains. (a) A and B chains; (b) C chain.

4. Intermolecular Interactions

In crystalline α-chymotrypsin, long chains of molecules in close contact run in the z direction. The contacts between adjacent molecules in this direction are arranged about two types of parallel local twofold axes, which alternate along the chains (Blow et al., 1964). The "averaged" electron density map shows clearly only those features of the two molecules in the asymmetric unit which are exactly related by the local symmetry. Since the environments of these molecules are different, many details of the intermolecular interactions are obscured in the map. However, the interactions between pairs of dyad-related molecules are the same on each side of the local twofold axes which pass between them and are thus seen clearly in the averaged map.

One of these two types of interaction brings the active centres of two molecules close to one another, so that the sulphonyl groups of adjacent tosylated molecules are only 11·9 Å apart (Sigler et al., 1966). The intermolecular interactions which occur about this axis (dyad A) appear to be the most well-developed in the crystal structure (Table 1).

TABLE 1

Molecular interactions about local twofold axes†

Interactions about dyad A

α-NH$_3^+$ (alanine 149)	β-COO$^-$ (aspartate 64)
α-Carboxyl (tyrosine 146)	peptide bond 57–58
Phenol (tyrosine 146)	imidazole (histidine 57)
Side chain (methionine 192) . . .	side chain (methionine 192)

Interactions about dyad B

Amide (asparagine 204)	amide (glutamine 240)
Hydroxyl (serine 125)	carboxyl (aspartate 128)

† The interacting residues are in close proximity. The exact nature of the interactions is not always clear.

Two residues which contribute greatly to this intermolecular interaction are the carboxyl terminus of the B chain (tyrosine 146), and the amino terminus of the C chain (alanine 149). Since these termini do not exist in δ-chymotrypsin or in the zymogen, neither the tetragonal crystal form of the enzyme (in which π-, δ- and γ-chymotrypsin are found) nor the crystalline zymogen could be expected to exhibit this type of molecular interaction (Fig. 3).

The intermolecular contact about dyad A is of particular interest because of the proximity of the active centres to the interacting groups. From Figure 4 it can be seen that the phenolic ring of tyrosine 146 is inserted "below" the imidazole substituent histidine 57 in the adjacent molecule. This situation explains the observation that in *solution*, tyrosine 146 is sufficiently exposed to be rapidly attacked by carboxypeptidase (Gladner & Neurath, 1954), and is the most rapidly iodinated residue (Glazer & Sanger, 1964; Dube, Roholt & Pressman, 1964), whereas in *crystals* of α-chymotrypsin tyrosine 146 is completely unreactive towards I$_3^-$ under conditions where tyrosine 171 is completely di-iodinated (P. B. Sigler, manuscript in preparation). Tyrosine 146 of

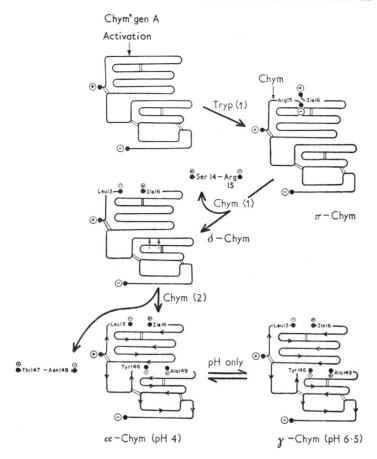

FIG. 3. Activation scheme for the chymotrypsins (Desnuelle, 1960; Corey, Battfay, Brueckner & Mark, 1965).

the dyad-related molecule is in a position which might be expected to interfere with the enzymic activity of its neighbour (Fig. 4). However, di-isopropyl fluorophosphate and sulphonyl fluorides readily acylate serine 195 in crystalline α-chymotrypsin (Sigler *et al.*, 1966) and preliminary experiments (in collaboration with Mrs S. Banks) have shown that the crystalline enzyme catalyses the hydrolysis of *N*-acetyl-L-tyrosine hydrazide in the crystal at about 7% of the rate observed in solution (see also Kallos, 1964). It follows that the interaction of tyrosine 146 with histidine 57 neither prevents the access of these bulky molecules to the catalytic centre nor disrupts the catalytic mechanism. These substrate and inhibitor molecules are considerably bulkier than the I_3^- ion, which is apparently excluded from access to tyrosine 146. This implies that the region occupied by tyrosine 146 is neither a binding site nor a necessary access route for a substrate molecule. In this way the crystal packing arrangement may provide a valuable clue to the route of the substrate to the active site.

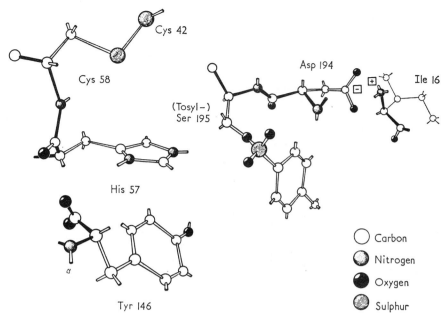

FIG. 4. Probable stereochemistry of selected residues in the vicinity of the active site.

The polypeptide backbone is indicated by a solid black line. The tyrosine residue at lower left belongs to a different molecule and it is not known whether its carboxyl group is charged under the conditions of crystallization.

Note added in proof: Further study of the electron density distribution has indicated that the phenolic part of tyrosine 146 lies in a plane parallel to the imidazole ring of the adjacent histidine 57, rather than the titled orientation shown in this Figure.

5. Difference between Native and Inhibited Enzyme

A three-dimensional map at 2 Å resolution of the differences in electron density between tosyl-α-chymotrypsin and the native enzyme shows significant features only in the vicinity of the active centre. None of the movements are more than 1 Å. In Figure 5 we show the important features of the difference map, superimposed on the relevant sections of the original map. These features may be interpreted as follows:

(a) the sulphonyl group of the inhibitor;

(b) movement of the β-oxygen of serine 195 into a position consistent with the formation of a sulphonyl ester;

(c) movement of histidine 57. (When the inhibitor is bound there is some rotation about the α–β bond, so that in the Figure the lower side of the imidazole ring is moving towards the reader, while the upper side moves into the paper.)

(d) movement of methionine 192.

It will be noted that the largest side-chain movement is that of methionine 192.

6. Active Centre Interactions

In tosyl-α-chymotrypsin the distance between the centre of the imidazole ring of histidine 57 and the centre of the sulphonyl group is 4·7 Å. In the native enzyme, the centre of the imidazole ring is less than 4 Å from the oxygen of serine 195. These

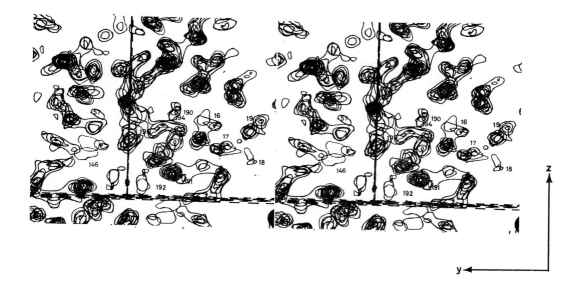

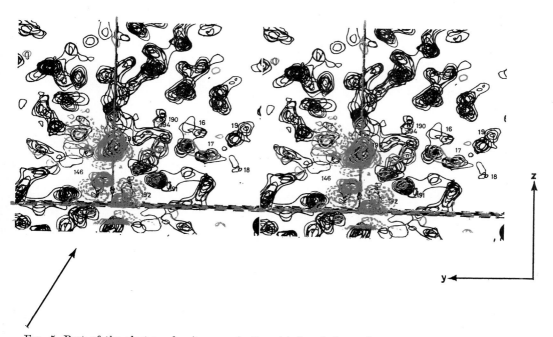

Fig. 5. Part of the electron density map in the vicinity of the active centre, including the sections $x = 17$ to 24. Most of the residues in Fig. 4 are included. In blue is shown the difference of density between the tosyl- and the native enzyme. The contours of the difference map are separated by about 0.04 el/Å³; the zero contour is omitted, and negative contours are dashed. For clarity, in the vicinity of the sulphonyl group, five higher contours of difference density are omitted. The arrow indicates the approximate view-point from which Figs 4 and 6 are drawn.

conformations are clearly compatible with the existence of a hydrogen bond between $N^{\epsilon 2}$ of histidine 57† and the oxygen of serine 195 in the native enzyme; or between $N^{\epsilon 2}$ of histidine 57 and the sulphonyl group in the inhibited enzyme.

In tosyl-α-chymotrypsin the region "behind" this residue (as seen in Fig. 4) seems rather loosely packed, and the region "above" it is mostly exposed to solvent, apart from an interaction with the sulphur of cystine 58. The "lower" side of histidine 57 is also on the surface of the molecule; but, as already mentioned, this region is largely filled by the side chain of tyrosine 146 from an adjacent molecule.‡

7. Homology between Chymotrypsin and Trypsin

Implicit in any discussion involving sequence homology is the doctrine that extensive similarity in amino acid sequence reflects a comparable similarity in tertiary structure (Perutz, Kendrew & Watson, 1965). The homology which exists between chymotrypsin and trypsin (Walsh & Neurath, 1964; Hartley, Brown, Kauffman & Smillie, 1965) affords an opportunity to check the degree of similarity at atomic resolution between the tertiary structures of two highly homologous proteins.

There are two disulphide bridges in trypsin which are not present in chymotrypsin. The two pairs of residues in chymotrypsin which, by homology, are replaced by cystine in trypsin are residues 22–157 and 127–232. The positions of these residues in the model of tosyl-α-chymotrypsin are represented by the ends of the thick arrows in Figure 6. In our rough model the distance between the α-carbon atoms of residues 22 and 157 is 5 Å; and between residues 127 and 232 is 7 Å. These values may be compared with the average distance 5·2 Å between the α-carbon atoms of the five cystine residues in the same rough model. In other words, the disulphide bridges which are present only in trypsin could be built into our model of tosyl-α-chymotrypsin with scarcely any movement of the main-chain atoms. This indicates not only that trypsin and chymotrypsin have nearly identical tertiary structures, but also demonstrates that the structure is independent of the presence of the disulphide bridges.

The direct observation of this structural homology is consistent with the classic experiments of Anfinsen and his colleagues (e.g. White, 1960,1961) on the renaturation of ribonuclease, which showed that the pairing of the correct half-cystines is a consequence of the equilibrium conformation of the protein (Epstein, Goldberger & Anfinsen, 1963).

In the cell, where conditions favour sulphydryl exchange, these disulphide bridges are doubtless highly reversible, allowing the protein to assume an optimal conformation. The disulphide bridges evidently do not "determine" the structure in the thermodynamic sense, but serve a kinetic role as chemical "staples" which stabilize the structure in an oxidative, highly variable extracellular environment.This appears to be the functional basis of the observation, first suggested to us by Dr B. S. Hartley, that the cystine bridge seems to be a characteristic structural feature of the *extracellular* proteins of animals.

---

† $N^{\epsilon 2}$ refers to the nitrogen of the imidazole ring further from the β-carbon (Edsall *et al.*, 1966).

‡ Since bond 146–147 is one of those hydrolysed by chymotrypsin in the transition from δ- to α-chymotrypsin (Fig. 3), it is tempting to suppose that this type of interaction might represent an enzyme–product complex. This seems impossible since the carboxyl group representing the site of bond cleavage is fully 8 Å from the oxygen of serine 195. Thus, although the cleavage of peptide bond 146–147 of δ-chymotrypsin must be accompanied by the binding of tyrosine 146 by another chymotrypsin molecule as an enzyme–substrate complex, such an interaction must be of a different type from that observed here.

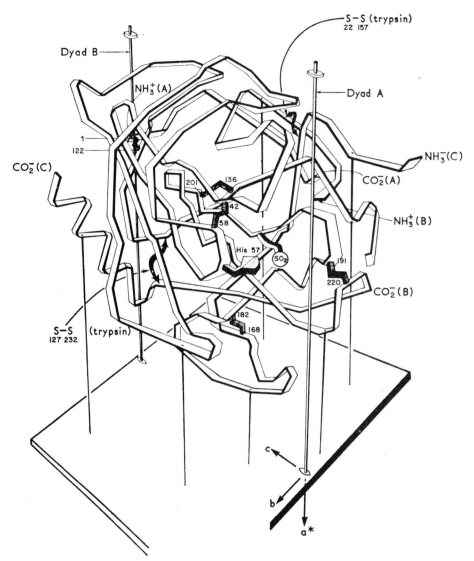

Fig. 6. A schematic drawing representing the conformation of the polypeptide chains in α-chymotrypsin. The positions of amino acids in positions homologous with the two additional cystines present in trypsin are shown. This drawing has been revised from the similar drawing in paper I (Matthews *et al.*, 1967) to improve its clarity and accuracy.

8. Stereochemistry of the Activation Process

The activation of chymotrypsinogen to chymotrypsin is triggered by the tryptic cleavage of the arginyl(15)–isoleucine(16) bond (Rovery & Desnuelle, 1953; Dreyer & Neurath, 1955) (Fig. 3). From the work of Hess and his collaborators (especially Oppenheimer, Labouesse & Hess, 1966) it appears that it is specifically the formation of a new terminal amino group, rather than any "unblocking" reaction, which is responsible for promoting enzymic activity. The role of the newly formed α-amino group of isoleucine 16 in stabilizing the enzymically active conformation was indicated briefly in paper I (Matthews *et al.*, 1967) where it was shown that the negatively charged β-carboxylate group of aspartate 194 and the positively charged α-amino group of isoleucine 16 form an ion-pair in a region of presumably low dielectric constant (Fig. 4).

We feel there is now sufficient structural information to suggest the following hypothesis for the stereochemistry of the activation process:

(i) *The over-all tertiary structure of the zymogen is similar to that of the active enzyme: there is no gross reorganization in the folding of the main chain nor a significant helix–coil transition.*

As noted in paper I (Matthews *et al.*, 1967), the positions of the chain termini created during the activation and subsequent autolysis are consistent with the cleavage of two di-peptides from the surface of a zymogen with fundamentally the same structure. Kraut *et al.* (1967) carried out a direct comparison of the low-resolution electron-density maps of chymotrypsinogen and δ-chymotrypsin. They found a close correspondence between almost all prominent features of these maps. By correlating these maps with the high-resolution map of α-chymotrypsin, Wright, Kraut & Wilcox (1968) were able to confirm the two regions where changes were observed as (A) the region of tryptic cleavage and (B) the general region of threonine 147 and asparagine 148.

Since the initial observation of Neurath, Rupley & Dreyer (1956) that a change in specific rotation parallels the activation process, there have been many optical rotatory dispersion studies. Raval & Schellman (1965) and Biltonen, Lumry, Madison & Parker (1965) ascribed changes in rotation to additional Cotton effects not associated with α-helices and concluded that the structural transition from zymogen to enzyme entailed little or no change in α-helix content. These conclusions are consistent with our interpretation. Various authors, however, have reached different conclusions, and most recently Fasman, Foster & Beychok (1966) have suggested, principally from measurements of circular dichroism, that the enzyme gains at least two turns of α-helix on conversion from the zymogen. This conclusion is not consistent with the interpretation of the X-ray diffraction results.

In the high-resolution map of α-chymotrypsin, the only region identified as an α-helix of more than one turn is the carboxyl-terminus of the C chain. This is the part of the molecule farthest from the chain termini formed on activation (Fig. 2). In the corresponding region of the low-resolution map of chymotrypsinogen (Kraut *et al.*, 1962) there is a strong feature consistent with the existence of a comparable length of α-helix. In the absence of a dramatic change in the over-all shape of the molecule, it is difficult to envisage any mechanism whereby this α-helix region could be significantly increased during the activation process.

We conclude that these changes in optical properties arise from alterations in the local stereochemistry of individual chromophores rather than changes of secondary structure.

(ii) *The stereochemistry of the activation of the zymogen is paralleled in the enzyme by a pH-dependent structural transition. In both transitions activity depends upon the integrity of the ion-pair between isoleucine 16 and aspartate 194.*

Hess and his collaborators have identified a pH-dependent structural transition in the native enzyme between an "active" and an "inactive" form. The active form dominates the equilibrium below pH 8 and is characterized functionally as the form which binds specific substrates in a productive mode (Himoe, Parks & Hess, 1967). This form can be recognized physically by its characteristic optical rotation. The other conformational state, the "inactive" form, is favoured by high pH and is unable to bind specific substrates and inhibitors in a productive mode. This inability is considered to be the basis for the fall-off in enzymic activity at pH > 8. Since the optical rotation is different from that of the active form, but similar to that of the zymogen, the inactive form is felt to resemble chymotrypsinogen in its tertiary structure.

The group with pK_a 8 to 9 which controls this transition has been identified as the α-amino group of isoleucine 16 (Oppenheimer et al., 1966; Ghelis, Labouesse & Labouesse, 1967). By treating the zymogen with acetic anhydride prior to activation, these workers prepared a derivative, acetyl-δ-chymotrypsin, in which all amino groups except this one were acetylated. This derivative was shown to be fully active at neutral pH, and to undergo the transition to an inactive form at high pH. Re-acetylation of the derivative, so as to block the α-amino group, inactivated the enzyme.

The di-isopropylphosphoryl-enzyme, which resembles the "active" form in its optical rotation, does not exhibit a structural transition at high pH. Titration studies indicate that the titratable proton of the α-amino group of isoleucine 16 is locked into the structure following esterification of serine 195 with di-isopropyl fluorophosphate (Oppenheimer et al., 1966, and references therein).

The stereochemical basis of this active–inactive equilibrium and its relationship to the activation process is schematically summarized in Figure 7. The relative positions of the relevant groups in tosyl-α-chymotrypsin are depicted in Figures 4 and 5. The active conformation is stabilized by the ion-pair formed by aspartate 194 and isoleucine 16. The existence of the ion-pair accounts for the anomalously high pK_a of the α-amino group of isoleucine 16, by comparison with the pK_a of less than 8 which is usual for peptide α-amino groups (Edsall & Wyman, 1958, p. 470). If the positive charge is removed from the amino group, the high potential created by an isolated negative charge in a medium of low dielectric constant within the molecule would require the carboxylate ion to seek an alternative orientation, in which it could point into the solvent. The model suggests that this could be accomplished by a movement of the carboxylate group into the vicinity of the side groups of serine 195 and histidine 57. We cannot predict the exact nature of the disruption of the active site which would follow from this, but we suggest that at high pH the conformation in the region of the active centre reverts to that of the zymogen. This accounts neatly for similarity in the optical properties and absence of enzymic activity in both proteins. It would follow that the zymogen and the inactive form lack activity for the same stereochemical reasons.†

† Several studies have been made of the binding of substrates and inhibitors to the enzyme as a function of pH (Deranleau & Neurath, 1966; Johnson & Knowles, 1967); and of binding to the zymogen (Vaslow & Doherty, 1953; Weiner, White, Hoare & Koshland, 1966). The evidence drawn from these studies is conflicting, and inconclusive. This is largely because in any study of an inactive analogue of the enzyme there is no direct way of defining "specific" binding, to distinguish it from other forms of molecular association.

Ghelis et al. (1967) recently reported preliminary experiments on δ-chymotrypsin in which all α-amino groups including that of isoleucine 16 have been acetylated, indicating that this material specifically hydrolyses nitrophenyl acetate. They point out that this finding disagrees with that of Scrimger & Hofmann (1967), who found that trypsin, after the α-amino group has been destroyed with nitrous acid, is unable to react with di-isopropyl fluorophosphate.

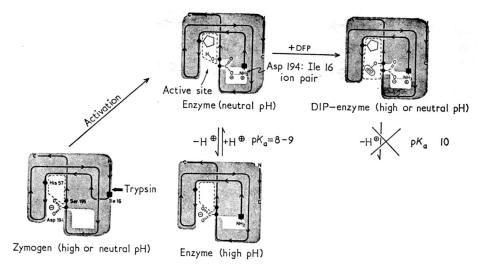

FIG. 7. Schematic diagram for the activation hypothesis.

The diagram at upper centre represents active chymotrypsin, in which the active site, including serine 195 and histidine 57, is intact. Deprotonation of the α-amino group of isoleucine 16 causes the aspartate 194 side chain to swing out to the surface (below). This disrupts the activity in an unknown way, represented by leaving part of the active site blank. The active site region is now the same as it is in the zymogen.

Attachment of a DIP-group to serine 195 blocks the aspartate 194 side chain from reaching the surface, locking the enzyme into the active conformation, and making deprotonation of the α-amino group much more difficult.

Himoe *et al.* (1967) have presented clear evidence that k_{cat}, which expresses the rate of catalysis when the enzyme is saturated with substrate, is effectively constant in the pH range 7 to 10. They find that the change of binding constant, K_m (apparent), for a specific amide substrate, is responsible for the loss of enzymic activity at high pH. Our hypothesis is that the inactive form of the enzyme is stereochemically like the zymogen, which possesses no catalytic activity. The above findings are consistent with the hypothesis if the zymogen is *totally* unable to bind substrate at the catalytic site. The following kinetic mechanism would then exist:

$$\begin{array}{c} I + S \\ K_I \Updownarrow \\ E + S \underset{K_s}{\overset{}{\rightleftharpoons}} ES \overset{k_0}{\longrightarrow} E + P \end{array}$$

In the pH range 7 to 10, the equilibrium constant K_I will be such that finite concentrations of the active enzyme E and the inactive form I are in equilibrium. However, as the substrate concentration $[S]$ is increased to $\gg K_m$, the ES form is favoured, so that even the I form is reduced to a negligible concentration. Thus

$$k_{cat} \left(\text{defined as} \lim_{[S] \to \infty} \frac{k_0 [S]}{K_m + [S]} \right) = k_0,$$

just as in the simple Michaelis–Menten case, but the apparent K_m is modified by the existence of the inactive form.

If the inactive form possesses a substrate-binding site the affinity of which for the substrate is K_s, but which is unable to catalyse the ES complex to products, a very different result follows. In this case *both* the apparent K_m, and k_{cat} are altered, as the available substrate is distributed between the two forms.

These simplified cases are both special cases of the scheme presented by Himoe *et al.* (1967).

Figure 7 also indicates the probable reason that the titratable proton becomes sequestered upon irreversible phosphorylation of serine 195. We suggest that the fixing of a bulky group to serine 195 sterically blocks the rotation required to bring the isolated carboxylate group to the surface upon deprotonation of the amino group of the ion pair (Fig. 4). In effect, the phosphoryl ester acts as a bulky "pin" to keep the carboxylate group pointed inward, thereby providing an electrostatic clamp to hold the proton, if not the protonated amino group, within the native protein.

The other part of the structural transition which occurs on activation must be the entry of the N-terminal isoleucine from a surface position into the environment where the ion-pair is formed. In α-chymotrypsin, isoleucine 16 and valine 17 make several hydrophobic interactions with threonine 138, alanine 158, leucine 160 and valine 188, and presumably this interaction energy provides part of the driving force for the activation transition. The homologous residues in trypsin (Ile-Val) and elastase (Val-Val) are also hydrophobic (Walsh & Neurath, 1964; Brown, Kauffman & Hartley, 1967). The transition of chymotrypsin at high pH does not necessarily parallel this part of the activation transition, since once the charge on the terminal amino group is abolished it may well remain internal, as indicated in Figure 7.

The electron density map presented in Fig. 1 depends on contributions to earlier stages of the work made by Dr M. G. Rossmann and Miss B. A. Jeffery. Invaluable technical assistance was given by Mrs J. Dawes, Miss P. Masters, Miss D. Singleton, Mrs S. Simpson and Mrs S. Wickham, and computer programs by R. A. Crowther, T. H. Gossling and Dr Hilary Muirhead were used. One of us (P. B. S.) was a Helen Hay Whitney Fellow and another (R. H.) is a Medical Research Council Scholar.

We are grateful to Dr B. S. Hartley, Professor G. P. Hess and Dr J. R. Knowles for discussion, and to Dr M. F. Perutz for his comments on the manuscript.

REFERENCES

Bernal, J. D., Fankuchen, I. & Perutz, M. F. (1938). *Nature*, **141**, 523.

Biltonen, R., Lumry, R., Madison, V. & Parker, H. (1965). *Proc. Nat. Acad. Sci., Wash.* **54**, 1412.

Blow, D. M., Rossmann, M. G. & Jeffery, B. A. (1964). *J. Mol. Biol.* **8**, 65.

Brown, J. R., Kauffman, D. L. & Hartley, B. S. (1967). *Biochem. J.* **103**, 497.

Corey, R. B., Battfay, O., Brueckner, D. A. & Mark, F. G. (1965). *Biochim. biophys. Acta*, **94**, 535.

Deranleau, D. & Neurath, H. (1966). *Biochemistry*, **15**, 1413.

Desnuelle, P. (1960). In *The Enzymes*, ed. by P. Boyer, H. Lardy & K. Myrbäck, vol. 4, chap. 5. New York: Academic Press.

Dreyer, W. J. & Neurath, H. (1955). *J. Biol. Chem.* **217**, 527.

Dube, S. K., Roholt, O. A. & Pressman, D. (1964). *J. Biol. Chem.* **239**, 3347.

Edsall, J. T., Flory, P. J., Kendrew, J. C., Liquori, A. M., Némethy, G., Ramachandran, G. N. & Scheraga, H. A. (1966). *J. Mol. Biol.* **15**, 395.

Edsall, J. T. & Wyman, J. (1958). *Biophysical Chemistry*, vol. 1. New York: Academic Press.

Epstein, C. J., Goldberger, R. F. & Anfinsen, C. B. (1963). *Cold Spr. Harb. Symp. Quant. Biol.* **28**, 439.

Fahrney, D. E. & Gold, A. M. (1963). *J. Amer. Chem. Soc.* **85**, 997.

Fasman, G., Foster, R. J. & Beychok, S. (1966). *J. Mol. Biol.* **19**, 240.

Ghelis, C., Labouesse, J. & Labouesse, B. (1967). *Biochem. Biophys. Res. Comm.* **29**, 101.

Gladner, J. A. & Neurath, H. (1954). *J. Biol. Chem.* **206**, 911.

Glazer, A. S. & Sanger, F. (1964). *Biochem. J.* **90**, 92.

Haake, P. & Turley, P. C. (1967). *J. Amer. Chem. Soc.* **89**, 4611.

Hartley, B. S. (1964). *Nature*, **201**, 1284.

Hartley, B. S., Brown, J. R., Kauffman, D. L. & Smillie, L. B. (1965). *Nature*, **207**, 1157.

Hartley, B. S. & Kauffman, D. L. (1966). *Biochem. J.* **101**, 229.

Himoe, A., Parks, P. C. & Hess, G. P. (1967). *J. Biol. Chem.* **242**, 919.

Johnson, C. H. & Knowles, J. R. (1967). *Biochem. J.* **103**, 428.

Kallos, J. (1964). *Biochim. biophys. Acta*, **89**, 364.

Kallos, J. & Rizok, D. (1964). *J. Mol. Biol.* **9**, 255.

Kraut, J., Sieker, L. C., High, D. F. & Freer, S. T. (1962). *Proc. Nat. Acad. Sci., Wash.* **48**, 1417.

Kraut, J., Wright, H. T., Kellerman, M. & Freer, S. T. (1967). *Proc. Nat. Acad. Sci., Wash.* **58**, 304.

McAuliffe, C. A. (1967). *J. Chem. Soc.* (A), no. 4, p. 641.

Matthews, B. W., Sigler, P. B., Henderson, R. & Blow, D. M. (1967). *Nature*, **214**, 652.

Neurath, H., Rupley, J. A. & Dreyer, W. J. (1956). *Arch. Biochem. Biophys.* **65**, 243.

Oppenheimer, H. L., Labouesse, B. & Hess, G. P. (1966). *J. Biol. Chem.* **241**, 2720.

Perutz, M. F., Kendrew, J. C. & Watson, H. C. (1965). *J. Mol. Biol.* **13**, 669.

Raval, D. N. & Schellman, J. A. (1965). *Biochim. biophys. Acta*, **107**, 463.

Rovery, M. & Desnuelle, P. (1953). *Biochim. biophys. Acta*, **13**, 300.

Scrimger, S. T. & Hofmann, T. (1967). *J. Biol. Chem.* **242**, 2528.

Sigler, P. B., Jeffery, B. A., Matthews, B. W. & Blow, D. M. (1966). *J. Mol. Biol.* **15**, 175.

Strumeyer, D. H., White, W. N. & Koshland, D. E. (1963). *Proc. Nat. Acad. Sci., Wash.* **50**, 931.

Vaslow, F. & Doherty, D. G. (1953). *J. Amer. Chem. Soc.* **75**, 928.

Walsh, K. A. & Neurath, H. (1964). *Proc. Nat. Acad. Sci., Wash.* **52**, 884.

Weiner, H., White, W. N., Hoare, D. G. & Koshland, D. E. (1966). *J. Amer. Chem. Soc.* **88**, 3851.

White, F. H. (1960). *J. Biol. Chem.* **235**, 383.

White, F. H. (1961). *J. Biol. Chem.* **236**, 1353.

Wright, H. T., Kraut, J. & Wilcox, P. E. (1968). *J. Mol. Biol.*, in the press.

Wyckoff, H. W., Hardman, K. D., Allewell, N. M., Inagami, T., Johnson, L. N. & Richards, F. M. (1967). *J. Biol. Chem.* **242**, 3984.

J. Mol. Biol. (1970) **47**, 15–28

Mutant Tyrosine Transfer Ribonucleic Acids

J. N. Abelson†, M. L. Gefter‡, L. Barnett, A. Landy§,

R. L. Russell and J. D. Smith

Medical Research Council,
Laboratory of Molecular Biology
Hills Road, Cambridge, England

(*Received 27 June 1969*)

Three independent mutants of the su_{III} tyrosine suppressor transfer RNA gene have been isolated. These mutants are shown to produce mutant tRNA's differing from the wild-type su_{III}^+ tRNA molecule in each case by a single base change. A mutant tRNA containing an A residue in place of a G in the "dihydrouracil loop" appears to be defective in a step in protein synthesis occurring after the acylated tRNA is bound to the ribosome. A mutant tRNA having an A residue in place of a G in the "anticodon stem" appears to be defective exclusively in its apparent affinity for the tyrosyl tRNA synthetase. The isolation of mutant tRNA's as well as their biological properties are discussed.

1. Introduction

Although the primary sequences of a variety of transfer ribonucleic acids from several different organisms are known, comparisons of these have not yet suggested a correlation between primary structure and function in the tRNA molecule, with the exception of the location of the anticodon. We have approached this problem by studying the effects of single base changes, induced by mutation, on the functional properties of the su_{III} tyrosine tRNA of *Escherichia coli*, whose complete nucleotide sequence is known (Goodman, Abelson, Landy, Brenner & Smith, 1968).

The wild-type su_{III} gene (su_0^-) is the structural gene for one of the tyrosine tRNA's of *E. coli* which recognizes the codons UAU and UAC. The su_{III}^+ amber suppressor results from a mutation changing the anticodon from GUA to CUA so that su_{III}^+ tRNA recognizes only the amber codon UAG. Mutation to su^- can occur not only by reversion to su_0^- but also by any change in the tRNA sequence which gives a defective tRNA. Although some of these might produce gross changes in the tRNA structure affecting all functions of the molecule, it is reasonable to expect that there may be base substitutions that affect specific recognition sites (e.g. those for the synthetase, transfer factor, etc.), analogous to the anticodon change which specifically alters codon recognition.

We have isolated a large number of mutants of the su_{III}^+ tyrosine transfer RNA

† Present address: Department of Chemistry, University of California (San Diego), La Jolla, Calif. 92037, U.S.A.

‡ Present address: Department of Biological Sciences, Columbia University, New York, N.Y. 10027, U.S.A.

§ Present address: Division of Biological and Medical Sciences, Brown University, Providence, R.I. 02912, U.S.A.

gene. Surprisingly, most of the absolute su^- mutants produce very little transfer RNA and even partially defective genes produce reduced amounts: the sequences and properties of three partial su^- mutants have been studied, and are described in this paper.

2. Materials and Methods

(a) Media and strains

B broth, low phosphate medium, and Tris maleate have been described previously (Landy, Abelson, Goodman & Smith, 1967; Russell et al., 1969). CA274 carries a lac^-_{amber} mutation, MB100 is a derivative of MB0 which carries a single, su^+_{III} minor tyrosine tRNA gene, and the transducing phage $\phi80psu^+_{III}-1$ (derived from $\phi80psu^+_{III}$) also carries a single, su^+_{III} tyrosine tRNA gene. These strains are described in more detail by Russell et al. (1969).

(b) Mutagenesis and mutant selection

MB100 was grown to 2×10^8 cells/ml. in B broth, concentrated to 4×10^8 cells/ml. in 10^{-2} M-MgSO$_4$ and infected with $\phi80psu^+_{III}-1$ at a multiplicity of 5. After 15 min at 37°C, the infected cells were washed once with Tris maleate and exposed to N-methyl-N'-nitro-N-nitrosoguanidine at 100 μg/ml. in Tris maleate for 30 min at 37°C. The mutagen-treated infected cells were washed once with B broth, and aerated in B broth at 37°C until lysis.

The progeny phages were plated at 2×10^4 plaques per plate on CA274 indicator bacteria, in the presence of 0·8 mg of 5-bromo-4-chloro-3 indolyl-β-D-galactoside and 0·24 mg of isopropyl-β-D-thiogalactoside per plate. (Su^+ phages suppress the lac^-_{amber} mutation of CA274, and the resulting, isopropyl-β-D-thiogalactoside-induced galactosidase converts the colourless 5-bromo-4-chloro-3 indolyl-β-D-galactoside to an insoluble blue residue which colours the phage plaque; su^- phage plaques remain colourless.) About 0·1% of the phages formed colourless or light blue plaques in distinction to the full blue plaques of the remainder. (Only about one-twentieth as many colourless or light blue plaques appeared among control, non-mutagenized, phages.) These presumptive su^-_{III} mutants were picked and purified by two more replatings. Standard plate stocks were then prepared, using 2×10^5 phages per plate and MB100 as indicator bacteria. Usually these stocks contained 5×10^{10} to 2×10^{11} phages/ml.

(c) Transfer RNA from bacteriophage-infected CA274

CA274 was grown in low phosphate medium to 2×10^8 cells/ml., centrifuged and the bacteria resuspended in 0·1 vol. of the supernatant low phosphate medium. They were irradiated for 90 sec with a germicidal lamp (30% survivors) and MgSO$_4$ added to give 0·02 M. The bacteria were infected with the $\phi80psu$ phage at a multiplicity of 10 or 20, and 10 min at 37°C (without aeration) allowed for adsorption. The infected bacteria were then diluted with low phosphate medium to give 2×10^8 cells/ml. and aerated at 37°C ($t = 0$ min). If the RNA was to be labelled, 0·05 to 0·1 mc [^{32}P]orthophosphate/ml. was added at 5 min. At 25 min chloramphenicol (50 μg/ml.) was added and the cells were harvested at 150 min. The culture volumes used were 30 to 200 ml. for ^{32}P-labelled cells and 4 to 8 l. for unlabelled cells. Transfer RNA was extracted with phenol (Brubaker & McCorquodale, 1963) and precipitated with ethanol from 0·2 M-sodium acetate pH 5. This is referred to as stage 1 tRNA. An initial purification by stepwise elution from DEAE-cellulose (Landy et al., 1967) gave stage 2 tRNA.

^{32}P-labelled tRNA was stripped of amino acids by incubation in 2 M-Tris chloride pH 9·1 at 37°C for 30 min. It was fractionated on DEAE-Sephadex using an NaCl gradient pH 5 containing 7 M-urea, as described previously (Gefter & Russell, 1969). Tyrosine tRNA was located by the aminoacylation assay using [^{3}H]tyrosine. It eluted behind the major ^{32}P peak. This fraction was pooled, aminoacylated with [^{3}H]tyrosine in the absence of other amino acids and purified on a benzoylated DEAE-cellulose column according to Gillam et al. (1967). A column of benzoylated DEAE-cellulose 0·4 cm diameter $\times$ 4 cm was equilibrated with a solution containing 0·3 M-NaCl, 0·01 M-potassium acetate pH 4·8, 0·01 M-MgCl$_2$, 0·002 M-mercaptoethanol and the tRNA applied in the buffer. The column was washed with 8 vol. of the same buffer containing 0·8 M-NaCl and then tyrosyl-tRNA eluted with the same buffer containing 1·0 M-NaCl, 10% ethanol (v/v). All tyrosine tRNA with the fully modified 2-methylthio-$^6N(\gamma,\gamma$, dimethylallyl)-adenosine and that with partially modified ^6N (γ,γ, dimethylallyl)-adenosine at residue 38 eluted in this fraction. This material was precipitated from ethanol, reprecipitated from 0·2 M-sodium acetate pH 5 with ethanol, dried *in vacuo* and used for the sequence determinations.

Unlabelled tyrosine tRNA was fractionated using the following three procedures in the order stated in the Results section: (1) chromatography on DEAE-Sephadex with an NaCl gradient at pH 5 in 7 M-urea (Gefter & Russell, 1969); (2) reversed phase partition chromatography as described by Kelmers, Novelli & Stulberg (1965) and adapted by Gefter & Russell (1969); (3) chromatography of the tyrosyl-tRNA on benzoylated DEAE-cellulose as described above. The column separations were scaled up in proportion to the increased amounts of tRNA.

Assays for tyrosine and phenylalanine accepting tRNA's were according to Smith, Abelson, Clark, Goodman & Brenner, (1966).

(d) *Nucleotide sequence determination*

The techniques were those described by Sanger, Brownlee & Barrell (1965) and Brownlee & Sanger (1967) as used in the sequence determination of tyrosine tRNA (Goodman *et al.*, 1968).

(e) *Hybridization of ^{32}P-labelled transfer RNA*

Hybridization of ^{32}P-labelled tRNA to ϕ80psu^+ DNA was carried out as described by Landy *et al.* (1967).

(f) *Ribosome binding measurements*

The trinucleoside diphosphates UAU and UAG were prepared according to Thach & Doty (1965). The polynucleotide phosphorylase was a gift from Dr M. Grunberg-Manago. Ribosomes were prepared according to Anderson, Bretscher, Clark & Marcker (1967) and were kindly supplied by Dr J. A. Steitz of this laboratory. Poly UAG (1:1:1) was given by Dr B. F. C. Clark of this laboratory. The supernatant transfer factor was prepared according to Ravel (1967). The preparation was only taken through the first ammonium sulphate fractionation of the S-150 fraction. This preparation stimulated the binding of phenylalanine tRNA to ribosomes in the presence of poly U and was assayed by the stimulation of binding of GTP to a Millipore filter. The preparation was capable of binding 0·4 m-mole of GTP/mg of protein.

(g) *Tyrosyl-tRNA synthetase*

The preparation of crude tyrosyl-tRNA synthetase was essentially as described previously (Smith *et al.*, 1966). The enzyme was then further purified by adsorption and elution from calcium phosphate gel and hydroxyapatite. The partially purified enzyme was stored at $-20°$C in 50% glycerol without significant loss in activity for several months. The preparation was capable of acylating 6 mμmoles of tyrosine tRNA/min/mg of protein.

Assays of tyrosine accepting activity and of *in vitro* suppression were carried out as previously described (Gefter & Russell, 1969).

3. Results

To examine the properties of mutant tRNA's, su^- mutants must be obtained on ϕ80psu transducing phages (see Smith *et al.*, 1966). Two methods are available; mutants can be selected in the bacterium and then transferred to ϕ80psu by lysogenization and recombination (see Russell *et al.*, 1970), or the mutants can be generated directly on the transducing phage. Both methods have been used: however, the mutants to be described here have been obtained by the latter procedure. The phenotypes of these mutants in the bacterium have not been studied. Of the mutant phages studied, some form entirely colourless plaques in the dye indicator test (see Materials and Methods) and are presumed to carry mutations which entirely eliminate function of the su_{III}^{+} tRNA; the rest form partially coloured plaques, indicating that the mutant tRNA they produce retains some suppressor activity.

(a) Synthesis of tyrosine tRNA in CA274 infected with ϕ80psu mutants

The ability of the phages carrying mutant su genes to promote the synthesis of tyrosine tRNA in infected CA274 was measured in two ways.

(1) The synthesis of tRNA capable of accepting tyrosine was measured. Crude tRNA (stage (1) in Materials and Methods) was isolated from infected cells and the relative amounts of tyrosine and phenylalanine tRNA's were compared with those of tRNA from a control culture infected with ϕ80.

When normalised to their phenylalanine tRNA content, ϕ80psu^{+}-infected cells contained 7 to 16 times the tyrosine tRNA of uninfected or ϕ80-infected cells. In contrast, five su^{-} mutants and two classified as weak su^{+} mutants, failed to show any increase in tyrosine tRNA. Six weak su^{+} mutants showed increases in tyrosine tRNA ranging from 1·5 to 3 times that of ϕ80-infected cells. The possibility that the failure of mutants to show any increase in tyrosine tRNA was due to the inability of their tRNA to be acylated was eliminated by the second test for tRNA synthesis.

(2) The synthesis of the tyrosine tRNA sequence, whether functional or not, was measured by hybridization of tRNA from infected cells to ϕ80psu^{+} DNA. Cells infected with mutant phage were labelled with [^{32}P]orthophosphate and the tRNA was isolated and purified to stage (2) (see Materials and Methods). The proportion of tRNA hybridizing to ϕ80psu^{+} DNA was determined. Transfer RNA isolated from ϕ80-infected cells served as control RNA. Thirty to forty per cent of tRNA isolated from ϕ80psu^{+} infected cells was shown to hybridize to ϕ80psu^{+} DNA, whereas 2 to 5% of control RNA hybridized. Four su^{-} mutant phages examined did not cause an accumulation of hybridizable tRNA, thus indicating that the failure of these mutants to stimulate accumulation of acylateable tyrosine tRNA was due to a lack of excess tyrosine tRNA in the preparation.

The weak su^{+} mutant phages that caused a slight increase in acylateable tyrosine tRNA also caused an increase in hybridizable tRNA. The increase resulting from infection with these mutants was 7 to 30% of that observed for the parental psu^{+} phage.

Approximate values for the amount of [^{32}P]tyrosine tRNA could also be obtained by measuring the amount of ^{32}P eluting in the region of tyrosine tRNA on chromatography on DEAE-Sephadex columns. Such measurements were qualitatively in agreement with the hybridization results.

As the su^{-} mutants produced no detectable tRNA, it was not possible to study the completely defective mutant tRNA's. Five of the weak su^{+} mutants produced sufficient tRNA to allow their characterization. Three of these were found by sequence analysis to be the same mutant (mutant 15). This and the other two mutants designated 12 and 24, will be discussed.

(b) Nucleotide sequences of the mutant tRNA's

The sequence of the su_{III}^{+} tyrosine tRNA is shown in Figure 1. The three su mutant tRNA's, 12, 15 and 24 each differ from su^{+} by a single $G \rightarrow A$ base change. The positions of these are indicated on the sequence in Figure 1.

For sequence analysis, both the T_1 and the pancreatic ribonuclease products from the purified mutant ^{32}P-labelled tRNA were separated in the two-dimensional electrophoretic system and compared with those from su_{III}^{+} tRNA. The sequence of the

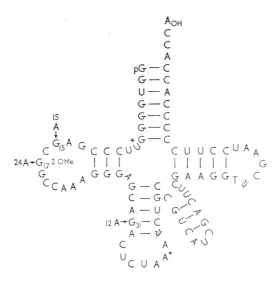

FIG. 1. Nucleotide sequence of the su_{III}^+ tyrosine suppressor tRNA. The anticodon (CUA) is at the bottom. U* is 4-thiouracil and A* is 2-methylthio-$^6N(\gamma,\gamma$, dimethylallyl) adenosine. The arrows indicate the positions where A has replaced G in the mutants isolated. The mutant with a base change at residue 15 is designated su^-15. The mutant with a base change at residue 17 (2'-O-methylguanosine) is designated su^-24 and the mutant with a base change at residue 31 (anticodon stem) is designated su^-12.

additional T_1 and pancreatic ribonuclease products from the mutant tRNA's were determined as described below. To confirm that there were no other sequence changes in the mutant tRNA's, all of the T_1 and pancreatic ribonuclease products were digested with pancreatic or T_1 ribonuclease, respectively, and the products separated by electrophoresis at pH 3·5 or on DEAE paper in 7% formic acid. These products were compared with those from the corresponding su_{III}^+ oligonucleotides. Certain of the di- and trinucleotides were simply characterized by their nucleotide composition.

In bacteria infected with $\phi80psu_{III}^+$ the amount of su_{III}^+ tyrosine tRNA synthesized is so great that the small proportion of tyrosine tRNA transcribed from the host genes is negligible in comparison. This is not so in cells infected with phages carrying the mutant tRNA genes because the amount of su tRNA synthesised after infection is one-fifth to one-quarter that made in $\phi80psu_{III}^+$ infected cells. Consequently all the ^{32}P-labelled su mutant tyrosine tRNA preparations contained a small but significant amount of the bacterial specified tyrosine tRNA's I and II, and where nucleotide replacements occurred in the mutant tRNA's the sequences missing in the mutant were not entirely absent from the tRNA fingerprint.

(i) su⁻ *15 tRNA*

Sequence analysis indicates that su^-15 tRNA has an A residue in place of G_{15} (see Fig. 1). The chart below shows the sequence around residue 15 in the su^+ tRNA and su^-15 tRNA. The oligonucleotides expected for T_1 ribonuclease and pancreatic ribonuclease digestions of these tRNA's are shown. The sequences in brackets are those from tRNA molecules containing an unmethylated ribose at G_{17}. This residue is incompletely methylated in phage 80 infected cells.

tRNA	su^+	su^-15
Sequence	.CGAGCGmGC. (.CGAGCGGC.) ↑ 15	.CGAACGmGC. (.CGAACGGC.) ↑ 15
T_1 Ribonuclease products	AG+CGmG (AG+CG+G)	AACGmG (AACG+G)
Pancreatic ribonuclease products	GAGC+GmGC (GAGC+GGC)	GAAC+GmGC (GAAC+GGC)

Comparison of the separated T_1 ribonuclease digestion products from su_{15} and su^+ tRNA (Plate I) shows the following differences: in su_{15} tRNA, CGmG is absent, CG is present in small amounts (0·32 mole) and 1 mole of AG is found instead of the normal yield of 2 moles. Two additional T_1 products are obtained from su_{15}^- tRNA and these were identified as AACGmG and AACG. AACGmG was hydrolysed with pancreatic ribonuclease and separated on electrophoresis at pH 3·5 giving two products with mobilities of AAC and GmG. These were identified by alkaline hydrolysis followed by electrophoresis at pH 3·5. AAC gives A(1·8 moles) and C(1 mole) while GmG was unchanged. The pancreatic products AAC and GmG were in approximately equimolar yield.

AACG after pancreatic ribonuclease hydrolysis and electrophoresis at pH 3·5 gave equimolar yields of AAC and G. AAC was identified by alkaline hydrolysis and gave 1·8 moles A/1 mole C. The combined yield of AACGmG and AACG was 0·7.

The pancreatic ribonuclease digestion products of su_{15} and su^+ show a single difference. In su_{15} GAGC is missing and is replaced by GAAC (Plate II). GAAC was identified by T_1 ribonuclease digestion and electrophoresis at pH 3·5 to give AAC and G. AAC was again identified by alkaline hydrolysis and gave 2·06 moles A/1 mole C. The yield of GAAC from su_{15}^- was 0·85 mole. This change is expected from the substitution $G_{15} \rightarrow A$ (Fig. 1).

(ii) su$^-$ 24 tRNA

The sequence analysis of su^-24 tRNA is compatible only with a change of 2'-O-methyl G_{17} to an A. The sequences around residue 17 are shown in the chart below. The expected products from T_1 ribonuclease and pancreatic ribonuclease digests of su_{III}^+ and su^-24 tRNA's are compared.

tRNA	su^+	su^-24
Sequence	.GCGmGC. (.GCGGC.) ↑ 17	.GCAGC. ↑ 17
T_1 Ribonuclease products	CGmG (CG+G)	CAG
Pancreatic ribonuclease products	GmGC (GGC)	AGC

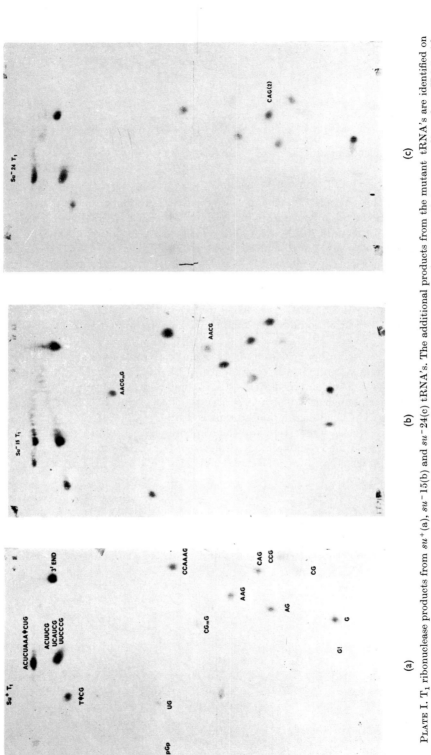

PLATE I. T_1 ribonuclease products from su^+(a), su^- 15(b) and su^- 24(c) tRNA's. The additional products from the mutant tRNA's are identified on the autoradiogram. The products were separated by electrophoresis on (1) cellulose acetate in 5% (w/v) acetic acid, 7 M-urea, adjusted to pH 3·5 with pyridine (origin on the right); (2) DEAE-paper in 7% aqueous formic acid (v/v) (origin at the top). The ^{32}P-labelled products were autoradiographed.

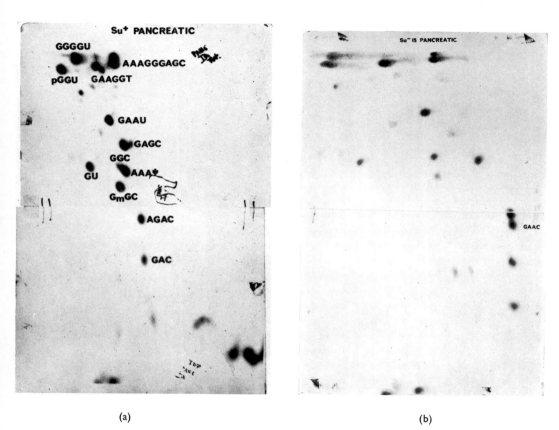

(a)   (b)

PLATE II. Pancreatic ribonuclease products from su^+(a) and su^-15(b) tRNA's. The additional products from the su^-15 tRNA are identified on the autoradiogram. Separation is as described in Plate I. The su^+ tRNA digest was run for a shorter time at pH 3·5.

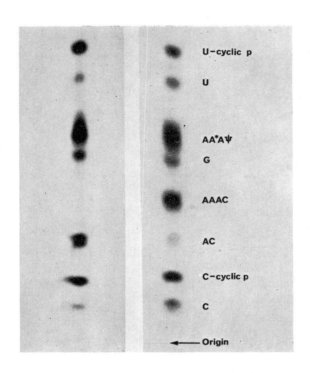

PLATE III. Analysis by pancreatic ribonuclease digestion of the anticodon-containing T_1 ribonuclease products from su^-12 and su^+ tRNA's; (left) ACUCUAA*AψCUG from su^+ tRNA. (right) CAAACUCUAA*AψCUG from su^-12 tRNA;

Electrophoresis was in pyridine acetate buffer pH 3·5 on Whatman 52 paper and the ^{32}P-labelled products were located by autoradiography.

T_1 ribonuclease fingerprints of su^-24 tRNA (Plate I) showed:
(1) only a trace of CGmG and a very low yield of CG,
(2) an increase of CAG from 1 mole to nearly 2 moles. (CAG was identified by the isolation of C and AG following pancreatic ribonuclease digestion.)

Relative molar yields (CCG = 1) of T_1 ribonuclease products from su^-24 tRNA were: CGmG, 0·09; CG, 0·3; CCG, 1·0; CAG, 1·8; AAG, 1·1; AG, 2·0. The molar yields of these nucleotides from su^+ tRNA were CGmG, 0·37; CG, 0·95; CCG, 1·0; CAG, 1·0; AAG, 1·05; AG, 2·2. The expected molar yields of (CGmG + CG) are 0 (su^-24) and 1·0 (su^+). For reasons already discussed above, the su^-24 tRNA preparation contains small amounts of the bacterial specified tyrosine tRNA's I and II accounting for the presence of small amounts of the nucleotides CGmG and CG.

In the pancreatic ribonuclease fingerprint GGC and GmGC are in greatly decreased yield and an additional product AGC appears in approximately molar yield. In the two-dimensional fingerprint AGC was incompletely separated from the pancreatic ribonuclease product GAC derived from the anticodon region. The combined spots were treated with T_1 ribonuclease and on electrophoresis at pH 3·5 gave approximately equimolar amounts of C, AC, AG and G. When the upper and lower portions of this spot were separately hydrolysed with T_1 ribonuclease, the upper region gave mainly G and AC and thus was largely GAC while the lower region gave principally AG and C and consequently was largely AGC.

(iii) su^-12 tRNA

Sequence analysis of this tRNA shows that residue G31 has been changed to an A. The sequences of the su_{III}^+ and su^-12 tRNA's around residue 31 are shown below. The products of ribonuclease T_1 and pancreatic ribonuclease are shown for comparison.

	su^+	su
	GCAGACUCUAA*AψCUG	GCAAACUCUAA*AψCUG
	↑	↑
T_1 ribonuclease products	31	31
	CAG+ACUCUAA*AψCUG	CAAACUCUAA*AψCUG
Pancreatic ribonuclease products	AGAC,U,C,U,AA*Aψ,C,U	AAAC,U,C,U,AA*AψC,U

The T_1 ribonuclease fingerprint shows two changes:
(1) CAG is almost absent; (2) the oligonucleotide ACUCUAAAψCUG derived from the anticodon loop is replaced by a product moving slightly slower in the first dimension (pH 3·5). Plate III shows the pancreatic ribonuclease products from this oligonucleotide and those from ACUCUAAAψCUG derived from su_{III}^+ for comparison. ACUCUAAAψCUG yields AC, AAAψ, C, U and G, while the product from su^-12 tRNA gives AAAC, AAAψ, C, U and G. AAAC was identified by nucleotide composition after alkaline hydrolysis and yielded 3 moles A/1 mole C. The molar proportions of C/U/-AAAC in the su^-12 product were 3:3:1.

These changes result from the substitution $G_{31} \rightarrow A$ as shown in Figure 1 and the new su^-12 product is CAAACUCUAAAψCUG.

The pancreatic ribonuclease fingerprint of su^-12 tRNA confirmed this by showing the absence of AGAC and its replacement by AAAC. The latter was identified by base composition.

(c) *Functional analysis of mutant transfer RNA's* su⁻12 *and* su⁻15

The colour of the plaques produced on the indicator plates by the transducing phages carrying *su⁻*12 and *su⁻*15 showed these to have a very weak suppressor activity, suggesting that the mutant tRNA's were partially defective in one or more functions. These two mutant tRNA's and *su⁺*tRNA were compared in their reactions with tyrosyl-tRNA synthetase, UAG-dependent binding to ribosomes and ability to suppress an amber mutant of f2 bacteriophage RNA *in vitro*.

Suppressor tRNA synthesized in the phage-infected cells can be separated into three species by reverse phase partition chromatography. These differ according to the extent of modification of the 2-methylthio-$^6N(\gamma,\gamma,$ dimethylallyl) adenosine residue 38, adjacent to the anticodon (Gefter & Russell, 1969). We chose the form in which residue 38 is $^6N(\gamma,\gamma,$ dimethylallyl) adenosine, lacking the 2-methylthio group (peak II on the reversed phase separation system) for comparisons of the mutant and *su⁺* tRNA's. This partially modified form of the suppressor tRNA is aminoacylated with the same kinetics as the fully modified form and only differs from the latter in being 50% as efficient in protein synthesis (Gefter & Russell, 1969). Peak II was chosen because the fully modified suppressor tRNA cannot be separated easily from the host-specified tyrosine tRNA present in the cells before infection. Only the tRNA's synthesized after infection are partially modified. Contamination of the mutant tRNA's with large amounts of host-specified tyrosine tRNA would complicate activity measurements in assays not distinguishing suppressor tRNA (UAG recognizing) from the UAU/UAC recognising host-specified tRNA. As will be discussed later the preparations of *su⁺* and *su⁻*12 tRNA's are essentially free of host-specified tRNA, whereas the *su⁻*15 tyrosine tRNA preparation contains 35% host specified tRNA.

The tRNA's were obtained from cells infected with the transducing phages and purified by either of two procedures (see Materials and Methods): (1) the total tyrosine tRNA was separated by chromatography on DEAE-Sephadex and then fractionated by reverse phase chromatography to isolate peak II; (2) peak II tyrosine tRNA was separated by reverse phase chromatography and further purified on benzoylated DEAE-cellulose. Both procedures yielded preparations of equivalent relative purity (approximately 30% of the tRNA could be acylated with tyrosine).

(i) *Kinetics of acylation of transfer RNA's*

The amount of tyrosine acceptor activity in the tRNA preparations was determined using a partially purified preparation of tyrosyl-tRNA synthetase (see Materials and Methods). The apparent K_m and V_{max} values were then determined for each tRNA by the method of reciprocal plots. The results obtained for two independent preparations of tRNA are summarized in Table 1. It is evident from the data from the first preparation of tRNA's, that the apparent K_m and V_{max} for *su⁻*15 tRNA is the same as the parental *su⁺* tRNA whereas the apparent K_m for *su⁻*12 tRNA is ten times higher and the V_{max} is three times higher than the parental tRNA. The results obtained with the second preparation indicate that the apparent K_m for *su⁻*12 tRNA is 16 times higher than the parental *su⁺* tRNA. The V_{max} for *su⁻*12 and *su⁺* tRNA's in these experiments appear to be the same. The results in Table 1 also demonstrate that the change in anticodon from su_0^- (G*UA) to *su⁺* (CUA) has no effect on the apparent affinity or velocity constants. Similar values were obtained for the apparent K_m of tyrosine tRNA present in uninfected cells.

The change of a G to an A at residue 31 in the anticodon stem thus considerably

TABLE 1

Kinetics of acylation of transfer RNA's

tRNA	Preparation 1		Preparation 2	
	Apparent K_m (μM)	Apparent V_{max} (mμmoles/min/mg)	Apparent K_m (μM)	Apparent V_{max} (mμmoles/1 min/mg)
su^+	0·066	1·6	0·025	2·4
su^-0	—	—	0·020	2·2
su^-15	0·052	1·6	—	—
su^-12	0·67	5·3	0·40	2·4

The reaction mixture, 0·1 ml., contained (in mM concentrations) ammonium cacodylate buffer (pH 7·1), 200; magnesium acetate 20; KCl, 40; ATP, 2; CTP, 0·4; β-mercaptoethanol, 10; [^{3}H]-tyrosine ($3\cdot6 \times 10^3$ cts/min/$\mu\mu$mole), $3\cdot7 \times 10^{-2}$; 19 other unlabelled amino acids each, $2\cdot5 \times 10^{-1}$; tyrosyl-tRNA synthetase, 0·15 μg and tRNA. The reaction mixture was incubated at 37°C for 10 min and the reaction stopped by the addition of trichloroacetic acid. The amount of tRNA acylated as a function of tRNA concentration was determined. Extrapolation of the reciprocal plots gave the results presented in the Table.

reduces the apparent affinity of tyrosyl tRNA for the activating enzyme, while a G to A change in the dihydrouracil loop at residue 15 (su^-15 tRNA)has no effect on this recognition.

(ii) *UAG-dependent ribosome binding of the mutant transfer RNA's*

The ability of the mutant-tRNA's to recognize the codon UAG was measured using the trinucleotide-dependent ribosome binding assay of Nirenberg & Leder (1964). This assay also gives a measure of the stability of the tRNA–ribosome–trinucleotide complex. In these assays both ribosomes and triplet were in excess, and conditions were established such that the amount of ^{3}H-labelled tyrosyl-tRNA bound was directly proportional to the input tRNA. In the experiments shown in Table 2 ribosome binding was measured at 20 mM-Mg^{2+} in the absence of transfer factors. Measurements were made after incubation for 15 min at 25°C when the reaction was complete. Similar measurements of the binding in the presence of UAU allowed calculation of the proportion of host-specified tyrosine tRNA in the preparations (Table 2). This is tRNA synthesized from the host tyrosine tRNA genes after infection. Su^-15 tyrosine tRNA contained 35% UAU recognising tRNA while su^-12 tRNA contained none.

Su^-15 tRNA bound to ribosomes with UAG to the same extent as su^+ tRNA but in contrast su^-12 tRNA hardly bound at all under these conditions (Table 2).

(iii) *Supernatant factor-dependent binding of mutant transfer RNA's to ribosomes and poly UAG*

The conditions of the assay described in Table 2 do not correspond to those required for protein synthesis where recognition of the messenger–ribosome complex by tRNA involves the interaction of the tRNA with transfer (T) factor and GTP at a lower Mg^{2+} concentration. This mode of recognition was assayed using the Nirenberg & Leder technique but at 4 mM-Mg^{2+}, with a random UAG polymer, in the presence of ribosomes, a supernatant preparation containing transfer factor, and GTP. The binding of tRNA to ribosomes is then dependent both on polymer and supernatant factor. GTP gave variable stimulation and was included in all reaction mixtures. Its failure

TABLE 2

Ribosome binding in the absence of transfer factors

tRNA	Input	Bound −triplet (cts/min)	Bound +UAU (cts/min)	% Bound specifically	Bound +UAG (cts/min)	% Bound specifically	% Bound specifically UAU +UAG (cts/min)
su^+	17,128	3724	4469	4·4	12,544	73	77·4
su_0^-	18,050	7728	15,409	42·5	8934	6·7	49·2
su^-15	19,670	2350	6881	23	10,701	42·6	65·6
su^-12	7500	869	1417	7·3	1531	8·8	16·1

The binding assay was done according to Nirenberg & Leder (1964). The reaction mixtures (0·05 ml.) contained (in μmoles) ammonium cacodylate buffer (pH 7·1) 2·5; KCl, 2·5; $MgCl_2$, 1·0; 1·0% ribosomes (0·43 A_{260} unit); trinucleotide UAU 0·2, A_{260} unit or UAG, 0·28 A_{260} unit; and tRNA charged with [^{3}H]tyrosine (2×10^4 cts/min/$\mu\mu$mole).

to give consistent stimulation was probably because no attempt was made to remove nucleotides from the charged tRNA preparations. The preparation and assay of the transfer factor are described in Materials and Methods. Under these conditions the amount of tRNA bound to ribosomes is proportional to input tRNA and both UAU- and UAG-recognizing tRNA's bind with the same efficiency; su_0^- and su_{III}^+ tRNA's do not compete for binding. The results summarized in Table 3 show that su^-15 tRNA binds as efficiently as the su^+ tRNA.

Surprisingly, su^-12 tRNA binds efficiently under these conditions, although it was almost inactive in non-factor dependent binding at 20 mM-Mg^{2+}.

Although we have not measured directly the interaction of transfer factor and the tRNA's, we conclude from these experiments that both su^-15 and su^-12 tRNA's can be recognized by the transfer factor and can form a stable complex with ribosomes and the polymer at 4 mM-Mg^{2+}.

TABLE 3

Supernatant factor-dependent binding of mutant transfer RNA's to ribosomes and UAG

tRNA	Input (cts/min)	Bound complete-polymer (cts/min)	Bound complete-factor (cts/min)	Complete (cts/min)	% Bound specifically
su^+	8564	439	549	2102	19·3
	17,128	881	884	5188	25
su^+	17,128	778	1007	4695	22·8
su^-	12,025	707	727	3586	23·8
su^+, su^-	29,153	1357	—	6713	18·4
su^-15	9800	608	472	2472	19·0
su^-12	7500	515	—	1870	18·0

The binding assays were done exactly as described in the legend to Table 2 except that the $MgCl_2$ concentration was lowered to 4 mM. In addition where indicated, poly UAG (0·03 A_{260} unit) was substituted for triplet, and supernatant factor (0·25 μg) and GTP (4·8 mμmoles) were added.

(iv) *Translation* in vitro *of the UAG codon*

The ability of the mutant tRNA's to translate a UAG codon was assayed by a modification of the procedure of Engelhardt, Webster, Wilhelm & Zinder (1965) (Gefter & Russell, 1969). This assay measures the frequency of translation of the amber mutant *sus-4A* in bacteriophage f2 RNA. The mutant UAG codon replaces the seventh codon in the phage coat protein gene. When f2 *sus-4A* RNA is a messenger in an *in vitro* protein synthesizing system from su^- *E. coli* a hexapeptide, comprising the first six amino acids of the coat protein (including formylmethionine), is synthesized and released. Addition of su_{III}^+ tRNA results in translation of the UAG codon as tyrosine and synthesis of longer polypeptides. In the assay both hexapeptide and longer polypeptides are equivalently labelled with [³⁵S]formylmethionine derived from [³⁵S]formylmethionyl tRNA and separated by precipitation of the longer peptides in 68% ethanol. The frequency of translation of the UAG codon (percentage transmission) is the ratios ([³⁵S]formylmethionine in ethanol insoluble peptides)/(³⁵S in total peptide synthesized).

At low suppressor tRNA concentrations the transmission is proportional to added tRNA and the assay is expected to measure the specific activity of a suppressor tRNA (transmission/unit amount of uncharged tRNA added). It would differ from an *in vivo* assay of suppression which is dependent both on the specific activity of the tRNA and the amount present in the cell. This is important because in transducing phage-infected cells, su^-12 and su^-15 tRNA's are synthesized at about one-fifth of the rate of su^+ or su_0^- tRNA's and this fact alone must contribute to the level of *in vivo* suppression as qualitatively judged on the indicator plates.

Figure 2 shows the relation between suppression and tRNA concentration for su^+, su^-12 and su^-15 tRNA's. The tRNA concentrations are those of the UAG recognizing tRNA, calculated for each preparation from the proportions of UAU and UAG-recognizing tRNA determined in Table 2. Mutant 12 tRNA suppressed about 50% as well as su^+ tRNA. It is quite possible that this difference is solely due to its higher K_m in the aminoacylation step; the concentrations of su^-12 tRNA in the assay mixture are below its apparent K_m. This could be decided in experiments using pre-charged

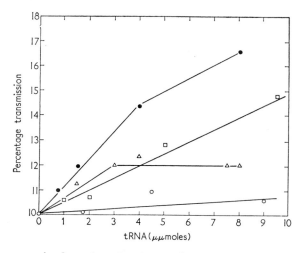

FIG. 2. *In vitro* suppression by su^+, —●—●—; su_0^-, —○—○—; su^-12, —□—□—; su^-15, —△—△— tRNA's. The details of the assay are described in the text and Gefter & Russell (1969).

tRNA. Su^-15 tRNA shows about 50% of the su^+ tRNA suppressor activity at low concentrations but there is an abrupt halt in increase in suppressor activity at higher tRNA concentrations. This anomalous behaviour has been consistently observed with two independent preparations of this tRNA. Although su^-15 tRNA is indistinguishable from su^+ tRNA in its rate of aminoacylation and in its ability to bind to ribosomes in the presence of the transfer factor and UAG codon, it suppresses poorly and anomalously. Thus this tRNA appears to be defective in a step in protein synthesis occurring after the aminoacylated tRNA is bound to the ribosome.

4. Discussion

An unexpected property of the su_{III} mutants is their reduced synthesis of tRNA compared with wild-type su^+ or su_0^-. These comparisons are based on the amount of su_{III} tRNA synthesized in CA274 infected with $\phi80psu$ phages. Mutants 12, 15 and 24 and other partly defective su_{III} mutants promote the synthesis of between 5 and 30% of the amount of tRNA in $\phi80psu^+$ or $\phi80psu_0^-$ cells. All the totally defective su^- mutants produce tRNA in even smaller amounts, in most cases undetectable. In several instances we have shown that the latter are not deletion mutants. The reduced synthesis is not generally due to additional mutations in bacteriophage genes resulting in poor phage replication since (1) $\phi80psu^-12$, 15 and 24 all have normal burst sizes, and (2) the reduced tRNA synthesis is also observed with su^- mutants isolated in the bacteria and subsequently obtained as $\phi80psu$ derivatives by transduction. Several different explanations could be suggested for the reduced net tRNA synthesis from the mutant su genes, for example: (a) the tRNA is synthesized at a reduced rate due to a second mutation in a specific region outside of the structural tRNA gene which controls its rate of transcription, and we have evidence for this type of mutation. Mutants 12, 15 and 24 would then have to be mutant both in the structural gene and in the region controlling transcription; (b) the possibility that all mutations affecting the conformation of tRNA result in a reduced rate of maturation (i.e. removal from the template, conversion from a precursor molecule) has not been excluded; (c) the tRNA is synthesized at the normal rate but is unstable and degraded in the cell. We have recently isolated some temperature-sensitive su^+ mutants which appear to have this property. However, the unexpected finding for which we cannot offer a unique explanation is that of some 40 mutants examined all show a greatly diminished amount of tRNA synthesis.

The partially defective su_{III} mutants we have described were obtained by mutagenesis of the transducing phage $80psu_{III}^+$. Stable lysogens of these were not isolated. Therefore, the phenotypes of these mutants when integrated in the bacterial chromosome is not known. The three mutants described in detail are certainly very weak suppressors. However, quantitative comparison of the specific suppressor activities of these mutant tRNA's *in vivo* cannot easily be made because the amounts of tRNA synthesized from the mutant genes in the phage-infected cells are different and markedly less than from the wild type su^+.

These three su^+ mutants specify mutant tRNA's each differing from the wild-type su^+ tRNA by a single G $\rightarrow$ A substitution. G $\rightarrow$ A substitutions would be expected from the specificity of the mutagen, N-methyl-N'-nitro-N-nitrosoguanidine, which gives predominantly the transitions G $\rightarrow$ A and C $\rightarrow$ T. Two of the mutants have base changes in sequences common to many tRNA's, the AG and GG sequences in the dihydrouracil loop.

In mutant 12 the base change is in the hydrogen-bonded arm of the anticodon loop resulting in an A·C pair in place of the wild-type G·C pair. One might expect this to produce some conformational change in this part of the molecule. Su^-12 tRNA shows a markedly decreased affinity for tyrosyl-tRNA synthetase. This suggests either that the anticodon arm interacts directly with the enzyme, or that correct hydrogen bonding of this arm is required to maintain the conformation of another part of the molecule involved in the specific recognition. Although we favour the former explanation, to distinguish between the two possibilities will require isolation of other mutants with similar phenotypes. Another possible approach is to look for a second site revertant of mutant 12 having a normal base pair restored. Su^-12 tRNA functions about 50% as well as su^+ tRNA in the suppressor activity assay (and this difference may solely result from the increased K_m for synthetase). This fact strongly suggests that su^-12 tRNA does not have a grossly changed conformation.

The codon-dependent ribosome binding assays on su^-12 tRNA show efficient binding at 4 mM-Mg^{2+} with transfer factor and GTP, but insignificant binding in the absence of transfer factor at 20 mM-Mg^{2+}. It would appear that the conformation of the anticodon loop at 20 mM-Mg^{2+} is unfavourable for codon-specified ribosome interaction, but that in the conditions of the factor dependent binding either the T factor or the lower Mg^{2+} concentration allows the anticodon loop to assume a correct conformation.

In the clover leaf arrangement of the tRNA sequence (Fig. 1) pseudouridine residue 40 is in the A·ψ pair adjacent to the A·C pair of mutant 12. In the su^+ tRNA this residue is entirely modified as pseudouridine but in some preparations of su^-12 tRNA we find a small proportion (up to 20%) of the tRNA molecules have uracil at this position. This implies that this pseudouridine residue is derived by modification of uridine and that the sequence change in mutant 12 reduces the efficiency of its modification.

In su^-15 tRNA, residue 15 is changed from G to A. This residue is part of the AG sequence in the dihydrouracil loop common to many tRNA's (although possibly not all). This change affects neither the kinetics of aminoacylation nor the ribosome-binding of the tRNA (whether factor dependent or not). However su^-15 tRNA does not function normally in the in vitro suppression assay. The curve relating suppressor activity to tRNA concentration shows no further increase in suppressor activity above quite low tRNA concentrations (Plate IV). Su^-12 tRNA thus appears to be defective in a step in protein synthesis subsequent to binding of the tRNA to the ribosome. This could be for example either peptide transfer, translocation or tRNA ejection, and analysis of these steps is in progress.

Our aim in this work is to relate sequence changes to functional defects in the tRNA molecule. The first two single base substitution mutants analysed produce effects on different functions of the tRNA.

The authors thank Miss E. Higgins and Mr T. V. Smith for their expert technical assistance. One of us (M.L.G.) is a fellow of the Jane Coffin Childs Memorial Fund for Medical Research. This investigation has been aided by a grant from the Jane Coffin Childs Memorial Fund for Medical Research.

REFERENCES

Anderson, J. S., Bretscher, M. S., Clark, B. F. C. & Marcker, K. A. (1967). *Nature*, **215**, 490.
Brownlee, G. G. & Sanger, F. (1967). *J. Mol. Biol.* **23**, 338.

Brubaker, L. H. & McCorquodale, D. J. (1963). *Biochim. biophys. Acta*, **76**, 48.

Engelhardt, D. L., Webster, R. E., Wilhelm, R. C. & Zinder, N. D. (1965). *Proc. Nat. Acad. Sci., Wash.* **54**, 1791.

Gefter, M. L. & Russell, R. L. (1969). *J. Mol. Biol.* **39**, 145.

Gillam, J., Millward, S., Blew, D., von Tigerstrom, M., Wimmer, E. & Tener, J. M. (1967). *Biochemistry*, **6**, 3043.

Goodman, H. M., Abelson, J., Landy, A., Brenner, S. & Smith, J. D. (1968). *Nature*, **217**, 1019.

Kelmers, A. D., Novelli, G. D. & Stulberg, M. P. (1965). *J. Biol. Chem.* **240**, 3979.

Landy, A., Abelson, J., Goodman, H. M. & Smith, J. D. (1967). *J. Mol. Biol.* **29**, 457.

Nirenberg, M. & Leder, P. (1964). *Science*, **145**, 1399.

Ravel, J. M. (1967). *Proc. Nat. Acad. Sci., Wash.* **57**, 1811.

Russell, R. L., Abelson, J. N., Landy, A., Gefter, M. L., Brenner, S. & Smith, J. D. (1970). *J. Mol. Biol.* **47**, 1.

Sanger, F., Brownlee, G. G. & Barrell, B. G. (1965). *J. Mol. Biol.* **13**, 373.

Smith, J. D., Abelson, J., Clark, B. F. C., Goodman, H. M. & Brenner, S. (1966). *Cold Spr. Harb. Symp. Quant. Biol.* **31**, 479.

Thach, R. E. & Doty, P. (1965). *Science*, **147**, 1310.

J. Mol. Biol. (1970) **48**, 443–453

A General Method Applicable to the Search for Similarities in the Amino Acid Sequence of Two Proteins

Saul B. Needleman and Christian D. Wunsch

*Department of Biochemistry, Northwestern University, and
Nuclear Medicine Service, V. A. Research Hospital
Chicago, Ill. 60611, U.S.A.*

(Received 21 July 1969)

A computer adaptable method for finding similarities in the amino acid sequences of two proteins has been developed. From these findings it is possible to determine whether significant homology exists between the proteins. This information is used to trace their possible evolutionary development.

The maximum match is a number dependent upon the similarity of the sequences. One of its definitions is the largest number of amino acids of one protein that can be matched with those of a second protein allowing for all possible interruptions in either of the sequences. While the interruptions give rise to a very large number of comparisons, the method efficiently excludes from consideration those comparisons that cannot contribute to the maximum match.

Comparisons are made from the smallest unit of significance, a pair of amino acids, one from each protein. All possible pairs are represented by a two-dimensional array, and all possible comparisons are represented by pathways through the array. For this maximum match only certain of the possible pathways must be evaluated. A numerical value, one in this case, is assigned to every cell in the array representing like amino acids. The maximum match is the largest number that would result from summing the cell values of every pathway.

1. Introduction

The amino acid sequences of a number of proteins have been compared to determine whether the relationships existing between them could have occurred by chance. Generally, these sequences are from proteins having closely related functions and are so similar that simple visual comparisons can reveal sequence coincidence. Because the method of visual comparison is tedious and because the determination of the significance of a given result usually is left to intuitive rationalization, computer-based statistical approaches have been proposed (Fitch, 1966; Needleman & Blair, 1969).

Direct comparison of two sequences, based on the presence in both of corresponding amino acids in an identical array, is insufficient to establish the full genetic relationships between the two proteins. Allowance for gaps (Braunitzer, 1965) greatly multiplies the number of comparisons that can be made but introduces unnecessary and partial comparisons.

2. A General Method for Sequence Comparison

The smallest unit of comparison is a pair of amino acids, one from each protein. The maximum match can be defined as the largest number of amino acids of one protein that can be matched with those of another protein while allowing for all possible deletions.

The maximum match can be determined by representing in a two-dimensional array, all possible pair combinations that can be constructed from the amino acid sequences of the proteins, A and B, being compared. If the amino acids are numbered from the N-terminal end, Aj is the jth amino acid of protein A and Bi is the ith amino acid of protein B. The Aj represent the columns and the Bi the rows of the two-dimensional array, MAT. Then the cell, MATij, represents a pair combination that contains Aj and Bi.

Every possible comparison can now be represented by pathways through the array. An i or j can occur only once in a pathway because a particular amino acid cannot occupy more than one position at one time. Furthermore, if MATmn is part of a pathway including MATij, the only permissible relationships of their indices are $m > i, n > j$ or $m < i, n < j$. Any other relationships represent permutations of one or both amino acid sequences which cannot be allowed since this destroys the significance of a sequence. Then any pathway can be represented by MATab ... MATyz, where $a \geqslant 1$, $b \geqslant 1$, the i and j of all subsequent cells of MAT are larger than the running indices of the previous cell and $y \leqslant K, z \leqslant M$, the total number of amino acids comprising the sequences of proteins A and B, respectively. A pathway is signified by a line connecting cells of the array. Complete diagonals of the array contain no gaps. When MATij and MATmn are part of a pathway, $i - m \neq j - n$ is a sufficient, but not necessary condition for a gap to occur. A necessary pathway through MAT is defined as one which begins at a cell in the first column or the first row. Both i and j must increase in value; either i or j must increase by only one but the other index may increase by one or more. This leads to the next cell in a MAT pathway. This procedure is repeated until either i or j, or both, equal their limiting values, K and M, respectively. Every partial or unnecessary pathway will be contained in at least one necessary pathway.

In the simplest method, MATij is assigned the value, one, if Aj is the same kind of amino acid as Bi; if they are different amino acids, MATij is assigned the value, zero. The sophistication of the comparison is increased if, instead of zero or one, each cell value is made a function of the composition of the proteins, the genetic code triplets representing the amino acids, the neighboring cells in the array, or any theory concerned with the significance of a pair of amino acids. A penalty factor, a number subtracted for every gap made, may be assessed as a barrier to allowing the gap. The penalty factor could be a function of the size and/or direction of the gap. No gap would be allowed in the operation unless the benefit from allowing that gap would exceed the barrier. The maximum-match pathway then, is that pathway for which the sum of the assigned cell values (less any penalty factors) is largest. MAT can be broken up into subsections operated upon independently. The method also can be expanded to allow simultaneous comparison of several proteins using the amino acid sequences of n proteins to generate an n-dimensional array whose cells represent all possible combinations of n amino acids, one from each protein.

The maximum-match pathway can be obtained by beginning at the terminals of the sequences ($i = y, j = z$) and proceeding toward the origins, first by adding to the value of each cell possessing indices $i = y - 1$ and/or $j = z - 1$, the maximum value from among all the cells which lie on a pathway to it. The process is repeated for indices $i = y - 2$ and/or $j = z - 2$. This increment in the indices is continued until all cells in the matrix have been operated upon. Each cell in this outer row or column will contain the maximum number of matches that can be obtained by originating any pathway at that cell and the largest number in that row or column is equal to the maximum match; the maximum-match pathway in any row or column must begin at this number. The operation of successive summations of cell values is illustrated in Figures 1 and 2.

	A	B	C	N	J	R	Q	C	L	C	R	P	M
A	1												
J					1								
C			1					1	1				
J					1								
N				1									
R							1	4	3	3	2	2	0
C	3	3	4	3	3	3	3	4	3	3	1	0	0
K	3	3	3	3	3	3	3	3	3	2	1	0	0
C	2	2	3	2	2	2	2	3	2	3	1	0	0
R	2	1	1	1	1	2	1	1	1	1	2	0	0
B	1	2	1	1	1	1	1	1	1	1	1	0	0
P	0	0	0	0	0	0	0	0	0	0	0	1	0

FIG. 1. The maximum-match operation for necessary pathways.
The number contained in each cell of the array is the largest number of identical pairs that can be found if that cell is the origin for a pathway which proceeds with increases in running indices. Identical pairs of amino acids were given the value of one. Blank cells which represent non-identical pairs have the value, zero. The operation of successive summations was begun at the last row of the array and proceeded row-by-row towards the first row. The operation has been partially completed in the R row. The enclosed cell in this row is the site of the cell operation which consists of a search along the subrow and subcolumn indicated by borders for the largest value, 4 in subrow C. This value is added to the cell from which the search began.

	A	B	C	N	J	R	Q	C	L	C	R	P	M
A	8	7	6	6	5	4	4	3	3	2	1	0	0
J	7	7	6	6	6	4	4	3	3	2	1	0	0
C	6	6	7	6	5	4	4	4	3	3	1	0	0
J	6	6	6	5	6	4	4	3	3	2	1	0	0
N	5	5	5	6	5	4	4	3	3	2	1	0	0
R	4	4	4	4	4	5	4	3	3	2	2	0	0
C	3	3	4	3	3	3	3	4	3	3	1	0	0
K	3	3	3	3	3	3	3	3	3	2	1	0	0
C	2	2	3	2	2	2	2	3	2	3	1	0	0
R	2	1	1	1	1	2	1	1	1	1	2	0	0
B	1	2	1	1	1	1	1	1	1	1	1	0	0
P	0	0	0	0	0	0	0	0	0	0	0	1	0

FIG. 2. Contributors to the maximum match in the completed array.
The alternative pathways that could form the maximum match are illustrated. The maximum match terminates at the largest number in the first row or first column, 8 in this case.

It is apparent that the above array operation can begin at any of a number of points along the borders of the array, which is equivalent to a comparison of N-terminal residues or C-terminal residues only. As long as the appropriate rules for pathways are followed, the maximum match will be the same. The cells of the array which contributed to the maximum match, may be determined by recording the origin of the number that was added to each cell when the array was operated upon.

3. Evaluating the Significance of the Maximum Match

A given maximum match may represent the maximum number of amino acids matched, or it may just be a number that is a complex function of the relationship between sequences. It will, however, always be a function of both the amino acid compositions of the proteins and the relationship between their sequences. One may ask whether a particular result found differs significantly from a fortuitous match between two random sequences. Ideally,one would prefer to know the exact probability of obtaining the result found from a pair of random sequences and what fraction of the total possibilities are less probable, but that is prohibitively difficult, especially if a complex function were used for assigning a value to the cells.

As an alternative to determining the exact probabilities, it is possible to estimate the probabilities experimentally. To accomplish the estimate one can construct two sets of random sequences, a set from the amino acid composition of each of the proteins compared. Pairs of random sequences can then be formed by randomly drawing one member from each set. Determining the maximum match for each pair selected will yield a set of random values. If the value found for the real proteins is significantly different from the values found for the random sequences, the difference is a function of the sequences alone and not of the compositions. Alternatively, one can construct random sequences from only one of the proteins and compare them with the other to determine a set of random values. The two procedures measure different probabilities. The first procedure determines whether a significant relationship exists *between* the real sequences. The second procedure determines whether the relationship *of* the protein used to form the random sequences *to* the other proteins is significant. It bears reiterating that the integral amino acid composition of each random sequence must be equal to that of the protein it represents.

The amino acid sequence of each protein compared belongs to a set of sequences which are permutations. Sequences drawn randomly from one or both of these sets are used to establish a distribution of random maximum-match values which would include all possible values if enough comparisons were made. The null hypothesis, that any sequence relationship manifested by the two proteins is a random one, is tested. If the distribution of random values indicates a small probability that a maximum match equal to, or greater than, that found for the two proteins could be drawn from the random set, the hypothesis is rejected.

4. Cell Values and Weighting Factors

To provide a theoretical framework for experiments, amino acid pairs may be classified into two broad types, identical and non-identical pairs. From 20 different amino acids one can construct 180 possible non-identical pairs. Of these, 75 pairs of amino acids have codons (Marshall, Caskey & Nirenberg, 1967) whose bases differ at only one position (Eck & Dayhoff, 1966). Each change is presumably the result of a single-point mutation. The majority of non-identical pairs have a maximum of only one or zero corresponding bases. Due to the degeneracy of the genetic code, pair differences representing amino acids with no possible corresponding bases are uncommon even in randomly selected pairs. If cells are weighted in accordance with the maximum number of corresponding bases in codons of the represented amino acids, the maximum match will be a function of identical and non-identical pairs. For comparisons in general, the cell weights can be chosen on any basis.

If every possible sequence gap is allowed in forming the maximum match, the significance of the maximum match is enhanced by decreasing the weight of those pathways containing a large number of gaps. A simple way to accomplish this is to assign a penalty factor, a number which is subtracted from the maximum match for each gap used to form it. The penalty is assigned before the maximum match is formed. Thus the pathways will be weighted according to the number of gaps they contain, but the nature of the contributors to the maximum match will be affected as well. In proceeding from one cell to the next in a maximum-match pathway, it is necessary that the difference between each cell value and the penalty, be greater than the value for a cell in a pathway that contains no gap. If the value of the penalty were zero, all possible gaps could be allowed. If the value were equal to the theoretical value for the maximum match between two proteins, it would be impossible to allow a gap and the maximum match would be the largest of the values found by simply summing along the diagonals of the array; this is the simple frame-shift method.

5. Application of the Method

To illustrate the role of weighting factors in evaluating a maximum match, two proteins expected to show homology, whale myoglobin (Edmundson, 1965) and human β-hemoglobin (Konigsberg, Goldstein & Hill, 1963), and two proteins not expected to exhibit homology, bovine pancreatic ribonuclease (Smyth, Stein & Moore, 1963) and hen's egg lysozyme (Canfield, 1963) were chosen for comparisons.

The FORTRAN programs used in this study were written for the CDC3400 computer. The operations employed in forming the maximum match are those for the special case when none of the cells of the array have a value less than zero. Four types of amino acid pairs were distinguished and variable sets consisting of values to be assigned to each type of pair and a value for the penalty were established. The pair types are as follows:

Type 3. Identical pairs: those having a maximum of three corresponding bases in their codons.

Type 2. Pairs having a maximum of two corresponding bases in their codons.

Type 1. Pairs having a maximum of one corresponding base in their codons.

Type 0. Pairs having no possible corresponding base in their codons.

The value for type 3 pairs was 1·0 and the value for type 0 pairs was zero for all variable sets.

At program execution time, the amino acids (coded by two-digit numbers) of the sequences to be compared were read into the computer, and were followed by a twenty-by-twenty symmetrical array, the maximum correspondence array, analogous to one used by Fitch (1966), that contained all possible pairs of amino acids and identified each pair as to type. The RNA codons for amino acids used to construct the maximum-correspondence array were taken from a single Table (Marshall *et al.*,

1967). The UGA, UAA and UAG codons were not used, but UUG was used as a codon for leucine. The subsequent data cards indicated the numerical values for a variable set.

The two-dimensional comparison array was generated row-by-row. The amino acid code numbers for Ai and Bj referenced the correspondence array to determine the type of amino acid pair constituted by Ai and Bj. The type number referenced a short array, the variable set, containing the type values, and the appropriate value from that set was assigned to the appropriate cell of the comparison array. The maximum match was then determined by the procedure of successive summations.

Following the determination of the maximum match for the real proteins, the amino acid sequence of only one member of the protein pair was randomized and the match was repeated. The sequences of β-hemoglobin and ribonuclease were the ones randomized. The randomization procedure was a sequence shuffling routine based on computer-generated random numbers. A cycle of sequence randomization–maximum-match determination was repeated ten times in all of the experiments in this report, giving the random values used for comparison with the real maximum-match. The average and standard deviation for the random values of each variable set was estimated.

6. Results and Discussion

The use of a small random sample size (ten) was necessary to hold the computer time to a reasonable level. The maximum probable error in a standard deviation estimate for a sample this small is quite large and the results should be judged with this fact in mind. For each set of variables, it was assumed that the random values would be distributed in the fashion of the normal-error curve; therefore, the values of the first six random sets in the β-hemoglobin–myoglobin comparison were converted to standard measure, five was added to the result, and these values were plotted as *one* group against their calculated probit. The results of the plot are shown in Figure 3. The fit is good indicating the probable adequacy of the measured standard deviations for these variable sets in estimating distribution functions for random values through two standard deviations. The above fit indicates no bias in the randomization procedure. In other words, randomization of the sequence was complete before the maximum match was determined for any sequence in a random set.

The results obtained in the comparison of β-hemoglobin with myoglobin are summarized in Table 1 and the results for the ribonuclease–lysozyme comparison are in Table 2. These Tables indicate the values assigned to the pair types, the penalty factor used in forming each of the maximum matches, and the statistical results obtained. The number of gaps roughly characterizes the nature of the pathway that formed the maximum match. A large number is indicative of a devious pathway through the array. One gap means that all of the pathway may be found on only two partial diagonals of the array.

The most important information is obtained from the standardized value of the maximum match for the real proteins, the difference from the mean in standard-deviation units. For this sample size all deviations greater than 3·0 were assumed to include less than 1% of the true random population and to indicate a significant difference. As might be expected, all matches of myoglobin and β-hemoglobin show a significant deviation. Among the sets of variables, set 1, which results in a search for identical amino acid pairs while allowing for all deletions, indicates that 63

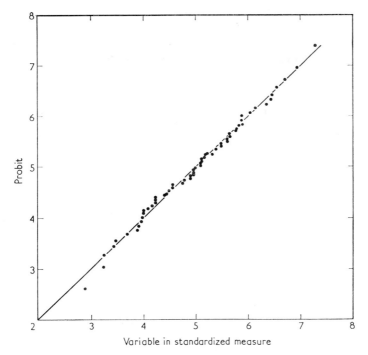

FIG. 3. Probit plot for six grouped random samples.

The solid line indicates the plot that would result from a probit analysis on an infinite number of samples from a normally-distributed population. The points represent the results of probit calculations on 60 random maximum match values that were assumed to have come from one population.

TABLE 1

β-Hemoglobin–myoglobin maximum matches

Variable set	Match values for pair types		Penalty	Maximum-match value sum		s	Real X	Minimum deletions	
	2	1		Real	Random†			Real	Random†
1	0	0	0	63·00	55·60	1·80	4·11	35	36·2
2	0	0	1·00	38·00	27·80	2·09	4·88	4	5·5
3	0·67	0·33	0	97·00	91·47	1·55	3·57	18	24·3
4	0·67	0·33	1·03	89·63	80·25	1·11	8·46	1	3·6
5	0·25	0·05	0	71·55	64·78	1·59	4·27	46	45·0
6	0·25	0·05	1·05	51·95	40·54	1·46	7·80	3	7·5
7	0·25	0·05	25	47·30	33·80	1·52	8·87	0	0

s is the estimated standard deviation; X, the standardized value, (real − random)/s, of the maximum match of the real proteins. The values for type 3 and type 0 pairs were 1·0 and 0, respectively, in each variable set.

† An average value from 10 samples.

TABLE 2

Ribonuclease–lysozyme maximum matches

Variable set	Match values for pair types		Penalty	Maximum-match value sum		s	Real X	Minimum deletions	
	2	1		Real	Random†			Real	Random†
1	0	0	0	48·00	44·20	2·56	1·48	34	29·2
2	0	0	1·00	23·00	22·00	1·73	0·58	5	5·2
3	0·67	0·33	0	78·33	76·17	0·82	2·64	21	18·8
4	0·67	0·33	1·03	67·93	67·37	1·27	0·43	2	2·2
5	0·25	0·05	0	56·00	52·26	2·12	1·77	35	35·5
6	0·25	0·05	1·05	33·70	33·02	1·66	0·41	8	6·8
7	0·25	0·05	25	28·15	27·67	1·75	0·22	0	0

s is the estimated standard deviation; X, the standardized value, (real−random)/s, of the maximum match of the real proteins. The values for type 3 and type 0 pairs were 1·0 and 0, respectively in each variable set.

† An average value from 10 samples.

amino acids in β-hemoglobin and myoglobin can be matched. To attain this match, however, it is necessary to permit at least 35 gaps. In contrast, when two gaps are allowed according to Braunitzer (1965), it is possible to match only 37 of the amino acids. Curiously, when this variable set was used for comparing *human* myoglobin (Hill, personal communication) with human β-hemoglobin, the maximum match obtained was not significant. Differences between real and random values were highly significant, however, when other variable sets were used.

Variable set 2 attaches a penalty equal to the value of one identical amino acid pair to the search for identical amino acid pairs. This penalty will exclude from consideration any possible pathway that leaves and returns to a principal diagonal, thereby needing two gaps, in order to add only one or two amino acids to the maximum match. This set results in a total of $30 + 4 = 42$ amino acids matched (the maximum-match value plus the number of gaps is reduced to four) and the significance of the result relative to set 1 appears to be increased. Braunitzer's comparison would have a value of $37 − 2 = 35$ using this variable set, hence it was not selected by the method.

Variable sets 3 and 4 have an interesting property. Their maximum-match values can be related to the minimum number of mutations needed to convert the selected parts of one amino acid sequence into the selected parts of the other. The minimum number of mutations concept in protein comparisons was first advanced by Fitch (1966). If the type values for these sets are multiplied by three, they become equal to their pair type and directly represent the maximum number of corresponding bases in the codons for a given amino acid pair. Thus the maximum match and penalty factors may be multiplied by three, making it possible to calculate the maximum number of bases matched in the combination of amino acid pairs selected by the maximum-match operation.

β-Hemoglobin, the smaller of the two proteins, contains 146 amino acids; consequently the highest possible maximum match (disregarding integral amino-acid composition data) with myoglobin is $146 \times 3 = 438$. Insufficient data are available to analyze the result from set 3 on the basis of mutations. If it is assumed that the gap in set 4 does not exclude any part of β-hemoglobin from the comparison, this set has a maximum of $3(89\cdot63 + 1\cdot03) = 272$ bases matched, indicating a minimum of $438 - 272 = 166$ point mutations in this combination. Using this variable set and placing gaps according to Braunitzer, a score of $88\cdot6$ was obtained, thus his match was not selected. Again it may be observed that the penalty greatly enhanced the significance of the maximum match.

Variable sets 5 and 6 have no intrinsic meaning and were chosen because the weight attached to type 2 and type 1 pairs is intermediate in value with respect to sets 1 and 2 and sets 3 and 4. The maximum match for set 6 is seen to have a highly significant value.

The data of set 7 are results that would be obtained from using the frame-shift method to select a maximum match; the penalty was large enough to prevent any gaps in the comparisons. The slight differences in significance found among the maximum-match values of β-hemoglobin and myoglobin resulting from use of sets 4, 6 and 7 are probably meaningless due to small sample size and errors introduced by the assumptions about the distribution functions of random values. Finding a value in set 7 that is approximately equal to those from sets 4 and 6 in significance is not surprising. A larger penalty factor would have increased the difference from the mean in sets 4 and 6 because almost every random value in each set was the result of more gaps than were required to form the real maximum match. Further, the gaps that are allowed are at the N-terminal ends so that about 85% of the comparison can be made without gaps. If an actual gap were present near the middle of one of the sequences, it would have caused a sharp reduction in the significance of the frame-shift type of match.

Set 3 is the only variable set in Table 2 that shows a possible difference. Assuming the value is accurate, other than chance, there is no simple explanation for the difference. A small but meaningful difference in any comparison could represent evolutionary divergence or convergence. It is generally accepted that the primary structure of proteins is the chief determinant of the tertiary structure. Because certain features of tertiary structure are common to proteins, it is reasonable to suppose that proteins will exhibit similarities in their sequences, and that these similarities will be sufficient to cause a significant difference between most protein pairs and their corresponding randomized sequences, being an example of submolecular evolutionary convergence. Further, the interactions of the protein backbone, side chains, and the solvent that determine tertiary structure are, in large measure, forces arising from the polarity and steric nature of the protein side-chains. There are conspicuous correlations in the polarity and steric nature of type 2 pairs. Heavy weighting of these pairs would be expected to enhance the significance of real maximum-match values if common structural features are present in proteins that are compared. The presence of sequence similarities does not always imply common ancestry in proteins. More experimentation will be required before a choice among the possibilities suggested for the result from set 3 can be made. If several short sequences of amino acids are common to all proteins, it seems remarkable that the relationship of ribonuclease to lysozyme in six of the seven variable sets appears to be truly a random one. It should be noted, however, that the standard value of the real maximum-match is positive in each variable set in this comparison.

This method was designed for the purpose of detecting homology and defining its nature when it can be shown to exist. Its usefulness for the above purposes depends in part upon assumptions related to the genetic events that could have occurred in the evolution of proteins. Starting with the assumption that homologous proteins are the result of gene duplication and subsequent mutations, it is possible to construct several hypothetical amino-acid sequences that would be expected to show homology. If one assumes that following the duplication, point mutations occur at a constant or variable rate, but randomly, along the genes of the two proteins, after a relatively short period of time the protein pairs will have nearly identical sequences. Detection of the high degree of homology present can be accomplished by several means. The use of values for non-identical pairs will do little to improve the significance of the results. If no, or very few, deletions (insertions) have occurred, one could expect to enhance the significance of the match by assigning a relatively high penalty for gaps. Later on in time the hypothetical proteins may have a sizable fraction of their codons changed by point mutations, the result being that an attempt to increase the significance of the maximum match will probably require attaching substantial weight to those pairs representing amino acids still having two of the three original bases in their codons. Further, if a few more gaps have occurred, the penalty should be reduced to a small enough value to allow areas of homology to be linked to one another. At a still later date in time more emphasis must be placed on non-identical pairs, and perhaps a very small or even negative penalty factor must be assessed. Eventually, it will be impossible to detect the remaining homology in the hypothetical example by using the approach detailed here.

From consideration of this simple model of protein evolution one may deduce that the variables which maximize the significance of the difference between real and random proteins gives an indication of the nature of the homology. In the comparison of human β-hemoglobin to whale myoglobin, the assignment of some weight to type 2 pairs considerably enhances the significance of the result, indicating substantial evolutionary divergence. Further, few deletions (additions) have apparently occurred.

It is known that the evolutionary divergence manifested by cytochrome (Margoliash, Needleman & Stewart, 1963) and other heme proteins (Zuckerkandl & Pauling, 1965) did not follow the sample model outlined above. Their divergence is the result of *non-random* mutations along the genes. The degree and type of homology can be expected to differ between protein pairs. As a consequence of the difference there is no *a priori* best set of cell and operation values for maximizing the significance of a maximum-match value of homologous proteins, and as a corollary to this fact, there is no best set of values for the purpose of detecting only slight homology. This is an important consideration, because whether the sequence relationship between proteins is significant depends solely upon the cell and operation values chosen. If it is found that the divergence of proteins follows one or two simple models, it may be possible to derive a set of values that will be most useful in detecting and defining homology.

The most common method for determining the degree of homology between protein pairs has been to count the number of non-identical pairs (amino acid replacements) in the homologous comparison and to use this number as a measure of evolutionary distance between the amino acid sequences. A second, more recent concept has been to count the minimum number of mutations represented by the non-identical pairs. This number is probably a more adequate measure of evolutionary distance because it utilizes more of the available information and theory to give some measure of the number of genetic events that have occurred in the evolution of the proteins. The approach outlined in this paper supplies either of these numbers.

This work was supported in part by grants to one of us (S.B.N.) from the U.S. Public Health Service (1 501 FR 05370 02) and from Merck Sharp & Dohme.

REFERENCES

Braunitzer, G. (1965). In *Evolving Genes and Proteins*, ed. by V. Bryson & H. J. Vogel, p. 183. New York: Academic Press.

Canfield, R. (1963). *J. Biol. Chem.* **238**, 2698.

Eck, R. V. & Dayhoff, M. O. (1966). *Atlas of Protein Sequence and Structure*. Silver Spring, Maryland: National Biomedical Research Foundation.

Edmundson, A. B. (1965). *Nature*, **205**, 883.

Fitch, W. (1966). *J. Mol. Biol.* **16**, 9.

Konigsberg, W., Goldstein, J. & Hill, R. J. (1963). *J. Biol. Chem.* **238**, 2028.

Margoliash, E., Needleman, S. B. & Stewart, J. W. (1963). *Acta Chem. Scand.* **17**, S 250.

Marshall, R. E., Caskey, C. T. & Nirenberg, M. (1967). *Science*, **155**, 820.

Needleman, S. B. & Blair, T. H. (1969). *Proc. Nat. Acad. Sci., Wash.* **63**, 1127.

Smyth, D. G., Stein, W. G. & Moore, S. (1963). *J. Biol. Chem.* **238**, 227.

Zuckerkandl, E. & Pauling, L. (1965). In *Evolving Genes and Proteins*, ed. by V. Bryson & H. J. Vogel, p. 97. New York: Academic Press.

J. Mol. Biol. (1970) **51**, 393–409

A Restriction Enzyme from *Hemophilus influenzae*

II. Base Sequence of the Recognition Site

Thomas J. Kelly, Jr.† and Hamilton O. Smith

Department of Microbiology
Johns Hopkins University, School of Medicine
Baltimore, Md. 21205, U.S.A.

(Received 18 February 1970)

Hemophilus influenzae strain Rd contains an enzyme, endonuclease R, which specifically degrades foreign DNA. With phage T7 DNA as substrate the endonuclease introduces a limited number (about 40) double-strand breaks (5′-phosphoryl, 3′-hydroxyl). The limit product has an average length of about 1000 nucleotide pairs and contains no single-strand breaks. We have explored the nucleotide sequences at the 5′-ends of the limit product by labeling the 5′- phosphoryl groups (using polynucleotide kinase) and characterizing the labeled fragments released by various nucleases. Two classes of 5′-terminal sequences were obtained: pApApCpNp . . . (60%) and pGpApCpNp . . . (40%), where N indicates that the base in the 4th position is not unique. The dinucleoside monophosphates at the 3′-ends were isolated after micrococcal nuclease digestion of the limit product and identified as TpT(60%) and TpC (40%). We conclude that endonuclease R of *H. influenzae* recognizes the following specific nucleotide sequence:

$$5' \ldots \text{pGpTpPy} | \text{pPupApCp} \ldots 3'$$
$$3' \ldots \text{pCpApPup} | \text{PypTpGp} \ldots 5'$$

The implications of the twofold rotational symmetry of this sequence are discussed.

1. Introduction

The accompanying paper (Smith & Wilcox, 1970) describes the purification and properties of endonuclease R of *Hemophilus influenzae*, an enzyme which appears to be capable of recognizing and degrading foreign DNA molecules. The purified enzyme has no detectable nucleolytic activity against either duplex or single-stranded *H. influenzae* DNA molecules but produces a limited number of duplex cleavages (5′-phosphoryl, 3′-hydroxyl) in a variety of native foreign DNA molecules. The limit product of endonuclease R digestion has an average chain length of the order of 1000 nucleotide pairs and contains no single-strand breaks. These properties are qualitatively similar to those of the *Escherichia coli* restriction enzymes studied by Meselson & Yuan (1968) and Linn & Arber (1968).

The fact that endonuclease R hydrolyzes only about 0·1% of the phosphodiester linkages potentially available to it strongly suggests that the enzyme "recognizes" a relatively small number of specific sites within a foreign DNA molecule. The experiments described in this paper were motivated by the assumption that the specificity of this recognition process is determined by the local nucleotide sequence at these

† Present address: National Institute of Allergy and Infectious Diseases, National Institutes of Health, Bethesda, Md. 20014, U.S.A.

sites. Using end-labeling techniques, we have explored the sequences at the ends of the fragments released by digestion with endonuclease R and succeeded in identifying the complete base sequence of its recognition site.

2. Materials and Methods

(a) Enzymes

Electrophoretically pure pancreatic DNase (2000 units/mg) was obtained from Sigma Chemical Co. Bacterial alkaline phosphatase (20 units/mg), snake venom phosphodiesterase (potency 0·3), micrococcal nuclease (1000 units/mg) and spleen phosphodiesterase were obtained from Worthington Biochemical Corp.

Polynucleotide kinase (fraction VI of Richardson (1965), (5000 units/ml.) and rechromatographed bacterial alkaline phosphatase, 20 units/ml.(Weiss, Live & Richardson, 1968). were supplied by Dr Bernard Weiss.

Exonuclease I from *E. coli* was received from Dr Paul Englund as the ammonium sulfate I fraction of a DNA polymerase purification and was reprecipitated with ammonium sulfate and chromatographed on DEAE-cellulose as described by Lehman & Nussbaum (1964). The final specific activity was 20,200 units/mg.

The purification of endonuclease R of *H. influenzae* has been described (Smith & Wilcox 1970). The final 200-fold purified preparation contained 16 units/ml.

(b) Preparation of ^{33}P-labeled phage T7 DNA

Phage T7 (from B. Weiss) was grown on *E. coli* B in a synthetic medium containing 1·5 μg phosphorus/ml. (Smith, 1968) and 2·5 μc ^{33}P/ml. (New England Nuclear Corp.; supplied as ^{33}P-labeled phosphoric acid containing 0·02 N-HCl). Following lysis the phage were purified by differential centrifugation and by banding twice in a preformed CsCl gradient (Thomas & Abelson, 1966). Phenol extraction was carried out as described by Kelly & Thomas (1969). After phenol extraction the DNA solution was exhaustively dialyzed against 0·05 M-NaCl, 0·01 M-Tris, pH 7·4. The preparation of ^{32}P-labeled T7 DNA was carried out by the same methods used for ^{33}P-labeled T7 DNA except that 5μc of carrier-free [^{32}P]orthophosphate (New England Nuclear Corp.) /ml. was substituted for the [^{33}P]phosphoric acid in the growth medium.

(c) Fractionation of nucleotides and oligonucleotides

Electrophoresis of nucleotides and oligonucleotides was performed on cellulose thin-layer sheets (Eastman Kodak Co.) at a potential gradient of 25 v/cm in a Warner–Chilcott model E-800-2B migration chamber cooled to 5°C. For routine separations 0·075 M-ammonium formate buffer, pH 3·55, was used. When a second round of electrophoretic fractionation was required, 8·7% acetic acid, 2·5% formic acid (v/v), pH 1·9, (Sanger, Brownlee & Barrell, 1965) was used.

Two-dimensional chromatographic separation of nucleotides on 6 cm × 6 cm PEI-cellulose thin-layers (Brinkman Instruments, Inc., Cel MN 300 PEI) was carried out by methods similar to those described by Randerath & Randerath (1967). The chromatograms were developed in the first dimension with 0·5 M-LiCl and in the second dimension with 0·5 M-acetic acid. Between dimensions the chromatograms were air dried and then desalted by immersion in anhydrous methanol for 15 min. Following chromatography the nucleotide spots were located with an ultraviolet lamp and cut from the thin layer for assay of radioactivity in a scintillation counter. With this procedure a nucleotide analysis can be performed in about 1 hr.

Fractionation of oligonucleotides according to net charge was carried out by the method of Tomlinson & Tener (1963). The sample containing 0·02 M-Tris, pH 7·6, and 7 M-urea was loaded onto a 1 cm × 10 cm column of DEAE-cellulose (Serva; in bicarbonate form) which had been equilibrated with the same solvent. The column was then developed with a linear gradient of NaCl (0 to 0·3 M) in 0·02 M-Tris, pH 7·6, 7 M-urea. The total gradient volume was generally 1000 ml. and the flow rate was 50 ml./hr. The oligonucleotides were subsequently freed from salt and urea by adsorption to DEAE-cellulose (Rushizsky & Sober, 1962) or partially inactivated Norit (Threlfall, 1957; Josse & Moffatt, 1965).

(d) Autoradiography

The distribution of radioactivity on a thin-layer sheet following fractionation of a collection of oligonucleotides by electrophoresis was generally determined by autoradiography. In a number of the experiments to be described in this paper the oligonucleotides were uniformly labeled with ^{33}P. However, those oligonucleotide species which originated from the 5′-termini of an endonuclease R limit product also contained ^{32}P (see section (e) below). In order to monitor the distribution of both labels the usual autoradiographic procedure was modified as follows: a sheet of X-ray film (Eastman Kodak Co., NS-54T) was placed in direct contact with the thin-layer matrix (unscreened autoradiograph). Both ^{33}P and ^{32}P disintegrations contributed to grain exposure on this film. A second sheet of X-ray film (screened autoradiograph) was separated from the thin-layer matrix by a thin screen. (The Mylar backing of commercial thin layers was found to be suitable.) Because of absorption of the lower energy ^{33}P disintegrations by the screen, grain exposure on this film was primarily due to ^{32}P disintegrations. Using this method it was possible to tell at a glance which of a number of oligonucleotide species contained ^{32}P and therefore originated from the 5′-ends of an endonuclease R limit product. Densitometer tracings of the autoradiographs were made with a Joyce–Loebl recording microdensitometer.

(e) ^{32}P-labeling of the 5′-end of uniformly ^{33}P-labeled endonuclease R limit product

A preparation of uniformly ^{33}P-labeled phage T7 DNA was digested to a limit product with endonuclease R (Smith & Wilcox, 1970). The reaction mixture (1·2 ml.) contained 500 mμmoles of ^{33}P-labeled T7 DNA (spec. act. of the order of 10^4 cts/min/mμmole), 50 mM-NaCl, 40 mM-Tris, pH 7·4, 7 mM-MgCl$_2$, 10 mM-mercaptoethanol and 0·6 unit of endonuclease R. After incubation for 15 min at 37°C an additional 0·6 unit of endonuclease R was added to the reaction mixture and the incubation was continued for another 15 min at 37°C. The resulting digest was then dialyzed twice against 1 liter of 0·05 M-NaCl, 0·01 M-Tris, pH 8·0.

The dialyzed preparation of endonuclease R limit product (1·0 ml.) was incubated for 30 min at 37°C with 2 units of rechromatographed alkaline phosphatase. One additional unit of phosphatase was then added and incubation was continued for 30 min at 37°C. The resulting digest was adjusted to 0·05 M in EDTA, extracted twice with redistilled phenol (saturated with 0·1 M-Tris, pH 7·4), and exhaustively dialyzed against 0·05 M-NaCl, 0·01 M-Tris, pH 7·4.

The resulting preparation of dephosphorylated endonuclease R limit product was rephosphorylated using [γ-^{32}P]ATP in the polynucleotide kinase reaction as described by Weiss et al. (1968). The reaction mixture (1·1 ml.) contained 500 mμmoles of dephosphorylated ^{33}P-labeled limit product, 50 mM-NaCl, 50 mM-Tris (pH 7·4), 10 mM-MgCl$_2$, 15 mM-mercaptoethanol, 1·5 mM-potassium phosphate buffer (pH 7·4) 15 μM-[γ-^{32}P]ATP, and 150 units of polynucleotide kinase. The reaction mixture was incubated for 1 hr at 37°C with the addition of another 75 units of kinase at 30 min. Precipitation of the 5′-^{32}P-end labeled limit product was accomplished by addition of 1 vol. of 0·1 M-sodium pyrophosphate and 1·33 vol. of 12·5% trichloroacetic acid. The precipitate was collected on a glass filter (Whatman GF/C) and washed extensively with 6% trichloroacetic acid, 0·1 M-sodium pyrophosphate followed by 95% ethanol. The DNA was recovered by macerating the filter in 1 ml. of 0·2 N-NH$_4$OH and then filtering the resulting suspension through a semi-micro, medium-sintered glass funnel. After washing the funnel twice with 1 ml. of 0·2 N-NH$_4$OH the filtrate containing the dissolved DNA was evaporated to the desired volume under a warm air jet.

The [γ-^{32}P]ATP used in this labeling procedure was provided by B. Weiss and by P. Englund. The specific activities used were in the range of 10^5 to 10^6 cts/min/mμmole.

(f) Preparation of 5′-terminal oligonucleotides from an endonuclease R limit digest of phage T7 DNA

(i) Mononucleotides

Digestion of 5′-^{32}P-terminally labeled endonuclease R limit product to mononucleotides was carried out by consecutive hydrolysis with pancreatic DNase and snake venom phosphodiesterase. A reaction mixture (0·1 ml.) containing 150 μM-DNA (prepared as described in section (e)), 10 mM-Tris (pH 7·4), 5 mM-MgCl$_2$ and 20 μg pancreatic DNase/ml. was incubated at 37°C for 30 min to yield oligonucleotides. The pancreatic DNase was then inactivated by heating to 95°C for 5 min. A 10-μl. sample of the digest was digested to mononucleotides by adding 1 μl. of 1 M-glycine buffer, pH 9·2, and 1 μl. of snake venom phosphodiesterase (4 mg/ml.) and incubating at 37°C for 30 min.

(ii) Dinucleotides

A 10-μl. sample of the above pancreatic DNase digest was adjusted to 0·1 M in glycine buffer, pH 9·2. Exonuclease I was added to a final concentration of 4×10^3 units/ml. and the reaction mixture incubated at 37°C for 30 min. This procedure is similar to that described by Weiss & Richardson (1967).

(iii) Trinucleotides

A portion of 5'-^{32}P-terminally labeled endonuclease R limit product was digested to an average chain length of 5 residues with pancreatic DNase. The reaction mixture (1 ml.) contained 300 μM-DNA, 10 mM-Tris (pH 7·4), 5 mM-MgCl$_2$, 50 mM-NaCl and 20 μg pancreatic DNase/ml. After incubation at 37°C for 45 min the reaction mixture was heated to 95°C for 5 min. An average of 1 to 2 nucleotides was then removed from the 3'-ends of the oligonucleotides in this digest by partial digestion with snake venom phosphodiesterase (Holley, Madison & Zamir, 1964). The pancreatic DNase digest was equilibrated at 20°C and 0·25 ml. of snake venom phosphodiesterase (40 μg/ml. in 0·06 M-MgCl$_2$) was added. Incubation at 20°C was continued for 15 min and then the digest was quickly frozen in a salt–ice bath. The average chain length of the final product was 2·1 residues. Preparatory to fractionation by DEAE-cellulose chromatography, 0·84 g of urea was added to the frozen oligonucleotide mixture, which was then allowed to thaw at room temperature. After addition of 0·04 ml. of 1 M-Tris, pH 7·6, the volume of the solution was adjusted to 2·0 ml. Fractionation according to chain length was carried out as described in section (c) above. The trinucleotide fraction was desalted and concentrated to 20 μl. for electrophoretic analysis.

(iv) Tetranucleotides

A solution containing 150 mμmoles of 5'-^{32}P-terminally labeled endonuclease R limit product was digested with pancreatic DNase under the conditions described in section (i) above. The tetranucleotides were isolated from the resulting digest by methods similar to those described for the isolation of trinucleotides.

(g) Determination of oligonucleotide chain length

The number-average chain length of a given collection of ^{33}P- or ^{32}P-labeled oligonucleotides was determined by measuring the ratio of total radioactivity to radioactivity released as P$_i$ by alkaline phosphatase treatment. The reaction mixture for phosphatase treatment (0·04 ml.) contained less than 1 mμmole of oligonucleotide, 0·125 M-Tris (pH 8·0) and 3 units of alkaline phosphatase/ml. Incubation was carried out at 37°C for 30 min. A sample of the reaction mixture was removed for determination of total radioactivity (T). A second sample was applied 1 cm from the bottom edge of a 1 cm × 5 cm strip of PEI-cellulose thin layer which had been banded with a Norit suspension (20% packed volume) in a 1-cm width, 2 cm from the bottom edge. The chromatographic strip was dried and then developed with 1 N-HCl for 15 min. In this procedure acid-insoluble DNA is bound at the origin, soluble oligonucleotides are quantitatively absorbed by the Norit band, and only inorganic phosphate migrates to the upper 1·5 cm of the strip. This upper 1·5 cm segment was dried, cut from the strip, and assayed for radioactivity (P). The average chain length is given by T/P.

(h) Isolation of 3'-terminal dinucleotides of endonuclease R limit product

Uniformly ^{32}P-labeled phage T7 DNA was digested to a limit product with endonuclease R in a reaction mixture (1·7 ml.) containing 375 mμmoles of DNA, 6 mM-MgCl$_2$, 6 mM-mercaptoethanol, 50 mM-NaCl, 10 mM-Tris, pH 7·4, and 0·4 unit of endonuclease R. After incubation at 37°C for 20 min an additional 0·4 unit of endonuclease R was added and the incubation was continued for another 10 min. The limit digest was extracted once with phenol (saturated with 0·1 M-Tris, pH 7·4) and then exhaustively extracted with ether to remove the phenol. The endonuclease R limit product was then digested to completion with micrococcal nuclease. The reaction mixture was constructed by addition of 0·16 ml. of dGMP (20 mM), 0·05 ml. CaCl$_2$ (0·1 M), and 0·015 ml. micrococcal nuclease (7500 units/ml.) to the phenol-extracted endonuclease R limit digest. After incubation at 37°C for 15 min an additional 5 μl. of micrococcal nuclease was added and incubation was continued for another 15 min at 37°C. The dinucleoside monophosphates from the 3'-ends of the endonuclease R limit product were the only species in the micrococcal digest which possessed a single negative charge at pH 7·6. They were recovered by fractionating the digest according to net charge by the method described in section (c). (Note: the dGMP was included in the micrococcal nuclease reaction mixture to reduce the possibility of production of spurious dinucleoside monophosphate species by any residual phosphatase activity.)

3. Results

(a) Nucleotide sequences at the 5'-ends of the limit product of endonuclease R digestion of T7 DNA

As described in Smith & Wilcox (1970), endonuclease R of H. influenzae introduces approximately 40 double-strand breaks in phage T7 DNA. The cleavage is such as to produce 5'-phosphoryl and 3'-hydroxyl end groups. Figure 1 shows the general scheme used in analyzing the nucleotide sequences at the 5'-termini of the limit product of endonuclease R digestion of T7 DNA. Uniformly ^{33}P-labeled T7 DNA was

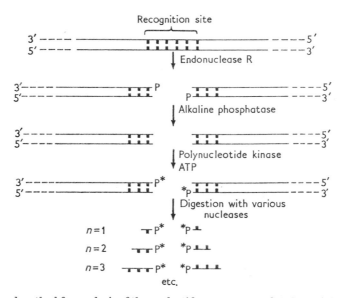

FIG. 1. General method for analysis of the nucleotide sequences at the 5'-termini of endonuclease R limit product.

Uniformly ^{33}P-labeled phage T7 DNA was digested to a limit product with endonuclease R of H. influenzae. The 5'-phosphoryl end-groups produced during this digestion were removed with alkaline phosphatase. ^{32}P-labeled phosphoryl groups were then transferred from [γ-^{32}P]ATP to the dephosphorylated 5'-termini of the limit product using polynucleotide kinase (Richardson, 1965). The 5'-terminal mono-, di- and trinucleotides were isolated following degradation of the terminally labeled limit product with various nucleases and identified as described in the text. Note: in the above Figure endonuclease R is pictured as producing an "even" duplex break at the exact center of its recognition site. It is possible to imagine a number of other possible patterns of cleavage. This particular pattern is shown only to simplify the illustration and does not imply any a priori assumptions concerning the mechanism of action of the enzyme.

digested to completion with endonuclease R. The 5'-phosphoryl groups produced during this digestion were removed with alkaline phosphatase. ^{32}P-labeled phosphoryl groups were then transferred from [γ-^{32}P]ATP to the dephosphorylated 5'-termini using polynucleotide kinase (Richardson, 1965). The resulting product (referred to below as 5'-^{32}P-terminally labeled endonuclease R limit product) was digested with various nucleases and the ^{32}P-labeled fragments released were isolated and characterized.

(i) 5'-Terminal nucleotides

A preparation of ^{33}P-labeled T7 DNA was digested to a limit product with endonuclease R. The 5'-ends were labeled with ^{32}P-phosphoryl groups as described above. A portion of this terminally labeled DNA was hydrolyzed to mononucleotides by sequential digestion with pancreatic DNase and snake venom phosphodiesterase, and the resulting digest was analyzed by electrophoresis on thin-layer cellulose at pH 3·55. Figure 2 shows densitometer tracings of unscreened (Fig. 2(a)) and screened (Fig. 2(b))

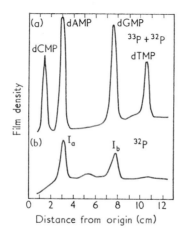

Fig. 2. Electrophoresis of the 5'-terminal mononucleotides of endonuclease R limit product. Endonuclease R limit product labeled at its 5'-termini with ^{32}P and internally with ^{33}P was hydrolyzed to mononucleotides by sequential digestion with pancreatic DNase and snake venom phosphodiesterase as described in Materials and Methods. The mononucleotides were separated by electrophoresis on a cellulose thin-layer sheet in 0·075 M-ammonium formate, pH 3·55, at a potential gradient of 25 v/cm. Following electrophoresis, screened and unscreened autoradiographs of the thin-layer sheet were prepared as described in Materials and Methods section (d).

(a) Microdensitometer tracing of unscreened autoradiograph (^{32}P and ^{33}P); (b) microdensitometer tracing of screened autoradiograph (^{32}P).

autoradiographs of the thin-layer sheet prepared following electrophoresis (see Materials and Methods). Examination of the screened autoradiograph revealed that dAMP and dGMP were the only mononucleotides which contained ^{32}P. However, the radioactivity was not divided equally betweeen the two mononucleotides, dAMP containing about 63% of the total ^{32}P radioactivity and dGMP containing about 37% (see Table 1). This result appears to reflect a real difference in the relative frequency of dAMP terminals *versus* dGMP terminals rather than a difference in the recovery of the two nucleotides, since the distribution of ^{33}P radioactivity closely paralleled the known base composition of T7 DNA. Furthermore, the relative frequency of dAMP and dGMP seems to be a function of the particular DNA substrate

TABLE 1

Analysis of 5′-terminal nucleotides

Nucleotide	[33]P (cts/min)	[32]P (cts/min)
d(pC)	720	0
d(pA)	710	170
d(pG)	690	100
d(pT)	700	5

The areas containing the four 5′-nucleoside monophosphates were cut from the cellulose thin-layer strip of Fig. 2. and assayed for [32]P- and ^{33}P-radioactivity in a liquid-scintillation spectrometer.

used, since analysis of the 5′-terminal nucleotides of an endonuclease R limit digest of phage P22 DNA yielded 85% dAMP and 15% dGMP. The simplest interpretation of these facts is that the nucleotide immediately adjacent to a site of phosphodiester bond cleavage by endonuclease R is not absolutely specified, i.e. the enzyme cleaves both strands of the DNA duplex on the 5′-side of a purine nucleotide but does not unambiguously distinguish between the two possible purine nucleotides, dAMP and dGMP.

(ii) 5′-Terminal dinucleotides

A portion of 5′-^{32}P-terminally labeled endonuclease R limit product was digested with pancreatic DNase. The resulting collection of oligonucleotides was then digested with exonuclease I. Exonuclease I acts from the 3′-hydroxyl end of a single-stranded

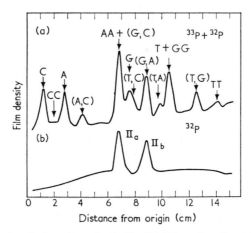

Fig. 3. Electrophoresis of the 5′-terminal dinucleotides of endonuclease R limit product. Endonuclease R limit product labeled at its 5′-termini with ^{32}P and internally with ^{33}P was digested sequentially with pancreatic DNase and exonuclease I as described in Materials and Methods. The resulting collection of mono- and dinucleotides was fractionated by electrophoresis at pH 3·55 as described in the legend to Fig. 2. The known positions of the various dinucleotides are indicated on the tracing of the unscreened autoradiograph (a). Since isomeric dinucleotides were not resolved under the conditions used for electrophoresis, the symbol (X, Y) is used to indicate the position of both pXpY and pYpX.

(a) Microdensitometer tracing of unscreened autoradiograph (^{32}P and ^{33}P); (b) microdensitometer tracing of screened autoradiograph (^{32}P).

TABLE 2

Nucleotide analysis of 5'-terminal dinucleotides

Nucleotide	Species IIa ^{32}P (cts/min)	Speeies IIb ^{32}P (cts/min)
d(pC)	2	1
d(pA)	1485	7
d(pG)	9	857
d(pT)	4	0

The areas containing species IIa and IIb of Fig. 3 were cut from the thin layer, and the radioactive dinucleotides eluted with distilled water. A portion of each was hydrolyzed to mononucleotides with snake venom phosphodiesterase. The reaction mixtures contained 0·1 M-glycine buffer (pH 9·2), 0·01 M-MgCl$_2$ and 0·2 mg snake venom phosphodiesterase/ml. After incubation at 37°C for 30 min, each reaction mixture was fractionated by electrophoresis at pH 3·55. The ^{32}P radioactivity in each mononucleotide was determined by liquid-scintillation counting.

DNA moledule releasing 5'-mononucleotides in stepwise fashion, but is unable to hydrolyze the 5'-terminal dinucleotide (Lehman, 1960). Electrophoretic analysis (pH 3·55) of the exonuclease I digest revealed two distinct ^{32}P-containing dinucleotide species (Fig. 3(b)). Species IIa contained about 62% of the total ^{32}P radioactivity and had an electrophoretic mobility compatible with three possible dinucleotides: pApA, pGpC and pCpG (Fig. 3(a)). Species IIa was eluted from the thin layer and hydrolyzed to 5'-mononucleotides with snake venom phosphodiesterase. The ^{32}P radioactivity was found only in dAMP (Table 2) indicating that species IIa was pApA. Species IIb contained about 38% of the total ^{32}P radioactivity and migrated at a rate compatible with either pGpA or pApG. Upon digestion to mononucleotides the ^{32}P radioactivity was found associated exclusively with dGMP (Table 2) indicating that species IIb was pGpA. This identification of the 5'-terminal dinucleotides of an endonuclease R limit digest of T7 DNA as pApAp and pGpA was confirmed by thin-layer cellulose chromatography against known marker dinucleotides in the solvent system of Markham & Smith (1952) as modified by Weiss & Richardson (1967) in experiments not reported here. On the basis of these results it was concluded that endonuclease R cleaves both strands of the DNA duplex on the 5'-side of the sequence d(pPupA).

(iii) 5'-Terminal trinucleotitdes

Extending the sequence analysis of the 5'-ends of endonuclease R limit product beyond the dinucleotide level presented certain special problems. In particular, no method has been described for obtaining quantitatively higher order terminal oligonucleotides. The following method was employed to obtain a reasonably representative sample of the 5'-terminal trinucleotides. A preparation of 5'-^{32}P-terminally labeled endonuclease R limit product was digested with pancreatic DNase to an average chain length of about five residues/oligonucleotide. An average of about 1·4 residues was then removed from the 3'-end of each chain by partial digestion with snake venom phosphodiesterase. The resulting oligonucleotide mixture was fractionated according to chain length by DEAE-cellulose chromatography in the presence of 7 M-urea (Fig. 4) (Tomlinson & Tener, 1963). The trinucleotide fraction, containing about 20% of the original ^{32}P radioactivity, was collected and analyzed by electrophoresis on cellulose thin-layer at pH 3·55. Two ^{32}P-containing trinucleotide species

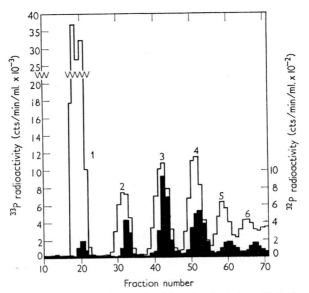

Fig. 4. Fractionation of 5′-terminal oligonucleotides of endonuclease R limit product according to chain length by DEAE-cellulose chromatography in the presence of 7 M-urea.

A preparation of endonuclease R limit product labeled at its 5′-termini with ^{32}P and internally with ^{33}P was digested to an average chain length of 5 residues with pancreatic DNase. The oligonucleotides thus formed were further degraded from the 3′-end with snake venom phosphodiesterase to a final average chain length of 2·1 residues (see Materials and Methods). The resulting digest was loaded onto a 1 cm × 10 cm DEAE-cellulose column, and the oligonucleotides were eluted according to chain length with a linear gradient of NaCl (0 to 0·3 M) in 0·02 M-Tris (pH 7·6) and 7 M-urea. The total gradient volume was 1 l. and the fraction volume was 8 ml. The chain lengths corresponding to the various oligonucleotide fractions are indicated in the Figure.

Shaded area, ^{32}P radioactivity; unshaded area ^{33}P radioactivity.

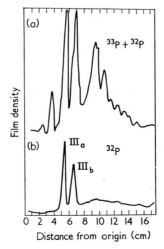

Fig 5. Electrophoresis of the 5′-terminal trinucleotides of endonuclease R limit product.

The trinucleotide fractions of Fig. 4 were pooled, desalted and concentrated as described in Materials and Methods. Fractionation by electrophoresis was carried out under the conditions described in the legend to Fig. 2.

(a) Microdensitometer tracing of unscreened autoradiograph (^{32}P and ^{33}P); (b) microdensitometer tracing of screened autoradiograph (^{32}P).

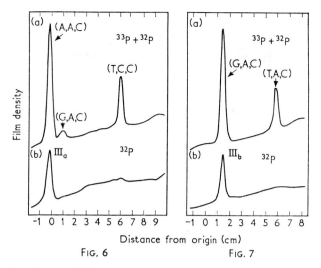

FIG. 6.   FIG. 7.

FIG. 6. Separation of species IIIa from contaminating trinucleotides by electrophoresis at pH 1·9.

The area of the thin-layer cellulose sheet corresponding to species IIIa (Fig. 5) was cut out and the labeled trinucleotides were eluted with 0·02 M-ammonium bicarbonate. Electrophoretic fractionation of this material was carried out on cellulose thin-layer in 8·7% acetic acid, 2·5% formic acid (pH 1·9) at a potential gradient of 25 v/cm. Autoradiography of the thin layer was performed in the usual manner (see Materials and Methods section (d)). The base composition of the various trinucleotides in the eluate is indicated on the tracing of the unscreened autoradiograph.

(a) Microdensitometer tracing of unscreened autoradiograph (^{32}P and ^{33}P); (b) microdensitometer tracing of screened autoradiograph (^{32}P).

FIG. 7. Separation of species IIIb from contaminating trinucleotides by eletrophoresis at pH 1·9.

The area containing species IIIb of Fig. 5 was cut from the cellulose thin-layer sheet and the labeled trinucleotides were eluted with 0·02 M-ammonium bicarbonate. Electrophoretic fractionation was carried out as described in the legend to Fig. 6.

(a) Microdensitometer tracing of unscreened autoradiograph (^{32}P and ^{33}P); (b) microdensitometer tracing of screened autoradiograph (^{32}P).

TABLE 3

Nucleotide analysis of 5′-terminal trinucleotides

| Nucleotide | Species IIIa | | Species IIIb | |
	^{33}P (cts/hr)	^{32}P (cts/hr)	^{33}P (cts/hr)	^{32}P (cts/hr)
d(pC)	3496	14	3410	0
d(pA)	6371	3020	2950	0
d(pG)	98	45	2783	2050
d(pT)	0	14	0	0

Species IIIa and IIIb of Figs 6 and 7, respectively, were eluted from the cellulose thin-layer with 0·02 M-ammonium bicarbonate. Separate samples of each species were hydrolyzed to mononucleotides as described in the legend to Table 2. The mononucleotides were separated by two-dimensional chromatography on a PEI-cellulose thin-layer sheet as described in Materials and Methods section (c). The ^{32}P- and ^{33}P-radioactivity in each of the four deoxyribonucleotides was assayed in a liquid-scintillation spectrometer.

were observed (Fig. 5). These species were individually eluted from the cellulose thin-layer and separated from contaminating trinucleotides by a second dimension of electrophoresis at pH 1·9 (Figs 6 and 7). Nucleotide analyses of the isolated 5'-terminal trinucleotides are shown in Table 3. In the case of species IIIa the ^{32}P radioactivity was associated exclusively with dAMP. The ^{33}P radioactivity was associated with dAMP and dCMP in a ratio of about 2:1. In the case of species IIIb the ^{32}P radio-activity was associated with dGMP. The ^{33}P radioactivity was found in dGMP, dAMP and dCMP in roughly equal amounts. On the basis of these results together with those described in the preceding section, species IIIa was identified as d(pApApC) and species IIIb was identified as d(pGpApC). It was concluded that endonuclease R cleaves both strands of the DNA duplex on the 5'-side of the sequence d(pPupApC).

(b) Nucleotide sequences at the 3'-ends of the limit product

Endonuclease R cleaves the two strands of the T7 DNA duplex at points in close enough proximity to one another to allow the molecule to fall apart into two essentially duplex fragments (Smith & Wilcox, 1970). During this process no "extra" nicks are introduced into either strand nor is there any detectable release of acid-soluble material (Smith & Wilcox, 1970). As indicated in the preceding section, both strands are cleaved on the 5'-side of the sequence pPupApC. These facts are compatible with three possible structures for the endonuclease R substrate region (Fig. 8). In structure (a) the two strands of the substrate are broken at exactly opposite points ("even"

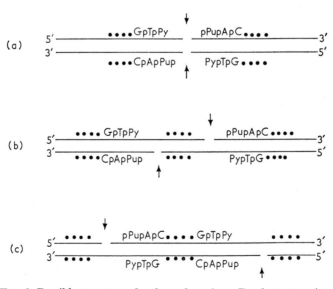

Fig. 8. Possible structures for the endonuclease R substrate region.
Analysis of the sequences at the 5'-ends of endonuclease R limit product has indicated that the enzyme nicks both strands of the DNA duplex on the 5'-side of the sequence pPupApC. The Figure shows 3 structures for the endonuclease R substrate region which are consistent with this finding. In structure (a) the nicks (shown by arrows) are exactly opposite one another so that the 3'-ends of the fragments produced are complementary to the 5'-ends. In structures (b) and (c) the nicks are staggered so that either the 3'-ends (b) or the 5'-ends (c) of the product fragments extend beyond their complementary partner strands as short single-stranded tails. In both of these structures the 3'-ends of the product fragments are not necessarily complementary to the 5'-ends.

break), while in structures (b) and (c) a short stretch of nucleotide pairs intervenes between the two breaks ("staggered" break). Structure (a) can easily be distinguished from the others because only in the case of an "even" break are the 3'-ends of the limit product necessarily complementary to the 5'-ends. To investigate this latter possibility an analysis of the 3'-dinucleotides of the endonuclease R limit product was undertaken.

Uniformly ^{32}P-labeled T7 DNA was digested to a limit product with endonuclease R. The resulting digest was then incubated with micrococcal nuclease. Micrococcal nuclease produces 3'-phosphoryl, 5'-hydroxyl endonucleolytic cleavages and reduces DNA molecules (native or denatured) to a limit product consisting of about 50% each of nucleoside-3'-phosphates (Xp) and dinucleoside-3'-phosphates (XpYp) (Sulkowski & Laskowski, 1962). The ends of the DNA molecules yield unique nucleotide species. The 5'-ends yield pXp and pXpYp-type compounds, while the 3' ends yield XpY's and nucleosides. In the present case the dinucleoside monophosphates (XpY's) from the 3'-ends of the endonuclease R limit digest were separated from all other species in the micrococcal digest by chromatography on DEAE-cellulose in the presence of 7 M-urea (Tomlinson & Tener, 1963). The isolated dinucleoside monophosphates were fractionated by electrophoresis at pH 3·55. Two major radioactive species were observed (Fig. 9). Species I released pC when digested with snake venom

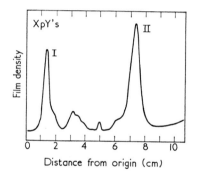

Fig. 9. Electrophoresis of the 3'-terminal dinucleoside monophosphates of endonuclease R limit product.

Uniformly ^{32}P-labeled endonuclease R limit product was digested to completion with micrococcal nuclease and the 3'-terminal dinucleoside monophosphates were isolated as described in Materials and Methods. Electrophoretic fractionation on a cellulose thin layer-sheet at pH 3·55 was performed as described in the legend to Fig. 2. The Figure shows a densitometer tracing of an unscreened autoradiograph of the thin-layer sheet.

phosphodiesterase and Tp when digested with spleen phosphodiesterase (Fig. 10) indicating that it was TpC. Species II yielded pT when digested with snake venom phosphodiesterase and Tp when digested with spleen phosphodiesterase indicating that it was TpT. This was the expected result for an "even" break. It was concluded that structure (a) of Figure 8 most closely describes the endonuclease R substrate region.

(c) Size of the recognition sequence

Assuming for simplicity that the base sequence of T7 DNA is random, the sequence of six nucleotide pairs depicted in Figure 8(a) would be expected to occur about once

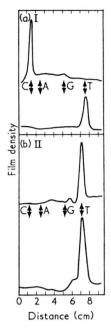

FIG. 10. Sequence analysis of the 3′-terminal dinucleoside monophosphates of endonuclease R limit product.

Species I and II of Fig. 9 were eluted from the cellulose thin-layer with 0·02 M-ammonium bicarbonate. Separate portions of each species were hydrolyzed (37°C for 30 min) with snake venom phosphodiesterase in reaction mixtures containing 0·1 M-glycine buffer, pH 9·2, 0·01 M-MgCl₂, and 0·2 mg phosphodiesterase/ml. Separate fractions of each were also hydrolyzed (37°C for 30 min) with spleen phosphodiesterase in reaction mixtures containing 0·1 M-Tris (pH 7·1), 0·02, M-MgCl₂, and 0·5 unit phosphodiesterase/ml. (Snake venom phosphodiesterase produces 5′-mononucleotides and spleen phosphodiesterase produces 3′-mononucleotides.) Marker mononucleotides were added to the 4 digests and electrophoretic fractionation was carried out on a cellulose thin-layer sheet in 0·075 M-ammonium formate, pH 3·55, at a potential gradient of 25 v/cm. Following electrophoresis a single unscreened autoradiograph of the thin-layer sheet was prepared.

(a) Nucleotides released by digestion of species I with snake venom phosphodiesterase (upper curve) and spleen phosphodiesterase (lower curve).

(b) Nucleotides released by digestion of species II with snake venom phosphodiesterase (upper curve) and spleen phosphodiesterase (lower curve.)

in every 1024 nucleotide pairs. The average length of the limit product of endonuclease R digestion of T7 DNA is about 1000 nucleotide pairs (Smith & Wilcox, 1970). Similarly, for phage P22 DNA the average length of the limit product is about 1300 nucleotide pairs. These facts suggest that the sequence of Figure 8(a) contains sufficient information to account fully for the observed degree of specificity of endonuclease R. However, since strictly speaking neither the T7 nor the P22 DNA molecule represents a truly random collection of nucleotides, the possibility remains that the recognition region for endonuclease R might be larger than the above calculations suggest. In order to rule out this possibility an analysis of the 5′-terminal tetranucleotides of the endonuclease R limit product was undertaken.

A 5′-³²P-terminally labeled limit digest of T7 DNA was hydrolyzed with pancreatic DNase to an average chain length of about 4·4. The digest was then fractionated

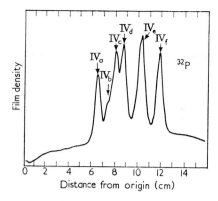

FIG. 11. Electrophoresis of 5'-terminal tetranucleotides of endonuclease R limit product. Endonuclease R limit product labeled at its 5'-termini with ^{32}P was digested to an average chain length of 4·4 with pancreatic DNase. The tetranucleotide fraction was isolated from this digest by ion-exchange chromatography in the presence of 7 M-urea (see Fig. 4) and fractionated by electrophoresis at pH 3·55 under the conditions described in the legend to Fig. 2. A screened auto-radiograph of the thin-layer sheet was prepared as described in Materials and Methods section (c). The Figure shows a microdensitometer tracing of this autoradiograph.

according to chain length by ion-exchange chromatography in the presence of 7 M-urea (Tomlinson & Tener, 1963). The tetranucleotide fraction, containing about 30% of the original ^{32}P radioactivity, was collected and analyzed by electrophoresis at pH 3·55 (Fig. 11). Six distinct ^{32}P-containing tetranucleotides were observed, clearly indicating that the fourth nucleotide from the 5'-ends of the limit product is not uniquely specified. It was concluded that the sequence of six nucleotide pairs illustrated in Figure 8(a) represents the complete recognition region for endonuclease R.

It should be noted that if the fourth position from the cleavage point can be occupied by any nucleotide pair, then eight tetranucleotides (rather than six) would be expected. It is possible that the two tetranucleotides not accounted for in the present experiment occur at such a low frequency in the T7 DNA molecule as to escape detection, or that two of the bands observed during electrophoresis contained more than one species.

4. Discussion

Endonuclease R of *H. influenzae* and other restriction enzymes belong to a class of proteins which act at a limited number of specific sites within a polynucleotide. Other members of this class (e.g. repressors, RNA polymerases, aminoacyl-tRNA synthetases) appear to have important roles in a number of basic cellular processes, including transcription, translation and the control of gene expression (for a recent review see Yarus, 1969). It is generally agreed that the site-specific action of these proteins is a consequence of their ability to recognize specific nucleotide sequences. However, at present the nature of these sequences is largely unknown.

The experiments described in this paper represent an attempt to reconstruct the complete nucleotide sequence of the recognition site of endonuclease R by studying the nucleotide sequences at the ends of the fragments produced by the action of the enzyme. The results indicate that endonuclease R recognizes the following nucleotide sequence:

$$
\begin{array}{c}
\downarrow \\
5'\ldots \text{GpTpPypPupApC}\ldots 3' \\
3'\ldots \text{CpApPupPypTpG}\ldots 5' \\
\uparrow
\end{array}
$$

where the arrows indicate the sites of phosphodiester bond hydrolysis.

Since the opportunities for interaction between the enzyme and the bases of this sequence are rather limited when DNA is in the double helical configuration, it seems reasonable to suppose that the helix is disrupted locally during the recognition process. It is not known at present which of the 12 bases in the sequence are involved in specific interaction with the enzyme. However, the fact that endonuclease R does not degrade single-stranded DNA suggests the possibility that bases on both strands are involved.

Reasoning by analogy to the *E. coli* restriction system, we assume that endonuclease R does not degrade *H. influenzae* DNA because this recognition sequence is modified in some way. Nevertheless, since we do not yet have direct evidence for the existence of a modification activity in *H. influenzae*, we must, for the sake of logical completeness, consider the possibility that *H. influenzae* DNA does not contain the sequence. However, the widespread occurrence in bacteria of host-controlled modification renders this latter alternative unlikely (Arber, 1965).

The most striking feature of the recognition site of endonuclease R is its symmetry. Disregarding for the moment the central ambiguity, the sequence possesses a 2-fold rotational axis of symmetry perpendicular to the helix axis, i.e. when read with the same polarity, the sequence of bases on one strand is the same as that on the other. Of course, the presence of the ambiguity in the two central positions means that the symmetry of the sequence may not always be exact in the geometric sense, i.e. one purine site may be occupied by A and the other by G. From an operational point of view, however, this distinction is probably not important since the enzyme does not distinguish between A and G in these positions. From the enzyme's point of view the sequence is always symmetrical and the molecular environment about each of the two cleavage points is the same.

It is unlikely that the symmetry of this sequence is fortuitous, since the number of possible asymmetrical sequences of this type is about 30 times the number of possible symmetrical sequences (See Arber & Linn (1969) for a general discussion of the expected properties of symmetrical and asymmetrical recognition sequences.) It is probable, therefore, that the observed symmetry is a reflection of some fundamental feature of the mechanism of action of endonuclease R. Figure 12 presents two possible models for the action of the enzyme which take advantage of symmetry.

In model I a single enzyme molecule recognizes six bases on one strand and introduces a single "nick". The presence of a symmetrical recognition site allows the same enzyme to recognize and nick the opposite strand in exactly the same place to produce a duplex break. If the sequence were asymmetrical and both the recognition and the nucleolytic processes involved only one strand as depicted in model I, two enzymes with different recognition specificities would be required.

A more interesting possibility is that the symmetry of the recognition site may reflect underlying symmetry in the enzyme. There is good reason to believe that closed oligomeric proteins composed of identical or nearly identical subunits display

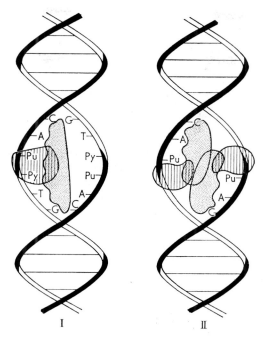

I   II

Fig. 12. Recognition of a symmetrical nucleotide sequence by endonuclease R. Two possible models.

Model I. A single enzyme molecule specifically binds to the base sequence GpTpPypPupApC. The active site of the enzyme (shaded area) catalyzes 5′-phosphoryl, 3′-hydroxyl cleavage of the phosphodiester bond between Py and Pu. The enzyme then detaches and attacks the other strand which (because of symmetry) contains the same base sequence. An even duplex break results. (Note: since endonuclease R is inactive on single-stranded DNA, it is necessary to assume in this model that the enzyme in some way "senses" the bihelical configuration of the substrate).

Model II. The enzyme is composed of two identical subunits (related by a 2-fold rotational axis of symmetry) which bind to the sequence pPupApC on opposite strands of the DNA duplex. Each subunit is actually a dimer constructed from a "recognition" subunit (stippled) and a "nuclease" subunit (shaded). The recognition subunits bind to A and C, while the nuclease subunits bind to Pu. Hydrolysis of the phosphodiester bonds between Py and Pu by the nuclease subunits results in a duplex break.

rotational symmetry (for a discussion of this point see Monod, Wyman & Changeux, 1965). In particular, a closed dimer is expected to possess a 2-fold rotational axis. These considerations suggest the possibility, depicted in model II, that recognition is accomplished by two identical subunits related by a 2-fold axis of symmetry. In this model each subunit recognizes the same sequence of three bases on opposite strands of the DNA duplex. From the standpoint of economy of genetic information it is much cheaper to specify two identical subunits each capable of recognizing three bases in a symmetrical sequence of six than to specify a larger protein capable of recognizing the entire sequence.

In model II each of the two identical subunits is constructed from two non-identical subunits, a "recognition" subunit and a "nuclease" subunit. We suggest the possibility that this division of labor occurs for two reasons. In the first place there is genetic evidence in the *E. coli* system that at least three gene products are involved in restriction and modification. One product carries out the restriction function, another the modification function, and the third determines the site specificity (Arber & Linn, 1969). In the second place the recognition sequence of endonuclease R contains two distinct levels of specificity. The two central base pairs are only partially specified, whereas the remainder of the sequence is absolutely specified. Many general nucleases are known to possess the limited type of specificity displayed in the central part of the sequence, e.g. pancreatic DNase preferentially cleaves phosphodiester linkages of the type PupPy (Laskowski, 1959). Thus, the partial specificity of endonuclease R at the cleavage point could be determined by a nuclease subunit which binds in this region, while the absolute specificity of the enzyme for the remainder of the sequence could be determined by a recognition subunit. This would mean that the recognition subunit need recognize only two bases. Of course, confirmation of these speculations will depend upon further structural studies on the enzyme.

In retrospect, it is probably not very surprising that endonuclease R recognizes a symmetrical sequence since the enzyme carries out an essentially symmetrical operation, namely the cleavage at equivalent points of two DNA strands of opposite polarity. This consideration suggests that symmetrical recognition sequences may be the rule for restriction enzymes in general. On the other hand, many of the proteins that are known to possess the ability to recognize a specific nucleotide sequence carry out fundamentally asymmetrical operations (e.g. RNA polymerase). The recognition sequences for these proteins would be expected to be asymmetrical.

This work was supported by U.S. Public Health Service grant number AI-07875. One of us (H.O.S.) was the recipient of U.S. Public Health Service Career Development Award no. AI 17902.

REFERENCES

Arber, W. (1965). *Ann. Rev. Microbiol.* **19**, 365.
Arber, W. & Linn, S. (1969). *Ann. Rev. Biochem.* **38**, 467.
Holley, R. W., Madison, J. T. & Zamir, A. (1964). *Biochem. Biophys. Res. Comm.* **17**, 389
Josse, J. & Moffatt, J. G. (1965). *Biochemistry*, **4**, 2825.
Kelly, T. J. & Thomas, C. A. (1969). *J. Mol. Biol.* **44**, 459.
Laskowski, M., Sr. (1959). *Ann. N.Y. Acad. Sci.* **81**, 776.
Lehman, I. R. (1960). *J. Biol. Chem.* **235**, 1479.
Lehman, I. R. & Nussbaum, A. L. (1964). *J. Biol. Chem.* **239**, 2628.
Linn, S. & Arber, W. (1968). *Proc. Nat. Acad. Sci., Wash.* **59**, 1300.
Markham, R. & Smith, J. D. (1952). *Biochem. J.* **52**, 552.
Meselson, M. & Yuan, R. (1968). *Nature*, **217**, 1110.
Monod, J., Wyman, J. & Changeux, J. (1965). *J. Mol. Biol.* **12**, 88.
Randerath, K. & Randerath, E. (1967). In *Methods in Enzymology*, ed. by L. Grossman & K. Moldave, vol. 12, part A, p. 323. New York: Academic Press.
Richardson, C. C. (1965). *Proc. Nat. Acad. Sci., Wash.* **54**, 158.
Rushizski, G. W. & Sober, H. A. (1962). *Biochem. biophys. Acta*, **55**, 217.
Sanger, F., Brownlee, G. G. & Barrell, B. G. (1965). *J. Mol. Biol.* **13**, 373.
Smith, H. O. (1968). *Virology*, **34**, 203.
Smith, H. O. & Wilcox, K. (1970). *J. Mol. Biol.* **51**, 379.
Sulkowski, E. & Laskowski, M., Sr. (1962). *J. Biol. Chem.* **237**, 2620.
Thomas, C. A. & Abelson, J. A. (1966). In *Procedures in Nucleic Acid Research*, ed. by G. Cantoni & D. Davies, p. 553. New York: Harper & Row.
Threlfall, C. J. (1957). *Biochem. J.* **65**, 694.
Tomlinson, R. V. & Tener, G. M. (1963). *Biochemistry*, **2**, 697.
Weiss, B., Live, T. R. & Richardson, C. C. (1968). *J. Biol. Chem.* **243**, 4530.
Weiss, B. & Richardson, C. C. (1967). *J. Mol. Biol.* **23**, 405.
Yarus, M. (1969). *Ann. Rev. Biochem.* **38**, 841.

J. Mol. Biol. (1970) **53**, 159–162

Calcium-dependent Bacteriophage DNA Infection

Escherichia coli cells of strain K12 and C can be made competent to take up temperate phage DNA without the use of "helper phage". This competence is dependent on the presence of calcium ions and is effective for both linear and circular DNA molecules.

It has been known that DNA extracted by phenol treatment from temperate coliphages such as λ, 434, 186 or P2 can infect sensitive *Escherichia coli* cells in the presence of "helper phage" (Kaiser & Hogness, 1960; Mandel, 1967). However, the exact role of the helper phage is still unknown. It seems that injection of the DNA of the helper phage and the presence of the intact helper phage DNA in a cell (Takano & Watanabe, 1967) are required for the cell to become competent in incorporating free DNA.

To be infective the DNA molecule must possess at least one free cohesive end (Strack & Kaiser, 1965; Kaiser & Inman, 1965). Moreover, there is a correlation between the specificity of the cohesive ends of the helper-phage DNA and the infectious DNA and the capacity of the phage to serve as a helper for DNA infectivity (Mandel & Berg, 1968; Kaiser & Wu, 1968). The DNA infection seems to depend on the homology between cohesive ends of the infecting DNA and of the DNA of the helper phage.

Since previous work by one of the authors (Mandel, 1967) had shown that changes in cell wall permeability occurred in *E. coli* (strain C600) when made competent by infection with helper phage, we became interested in the effects of both monovalent and divalent ions on *E. coli* cell wall permeability and its correlation with DNA uptake.

During the course of this investigation we found that the DNA of temperate phages P2 and λ could infect a sensitive host in the absence of helper phage and that DNA uptake depended on the presence of calcium ions.

Bacterial strains used were *E. coli* K12 strain C600 (designated K38) as a host for λi⁴³⁴ DNA and *E. coli* C1a as a host for P2 DNA. Phage λi⁴³⁴ was obtained by ultraviolet induction of a λi⁴³⁴ lysogen and P2 by infection of sensitive cells. Phages were purified by differential centrifugation and phage DNA extracted with buffer-saturated phenol (0·01 M-Tris, pH 8·0). DNA of a streptomycin-resistant mutant of K38 was extracted by the methods suggested by Smith (1968) and Avadhani, Mehta & Rege (1969). Competent cells were prepared by inoculating supplemented P medium (Radding & Kaiser, 1963) with a 1 to 500 dilution of an overnight culture of K38 or C1a and grown with aeration at 37°C until an optical density of 0·6 was reached (1×10^9 cells/ml.). The cells were then quickly chilled, centrifuged and resuspended in 0·5 volume $CaCl_2$, kept cold for 20 minutes, then centrifuged and resuspended in 0·1 volume of cold $CaCl_2$. Chilled DNA samples, 0·1 ml. in volume, in standard saline citrate (0·15 M-NaCl, 0·015 M-sodium citrate, pH 7·0) were added to 0·2 ml. of competent cells, further chilled for 15 minutes and incubated for 20 minutes at 37°C. At the end of the incubation period, the reaction mixture was either chilled or treated with DNase for five minutes at 37°C. Dilutions of the mixture were made and plated on appropriate indicators. Under those assay conditions we obtained approximately 10^5 to 10^6 plaques per μg of DNA.

Our work shows that *E. coli* K12 and *E. coli* C grown in P medium can take up phage DNA quite readily in the presence of calcium ions. In Figure 1 we see the

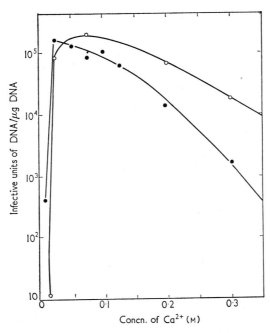

Fig. 1. DNA infectivity as a function of Ca^{2+} concentration. Experimental procedure as described in text. —○—○—, P2 DNA; —●—●—, λi^{434} DNA.

extremely rapid rise in competence of both K38 and C1a in going from 0·01 to 0·025 m-Ca^{2+} and the much slower decline in competence at molarities above 0·1. At 0·5 m-Ca^{2+} concentration the competence of K38 is reduced to practically zero while that of C1a is reduced by a factor of about 100 relative to its peak competence. This may be related to the survival of K38 and C1a cells which when incubated for 20 minutes in 0·5 m-Ca^{2+} give 0·1 and 25% survival, respectively. As a further test of changes in cell wall permeability we incubated competent K38 cells (0·1 m-Ca^{2+}) in the presence of 10 μg actinomycin D/ml. and found 25% survival compared to 100% survival for K38 in 0·01 m-Tris–0·01 m-Mg^{2+} in the presence of actinomycin.

The time-course of the interaction between λi^{434} DNA and K38 at 37°C is shown in Figure 2. The number of plaques increases very rapidly with time, reaching a maximum at about 20 minutes under the conditions of this assay. Since the competent cells and DNA are pre-chilled then mixed at cold room temperatures, and kept at 0°C for 15 minutes before starting the experiment, the plaques obtained at zero time represents DNA taken up by the cells at 0°C and protected from DNase. When the experiment is done with cells and DNA mixed at room temperature and assayed immediately, we find no DNA infectivity at zero time. In both cases the major portion of the interaction is completed in two minutes, which is quite rapid compared to the kinetics of the

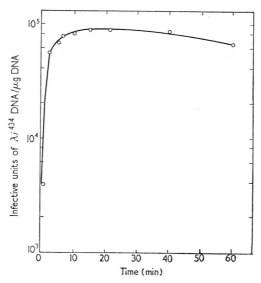

FIG. 2. Time-course of the reaction between λi^{434} DNA at a concentration of 0·1 μg/ml. and K38 cells at a concentration of 2×10^{10}/ml. in the presence of 0·05 M-Ca^{2+}. Procedure was the same as that described in the text except for proportionately larger volumes. At the times indicated on the abscissa, samples of 0·3 ml. each were removed from the incubation mixture and added to 0·1 ml. of 20 μg pancreatic DNase/ml., incubated 5 min at 37°C and plated for assay.

helper-phage DNA infection under similar conditions. The decrease of infective units after 20 minutes reflects the survival rate of K38 incubated in 0·05 M-Ca^{2+}.

The number of infectious centers obtained is linearly proportional to the concentration of phage DNA and concentration of cells. The saturation level of DNA is dependent on the cell concentration and under conditions of our assay (cell concentration$\sim 2 \times 10^{10}$ml.), 10 μg DNA/ml. was not saturating.

In contrast to helper-phage DNA infectivity, cells, competent in the presence of calcium ions, take up both linear and hydrogen-bonded circular DNA (Hershey & Burgi, 1965) with equal efficiency, as shown in Table 1.

Our attempts to transform K38 streptomycin-sensitive cells to streptomycin resistance with DNA extracted from K38 streptomycin-resistant cells met with failure.

TABLE 1

Effect of DNA form on infectivity

| | P2 DNA infectivity per μg DNA | |
	Linear DNA† (heated)	Circular DNA† (not heated)
Ca^{2+} assay	$1\cdot1 \times 10^6$	$1\cdot0 \times 10^6$
Helper-phage assay	$6\cdot0 \times 10^5$	$1\cdot5 \times 10^5$

† Circular DNA prepared as described in Hershey & Burgi (1965). The percentage of circles in DNA preparations depends on salt concentration and duration of storage. Heating at 75°C for 5 min and quick cooling to 0°C converts the closed, circular form to the open linear form.

However, since the two DNA extracts which we carefully prepared showed no hyperchromic shift on melting, the problem may well be in the technique of extracting undegraded DNA from *E. coli*.

We would like to thank Professor R. Calendar for first pointing out to us that a calcium-rich salt solution improved the helper-phage P2 DNA infectivity assay, and Leslie Jensen for her technical assistance.

This work was supported by research grant no. AI–07919 from the National Institutes of Health to one of us (M.M.).

Department of Biochemistry and Biophysics   M. MANDEL
University of Hawaii   A. HIGA†
Honolulu, Hawaii, U.S.A.

Received 26 May 1970

REFERENCES

Avadhani, N.-G., Mehta, B. M. & Rege, D. V. (1969). *J. Mol. Biol.* **42**, 413.
Hershey, A. D. & Burgi, E. (1965). *Proc. Nat. Acad. Sci., Wash.* **53**, 325.
Kaiser, A. D. & Hogness, D. S. (1960). *J. Mol. Biol.* **2**, 392.
Kaiser, A. D. & Inman, R. B. (1965). *J. Mol. Biol.* **13**, 78.
Kaiser, A. D. & Wu, R. (1968). *Cold Spr. Harb. Symp. Quant. Biol.* **33**, 729.
Mandel, M. (1967). *Molec. Gen. Genetics,* **99**, 88.
Mandel, M. & Berg, A. (1968). *Proc. Nat. Acad. Sci., Wash.* **60**, 265.
Radding, C. M. & Kaiser, A. D. (1963). *J. Mol. Biol.* **7**, 225.
Smith, H. O. (1968). *Virology,* **34**, 203.
Strack, H. B. & Kaiser, A. D. (1965). *J. Mol. Biol.* **12**, 36.
Takano, T. & Watanabe, T. (1967). *Virology,* **31**, 722.

† Present address: Hawaiian Sugar Planters' Association, Honolulu, Hawaii, U.S.A.

J. Mol. Biol. (1972) **64**, 417–437

Repeating Sequences and Gene Duplication in Proteins

A. D. McLachlan

Medical Research Council
Laboratory of Molecular Biology
Hills Road, Cambridge
England

(Received 23 October 1970, and in revised form 17 November 1971)

The theory that proteins have evolved by repeated internal duplication of short segments of polypeptide chains has been tested by looking for repeats and near repeats in over 50 different proteins, many of them of known structure. The probability that the observed repeats could arise by chance has been calculated.

The search does not yield a single new example where the evidence for gene duplication is compelling. No protein shows a unique internally consistent pattern of repeats which both correlates with repeats in the structure and cannot be explained by chance. The evidence is discussed in detail for haemoglobin, chymotrypsin, subtilisin and carboxypeptidase. The evolution of complex large proteins from simple small ones has probably been a process of continuous growth in which chains have been gradually added to the outer surface surrounding an invariable core near the active centre.

1. Introduction

The idea has often been put forward that in the earliest stages of evolution of primitive forms of life the primitive proteins began with the production of repeating amino-acid sequences composed of a number of short similar segments. Thus, a large protein could be built up by repeating a small number of basic sequences which might form structural building blocks for the whole large structure. It is assumed that the repeating amino-acid sequence is coded for by a repeating DNA sequence, which is formed by duplicating again and again short pieces from the original sequence. Bacterial proteins then are expected to be more primitive and repetitive than those from mammals.

The bacterial ferredoxins (Benson, Mower & Yasunobu, 1967) have been cited as evidence for this theory (Jukes, 1966; Eck & Dayhoff, 1966), since the second half of the amino-acid sequence is an almost exact duplicate of the first, and each half has a regular periodic arrangement of cysteines. More recently Dus, Sletten & Kamen (1968) noticed repetitive pieces in the sequence of cytochrome c_2 from *Rhodospirillum rubrum*, and suggested that these features were survivals of a primitive ancestral repeating sequence. Other repeats have been noted in clupeine Z from the pacific herring (Black & Dixon, 1967), subtilisins BPN' and Carlsberg (Markland & Smith, 1967; Wright, Alden & Kraut, 1969; Smith, DeLange, Evans, Landon & Markland, 1968), the haemoglobins (Cantor & Jukes, 1966; Fitch, 1966), and carboxypeptidase A (Bradshaw, Neurath & Walsh, 1969; Neurath, Bradshaw & Arnon, 1969). Thus, there appears to be a considerable body of circumstantial evidence in favour of the theory.

The object of the work described in this paper was to test the sequences of a large number of proteins, particularly those the structures of which are already known, for evidence that they could have been formed originally from a mosaic of repeating sequences. It is fairly straightforward to test whether two long halves of a long protein sequence have evolved from a common ancestor, by calculating the minimum number of base changes required, according to the genetic code (Fitch, 1966; Nolan & Margoliash, 1968; Needleman & Blair, 1969), or by looking for sequences of chemically similar amino acids (Haber & Koshland, 1970). To test whether a given sequence can have been formed from a relatively *large* number of *short* repeating pieces is much more difficult, but I believe that the methods which I have used are sufficiently sensitive to detect any underlying pattern if it exists. I have already (McLachlan, 1971) given reasons why the repeats in cytochrome c are not significant. We shall now see that this further investigation fails to show any convincing new examples of gene duplication or ancestral repetition, and supports Haber & Koshland's conclusion (1970) that many of the repeats noted in protein sequences may easily have occurred by chance.

If these proteins ever did arise from ancestral repeating sequences, no regularity in sequence or structure now remains which cannot reasonably be accounted for in other ways.

A second reason for searching for similar short segments in protein sequences is to see whether chemically similar segments have a similar three-dimensional structure (Low, Lovell & Rudko, 1968). We shall see that the structures of even very similar segments are often quite different, so that there is no simple relationship.

2. Gene Duplication

Parallel duplication of a gene to produce two distinct new genes is a well-known event in evolution (Bridges, 1936; Stephens, 1951; Jukes, 1966; Dixon, 1966). An example is the formation of the haemoglobin γ chains from the normal β chain (Ingram & Stretton, 1961). Series duplication, in which the new gene codes for a single polypeptide chain of twice the original length, has also occurred sometimes. The constant parts of the antibody heavy chains are good examples (Edelman *et al.*, 1969; Cohen & Milstein, 1967*a,b*; Singer & Doolittle, 1966; Milstein & Pink, 1970). Duplication has also occurred in ferredoxin (Tanaka, Nakashima, Benson, Mower & Yasunobu, 1966; Benson *et al.*, 1967; Eck & Dayhoff, 1966), haptoglobin (Smithies, Connell & Dixon, 1962; Dixon, 1966; Black & Dixon, 1968) and probably in the diheme cytochrome c_3 from *Desulphovibrio vulgaris* (Ambler, 1968). However, in spite of these examples, series duplication is a rare genetic event, and when duplication does begin to occur the extra piece may often be eliminated by looping out a portion of the DNA helix, so that it is cut out and the defect is repaired (Russell *et al.*, 1970). Because duplication is so rare one needs to look critically at the evidence in any case where it is postulated to explain a repeat in a protein sequence.

3. Tests for Significance

The usual method for testing whether two sections of a protein sequence are ancestrally related is to calculate the minimum number of base changes needed to convert one into another, according to the genetic code. This quantity is called the minimum mutation distance (Fitch, 1966,1970; Dayhoff, 1969). Two long proteins can be tested for homology by taking a long span of, say, twenty amino acids and comparing every possible segment of one protein with every possible segment of the other. If the proteins are unrelated the statistical distribution of the minimum mutation distances is approximately Gaussian. For related sequences there is an abnormally high number of short mutation distances, and Fitch's tests (Fitch & Margoliash, 1967; Fitch, 1970) can be used to detect deviations from a Gaussian curve.

For comparing *short* segments of sequence the minimum mutation distance is a poor test unless the pieces are closely related. The actual mutation distance can differ significantly from the minimum. Single-step mutations are not particularly significant by themselves because the genetic code (Crick, 1968) allows a large number of mutations to occur between amino acids which are completely dissimilar in size, shape and charge. Thus it is possible to pair off sections of sequences which are connected by many potential single-step mutations and identities, but could never conceivably have the same three-dimensional structure. Now the structure of a protein evolves much more slowly than the amino-acid sequence. For instance, in the haemoglobins even very distant species such as insect (Huber, Formanek & Epp, 1968), marine worm (Padlan & Love, 1968), lamprey (Braunitzer & Fujiki, 1969), carp and mammals (Dayhoff, 1969) share a common molecular architecture (Perutz, Muirhead, Cox & Goaman, 1968; Perutz, Kendrew & Watson, 1965). Hence a more discriminating test of distant evolutionary relationship is to ask whether two sequences could conceivably have a common structure; or whether corresponding pairs of amino acids are structurally similar in size, shape, polarity and so on (Thiebaux & Pattee, 1967). A good match between two sequences, either through mutation distances or chemical properties, only points to common ancestry if it has a low probability of occurring for other reasons; the degeneracy of the genetic code is so great that approximately 3^n nucleotide sequences can code for a given set of n amino acids, and the probability that even two short identical peptides correspond to the same DNA is very low.

The methods used in this paper have been described already (McLachlan, 1971). They are developed from those used by Fitch (1966), Cantor & Jukes (1966), Needleman & Blair (1969) and Haber & Koshland (1970).

The first step is to set up a measure of similarity for each pair of amino acids. In McLachlan (1971) this was based on the observed frequencies of amino-acid replacements in homologous proteins. However, in this work, which was done earlier, we used a more intuitive scoring scheme† based on polar or non-polar character, size, shape and charge (Sneath, 1966). To each pair of amino acids i, j we assign a similarity score $m(i, j)$ which ranges from 0 to 6 (see Table 1). The score $m(i, i)$ for matching an amino acid with itself is normally 5, but rises to 6 for the less common ones.

To compare two sequences A and B, in which the amino acids at positions p and q are a_{pA} and a_{qB}, one first sets up a score matrix $M(p,q) = m(a_{pA}, a_{qB})$ in which $m(a_{pA}, a_{qB})$ is the similarity between a_{pA} and a_{qB}. The next step is to assign a score to the match between two segments of protein centred on positions p and q. Here the

---

† The scores assigned to the repeating segments observed in this work are a little different if one uses the more objective scoring scheme of McLachlan (1971), and the matching probabilities are also affected, but the alterations do not affect any of the conclusions which we reach.

TABLE 1

Chemical similarity scores for the amino acids

Score				Pairs				
6	FF	MM	YY	HH	CC	WW	RR	GG
5	LL	II	VV	SS	PP	TT	AA	QQ
	NN	KK	DD	EE				
3	FY	FW	LI	LM	IM	ST	AG	QE
	ND	KR						
2	FL	FM	FH	IV	YH	YW	SC	HQ
	QN	DE						
1	FI	FV	LV	LP	LY	LW	IT	IY
	IW	MV	MY	MW	VP	SA	SN	SQ
	PT	PA	TA	TN	HN	HW	QK	QD
	NE							
0	All others, including unknowns and deletions							

One-letter code. F(Phe), L(Leu), I(Ile), M(Met), V(Val), S(Ser), P(Pro), T(Thr), A(Ala), Y(Tyr), H(His), Q(Gln), N(Asn), K(Lys), D(Asp), E(Glu), C(Cys), W(Trp), R(Arg), G(Gly), B(Asx), Z(Glx).

score for two segments of length s is taken to be a weighted sum of the successive M values, with weights W_h:

$$C(p,q) = \sum W_h M(p + h, q + h), \qquad h = -g, \ldots, + g. \tag{1}$$

It is convenient to take s an odd number, $s = 2g + 1$. s should be large enough to make a high score statistically significant but short enough to avoid missing gaps. The weights W_h can be chosen at will. The matrix of weighted sums $C(p,q)$ is called the *comparison matrix* for the two sequences. If two sections of sequence are similar the comparison matrix shows a line of high scores running parallel to the main diagonal. A computer can be set to construct the matrix and print out suitable symbols to indicate scores which have different levels of significance, giving a correlation diagram for the sequences (Gibbs & McIntyre, 1970).

To judge the significance of a high score $C(p,q)$ one needs to know the probability distribution of the scores for a pair of random sequences of given compositions. The scores $C(p,q)$ and $C(p + r, q + r)$ will in fact be highly correlated if $r < s$. An exact calculation of probabilities is difficult, but we can calculate a related distribution exactly.

Consider the following experiment. Two infinite packs of cards A and B are shuffled. The cards represent amino acids, and the composition of each pack is in the same proportions as the protein A or B. A set of s cards $a_1, a_2, \ldots, a_s$ are drawn in order from A and compared in turn with another set $b_1, b_2, \ldots, b_s$ drawn from pack B, the score for each pair being $m_r = m(a_r, b_r)$. The weighted score for the entire match is defined as:

$$M = \sum_r W_r m_r. \tag{2}$$

Then the *double matching probability* Q_{AB} (M), that the observed score in this experiment is greater than or equal to M, can be calculated (see McLachlan, 1971); so can the mean score and the standard deviation. If the number of cards is large the central part of the distribution is approximately Gaussian.

As an example, consider the comparison matrix for a haemoglobin α chain against the horse β chain, with span 5 and weights 1, 2, 3, 2, 1. Here the calculated mean and standard deviation are 6·12 and 6·22, respectively. In this case scores of 40, 30, 24 and 18 correspond to double matching probabilities of 9×10^{-5}, 3×10^{-3}, 2×10^{-2}, and 3×10^{-3}.

Another useful probability is the *single matching probability*. Suppose that we place a *given* set of s cards $a_1, a_2, \ldots, a_s$ in order on a table. Now draw cards $b_1, b_2, \ldots, b_s$ in turn from an infinite shuffled pack with the composition of protein B, and record each score $m_r = m(a_r, b_r)$. The single matching probability $R_{aB}(M)$ is the probability that the sum $m_1 + m_2 + \cdots + m_s$ is greater than, or equal to, M. It depends on the compositions of the peptide α and of the entire protein B.

For a match to be statistically significant one requires a low value for the probabilities Q_{AB} or R_{aB}. Since a comparison of every pair of short segments from two proteins of lengths n_A, n_B entails about $n_A n_B$ potential matches there are often as many as 10^4 to 10^5 entries in the comparison matrix. Thus, events with probabilities R_{aB} in the range 10^{-4} to 10^{-6} are likely to be observed reasonably often in pairs of random sequences.

So far we have considered only a single pair of sequences. Often one has families of proteins, such as the haemoglobins, in which the sequences from related animals form a homologous series. With such a family it is possible that remnants of an ancestral repeating pattern might persist in some members, but not in others, and yet still be detected by analysis of the entire family as a whole. With this object in mind we have set up a *family comparison matrix* $C_{max}(p,q)$. Let $S_{1A}, S_{2A} \ldots$ be a set of j_A homologous sequences from a family A, and let $S_{1B}, S_{2B}, \ldots$ be a set of j_B sequences which form a second family B (A and B could be the same). Suppose that $C_{xy}(p,q)$ is the p,q element of the comparison matrix for S_{xA} and S_{yB}. Then $C_{max}(p,q)$ is defined to be the maximum of the $C_{xy}(p,q)$ for all the pairs x,y (p,q being fixed). The family comparison matrix displays all the strongest matches between the two groups, and one can then search for any underlying pattern.

If an interesting match is found one can first try to calculate the probability that the observed regularity could have arisen by chance (Šorm & Knichal, 1958; Fitch, 1970). One fundamental difficulty is that any amino-acid sequence can be tested for an infinite number of different special features, and every sequence will therefore possess some unique but insignificant coincidental feature which may be exceedingly improbable (Šorm & Keil, 1962). By fastening on such features, which are suggested by the particular sequence in question, one can easily be persuaded that some subtle pattern of repeats exists (Urbain, 1969; Williams, Clegg & Mutch, 1961).

To avoid this danger we have restricted ourselves to one well-defined class of regularity, expressed as a high score on the comparison matrix. We have also done control experiments by comparing totally unrelated proteins such as carboxypeptidase and haemoglobin, or one protein sequence with another written in reverse order. These experiments confirm the results of the statistical calculations.

Suppose that several matches of various lengths and strengths have been found. If they are really remnants of a set of ancestral repeats, which may be interrupted by insertions or deletions, they must not merely be statistically improbable. They must satisfy the test of *mutual consistency*. Suppose for example that a segment *abcdefg* is related to two pieces $a'b'c'd'e'$ and $c''d''e''f''g''$. There should then be a match between $c'd'e'$ and $c''d''e''$. Often, however, such a match is absent, and there may even be a weak match out of phase in which, say $a'b'c'd'e'$ resembles $c''d''e''f''g''$.

Once a mutually consistent set of repeats has been found a further test can be applied. This is to see whether the related segments share a common structure. An underlying regularity of structure would be good evidence that a protein had begun as a repeating sequence. On the other hand, similar sequences do not as a rule have similar conformations in different parts of a protein, and the structure may change in the course of evolution. Absence of a regular structure cannot therefore rule out the possibility that a sequence repeat is ancient, but the possibility becomes more remote.

The main features to be looked for in related segments of a protein which has evolved by repeated internal duplication of its sequence are therefore as follows:

(1) identical or chemically similar amino acids in weakly matched pieces; short mutation distances in strongly matched pieces;
(2) low probability that the observed repeats could occur by chance;
(3) mutually consistent sets of repeats;
(4) similar structures;
(5) persistence of a repeat in many members of a family of homologous proteins.

The next sections describe the correlations found in a variety of proteins. We use the word *correlation* to describe any repeating feature which appears to be interesting, without implying by this word that it is statistically significant in any way.

4. Haemoglobin and Myoglobin

The haemoglobins are a suitable family for testing because the structure is known and a great many sequences are available (Perutz *et al.*, 1965). The fact that the structure is nearly all built of α helices, many of which are of similar lengths, makes it seem possible that if any proteins have evolved in a repetitive way then haemoglobin could be one of them.

A preliminary study of myoglobin showed some promising similarities between helices A and B, B and G, and B and E. Also the C helix and the FG corner appeared to be related. The score matrix based on the genetic code showed several long runs of single-step mutations. Individual haemoglobin chains showed various short repeats, including some quite long correlations.

The next step was to make a family comparison matrix for ten homologous haemoglobin chains. One matrix, with span of 5, weights 1, 2, 3, 2, 1 and contour levels of 40, 30, 24, 18 showed all the short repeats with tripeptides and other fragments. Another, with a span of 11, gave the longer correlations which are listed in Table 2.

These correlations are found by comparing each of the ten sequences with each of the others in every possible registration. A correlation which is strong between one pair of species may be weak between others, because the sequences vary considerably. For each correlation Table 2 also lists the shift. This is the number of spaces which one has to slide the sequences past one another in order to align the two matched segments. The shift is used to test for consistency. For example, in Table 2 the shifts for m (relating helices E and H) and c (relating A and H) are 66 and 117. There should, therefore, be some relation between helices A and E with a shift of $(117-66) = 51$. The comparison matrix does show such a relation, but it is very weak, and insignificant compared with the other observed correlation (shift of 58).

The individual correlations, though never very long, are quite strong, in the sense that each pair of spans is at least as similar as the majority of diagonal spans in the comparison of two homologous but distantly related proteins. For example, the correlation between the E helices of sperm whale myoglobin and horse haemoglobin β is weaker than that between the two segments m in Table 2. But although the individual correlations are strong it proves to be impossible to relate the different fragments to one another in a meaningful way according to any evolutionary family tree. The correlations b, c and m discussed above illustrate the difficulty; one cannot reconcile them simultaneously and thus demonstrate a single common ancestral sequence for helices A, E and H. A search through the 19 correlations of Table 2, taken in threes, shows that there is no underlying pattern of repeats. Instead one must regard the correlations as the chance result of unrelated variations in different parts of the molecule.

Are these correlations related to the structure? It would be interesting if they brought out some general similarity between the different helical regions, or some common features of the corners. The most permanent feature of the haemoglobin sequences (apart from the haem-linked histidines) is the persistence of non-polar side chains at certain internal positions (Perutz et al., 1965). The long correlation m between helices E and H, which has already been commented on by Fitch (1966) and Cantor & Jukes (1966), makes use of the fact that both helices have a similar distribution of internal side chains, with non-polar sites at positions 4, 8, 11, 12, 15 and 19. The distribution of non-polar sites in the other helices is less regular and cannot be matched to that of E or H except over short sections. In fact there is little or no relation between the structure and the other correlations. For example, in the correlation g, helix B matches the GH corner and the first half of the H helix. The most that can be said is that since haemoglobin has such a high helix content (even the corners contain helical fragments), many pairs of short segments have some local structural similarity. As a further test we compared the haemoglobin sequences with carboxypeptidase and chymotrypsin to see whether the helical regions in these latter proteins would be picked out. No significant relationship was detected. Haemoglobin was also tested against the sequence of horse α chain in reverse order, and gave several quite long correlations. This reinforces the view that most of the correlations in Table 2 are random.

These tests cannot, of course, prove that haemoglobin did not originally arise from a primitive repeating sequence. But they do demonstrate that no trace of such a sequence remains detectable beneath the correlations which naturally arise by chance in any long protein chain.

5. Chymotrypsin

The X-ray analyses of α-chymotrypsin and elastase (Birktoft, Blow, Henderson & Steitz, 1970; Matthews, Sigler, Henderson & Blow, 1967; Shotton & Watson, 1970; Watson, Shotton, Cox & Muirhead, 1970) show that both molecules share a common structural framework. This consists principally of two independent hydrogen-bonded substructures which rest upon one another, forming two sides of the active site. The substructures have a similar pattern of antiparallel pleated sheets linked by hydrogen bonds, with loops closed by disulphide bridges. There is therefore a possibility that the two substructures might have been formed by gene duplication.

Birktoft & Blow examined the amino-acid sequence to see whether there was any

TABLE 2

Strongest repeats in haemoglobin and myoglobin

Pair	Shift	Regions	Sequences	Species
a	42	A3-A13	A D K T N V K A A W G	Human α
		CD5-D7	G D L S N A K A V M A	Horse β
b	58	A8-B1	V L H V W G K K V G A H	Myoglobin/human α
		E11-EF1	V L H S F G K A V G H	Horse β/rabbit α
c	117	A7-B2	A V L A L W D K V E A D V	Horse β/myoglobin
		H3-H15	A V H A S L D K F L A D V	Rabbit α
d	72	A7-AB1	N L K G T F A K L S	Kangaroo β
		EF4-F5	N V K A A W S K V G	Horse α
e	44	A13-B3	G K V G A H A G	Human α
		E2-E9	P K V L A H G A	Kangaroo β
f	40	B1-B9	D V A G H G Q D I	Human α
		E3-E11	Q V K A H G K K V	Horse α
g	99	B1-B13	H A G E Y G A E A L E R M	Human α
		GH1-H8	H P G N F G A D A Q G A M	Myoglobin
h	37	CD2-D6	D H F G D L S N A K V M	Kangaroo β
		EF4-F7	D L P G A L S D L S N L	Horse α
i	40	CD3-D3	S F G D L S D P	Horse β
		EF8-F7	T F A Q L S E L	Human γ
j	30	CD4-E1	F K H L S N A K A V M A N	Myoglobin/kangaroo β
		E19-F3	I K H L D D L K G T F A Q	Human γ

			Sequence	
k	33	CD4-D7	F G D L S S A D A I L	Human γ
l	46	EF2-F4	L D D L K G A F A S L	Cow γ
		E8-E16	G V T V L H S F G E G	Myoglobin/horse β
		G12-GH2	L V T V L H S R H P G	Human γ/myoglobin
m	66	E2-EF1	P K V K A H G A F S D G L A H	Horse β
		H2-H21	P E L Q A S Y Q K V V A G V A N A L A H	Human β
n	17	E12-EF1	L T S L G D A I K H	Human γ
		F1-FG1	L S T L S D L H A H	Rabbit α
o	10	E15-EF6	L G D A I K N L D N L K	Kangaroo β
		E5-F8	M P N A L S A L S D L H	Human α
p	37	E19-F3	V G H L D D D L P G A L S T	Rabbit α
		G14-H2	V G I M F Y L P G D F P P	Carp α
q	3	EF6-F4	P G A A L S D L	Horse α
		F1-F7	F A A L S E L	Horse β
r	28	G6-G17	K L L G N V L V T V L A	Human γ
		H10-H21	K F L A D V S T V L T S	Rabbit α
s	4	G7-GH1	L L G N V L A L V V A R H F	Horse β
		G11-GH5	V L V T V L A I H F G K E F	Human γ

These 19 repeats are the strongest selected from a family comparison matrix, using a span of 11 with weights of 1, 2, 2, 3, 3, 3, 3, 3, 2, 2, 1. Those correlations were selected in which there were 7 or more consecutive spans with a score of at least 60. Ten sequences were used. Haemoglobin α: horse, carp, rabbit, human. Haemoglobin β: horse, human, kangaroo. Haemoglobin γ: cow, human. Myoglobin: sperm whale. The total number of spans compared was approximately 1,000,000. Sequences were taken from Dayhoff (1969) except for the kangaroo β chain (Air, Thompson, Richardson & Sharman, 1971). The sequences of lamprey and insect haemoglobins show no additional correlations.

TABLE 3

Repeats in the chymotrypsin family

a	19–27	e	128–135
	39–47		179–185
b	12–20	f	132–139
	142–150		178–185
c	15–24	g	138–145
	201–230		170–175
d	39–46	h	141–148
	179–185		192–199

Positions are numbered as in chymotrypsinogen (Blow, Birktoft & Hartley, 1969; Shotton & Hartley, 1970; Brown & Hartley, 1966). Correlations are taken from a comparison matrix with span of 11, weights 1, 2, 2, 3, 3, 3, 3, 3, 2, 2, 1. They have at least 7 consecutive spans with a score $\geq$ 60, using the scores of Table 1. Five sequences were used: chymotrypsinogen A, B (cow), elastase (pig), trypsin (cow) and fragments of trypsinogen (dogfish). (References: Shotton & Hartley, 1970; Bradshaw, Neurath, Tye, Walsh & Winter, 1970.)

repeat, but found none (Birktoft *et al.*, 1970). We searched for repeats, using the family comparison matrix for five sequences (Table 3). The matrix shows eight repeats of similar strength to those in haemoglobin, but they bear no relation to the pattern of hydrogen bonding or the disulphide bridges. The strongest repeat:

G L S R I V N G E   (12–20) chymotrypsin
G L T R – T N G Q   (142–150) elastase

relates the junctions between sections A, B and sections B, C of the chain, which are cut when the enzyme is activated. It is not statistically significant, since the matching probability for these two segments is of the order of 1 in 100,000.

The active sites of chymotrypsin and subtilisin are similar, since they each contain histidine, serine and aspartic acid residues arranged in almost the same relative orientations in space. These amino acids do not occur in the same order in the two sequences, and the structural organization of the two proteins is quite different (Wright *et al.*, 1969), so that there can be no evolutionary connection between them. Instead these enzymes are a good example of convergent evolution. The comparison matrix shows a few weak correlations which bear no relation to the active site. There is also one strong repeat:

A N T V P Y Q V S   (24–32) trypsin
A Q S V P Y G V S   (1–9) subtilisin BPN′.

The matching probability is approximately 3×10^{-6}, which is not very low. It is interesting that two such similar fragments have entirely different structures. The first contains a corner between two pieces of β structure. The second lies within the first two helices of subtilisin.

6. Lysozyme and Ribonuclease

The structures of lysozyme (Blake, Mair, North, Phillips & Sarma, 1967) and ribonuclease (Wyckoff *et al.* 1967, 1970; Kartha, Bello & Harker, 1967) are broadly similar: both have two wings and a central cleft; both have an irregular structure made up of α helices and β structure; both have an active site in the cleft which binds a sugar molecule; the sequences are almost the same length. In 1967, Manwell (1967) noticed some correlations between their sequences which he considered to be statistically significant in the light of the genetic code. Hence he argued that there might be a distant evolutionary link between the two enzymes. Later investigations by Needleman & Blair (1969) and Haber & Koshland (1970) have shown that there is no statistically significant relation between the amino-acid sequences, even if one introduces several deletions, and our comparison matrix confirms their conclusions. Nevertheless there are a number of short peptides which are identical in both proteins (Low et al., 1968). These suggested the possibility that if both proteins arose by repeating a number of short pieces of sequence, then lysozyme and ribonuclease might be formed from the same set of pieces, put together in a different pattern.

A family comparison matrix for three lysozyme sequences and three ribonucleases shows several short correlations, of which the strongest are shown in Table 4.

TABLE 4

Similarity between lysozyme and ribonuclease

a	A	A	K	F	E	S	N	F	L(31- 38)	Hen
	A	A	K	F	E	R	Q	H	R(8–15)	Cow
b	M	K	R	H	G	L			L(11–16)	Hen
	M	K	R	Q	G	M			R(33–38)	Cow
c	D	V	Q	A					L(121–124)	Hen
	D	V	Q	A					R(56–59)	Rat
d	A	L	C	S	E	K			L(114–119)	Lactalbumin
	A	I	C	S	Q	K			R(59–64)	Rat
e	S	S	N	I	C	N			L(72–77)	Lactalbumin
	S	S	N	Y	C	N			R(25–30)	Cow

Lysozymes: hen (Canfield, 1963), part of duck (Jollès, 1969), lactalbumin (Brew, Vanaman & Hill, 1967). Ribonucleases: cow (Smyth, Stein & Moore, 1963), rat, part of horse (Beintema & Gruber, 1967).

Each has a matching probability between 10^{-5} and 10^{-6}. However, these fragments cannot be fitted into any consistent pattern, and one is forced to conclude that they arise by chance.

Some of the pairs of segments in Table 4 have similar structures. Both segments of a, c and d are α-helical. The partners in b and c fold differently.

Comparisons between hen lysozyme and phage T4 lysozyme (Tsugita & Inouye, 1968), and between rat ribonuclease and nuclease T1 (Takahashi, 1965) showed no significant correlations.

7. Subtilisin

The sequences of subtilisins BPN' (Markland & Smith, 1967) and Carlsberg (Smith et al., 1968) are much more repetitive than haemoglobin or chymotrypsin, and the most prominent repeats have already been noted (Smith et al., 1968; Wright et al., 1969). Table 5 shows the longest segments which are picked out on a comparison matrix which uses the genetic code. Almost the same pieces are picked out by using the chemical similarity test. Here a_2, a_3 and a_4 are closely related to a_1, with chemical matching probabilities of the order 10^{-4}, 10^{-5} and 10^{-5}, respectively. Segments a_3 and a_4 are also related, with a probability of 10^{-4}. The probabilities for matching

TABLE 5

Repeats in subtilisin

Section	Sequence	Position
a_5	S I G V L G V A P S S A L Y	(78–91)
a_4	N V K V A V I D S G I D S S H P N L	(26–42)
a_3	N L K V A G G A S M V P P S E T P N F	(41–58)
a_2	H V A A G T V A A L	(67–75)
a_1	M A S P H V A G A A A A L I L K S H P N W	(221–241)
b_1	V A S G V V V A A A A G N Q G G S T G S S S	(143–163)
b_2	A L H S Q G G Y T G S	(15–25)
b_3	(P S S A L) Y A V K V L G N A G S (Q G Y S)	(91–101)
b_4	(S V I A) V G A V D S S N Q R A S F S	(177–190)

These segments are taken from a family comparison matrix, span 19, using a scoring system based on the genetic code: 5 for an identity, 1 for a single-step mutation. Regions are selected if the sum of the scores is greater than 36 in at least 8 consecutive spans. Segment a_5 is related to a_1 rather more weakly. a_2 is selected because of its chemical similarity to a_1. In the matrix a_1 is related to the first 11 amino acids of b_1 and to each of a_2–a_5. a_3 is also related to a_4. b_1 is related to b_2, b_3 and b_4.

each of a_1 and b_2 to b_1 are also about 10^{-4}, while b_3 and b_4 are more weakly related to b_1.

These repeats are longer and show a greater degree of internal consistency than those in other proteins, but matching probabilities as high as 10^{-5} are not very significant in a comparison of about 100,000 spans. Many of the repeats depend on local concentrations of alanine, valine and glycine residues. We have examined their structures on a model of subtilisin BPN' built at Cambridge by Dr C. Wright.

Segments a_1 and a_2 are interesting because both contain histidines close to the active site (Wright et al., 1969). Also, both histidines are near the beginning of two long helices (63–73) and (223–238) which rest upon one another and run parallel to one another through the centre of the molecule. On the other hand, segment b_1 is totally unlike a_1: it consists of the end of a helix (143–145) followed by a long internal piece of extended chain (146–154) which then runs along the surface (155–159) and leads into an irregular loop (160–164). It is also interesting that several of the other segments include long pieces of internal extended chain: a_3 (45–50), a_4 (26–32), a_5 (80–85) and (89–95), b_3 (89–95) and b_4 (137–180).

The statistical evidence for gene duplication is weak, since the probability that the tetrapeptide HVAG should be repeated somewhere in the sequence is greater than 10^{-2}. The structural similarity of the two helices (a_1 and a_2) and the pieces of extended chain is suggestive but not compelling. There is no sign of regularity in the molecular structure as a whole. Of the two long correlations noted by Haber & Koshland (1970), one (133–164 with 212–243) includes the relation between a_1 and b_1 above; the other (5–50 with 113–158) embraces no structurally similar regions and does not appear prominently in our comparison matrix.

8. Carboxypeptidase

The most interesting repeat in carboxypeptidase A is the tripeptide IHS at 68–70 and 195–197. Both histidines are zinc ligands at the active site (Bradshaw, Ericsson, Walsh & Neurath, 1969; Lipscomb et al., 1968; Lipscomb, Reeke, Hartsuck, Quiocho & Bethge, 1970), and both occur at the ends of two parallel strands of extended chain which run from 61–68 and 190–196 through the core of the structure. There is, however, no continuing correspondence of either structure or sequence on either side of this pair of regions. In carboxypeptidase B the first histidine tripeptide becomes FHA (Bradshaw, Neurath & Walsh, 1969). The matching probability for the tripeptide IHS to be repeated is of the order of 10^{-3} and is not statistically significant. If gene duplication has occurred in this region it must have been overlaid by very extensive later changes in the three-dimensional structure. Neurath et al. (1969) have also proposed that a repeated sequence near arginine 145 has duplicated:

N R L W K T R S – – – V T S S S L C   (123–138)
N R N W D A G F G K A G A S S S P C   (144–161).

Here there is no discernible structural similarity in the two pieces, and the correlation is again not statistically significant. Finally there is a weak correlation linking two helical regions 18–26 and 99–107.

9. Papain

Papain (Drenth, Jansonius, Koekoek, Swen & Wolthers, 1968; Husain & Lowe, 1969; Light, Frater, Kimmel & Smith, 1964) contains a repeated tetrapeptide PVKN at 15–18 and 209–212 which occurs in both places as a piece of extended chain in the surface. There are also three interesting strongly related segments, with matching probabilities of about 10^{-5}:

```
G  I  I  K  I  R  T  G  N  L  N  Q  Y        (36–48)
G  Y  I  L  I  K  N  S  W  G                  (169–178)
G  Y  I  R  I  K  R  G  T  G  N  S  Y        (185–197).
```

The structure of the first, which includes a section of the helical core (26–40) of the second lobe of the structure, is totally unlike the latter pair. These include two antiparallel strands of β structure (169–175 and 187–190) which are linked to one another in the surface of the first lobe.

10. Histones

The sequence of calf thymus histone IV has several short repeated sections (DeLange, Famborough, Smith & Bonner, 1969). The longest of these:

```
L  G  K  G  G  A  K  R  H          (10–18)
L  A  R  G  G  V  K  R  R          (37–44)
```

has a matching probability of about 10^{-4}, so it is not statistically significant. There is no evidence of any regular pattern of repeats. The lysine-rich histone (Iwai, Ishikawa & Hayashi, 1970) appears to be quite unrelated in sequence and contains a repeat (13–19 with 114–120) which is too short to be statistically significant.

11. The Significance of Repeats

The proteins which we have studied, including several others which do not merit special mention (tobacco mosaic virus coat, phage f2 coat, lysozyme T4, azurins, penicillinases, staphylococcal nuclease, tryptophan synthetase alpha and glyceraldehyde 3-phosphate dehydrogenases), do not yield a single example of long mutually consistent regular repeats which are strong enough to be statistically significant evidence for an ancestral repeating sequence. In those examples where a few short repeating segments are found to be well correlated, the structures of corresponding pieces are, as often as not, quite different. There is no sign yet that bacterial proteins, or ancient proteins such as the histones, are consistently more repetitive than proteins which have evolved more recently. The only two examples where a sequence repeat appears suggestive are the pair of central helices in subtilisin and the two central strands of β structure in carboxypeptidase. Here, however, the evidence is not statistically significant, and one has to rely on the fact that the structural core of each protein adjacent to the active site contains two similar segments which each carry a chemically active histidine.

None of the proteins studied shows a large-scale repeat in its structural organization, accompanied by a repeat in the sequence, which cannot be accounted for by chance.

There is, therefore, no necessity to postulate that gene duplication, either in the form of regular repeating sequences, or a mosaic of short repeated pieces of various kinds, has been a dominant influence in the recent evolution of proteins. If events of this kind did occur in the earliest stages of evolution of the most primitive proteins, their traces have been almost completely obscured by later changes in the sequence and structure.

Chromosomes of many higher animals (e.g. mice) contain large amounts of highly repetitive satellite DNA near the centromeres. This DNA contains repeat periods as short as six nucleotides (Southern, 1970) and does not code for any known protein. Even if satellite DNA is a source for the random evolution of new proteins, it can bear little relevance to the processes of evolution in bacteria.

It is worth emphasising that surprisingly strong short repeats arise fairly often by chance. They can easily appear significant when taken by themselves, especially if the three-dimensional structure of the protein is unknown. The case against the significance of repeats relies more on the careful examination of the mutual consistency of the observed correlations and their relationship with the folding of the protein than on statistical arguments. The early work of Cantor & Jukes (1966) and Fitch (1966) thus suggested possibilities that were of great interest and potential importance, which could not be dismissed on the evidence then available to them.

There is an apparent weakness in our arguments because it has been assumed throughout that two sections of protein sequence which share a common ancestor must fold similarly. At the same time many examples have been pointed out where similar short sequences have quite different structures. Why should there not be duplication of short pieces of sequences which then form different structures? To this there are two answers. One is that no statistically significant evidence for this kind of duplication event yet exists. The other is that ancestral similarities in sequence only persist for a long time during evolution if the structures to which they belong remain the same. Hence any weak repeat which does not correspond to a structural repeat is far less likely to be a long-standing ancestral feature than one which does.

The occurrence of very similar segments with quite different structures in so many proteins suggests that local sequence may be even less important in protein folding than is usually thought, and that the balance of molecular forces is exceedingly delicate.

12. Evolution of Large Proteins from Smaller Ones

During the course of evolution, proteins have tended to acquire successively more complicated and sophisticated functions. These new functions often make use of larger structures, either with longer chains or with several interacting subunits. Thus, many protein chains which now possess 200 to 400 amino acids have probably evolved from shorter proteins of 50 to 100 amino acids. There is probably also a critical length of about 50 amino acids below which it is difficult to form a protein structure which is stable in solution under normal conditions.

In the very earliest stages of evolution, after the setting up of the genetic code, the rate of error in protein synthesis would tend to control the length of a protein. Two extremes are conceivable: large imprecisely folded molecules which were relatively inefficient, but which could tolerate many sequence errors without losing activity; or shorter, precisely formed, and highly intolerant structures which could function very effectively provided they were free of error. For example, if we require that the protein synthesis apparatus must give a 90% yield of perfect sequences, the tolerable error rates per amino acid for chains of lengths 20, 100 and 400 are 0·002, 0·0004 and 0·0001, respectively. If the *proportion* of errors, rather than their total number, determined whether primitive proteins were acceptable, the argument for short proteins disappears, since long chains would have a better chance of forming stable structures.

In any case there are certainly a large number of proteins of the second, intolerant type which have increased in size during more recent evolution; and one can ask what processes may have led to their growth.

There are three principal possibilities.

(a) *Multiple repeats followed by consolidation of the whole*

A single short sequence of, say, 10 to 15 amino acids was exactly copied n times over at one time to produce a primitive repeating sequence having approximately the same length as the final protein (Eck & Dayhoff, 1966). The fundamental sequence was capable of forming some simple kind of stable structural unit—a helix or a loop of β structure—which was repeated several times. The units might then aggregate together to form a stable whole, forming a protein with a repetitive secondary structure, folded into a tertiary structure the chief features of which were determined once and for all. Proteins which happened to be able to bind other molecules might then later acquire some activity as an enzyme without undergoing large changes in tertiary structure. On this view the primitive repeating protein would have no biological function until after its over-all structure was decided.

(b) *Random duplication and consolidation of segments*

The initial short sequence, which has a primitive biological function, grows by adding a segment of 10 to 15 amino acids at long intervals of time. Each segment is a copy of some existing portion of the chain, and may be added anywhere in the sequence. After a piece is added there is a rapid series of small evolutionary changes to consolidate its position in the new structure. The protein would maintain some biological function which changes in small steps. A rarer process would be the duplication of an entire protein chain, to produce a molecule built out of two similar substructures in contact. This event could lead to a discontinuous change of function or a large improvement in biological activity. Two identical substructures tend to fit together well, because pieces which are of similar structure are more likely to fit one another in complementary fashion than are two dissimilar pieces (Monod, Wyman & Changeux, 1965).

The sequences and structures studied in this paper give no support at all to the idea of multiple repeats, and little, if any, to the idea of limited duplication. The theory of multiple repeats also has the drawback that it requires a large structure to evolve before it acquires any well-defined biological function. Thus we are led to a third possibility.

(c) *Piecemeal growth*

The protein would begin as a short chain with some biological activity, centred on an active site, and then alter its structure, gradually for the most part, inserting or deleting one amino acid at a time at points on the outer surface in such a way as to conserve the structural core round the active site. In this way successively more complicated surface loops and supporting structures could be added to strengthen the original molecular framework and improve the activity. During these changes the biological function would always exist, although it might change its character, as happens in going from chymotrypsin to thrombin (Magnusson, 1968,1970) or from lysozyme to α-lactalbumin (Brew & Campbell, 1967; Brew *et al.*, 1967; Browne *et al.*, 1969). The sequences of cytochrome *c* (Nolan & Margoliash, 1968) or of myoglobin, lamprey haemoglobin and mammalian haemoglobins (Dayhoff, 1969; Braunitzer & Fujiki, 1969) show that protein chains can easily lengthen or contract at either end and at corners. Elastase and thrombin yield examples where growth occurs at bends in the middle of a chain (Shotton & Hartley, 1970) while haemoglobin Gun Hill (Bradley, Wohl & Rieder, 1967) provides one where several deletions may have occurred at the same time. It is more difficult to imagine several simultaneous insertions, because the new amino acids are unlikely to fit well into the existing structure. Gene duplication would only occur as a rare and atypical event.

Two important features are common to many enzyme structures (Blow & Steitz, 1970): the presence of buried polar groups in the active site, and the way in which these polar groups are attached to different lobes or sub-assemblies of the structure which come together on different sides of the active centre. One reason for this, suggested by Blow & Steitz, is that the folding of the protein must supply sufficient free energy to compensate for the *electronic strain* energy which is used to abstract the polar groups from the surrounding solution: one simple way is to use the free energy of adhesion of several large rigid substructures. A second reason may be that small changes in the mutual packing of the substructures during evolution allow very precise adjustment of the positions of the active groups to optimise the catalytic activity. Figure 1 illustrates a typical scheme of gradual piecemeal evolution which

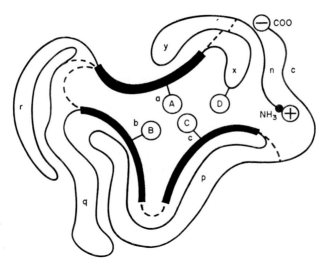

FIG. 1. Scheme for gradual growth of an enzyme chain. Initially the active site consists of three lobes, a, b and c, which support the active groups A, B, C. Later the closed loops p, q and r grow successively on the outer surface. A new loop x, later supported by the section y, introduces a new group D into the active centre, modifying the activity. The amino and carboxyl ends grow at n and c to link together the ends of the chain.

embodies these features. Here the active site begins with three lobes which gradually grow and become more extensively interlinked. Later a fourth outer lobe is added with a new active group which modifies the catalytic function.

According to this type of scheme the final large structure gradually builds up from the interior in successive layers centred on the active site. The path by which the enzyme folds would evolve too, but if there is any orderly series of stages there should be a tendency for the folding of the parts which evolved earliest to guide the folding of the later and more peripheral sections. For example, in the final structure of the protein in Figure 1 the sections p, q and r could not achieve their final conformation until a, b and c were assembled. This idea of a hierarchy of folding events set up during evolution of a large chain, and each dominating the folding of successive portions of the structure, is related to the idea that a protein contains nucleation points about which it folds. However, it goes further, since it suggests a way in which a very large structure can "learn" to fold as it evolves, and suggests that distantly related proteins of similar structure but widely different chain lengths should have whole loops of peripheral structure inserted or deleted, while conserving other parts of the core almost unchanged.

I thank Mrs P. M. E. Altham, Dr F. H. C. Crick, Dr M. F. Perutz, Dr L. E. Orgel and Dr D. M. Blow for discussions.

APPENDIX

Substitution Frequencies in Proteins

Here are the data used in Figure 1 of the preceding paper (McLachlan, 1971). They are taken from 17 homologous families of proteins: 2 subtilisins, 13 haemoglobins, 8 cytochromes c, 2 penicillinases, 14 antibody light-chain variable regions, 12 antibody constant regions, 5 tobacco mosaic viruses, 6 azurins, 3 glyceraldehyde-3-phosphate dehydrogenases, 6 chymotrypsin enzymes, 2 cytochromes c_3, 3 cytochromes c_{551}, 4 lysozymes, 3 ribonucleases, 17 insulins, 3 bacterial ferredoxins, and 3 plant ferredoxins.

In each family we have counted $N(i,j)$, the number of *positions* at which amino acids i and j occur at least once as alternatives, and $n(i)$, the number of positions at which amino acid i ever occurs. The total number of substitutions for each amino acid is defined as:

$$N(i) = \sum N(i,j), \qquad j \neq i. \tag{A1}$$

If the substitutions were all equally probable, the expected values of $N(i,j)$ and $N(i)$ would be:

$$E(i,j) = \alpha n(i)n(j), \tag{A2}$$

$$E(i) = \sum E(i,j), \qquad j \neq i, \tag{A3}$$

where

$$\alpha = N_1/N_2; \; N_1 = \sum N(i,j), \; N_2 = \sum n(i)n(j), \; i \neq j. \tag{A4}$$

The relative frequencies of the substitutions are defined to be:

$$f(i,j) = N(i,j)/E(i,j) \quad \text{and} \quad f(i) = N(i)/E(i).$$

In Table A1 the diagonal elements are $N(i)$ and $f(i)$. The off-diagonal elements are $N(i,j)$ and $f(i,j)$. The row below the main body of the Table gives $n(i)$.

TABLE A1

Frequencies of amino-acid substitutions in proteins

AMINO ACID REPLACEMENTS IN 17 FAMILIES OF PROTEINS

	V	L	I	M	F	W	Y	G	A	P	S	T	C	H	R	K	Q	E	N	D
VAL	662	82	102	28	27	5	19	35	79	25	50	46	6	12	18	41	21	27	18	21
	0.94	1.74	2.77	1.66	1.19	0.64	0.81	0.67	1.09	0.84	0.68	0.85	0.46	0.67	0.71	0.73	0.66	0.72	0.43	0.47
LEU	82	545	72	44	43	7	18	18	43	12	38	35	2	11	17	34	21	12	18	18
	1.74	0.93	2.40	3.19	2.33	1.10	0.95	0.43	0.73	0.49	0.63	0.79	0.19	0.75	0.82	0.74	0.80	0.39	0.53	0.49
ILE	102	72	411	19	18	5	14	13	31	11	31	33	2	7	7	13	6	10	10	7
	2.77	2.40	0.89	1.77	1.25	1.01	0.94	0.39	0.67	0.58	0.66	0.96	0.24	0.61	0.43	0.36	0.30	0.42	0.38	0.25
MET	28	44	19	227	12	1	5	5	20	3	17	20	3	6	3	10	9	6	8	8
	1.66	3.19	1.77	1.04	1.82	0.44	0.73	0.33	0.95	0.34	0.79	1.26	0.79	1.15	0.40	0.61	0.96	0.55	0.66	0.61
PHE	27	43	18	12	249	10	49	2	17	5	18	12	0	10	5	6	1	4	4	6
	1.19	2.33	1.25	1.82	0.86	3.28	5.36	0.10	0.60	0.43	0.62	0.56	0.0	1.43	0.50	0.27	0.08	0.27	0.25	0.34
TRP	5	7	5	1	10	85	16	4	4	0	9	5	1	2	3	4	3	3	1	2
	0.64	1.10	1.01	0.44	3.28	0.84	5.09	0.57	0.41	0.0	0.91	0.68	0.57	0.83	0.88	0.53	0.70	0.59	0.18	0.33
TYR	19	18	14	5	49	16	247	6	13	3	24	13	2	10	7	13	5	9	11	10
	0.81	0.95	0.94	0.73	5.36	5.09	0.83	0.29	0.44	0.25	0.81	0.59	0.38	1.38	0.68	0.57	0.39	0.59	0.66	0.55
GLY	35	18	13	5	2	4	6	535	78	26	73	37	5	11	20	46	21	41	46	48
	0.67	0.43	0.39	0.33	0.10	0.57	0.29	0.84	1.20	0.97	1.10	0.76	0.43	0.68	0.88	0.91	0.73	1.22	1.24	1.20
ALA	79	43	31	20	17	4	13	78	878	52	135	87	6	19	24	68	43	68	44	47
	1.09	0.73	0.67	0.95	0.60	0.41	0.44	1.20	1.02	1.39	1.46	1.28	0.37	0.85	0.75	0.97	1.07	1.45	0.85	0.84
PRO	25	12	11	3	5	0	3	26	52	331	46	27	2	8	12	27	16	28	8	20
	0.84	0.49	0.58	0.34	0.43	0.0	0.25	0.97	1.39	0.88	1.21	0.96	0.30	0.87	0.92	0.93	0.97	1.45	0.37	0.87
SER	50	38	31	17	18	9	24	73	135	46	1039	144	12	25	48	87	53	62	98	69
	0.68	0.63	0.66	0.79	0.62	0.91	0.81	1.10	1.46	1.21	1.18	2.08	0.72	1.09	1.48	1.22	1.30	1.30	1.86	1.21
THR	46	35	33	20	12	5	13	37	87	27	144	751	8	22	25	67	37	47	43	43
	0.85	0.79	0.96	1.26	0.56	0.68	0.59	0.76	1.28	0.96	2.08	1.13	0.65	1.31	1.05	1.27	1.23	1.33	1.10	1.02
CYS	6	2	3	3	0	1	2	5	6	2	12	8	68	3	3	2	1	1	4	5
	0.46	0.19	0.24	0.79	0.0	0.57	0.38	0.43	0.37	0.30	0.72	0.65	0.40	0.74	0.52	0.16	0.14	0.12	0.43	0.50
HIS	12	11	7	6	10	2	10	11	19	8	25	22	3	249	18	25	14	8	20	18
	0.67	0.75	0.61	1.15	1.43	0.83	1.38	0.68	0.85	0.87	1.09	1.31	0.74	1.08	2.29	1.44	1.41	0.69	1.56	1.30
ARG	18	17	7	3	5	3	7	20	24	12	48	25	3	18	359	65	30	20	21	13
	0.71	0.82	0.43	0.40	0.50	0.88	0.68	0.88	0.75	0.92	1.48	1.05	0.52	2.29	1.11	2.64	2.13	1.21	1.15	0.66
LYS	41	34	13	10	6	4	13	46	68	27	87	67	2	25	65	701	46	52	53	42
	0.73	0.74	0.36	0.61	0.27	0.53	0.57	0.91	0.97	0.93	1.22	1.27	0.16	1.44	2.64	1.02	1.48	1.43	1.32	0.97
GLN	21	21	6	9	1	3	5	21	43	16	53	37	1	14	30	46	450	56	33	34
	0.66	0.80	0.30	0.96	0.08	0.70	0.39	0.73	1.07	0.97	1.30	1.23	0.14	2.13	1.48	1.11		2.70	1.44	1.37
GLU	27	12	10	6	4	3	9	41	68	28	62	47	1	8	20	52	56	568	35	79
	0.72	0.39	0.42	0.55	0.27	0.59	0.59	1.22	1.45	1.45	1.30	1.33	0.12	0.69	1.21	1.43	2.70	1.21	1.30	2.73
ASN	18	18	10	8	4	1	11	46	44	8	98	43	4	20	21	53	33	35	534	59
	0.43	0.53	0.38	0.66	0.25	0.18	0.66	1.24	0.85	0.37	1.86	1.10	0.43	1.56	1.15	1.32	1.44	1.30	1.03	1.84
ASP	21	18	7	8	6	2	10	48	47	20	69	43	5	18	13	42	34	79	59	549
	0.47	0.49	0.25	0.61	0.34	0.33	0.55	1.20	0.84	0.87	1.21	1.02	0.50	1.30	0.66	0.97	1.37	2.73	1.84	0.99
POSN	394	321	250	115	154	53	159	353	493	203	502	370	89	122	173	382	218	255	282	304

N1 = 9438   N2 = 25290138   ALPHA = 0.00037

REFERENCES

Air, G. M., Thompson, E. O. P., Richardson, B. J. & Sharman, G. B. (1971). *Nature*, **229**, 391.

Ambler, R. P. (1968). *Biochem. J.* **109**, 47.

Beintema, J. J. & Gruber, M. (1967). *Biochim. biophys. Acta*, **147**, 612.

Benson, A. M., Mower, H. F. & Yasunobu, K. T. (1967). *Arch. Biochem. Biophys.* **121**, 563.

Birktoft, J. J., Blow, D. M., Henderson, R. & Steitz, T. A. (1970). *Phil. Trans. Roy. Soc. Lond.* B, **257**, 67.

Black, J. A. & Dixon, G. H. (1967). *Nature*, **216**, 152.

Black, J. A. & Dixon, G. H. (1968). *Nature,* **218,** 736.

Blake, C. C. F., Mair, G. A., North, A. C. T., Phillips, D. C. & Sarma, V. R. (1967). *Proc. Roy. Soc. Lond.* B, **167,** 365.

Blow, D. M., Birktoft, J. J. & Hartley, B. S. (1969). *Nature,* **221,** 337.

Blow, D. M. & Steitz, T. A. (1970). *Ann. Rev. Protein Chem.* **39,** 63.

Bradley, T. B., Wohl, R. C. & Rieder, R. F. (1967). *Science,* **157,** 1581.

Bradshaw, R. A., Ericsson, L. H., Walsh, K. A. & Neurath, H. (1969). *Proc. Nat. Acad. Sci., Wash.* **63,** 1389.

Bradshaw, R. A., Neurath, H., Tye, R. W., Walsh, K. A. & Winter, W. P. (1970). *Nature,* **226,** 237.

Bradshaw, R. A., Neurath, H. & Walsh, K. A. (1969). *Proc. Nat. Acad. Sci., Wash.* **63,** 406.

Braunitzer, G. & Fujiki, H. (1969). *Naturwiss.* **56,** 322.

Brew, K. & Campbell, P. N. (1967). *Biochem. J.* **102,** 258.

Brew, K., Vanaman, T. C. & Hill, R. L. (1967). *J. Biol. Chem.* **242,** 3747.

Bridges, C. B. (1936). *Science,* **83,** 210.

Brown, J. R. & Hartley, B. S. (1966). *Biochem. J.* **101,** 214, 229.

Browne, W. J., North, A. C. T., Phillips, D. C., Brew, K., Vanaman, T. C. & Hill, R. L. (1969). *J. Mol. Biol.* **42,** 65.

Canfield, R. E. (1963). *J. Biol. Chem.* **238,** 2698.

Cantor, C. & Jukes, T. H. (1966). *Proc. Nat. Acad. Sci., Wash.* **56,** 172.

Cohen, S. & Milstein, C. (1967a). *Advanc. Immunology,* **7,** 1.

Cohen, S. & Milstein, C. (1967b). *Nature,* **214,** 449.

Crick, F. H. C. (1968). *J. Mol. Biol.* **38,** 367.

Dayhoff, M. O. (1969). *Atlas of Protein Sequence and Structure 1969.* Silver Spring, Maryland: National Biochemical Research Foundation.

DeLange, R. J., Famborough, D. M., Smith, E. L. & Bonner, J. (1969). *J. Biol. Chem.* **244,** 319.

Dixon, G. H. (1966). *Essays in Biochemistry,* vol. 2, ed. by P. N. Campbell & G. D. Greville. New York: Academic Press.

Drenth, J., Jansonius, J. N., Koekoek, R., Swen, H. M. & Wolthers, B. G. (1968). *Nature,* **218,** 929.

Dus, K., Sletten, K. & Kamen, M. (1968). *J. Biol. Chem.* **243,** 5507.

Eck, R. V. & Dayhoff, M. O. (1966). *Science,* **152,** 363.

Edelman, G. M., Cunningham, B. A., Gall, W. E., Gottlieb, P. D., Rutishauser, U. & Waxdal, M. J. (1969). *Proc. Nat. Acad. Sci., Wash.* **63,** 78.

Fitch, W. M. (1966). *J. Mol. Biol.* **16,** 1, 8, 17.

Fitch, W. M. (1970). *J. Mol. Biol.* **49,** 1, 15.

Fitch, W. M. & Margoliash, E. (1967). *Science,* **155,** 279.

Gibbs, A. J. & McIntyre, G. A. (1970). *Europ. J. Biochem.* **16,** 1.

Haber, J. E. & Koshland, D. (1970). *J. Mol. Biol.* **50,** 617.

Huber, R., Formanek, H. & Epp, O. (1968). *Naturwiss.* **2,** 75.

Husain, S. S. & Lowe, G. (1969). *Biochem. J.* **114,** 279.

Ingram, V. M. & Stretton, A. O. W. (1961). *Nature,* **190,** 1079.

Iwai, K., Ishikawa, K. & Hayashi, H. (1970). *Nature,* **226,** 1057.

Jollès, P. (1969). *Angewandte Chemie* (Internat. edn.), **8,** 227.

Jukes, T. H. (1966). *Molecules and Evolution.* New York: Columbia University Press.

Kartha, G., Bello, J. & Harker, D. (1967). *Nature,* **213,** 862.

Light, A., Frater, R., Kimmel, J. R. & Smith, E. L. (1964). *Proc. Nat. Acad. Sci., Wash.* **52,** 1276.

Lipscomb, W. N., Hartsuck, J. A., Reeke, G. N., Quiocho, F. A., Bethge, P. H., Ludwig, M., Steitz, T. A., Muirhead, H. & Coppola, J. C. (1968). *Structure, Function and Evolution in Proteins,* Brookhaven Symposia in Biology. **21,** 23.

Lipscomb, W. N., Reeke, G. N., Hartsuck, J. A., Quiocho, F. A. & Bethge, P. H. (1970). *Phil. Trans. Roy. Soc. Lond.* B, **257,** 177.

Low, B. W., Lovell, F. M. & Rudko, A. D. (1968). *Proc. Nat. Acad. Sci., Wash.* **60,** 1515.

McLachlan, A. D. (1971). *J. Mol. Biol.* **61,** 409.

Magnusson, S. (1968). *Biochem. J.* **110**, 25.

Magnusson, S. (1970). In *Structure–Function Relationships of Proteolytic Enzymes*, p. 138, ed. by P. Desnuelle, H. Neurath & M. Ottesen. Copenhagen: Munksgaard.

Manwell, J. (1967). *J. Comp. Biochem. Physiol.* **23**, 383.

Markland, F. S. & Smith, E. L. (1967). *J. Biol. Chem.* **242**, 5198.

Matthews, B. W., Sigler, P. B., Henderson, R. & Blow, D. M. (1967). *Nature*, **214**, 652.

Milstein, C. & Pink, J. R. L. (1970). *Progress in Biophysics & Molecular Biology*, **21**, 209, ed. by J. A. V. Butler & D. Noble. Oxford: Pergamon Press.

Monod, J., Wyman, J. & Changeux, J. P. (1965). *J. Mol. Biol.* **12**, 88.

Needleman, S. B. & Blair, T. T. (1969). *Proc. Nat. Acad. Sci., Wash.* **63**, 1227.

Neurath, H., Bradshaw, R. A. & Arnon, R. (1969). *International Symposium on Structure–Function Relationships of Proteolytic Enzymes*. Copenhagen: Munksgaard.

Nolan, C. & Margoliash, E. (1968). *Ann. Rev. Biochem.* **37**, 727.

Padlan, E. A. & Love, W. E. (1968). *Nature*, **220**, 376.

Perutz, M. F., Kendrew, J. C. & Watson, H. C. (1965). *J. Mol. Biol.* **13**, 669.

Perutz, M. F., Muirhead, H., Cox, J. M. & Goaman, L. G. C. (1968). *Nature*, **219**, 139.

Russell, R. L., Abelson, J. N., Landy, A., Gefter, M. L., Brenner, S. & Smith, J. D. (1970). *J. Mol. Biol.* **47**, 1.

Shotton, D. M. & Hartley, B. S. (1970). *Nature*, **225**, 802.

Shotton, D. M. & Watson, H. C. (1970). *Nature*, **225**, 811.

Singer, S. J. & Doolittle, R. F. (1966). *Science*, **153**, 13.

Smith, E. L., DeLange, R. J., Evans, W. H., Landon, M. & Markland, F. S. (1968). *J. Biol. Chem.* **243**, 2184.

Smithies, O., Connell, G. E. & Dixon, G. H. (1962). *Nature*, **196**, 232.

Smyth, D. G., Stein, W. H. & Moore, S. (1963). *J. Biol. Chem.* **238**, 227.

Sneath, P. H. A. (1966). *J. Theoret. Biol.* **12**, 157.

Šorm, F. & Keil, B. (1962). *Advanc. Protein Chem.* **17**, 167.

Šorm, F. & Knichal, V. (1958). *Collection Czech. Chem. Commun.* **23**, 1575.

Southern, E. M. (1970). *Nature*, **227**, 794.

Stephens, S. G. (1951). *Advanc. Genetics*, **4**, 247.

Takahashi, K. (1965). *J. Biol. Chem.* **240**, pc 4117.

Tanaka, M., Nakashima, T., Benson, A., Mower, H. & Yasunobu, K. T. (1966). *Biochemistry*, **5**, 1666.

Thiebaux, H. J. & Pattee, H. H. (1967). *J. Theoret. Biol.* **17**, 121.

Tsugita, A. & Inouye, M. (1968). *J. Mol. Biol.* **37**, 201.

Urbain, J. (1969). *Biochemical Genetics*, **3**, 249.

Watson, H. C., Shotton, D. M., Cox, J. M. & Muirhead, H. (1970). *Nature*, **225**, 806.

Williams, J., Clegg, J. B. & Mutch, M. O. (1961). *J. Mol. Biol.* **3**, 532.

Wright, C. S., Alden, R. A. & Kraut, J. (1969). *Nature*, **221**, 235.

Wyckoff, H. W., Hardman, K. D., Allewell, N. M., Inagami, T., Johnson, L. N. & Richards, F. M. (1967). *J. Biol. Chem.* **242**, 3749.

Wyckoff, H. W., Tsernoglu, D., Hanson, A. W., Knox, J. R., Lee, B. & Richards, F. M. (1970). *J. Biol. Chem.* **245**, 305.

J. Mol. Biol. (1972) **72**, 209–217

Studies on Polynucleotides†

CIII.‡ Total Synthesis of the Structural Gene for an Alanine Transfer Ribonucleic Acid from Yeast

H. G. Khorana[a], K. L. Agarwal[a], H. Büchi[b], M. H. Caruthers[a],
N. K. Gupta[c], K. Kleppe[d], A. Kumar[e], E. Ohtsuka[f],
U. L. RajBhandary[a], J. H. van de Sande[a], V. Sgaramella[g],
T. Terao[h], H. Weber[i] and T. Yamada[j]

*Institute for Enzyme Research of the University of Wisconsin and the
Departments of Biology and Chemistry, Massachusetts Institute of
Technology, Cambridge, Mass. 02139, U.S.A.*

(*Received 9 December 1971*)

A plan for the total synthesis of the DNA duplex, 77 nucleotide units long, corresponding in sequence to the major yeast alanine transfer RNA, is formulated. The plan involves: (a) the chemical synthesis of 15 polydeoxynucleotide segments ranging in length from five to 20 nucleotide units and (b) ligase-catalyzed covalent joining of several segments to form three parts of the duplex, followed by joining of the three parts to construct the entire duplex. Twelve accompanying papers describe the experimental realization of this objective.

1. Introduction

Methods have been developed in recent years for the chemical synthesis of polydeoxynucleotides of defined nucleotide sequence (Khorana, 1968a), but there is a severe practical limit on the length of the polynucleotide chains which can be assembled unambiguously by purely chemical methods. On the other hand, for biological studies

† The work reported in this series (papers CIII to CXV) was supported by grants from the National Cancer Institute of the National Institutes of Health, U.S. Public Health Service (grants nos 72576 and CA05178), The National Science Foundation, Washington, D.C. (grants nos 73078 and GB-7484X) and The Life Insurance Medical Research Fund.

‡ Paper CII in this series is by Agarwal & Khorana, 1972.

[a] Present address: Departments of Biology and Chemistry, Massachusetts Institute of Technology, Cambridge, Mass. 02139, U.S.A.

[b] Present address: Neue Wangerstrasse, Haus Aurora, 7320 Sargans, Switzerland.

[c] Present address: Department of Chemistry, University of Nebraska, Lincoln, Nebraska, U.S.A.

[d] Present address: Department of Biochemistry, University of Bergen, Bergen, Norway.

[e] Present address: All India Institute of Medical Science, Dept. of Biochemistry, Ansari Nagar, New Delhi, India.

[f] Present address: Faculty of Pharmaceutical Sciences, Osaka University, Toyonaka, Osaka-fu, Japan.

[g] Present address: Department of Genetics, Stanford University Medical School, Palo Alto, Calif., U.S.A.

[h] Present address: Department of Pharmaceutical Sciences, Tokyo University, Tokyo, Japan.

[i] Present address: Institut für Molekularbiologie, Universität Zürich, Winterthurerstrasse 260, Zürich, Switzerland.

[j] Present address: Research Institute for Microbial Diseases, 3 Dojima Nishimachi, Kitaku, Osaka, Japan.

14

of the nucleic acids, it is often the high molecular weight nucleic acids which are the most useful. It is therefore desirable, or even necessary, to couple methods of organic chemistry which alone can afford oligonucleotides of defined sequence, with other concepts or methods, in order to prepare nucleic acids of known sequences. In earlier work reported from this laboratory, it was possible to prepare double-stranded DNA-like polymers of known repeating sequences by using short synthetic polydeoxy-nucleotides as templates for the DNA polymerases. The availability of the resulting polymers permitted extensive studies of the cell-free protein synthesis and of the genetic code (Khorana, 1967,1968b,c). Clearly, future work in this field must aim at the synthesis of bihelical DNA's with specific non-repeating nucleotide sequences. As a first objective, the total synthesis of a double-stranded DNA coding for a transfer RNA was undertaken. We now report the total synthesis of the DNA corresponding to the principal yeast alanine transfer RNA (Fig. 1). This was the only tRNA the

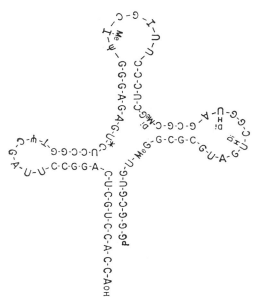

Fig. 1. Structure of the major yeast alanine transfer RNA as deduced by Holley and his co-workers (1965). The structure is shown in the familiar cloverleaf model for the secondary structure of tRNA's. Ψ, pseudouridine; T, ribothymidine; I, inosine, Me-I, methylinosine; DiHU, 5,6-dihydrouridine.

sequence of which was known at the time this work was started (Holley et al., 1965). The present paper serves to introduce the experimental plan while the practical realization of this objective is described in accompanying papers. Brief reports of the work have appeared previously (Agarwal et al., 1970; Khorana, 1971).

   The decision to synthesize the gene for a transfer RNA followed from a variety of considerations. The first clear argument in its favor was that the sequence of the deoxynucleotides in the gene can be derived directly from the sequence of the transfer RNA†. Moreover, the general functions of the tRNA's are clearly established. These

† In deriving the DNA sequence from the tRNA sequence the general assumption has been made that all the minor bases are produced by modification of the four parent bases and that these modifications occur after transcription of the DNA gene with the four standard bases. Thus, inosine is formed by deamination of adenosine, and hence it originates from an A·T base pair in

molecules have to be recognized by a rather large number of components of the protein-synthesizing machinery, such as by the aminoacyl-tRNA synthetases, by the nucleotidyl transferase which repairs the CCA end, by the ribosomes and by several protein factors involved in protein chain initiation, elongation and termination, and finally by messenger RNA. Also, tRNA molecules abound in minor bases and the nascent tRNA molecules have to be recognized by several modifying enzymes. The tRNA's are a unique class of molecules, possessing attributes of both nucleic acids and protein. There is considerable evidence now to suggest that, in addition to a common secondary structure (cloverleaf model, Fig. 1), these molecules possess a tertiary structure. It is possible that a part of the evolution of the genetic code is synonymous with the evolution of tRNA molecules. The total area of the structure–function relations in these molecules is an open field, despite intense activity. It is clear that chemical synthesis could, in principle, offer a definitive approach of wide scope. Different parts of the tRNA structure could be systematically modified at the gene level. The modifications could involve additions, deletions or substitutions of single or a few bases, or could be more extensive, such as the replacements of loops and stems by those present in a different tRNA.

2. Early Work

In developing step-by-step synthesis of macromolecular DNA, the central concept which we have wanted to exploit is the inherent ability of polynucleotide chains to form ordered bihelical complexes by virtue of base pairing. In the work carried out between 1965 and 1967, chemical syntheses of the two icosadeoxynucleotides shown in Figure 2 were accomplished. The two icosanucleotides together span nucleotides

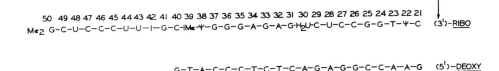

FIG. 2. Chemically synthesized icosadeoxynucleotides corresponding to sequences 21 to 50 of yeast alanine tRNA. I, inosine; Ψ, pseudouridine.

21 to 50 of the tRNA gene, belong to the complementary strands and overlap through halves of their length in the required antiparallel manner. The polynucleotides were in the size range which is the upper limit for the available methodology of organic synthesis. From previous experience, it seemed reasonable to believe that the overlap of ten nucleotides would provide sufficient stability for the duplex composed of the two icosanucleotides. The duplex could be subjected to a repair reaction by the DNA polymerase I of *Escherichia coli*, the repaired strands separated and the separated strands could again be annealed with a short, partly complementary polydeoxy-

DNA. Similarly, pseudouridine (5-ribosyluracil) and dihydrouridine both originate from uridine and hence correspond to an A·T base pair. There seems to be considerable support for this point of view (see e.g. Söll, 1971).

nucleotide and the repair reaction repeated. A device could thus be available for chain elongation by using only short chemically synthesized polydeoxynucleotides (Khorana, 1967). An alternative important possibility considered was as follows. Icosanucleotides (or shorter polynucleotides) corresponding to the entire two strands of the DNA duplex would be chemically synthesized and three or more such segments, when aligned to form an ordered duplex, could be joined end-to-end in aqueous solution. Such joining by water-soluble carbodiimide had been studied in this laboratory several years ago by P. T. Gilham (Naylor & Gilham, 1966).

The discovery of the DNA-joining enzymes (Zimmerman, Little, Oshinsky & Gellert, 1967; Weiss & Richardson, 1967; Olivera & Lehman, 1967; Gefter, Becker & Hurwitz, 1967; Cozzarelli, Melechen, Jovin & Kornberg, 1967) opened up the important possibility of the use of these enzymes in end-to-end joining of chemically synthesized polydeoxynucleotides. Studies were therefore carried out to determine the minimum length of the oligodeoxynucleotide chains which these enzymes require to bring about the joining reactions. Two sets of studies, the first with repeating polymers (Gupta, Ohtsuka, Weber, Chang & Khorana, 1968b) and the second with the polydeoxynucleotides of Figure 3 (Gupta et al., 1968a) showed that the chain lengths necessary

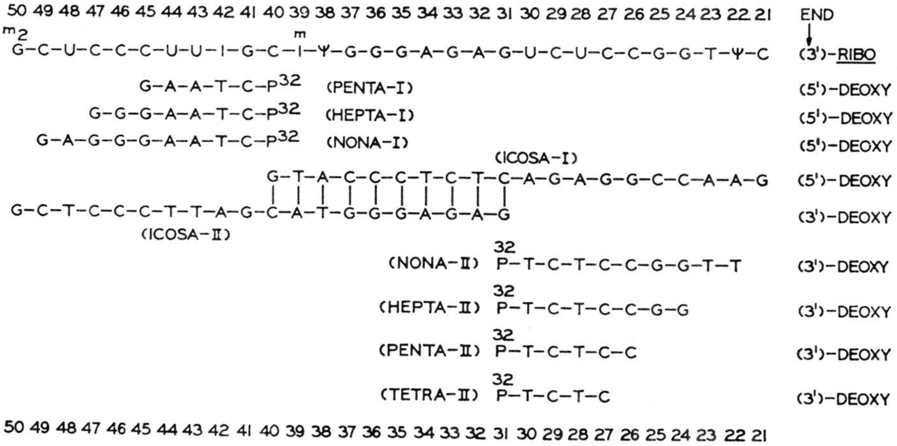

FIG. 3. Chemically synthesized polydeoxynucleotides corresponding to sequences 21 to 50 of yeast alanine tRNA. The icosanucleotide-I (Icosa-I) represents a sequence complementary to nucleotides 21 to 40 of the tRNA and has polarity opposite to that of the tRNA; the nona-, hepta- and pentanucleotides (Nona-I, Hepta-I, and so on) similarly contain sequences complementary to nucleotides 41 to 49 or less, and again have polarity opposite to that of the tRNA. The poly-deoxynucleotides (Icosa-II, Nona-II, Hepta-II, Penta-II and Tetra-II) are segments, complements of the complement, and therefore contain the same sequences and polarity as the tRNA itself. [32]P represents the radioactively labeled 5′-phosphate end-group wherever shown.

were quite small. Thus, in the system shown in Figure 3, which consists of the two icosanucleotides and the appropriate oligonucleotides, the pentanucleotides C-T-A-A-G (Penta-I) and T-C-T-C-C (Penta-II) joined quite well and even the tetranucleotide, T-C-T-C, could indeed be joined to the appropriate icosanucleotide under suitable conditions of temperature, magnesium ion and enzyme concentration. The results seemed to promise a remarkably simple strategy for the synthesis of bihelical DNA which would comprise the following three steps: (1) chemical synthesis

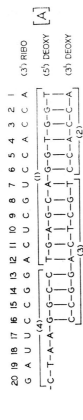

Fig. 4. Total plan for the synthesis of the yeast alanine tRNA gene. The chemically synthesized segments are in brackets, the serial number of the segment being shown inserted in the brackets in parentheses. A total of 17 segments (including 10′ and 12′) varying in chain length from penta- to icosanucleotides was synthesized.

of polydeoxynucleotide segments of chain length in the range of 8 to 12 units with 3′ and 5′-hydroxyl end-groups free; the segments would represent the entire two strands of the intended DNA and those belonging to the complementary strands would have an overlap of four to five nucleotides; (2) the phosphorylation of the 5′-hydroxyl group with ATP carrying a suitable label in the γ-phosphoryl group using the T4 polynucleotide kinase; and (3) the head-to-tail joining of the appropriate segments when they are aligned to form bihelical complexes using the T4 polynucleotide ligase.

3. The Total Plan

Being already in possession of the two icosanucleotides and the corresponding shorter segments, which had served as intermediates in chemical synthesis, the plan shown in Figure 4 was formulated for the total synthesis of the transfer RNA gene. Thus, the gene was divided into three parts shown as [A], [B] and [C] (or [C′]) and each part was to consist of several chemically synthesized segments. The segments are indicated in Figure 4 by brackets, the serial number being inserted into the bracket in parentheses. Additional considerations which led to the derivation of this plan were as follows. Starting with the [B] part (bottom strand) and proceeding to the right, segment 8 consisted of the hexadecanucleotide rather than the icosanucleotide shown in Figure 3. The former was preferred because (1) the heptanucleotide (segment 7) could be prepared readily, and (2) the starting tetranucleotide block (C-C-G-G) in the chemical synthesis of segment 5 could be used, in addition, for the synthesis of both segments 3 and 4 (part [A] of the gene). This left only the dodecanucleotide (segment 1) and the hexanucleotide, C-C-A-C-C-A (segment 2), to complete the [A] part of the gene†.

Adoption of the above plan for the [A] part of the gene required some experimental information on the feasibility of joining the [A] part to the [B] part through the weak tetranucleotide (C-T-A-A-) overlap which contains three A·T base pairs. Model experi-
                    (-G-A-T-T)
ments were therefore performed using the system shown in Figure 5, the decanucleotide, d-G-A-A-C-C-G-G-A-G-A, used being available from synthesis of the icosanucleo-

$$^{32}\text{P} \quad \text{OH}$$

A-G-A-G-G-C-C-A-A-G   C-T-A-A-G-G-C-C
C-C-G-G-T-T-C — G-A-T-T

FIG. 5. A three-component joining system corresponding to nucleotide sequence 13 to 30 of yeast tRNA and containing the weak tetranucleotide overlap (C-T-A-A and its complement).

tide I (Fig. 2). The joining reaction proceeded at 5°C and 10 mM-Mg^{2+} concentration to at least 30% and, therefore, the plan adopted for the [A] and [B] parts of the DNA appeared feasible.

Proceeding to the left of segment 6 (Fig. 4), it is seen that this region predominates in purine nucleotides, which cause greater difficulty in chemical synthesis. It was

† As seen later (paper CXII, Sgaramella & Khorana, 1972), this plan for the [A] part of the DNA presented an interesting and instructive difficulty in the enzymic work. The self-complementary tetranucleotide sequence, C-C-G-G, in segment 3 led to dimerization of the product formed from segments 1, 2 and 3. This precluded the formation of [A] by straightforward one-step joining of segments 1 to 4.

therefore decided to use a pentanucleotide as segment 9 and then to go on to divide the remainder of the gene (part [C]) arbitrarily into decanucleotides (segments 10, 11, 12 and 13) and, finally, to provide the terminal sequences in the form of a dodeca-nucleotide (segment 14) and a heptanucleotide (segment 15). (The latter again predominates in the guanine nucleotide.) Because of the high (G + C) content in part [C], problems of intramolecular secondary structure and possible self-aggregation were anticipated and, in particular, the self-complementarity of segment 10 was obvious. Thus, as seen in Figure 6(a), this segment can form a tight duplexed structure by itself. (Absorbance–temperature profiles described in paper CIX (Agarwal, Kumar & Khorana, 1972) have borne this out, the T_m in 5 mM-Mg^{2+} + 0·01 M-Tris·HCl buffer being 62°C.) To guard against the possibility of failure in joining of segment 10 to [B] or [C], or both simultaneously, a modified scheme ([C'] of Fig. 4) for this part of the gene was also adopted. In this, the length of the segment 10' was reduced to eight nucleotides (probable secondary structure in Fig. 6(b)) while the adjoining segment 12' was increased to a dodecanucleotide. In the work described, no difficulty has in fact been encountered in realizing synthesis according to plan [C].

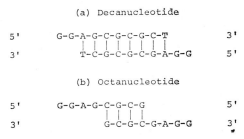

FIG. 6. Self-complementarity of segments 10 and 10'. For numbering of the segments see Fig. 4.

Eight succeeding papers (CIV to CXI) (Weber & Khorana, 1972; Büchi & Khorana, 1972; Kumar, Ohtsuka & Khorana, 1972; Ohtsuka, Kumar & Khorana, 1972; Kumar & Khorana, 1972; Agarwal *et al.*, 1972; Caruthers, van de Sande & Khorana, 1972; Caruthers & Khorana, 1972) describe the synthesis of the fifteen segments including the icosanucleotide corresponding to the nucleotide sequence 31 to 50, which has not been used in the present work. Four succeeding papers (CXII to CXV) describe the enzymic work leading to the synthesis of the duplex segment [A] (Sgaramella & Khorana, 1972), of the duplex [B] (Sgaramella, Kleppe, Terao, Gupta & Khorana, 1972), of the duplex [C] (van de Sande, Caruthers, Sgaramella, Yamada & Khorana, 1972), and of the entire 77 nucleotide long duplex (Caruthers, Kleppe *et al.*, 1972).

Following the synthesis, the first task is the replication of the synthetic DNA, for only then can one ensure adequate future supplies for the many studies that suggest themselves. Initial work on the problem of repair replication by using DNA poly-merases has already been published (Gupta & Khorana, 1968; Kleppe, Ohtsuka, Kleppe, Molineux & Khorana, 1971). Further work along these lines is in progress. Transcription of the appropriate strand to produce the tRNA will be the next step and because the gene lacks the natural start signal for the DNA-dependent RNA poly-merase, the initiation and termination will have to be artificially controlled. Model experiments using parts of the present DNA will be reported separately (Terao *et al.*, manuscript in preparation).

Merrill (1968) has suggested that one G residue be added to the Holley sequence (between nucleotides nos 47 and 48). This possible alteration is not of serious concern, from the standpoint of present work. Indeed, as pointed out above, a main aim of the synthetic approach is to study the structure–function relations by deliberate base substitutions, deletions and additions. It should also be pointed out that because, from many points of view, it seems advantageous to study an *E. coli* rather than a yeast tRNA by the present approach, the synthesis of *E. coli* tyrosine suppressor tRNA is also in progress in this laboratory (Besmer *et al.*, 1971). In the meantime, the DNA corresponding to the yeast alanine tRNA, of course, provides opportunities for study of DNA replication *in vitro* and of transcription.

REFERENCES

Agarwal, K. L., Büchi, H., Caruthers, M. H., Gupta, N., Kleppe, K., Kumar, A., Ohtsuka, E., RajBhandary, U. L., van de Sande, J. H., Sgaramella, V., Weber, H., Yamada, T. & Khorana, H. G. (1970). *Nature*, **227**, 27.

Agarwal, K. L. & Khorana, H. G. (1972). *J. Amer. Chem. Soc.*, in the press.

Agarwal, K. L., Kumar, A. & Khorana, H. G. (1972). *J. Mol. Biol.* **72**, 351.

Besmer, P., Agarwal, K. L., Caruthers, M. H., Cashion, P. J., Fridkin, M., Jay, E., Kumar, A., Loewen, P. C., Ohtsuka, E., van de Sande, J. H., Siderova, N. & RajBhandary, U. L. (1971). *Fed. Proc.* **30**, 1314.

Büchi, H. & Khorana, H. G. (1972). *J. Mol. Biol.*, **72**, 251.

Caruthers, M. H., Kleppe, K., van de Sande, J. H., Sgaramella, V., Agarwal, K. L., Büchi, H., Gupta, N., Kumar, A., Ohtsuka, E., RajBhandary, U. L., Terao, T., Weber, H., Yamada, T. & Khorana, H. G. (1972) *J. Mol. Biol.* **72**, 475.

Caruthers, M. H. & Khorana, H. G. (1972). *J. Mol. Biol.* **72**, 407.

Caruthers, M. H., van de Sande, J. H. & Khorana, H. G. (1972). *J. Mol. Biol.* **72**, 375.

Cozzarelli, N. R., Melechen, N. E., Jovin, T. M. & Kornberg, A. (1967). *Biochem. Biophys. Res. Comm.* **28**, 578.

Gefter, M. L., Becker, A. & Hurwitz, J. (1967). *Proc. Nat. Acad. Sci., Wash.* **58**, 240.

Gupta, N. K. & Khorana, H. G. (1968). *Proc. Nat. Acad. Sci., Wash.* **61**, 215.

Gupta, N. K., Ohtsuka, E., Sgaramella, V., Büchi, H., Kumar, A., Weber, H. & Khorana, H. G. (1968a). *Proc. Nat. Acad. Sci., Wash.* **60**, 1338.

Gupta, N. K., Ohtsuka, E., Weber, H., Chang, S. H. & Khorana, H. G. (1968b). *Proc. Nat. Acad. Sci., Wash.* **60**, 285.

Holley, R. W., Apgar, J., Everett, G. A., Madison, J. T., Marquisee, M., Merrill, S. H., Penswick, J. R. & Zamir, A. (1965). *Science*, **147**, 1462.

Khorana, H. G. (1967). In *Proc. 7th Intern. Cong. Biochem.*, Tokyo. The International Union of Biochem. Vol. 36, p. 17.

Khorana, H. G. (1968a). *Pure Appl. Chem.* **17**, 349.

Khorana, H. G. (1968b). *Biochem. J.* **109**, 709.

Khorana, H. G. (1968c). *The Harvey Lectures*, series 62, p. 79.

Khorana, H. G. (1971). *Pure Appl. Chem.* **25**, 91.

Kleppe, K., Ohtsuka, E., Kleppe, R., Molineux, I. J. & Khorana, H. G. (1971). *J. Mol. Biol.* **56**, 341.

Kumar, A. & Khorana, H. G. (1972). *J. Mol. Biol.* **72**, 329.

Kumar, A., Ohtsuka, E. & Khorana, H. G. (1972). *J. Mol. Biol.* **72**, 289.

Merrill, C. R. (1968). *Biopolymers*, **6**, 1727.

Naylor, R. & Gilham, P. T. (1966). *Biochemistry*, **5**, 2722.

Ohtsuka, E., Kumar, A. & Khorana, H. G. (1972). *J. Mol. Biol.* **72**, 309.

Olivera, B. M. & Lehman, I. R. (1967). *Proc. Nat. Acad. Sci., Wash.* **57**, 1426.

van de Sande, J. H., Caruthers, M. H., Sgaramella, V., Yamada, T. & Khorana, H. G. (1972). *J. Mol. Biol.* **72**, 457.

Sgaramella, V., Kleppe, K., Terao, T., Gupta, N. & Khorana, H. G. (1972). *J. Mol. Biol.* **72**, 445.

Sgaramella, V. & Khorana, H. G. (1972). *J. Mol. Biol.* **72,** 427.

Söll, D. (1971). *Science*, **173**, 293.

Weber, H. & Khorana, H. G. (1972). *J. Mol. Biol.* **72,** 219.

Weiss, B. & Richardson, C. C. (1967). *Proc. Nat. Acad. Sci.*, *Wash.* **57,** 1021.

Zimmerman, S. B., Little, J. W., Oshinsky, C. K. & Gellert, M. (1967). *Proc. Nat. Acad. Sci.*, *Wash.* **57,** 1841.

J. Mol. Biol. (1973) **78**, 363–376

Studies of Simian Virus 40 DNA

VII.† A Cleavage Map of the SV40 Genome

Kathleen J. Danna‡, George H. Sack, Jr and Daniel Nathans

Department of Microbiology, The Johns Hopkins University
School of Medicine, Baltimore, Md 21205, U.S.A.

(Received 28 December 1972)

A physical map of the Simian virus 40 genome has been constructed on the basis of specific cleavage of Simian virus 40 DNA by bacterial restriction endonucleases. The 11 fragments produced by enzyme from *Hemophilus influenzae* have been ordered by analysis of partial digest products and by analysis of an overlapping set of fragments produced by enzyme from *Hemophilus parainfluenzae*. In addition, the single site in SV40 DNA cleaved by the *Escherichia coli* R_I restriction endonuclease has been located. With this site as a reference point, the *H. influenzae* cleavage sites and the *H. parainfluenzae* cleavage sites have been localized on the map.

1. Introduction

The genome of the oncogenic Simian virus 40 is a double-stranded, covalently closed DNA molecule with a molecular weight of about 3×10^6. We have been studying the structure and function of this molecule by specific cleavage with bacterial restriction endonucleases (Adler & Nathans, 1970; Danna & Nathans, 1971,1972; Nathans & Danna, 1972; Sack & Nathans, 1973). Such enzymes make double strand breaks in DNA, generally at specific sites (Kelly & Smith, 1970), thus providing reference points as well as products derived from particular parts of the genome. We reported earlier that restriction endonuclease from *Hemophilus influenzae* produces eleven fragments from SV40§ DNA (Danna & Nathans, 1971) and that the enzyme from *Hemophilus parainfluenzae* produces three major fragments separable by electrophoresis (Sack & Nathans, 1973). In order to use these fragments and enyzme cleavage sites to localize genes and template functions of the SV40 genome, it is necessary to order the fragments in the molecule. By analysis of partial digest products and overlapping sets of fragments, we have ordered the SV40 DNA pieces produced by the restriction endonucleases from *H. influenzae* and *H. parainfluenzae*. In addition, the site cleaved by the *Escherichia coli* R_I restriction endonuclease (Morrow & Berg, 1972; Mulder & Delius, 1972) has also been localized. On the basis of these data and estimates of the size of each DNA fragment, we have constructed a physical map of the SV40 genome.

---

† Paper VI in this series is Sack & Nathans, 1973.

‡ Present address: Laboratorium voor Moleculaire Biologie, K.L. Ledeganckstraat 35, Rijksuniversiteit-Gent, Gent 9000, Belgium.

§ Abbreviation used: SV40, simian virus.

2. Nomenclature for DNA fragments

In this paper we have used a simplified nomenclature to designate fragments of SV40 DNA produced by restriction endonucleases. Each fragment present in a complete digest is assigned a capital letter in order of fragment size (A, B, C ... where A is largest) with an italicized prefix designating the enzyme used to produce the fragment. The restriction endonuclease of *H. influenzae* (Endo R of Smith & Wilcox, 1970) is denoted as *Hin*, that from *H. parainfluenzae* (Gromkova & Goodgal, 1972) as *Hpa*, and the *E. coli* R_I enzyme (Yoshimoro, 1971) as Eco_{RI}. Thus the SV40 DNA fragments found in an *H. influenzae* enzyme digest are *Hin*-A, *Hin*-B, etc., and those found in an *H. parainfluenzae* enzyme digest are *Hpa*-A, *Hpa*-B, etc. When a new fragment results from sequential digestion by two enzymes, it is denoted by combining the designations of the single digest fragments from which it originates. For example, as will be shown below, sequential digestion of SV40 DNA with restriction endonucleases of *H. parainfluenzae* and *H. influenzae* yields a new fragment, not present in either single digest, which represents the overlap between fragments *Hpa*-B and *Hin*-C. The new fragment is designated *Hpa*-B *Hin*-C or *Hin*-C *Hpa*-B (Fig. 1(a)).

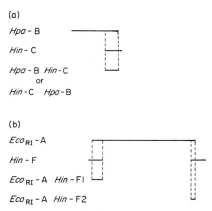

Fig. 1. Examples of nomenclature for new fragments resulting from sequential digestion of SV40 DNA by 2 restriction endonucleases. Horizontal lines indicate fragments and vertical lines indicate cleavage sites.

In those cases in which one of the restriction enzymes cleaves a circular molecule once, or, more generally, only between two adjacent cleavage sites of a second enzyme used subsequently, the above nomenclature is ambiguous; sequential cleavage by the second enzyme yields a new fragment from each end of the first product. We have labeled these fragments 1 and 2, fragment 1 being the larger. For example, the R_I restriction endonuclease from *E. coli* cleaves SV40 DNA at a single unique site within *Hin*-F (see below). Subsequent cleavage of this full-length linear product (Eco_{RI}-A) by the *H. influenzae* enzyme would be expected to yield two new products consisting of the left and right parts of *Hin*-F. These are designated Eco_{RI}-A *Hin*-F1 for the larger product and Eco_{RI}-A *Hin*-F2 for the smaller product (Fig. 1(b)).

In some instances, an isolated fragment produced by one enzyme may be further cleaved by a second enzyme into two or more products. Until the origin of each final product is known it is not possible to designate these fragments by the above nomenclature. In this instance, we suggest a less specific designation indicating the origin of the isolated fragment and the second enzyme used. For example, the *H. parainfluenzae* enzyme cleaves the isolated *Hin*-C fragment into two products. Before identifying the *Hpa* fragments from which the two products are derived, we designate them *Hpa Hin*-C1 and *Hpa Hin*-C2, respectively, *Hpa Hin*-C1 being the larger.

In addition to this nomenclature for DNA fragments, we have used arabic numerals to designate restriction enzyme cleavage *sites* in the DNA, again with the prefix indicating the enzyme used. For example, *H. influenzae* cleavage sites are *Hin*-1, *Hin*-2, etc.

As many more restriction endonucleases from a single species come into use, it will be necessary to add a qualifying term to the enzyme designation, as is already the case for *E. coli* enzymes. We suggest a subscript be used, as done in this paper for the *E. coli* R_I enzyme (Eco_{RI}).

3. Materials and Methods

(a) Cell lines and virus

Small-plaque SV40 (isolated from strain 776 by K. Takemoto) was grown from a cloned stock in the BSC-1 line of African green monkey kidney cells, as described previously (Danna & Nathans, 1971). Stocks were made by infecting at a multiplicity of 0·001 plaque-forming units/ml.

(b) Preparation of ^{32}P-labeled SV40 covalently closed circular DNA

Procedures for infection of cells with SV40, and for labeling and purification of covalently closed SV40 DNA, have been given elsewhere (Danna & Nathans, 1971). Briefly, BSC-1 cells were infected at a multiplicity of 10 to 20 plaque-forming units/ml, labeled with [^{32}P]orthophosphate, and lysed by the method of Hirt (1967). Virus DNA was purified from the supernatant fluid by incubation with heated ribonuclease A, phenol extraction, ethanol precipitation, equilibrium centrifugation in CsCl/ethidium bromide, and sedimentation in a neutral sucrose gradient. Final spec. act. of ^{32}P-labeled covalently closed circular SV40 DNA was 2×10^5 to 3×10^5 cts/min/μg.

(c) Cleavage of SV40 DNA with H. influenzae restriction endonuclease

Two enzyme preparations purified by the method of Smith & Wilcox (1970) were used for most of the experiments to be described. A third preparation of enzyme used in some of the later experiments was supplied by H. O. Smith. Enzyme units have been defined by Smith & Wilcox (1970). For preparation of complete digests, SV40 covalently closed circular DNA was incubated with *H. influenzae* restriction endonuclease at 37°C in a reaction mixture containing 6·6 mM-Tris (pH 7·5), 6·6 mM-MgCl$_2$, 50 mM-NaCl. With each new preparation of enzyme, a series of enzyme concentrations was used to determine the optimum conditions for complete digestion of SV40 DNA. Some variation was found with different preparations and often high enzyme concentrations were inhibitory. Generally, incubation of 50 μg of covalently closed circular SV40 DNA with 0·066 unit of restriction endonuclease in a volume of 0·12 ml for 1 h resulted in complete digestion, as determined by electrophoresis of the products (see below). All reactions were terminated by addition of EDTA to 0·02 M.

To obtain incomplete digests of ^{32}P-labeled SV40 DNA, both the ratio of enzyme to DNA and the time of incubation were varied, but in all cases incubation was carried out at 37°C in the standard buffer described above. As an example, 0·004 unit of enzyme incubated with 2·3 μg of DNA in a volume of 0·12 ml for 30 min yielded at least 29 intermediates in addition to some limit products of the reaction. The legend to Plate I describes other conditions used for partial digestion. Individual, partially digested fragments (at concentrations ranging from 0·5 to 300 μg/ml) were completely digested by incubation with excess enzyme under standard conditions in volumes of 20 to 30 μl.

(d) *Cleavage of SV40 DNA with* H. parainfluenzae *restriction endonuclease*

The enzyme preparation used for these experiments was purified as described earlier (Gromkova & Goodgal, 1972; Sack & Nathans, 1973) and stored at 4°C in 25% glycerol. It contained 8·5 units/ml. Incubations were carried out for 90 min at 30°C in buffer containing 13 mM-Tris·HCl (pH 7·5), 20 mM-KCl, 5 mM-MgCl$_2$, 13 mM-mercaptoethanol, 3% bovine serum albumin and 0·024 unit of enzyme/μg of DNA. Under these conditions the DNA was completely digested, as judged by electrophoresis and radioautography.

(e) *Cleavage of SV40 DNA with* E. coli R_I *restriction endonuclease*

The enzyme preparation used was generously provided by H. Boyer. It was stored at −15°C in 30% glycerol. Incubations were carried out for 2 to 16 h at 30°C in 0·1 M-Tris·HCl (pH 7·6), 6 mM-gMCl$_2$, 6 mM-mercaptoethanol, 200 μg bovine serum albumin/ml and 20 μg of enzyme/μg of DNA. Conversion of covalently closed circular DNA to full length linear molecules was confirmed by centrifugation of the product together with marker SV40 DNA in 5% to 20% sucrose gradients containing 1 × SSC (SSC is 0·15 M-NaCl, 0·015 M-sodium citrate), 10 mM-Tris·HCl (pH 7·4) at 49,000 revs/min in a SW50 Spinco rotor at 20°C for 3 h.

(f) *Vertical slab gel electrophoresis*

Before electrophoresis, samples of digested DNA were made 1% in sodium dodecyl sulfate, incubated at 37°C for 30 min and 0·25 vol. of 75% sucrose, 1% bromphenol blue was added. Electrophoresis was carried out at a constant voltage of 160 V for 16 to 20 h in 4% polyacrylamide vertical slab gels (15 cm × 40 cm × 0·16 cm) in a buffer containing 40 mM-Tris, 20 mM-sodium acetate, 2 mM-sodium EDTA, pH 7·8. A gel chamber similar to those described by DeWachter & Fiers (1971) and by Reid & Bieleski (1968) was used. Preparative electrophoresis of digestion products was followed by radioautography of the wet gel to locate [^{32}P]DNA bands. Gel segments corresponding to these bands were excised at room temperature and eluted with 2 successive vol. of 0·4 ml of 0·1 × SSC, pH 7·4. In preparation for redigestion, eluates were concentrated to dryness by evaporation, dissolved in 20 to 40 μl of sterile deionized water, and dialyzed against 0·1 × SSC, pH 7·4. For radioautographic analysis of DNA fragments, gels were dried by Maizel's modification (1971) of the method described by Fairbanks *et al.* (1965) and then placed in direct contact with Kodak blue medical X-ray film. Satisfactory radioautograms were obtained with as little as 100 cts/min of a ^{32}P-labeled fragment after exposure for 5 days, the band area being about 6 mm^2.

4. Results

(a) *Order of SV40 DNA fragments produced by* H. influenzae *restriction endonuclease*

We have reported earlier that cleavage of SV40 DNA by *H. influenzae* restriction endonuclease yields 11 fragments (designated *Hin*-A to *Hin*-K) separable by polyacrylamide gel electrophoresis (Plate I(a)). On the basis of the relative yield of each fragment from uniformly labeled [^{32}P]DNA, their molecular weights have been estimated (Table 1), and for fragments A to F, these values have been confirmed by electron micrographic length measurements (Danna & Nathans, 1971). Although it has not been established directly that the smaller fragments (G to K) are present in unimolar amount, we shall assume that this is the case; as shown later, none of the fragments is present in more than one location in the SV40 DNA molecule.

The general approach for ordering DNA fragments to be presented in this section was the separation by polyacrylamide gel electrophoresis and localization by radioautography of individual ^{32}P-labeled fragments incompletely digested with *H. influenzae* restriction enzyme; subsequently, each partial product was eluted and

TABLE 1

Molecular weight estimates of SV40 DNA fragments produced by cleavage
with H. *influenzae restriction endonuclease*

Hin fragment	Relative molecular weight (% of SV40 DNA)
A	22·5
B	15·0
C	10·5
D	10·0
E	8·5
F	7·5
G	7·0
H	5·5
I	5·0
J	4·5
K	4·0

Estimates are based on the distribution of radiolabel in a complete *Hin* digest of ^{32}P-labeled SV40 DNA. For A to F, the values were confirmed by electron micrographic measurements (Danna & Nathans, 1971).

digested with excess enzyme and the ultimate products identified by their electrophoretic mobilities. In this way, several overlapping fragments were analyzed and the order of nearly all 11 of the *Hin* fragments was deduced.

Partial digestion products of ^{32}P-labeled SV40 DNA were prepared by incubating the DNA with *H. influenzae* enzyme under standard conditions for short periods or with small quantities of enzyme for longer periods. Either procedure led to the accumulation of numerous intermediate products, identified by electrophoresis and radioautography, the average size of which depended on the time of incubation and enzyme concentration. Typical results, illustrating the effect of incubation time, are shown in the radioautogram in Plate I(b). As seen in the Plate (samples 1, 2 and 3) several discrete fragments are present in addition to the limit digestion products. Also, it is evident from these experiments and numerous others that some of the limit products (fragments A, E and K) appeared quite early during the course of the reaction, indicating that certain of the susceptible sites in SV40 DNA are cleaved preferentially under these conditions. (With some preparations of enzyme, a different set of fragments (G and J) consistently appeared early, suggesting that the *H. influenzae* endonuclease preparation contained enzymes of more than one specificity (see Discussion).)

Several large-scale partial digests of ^{32}P-labeled SV40 DNA were prepared under different incubation conditions and the products purified and redigested as described in Materials and Methods and in the legend to Plate I. For radioautographic analysis, each redigested partial product was electrophoresed in a slab gel in parallel with a sample of the original undigested partial product and a complete digest marker. The R_F for each undigested partial product was determined relative to fragment G, and the limit products derived after redigestion were identified by visual comparison with the marker. Several examples of these analyses are presented in Plate I ((c), (d) and (e)). In the examples of Plate I(c), sample 2 yielded fragments B, F, G, J and K upon redigestion (sample 2d); samples 3 and 4 each yielded fragments B, F, G and

J after redigestion (samples 3d and 4d); sample 6 produced fragments C and D upon redigestion (sample 6d); and sample 7 produced fragments H and I (7d). The sample shown in Plate I(d) yielded fragments B, G and J; sample 1 of Plate I(e) yielded fragments E and K; and sample 2 of Plate I(e) produced fragments G and J after redigestion. In some instances, the putative partial products isolated from preparative gels were actually limit products. For example, in Plate I(c) sample 5 from the partial digest had the same mobility as the marker fragment A, and after redigestion, yielded no other fragments. The same result was obtained with putative partial products that had mobilities identical to those of fragments B, C, D, E, G, H and K, confirming that these fragments do not contain additional cleavage sites for the *H. influenzae* restriction endonuclease.

Some undigested partial products were clearly a mixture of two fragments of approximately equal length, as shown by electrophoresis of the limit products, and in some cases by re-electrophoresis of the partial product. Upon redigestion, two groups of limit products were formed, readily distinguished by the intensity of bands in the radioautogram. In the example shown in Plate I(c), sample 1, fragments A, C and D constitute one group; fragments B, F, G, H, I and J form the other.

As a further check on the composition of the partial products, we compared the molecular weight of each partially digested fragment with the sum of the molecular weights of the products derived from it. For these calculations we used the molecular weight values for the limit products shown in Table 1. Molecular weights of the partial products were estimated graphically from a plot relating log molecular weight

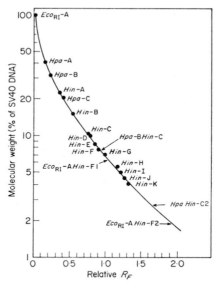

Fig. 2. Log molecular weight *versus* electrophoretic mobility of DNA fragments in 4% polyacrylamide gel (5% cross-linking). Mol. wt is expressed as a percentage of the mol. wt of SV40 DNA, and mobility is given relative to fragment *Hin*-G

$$\left(\frac{R_F \text{ of fragment}}{R_F \text{ of } Hin\text{-G}}\right),$$

For each fragment, the mol. wt has been estimated by the yield of each fragment in a total *Hin* or *Hpa* digest. The arrows indicate the mobility of new fragments resulting from sequential cleavage by 2 restriction endonucleases (see text).

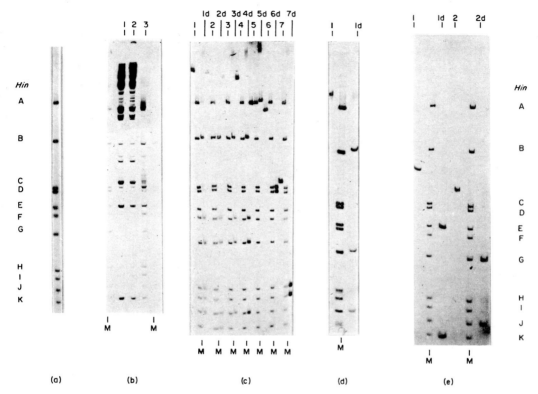

PLATE I. Examples of partial and completed digests of ³²P-labeled SV40 DNA with restriction endonuclease from *H. influenzae*. Each plate is a radioautogram of a single gel slab following electrophoresis. The origin is at the top.

(a) A complete digest of SV40 DNA. Conditions for digestion, electrophoresis and radioautography are described in Materials and Methods. Electrophoresis was for 17·5 h at 150 V.

(b) Partial digests of SV40 DNA. 25 μg of ³²P-labeled SV40 DNA I (10⁵ cts/min/μg) were incubated in a vol. of 0·1 ml with 0·016 of a unit of enzyme under standard conditions. Samples were removed after 20 min digestion (sample 2) and after 30 min digestion (sample 1). For sample 3, 23 μg of ³²P-labeled SV40 DNA I (4·5 × 10⁴ cts/min/μg) were incubated for 1 h with 0·055 of a unit of enzyme in a vol. of 0·32 ml. Samples of the partial digests (1, 2 and 3) and of a complete digest marker (M) were electrophoresed for 16 h at 160 V.

(c), (d) and (e) Redigestion of partial digest products eluted from electrophoresis gels. For each sample, the partial product (1, 2 etc.) and the redigested partial product (1d, 2d etc.) were electrophoresed in the sample gel slab with a complete digest of SV40 DNA as marker (M). (c) Electrophoresis was carried out at 165 V for 17 h; (d) 130 V for 21 h and 180 V for an additional 3 h; (e) 150 V for 17·5 h.

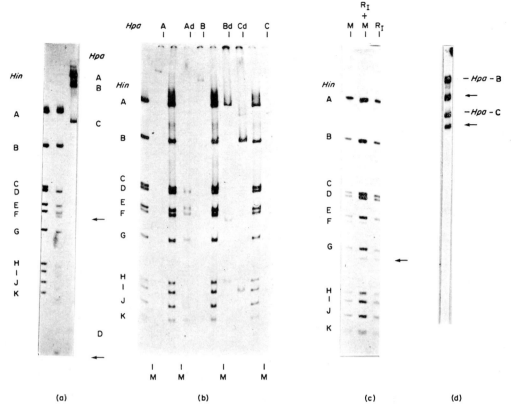

PLATE II. Analysis of ^{32}P-labeled SV40 DNA digests produced by restriction endonuclease from *H. parainfluenzae* and from *E. coli* R_I.

(a) Digestion of SV40 DNA with *H. influenzae* and *H. parainfluenzae* enzymes. The column on the left is a *Hin* digest, that on the right is a *Hpa* digest, and the middle column is a digest produced by sequential incubation with *Hpa* and *Hin* enzymes. Conditions of electrophoresis are given in the legend to Plate I(a). (The slightly faster mobility of some of the double digest products is probably related to the higher salt concentration in this sample—see Plate II(b).) The arrows indicate fragments *Hpa Hin*-C1 (upper) and *Hpa Hin*-C2 (lower).

(b) Digestion of isolated *Hpa* fragments with *H. influenzae* enzyme. The undigested *Hpa* fragments (A, B and C) and the digested *Hpa* fragments (Ad, Bd and Cd) were electrophoresed in a 4% slab gel with complete *Hin* digest marker (M) for 20 h at 145 V.

(c) Digestion of *Eco*$_{RI}$-A with *H. influenzae* enzyme. Samples were electrophoresed for 18 h at 155 V. The column on the left (M) is a digest of covalently closed circular SV40 DNA with *H. influenzae* enzyme; the column on the right (R_I) is a digest of *Eco*$_{RI}$-A with *H. influenzae* enzyme; and the middle column (R_I + M) is a mixture of the 2 digests. The arrow designates a new fragment seen only in the R_I sample. (The expected second new fragment was not observed; see text.) Conditions of electrophoresis as in (b).

(d) Digestion of *Eco*$_{RI}$-A with *H. parainfluenzae* enzyme. The location of the *Hpa* fragments in another part of the gel is indicated. Conditions of electrophoresis as in (b). The arrows indicate the 2 new fragments *Eco*$_{RI}$-A *Hpa*-A1 (top) and *Eco*$_{RI}$-A *Hpa*-A2 (bottom).

to R_F in 4% polyacrylamide gels, using as standards of known molecular weight the *Hin* limit products A to D (Danna & Nathans, 1971); unit-length linear SV40 DNA produced by cleavage with the *E. coli* R_I restriction endonuclease (Morrow & Berg, 1972; Mulder & Delius, 1972); and two fragments produced by the restriction endonuclease from *H. parainfluenzae*, fragments *Hpa*-A (relative mol. wt, 42%) and *Hpa*-B (relative mol. wt, 34%) (Sack & Nathans, 1973) (Fig. 2). Although the curve increases markedly in slope above a relative molecular weight of about 30% of the SV40 DNA, it was nonetheless possible to estimate the sizes of large partial products in this way.

Table 2 summarizes the data from all partial products analyzed. The results were generally straightforward, the molecular weight of each partial digestion product being identical to or close to the sum of the molecular weights of the products derived from it. Slight discrepancies might be due to the non-linearity of the relationship between log molecular weight and R_F, particularly for larger fragments, and the consequent difficulty in estimating molecular weights.

To deduce the physical order of the *Hin* products from the data in Table 2, we have arranged the limit products derived from each incompletely digested fragment

TABLE 2

Redigestion of partial digestion products with H. influenzae *restriction endonuclease*

	Undigested partial product		Redigested partial product	
R_F	Estimated molecular weight (% of SV40 DNA)	Products	Sum of product molecular weights (% of SV40 DNA)	Overlapping orders
0·69	12·0†	G,J	12·0	G J
0·68	11·5†	H,I	11·5	(H,I)
0·66	12·0†	F,K	11·5	F K
0·62	13·0†	E,K	12·5	K E
0·59	13·5†	F,J	12·5	J F
0·39	20·3†	F G,J	19·5	G J F
0·34	22·0†	C,D	20·5	(C,D)
	22·0†	B,G	22·0	B G
0·27	24·5†	B,G,J	27·0	B G J
0·19	36·0‡	B,F,G,J	34·5	B G J F
0·17	40·0‡	B,F,G,J,K	38·5	B G J F K
0·17	40·0‡	A,C,D	43·0	(C,D) A
0·16	43·0‡	A,C,D	43·0	(C,D) A
	43·0‡	B,F,G,H,I,J	45·0	(H,I) B G J F
0·14	48·0‡	B,F,G,H,I,J,K	49·0	(H,I) B G J F K
0·13	51·0‡	A,C,D,E	51·5	E (C,D) A
0·11	57·0‡	A,B,G,H,I,J	60·0	A (H,I) B G J

The R_F for each partial product was determined relative to fragment G in the complete digest in the same gel and is expressed as

$$\frac{\text{distance of partial product from origin}}{\text{distance of G from origin}}.$$

When 2 readily distinguishable groups of final products were derived from a partial product, the groups are shown on 2 separate lines in the Table (R_F 0·16 and R_F 0·34).

† The *Hin* products A to D were the standards of known mol. wt in the plot relating log mol. wt to R_F used to estimate these values.

‡ The *Hpa* products A and B and full length linear SV40 DNA were the standards of known mol. wt in the plot relating log mol. wt to R_F used to estimate these values.

in overlapping positions, as shown in the last column of the Table. In parentheses are two groups, namely the pairs (C,D) and (H,I) whose positions could not be uniquely determined from these data. The order of the *Hin* products established from analysis of partial digestion products is therefore E (C,D) A (H,I) B G J F K, with E and K being contiguous in the circular molecule.

(b) *Order of SV40 DNA fragments produced by* H. parainfluenzae *restriction endonuclease*

We have reported recently (Sack & Nathans, 1973) that a restriction endonuclease from *H. parainfluenzae* (Gromkova & Goodgal, 1972) cleaves SV40 DNA into three predominant fragments: *Hpa*-A (about 42% of the length of SV40 DNA), *Hpa*-B (34%) and *Hpa*-C (20%) (Plate II(a)). To orient these large fragments in the SV40 DNA molecule, we have determined which *Hin* fragments are present in each of the *Hpa* fragments. At the same time, the results of this analysis have resolved the ambiguities in the order of *Hin* fragments.

When ^{32}P-labeled SV40 DNA I was digested successively with restriction enzymes from *H. parainfluenzae* and from *H. influenzae* and the double digestion products separated by electrophoresis, fragments corresponding to all of the *Hin* fragments except *Hin*-C were observed (Plate II(a)). In addition, two new fragments appeared, one between *Hin*-F and *Hin*-G with a mobility of 0·80 relative to *Hin*-G and the other with a mobility of 1·65 relative to *Hin*-G. It appeared, therefore, that one of the *Hpa* sites was within fragment *Hin*-C, and the two new fragments were derived by cleavage of *Hin*-C. Direct evidence for this conclusion was obtained by digesting electrophoretically purified *Hin*-C with the *H. parainfluenzae* enzyme, which resulted in the appearance of both of the new fragments just described. On the basis of electrophoretic mobilities (Fig. 2) we estimate the size of the larger product (*Hpa Hin*-C1) as 8·0% and the smaller product (*Hpa Hin*-C2) as 2·5% of the SV40 DNA molecule.

In order to determine which *Hin* fragments were contained in particular *Hpa* fragments, we isolated fragments *Hpa*-A, *Hpa*-B and *Hpa*-C following electrophoresis in 3% acrylamide gel and digested them individually with *H. influenzae* enzyme. Fragment *Hpa*-A yielded *Hin*-D, E, F, G, J, K; fragment *Hpa*-B yielded *Hin*-A and H and the new fragment *Hpa Hin*-C1; and fragment *Hpa*-C yielded *Hin*-B and I (Plate II(b) and Table 3). Therefore, we can conclude that one of the *Hpa* sites is within fragment *Hin*-C and the other two sites are identical to, or very close to, the *Hin* sites between *Hin*-H and I and between *Hin*-B and G (Table 3). (As noted in Table 3, although we expected to detect the small fragment derived from *Hin*-C (*Hpa Hin*-C2) in the *H. influenzae* enzyme digest of *Hpa*-A, we did not observe this product. This could be due to the presence of a fourth *Hpa* site near the *Hin*-C–*Hin*-D junction: see Discussion.)

By combining the data in Table 3 with those in Table 2, the orientation of the large *Hpa* fragments in the SV40 DNA molecule can be deduced and the uncertainties in the order of *Hin* fragments resolved. Ignoring the smaller part of fragment *Hin*-C that is unaccounted for, we can relate the two sets of fragments as follows:

Hin E D C A H I B G J F K

Hpa A B C A

TABLE 3

Digestion of H. parainfluenzae *restriction endonuclease products with* H. influenzae *restriction endonuclease*

Product	R_F	Estimated molecular weight (% of SV40 DNA)	Products	Sum of product molecular weights (% of SV40 DNA)	Order
Hpa-A	0·17	42	D,E,F,G,J,K	41·5	G J F K E D
Hpa-B	0·21	34	A,H,C1	34	C A H
Hpa-C	0·39	20	B,I	20	I B
					C A H I B G J F K E D

The R_F was determined relative to fragment *Hin*-G in the complete *Hin* digest in the same gel and is expressed as

$$\frac{\text{distance of } Hpa \text{ fragment from origin}}{\text{distance of } Hin\text{-G from origin}}.$$

Molecular weights of *Hpa* fragments were estimated by relative yield (Sack & Nathans, 1973). Although we expected to find the product *Hpa Hin*-C2 in the digest of *Hpa*-A, this small fragment was not observed.

(c) *The site of cleavage of SV40 DNA by* E. coli *R_I restriction endonuclease*

The *E. coli* R_I restriction endonuclease (Yoshimoro, 1971) has been shown to produce one specific double-strand break in SV40 DNA (Morrow & Berg, 1972; Mulder & Delius, 1972; Fareed *et al.*, 1972) which is valuable as a reference point in the circular molecule. For this reason, we thought it worthwhile to localize this site on the cleavage map. We first determined which *Hin* fragment contains the Eco_{RI} site by comparing an *H. influenzae* enzyme digest of covalently closed circular SV40 DNA with a digest of Eco_{RI}-A (i.e. the full length linear product obtained by cleavage of covalently closed circular DNA with the R_I enzyme). Electrophoretograms of the two digests are shown in Plate II(c). The only difference in the two digests was the absence of *Hin*-F among the Eco_{RI}-A products and the appearance of a new fragment, between *Hin*-G and *Hin*-H. It is evident, therefore, that the R_I site is within *Hin*-F. This has been confirmed by direct cleavage of *Hin*-F with the Eco_{RI} enzyme, the two products (Eco_{RI}-A *Hin*-F1 and F2) being identified electrophoretically. From their electrophoretic mobilities of 1·05 and 1·93, relative to *Hin*-G (Fig. 1), we can estimate the size of Eco_{RI}-A *Hin*-F2 as 6·0 to 6·5% and Eco_{RI}-A *Hin*-F2 as 1·5 to 2·0% of the SV40 DNA molecule.

To determine which end of fragment *Hin*-F is closest to the R_I site, we needed to isolate larger pieces of DNA from Eco_{RI}-A and determine whether fragment Eco_{RI}-A *Hin*-F1 is contiguous to fragment J or fragment K, the two *Hin* fragments adjacent to *Hin*-F. This was done by cleaving Eco_{RI}-A with *H. parainfluenzae* enzyme (Plate II(d)), isolating the two new pieces (derived from *Hpa*-A and designated Eco_{RI}-A *Hpa*-A1 and A2), and cleaving each of the latter pieces with *H. influenzae* enzyme. As shown in Table 4, fragment Eco_{RI}-A *Hpa*-A1 yielded *Hin*-K, E and D, whereas fragment Eco_{RI}-A *Hpa*-A2 yielded *Hin*-G, J, and F1. Therefore, fragment Eco_{RI}-A *Hin*-F1 is next to *Hin*-J and the SV40 cleavage site of the R_I enzyme is 6·0 to 6·5% of the length of SV40 DNA from the F–J junction within fragment *Hin*-F.

It should be noted that in the experiments involving cleavage of Eco_{RI} products by *H. influenzae* enzyme, the expected smaller *Hin*-F2 fragment was not observed, although cleavage of *Hin*-F by the Eco_{RI} enzyme did yield both pieces of *Hin*-F. At present we have no explanation for the failure to observe the *Hin*-F2 fragment in *H. influenzae* digests of Eco_{RI}-A *Hpa*-A1.

TABLE 4

Analysis of fragments Eco_{RI}-A Hpa-*A1 and* Eco_{RI}-A Hpa-*A2 with*
H. influenzae *restriction endonuclease*

Eco_{RI}-A *Hpa* product	R_F	Estimated[†] molecular weight (% of SV40 DNA)	*Hin* products	Sum of product molecular weights (% of SV40 DNA)
Eco_{RI}-A *Hpa*-A1	0·29	26·0	K E D	(22·5 to 26·5)[‡]
Eco_{RI}-A *Hpa*-A2	0·46	17·0	G J F1	17·5

The R_F for each product was determined relative to fragment *Hin*-G in the same gel.

† The mol. wts were estimated graphically from a plot relating log mol. wt and R_F (Fig. 1).
‡ In addition to *Hin*-K, E and D (mol. wt 22·5%) Eco_{RI}-A *Hpa*-A1 should have a small part of *Hin*-F (equal to 1·5% of SV40 DNA) and possibly a small part of *Hin*-C (equal to 2·5% of SV40 DNA), neither of which was detected in this experiment.

(d) *A cleavage map of the SV40 genome*

We can usefully summarize all the foregoing data in a single cleavage map of the SV40 genome, incorporating the various cleavage sites and molecular weight estimates of the fragments (Fig. 3). Since the *E. coli* R_I enzyme site appears to be unique, this site has been designated the zero point, and measurements have been made arbitrarily in the direction F-J-G-B- . . . from this point. Map distances are shown in Figure 3 as fractions of the length of SV40 DNA, and the position of each cleavage site is given in these units in Table 5.

TABLE 5

Map distances for Hin *and* Hpa *cleavage sites*

Cleavage site	Distance from Eco_{RI} site (map units or fractional length of SV40 DNA)
Eco_{RI}	0
Hin 1	0·060
Hin 2	0·105
Hin 3	0·175
Hin 4	0·325
Hin 5	0·375
Hin 6	0·430
Hin 7	0·655
Hin 8	0·760
Hin 9	0·860
Hin 10	0·945
Hin 11	0·985
Hpa 1	0·175
Hpa 2	0·375
Hpa 3	0·735
(*Hpa* 4	0·760)

5. Discussion

A cleavage map of a DNA molecule, such as that shown in Figure 3, is a physical map based on specific sites susceptible to endonucleases. The particular map of the SV40 genome shown in the Figure is based on the sites of cleavage of SV40 DNA by three bacterial restriction endonucleases: the enzymes from *H. influenzae*, *H. parainfluenzae*, and *E. coli* R_I. As other specific restriction endonucleases are used on SV40 DNA, it will be possible to map new fragments in relation to these sites.

The methods we have used to order DNA fragments are analogous to those used for ordering peptides in a protein molecule or in sequencing RNA, i.e. analysis of overlapping fragments obtained either by partial digestion with a specific cleaving enzyme or by digestion with a second enzyme of different specificity. Although in the case of proteins and RNA, amino acid or nucleotide sequence is generally used to identify overlapping oligomers, we were able to identify overlapping SV40 DNA fragments by the distinctive electrophoretic mobility of their final digestion products. In the case of more complex digests, it will be necessary to cross-hybridize DNA fragments or to determine partial nucleotide sequences in order to localize the fragments in the cleavage map. Alternatively, DNA fragments can be ordered by electron microscopic examination of heteroduplexes formed between specific fragments or between a fragment and the full-length linear molecule. The ability to visualize the duplex region, however, sets a lower limit to the length of a fragment which can be ordered in this way. In the case of SV40 DNA, specific heteroduplexes between *H. influenzae* fragments and the linear Eco_{RI} product have been observed (T. J. Kelly, Jr, personal communication). Hybridization of denatured *Hin*-F and Eco_{RI} product yielded single-stranded circular molecules, thus confirming the localization of the Eco_{RI} site within *Hin* F.

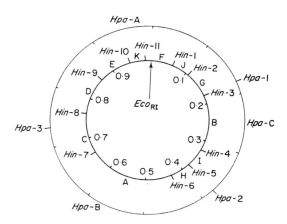

FIG. 3. A cleavage map of the SV40 genome. Map units are given as

$$\frac{\text{distance from } Eco_{RI} \text{ site}}{\text{length of SV40 DNA}}$$

in the direction F-J-G-B . . ., as given in Table 5. A fourth *Hpa* site (*Hpa*-4) probably corresponds to *Hin*-8 (see text).

Several features of the SV40 cleavage map presented in Figure 3 and Table 5 deserve comment. First, the Eco_{RI} site has been selected as the zero point of the map, since the unique full-length linear SV40 DNA molecule produced by the R_I enzyme is being used in mapping by electron microscopy. This site has been localized by identifying the *Hin* fragment in which the cleavage occurs (*Hin*-F), determining the nearest neighbor of Eco_{RI}-A *Hin*-F1 (the larger double digest product), and then estimating the molecular weight of this fragment by electrophoretic mobility.

Second, the *Hin* fragments, which have all been localized, are each present at only one position in the molecule. Earlier, we had suggested that fragments *Hin*-I, J and K appeared to be multiple on the basis of their relative electrophoretic mobility and yield (Danna & Nathans, 1971). However, with better separation of the fragments on long gel slabs, a more accurate yield of each small fragment could be determined than previously reported, and no sharp discontinuities in the yield *versus* mobility plot were found (Fig. 2). Moreover, in the absence of independent estimates of molecular weights, it is not clear that the relationship between log molecular weight and electrophoretic mobility is uniform for small pieces of DNA. In view of the unique map position for each *Hin* fragment, we now assume that the smaller fragments are actually equimolar with the other fragments. Definite proof for this assumption will require independent determinations of molecular weights, which are now being carried out.

A third feature of the map we should like to note is the coincidence of some of the *Hpa* sites with *Hin* sites. Although we cannot now exclude the possibility that these apparently coincident sites are merely very close, it seems more likely that the nucleotide sequences recognized by the *H. parainfluenzae* enzyme may include a subset of those sequences recognized by the *H. influenzae* enzyme(s) (Kelly & Smith, 1970). As pointed out earlier, in addition to the three *Hpa* sites shown in Figure 3, there may be a fourth *Hpa* site near *Hin*-8, which could account for the failure to detect fragment *Hpa Hin*-C2 from *Hpa*-A and also account for the small fragment (*Hpa*-D) present in the *H. parainfluenzae* digest of SV40 DNA (Sack & Nathans, 1973). This possibility is being investigated with more highly purified *H. parainfluenzae* endonuclease.

We should also like to comment on the finding that SV40 DNA has certain sites that are preferentially cleaved by the *H. influenzae* enzyme. Recently, the phosphocellulose fraction of *H. influenzae* restriction endonuclease has been separated into two subfractions which appear to have different specificities (H. O. Smith, personal communication). One of these fractions cleaves SV40 DNA into six fragments including *Hin*-A, E and K, and the other produces five fragments including *Hin*-G and J (Lee, Danna, Smith & Nathans, unpublished observations). From the position of these fragments in the cleavage map shown in Figure 3 and further analysis of digest products, we can deduce that the first enzyme fraction (designated *Hind* III) cleaves SV40 DNA at *Hin* sites 4, 6, 7, 9, 10 and 11, whereas the second enzyme fraction (designated *Hind* II) cleaves at *Hin* sites 1, 2, 3, 5 and 8.

The availability of ordered sets of specific fragments of SV40 DNA is proving useful in analyzing functions of the genome. For example, the initiation site for DNA replication has been located within fragment *Hin*-C close to site *Hin*-7 (at about 0·67 of a map unit) and the termination of replication has been mapped within fragment *Hin*-G (at about 0·15 of a map unit), thus allowing the conclusion that SV40 DNA replication is bidirectional (Danna & Nathans, 1972) and proceeds at about equal rates in the two replication arms. (These values are in close agreement with those reported by Fareed *et al.* (1972) who used electron micrograph length measurements to localize the origin of replication.) *Hin* fragments have also been used to localize the genes in the SV40 chromosome that are transcribed early and late in infection and to determine the direction of transcription (Khoury *et al.*, 1973). Similarly, deletions in the SV40 chromosome (Yoshiike, 1968) and substitutions with cellular DNA (Lavi & Winocour, 1972; Tai *et al.*, 1972), that occur during serial passage of SV40 at high multiplicity, are being mapped in relation to the position of restriction endonuclease cleavage sites (Brockman *et al.*, 1973). Finally, the SV40 cleavage map provides a framework for relating the results of nucleotide sequence analyses of individual fragments to the over-all structure of the genome.

We thank Dr Theresa Lee for her help in some of the experiments reported.

This research was supported by grants from the National Cancer Institute, United States Public Health Service, and the Whitehall Foundation, Inc. One of us (G. H. S.) was supported by training grant no. GM00624 from the United States Public Health Service and another author (K. J. D.) was a predoctoral fellow of the United States Public Health Service.

The results reported in this paper were presented at the Cold Spring Harbor Symposium on Tumor Viruses in August, 1972.

REFERENCES

Adler, S. P. & Nathans, D. (1970). *Fed. Proc.* **29**, 725.

Brockman, W. W., Lee, T. N. H. & Nathans, D. (1973). *Virology*, in the press.

Danna, K. J. & Nathans, D. (1971). *Proc. Nat. Acad. Sci., U.S.A.* **68**, 2913.

Danna, K. J. & Nathans, D. (1972). *Proc. Nat. Acad. Sci., U.S.A.* **69**, 3097.

DeWachter, R. & Fiers, W. (1971). In *Methods in Enzymology* (Colowick, S. P. & Kaplan, N. O., eds), vol. 21, pp. 167–178, Academic Press, New York.

Fairbanks, G., Jr, Levinthal, C. & Reeder, R. H. (1965). *Biochem. Biophys. Res. Commun.* **20**, 393.

Fareed, G. C., Garon, C. F. & Salzman, N. P. (1972). *J. Virol.* **10**, 484.

Gromkova, R. & Goodgal, S. H. (1972). *J. Bact.* **109**, 987.

Hirt, B. (1967). *J. Mol. Biol.* **26**, 365.

Kelly, T. J., Jr & Smith, H. O. (1970). *J. Mol. Biol.* **51**, 393.

Khoury, G., Martin, M. A., Lee, T. N. H., Danna, K. J. & Nathans, D. (1973). *J. Mol. Biol.* **78**, 377.

Lavi, S. & Winocour, E. (1972). *J. Virol.* **9**, 309.

Maizel, J. V., Jr. (1971). In *Methods in Virology* (Maramorosch, K. & Koprowski, H., eds), vol. 5, pp. 180–247, Academic Press, New York.

Morrow, J. & Berg, P. (1972). *Proc. Nat. Acad. Sci., U.S.A.* **69**, 3365.

Mulder, C. & Delius, H. (1972). *Proc. Nat. Acad. Sci., U.S.A.* **69**, 3215.

Nathans, D. & Danna, K. J. (1972). *J. Mol. Biol.* **64**, 515.

Reid, M. S. & Bieleski, R. L. (1968). *Anal. Biochem.* **22**, 374.

Sack, G. H., Jr. & Nathans, D. (1973). *Virology*, **51**, 517.

Smith, H. O. & Wilcox, K. (1970). *J. Mol. Biol.* **51**, 379.

Tai, H. T., Smith, C. A., Sharp, P. A. & Vinograd, J. (1972). *J. Virol.* **9**, 317.

Yoshiike, K. (1968). *Virology*, **34**, 391.

Yoshimoro, R. N. (1971). Ph.D. Thesis, The University of California Medical Center, San Francisco, U.S.A.

J. Mol. Biol. (1973) **78**, 453–471

Enzymatic End-to-end Joining of DNA Molecules

Peter E. Lobban† and A. D. Kaiser

*Department of Biochemistry, Stanford University
School of Medicine, Palo Alto, Calif. 94305, U.S.A.*

(*Received 9 February 1973*)

A way to join naturally occurring DNA molecules, independent of their base sequence, is proposed, based upon the presumed ability of the calf thymus enzyme terminal deoxynucleotidyltransferase to add homopolymer blocks to the ends of double-stranded DNA. To test the proposal, covalently closed dimer circles of the DNA of bacteriophage P22 were produced from linear monomers. It is found that P22 DNA as isolated will prime the terminal transferase reaction, but not in a satisfactory manner. Pre-treatment of the DNA with λ exonuclease, however, improves its priming ability. Terminal transferase can then be used to add oligo(dA) blocks to the ends of one population of P22 DNA molecules and oligo(dT) blocks to the ends of a second population, which enables the two DNAs to anneal to one another to form dimer circles. Subsequent treatment with a system of DNA repair enzymes converts the circles to covalently closed molecules at high efficiency. It is demonstrated that the success of the joining system does not depend upon any obvious unique property of the P22 DNA.

The joining system yields several classes of by-products, among them closed circular molecules with branches. Their creation can be explained on the basis of the properties of terminal transferase and the DNA repair enzymes.

1. Introduction

Given a single deoxyribonucleoside triphosphate as a substrate, calf thymus terminal deoxynucleotidyltransferase (terminal transferase) will attach homopolymer blocks to pre-existing "primer" DNA molecules by catalyzing the step-wise addition of nucleotide residues to their 3′-termini (Kato *et al.*, 1967; Chang & Bollum, 1971). Primers commonly used with the transferase are single-stranded oligonucleotides and heat-denatured DNA (Yoneda & Bollum, 1965; Kato *et al.*, 1967). If the enzyme will also accept double-stranded primers, a method for the joining of naturally occurring DNA molecules end-to-end can be proposed. First, terminal transferase would be used to add homopolymer blocks to the ends of one population of DNA molecules and complementary blocks to the ends of the other population. Next, the two DNAs would be mixed and annealed. Finally, the joined molecules, held together by hydrogen bonds between the bases in the homopolymers, would be exposed to a DNA repair system to render the junctions covalently continuous.

The subject of this paper is a test of the proposed joining system: the creation of covalently closed circular dimers of the DNA of bacteriophage P22, chosen as a typical double-stranded molecule with homogeneous physical properties (Rhoades

† Present address: Department of Medical Cell Biology, Medical Sciences Building, University of Toronto, Toronto 181, Ontario, Canada.

531

et al., 1968). Oligo(dA) and oligo(dT) were selected as the homopolymer blocks to mediate joining because they form a double helix known to be recognized by *Escherichia coli* DNA ligase (Olivera & Lehman, 1968), which was to be a component of the DNA repair system. Circular dimers were preferred to the other possible products of joining, long linear and circular oligomers, because they would be uniform in size and readily detected by their characteristic buoyant density in the presence of ethidium bromide (Radloff *et al.*, 1967) or their rapid rate of sedimentation in alkaline media (Vinograd *et al.*, 1965). Also, the production of molecules having no remaining single-strand interruptions constitutes a rigorous test of the final step of joining.

Jackson *et al.* (1972) have tested a similar joining system with the DNA of an animal virus.

2. Materials and Methods

(a) *Reagents*

Reagents and their sources were CsCl, optical grade, Harshaw Chemical Co.; ethidium bromide, Calbiochem; deoxyribonucleosides, Calbiochem; deoxyribonucleoside triphosphates, P-L Biochemicals; and Sarkosyl, Geigy Chemical Corp. Deoxyribonucleoside triphosphates labeled with ^{32}P in the α-position were prepared by the methods of Symons (1968,1969) and checked for radiochemical purity according to Gilliland *et al.* (1966). Tris/EDTA buffer is 10 mm-Tris·HCl, 1 mm-EDTA, pH as indicated.

(b) *Scintillation counting*

Samples for scintillation counting were dried onto glass filters (Whatman GF/C, 2·4 cm) and counted in a scintillation fluid consisting of 4 g 2,5-diphenyloxazole/l and 50 mg 1,4-bis-2-(4-methyl-5-phenyloxazolyl)-benzene/l in toluene.

(c) *Acid-precipitable radioactivity*

A sample to be assayed for acid-precipitable radioactivity was pipetted in a volume of 0·075 ml or less to a chilled 0·2-ml sample of 0·1 mg carrier DNA/ml in 0·1 m-sodium pyrophosphate, and 1·2 ml of 0·4 m-trichloroacetic acid, 0·02 m-sodium pyrophosphate was added. After sitting 10 min at 0°C, the mixture was poured through a glass filter, which was then washed with 40 ml 1 m-HCl, 0·1 m-sodium pyrophosphate; 40 ml 1 m-HCl, 0·1 m-sodium phosphate; and 20 ml 1 m-HCl. The filter was then wet with ethanol, dried, and counted.

(d) *Acid-soluble radioactivity*

A sample to be assayed for acid-soluble radioactivity was pipetted in a volume of 0·2 ml into a centrifuge tube containing 0·075 ml of 2·5 mg carrier DNA/ml and 0·01 ml of 0·1 m-EDTA at 0°C. After addition of 0·015 ml concentrated HCl, the mixture was left at 0°C for 10 min and then centrifuged at 6000 *g* for 10 min. A sample of the supernate was transferred to a glass filter, and the filter was dried and counted as described.

(e) *Sucrose gradients*

Sedimentation of DNA through sucrose gradients was done in cellulose nitrate tubes in the Spinco SW40 rotor. Neutral gradients were made by pouring 5% to 20% sucrose gradients in 0·3 m-NaCl, 25 mm-Tris·HCl, 2 mm-EDTA, pH 7·5, into tubes containing 1-ml "cushions" of 60% by weight CsCl in 20% sucrose. Samples were layered on in 0·2 to 0·3 ml. Sedimentation conditions are described elsewhere. Fractions were collected directly onto glass filters, which were then dried, wet with 4 drops of 1 m-trichloroacetic acid, 0·05 m-sodium pyrophosphate, washed with 20 ml 1 m-HCl on a suction filter, washed with ethanol, dried, and counted. Recovery of label was quantitative.

Alkaline gradients were 5% to 20% sucrose gradients in 0·3 m-NaOH. Fractions were collected onto glass filters which had been wet with 0·1 ml of 1 m-NaH₂PO₄ and dried before use.

(f) *Extraction of ethidium bromide*

Ethidium bromide was removed from DNA solutions by extraction with 5 or more successive equal volumes of cold *n*-butanol, a method suggested by J. C. Wang (personal communication).

(g) *Phenol extraction*

Phenol extraction was used to terminate preparative terminal transferase and λ exonuclease reactions and to obtain P22 DNA from the phage. The phenol was redistilled, stored under nitrogen at −20°C, and equilibrated with buffer shortly before use.

(h) *Bacteriophage P22 DNA*

P22 $tsc_2{}^{29}$, a thermally-inducible mutant of P22 phage, was supplied by M. Levine. *thyA57*, a thymine-requiring strain of *Salmonella typhimurium* LT2, came from the strain collection of K. Sanderson *via* E. Lederberg. The lysogen *thyA57* (P22 $tsc_2{}^{29}$) was prepared in this laboratory.

Tritium-labeled P22 phage were obtained by thermal induction of *thyA57* (P22 $tsc_2{}^{29}$) in the presence of tritiated thymidine. Phage were concentrated from lysates by the method of Yamamoto *et al.* (1970) and purified by sedimenting them into pre-formed CsCl step gradients and then banding them twice to equilibrium in CsCl solutions.

Phage P22 DNA was obtained from purified phage by diluting them to an absorbance at 260 nm of 25 and phenol-extracting. The DNA was then dialyzed into Tris/EDTA buffer, pH 7·9. Its absorbancy ratio (A_{260}/A_{280}) was 1·95 to 1·99, and it had a specific activity of about 4000 cts/min/μg.

The molecular weight of the sodium salt of phage P22 DNA was taken as $28·0 \times 10^6$ by correcting the data of Rhoades *et al.* (1968) to the molecular weight of phage T7 DNA given by Freifelder (1970). The absorbance at 260 nm of a solution of P22 DNA that is 1 mM in DNA-phosphorus was assumed to be 6·75 as for phage T7 DNA (Richardson, 1966).

Tritiated P22 DNA with a ^{32}P label in its terminal 5′-phosphoryl groups was prepared according to Weiss *et al.* (1968).

(i) *Enzymes*

Lambda exonuclease, $8·8 \times 10^4$ units/mg, was prepared according to Little *et al.* (1967) and had no detectable endonucleolytic activity under the conditions of use. Calf thymus terminal deoxynucleotidyltransferase at 6700 dATP units/mg (Kato *et al.*, 1967; Yoneda & Bollum, 1965) was the gift of R. L. Ratliff and J. L. Hanners and was devoid of exonucleolytic activity. DNA polymerase I of *E. coli* was obtained from A. Kornberg at 12,000 units/mg as assayed with activated calf thymus DNA (Richardson *et al.*, 1964*b*; Jovin *et al.*, 1969). *E. coli* exonuclease III at $1·3 \times 10^5$ units/mg (Jovin *et al.*, 1969) and a homogeneous preparation of *E. coli* DNA ligase at 10^4 units/mg (Modrich & Lehman, 1970, and unpublished results) were the gifts of P. Modrich.

(j) *Electron microscopy*

DNA samples were prepared for electron microscopy by a modification of the aqueous technique of Davis *et al.* (1971), in which Tris·HCl buffers at the same concentrations and pH are substituted for ammonium acetate in both the spreading solution and the hypophase. Molecules were measured on micrographs by projecting them from behind onto a translucent screen and tracing them with a map measure. Preparations of the open circular form of bacteriophage PM2 DNA (Espejo *et al.*, 1969), used as a length standard, were the gifts of F. Schachat and J. Mertz.

3. Results

(a) *Addition of homopolymers to bacteriophage P22 DNA*

Figure 1 shows that P22 DNA will prime the synthesis of oligo(dA) and oligo(dT) homopolymers by terminal transferase. However, both reactions accelerate in the beginning, suggesting that the enzyme may be initiating asynchronously. Since asynchronous initiation would tend to broaden the size distribution of the homopolymer blocks made, a way was developed for modifying P22 DNA so that it will accept nucleotide residues without an acceleration phase. (Alternatively, the acceleration may be due to a progressive improvement in the priming capacity of the P22 strands as they are elongated. In that case, circumventing the acceleration phase would still allow the transferase to synthesize homopolymers of a given average length in a shorter time, thus limiting the exposure of the DNA to deleterious contaminants in the enzyme (see below).)

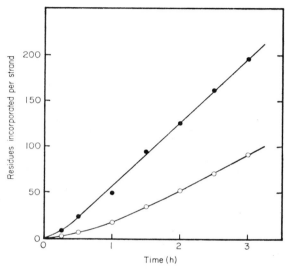

FIG. 1. The terminal transferase reaction with P22 DNA as a primer.
The reaction mixtures contained 74 μg P22 DNA/ml and 8·8 μg terminal transferase/ml in 0·1 M-potassium cacodylate, 7·5 mM-potassium phosphate, 8 mM-MgCl$_2$, 2·1 mM-β-mercaptoethanol, pH 7·0. The substrates were d-[α-^{32}P]ATP at 0·25 mM (—●—●—) and d-[α-^{32}P]TTP at 0·3 mM (—○—○—). The reactions were incubated at 37°C, and samples were withdrawn at intervals to assay for ^{32}P made acid-insoluble.

The modification chosen consisted of treating the DNA with λ exonuclease to remove a small number of residues from its 5'-ends (Little, 1967) so that the 3'-ends are no longer base-paired. The experiment of Figure 2 demonstrates that such a treatment is possible. Tritiated P22 DNA bearing a ^{32}P label in its terminal 5'-phosphoryl groups was exposed to an excess of the exonuclease at 0°C. The ^{32}P was removed rapidly and quantitatively, showing that all termini were attacked in synchrony. Subsequent digestion proceeded at the rate of 64 residues per strand per hour, slowly enough to allow the reaction to be terminated after only a limited number of residues have been removed. We designate P22 DNA treated with λ exonuclease to remove an average of x residues per strand by "P22$_{-x}$." Values of x ranged from 25 to 40.

When P22$_{-x}$ primes terminal transferase, there is no acceleration phase (Fig. 3). Therefore, treatment of the P22 DNA with λ exonuclease was incorporated into the

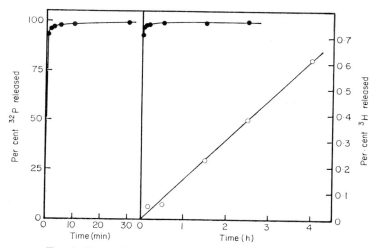

FIG. 2. The action of λ exonuclease on P22 DNA at 0°C.

The reaction mixture contained tritiated P22 DNA with a ^{32}P label in its 5′-phosphoryl groups at 60·3 μg/ml and λ exonuclease at 4·52 μg/ml in 0·067 M-potassium glycinate, 4 mM-MgCl$_2$, pH 9·4 (measured at room temperature). Incubation was at 0°C in an ice–water bath, and samples for the assay of acid-soluble radioactivity were taken with chilled pipettes. —●—●—, ^{32}P label and —○—○—, ^{3}H label, whose rate of release corresponds to the removal of 64 residues per strand/h.

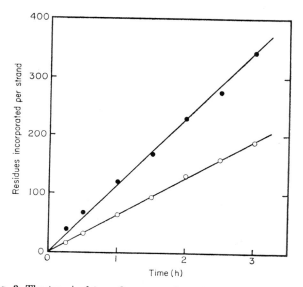

FIG. 3. The terminal transferase reaction with P22$_{-z}$ as a primer.

The reaction mixtures were identical to those of Fig. 1 except that P22$_{-29}$ at 67 μg/ml replaced the untreated P22 DNA. The symbols are the same as in Fig. 1.

joining procedure. Preparative terminal transferase reactions were done with $P22_{-x}$ by scaling up the reactions of Figure 3 and terminating by phenol extraction. The products are symbolized by "dA_y–$P22_{-x}$–dA_y" and "dT_y–$P22_{-x}$–dT_y," with y signifying the number of residues incorporated by the transferase per strand of $P22_{-x}$. The shorter symbols dA–P22–dA and dT–P22–dT serve when x and y are immaterial.

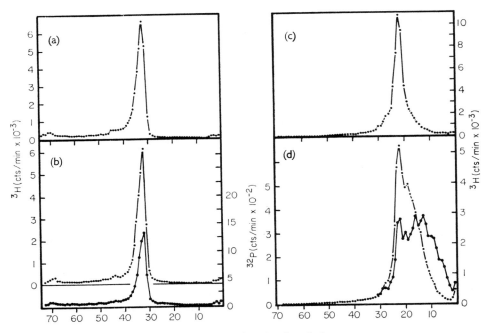

FIG. 4. Sedimentation of dA-P22-dA and dT-P22-dT.

(a) 5 μg of ³H labeled $P22_{-29}$ was layered onto a neutral sucrose gradient in 0·2 ml and sedimented at 20°C for 3 h at 37,000 revs/min.

(b) 5 μg of dT_{66}-$P22_{-29}$-dT_{66}, labeled with ³H in the $P22_{-29}$ (--●--●--) and with ³²P in the homopolymer (—●—●—), was sedimented as in (a).

(c) 9 μg of tritiated P22 DNA was layered onto an alkaline sucrose gradient in 0·2 ml and sedimented at 37,000 revs/min for 3 h at 20°C.

(d) 9·5 μg of dA_{72}-$P22_{-29}$-dA_{72}, labeled as for dT_{66}-$P22_{-29}$-dT_{66}, was sedimented as in (c). The symbols are those used in (b). ³²P counts from the lower portion of the gradient were near background and were not plotted.

The neutral and alkaline sucrose gradients of Figure 4 show that the homopolymer blocks (³²P label) synthesized by terminal transferase in the presence of $P22_{-x}$ cosediment with the DNA (³H label) and are therefore covalently attached to it. As some of the ³²P label sediments with full-length single strands of $P22_{-x}$ (gradient (d) Fig. 4), at least some of the homopolymers are attached at the DNA termini. However, comparison of gradients (c) and (d) in Figure 4 shows that $P22_{-x}$ is extensively nicked during treatment with terminal transferase, probably because of contaminating endonuclease activity. Moreover, the ratio of ³²P to ³H in the region of gradient (d) (Fig. 4) where short single strands appear is greater than would be expected if the transferase acted only at termini. Thus, addition of homopolymers to nicks may be occurring. Direct evidence for that possibility has been obtained by Drs David Jackson and Paul Berg (personal communication), who have shown that terminal transferase accepts nicked circular DNA as a primer, but not closed circles.

(b) *Annealing of dA–P22–dA to dT–P22–dT*

When either $P22_{-x}$, dA–P22–dA, or dT–P22–dT is self-annealed, it sediments in neutral sucrose gradients at the same rate as untreated P22 DNA (Fig. 5(b) to (e)): the small peak in fractions 23 and 24 of all four gradients is apparently artifactual,

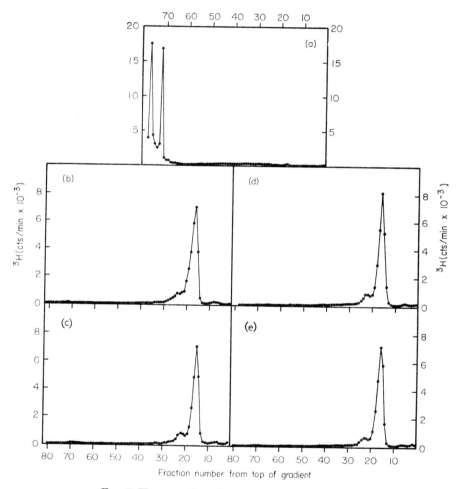

Fig. 5. The annealing of dA-P22-dA to dT-P22-dT.

The following DNA samples were sedimented through neutral sucrose gradients at 20°C for 90 min at 37,000 revs/min: (a) dA_{72}-$P22_{-29}$-dA_{72} * dT_{66}-$P22_{-29}$-dT_{66}, 10 μg; (b) P22 DNA, 6·8 μg; (c) $P22_{-29}$, 5 μg; (d) dA_{72}-$P22_{-29}$-dA_{72}, 5 μg; and (e) dT_{66}-$P22_{-29}$-dT_{66}, 5 μg. Before sedimentation, the DNAs for gradients (a) and (c) to (e) were annealed at 49 μg/ml in 0·5 M-NaCl, Tris/EDTA buffer, pH 7·8, for 10 min at 65°C, 45 min at 45°C at 45 min cooling from 45°C to 39°C. The sample for gradient (b) was heated at 34 μg/ml to 70°C for 10 min in 0·5 M-NaCl, Tris/EDTA buffer, pH 7·8, and quenched by swirling the incubation tube in an ice–water bath. The peak in fraction 72 of (a) is on the high-density cushion at the bottom of the gradient.

as it is not present in parallel gradients in which the 0·3 M-NaCl is omitted, and DNA from those fractions is all of unit length as judged by electron microscopy (data not shown). In contrast, when dA–P22–dA and dT–P22–dT are mixed in equimolar proportions and annealed (symbolized by dA–P22–dA * dT–P22–dT), most of the DNA sediments several times more rapidly than P22 DNA (Fig. 5(a)). We conclude that the homopolymer blocks allow molecules of dA–P22–dA to join to molecules of dT–P22–dT in a specific manner. As the aggregates in dA–P22–dA * dT–P22–dT must contain many P22 moieties in order to sediment so rapidly, it also follows that most molecules of dA–P22–dA and dT–P22–dT bear at least two sites capable of mediating joining.

The sedimentation properties of dA_y–$P22_{-x}$–dA_y * dT_z–$P22_{-x}$–dT_z are independent of the values of y and z so long as both exceed about 40 residues per strand. When y and z are less than 25, on the other hand, much of the DNA sediments as linear monomers and short aggregates (data not shown). Accordingly, values of y and z in the range of 50 to 80 residues per strand were used for joining.

TABLE 1

Structures of annealed DNA molecules

DNA preparation	Linear molecules	Branched linear molecules	Circular molecules	Branched circular molecules	Unscorable molecules
dA-P22-dA	60	0	0	0	13
dT-P22-dT	52	0	0	0	11
dA-P22-dA * dT-P22-dT	98	1	28	13	62

DNA samples were prepared for electron microscopy after annealing as described for Fig. 6. Molecules seen were scored according to structure. Unscorable molecules were those whose contours could not be followed unambiguously; in the case of dA-P22-dA * dT-P22-dT, that class included some very large, multiply branched forms. Linear fragments, which constituted about 5% of each DNA preparation, were not included in the scoring.

The structure of the aggregates in dA–P22–dA * dT–P22–dT was investigated using the electron microscope. For that purpose, the DNA was annealed at low concentration (5 µg/ml) to favor the production of dimer circles over longer forms (Wang & Davidson, 1966). Molecules were scored according to structure (Table 1), and measured (Fig. 6). The data support the following generalizations: (1) the predominant species in self-annealed dA–P22–dA or dT–P22–dT is the linear monomer of P22 DNA, and no circles are seen; (2) among the linear molecules in dA–P22–dA * dT–P22–dT are some which are dimeric or trimeric; and (3) there are circles in dA–P22–dA * dT–P22–dT, most of which are dimers. It follows that the homopolymer blocks attached to $P22_{-x}$ by terminal transferase can mediate an end-to-end joining reaction.

The joining of dA–P22–dA to dT–P22–dT also yields branched DNA molecules (Table 1). The most frequent types seen were "σ-forms", which were circles with a single branch, and "θ-forms", in which three DNA threads began at one point and converged on a second point (Plate I). There were also some more highly branched forms too complex for classification.

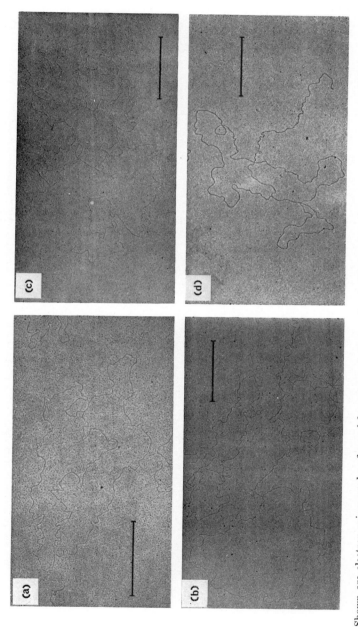

PLATE I. Shown are electron micrographs of one θ-form (a) and three σ-forms ((b) to (d)). The θ-form has arms of lengths 0·70, 0·56 and 0·71 P22 unit. The circular parts of the σ-forms are, respectively, 1·26, 2·01 and 1·49 units long, and the linear parts are of lengths 0·75, 1·02 and 0·49 units. Each calibration mark shows 2 μm.

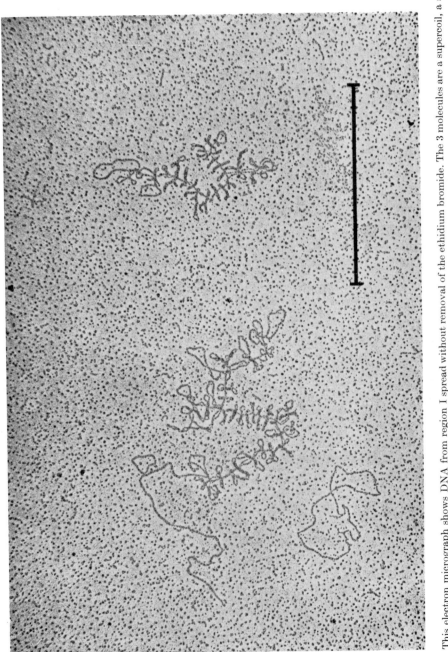

PLATE II. This electron micrograph shows DNA from region I spread without removal of the ethidium bromide. The 3 molecules are a supercoil, a super-coil with a branch and an open PM2 circle. The calibration mark is 2 μm long.

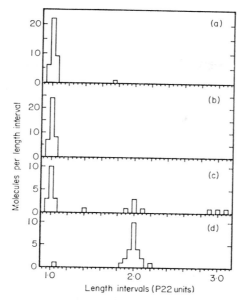

FIG. 6. Measurement of DNA molecules in annealed preparations.

DNA samples were annealed at 5 μg/ml in 0·1 M-NaCl, 25 mM-Tris·HCl, 0·5 mM-EDTA (pH 7·5), for 15 min at 65°C, 2 h at 45°C, and 3 h cooling from 45°C to 32°C. Electron micrographs were taken at random, and the molecules were measured using P22 monomers present as length standards. The samples were (a) dA-P22-dA; (b) dT-P22-dT; (c) dA-P22-dA * dT-P22-dT, linear molecules only; and (d) dA-P22-dA * dT-P22-dT, circular molecules only. One of the linear trimers in (c) was a Y-shaped molecule whose arm lengths were 0·69, 0·98 and 1·26 P22 units; all other linear forms were unbranched.

Most θ-forms are dimers with non-integral arm lengths (Table 2). The lengths of the circular parts of the majority of σ-forms are non-integral in the range from one to two P22 units (Fig. 7, filled and open circles), and most of those molecules are dimeric or trimeric in over-all length (Fig. 7, filled circles). Those data imply that most branches occur where the homopolymer block at an end of one DNA molecule

TABLE 2

Measurements of θ-forms

Arm 1	Arm 2	Arm 3	Total length
0·22	0·50	1·35	2·07
0·28	0·82	0·98	2·08
0·56	0·70	0·71	1·97
0·18	0·50	1·37	2·05
0·42	0·79	0·82	2·03
0·36	0·76	0·84	1·96
0·08	0·09	1·92	2·09
0·38	0·62	0·67	1·67

The arm lengths (in P22 units) of several θ-forms in dA-P22-dA * dT-P22-dT are tabulated here in order of increasing magnitude.

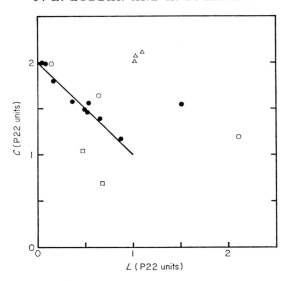

FIG. 7. Measurement of σ-forms.

The lengths L of the linear portions and C of the circular portions of several randomly selected σ-forms in dA-P22-dA * dT-P22-dT are plotted above. Four classes are distinguished: dimers and trimers with $1 < C < 2$ (●), molecules with $1 < C < 2$ whose over-all lengths are non-integral (○), molecules with $C = 2$ and $L = 1$ (△), and other types (□). The line shows the theoretical curve for dimers of the first class.

has annealed to an internal site on a second molecule, probably at a nick where ter-minal transferase has attached a complementary homopolymer. Only a minority of σ-forms, those consisting of a dimer circle with a monomeric branch (Fig. 7, triangles), can have originated from the annealing of two homopolymer blocks of one type to a single complementary block. Several of the σ-forms have non-integral over-all lengths, suggesting that DNA fragments were involved in their formation.

(c) Covalent closure of the joined molecules

The postulated structure of the dimer circles in dA–P22–dA * dT–P22–dT is shown diagrammatically in Figure 8. Each molecule contains four gaps bounded by 3′-hydroxyl and 5′-phosphoryl groups at the points where the P22 moieties are joined. One or more nicks acquired during the incubation with terminal transferase may also be present, and it has been shown that our preparations of the transferase con-tain an activity that attacks DNA in such a way as to leave 3′-phosphoryl groups (D. Brutlag, personal communication). Accordingly, the system chosen for the cova-lent closure of the dimer circles consisted of the *E. coli* enzymes DNA polymerase I, DNA ligase and exonuclease III. The former two enzymes can repair single-strand interruptions with either type of 5′-boundary so long as the 3′-boundaries bear free hydroxyl groups (Richardson *et al.*, 1963,1964*a*; Goulian & Kornberg, 1967; Masa-mune *et al.*, 1971). Since exonuclease III can remove 3′-phosphoryl groups from DNA (Richardson & Kornberg, 1964), its participation in the closure reaction should enable interruptions with phosphorylated 3′-boundaries to be repaired as well. In addition,

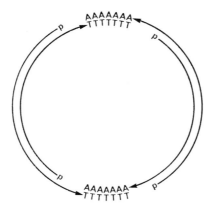

FIG. 8. The postulated structure of the dimer circle.

The strings of A's and T's represent the homopolymer blocks, the p's stand for phosphoryl groups, and the arrows are directed toward the 3'-ends of the respective strands. The end groups are shown as terminal transferase (Kato *et al.*, 1967) and λ exonuclease (Little, 1967) would leave them.

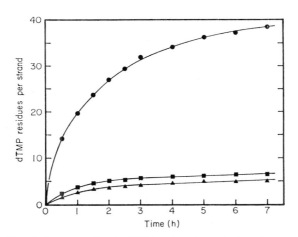

FIG. 9. Incorporation by DNA polymerase during closure.

DNA was prepared for closure by annealing it as for Fig. 6 and adding $(NH_4)_2SO_4$ to 10 mM; $MgCl_2$ to 5 mM; β-mercaptoethanol to 0·5 mM; dATP, dGTP, and dCTP to 45 μM each; NAD$^+$ to 75 μM; bovine plasma albumin to 5 μg/ml; and d-[α-^{32}P]TTP to 8 μM. The mixture was then chilled to 0°C, and the enzymes were added: ligase, 0·75 units/ml; polymerase, 1·2 μg/ml; and exonuclease III, 0·12 units/ml. The reaction was started by placing the mixture at 15°C, and samples were taken at intervals to measure acid-insoluble ^{32}P. A supplement of polymerase (0·8 μg/ml) was added at 135 min and of ligase (0·2 units/ml) at 165 min and at 330 min. The substrates for closure were dA-P22-dA (—▲—▲—), dT-P22-dT (—■—■—), and dA-P22-dA * dT-P22-dT (—●—●—).

the 3'-to-5' exonuclease of DNA polymerase I (Brutlag & Kornberg, 1972) will remove any residues added to the 3'-boundaries of nicks by terminal transferase, so
that repair can then occur; any unpaired residues that may be present at the 3'-
terminus of a homopolymer block which has annealed to a shorter complementary
block should be subject to removal in a similar manner.

Figure 9 shows the kinetics of incorporation of dTTP when dA–P22–dA, dT–
P22–dT, and dA–P22–dA * dT–P22–dT were each subjected to closure. As expected,
dA–P22–dA * dT–P22–dT supported much more incorporation than either of the
other two, indicating that repair was occurring at the gaps formed by the annealed
homopolymers. The over-all incorporation is not great in any of the three reactions,
so very little of the DNA present in the products of closure is newly synthesized.

Samples taken from closure reactions and sedimented to equilibrium in ethidium
bromide/CsCl gradients gave the profiles of Figure 10. When dA–P22–dA * dT–P22–
dT was the substrate for closure, the products appeared in four distinct regions of
the gradient: I, a discrete, high-density peak as expected for closed circles (Radloff
et al., 1967); II, a polydisperse population whose densities ranged from that of closed

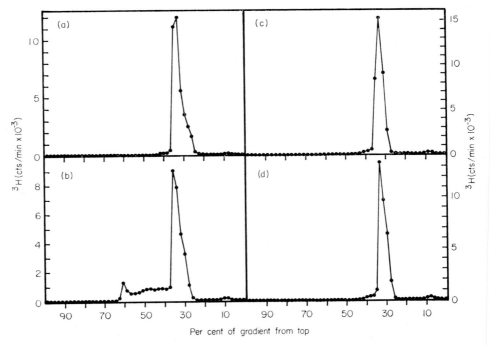

Fig. 10. The closure products in ethidium bromide/CsCl gradients.
Closure reactions were done as for Fig. 9 except that the dTTP was unlabeled. At specified times,
samples of 1·8 ml were removed, mixed with an ethidium bromide/CsCl solution (0·1 ml 10%
Sarkosyl; 0·25 ml 0·2 M-EDTA; 0·4 ml 1 M-Tris·HCl, pH 8·0; 2·88 ml water; 0·25 ml 10 mg
ethidium bromide/ml; and 5·117 g CsCl), and centrifuged at 7°C for 40 h at 40,000 revs/min in
the Spinco 65 rotor. Fractions were collected and treated as for sucrose gradients. The DNA
samples were (a) dA-P22-dA * dT-P22-dT after zero min of closure (enzymes present),
(b) dA-P22-dA * dT-P22-dA after 390 min of closure, (c) dA-P22-dA after 390 min of closure,
and (d) dT-P22-dT after 390 min of closure. Region I of gradient (b), centered at 60% from the
top, contained 6·0% of the DNA recovered from the gradient.

circles to that of linear molecules; III, a discrete peak at the density of the starting material; and IV, a small, broad peak to the low-density side of region III. In contrast, the products of exposing dA–P22–dA or dT–P22–dT to the closure enzymes banded only in regions III and IV, as did dA–P22–dA * dT–P22–dT taken from the closure reaction at zero time.

TABLE 3

Effects of various modifications to the closure reaction on the distribution of the product DNA in ethidium bromide/CsCl gradients

Sampling time (min)	Ligase present?	Exo III concentration (units/ml)	Region I (%)	Region II (%)	Region III (%)	Region IV (%)
0	+	0·12	<0·1	0·0	98·0	2·0
390	—	0·12	<0·1	0·0	97·8	2·2
390	+	0·0	0·9	6·9	89·8	2·4
390	+	0·012	1·3	8·2	88·0	2·5
390	+	0·12	4·0	13·0	80·2	2·8
390	+	0·36	5·3	16·2	75·3	3·2
390	+	1·2	7·3	15·9	73·6	3·2
90	+	0·36	5·0	15·3	76·1	3·6
210	+	0·36	5·2	20·1	71·6	3·1
390	+	0·36	7·6	20·4	69·0	3·0

The closure reaction was run with dA-P22-dA * dT-P22-dT as the substrate according to the methods of Fig. 10 except for the indicated modifications. Samples were taken at the times shown for analysis on ethidium bromide/CsCl gradients.

Table 3 shows how modifications to the closure system affect the nature of the product DNA. Ligase is absolutely required for material to appear in region I, and exonuclease III has a stimulatory effect, suggesting that most but not all of the dimer circles in dA–P22–dA * dT–P22–dT contain 3′-phosphoryl groups that block repair unless removed. The data also show that the closure reaction is rapid and that the closed molecules are stable in the reaction mixture once formed.

DNA molecules taken from regions I and II of an ethidium bromide/CsCl gradient were scored for structure after removal of the ethidium bromide and CsCl (Table 4).

TABLE 4

Structures of DNA molecules from an ethidium bromide/CsCl gradient

Region	Circles	θ-forms	σ-forms	Multiply branched circles
I	74(74%)	11(11%)	12(12%)	3(3%)
II	4(8%)	4(8%)	23(46%)	19(38%)

Fractions from a given region of a preparative ethidium bromide/CsCl gradient were pooled, freed of ethidium bromide and CsCl, and prepared for electron microscopy. Molecules seen were scored for structure. The most common type of multiply branched circle was like a σ-form except that it had more than one branch. There were also θ-forms with branches and molecules resembling two σ-forms joined at the distal ends of their branches. Region III was too heterogeneous to score.

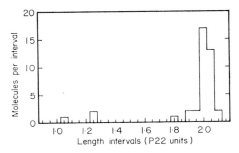

FIG. 11. Lengths of region I circles.

Forty randomly selected circles from region I of an ethidium bromide/CsCl gradient like that of Fig. 10(b) were measured with respect to the open-circular form of PM2 DNA as a length standard. The ratio of the length of P22 monomer to the length of PM2 DNA was determined separately to be 4·21 ($\pm$1·4%, 16 observations); the corresponding ratio for the dimer circles was 8·48 ($\pm$1·9%, 35 observations), making their average length 2·01 P22 units. The circle in length interval 1·05 was 1·06 units long and therefore not monomeric; and one of the circles in interval 1·90 was too short to be a dimer (1·91 units).

Approximately three-quarters of the molecules in region I were circles. Forty such circles were measured to give the histogram of Figure 11, which shows that 35 of the molecules were dimeric, with a mean length of 2·01 P22 units. The other five circles had non-integral lengths ranging from one to two units, suggesting that they resulted from the joining of one P22 molecule to a DNA fragment. Thus, closed dimer circles were produced by the closure reaction in an over-all yield of about 4%; 6% of the DNA banded in region I, 74% of that DNA was circular, and a fraction 35/40 of the circles were dimeric.

Closed circles could also be demonstrated among the products of closure by sedimentation of the DNA through alkaline sucrose gradients (Fig. 12). The peak in fractions 45 to 50 of gradient (c), displaying the characteristic rapid sedimentation rate of closed forms (Vinograd et al., 1965), constituted 4·4% of the DNA recovered, in good agreement with the amount of DNA from the same closure reaction that banded in region I of an ethidium bromide/CsCl gradient (6·1%, remembering that only three-quarters of the DNA in region I is circular). The slowly sedimenting peak in gradient (c) of Figure 12 appears three fractions further down the gradient than do full-length single strands of P22 DNA (gradient (a) Fig. 12), suggesting that efficient covalent joining of DNA strands has occurred even in those products of closure that are not closed circles.

(d) Closed branched molecules

The data of Table 4 show that among the products of closure are branched circular molecules appearing in regions I and II of ethidium bromide/CsCl gradients. The buoyant densities of those molecules indicate that they are "closed"; that is, a constraint operates in at least some portion of each molecule to prevent the two strands from rotating with respect to one another about the axis of the double helix, so that only as much ethidium can be bound there as in a closed circle of equivalent size (Radloff et al., 1967). We would like to suggest that closed σ and θ-forms have the structures shown diagrammatically in Figure 13. Structures (b) and (c) are closed by covalent bonds, while (a) is closed topologically: one strand of the circular portion

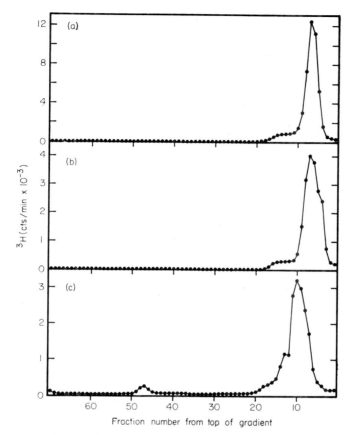

FIG. 12. Sedimentation of the closure products through alkaline sucrose.

Samples were taken from a closure reaction at zero time (b) and after 390 min of incubation (c). Each was made 0·38% in Sarkosyl and 12·5 mM in EDTA, concentrated by dialysis against solid polyethylene glycol 6000, and then dialyzed into 0·2% Sarkosyl, 20 mM-Tris·HCl, 1 mM-EDTA, pH 7·5. A sample of 0·3 ml of each was then layered onto an alkaline sucrose gradient and centrifuged at 20°C for 75 min at 31,000 revs/min. Gradient (a) shows a control sample of P22 DNA.

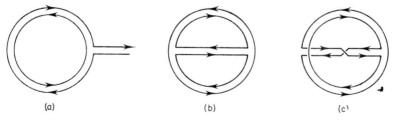

FIG. 13. The postulated structures of closed σ and θ-forms.

Shown are the structures attributed to closed σ-forms (a) and to closed θ-forms ((b) and (c)). The arrows indicate strand polarities.

cannot rotate with respect to the other without breaking all of the base pairs in the branch. Because the branch of a closed σ-form is under no such constraint, it should bind as much ethidium as a linear molecule of the same length; therefore, the whole molecule can have any buoyant density between that of a closed circle and that of an open form, depending upon the ratio of the lengths of its two parts. Similarly, a θ-form closed except for a nick in one arm should also have an intermediate density because only that arm will bind the full amount of ethidium. In agreement with this

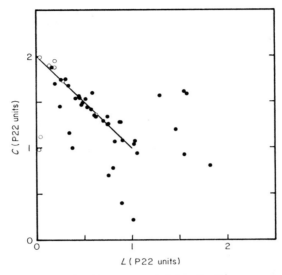

FIG. 14. Measurements of closed σ-forms.
The graph is plotted in the manner of Fig. 7 for σ-forms found in region I (○) and region II (●) of an ethidium bromide/CsCl gradient.

TABLE 5

Measurements of closed θ-forms

Arm 1	Arm 2	Arm 3	Total length
0·53	0·68	0·78	1·99
0·40	0·62	1·04	2·06
0·49	0·52	0·96	1·97
0·50	0·67	0·79	1·96
0·18	0·77	0·96	1·91
0·48	0·68	0·81	1·97
0·24	0·84	0·93	2·01
0·06	0·31	1·71	2·08
0·13	0·34	1·60	2·07
0·49	0·57	1·10	2·16
0·37	0·66	0·96	1·99
0·55	0·76	0·76	2·07
0·07	0·25	1·71	2·03
0·10	0·96	1·46	2·52

Closed θ-forms from regions I and II of a preparative ethidium bromide/CsCl gradient were measured and the results tabulated as for Table 2.

reasoning, the circular part of a closed σ-form mounted for electron microscopy in the presence of ethidium bromide displays the twisted configuration of a supercoil (Vinograd *et al.*, 1965), while its branch has the extended appearance typical of linear DNA (Plate II).

The most probable source of the closed branch forms is the population of open branched circles present in dA–P22–dA * dT–P22–dT before closure. Closed σ and θ-forms obey the same constraints on the lengths of their various parts as do the corresponding open forms (Fig. 14 and Table 5). Note that closed σ-forms from region I tend to have short branches, as expected.

4. Discussion

The experiments presented here show that P22 DNA molecules can be covalently joined to each other. Since the joining is observed only when molecules bearing both kinds of homopolymer blocks are present, and since the circular products of joining are exclusively dimeric, it follows that the reaction is mediated by the homopolymers and not by some unique structural feature of the P22 DNA (for example, its terminally repetitious regions (Rhoades *et al.*, 1968)). Thus it is likely that the joining method can be applied to any pair of double-stranded DNAs. Indeed, Jackson *et al.* (1972) have developed a similar joining system and have used it to create dimer circles of the DNA and to join Simian virus 40 and λ*dvgal* DNAs together. Also, Jensen *et al.* (1971) have reported progress toward the joining of phage T7 DNA molecules by analogous methods.

The creation of closed branched molecules by the closure enzymes merits further discussion. The probable structure of a σ-form in dA–P22–dA * dT–P22–dT is diagrammed in Figure 15. The molecule has three gaps which are identical to those in a dimer circle and therefore present no problem for repair. The fourth gap, whose 3′-boundary is the end of strand 1, is different: soon after initiating synthesis there, DNA polymerase will encounter the branch point rather than the 5′-end of another strand. At that time the enzyme can continue to use strand 2 as its template by displacing strand 3 (Kelly *et al.*, 1970); after each insertion of a base residue, the structure of the DNA will be the same as in the diagram except that the single-stranded portion of the strand being displaced will have grown. If at any time, however, the

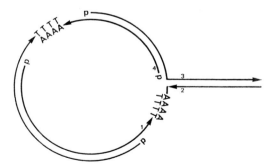

FIG. 15. The postulated structure of a σ-form in dA-P22-dA * dT-P22-dT.
The structure shown is based upon the evidence that most branch points form where terminal transferase has added a homopolymer block to the 3′-boundary of a nick. The symbols are the same as in Fig. 8.

enzyme begins to use strand 3 as its template, it will move toward the 5'-end of strand 4 and eventually create a structure in which ligase can join that end to the growing end of strand 1 to give a closed σ-form as depicted in Figure 13(a). Similar events could close a θ-form. The ability of DNA polymerase I to switch template strands during strand displacement has been previously implicated in the production of branched DNA (Schildkraut et al., 1964; Masamune & Richardson, 1971). It is also possible that the closing of branched forms could be catalyzed by ligase if that enzyme were able to join the 3'-end of strand 1 to the 5'-end of strand 4 as soon as the polymerase brought the former into juxtaposition with the latter at the branch point. We know of no precedent for such a reaction, however.

A different method for joining DNA molecules arises from the ability of the ligase of bacteriophage T4 to catalyze the formation of phosphodiester bonds between the ends of fully double-stranded DNA molecules (Sgaramella et al., 1970). Linear dimers and trimers of P22 DNA have been made with that enzyme (Sgaramella, 1972). Joining by T4 ligase differs from the method described here in that there is no apparent way to direct the enzyme to join molecules of one type only to molecules of another type and not to each other. The same is true of a third joining method dependent upon the ability of the R_I restriction endonuclease to generate small cohesive ends when it hydrolyzes DNA (Mertz & Davis, 1972; Hedgpeth et al., 1972).

The apparent generality of the method described here for the joining of any pair of double-stranded DNA molecules to each other may provide a new way to study eukaryotic genes and their controlling elements. Any block of genes from any organism could be inserted into the genome of a temperate bacteriophage to generate a specialized transducing phage. The genes borne by the phage could then perhaps be studied in bacteria, possibly even at the level of fine-structure genetic analysis. Alternatively, the phage could be used to produce many copies of the genes to be used to detect specific DNA-binding macromolecules in the donor organism. In short, it may become possible to apply the techniques of bacterial genetics and biochemistry to genes from eukaryotic cells.

We would like to acknowledge the fruitful discussions and free exchange of ideas we enjoyed with Drs David Jackson, Robert Symons and Paul Berg, who worked concurrently on a similar project. This work was supported by a research grant (AI04509) from the National Institutes of Health. One of us (P. E. L.) was a pre-doctoral fellow of the National Science Foundation while part of this work was in progress.

REFERENCES

Brutlag, D. & Kornberg, A. (1972). J. Biol. Chem. **247**, 241.
Chang, L. M. S. & Bollum, F. J. (1971). Biochemistry, **10**, 536.
Davis, R. W., Simon, M. & Davidson, N. (1971). In Methods in Enzymology (Grossman, L. & Moldave, K., eds), vol. 21, p. 413, Academic Press, New York.
Espejo, R. T., Canelo, E. S. & Sinsheimer, R. L. (1969). Proc. Nat. Acad. Sci., U.S.A. **63**, 1164.
Freifelder, D. (1970). J. Mol. Biol. **54**, 567.
Gilliland, J. M., Langman, R. E. & Symons, R. H. (1966). Virology, **30**, 716.
Goulian, M. & Kornberg, A. (1967). Proc. Nat. Acad. Sci., U.S.A. **58**, 1723.
Hedgpeth, J., Goodman, H. M. & Boyer, H. W. (1972). Proc. Nat. Acad. Sci., U.S.A. **69**, 3448.
Jackson, D. A., Symons, R. H. & Berg, P. (1972). Proc. Nat. Acad. Sci., U.S.A. **69**, 2904.

Jensen, R. H., Wodzinski, R. J. & Rogoff, M. H. (1971). *Biochem. Biophys. Res. Commun.* **43**, 384.

Jovin, T. M., Englund, P. T. & Bertsch, L. L. (1969). *J. Biol. Chem.* **244**, 2996.

Kato, K.-I., Gonçalves, J. M., Houts, G. E. & Bollum, F. J. (1967). *J. Biol. Chem.* **242**, 2780.

Kelly, R. B., Cozzarelli, N. R., Deutscher, M. P., Lehman, I. R. & Kornberg, A. (1970). *J. Biol. Chem.* **245**, 39.

Little, J. W. (1967). *J. Biol. Chem.* **242**, 679.

Little, J. W., Lehman, I. R. & Kaiser, A. D. (1967). *J. Biol. Chem.* **242**, 672.

Masamune, Y., Fleischman, R. A. & Richardson, C. C. (1971). *J. Biol. Chem.* **246**, 2680.

Masamune, Y. & Richardson, C. C. (1971). *J. Biol. Chem.* **246**, 2692.

Mertz, J. E. & Davis, R. W. (1972). *Proc. Nat. Acad. Sci., U.S.A.* **69**, 3370.

Modrich, P. & Lehman, I. R. (1970). *J. Biol. Chem.* **245**, 3626.

Olivera, B. M. & Lehman, I. R. (1968). *J. Mol. Biol.* **36**, 261.

Radloff, R., Bauer, W. & Vinograd, J. (1967). *Proc. Nat. Acad. Sci., U.S.A.* **57**, 1514.

Rhoades, M., MacHattie, L. A. & Thomas, C. A., Jr. (1968). *J. Mol. Biol.* **37**, 21.

Richardson, C. C. (1966). *J. Mol. Biol.* **15**, 49.

Richardson, C. C., Schildkraut, C. L., Aposhian, H. V., Kornberg, A., Bodmer, W. & Lederberg, J. (1963). In *Symposium on Informational Macromolecules* (Vogel, H., ed.), p. 13, Academic Press, New York.

Richardson, C. C. & Kornberg, A. (1964). *J. Biol. Chem.* **239**, 242.

Richardson, C. C., Inman, R. B. & Kornberg, A. (1964a). *J. Mol. Biol.* **9**, 46.

Richardson, C. C., Schildkraut, C. L., Aposhian, H. V. & Kornberg, A. (1964b). *J. Biol. Chem.* **239**, 222.

Schildkraut, C. L., Richardson, C. C. & Kornberg, A. (1964). *J. Mol. Biol.* **9**, 24.

Sgaramella, V. (1972). *Proc. Nat. Acad. Sci., U.S.A.* **69**, 3389.

Sgaramella, V., van de Sande, J. H. & Khorana, H. G. (1970). *Proc. Nat. Acad. Sci., U.S.A.* **67**, 1468.

Symons, R. H. (1968). *Biochim. Biophys. Acta*, **155**, 609.

Symons, R. H. (1969). *Biochim. Biophys. Acta*, **190**, 548.

Vinograd, J., Lebowitz, J., Radloff, R., Watson, R. & Laipis, P. (1965). *Proc. Nat. Acad. Sci., U.S.A.* **53**, 1104.

Wang, J. C. & Davidson, N. (1966). *J. Mol. Biol.* **19**, 469.

Weiss, B., Live, T. R. & Richardson, C. C. (1968). *J. Biol. Chem.* **243**, 4530.

Yamamoto, K. R., Alberts, B. M., Benzinger, R., Lawhorne, L. & Treiber, G. (1970). *Virology*, **40**, 734.

Yoneda, M. & Bollum, F. J. (1965). *J. Biol. Chem.* **240**, 3385.

J. Mol. Biol. (1974) **89**, 255–272

Site-directed Mutagenesis : Generation of an Extracistronic Mutation in Bacteriophage Qβ RNA

R. A. Flavell[†], D. L. Sabo, E. F. Bandle and C. Weissmann

Institut für Molekularbiologie I, Universität Zürich
Zürich, Switzerland

(Received 2 May 1974)

The preparation *in vitro* and chemical characterization of bacteriophage Qβ RNA with an extracistronic mutation, a G → A transition in the 16th position from the 3′-terminus, is described. The 5′-terminal region of the Qβ minus strand was synthesized *in vitro* up to position 14 (inclusive) by using ATP and GTP as the only substrates. The mutagenic nucleotide analog N^4-hydroxyCMP was then incorporated into position 15 instead of CMP. The minus strand was completed with the four standard ribonucleoside triphosphates, purified and used as a template for the synthesis of plus strands. Of the plus strand product, 33% had a G → A transition in the 16th position from the 3′-end (which corresponds to position 15 of the minus strand), as shown by nucleotide sequence analysis of the terminal T_1 oligonucleotide. The modified RNA was efficiently replicated by Qβ replicase and a preparation containing 55% of the mutant RNA was obtained.

The general approach to directed mutagenesis outlined above should allow the introduction of mutations into the 5′ and 3′-terminal regions of Qβ RNA as well as into the intercistronic sequences.

1. Introduction

Analysis of the RNA of small RNA-containing bacteriophages such as Qβ, R17, MS2 and f2 has shown that the cistrons specifying the viral proteins are flanked by non-translated nucleotide sequences. In Qβ RNA (cf. Fig. 1) a stretch of 61 residues, in R17, MS2 and f2 one of 129 nucleotides precedes the first cistron, and a segment of as yet unknown length (greater than 61 in Qβ, greater than 22 in R17 and MS2) follows the last one. Each of the intercistronic regions comprises about 20 to 40 nucleotides (cf. review by Weissmann *et al.*, 1973). A comparison of the nucleotide sequences of the 5′ and 3′-terminal extracistronic regions of the related phages R17 and MS2 showed that these are invariant (Adams & Cory, 1970; De Wachter *et al.*, 1971; Adams *et al.*, 1972; Robertson & Jeppesen, 1972), while the cistronic sequences differ in about 3% of the nucleotide positions (Nichols & Robertson, 1971; Min Jou *et al.*, 1972; Robertson & Jeppesen, 1972). This invariance suggests that the biological competence of the virus is dependent on the precise primary structure of the non-translated terminal RNA segments (Min Jou *et al.*, 1972). In order to test this hypothesis and at the same time explore the function of the extracistronic segments, it would be desirable to obtain mutants with base changes in these regions.

---

† Present address: Universiteit van Amsterdam, Laboratorium voor Biochemie, Amsterdam, The Netherlands.

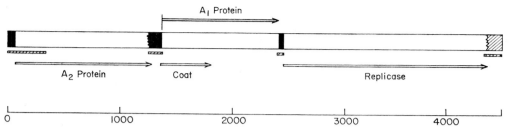

FIG. 1. Map of the Qβ genome.
Non-translated regions are dark; the locations of the cistrons are marked by arrows. The areas of known nucleotide sequence are indicated by the narrow bars under the map. (Based on Hindley et al., 1970; Staples et al., 1971; Billeter et al., 1969; cf. review by Weissmann et al., 1973.)

If only classical genetic techniques are considered, this would constitute an arduous task, since it would be necessary to generate mutations at random and then search for mutants with lesions in the region of interest. No simple criteria exist for selecting or identifying extracistronic mutants and ultimately only sequence analysis of a large number of phage clones could lead to success. However, if lesions in the extracistronic regions are unconditionally lethal, then all classical approaches are doomed to failure.

In view of these considerations it appeared desirable to synthesize viral RNAs with base substitutions in precisely defined sites within the regions of interest and then determine their biological properties.

In this paper we describe the preparation of Qβ RNA with a G to A transition in the 16th position from the 3′-terminus. The 5′-terminal region of the Qβ minus strand (cf. Fig. 2) was synthesized in vitro up to position 15 (exclusive) by using ATP and GTP as the only substrates (Billeter et al., 1969; Bandle & Weissmann, 1972). The mutagenic nucleotide analog N^4-hydroxyCMP (see for example Freese et al., 1961; Osborn et al., 1967; Janion & Shugar, 1968; Tessman, 1968; Poslovina et al., 1973; Popowska & Janion, 1974; and reviews by Phillips & Brown, 1967; Kochetkov & Budowsky, 1969; Singer & Fraenkel-Conrat, 1969) was then incorporated into position 15.† The minus strand was completed with the four standard ribonucleoside triphosphates, purified and used as template for the synthesis of plus strands. Of the plus strand product, 33% showed a G to A transition in the 16th position from the 3′-end.

2. Materials and Methods

Qβ replicase (spec. act. 625 units/mg; units as defined by Eoyang & August (1968)) and host factor I (spec. act. $8·6 \times 10^3$ units/mg unless stated otherwise; units as defined by Franze de Fernandez et al. (1972)) were prepared as described by Kamen et al. (1972). Labeled and unlabeled Qβ RNA were extracted from purified phage using previously published procedures (Weissmann et al., 1968). Qβ RNA fragments of about 11 S were prepared as described by Pollet et al. (1967). Yeast RNA used as carrier (BDH Chemicals, Ltd) was purified of RNAase contamination by incubating a 10 mg/ml solution in 20 mm-Tris·HCl (pH 7·5), 5 mm-EDTA, 0·1% sodium dodecyl sulfate with 0·4 mg

† Synthesis of the minus strand begins opposite the penultimate nucleotide (CMP) of the plus strand (Weith et al., 1968; Kamen, 1969; Rensing & August, 1969); therefore position 15 from the 5′-terminus of the minus strand corresponds to position 16 from the 3′-end of the plus strand (cf. Fig. 6).

pronase/ml (Calbiochem; self-digested in 0·15 M-EDTA (pH 7·5) for 30 min at 37°C) for 30 min at 37°C. The RNA was extracted 3 times with phenol, then 3 times with ether, twice precipitated with ethanol and finally dissolved in 1 mM-EDTA. Unlabeled ribonucleoside triphosphates were obtained from Boehringer GmbH (Mannheim, Germany). ATP and GTP used for substrate-limited synthesis were further purified to remove traces of other contaminating ribonucleoside triphosphates by chromatography on Dowex-1 formate. ATP or GTP were dissolved as 0·1 mmol in about 5 ml of 1/20 × FSS (FSS is 8 M-formic acid, 3·6 M-ammonium formate) and adsorbed to a 1·8 cm × 65 cm column of Dowex 1-X8 (formate), 200 to 400 mesh (AG grade, Calbiochem or Fluka, prewashed with 1 l of FSS followed by 1 l of 1/20 × FSS). In the case of ATP the column was washed with 90 ml of 1/8 × FSS and eluted with 2 l of a linear gradient of 1/8 to 5/16 × FSS. In the case of GTP the column was washed with 90 ml of 3/16 × FSS and eluted with 2 l of a linear gradient from 3/16 to 1/2 × FSS. The flow rate was 66 ml/h; chromatography was at 20°C. The peak fractions of the triphosphate were pooled (90 ml for ATP and 180 ml for GTP) and passed through a column (70 ml bed volume for ATP, 200 ml bed volume for GTP) of Dowex 50-WX8 (H$^+$), 50 to 100 mesh (Fluka, prewashed with 350 ml of 4 M-HCl and 1·5 l of water). The formic acid was removed by flash evaporation. α-^{32}P-labeled ribonucleoside triphosphates were the gifts of Dr M. A. Billeter and Dr L. Rymo of this Institute. N^4-hydroxyCTP and the corresponding 2′ (3)′-CMP derivative were prepared by reaction of CTP and 2′ (3′)-CMP, respectively with hydroxylamine under the conditions described by Budowsky *et al.* (1971*a*,1972) for the preparation of the triphosphate. The triphosphate was purified by chromatography on Dowex 1 (Cl$^-$) as described by Budowsky *et al.* (1972) and the monophosphate by chromatography on a column of Sephadex G10 (1 cm × 20 cm) using 10 mM-Tris·HCl (pH 8·6) as eluant. The spectral properties of both derivatives were identical to those reported previously (Budowsky *et al.*, 1971*a*). On electrophoresis in pyridine/acetate at pH 3·5 each of the compounds gave a single spot clearly separated from the corresponding U and C nucleotides (Table 1). N^4-hydroxyCTP has the same R_F value (0·18) as UTP on cellulose-coated thin-layer plates (Merck DC Alufolien cellulose, Art. 5552) with the solvent system isobutyric acid: 1 M-ammonium hydroxide: 0·5 M-EDTA (50:30:0·16).

TABLE 1

Electrophoretic mobilities of some nucleotides

2′(3′)-Monophosphate	R_B	5′-Triphosphate	R_B
CMP	0·40	CTP	2·14
AMP	0·69	HOCTP	2·39
GMP	1·23	UTP	2·61
HOCMP	1·41		
UMP	1·71		

Separations were carried out in a buffer containing 5% (v/v) acetic acid, 0·5% (v/v) pyridine and 0·5 mM-EDTA (final pH 3·5) in Savant-type electrophoresis tanks.

To achieve sufficient resolution, mononucleotides had to be separated on Whatman 52 paper (the blue marker, xylene cyanol FF, was run 20 cm from the origin) and triphosphates on Whatman 3MM paper (the blue marker was run 15 cm from the origin). Mobilities (R_B) are expressed relative to the blue marker.

The standard assay conditions for Qβ replicase are described in Kamen *et al.* (1972). Acid-insoluble radioactivity was determined by adding (per 25-μl assay) 200 μl of a solution containing 0·02 M-sodium pyrophosphate (pH 7), 0·15 M-NaCl, 0·015 M-sodium citrate, 10 μg bovine serum albumin/ml, mixing and adding 20 μl of 60% trichloroacetic acid. After a few minutes at 0°C the sample was filtered through glass-fiber filters

(Whatman GF/C) or Millipore HAWP filters. The filters were dried at 110°C and placed in vials containing 10 ml of a toluene solution containing 4 g butyl-PBD/ml (Ciba–Geigy AG, Basel) and 100 mg POPOP/l (Fluka AG, Buchs). Radioactivity was determined by scintillation counting.

(a) Preparation of minus strands with N^4-hydroxyCMP in position 15

All manipulations were carried out at 0 to 4°C unless stated otherwise. $Q\beta$ replicase (200 units), $Q\beta$ RNA (250 μg) and host factor I (200 units) were incubated in 1·25 ml of 0·5 mM-repurified GTP, 0·2 mM-repurified ATP, 80 mM-Tris·HCl (pH 7·5), 12 mM-MgCl$_2$ and 1 mM-EDTA for 5 min at 37°C. A 10-μl portion was removed for the subsequent determination of the enzymatic activity of the complex by the chain completion assay. For this purpose the portion was mixed with 20 μl of a solution containing 0·15 mM each of ATP, UTP and [α-^{32}P]CTP (1·5×10^5 cts/min per nmol), 0·5 mM-GTP, 50 μg polyethylene sulfonate/ml, 12 mM-MgCl$_2$, 80 mM-Tris·HCl (pH 7·5) and 1 mM-EDTA. After incubation for 10 min at 37°C the acid-insoluble radioactivity was determined. The total enzymatic activity, as measured by the chain completion assay was calculated to be 31 nmol of CMP for the main sample. The main sample (1·2 ml) was chromatographed through a 0·7 cm × 40 cm column of Sephadex G100 in 20 mM-Tris·HCl (pH 7·5) at a flow-rate of about 0·1 ml/min and 0·5-ml fractions were collected. The excluded fractions containing the replication complex were identified by protein determination (Schaffner & Weissmann, 1973) and pooled (2·9 ml). The enzymatic activity of the complex was determined by the chain completion assay as above: 110% of the activity originally measured was recovered. The main sample was adjusted to final concentrations of 10 μM-N^4-hydroxyCTP, 3 μM-[α-^{32}P]ATP (2·5×10^7 cts/min per nmol), 12 mM-MgCl$_2$, 100 mM-Tris·HCl (pH 7·5), 1 mM-EDTA and 50 μg polyethylene sulfonate/ml, and incubation was carried out for 10 min at 37°C. To the main sample 0·6 μmol of [^{3}H]GTP (3·3×10^4 cts/min per nmol), 8 μmol of ATP (an 800-fold excess over the [α-^{32}P]ATP), 6·2 μmol of CTP (a 200-fold excess over the N^4-hydroxyCTP) and 5 μmol of UTP were added, and incubation was continued for 15 min to complete the minus strands. The reaction mixture was chilled, 10-μl portions were removed for the determination of acid-insoluble radioactivity, sodium dodecyl sulfate was added (final concn 0·1%) and the sample was extracted once with 1 volume of distilled phenol. The aqueous phase was chromatographed through a column (0·7 cm diameter) consisting of 1 cm of Chelex-100 (Na$^+$) overlayered with 30 cm of Sephadex G100, equilibrated with 20 mM-Tris·HCl (pH 7·5). The excluded fractions were located by determining the Cerenkov radiation of each sample and pooled (4·3 ml). The solution was adjusted to 0·05 M-Tris·HCl (pH 7), 0·1 M-NaCl, 0·005 M-EDTA and the RNA precipitated with 2 volumes of ethanol. After 2 h or more at −20°C the precipitate was collected by centrifugation (10 min at 10,000 revs/min in the HB4 Sorvall rotor). Purification of the newly synthesized minus strands was carried out essentially as described by Pollet et al. (1967). The sample (containing about 250 μg of the original template and about 25 μg of product) in 0·25 ml of 0·5 mM-EDTA was heated in a boiling water bath for 30 s, a solution of 1·2 mg of 11 S $Q\beta$ RNA fragments in 0·75 ml of 0·5 mM-EDTA (heated to about 95°C) were added and the mixture was further heated for 60 s. After rapidly cooling in an ice bath, 20 × SSC (SSC is 0·15 M-NaCl, 0·015 M-sodium citrate, pH 7) was added to a final concentration of 0·5 × SSC and the mixture was annealed for 15 min at 55°C to convert the minus strands into a double-stranded form. Double-stranded RNA was separated from both plus strand RNA and RNA fragments by the general procedure of Erikson & Gordon (1966). The preparation was chromatographed on a 1 cm × 30 cm column of Sepharose 2B (Pharmacia, Uppsala) equilibrated with Tris/NaCl/EDTA buffer (0·05 M-Tris·HCl (pH 7), 0·1 M-NaCl, 0·005 M-EDTA), at a flow rate of 1·5 ml/h; 1-ml fractions were collected and their ^{32}P radioactivity was determined by measuring the Cerenkov radiation. The excluded material (double-stranded RNA) was pooled and the RNA was precipitated with 2 volumes of ethanol. The precipitate was collected by centrifugation, washed with 66% aqueous ethanol and dissolved in 0·2 ml of 0·5 mM-EDTA. The double-stranded RNA was denatured by heating for 90 s in a boiling water bath. The full-length minus strands were separated from plus strands by centrifuging the sample through a 5% to 23% sucrose density gradient in 50 mM-Tris·HCl (pH 7·5), 5 mM-EDTA for 120 min, at 60,000 revs/min and 15°C, in the Spinco SW65 rotor. A sample of authentic Qβ RNA was centrifuged as marker in a parallel tube. Portions (1 μl) of each fraction were spotted on glass filters and the ³H and ³²P-radioactivities were determined. The peak fractions sedimenting at 30 S were pooled, adjusted to 0·05 M-Tris·HCl (pH 7), 0·1 M-NaCl, 0·005 M-EDTA and 2 volumes of ethanol were added. The precipitate was collected by centrifugation, dissolved in 20 μl of 0·5 mM-EDTA and stored at −70°C. Table 2 shows the yield of radioactivity throughout the purification.

TABLE 2

Yield of minus strands substituted at position 15

Stage of purification	³²P Yield (cts/min × 10⁻⁵) (pmol RNA)†		³H (cts/min × 10⁻⁵) (pmol RNA)‡	
1 Crude product	22·5	(30)	8·1	(22·9)
2 Sephadex chromatography	16	(21·3)	6·7	(19·0)
3 Agarose chromatography	12·5	(16·7)	6·0	(17·0)
4 Sucrose gradient centrifugation	3·0	(4·0)	1·5	(4·2)

Minus strands substituted with N^4-hydroxyCMP in position 15 were synthesized and purified as described in Materials and Methods.

† ³²P-radioactivity is due to the incorporation of 3 molecules per chain of [α-³²P]AMP (2·5 × 10⁷ cts/min per nmol) in positions 16 to 18 during the second step of synthesis. The amount of RNA labeled (pmol of RNA chains) is therefore obtained by dividing total radioactivity by 7·5 × 10⁴.

‡ ³H-radioactivity is due to the incorporation of 1070 molecules per chain of [³H]GMP (3·3 × 10⁴ cts/min per nmol) in the last stage of synthesis (from position 19 to the end of the chain). The amount of RNA labeled (pmol of RNA chains) is obtained by dividing total radioactivity by 3·53 × 10⁴.

(b) *Preparation of wild-type Qβ minus strands*

A mixture of plus and minus strands was synthesized *in vitro* by incubating Qβ replicase (300 units), Qβ RNA (150 μg) and host factor I (300 units) in 2·5 ml of 0·4 mM-[³H]ATP (15,000 cts/min per nmol), 0·8 mM each of GTP, CTP and UTP, 0·2 M-KCl, 80 mM-Tris·HCl (pH 7·5), 12 mM-MgCl₂, 1 mM-EDTA for 70 min at 37°C. KCl was added to diminish the synthesis of 6 S RNA, which may otherwise arise after long-term incubations (R. Joho & C. Weissmann, unpublished observations). The incorporation into acid-insoluble material was 180 nmol of [³H]AMP (250 μg of [³H]RNA, of which about 100 μg were minus strands). The preparation was extracted 3 times with 1 volume of distilled phenol. The RNA was precipitated with ethanol and dissolved in 0·25 ml of 1 mM-EDTA. The solution was chromatographed through a 0·5 cm × 7 cm column of Sephadex G50 equilibrated with SSC, the excluded fractions were pooled and passed through a 0·1-ml column of Chelex-100 (Na⁺) equilibrated with SSC. The RNA was precipitated with ethanol, washed with 66% aqueous ethanol and dissolved in 200 μl of 1 mM-EDTA (recovery, 160 nmol of [³H]AMP). The minus strands were purified as described above; 13 nmol of [³H]AMP (15·6 μg of RNA), corresponding to an over-all yield of minus strands of about 13%, were recovered.

(c) *Synthesis of ³²P-labeled plus strands using minus strands as template*

(i) *Single-round synthesis*

Qβ replicase (3 units), 1 μg of minus strand substituted with N^4-hydroxyCMP in position 15, 3 units of factor I (spec. act. 1·3 × 10⁴ units/mg), 0·8 mM-ATP, 0·8 mM-GTP, 12 mM-MgCl₂, 80 mM-Tris·HCl (pH 7·5), 1 mM-EDTA were incubated in a volume of 20 μl for 5 min at 37°C. Polyethylene sulfonate (final concn 50 μg/ml), UTP (final concn 0·2 mM) and

[α-^{32}P]CTP (final concn 0·15 mM; spec. act. 3·5 × 10^7 cts/min per nmol) were added (final vol. 44 μl) and incubation was continued for 10 min. Samples of 1 μl were taken for the determination of acid-insoluble radioactivity. After the addition of 80 μl of Tris/NaCl/EDTA and sodium dodecyl sulfate to 0·1% the mixture was extracted with 2 volumes of phenol and chromatographed through a 0·5 cm × 6 cm column containing 5·5 cm of Sephadex G50 on a 0·5 cm layer of Chelex-100 (Na$^+$), equilibrated with 20 mM-Tris·HCl (pH 7·5), 2 mM-EDTA. The excluded fractions were pooled, 0·1 volume of 2 M-potassium acetate (pH 5·4) and 2 volumes of ethanol were added. The precipitate was collected as before and dissolved in 10 μl of 20 mM-Tris·HCl (pH 7·5), 2 mM-EDTA. The yield was 2·5 × 10^6 cts/min (about 0·1 μg of [^{32}P]RNA).

(ii) *Multiple-round synthesis*

In a typical experiment, Qβ replicase (5 units), 0·75 μg of N^4-hydroxyCMP-substituted minus strand, 0·8 mM of each standard ribonucleoside triphosphate (spec. act. of [α-^{32}P] CTP 2·1 × 10^5 cts/min per nmol), 0·2 M-KCl, 5 units of factor I, 12 mM-MgCl$_2$, 80 mM-Tris·HCl (pH 7·5), and 1 mM-EDTA (final vol. 50 μl) were incubated for 120 min at 37°C. The total incorporation was 17 nmol of CMP, corresponding to about 24 μg of RNA. The preparation was extracted once with 1 volume of phenol and chromatographed through a combined Sephadex G50 and Chelex-100 (Na$^+$) column as above. The RNA was precipitated with ethanol, dissolved in 200 μl of 0·5 mM-EDTA and centrifuged through a linear sucrose density gradient as above. The fractions containing the high molecular weight RNA were pooled and adjusted to 0·05 M-Tris·HCl, 0·1 M-NaCl, 0·005 M-EDTA. The RNA was precipitated with 2 volumes of ethanol, collected by centrifugation and dissolved in 25 μl of water. The yield was 14 μg of RNA.

In order to prepare highly radioactive plus strands for analysis, 3 μg of product were dissolved in 0·5 mM-EDTA, denatured for 90 s in a boiling water bath and used as template under the conditions of synthesis described above, except that the labeled nucleoside triphosphate(s) were of high specific activity (1 × 10^7 to 5 × 10^7 cts/min per nmol). The RNA was purified as above. To remove radioactive minus strands, the purified product was denatured in 0·3 ml of 0·5 mM-EDTA in the presence of an excess (100 μg) of unlabeled Qβ RNA. Following the addition of 20 × SSC to a final concentration of 0·5 × SSC, the mixture was annealed for 15 min at 55°C. Single-stranded RNA (consisting of labeled, newly synthesized plus strands and unlabeled Qβ RNA) was separated from double-stranded material by chromatography on a 0·7 cm × 10 cm column of cellulose CF-11 (Franklin, 1966) and precipitated with ethanol. In a typical experiment the final yield was 1·2 μg of labeled plus strands.

(d) *Analysis of ^{32}P-labeled RNA*

The minus strand content was determined by hybridizing one portion with a 250-fold excess of denatured Qβ replicative form RNA and another with a similar excess of Qβ RNA, and determining the RNAase-resistant radioactivity of both samples (cf. Weissmann *et al.*, 1968).

Degradation by T$_1$ and pancreatic RNAase A and 2-dimensional electrophoretic fractionation of oligonucleotides were carried out as described by Sanger & Brownlee (1967). Determination of ^{32}P-labeled nucleotide composition was carried out by adding 10 μg of purified yeast RNA as carrier and hydrolyzing the samples with 50 units of partially purified T$_2$ RNAase per ml (Hiramaru *et al.*, 1966) and 0·4 mg pancreatic RNAase A/ml in 7·5 mM-sodium acetate (pH 4·5), 0·5 mM-EDTA (final vol. 20 μl) and separating the nucleotides by electrophoresis on Whatman 52 paper in pyridine/acetate buffer at pH 3·5. Alkaline digestion was avoided, since N^4-hydroxyCMP is thereby partly degraded to UMP (unpublished observation).

Two-dimensional polyacrylamide gel electrophoretic separation of the T$_1$ oligonucleotides of Qβ RNA was carried out as follows. The preparation of purified ^{32}P-labeled plus strands (containing 2 × 10^6 to 1 × 10^7 cts/min and 100 μg of unlabeled Qβ RNA) was digested with 50 units of T$_1$ RNAase in 10 μl of 20 mM-Tris·HCl (pH 7·5), 2 mM-EDTA for 30 min at 37°C. Two-dimensional electrophoresis on polyacrylamide slab gels was carried out as described by De Wachter & Fiers (1972); the first dimension was run in

10% polyacrylamide (pH 3·5) containing 6 M-urea and the second dimension in a 20% gel (pH 8). Oligonucleotides were located by autoradiography and cut out with a cork-borer.

The gel disks were finely ground in a mortar and extracted twice with 0·5 ml of 0·5 M-NaCl containing 20 μg yeast RNA/ml and once with 0·5 ml of 0·5 M-NaCl. Precipitation was achieved by the addition of an equal volume of isopropanol. After 12 h at −20°C, the precipitate was collected and washed once with 50% aqueous isopropanol. In some cases the oligonucleotides were further purified to remove material extracted from the polyacrylamide gel, as follows. The gel extract was adjusted to 20 mM-Tris·HCl, 0·2 M-NaCl (pH 7·5) and filtered through a 0·7 cm × 1 cm column of Cellex D. After washing the column successively with 10 ml of 5 mM-Tris·HCl, 50 mM-NaCl (pH 7·5) and 0·5 ml of a 3% (v/v) solution of triethylamine adjusted to pH 9·5 with CO_2, the material was eluted with 0·6 ml of a 30% (v/v) solution of triethylamine adjusted to pH 9·5 with CO_2. The eluate was adjusted to pH 5·0 with glacial acetic acid. NaCl was added to a final concentration of 0·5 M and precipitation was achieved as described above.

3. Results

(a) *Incorporation of N^4-hydroxyCMP into position 15 of the Qβ minus strand*

We have shown previously (Goodman *et al.*, 1970; Bandle & Weissmann, 1972) that incubation of Qβ replicase with Qβ RNA as template and with GTP and ATP as the only substrates leads to the synthesis of minus strand segment comprising only the first 14 nucleotides of the chain, up to the position where CTP would be required for further elongation. The replication complex, consisting of enzyme, template and nascent RNA chain, was separated from the substrates by chromatography on Sephadex and recovered in an enzymatically active form in good yield by procedures described elsewhere (Bandle, 1973). As shown in Figure 2, the 14th nucleotide of the minus strand is followed by the sequence C-A-A-A-G . . . . Incubation of the purified replication complex with CTP and [α-^{32}P]ATP was expected to lead to the formation of a radioactive product, which, after hydrolysis to the mononucleotides, should yield one mole of ^{32}P-labeled 3'-CMP and two moles of ^{32}P-labeled 3'-AMP. Figure 3 shows the time-course of incorporation of [^{32}P]AMP at concentrations of 5 and 10 μM-CTP; the analysis of representative products (Table 3) gave a ratio of AMP to CMP of 2:1·2 in all cases. When N^4-hydroxyCTP was substituted for CTP

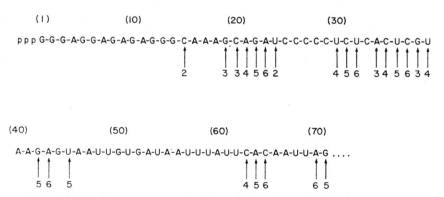

FIG. 2. Nucleotide sequence at the 5'-terminal region of the Qβ minus strand.
The arrows show the positions readily susceptible to mutagenesis. The number of stepwise elongations required to carry out incorporation of a nucleotide analog at a particular position are indicated. (Based on Goodman *et al.*, 1970 and Bandle, 1973).

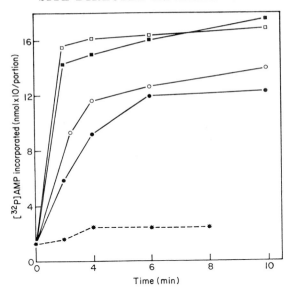

FIG. 3. Time-course of incorporation of [α-³²P]AMP into positions 16 to 18 of the minus strand using either N^4-hydroxyCTP or CTP as substrate for position 15.

Qβ replicase (200 units, spec. act. 625 units/mg), Qβ RNA (175 μg), host factor I (200 units, spec. act. 8.6×10^3 units/mg) were incubated in 1·25 ml of 0·5 mM-repurified GTP, 0·2 mM-repurified ATP, 80 mM-Tris·HCl (pH 7·5), 12 mM-MgCl₂ and 1 mM-EDTA for 5 min at 37°C. The mixture was chromatographed through a 1 cm × 30 cm column of Sephadex G100 in 20 mM-Tris·HCl (pH 7·5). The excluded fractions were pooled. Portions (100 μl) were adjusted to 12 mM-MgCl₂, 1 mM-EDTA, 80 mM-Tris·HCl (pH 7·5), 5·3 μM-[α-³²P]ATP, and either CTP or N^4-hydroxyCTP were added as required, giving a final volume of 150 to 160 μl. At the times indicated, samples (30 μl) were removed, the product was separated from unreacted nucleoside triphosphates by chromatography through a 1·4 ml column of Sephadex G50 and the acid-insoluble radioactivity of the excluded fractions was determined on portions corresponding to 0·28 units of replicase. —■—■—, 4·9 μM-CTP; —□—□—, 9·6 μM-CTP; —●—●—, 5·1 μM-N^4-hydroxyCTP; —○—○—, 9·9 μM-N^4-hydroxyCTP; --*--*--, neither CTP nor N^4-hydroxyCTP.

under otherwise identical reaction conditions, the rate of incorporation was reduced to about one half, while the extent of incorporation was 80% of the control. Representative labeled products contained 0·75 to 1 mole of N^4-hydroxyCMP and 0·24 to 0·27 mol of CMP per two moles of AMP (Table 3). The CMP is believed to be derived from traces of CTP contaminating the N^4-hydroxyCTP, since incubation of the replication complex with [α-³²P]ATP alone gave no incorporation (Fig. 4)†.

(b) Synthesis and purification of Qβ minus strands with N^4-hydroxyCMP in position 15

In order to substitute the CMP in position 15 by its N^4-hydroxy analog, a replicating complex with the product strand elongated up to position 14 (inclusive) was prepared on a large scale. Polyethylene sulfonate was added to prevent further initiation

† It should be noted that even though no CTP was detected in the N^4-hydroxyCTP preparation on electrophoretic analysis, enough CTP may have been present to allow substantial incorporation if the K_m for CTP is an order of magnitude or more lower than that for N^4-hydroxyCTP.

Table 3

Base composition of products labeled with [α-³²P]AMP in positions 16 to 18,
using CTP or N⁴-hydroxyCTP as substrate for position 15

N^4-hydroxyCTP	CTP	[α-³²P]ATP	Time of incubation (min)	N^4-hydroxyCMP	CMP	AMP
—	4·9	5·3	2	—	1·1	2
			16	—	1·2	2
—	9·6	5·3	2	—	1·2	2
			16	—	1·2	2
5·1	—	5·3	2	0·75	0·27	2
			8	0·77	0·24	2
			16	0·79	0·26	2
11·1	—	6·4	8	1·06	0·25	2
			16	1·02	0·25	2

Column group headers: Concentration of substrates for 2nd step of synthesis (μM) — N^4-hydroxyCTP, CTP, [α-³²P]ATP; Base composition (mol)† — N^4-hydroxyCMP, CMP, AMP.

The RNA samples (2400 cts/min or more) examined are those described in the legend to Fig. 3 (except for the experiment with 11·1 μM-N^4-hydroxyCTP, which was done separately). Nucleotide analysis was carried out as described in Materials and Methods.

† Molar values are calculated assigning the value 2 to AMP. The values for GMP and UMP were < 0·1 in all cases.

(Kondo & Weissmann, 1972) and the preparation was incubated with N^4-hydroxy-CTP and [α-³²P]ATP. The analog incorporated into position 15 would thus be labeled by its right-hand nearest neighbor. In order to complete the minus strands with the standard substrates, incubation was continued with a 200-fold excess of CTP over N^4-hydroxyCTP and an 800-fold excess of unlabeled over labeled ATP as well as with [³H]GTP and UTP. Under these conditions no significant incorporation of N^4-hydroxyCMP should occur, since even at a 1:1 ratio of CTP to analog, less than 2% (the detection limit of the assay) of the latter was incorporated (unpublished data). The minus strands were purified in order to remove the wild-type plus strands used as template. Figure 4 shows that the purified, substituted minus strands cosedimented with authentic Qβ RNA. Hybridization analysis showed that 99·4% of the radioactivity was in minus strand RNA.

(c) *Radiochemical analysis of the substituted minus strand*

In order to characterize the nature and extent of the nucleotide substitution, purified substituted minus strands were digested with RNAase T_1 and the products separated by two-dimensional electrophoresis. The autoradiogram of a typical preparation (Plate I) shows the presence of one major and one minor labeled T_1 fragment. Each of these fragments was eluted, digested with pancreatic RNAase and the products were separated by unidimensional electrophoresis. The major spot was

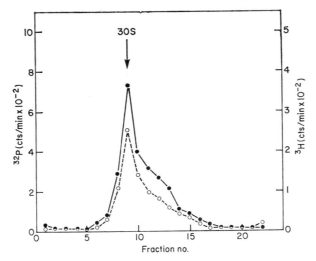

F<small>IG</small>. 4. Sedimentation profile of minus strands substituted with N^4-hydroxyCMP in position 15. Minus strands substituted with N^4-hydroxyCMP in position 15 from the 5′-end were labeled with [^{32}P]AMP in positions 16 to 18 and with [^{3}H]GMP from position 19 to the 3′-end, and purified as described in Materials and Methods. The Figure shows the sedimentation profile of the ^{32}P (—●—●—) and ^{3}H (--○--○--) labeled material in the final sucrose gradient centrifugation of the purification procedure. The arrow indicates the position of authentic 30 S ^{3}H-labeled Qβ RNA run in parallel.

T<small>ABLE</small> 4

Analysis of radioactive T_1 oligonucleotides from minus strands synthesized with N^4-hydroxyCTP and [α-^{32}P]ATP in the second step (positions 15 to 18)

Oligonuc-leotide	Yield (cts/min)	(%)	Products of pancreatic RNAase (mol ^{32}P)		^{32}P-labeled nucleotide composition of pancreatic RNAase products	Conclusion
1	3400	(70)	$\overline{\text{HOC}}$ A-A-A-G	(1) (2·3)	n.d. A	$\overline{\text{HOC}}$p*Ap*Ap*ApGp
2	950	(20)	C A-A-A-G	(1) (1·9)	n.d. A	Cp*Ap*Ap*ApGp
3†	150	(3)	$\overline{\text{HOC}}$ A-G		n.d. n.d.	$\overline{\text{HOC}}$p*ApGp*†
4†	370	(8)	C A-G		n.d. n.d.	Cp*ApG*†

Minus strands were prepared using N^4-hydroxyCTP and [α-^{32}P]ATP in the second step of synthesis as described in Materials and Methods. After digestion with T_1 RNAase the oligonucleotides were separated by 2-dimensional electrophoresis on cellulose-acetate and DEAE-paper and analyzed as described in the legend to Plate I and in Materials and Methods.

† These products were only occasionally present and are missing in the experiment of Plate I. They result when traces of CTP are present during the first step of synthesis, allowing some chains to be elongated up to position 19, where N^4-hydroxyCMP is inserted during the second step of synthesis (cf. Fig. 2). n.d., not determined,

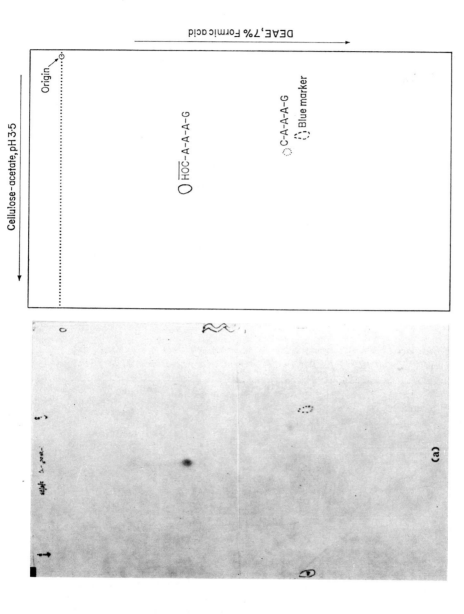

(b)

Cellulose-acetate, pH 3·5

DEAE, 7% Formic acid

Origin

$\overline{HOC}$–A–A–A–G

C–A–A–A–G

Blue marker

(a)

PLATE I. Fingerprint of purified minus strands substituted in position 15 with N^4-hydroxyCMP and labeled with [^{32}P]AMP in positions 16 to 18. A portion (3.3×10^4 cts/min) of the ^{32}P-labeled minus strands used for the analysis described in the legend to Fig. 4 was mixed with 10 μg of yeast RNA and treated with 25 units of RNAase T_1 for 30 min at 37°C. The digest was analyzed by 2-dimensional electrophoresis following the procedure of Sanger & Brownlee (1967). (a) Autoradiogram, (b) tracing with identification of spots (cf. Table 4).

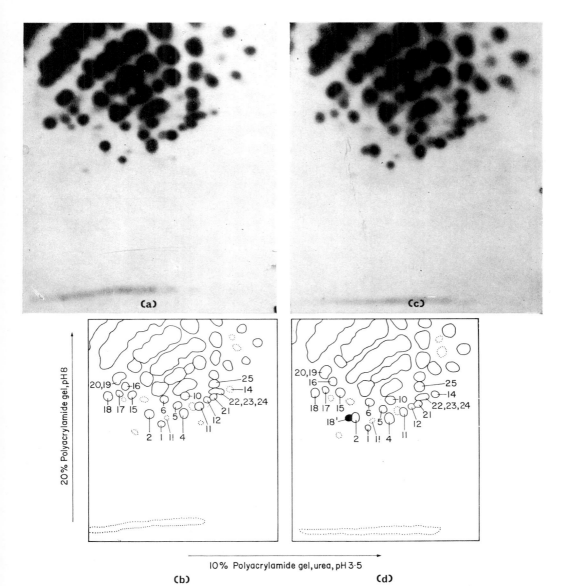

PLATE II. Two-dimensional polyacrylamide gel electrophoretic analysis of the T_1 oligonucleotides of plus strands synthesized from wild-type minus strands and from minus strands substituted in position 15 with N^4-hydroxyCMP.

Plus strands were prepared by multiple rounds of replication *in vitro* using wild type or substituted minus strands as template and $[\alpha\text{-}^{32}\text{P}]\text{CTP}$ ($8\cdot5 \times 10^7$ cts/min per nmol) as radioactive substrate and were purified by the general procedures described in Materials and Methods. Each preparation (20×10^6 cts/min wild type; 15×10^6 cts/min modified plus strands) contained 100 μg of $Q\beta$ RNA and was digested with 50 units of RNAase T_1 in 20 mM-Tris·HCl (pH 7·5), 2 mM-EDTA (final vol. 11 μl) for 30 min at 37°C. Then 8 μl of a solution containing 500 g of sucrose, 300 g of urea, 2 g of xylene cyanol FF and 2 g of bromophenol blue/l were added and 2-dimensional polyacrylamide gel electrophoresis was carried out as described by De Wachter & Fiers (1972), using a 10% gel in the first, and a 20% gel in the second dimension. (a) Autoradiogram of T_1 digest of wild-type plus strands; (b) tracing of (a); (c) autoradiogram of T_1 digest of plus strands synthesized using substituted minus strands as template; (d), tracing of (c).

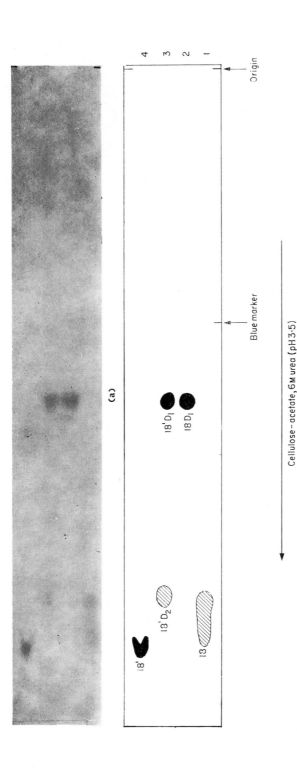

(a)

(b)

Cellulose-acetate, 6M urea (pH 3·5)

Origin

Blue marker

4

3

2

1

18'D₂

18'

18'D₁

18D₁

18

PLATE III. Electrophoretic analysis of the products obtained by digestion of oligonucleotides 18 and 18' with RNAase U₂.

A mixture of wild-type and modified plus strands was synthesized *in vitro* using all four α-³²P-labeled ribonucleoside triphosphates (spec. act. 5×10^6 cts/min per nmol) following the general procedures outlined in Materials and Methods. The purified plus strands (15×10^6 cts/min) were digested with RNAase T₁ and separated by 2-dimensional polyacrylamide gel electrophoresis as in the experiment described in the legend to Plate II. After autoradiographic localization, the gel areas corresponding to oligonucleotides 18 and 18' were cut out, extracted in the presence of 20 μg of yeast RNA, and purified by chromatography on Cellex D as described in Materials and Methods. Portions of oligonucleotides were mixed with 20 μg of yeast RNA. Incubations were carried out in 5 μl of 0·05 M-sodium-acetate, 0·002 M-EDTA (pH 4·5) containing 0·1 mg/ml of bovine serum albumin, for 3 h at 37°C. (1) 1000 cts/min oligonucleotide 18 unheated ; (2) 1000 cts/min oligonucleotide 18 heated with 0·06 units RNAase U₂; (3) 2000 cts/min oligonucleotide 18' heated with 0·06 units RNAase U₂; (4) 1000 cts/min oligonucleotide 18' unheated. (a) Autoradiogram; (b) tracing of (a).

identified as $\overline{\text{HOC}}$†-A-A-A-G and the minor one as C-A-A-A-G (cf. Table 4). Thus, 78% of the minus strands had the desired substitution at position 15. The T_1 fragment C-A-A-A-G arises from chains in which CMP was incorporated at position 15, due to CTP contamination of one of the components, probably the N^4-hydroxyCTP, used during stepwise synthesis (cf. section (a), above).

(d) *Synthesis of plus strands using substituted minus strands as template*

It has been shown earlier that non-infectious minus strands serve as template for the synthesis of infectious plus strands *in vitro* (Feix *et al.*, 1968). Purified minus strands substituted in position 15 with N^4-hydroxyCMP were incubated with $Q\beta$ replicase and the four standard ribonucleoside triphosphates, plus [α-^{32}P]CTP as labeled substrate. A similar synthesis was carried out using wild-type minus strands as template. With 0·5 μg of minus strands as template, 2·4 and 3·4 μg of plus strands were purified from the substituted and the control preparations, respectively. Figure 5 shows the sedimentation profile of the product synthesized on substituted minus strands, with authentic $Q\beta$ RNA as an internal marker.

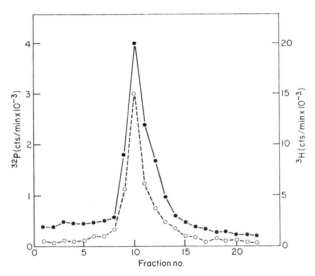

Fig. 5. Sedimentation profile of the product synthesized using minus strands substituted with N^4-hydroxyCMP in position 15 as template.

The ^{32}P-labeled RNA was synthesized by multiple rounds of replication and purified by sucrose gradient centrifugation as described in Materials and Methods. A portion (15 × 10^3 cts/min; 2·5 × 10^{-5} μg) was denatured for 90 s in a boiling water bath, mixed with ^{3}H-labeled $Q\beta$ RNA (3·6 × 10^4 cts/min; 1·2 μg) and centrifuged as described in the legend to Fig. 4. --○--○--, ^{3}H; —●—●—, ^{32}P.

(e) *Radiochemical analysis of plus strands synthesized from substituted minus strands*

In order to determine whether and if so to what degree a mutation had been generated in plus strands synthesized on minus strands containing N^4-hydroxyCMP

† Abbreviation used: $\overline{\text{HOC}}$, N^4-hydroxycytidine.

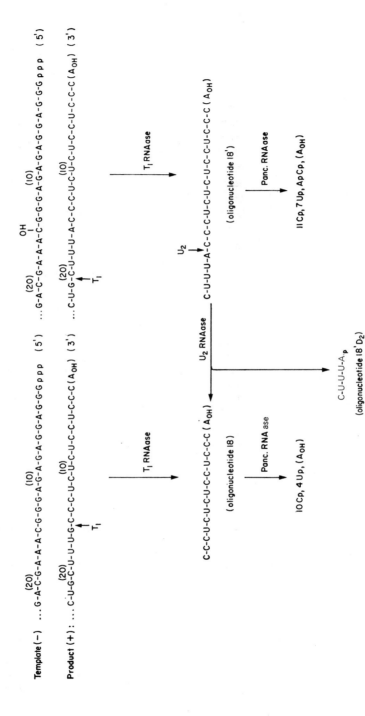

FIG. 6. Scheme for the analysis of the 3'-terminal T₁ fragment of wild-type Qβ RNA and Qβ RNA with a G → A transition in the 16th position from the 3'-terminus. Panc, pancreatic.

in position 15, it was necessary to isolate the complementary 3′-terminal portion of the plus strands and subject it to nucleotide sequence analysis. As shown in the scheme of Figure 6, digestion of a wild-type plus strand with RNAase T_1 leads to cleavage to the right of the GMP residue located at the 16th position from the 3′-end and releases a 3′-terminal fragment comprising 15 nucleotides. In the case of the mutant plus strand in which the G residue in position 16 has been substituted by an A residue, RNAase T_1 digestion should yield a 3′-terminal fragment extending up to the G residue at position 21. This oligonucleotide would be longer by five nucleotides than that derived from the wild-type strand and the two should be easily resolved on polyacrylamide gel electrophoresis. The two-dimensional polyacrylamide slab gel system described by De Wachter & Fiers (1972) is well-suited to separate large T_1 oligonucleotides, and most of the Qβ fragments resolved in this system (Plate II(a)) have been sequenced (M. A. Billeter, unpublished results). Fragment 18 was identi-fied as the 3′-terminal oligonucleotide of Qβ RNA (cf. Fig. 6) while fragment 2 (cf. legend to Table 5) is derived from the middle of the molecule (M. A. Billeter, un-published observations). Plate II allows a comparison of the fingerprints obtained from *in vitro* wild-type Qβ RNA and the plus strands made using the substituted minus strands as template in a multiple-round synthesis. In the fingerprint of the modified plus strand preparations (Plate II(b)) the molar quantity of oligonucleotide

TABLE 5

Yield of wild-type and modified 3′-terminal T_1 oligonucleotides from plus strands synthesized from wild-type minus strands or from minus strands substituted in position 15 with N^4-hydroxyCMP

Oligonucleotide[†]	Radioactivity in T_1 oligonucleotides of plus strand					
	Template: substituted minus strand (one-round synthesis)		substituted minus strand (multiple-round synthesis)		wild-type minus strand (multiple-round synthesis)	
	(cts/min)	(mol)[‡]	(cts/min)	(mol)[‡]	(cts/min)	(mol)[‡]
18	3550	0·28 (67%)	700	0·28 (45%)	2560	0·43 (100%)
18′	1910	0·14 (33%)	930	0·34 (55%)	—	—
2	7680	1·0	1490	1·0	3580	1·0

Wild-type minus strands and minus strands substituted in position 15 were used as templates for the synthesis of plus strands with [α-^{32}P]CTP as radioactive substrate. The purified plus strands were digested with RNAase T_1 and the resulting oligonucleotides were separated by 2-dimensional polyacrylamide gel electrophoresis. After localization by autoradiography, gel disks corresponding to the oligonucleotides were cut out and their Cerenkov radiation was determined. See Materials and Methods section for experimental details.

† The structures and origins of the oligonucleotides are as follows:

18	C-C-C-U-C-U-C-U-C-C-U-C-C-C(A)	3′-terminus of wild-type Qβ RNA (Goodman *et al.*, 1970)
18′	C-U-U-U-A-C-C-C-U-C-U-C-U-C-C-U-C-C-C(A)	3′-terminus of modified Qβ RNA (cf. Table 6)
2	C-C-A-C-U-C-A-A-A-U-C-U-U-U-C-A-U-A-A-A-A-G	from region at the end of A_2 cistron (Billeter, Goldfarb and Weissmann, unpublished data)

‡ Relative molar yields of an oligonucleotide were calculated by dividing its radioactivity by the number of ^{32}P atoms (introduced by [α-^{32}P]CTP) it contains and normalizing to the value obtained for oligonucleotide 2.

TABLE 6

Analysis of the 3'-terminal oligonucleotides 18 and 18'

Oligo-nucleotide	α-³²P-labeled ribonucleoside triphosphate	Total hydrolysis	Pancreatic RNAase products	U₂ RNAase products	Interpretation†
18	All four	10·0 Cp*, 3·8 Up*	n.d.	18D₁ (10·0 Cp*, 4·8 Up*)	Cp*Cp*Up*Cp*Up*Cp*Up*Cp*Up*Cp*Cp*(A)
	CTP	n.d.	5·0 Cp*, 3·8 Up*	n.d.	Cp*Cp* Up*Cp Up*Cp*Cp Up*Cp*Cp*Cp (A)
	UTP	Cp*	n.d.	18D₁ (Cp*)	Cp Cp Cp*Up Cp*Up Cp Cp*Up Cp Cp Cp (A)
18'	All four	11·0 Cp*, 1·2 Ap*, 7·5 Up*	10·0Cp*, 6·1Up*, 2·0 Ap*Cp*	14·0 18'D₁ (10·0 Cp* 4·3 Up*), 7·2 18'D₂ (1·0 Cp*, 1·2 Ap*, 3·0 Up*)	Cp*Up*Up*Up*Ap*Cp*Cp*Up*Up*Cp*Up*Cp*Cp*Up*Cp*Cp*(A) [18'D₂] [18'D₁]
	CTP	n.d.	4·0 Cp*, 4·0 Up*, 2·4 Ap*Cp*	n.d.	Cp Up Up Up Ap* Cp*Cp Up*Cp Up*Cp* CpUp*Cp*Cp*Cp (A)
	UTP	5·0 Cp*, 2·0 Up*	n.d.	4·0 18'D₁ (Cp*), 2·7 18'D₂ (1·0 Cp*, 2·0 Up*)	Cp*Up*Up*Up Ap Cp Cp*Up Cp*Up Cp Cp*Up Cp Cp Cp (A) [18'D₂] [18'D₁]

RNA was synthesized *in vitro* using as template wild-type Qβ RNA or RNA obtained by multiple rounds of synthesis using modified minus strands as template, as described in Materials and Methods. After purification, the radioactive plus strands were digested with T₁ RNAase and the fragments were separated by 2-dimensional polyacrylamide gel electrophoresis (cf. Plate II). Oligonucleotides 18 and 18' were located by autoradiography, extracted and analyzed as indicated. The products of pancreatic and U₂ RNAase digestion (other than mononucleotides) were enzymatically hydrolyzed with a mixture of T₂ and pancreatic RNAase; the resulting labeled mononucleotides are indicated by asterisks. Relative yields of products are based on moles of [³²P]phosphate; the underlined number was used as reference value. Values less than 0·3 are not given.

† The interpretations of the data on oligonucleotide 18 are based on prior knowledge of its sequence (Weith & Gilham, 1969; Goodman *et al.*, 1970). Treatment of oligonucleotide 18 with U₂ RNAase and electrophoresis on cellulose-acetate in pyridine/acetate buffer (pH 3·5), containing 6 M-urea gave a single spot (designated 18D₁), with a mobility 0·6 times that of the untreated oligonucleotide. This change in mobility is an artifact (cf. text). Digestion of 18' with RNAase U₂ yielded 2 fragments: 18'D₁, which had the same mobility and analytical values as 18D₁, and 18'D₂, which had a mobility 1·6 times that of 18D₁. The sequence of 18'D₂ can be uniquely derived from its base composition and nearest-neighbor data; the sequence of 18' is based on its pancreatic digestion products, on the identity of 18'D₁ with 18 and the structure of 18'D₂.

n.d., not determined.

18 (relative to oligonucleotide 2) is about half of that in the control (Table 5) and, moreover, a new spot has appeared, 18'. The sum of the relative molar quantities of 18 and 18' in the modified RNA is approximately that of 18 in the wild type control. The electrophoretic mobility in the second dimension (which is an inverse function of chain length) is less for oligonucleotide 18' than for 18 (14 nucleotides) and is slightly more than that of 2 (22 nucleotides), suggesting a chain length of 15 to 20 nucleotides. Oligonucleotides 18 and 18', labeled uniformly or with [α-^{32}P]CTP or [α-^{32}P]UTP only, were analyzed by digestion with pancreatic RNAase, RNAase U$_2$ and by total enzymatic hydrolysis (Table 6). The relationship of 18' to 18 was demonstrated by digesting 18' with RNAase U$_2$, which hydrolyses at purine residues. Two fragments, 18' D$_1$ and 18' D$_2$, were separated by electrophoresis on cellulose acetate. 18' D$_1$ is identical with 18 as judged by the following experiments. (a) The base composition, pancreatic RNAase digestion products and nearest-neighbor analysis are the same for 18' D$_1$ and 18 (Table 6); (b) 18' and 18 were digested with RNAase U$_2$ under the same conditions and electrophoresed side-by-side on cellulose-acetate at pH 3·5 in urea. Fragments 18' D$_1$ and 18 had the same mobility, while 18' D$_2$ ran faster (Plate III). Unexpectedly, the mobility of 18 was reduced after its exposure to RNAase U$_2$; since no cleavage by this enzyme is expected of an oligonucleotide devoid of purine residues, and since only a single spot was detected, we suspected that the modified mobility could be due to an association of the oligonucleotide with protein. The belief that 18 had not been cleaved was supported by the finding that 18 and 18' incubated with RNAase U$_2$ showed the same mobility on electrophoresis in 10% polyacrylamide gels in 6 M-urea (pH 3·5), and that after recovery from the gel the two preparations had the same mobility on cellulose-acetate, under the conditions described above (data not shown). Whatever the explanation for the aberrant mobility of 18 on cellulose-acetate after treatment with RNAase U$_2$, it co-electrophoreses with the larger cleavage product of 18'. Furthermore the larger cleavage product had the same electrophoretic mobility as 18 (either treated or untreated with enzyme) when run on a polyacrylamide gel as above. Fragment 18' D$_2$ has the structure C-U-U-U-A, as shown by the data of Table 6. Since digestion with pancreatic RNAase of 18' labeled with [α-^{32}P]CTP yielded Ap*Cp*, in addition to Cp* and Up*, the structure of 18' is unambiguously defined as C-U-U-U-A-C-C-C-U-C-U-C-U-C-C-U-C-C-C(A). This shows that indeed the G residue in position 16 from the 3'-terminus has been replaced by an A residue in the modified plus strands.

(f) *Efficiency of nucleotide transition induced by N⁴-hydroxyCMP*

The plus strand preparation obtained after multiple rounds of synthesis gave a molar ratio of 18' to 18 of about 1·22. Thus, about 55% of the plus strands contained the substitution in position 16. Under these conditions of synthesis the proportion of mutant RNA does not necessarily reflect the efficiency of mutagensis, since positive or negative selection may have occurred during replication. A plus strand preparation was therefore prepared by one round of synthesis using substituted minus strands as template. This was achieved by starting synthesis with GTP and ATP, adding polyethylene sulfonate to block further initiation and then allowing completion of the initiated strands by the addition of CTP and UTP. Analysis of the T$_1$ oligonucleotide composition of the product showed that the content of mutated RNA was 33% (Table 5).

4. Discussion

The approach to directed mutagenesis described in this paper is based on four main features of the replicase system. (1) It is possible to synthesize $Q\beta$ RNA minus strands enzymatically in discrete steps, controlled by the substrates made available during the successive incubations (Bandle & Weissmann, 1972; Bandle, 1973). (2) Minus strands can be purified free of plus strands and utilized as templates for the synthesis of infectious plus strands (Pollet *et al.*, 1967; Feix *et al.*, 1968). (3) The nucleotide analog N^4-hydroxyCTP is efficiently utilized by $Q\beta$ replicase in place of CTP and can be inserted into a single selected position. (4) N^4-hydroxyCMP incorporated into $Q\beta$ minus strands has template properties of UMP and CMP to about equal extent, thereby generating nucleotide transitions with high efficiency. The mutagenic properties of N^4-hydroxyCMP are ascribed to the fact that this compound and its derivatives can occur in two tautomeric forms, one of which (the amino form) is equivalent to CMP in its hydrogen-bonding properties, while the other (the imino form) is equivalent to UMP (see reviews by Phillips & Brown, 1967; Kochetkov & Budowsky, 1969; Singer & Fraenkel-Conrat, 1969). The ability of N^4-hydroxyCTP to serve as substrate for DNA-dependent RNA polymerase has been documented earlier, and the use of defined templates has shown that the analog can be incorporated in place of both UMP and CMP (Banks *et al.*, 1971; Budowsky *et al.*, 1971, 1972). Despite the fact that the tautomeric equilibrium in aqueous solution favors the imino form of N^4-hydroxyCMP over the amino form in a ratio of about 10:1 (Brown *et al.*, 1968; Brown & Hewlins, 1968*a,b*; Sverdlov *et al.*, 1971 (as quoted by Budowsky *et al.*, 1972)) the analog is incorporated by DNA-dependent RNA polymerase more readily in place of CMP than of UMP (Banks *et al.*, 1971). We have made analogous observations in the case of $Q\beta$ replicase (unpublished results). Once incorporated into a polymer, N^4-hydroxyCMP directs the incorporation of AMP residues by DNA-dependent RNA polymerase (Brown & Phillips, 1965; Phillips *et al.*, 1965,1966; Wilson & Caicuts, 1966; Singer & Fraenkel-Conrat, 1970; Banks *et al.*, 1971) in great preference to GMP residues (Banks *et al.*, 1971). This differs quantitatively from the results with $Q\beta$ replicase reported above; it may be that the environment of the analog has a strong influence on whether it behaves like a C or a U residue, as previously pointed out by Banks *et al.* (1971).

The methodology outlined in this paper has been used to obtain a $Q\beta$ RNA preparation in which the G residue in position 16 from the 3'-end has been substituted by an A residue in 33% of the molecules.† The mutant RNA was effective as template for further replication by $Q\beta$ replicase, as shown by the fact that after repeated replication its proportion increased to 55%. Apart from being of great use in obtaining mutant RNA in milligram quantities, these experiments show that the nucleotide transition in the 3'-terminal extracistronic region not only fails to diminish the template efficiency of the RNA but increases it slightly. It is obviously of great interest to determine whether the specific infectivity of $Q\beta$ RNA is affected by the substitution in position 16 from the 3'-end and the relevant experiments are in progress.

As mentioned above, it is also possible to incorporate N^4-hydroxyCMP into a minus strand in place of UMP; again the analog directs incorporation of either

---

† Since only 78% of the minus strands had a C → $\overline{\text{HOC}}$ replacement, the mutagenic efficiency of the analog in position 15 is 42%.

GMP or AMP in the corresponding position of the plus strand, this time leading to an A → G transition (unpublished results). Undoubtedly analogs causing efficient U ↔ C transitions can also be found. Since stepwise synthesis can be carried out up to about position 100 from the 5'-end, the approach to directed mutagenesis described above can be used to modify a large number of positions (Fig. 2). By using minus strands as template it will also be possible to insert analogs into the 5'-terminal region of the plus strand and thereby generate mutations in this region. Finally, resynchronization of minus strand synthesis at the ribosome binding sites (Kolakofsky *et al.*, 1973) should make the intercistronic regions and the beginnings of the coat and replicase cistrons amenable to site-directed transitions and permit the relationship between nucleotide sequence and ribosome binding capacity to be explored. The general approach to directed mutagenesis outlined in this paper might also be used to generate modified eukaryotic DNA segments if RNA-directed DNA polymerases could be brought to synthesize complete complements of a eukaryotic messenger RNA. Repair synthesis using DNA polymerase on specific DNA fragments obtained by the use of restriction enzymes and appropriate selection procedures could provide an alternative approach. Expression of modified DNA *in vivo* or *in vitro* might ultimately allow the synthesis of proteins with predetermined amino acid substitutions.

We thank Dr M. A. Billeter for a generous supply of α-^{32}P-labeled ribonucleoside triphosphates as well as for valuable advice and unpublished information. This project was supported by grants from the Schweizerische Nationalfonds nos 3506 and 3132 and the Jane Coffin Childs Fund no. 243 to one of us (C. W.), by fellowships of EMBO to another author (R. A. F.) and of the American Cancer Society to another (D. L. S.).

REFERENCES

Adams, J. M. & Cory, S. (1970). *Nature (London)*, **227**, 570–574.

Adams, J. M., Spahr, P. F. & Cory, S. (1972). *Biochemistry*, **11**, 976–988.

Bandle, E. (1973). Dissertation, University of Zürich.

Bandle, E. & Weissmann, C. (1972). *Experientia*, **28**, 743–744.

Banks, G. R., Brown, D. M., Streeter, D. G. & Grossman, L. (1971). *J. Mol. Biol.* **60**, 425–439.

Billeter, M. A., Dahlberg, J. E., Goodman, H. M., Hindley, J. & Weissmann, C. (1969). *Nature (London)*, **224**, 1083–1086.

Brown, D. M. & Hewlins, M. J. E. (1968a). *J. Chem. Soc.* (C), 1922–1924.

Brown, D. M. & Hewlins, M. J. E. (1968b). *J. Chem. Soc.* (C), 2050–2055.

Brown, D. M. & Phillips, J. H. (1965). *J. Mol. Biol.* **11**, 663–671.

Brown, D. M., Hewlins, M. J. E. & Schell, P. (1968). *J. Chem. Soc.* (C), 1925–1929.

Budowsky, E. I., Sverdlov, E. D., Shibaeva, R. P., Monastyrskaya, G. S. & Kochetkov, N. K. (1971a). *Biochim. Biophys. Acta*, **246**, 300–319.

Budowsky, E. I., Sverdlov, E. D. & Spasokukotskaya, T. N. (1971b). *FEBS Letters*, **17**, 336–338.

Budowsky, E. I., Sverdlov, E. D. & Spasokukotskaya, T. N. (1972). *Biochim. Biophys. Acta*, **287**, 195–210.

De Wachter, R. & Fiers, W. (1972). *Anal. Biochem.* **49**, 184–197.

De Wachter, R., Merregaert, J., Vandenberghe, A., Contreras, R. & Fiers, W. (1971). *Eur. J. Biochem.* **22**, 400–414.

Eoyang, L. & August, J. T. (1968). In *Methods in Enzymology* (Grossman, L. & Moldave, K., eds), vol. 12B, pp. 530–540, Academic Press, New York.

Erikson, R. L. & Gordon, J. A. (1966). *Biochem. Biophys. Res. Commun.* **23**, 422–428.

Feix, G., Pollet, R. & Weissmann, C. (1968). *Proc. Nat. Acad. Sci., U.S.A.* **59**, 145–152.

Franklin, R. M. (1966). *Proc. Nat. Acad. Sci., U.S.A.* **55**, 1504–1511.

Franze de Fernandez, M. T., Hayward, W. S. & August, J. T. (1972). *J. Biol. Chem.* **247**, 824–831.

Freese, E., Bautz-Freese, E. & Bautz, E. (1961). *J. Mol. Biol.* **3**, 133–143.

Goodman, H. M., Billeter, M. A., Hindley, J. & Weissmann, C. (1970). *Proc. Nat. Acad. Sci., U.S.A.* **67**, 921–928.

Hindley, J., Staples, D. H., Billeter, M. A. & Weissmann, C. (1970). *Proc. Nat. Acad. Sci., U.S.A.* **67**, 1180–1187.

Hiramaru, M., Uchida, T. & Egami, F. (1966). *Anal. Biochem.* **17**, 135–142.

Janion, C. & Shugar, D. (1968). *Acta Biochim. Polon.* **15**, 107–121.

Kamen, R. (1969). *Nature (London)*, **221**, 321–325.

Kamen, R., Kondo, M., Römer, W. & Weissmann, C. (1972). *Eur. J. Biochem.* **31**, 44–51.

Kochetkov, N. K. & Budowsky, E. I. (1969). In *Progress in Nucleic Acid Research and Molecular Biology* (Cohn, W. E. & Davidson, J. N., eds), vol. 9, pp. 403–438, Academic Press, New York.

Kolakofsky, D., Billeter, M. A., Weber, H. & Weissmann, C. (1973). *J. Mol. Biol.* **76**, 271–284.

Kondo, M. & Weissmann, C. (1972). *Biochim. Biophys. Acta*, **259**, 41–49.

Min Jou, W., Haegeman, G., Ysebaert, M. & Fiers, W. (1972). *Nature (London)*, **237**, 82–88.

Nichols, J. L. & Robertson, H. D. (1971). *Biochim. Biophys. Acta*, **228**, 676–681.

Osborn, M., Person, S., Philips, S. & Funk, F. (1967). *J. Mol. Biol.* **26**, 437–447.

Phillips, J. H. & Brown, D. M. (1967). In *Progress in Nucleic Acid Research and Molecular Biology* (Cohn, W. E. & Davidson, J. N., eds), vol. 7, pp. 349–368, Academic Press, New York.

Phillips, J. H., Brown, D. M., Adman, R. & Grossman, L. (1965). *J. Mol. Biol.* **12**, 816–828.

Phillips, J. H., Brown, D. M. & Grossman, L. (1966). *J. Mol. Biol.* **21**, 405–419.

Pollet, R., Knolle, P. & Weissmann, C. (1967). *Proc. Nat. Acad. Sci., U.S.A.* **58**, 766–773.

Popowska, E. & Janion, C. (1974). *Biochem. Biophys. Res. Commun.* **56**, 459–466.

Poslovina, A. S., Vasyunina, E. A., Andreeva, I. S. & Salganik, R. I. (1973). *Genetika*, **9**, 78–81.

Rensing, U. & August, J. T. (1969). *Nature (London)*, **224**, 853–856.

Robertson, H. D. & Jeppesen, P. G. N. (1972). *J. Mol. Biol.* **68**, 417–428.

Sanger, F. & Brownlee, G. G. (1967). In *Methods in Enzymology* (Grossman, L. & Moldave, K., eds), vol. 12A, pp. 361–381, Academic Press, New York.

Schaffner, W. & Weissmann, C. (1973). *Anal. Biochem.* **56**, 502–514.

Singer, B. & Fraenkel-Conrat, H. (1969). In *Progress in Nucleic Acid Research and Molecular Biology* (Davidson, J. N. & Cohn, W. E., eds), vol. 9, pp. 1–29, Academic Press, New York.

Singer, B. & Fraenkel-Conrat, H. (1970). *Biochemistry*, **9**, 3694–3701.

Staples, D. H., Hindley, J., Billeter, M. A. & Weissmann, C. (1971). *Nature New Biol.* **34**, 202–204.

Sverdlov, E. D., Krapevko, A. P. & Budowsky, E. I. (1971). *Khim. Geterotsikl. Soedin. (U.S.S.R.)*, **9**, 1264–1267.

Tessman, I. (1968). *Virology*, **35**, 330–333.

Weissmann, C., Colthart, L. & Libonati, M. (1968). *Biochemistry*, **7**, 865–874.

Weissmann, C., Billeter, M. A., Goodman, H. M., Hindley, J. & Weber, H. (1973). *Annu. Rev. Biochem.* **42**, 303–328.

Weith, H. L. & Gilham, P. T. (1969). *Science*, **166**, 1004–1005.

Weith, H. L., Asteriadis, G. T. & Gilham, P. T. (1968). *Science*, **160**, 1459–1460.

Wilson, R. G. & Caicuts, M. J. (1966). *J. Biol. Chem.* **241**, 1725–1731.

J. Mol. Biol. (1975) **94**, 425–440

Molecular Structure Determination by Electron Microscopy of Unstained Crystalline Specimens

P. N. T. Unwin and R. Henderson

Medical Research Council
Laboratory of Molecular Biology
Hills Road, Cambridge, England

(Received 15 November 1974)

The projected structures of two unstained periodic biological specimens, the purple membrane and catalase, have been determined by electron microscopy to resolutions of 7 Å and 9 Å, respectively. Glucose was used to facilitate their *in vacuo* preservation and extremely low electron doses were applied to avoid their destruction.

The information on which the projections are based was extracted from defocussed bright-field micrographs and electron diffraction patterns. Fourier analysis of the micrograph data provided the phases of the Fourier components of the structures; measurement of the electron diffraction patterns provided the amplitudes.

Large regions of the micrographs (3000 to 10,000 unit cells) were required for each analysis because of the inherently low image contrast ($<1\%$) and the statistical noise due to the low electron dose.

Our methods appear to be limited in resolution only by the performance of the microscope at the unusually low magnifications which were necessary. Resolutions close to 3 Å should ultimately be possible.

1. Introduction

The structure determination of unstained biological molecules by electron microscopy would be straightforward but for two principal factors. One is their high sensitivity to electron damage; typical proteins, for example, begin to disrupt at doses of the order of 1 electron/Å² (Stenn & Bahr, 1970), and such doses are far smaller than those used in routine observations. The other is the loss of three-dimensional order that generally takes place when the specimen's natural aqueous environment is replaced by the high vacuum of the microscope.

The high radiation sensitivity means that resolution in a "non-destructive" image of an isolated molecule is limited by electron noise to a figure of at least several tens of Ångstroms (Glaeser, 1973). A statistical barrier of this nature can be surmounted, however, if a highly ordered array of the molecules can be made, by utilizing the redundancy of information which is then present in the image (see e.g. McLachlan, 1958). A low dose image of this array will not display the projected structure of an *individual* molecule or unit cell directly; nevertheless, provided that the object is comprised of a large enough number of unit cells and that the periodicities are precisely maintained, all the information needed to reconstruct the "average" molecule

573

or unit cell will be available. There are a number of optical and computer methods for extracting such information (see e.g. Huxley & Klug, 1971).

The loss of order in the vacuum is, to some extent, prevented by conventional preparation methods in which the specimens are encased in negative stain, or fixed, embedded and sectioned; but these methods also restrict the information obtainable to comparatively gross features, such as the overall morphology of the specimen and the subunit structure. No information about the internal structure of the specimen is obtained. Elimination of such preparation procedures altogether by maintaining the specimen in a wet state with the aid of a hydration chamber (Parsons et al., 1974) or by embedding it in ice (Taylor & Glaeser, 1974) in theory provides a more satisfactory solution; however, the technical problems associated with these methods may make high *image* resolutions difficult to achieve in practice.

The experiments we describe here using thin crystalline platelets of beef liver catalase suggest that a more profitable approach towards *in vacuo* preservation is to replace the aqueous medium by another liquid which has similar chemical and physical properties, but is non-volatile in addition. A number of sugars and related compounds have these properties and X-ray evidence (Unwin, unpublished results) indicates that they can be successfully substituted without greatly disturbing the order inherent in the native hydrated crystals, at least to a resolution of 3 to 4 Å. The second object we have used in these experiments, the purple membrane from *Halobacterium halobium*, is unusual in being almost unaffected by drying (Blaurock & Stoeckenius, 1971), but it also requires the presence of a similar fluid medium to minimize fragmentation or cracking of the membrane when it is allowed to dry down on the microscope grid.

This paper describes the determination of the projected structures of both catalase and the purple membrane to resolutions of 9 and 7 Å, respectively, by making use of the principles and methods of preservation that we have just outlined. Quantitative computer processing methods have been used to extract the data from the electron micrographs (DeRosier & Klug, 1968; Erickson & Klug, 1971) and this enables us to give objective assessments of the accuracy of the final structures.

The resolution limits in the present case are not set by poor specimen preservation or by radiation sensitivity, but mainly, we believe, by electron–optical factors that become important at relatively low magnifications. We are confident that our methods will be applicable to biological structure determination at resolutions close to the limit set ultimately by the microscope.

2. Experimental Technique

(a) *Specimen preparation*

Purple membranes were prepared according to the procedure of Oesterhelt & Stoeckenius (1974) from cultures of *Halobacterium halobium* R_1. Beef liver catalase crystals were prepared from a twice crystallized aqueous suspension (Sigma Chemical Co.) according to the NaCl dialysis procedure of Sumner & Dounce (1955). This procedure gave large numbers of platelets of a suitable thickness (400 to 600 Å as determined by shadowing from a known angle) for imaging and diffraction.

Solutions of either specimen were applied to carbon-coated grids and washed with a 1% solution, usually of glucose (but see also Results section (a)), before being allowed to dry— the technique being essentially the same as one would use for negative staining (e.g. Huxley & Zubay, 1960).

The character of the grids we used was different in the two cases. For the membranes they were made to consist of a holey carbon film, coated with "gold islands", and overlaid with a thin carbon film; for the thicker catalase crystals they were made to consist simply of ~150 Å thick carbon film, coated very sparsely, but evenly, with the particles which are produced by evaporating gold in a vacuum of ~10^{-1} torr. Micrographs of the purple membrane sheets were taken over holes in the thick carbon films. The "islands" and sparsely populated particles served as focussing aids.

(b) Photographic materials

For recording either images under normal dosage conditions, or electron diffraction patterns, we used Ilford Special Lantern Contrasty plates in conjunction with PQ Universal Developer (diluted 1+4). For recording under low dosage conditions we required a thin, fine-grained, emulsion having a low unit density exposure and a high detective quantum efficiency. We chose Kodak Electron Image plates for this purpose and these were developed strictly according to the manufacturers' instructions for maximum speed development (concentrated D19 developer, 12 min at 20°C). Since it was required to record both low and high dose images in pairs (see Theoretical Background section), the two types of plate were alternated in the camera box (and developed separately).

Optical density measurements from the Electron Image plates, interpreted with the aid of calibration charts supplied by Kodak, were the sources of electron dose estimates. Although such estimates may not be exact on an *absolute* scale (Matricardi *et al.*, 1972) they are reproducible and simple to obtain.

(c) Electron microscopy

Electron microscopy was carried out at 100 kV with a Philips EM301 using a narrow coherent illuminating beam, which was produced by having thin 25 μm (imaging) and 12 μm (diffraction) second condenser apertures and a strongly excited first condenser lens. The diameter of the illuminated area at the object plane, when recording the images, was about 5 μm. Electron diffraction patterns and conventional bright-field images were taken by standard techniques. No objective aperture was used.

For recording under low-dosage conditions we found it desirable to have a shutter above the specimen so that at any stage the entire grid under observation could be protected from irradiation. Our early attempts to provide an appropriate shutter action by making use of the deflection facilities were unsuccessful because it was found impossible with these to eliminate movement of the specimen image at high magnification during the first fraction of a second the illumination was applied. The shutter finally constructed consisted of a very simple mechanism which cut off or let through the electron beam by rapidly translating the first condenser aperture to either of two fixed positions; it did not appear to suffer from the above defect.

Further minor modifications to the microscope involved the mounting of a second, high-resolution phosphor (Levy–West Laboratories) viewing screen about 7 cm off the optical axis in the direction of the observer, and also a second set of ×15 binoculars with which to view it.

Before attempting to produce micrographs under low-dosage conditions, several steps other than the standard ones for high-resolution imaging had to be taken. Thus: (i) it was checked that an image feature placed on the optical axis (or a marked point on the central viewing screen) did not change its position significantly on varying the magnification over the range to be used (~500 to 40,000×); (ii) the deflection controls were preadjusted so that at the upper magnification the focussed beam, which illuminated the on-axis viewing screen when they were switched off, would illuminate the off-axis viewing screen when they were activated; (iii) two settings of the second condenser lens were noted, one being that required to give a focussed beam and the other being that required to give an expanded beam of an intensity such as to deliver the required low dose to the photographic plate in a 4 to 8 s exposure.

These preliminary steps having been carried out, a grid was scanned at a magnification of 500 to 1500×, using an illumination level (~5×10^{-3} electrons/Å^2 per s in the plane of the object) just sufficient to detect specimen outlines (so that the dose delivered would be negligible in comparison with the dose needed for the photograph). On finding a suitable area, it was placed accurately on the optical axis (or the marked point on the viewing screen) and immediately isolated from the electron beam by closing the shutter. Appropriate adjustments were then made at leisure.

These were such that the shutter would be re-opened with the magnification at the level required for taking the photograph, the beam focussed on the off-axis viewing screen, and the region of interest (to be illuminated subsequently) centred on the optical axis.

The final procedure was then simply to open the shutter, focus using the off-axis screen, close the shutter once again, switch off the deflection controls, expand the beam, move the photographic plate into position and expose by opening the shutter once more for a suitable length of time.

(d) Data processing

Optical diffraction was used to select regions in the low-dose micrographs which would be appropriate for further analysis and as a means for determining the contrast transfer conditions (see Theoretical Background section (a)); it further provided a valuable means of assessing the microscope's performance at various stages of the work. Otherwise we resorted to numerical methods of analysis involving densitometer measurement of the optical densities on the photographic plates and calculations by digital computer.

Intensity data had to be collected both from the electron diffraction patterns and from the low-dose micrographs. The very sharp diffraction spots (<40 μm in diameter) were scanned with a Joyce-Loebl mark IIIc microdensitometer, using a very small sampling aperture. Carefully selected regions in the low-dose micrographs were scanned with a Perkin-Elmer model 1010A automatic microdensitometer (with the help of Dr J. Pilkington of the Royal Greenwich Observatory, Herstmonceux, Sussex). The step and sampling aperture sizes in this case were 10 μm and 11×11 μm, respectively, and the areas recorded consisted of 2048×2048 arrays of optical density readings. The optical densities were measured at intervals of 0·005, which is sufficiently accurate to record the variation due to statistical noise without significant error.

To establish the reciprocal space co-ordinates of a few of the stronger reflections in the Fourier transforms of each array, and hence the reciprocal lattice vectors, we initially processed a 1024×1024 region from the array on an IBM 370/165 computer, using a fast Fourier transform program written by L. Amos and L. Ten Eyck. Subsequently, to calculate the full transforms at the reciprocal space positions of all of the reflections, each 2048×2048 array was processed using the IBM 370/165 for carrying out the transform in one dimension and a PDP11/10 for carrying it out (on selected regions only) in the other. (The use of separate computers was a convenience rather than a necessity.) The final PDP11 output was in the form of 9×9 grids of amplitudes and phases, centred over each of the calculated positions of the reflections. An example of the form of the final amplitude output is given in Fig. 3.

3. Theoretical Background

We planned to combine data from bright-field images and electron diffraction patterns to determine the projected structures of the two specimens. Basically, the method proposed was to filter out statistical noise in Fourier transforms of selected low-dose micrographs by utilizing the phases at the reciprocal lattice points in the transforms, and to combine these phases with the amplitudes established by measurement of the intensities in the electron diffraction patterns. The electron diffraction intensities were to be used because they provide more accurate amplitude data than the image transforms (see Results section (c)). The resulting Fourier synthesis should give an undistorted map of the structure.

The validity of the maps obtained in this way depends largely on how faithfully the bright-field image records the projected structure; the general usefulness of the method depends to a great extent on the area of micrograph that needs to be processed to generate the necessary information. These aspects are of primary importance and it is appropriate now to discuss them in some detail.

(a) *Bright-field imaging of unstained periodic objects*

Diffraction of electrons by a periodic object is caused by periodic modulations in the potential field distribution associated with its constituent atoms. With unstained material (immersed in a liquid of similar density), these modulations are small, since the contributions due to the various small groups of like atoms placed more or less irregularly within the unit cell are similar and tend to cancel out. Therefore, diffraction is very weak in comparison, say, with that from thin, negatively stained biological specimens.

In order to describe this interaction, it is convenient to represent the incident electron beam by a coherent plane wave of amplitude unity. This, on passing through the potential field of the object is modified by a transmission function, $\exp(i\sigma\phi(x, y))$, where $\phi(x, y) = \int_{-\infty}^{\infty} \phi'(x, y, z)\mathrm{d}z$ is the projection in the beam direction of the three-dimensional potential distribution, $\phi'(x, y, z)$, in the object, and $\sigma = \pi/\lambda E$, E being the accelerating potential. The wave function immediately behind the object is therefore:

$$\psi(x_0, y_0) = \exp(i\sigma\phi(x_0, y_0)),$$
$$\simeq 1 + i\sigma\phi(x_0, y_0), \tag{1}$$

since the effect of the potential field on the incident electrons is small (weak phase object approximation).

A further term is sometimes included in equation (1) to account for attenuation of the coherent incident wave by, for example, inelastic scattering. This "amplitude" component is, however, weak in comparison with the phase component given above, and since its influence is undetectable in bright-field images of carbon films (Thon, 1971) it is reasonable to neglect it in a discussion of bright-field images of unstained specimens (see, however, footnote on p. 434).

The wave function at the diffraction plane is the Fourier transform of $\psi(x_0, y_0)$, multiplied by a phase factor, $\exp(i\chi)$, i.e.:

$$\Psi(h, k) = F(\psi(x_0, y_0))\exp(i\chi), \tag{2}$$

the integer variables h and k being used to indicate that we are dealing with a periodic object which produces discrete reflections on a reciprocal lattice. The factor, $\exp(i\chi)$, accounts for the modification of the phases of the various reflections as a result of defocussing and spherical aberration. The magnitude of this phase shift is given by:

$$\chi = 2\pi\lambda^{-1} (\delta f\theta^2/2 - C_s\theta^4/4), \tag{3}$$

where δf is the degree of underfocus, C_s is the spherical aberration coefficient and θ is the scattering angle.

Equation (2), using the approximation in equation (1), becomes:

$$\Psi(h, k) = \delta(0, 0) - \sigma\Phi(h, k)\sin\chi + i\sigma\Phi(h, k)\cos\chi, \tag{4}$$

where $\delta(0, 0)$ is a delta function representing the unscattered electron beam at the origin and $\Phi(h, k)$ is the Fourier transform of $\phi(x_0, y_0)$. The Fourier transform of $\Psi(h, k)$ gives the wave function, $\psi(x_i, y_i)$, at the image plane. However, since it is the intensity distribution, $|\psi(x_i, y_i)|^2$, that is recorded and since $\sigma\Phi(h, k) \ll \delta(0, 0)$, it is clear from equation (4) that the imaginary part of $\Psi(h, k)$ contributes to the image squared terms only in quantities which are small. Therefore, in considering the Fourier terms which make up the image, only the real part of equation (4) is relevant. Thus the Fourier transform of the intensity distribution in the image is proportional to:

$$\delta(0, 0) - 2\sigma\Phi(h, k)\sin \chi.$$

$-2 \sin \chi$, known as the *phase contrast transfer function*, is a term which modulates the amplitudes of $\Phi(h, k)$ and the signs of its phases. A plot of this function for a typical case is given in Figure 2(c). Its significance is of course that it relates the Fourier transform of the object to the Fourier transform of its image, and we have made use of this direct relation to establish the phases of the reflections originating from our specimens.

This approach, which was implicit in discussions by Hanszen & Morgenstern (1965) and by Hoppe (1970), has been shown to be applicable to thin, negatively stained specimens by Erickson & Klug (1970,1971).

The approach is valid, at least in the absence of effects due to curvature of the Ewald sphere (see Results section (a)), if (i) the second and higher-order terms in the expansion of $\exp(i\sigma\phi(x, y))$ are too small to introduce appreciable error, and (ii) the specimens are sufficiently thin that Fresnel diffraction effects (Cowley & Moodie, 1957) may be neglected.

The former source of error can be evaluated roughly from observed figures for $|F_{max}|^2$, the intensity of the strongest reflection in the electron diffraction pattern relative to the intensity of the unscattered beam. For either material under investigation (see Results section (a)), $|F_{max}|^2 < 5 \times 10^{-5}$†. Since the *phase* modulation of the incident wave by a Fourier component of the potential distribution in the object plane is directly related to its *amplitude* in the diffraction plane, this means that the phase modulations responsible for the strongest reflections are the order of 10^{-2} rad or less (and the contrast associated with them in the image, less than 1%). The greatest contribution from the second and higher-order terms arising from one Fourier component is therefore $\sim 0.5\%$ of the first-order term, i.e. certainly small enough to be neglected.

The effect of finite object thickness is to multiply the amplitude distribution in the diffraction plane by an additional phase factor, $\exp(i\pi Z\theta^2/2\lambda)$ (Cowley & Moodie, 1957), where Z is the specimen thickness. Clearly the higher the scattering angle, θ, the more significant does this source of error become. For the image resolutions we obtain however, the maximum value of the phase shift from this effect is only $\pi/6$ for the thicker specimen and $\pi/60$ for the thin one. These are still small shifts and consequently we do not expect the effects of finite object thickness to be very important.

(b) *Estimation of the required number of unit cells*

An estimate of the number of unit cells needed to provide sufficient information is useful for demonstrating the theoretical feasibility of our method. It is also important in practice that the required number of unit cells should fit easily onto a normal photographic plate.

† To appreciate the smallness of this figure it should be compared to the equivalent figure of 0·06 obtained by Gerchberg (1972), from the same catalase crystals, negatively stained.

A simple way of estimating this number is to divide the object into elements having the same projected area as a single unit cell, and assume each such element to be irradiated, on average, by aN electrons (N electrons/$Å^2$ incident on unit cell area, a $Å^2$). One then represents the electrons incident on n of the unit cells by a random array of delta functions. Now it is easily shown that in the Fourier transform of such an array, the amplitude at any general point is, on average, $(naN)^{\frac{1}{2}}$, compared to naN at the origin. These quantities would apply to the Fourier transform of the image intensity distribution, but for the fact that only a fraction (f) of the electrons incident on the object are actually detected on the photographic plate. Allowing for this, the "noise" in the Fourier transform of n unit cells in the image, expressed as a fraction of the amplitude at the origin, is $(fnaN)^{\frac{1}{2}}/(fnaN) = (fnaN)^{-\frac{1}{2}}$.

The strength of the weakest "signal" to be detected in the Fourier transform of the image intensity distribution, scaled in the same way, is $2|F_{min}|$ (i.e. twice the amplitude of the weakest relevant reflection in the *electron* diffraction pattern divided by the amplitude at the origin). If N' is taken as the maximum dose over which the periodic detail in the specimen remains intact (see also Results, section (a)), we have then: $2|F_{min}| > (fnaN')^{-\frac{1}{2}}$, or $2|F_{min}| = k(fnaN')^{-\frac{1}{2}}$ for a detectable signal to be obtained on Fourier transforming n unit cells. Alternatively, the minimum number of unit cells required to be present in the image is:

$$n = k^2/4faN'|F_{min}|^2. \tag{5}$$

The effect of noise from the substrate is not accounted for in this equation. However, provided that the actual dose given does not greatly exceed N' and that the substrate is not significantly more strongly scattering (i.e. thicker) than the specimen, it is likely to be small in comparison with noise of a statistical nature.

Since the "signal" in the Fourier transform of the image consists of discrete, regularly spaced peaks whose positions can be decided on with precision, the factor, k, in equation (5) (i.e. the ratio of the weakest peak to r.m.s. noise) can be quite small. Experimental measurements (Results, section (c)) show that a value of 1 would be satisfactory for perfect contrast transfer, but more generally we would expect a value of, say, 2 to be appropriate. The magnitude of f depends on the thickness of the specimen (since this determines the number of electrons which reach the image) and the detective quantum efficiency of the photographic plate; we estimate f to be about $\frac{1}{2}$ for purple membrane and $\frac{1}{4}$ for catalase. If we further anticipate our electron diffraction results (Results, section (a)) and put $N' = 0.5$ electrons/$Å^2$, $|F_{min}|^2 = 10^{-6}$, we obtain a required minimum number of about 1200 unit cells for purple membrane (unit cell area $\sim$3300 $Å^2$) and 700 unit cells for catalase (unit cell area $\sim$12,000 $Å^2$).

These numbers of unit cells can easily be accommodated on the normal photographic plate without having to resort to unduly low magnifications where high image resolution would be difficult to attain. Therefore, it should be possible to obtain single bright-field electron micrographs of such objects which contain all the required structural information. In practice, because of missing information at the zeros of the contrast transfer function, at least two would be required.

Finally, we draw attention to the fact that in contrast to the related expression for visualizing isolated molecules (e.g. Glaeser, 1973) equation (5) says nothing about resolution. A resolution limitation is however implied, since $|F_{min}|$ is, in general, the highest resolution reflection in the object's electron diffraction pattern. Of course, it might not be possible to attain the resolution predicted by equation (5) if the specimen is coherently ordered to a resolution which is greater than that of which the microscope is capable. It will become evident below that this is the case with our specimens.

4. Results

(a) Electron diffraction

Electron diffraction patterns (Plates I and II) demonstrate that crystalline order is preserved out to at least 3·5 Å in both the purple membrane sheets and the catalase platelets immersed in glucose. This resolution is close to that we obtain by X-ray diffraction and therefore refers to the inherent degree of order in the specimens.

Electron diffraction patterns of catalase having *low-angle* intensity distributions somewhat different from those shown in Plate I may be obtained by immersing the crystals in media composed of sucrose, ribose, inositol and other small hydrophilic molecules, which form reasonably stable non-volatile fluids. Such a variation in the low-angle intensity distribution is to be expected with media having slightly different densities, close to that of the protein (see e.g. Bragg & Perutz, 1952; Harrison, 1969). The high-angle intensities on the other hand are almost identical in the different media. This latter observation strongly suggests that the structure of the native hydrated protein is not greatly altered when the aqueous solution is replaced. Presumably, therefore, the properties of these sugars and related compounds are sufficiently close to those of water that they interact with the protein in a similar way.

The relative intensities of the reflections in the diffraction patterns shown in Plates I and II were sensitive, by differing degrees, to (i) the deviation of the normal of the plane of the specimen from the direction of the incident beam and (ii) the electron dose. The first factor is only of consequence with the catalase crystals and is due to the combined effect of their thickness (meaning that the reflections only extend a relatively short distance in reciprocal space) and the curvature of the Ewald sphere. It is easy to show that the 3·5 Å resolution reflections in the catalase pattern will only just touch the Ewald sphere when the crystals are exactly oriented. This means that deviations as little as $\frac{1}{2}°$ are sufficient to disturb the symmetry of the intensity distribution in the high-angle region quite noticeably. The effect will of course be evident in the image and needs to be minimal in the micrographs from which the projections are to be calculated, since it may lead to detail which is not representative of the detail that a true projection down the c-axis would show. Particular care therefore has to be taken with catalase in selecting micrographs which are appropriate, but this does not become a serious problem in practice until resolutions beyond our present 9 Å.

The behaviour of the two materials as a function of electron dose is illustrated in Figure 1. With purple membrane, the reflections decay exponentially (to a good approximation), at rates which are similar, although tending to be slightly more rapid at higher resolutions. With catalase on the other hand, while the higher resolution reflections tend to decay exponentially and at similar rates, the lower resolution reflections alter in a variable manner, a few actually increasing in intensity during the initial period of exposure.

Comparing the two sets of dose-response curves, it is interesting to note a good correspondence between the two materials in the decay rates of the highest resolution reflections. Since fading of diffraction patterns is due to loss of long-range order, the

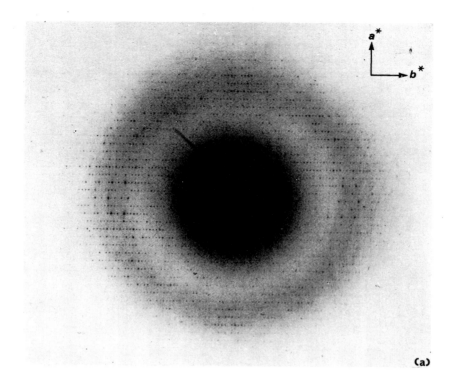

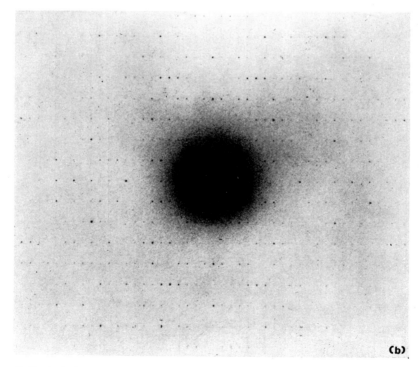

PLATE I. Typical electron diffraction patterns from nearly exactly oriented (see text) catalase crystals immersed in glucose; (a) high-angle and (b) low-angle pattern. The two patterns are shown separately since low-angle inelastic scattering makes it impossible to reproduce both in one picture. The fact that reflections from upper layer planes are not present in (a), means that the resolution limit does not extend much beyond 3·5 Å. Scales: (a) 1 cm: 0·068 Å^{-1}; (b) 1 cm: 0·026 Å^{-1}.

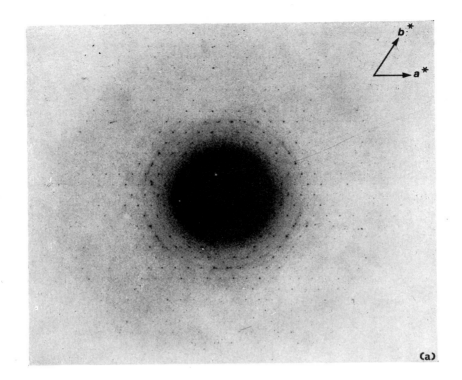

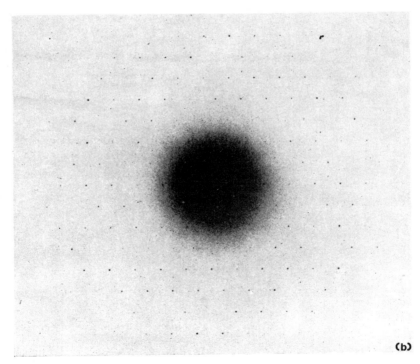

PLATE II. Typical electron diffraction patterns from the purple membrane; (a) high-angle and (b) low-angle pattern. They are from reconstituted membrane (Henderson, unpublished results), not the native one on which the micrographs are based. The distribution of intensities is the same in either case, but the reconstituted membrane has a better signal to noise ratio since it contains a much greater number of unit cells. Scales: (a) 1 cm: 0.056 Å^{-1}; (b) 1 cm: 0.028 Å^{-1}.

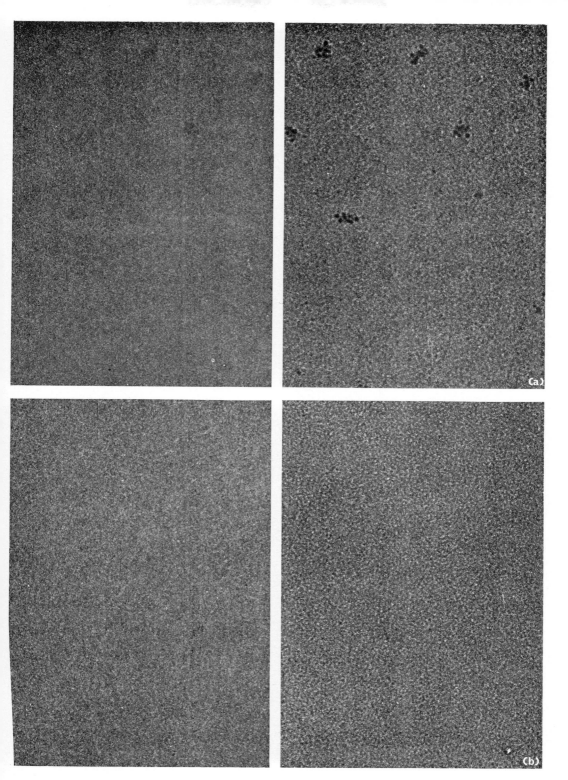

PLATE III. Typical low-dose and high-dose (longer exposure), bright-field micrographs of the same regions of (a) catalase and (b) purple membrane. The gold particles, used as focussing aids, can be seen in (a).

The two types of specimen are about equally sensitive to radiation damage and in both cases only the low-dose pictures contain significant detail about their structure; however, because of the low signal to noise ratio present in the low-dose pictures, such detail cannot be detected by eye. The high-dose pictures contain information on the contrast transfer conditions (see also Plates IV and V). Magnification: all at 500,000×.

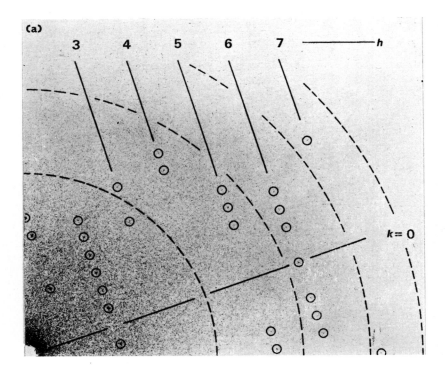

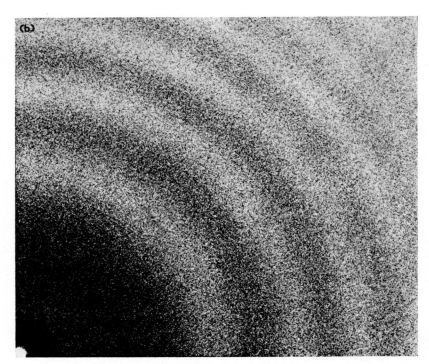

PLATE IV. Optical transform quadrants of the catalase micrographs containing the regions shown in Plate III; (a) of the low-dose micrograph, (b) of the high-dose micrograph. Peaks on the lattice corresponding to that in the electron diffraction pattern are evident in (a), their intensities being modulated according to the intensity distribution in (b).

Only the stronger peaks (circled) are readily observed in (a) (also in Plate V(a)) but this is partly because they are extremely sharp and hence difficult to reproduce photographically (the computed transforms, e.g. Fig. 3, indicate that their diameter should be less than 100 μm on the scale given). The positions of the minima in (b) are indicated by broken lines in (a). Scale: 1 cm: 0·011 Å^{-1}.

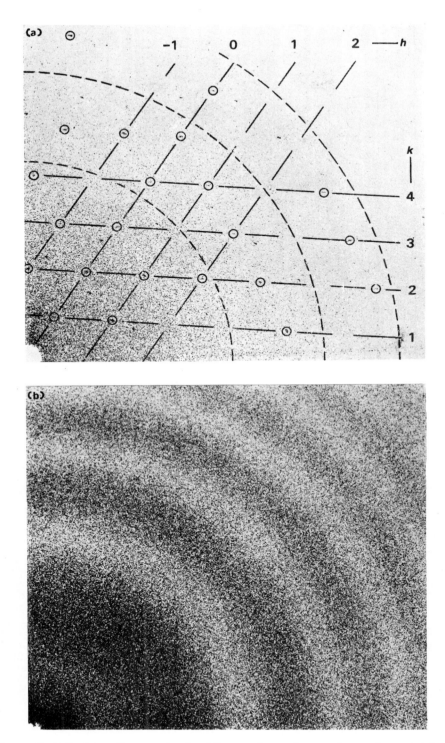

PLATE V. Optical transform quadrants of the purple membrane micrographs containing the regions shown in Plate III; (a) of the low-dose micrograph, (b) of the high-dose micrograph. There are two membranes in different orientations contributing to the peaks in (a); the lattice formed by one of them is indicated. Scale: 1 cm: 0·012 Å⁻¹.

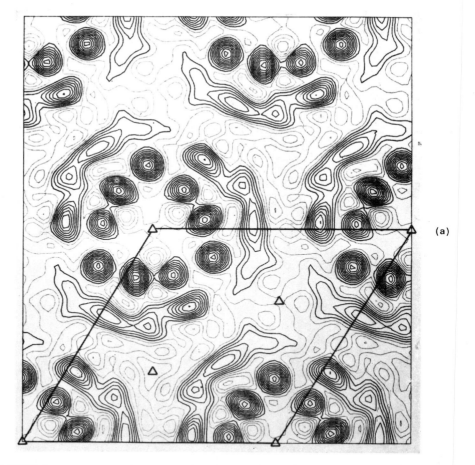

(a)

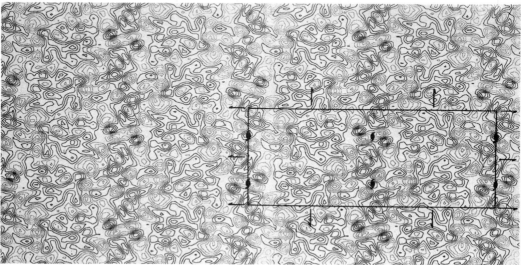

PLATE VI. Contour maps of the projected structures of (a) the purple membrane at 7 Å resolution and (b) catalase at 9 Å resolution, calculated according to the methods described in the text. The $F(0, 0)$ term was excluded from the syntheses. Positive contours are indicated by thicker lines; the positive peaks are due to high concentrations of scattering matter (such as dense regions in the protein). The unit cell dimensions shown in (a) are 62 Å × 62 Å and in (b) 69 Å × 173·5 Å.

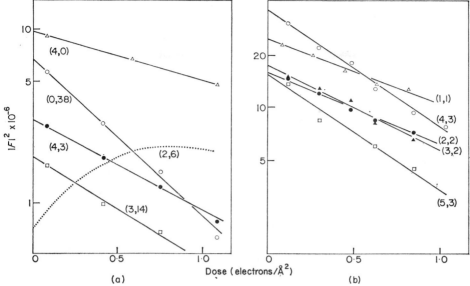

FIG. 1. The intensities (on a logarithmic scale) of some typical reflections in (a) the catalase and (b) the purple membrane electron diffraction pattern, plotted as a function of electron dose. The data for (a) were taken from "exactly oriented" single crystals (where the symmetry and Friedel-related reflections all have equal intensities), the dotted curve referring to one of the few reflections which actually increase in intensity during the initial exposure period. (Such reflections are at variance with the more typical behaviour, over the dose range shown, of exponential decay.) The data for (b) were taken from the "ring" diffraction patterns produced by a large number of native membranes several layers thick, so that the intensity of the (4, 3) ring is contributed by both the stronger (4, 3) and the weaker (3, 4) lattice reflection, and so on.

The intensities are scaled relative to unit intensity for the unscattered beam (where the area contributing to the diffraction pattern is smaller than that occupied by the specimen and there is only one specimen present). The absolute figures are more approximate in the case of catalase, where there is some variation in thickness from one crystal to the next.

correspondence probably reflects similarities in the strengths of the bonds stabilizing the two assemblies.

It is the highest decay rates that determine the characteristic dose, N', and hence the length of exposure required for the micrograph. Since the reflections should retain a good proportion of their original intensity over the total exposure period, and since the highest decay rates are approximately exponential, it seems reasonable to define the characteristic dose as that dose over which the *fastest fading peaks* have fallen to $1/e$ of their original intensity. This gives $N' = 0.5$ electrons/Å^2 for both specimens.

The strongest reflections are the (4, 3) in the purple membrane diffraction pattern and the (0, 4) in the catalase diffraction pattern. The intensities of these reflections, as fractions of the unscattered beam are 2.5×10^{-5} and 4×10^{-5}. 90% of the reflections from either structure are greater than 10^{-6}, i.e. 1/25 of the strongest intensity in the purple membrane case and 1/40 of the strongest intensity in the catalase case. Our aim is to include in the structure analysis all reflections above this limit. The exclusion of the remaining 10% of extremely weak reflections will have an insignificant effect on the final result. It is the strength of the weaker reflections which are to be included which determines the number of unit cells to be processed (see Theoretical Background section (b)).

(b) *Imaging*

The photographic factors which influence the choice of magnification for a low-dose image are (i) the resolution of the emulsion and (ii) the fog level. Our measurements on electron image plates gave figures of $\sim$10 μm and 0·03 optical density units for (i) and (ii), respectively, the magnitude of the former figure being a consequence of spread of electrons in the photographic emulsion (Hamilton & Marchant, 1967) rather than grain size. On this basis, the lower limit in magnification for, say, a 5 Å map is 20,000 $\times$. The upper limit is encountered when the optical density recorded on the plate produced by the dose required to disrupt the specimen (N') approaches the fog level. Allowing for attenuation of the incident beam by the specimen and assuming a unit density exposure of 0·46 electrons/μm^2 (estimated from data supplied by Kodak), we obtain an upper limit of about 40,000 $\times$ for purple membrane, and a somewhat lower figure for catalase, where the attenuation is greater.

The latter magnifications fortunately coincide with the practical requirement of having the size of the areas that need to be processed a fair amount smaller than the plate size; we accordingly chose magnifications of 40,000 $\times$ and 33,000 $\times$ for purple membrane and catalase, respectively, and exposures which gave optical density readings of about 0·1. (These exposures correspond to doses a little greater than N'.)

The typical low dose and high dose (longer exposure) micrographs we obtained for each of the two specimens are illustrated in Plate III. It is not possible to detect by eye any periodicities in the low-dose pictures,† although their presence is of course obvious from their Fourier (or optical) transforms (see below). The repeating structure rapidly disappears under irradiation and the second micrographs from each of the specimens show little or no evidence for it. These micrographs, being exposed by a dose approximately ten times that of the low-dose micrographs, are however less ⁻tatistically noisy and therefore display the usual "out of focus" granular phase detail distinctly.

(c) *Image analysis*

Computed and optical transforms of the low-dose electron micrographs (Fig. 3 and Plates III(a) and IV(a)) display discrete peaks on well-defined lattices, identifiable of course with those in the electron diffraction patterns; optical transforms of the high-dose micrographs (Plates III(b) and IV(b)) display concentric rings of intensity (Thon, 1966) deriving from the contrast transfer conditions prevailing. The computed transforms of the low-dose micrographs provided the phases at the lattice points of the transform of the specimen *image*. The optical transforms of the high-dose micrographs were then used to decide which of these image phases needed no modification and which required adjustment by 180° to produce the correct phases for the *object*‡.

The amplitudes of the structure factors were obtained by measurement of intensities

---

† With catalase the low resolution periodicities can be seen in over-exposed, low dose, micrographs. This is because they are associated with low-resolution diffraction peaks, some of which are relatively insensitive to electron dose.

‡ All the micrographs analysed were of under-, rather than over-focussed images. Therefore, in every case, the regions up to and including the first ring in the optical transforms, and every alternate ring thereafter, are associated with phase changes of 180° (cf. Fig. 2(c)).

Note that in the under-focussed image any small, low-frequency contribution due to amplitude contrast adds to the phase-contrast effect (see Erickson & Klug, 1970). The under-focussed image therefore avoids any ambiguity of sign in the low-angle region, which could arise in an over-focussed image with some amplitude contrast present.

in the electron diffraction patterns. We believe it is better to use the electron diffraction pattern rather than the Fourier transform of the image for this purpose because the electron diffraction pattern is insensitive to effects associated with the contrast transfer function and gives all the intensities on the same scale.

The ratios of the amplitudes computed from individual low-dose micrographs to the amplitudes from the electron diffraction patterns, when plotted against spatial frequency, form curves showing a series of well-defined maxima and minima (Fig. 2). The positions of these maxima and minima can be identified almost exactly with those in the optical transforms of the corresponding high-dose micrographs and are consistent with the theoretical contrast transfer functions calculated after estimating, from the high-dose micrograph, the degree of under-focus (see Fig. 2). We can thus confirm

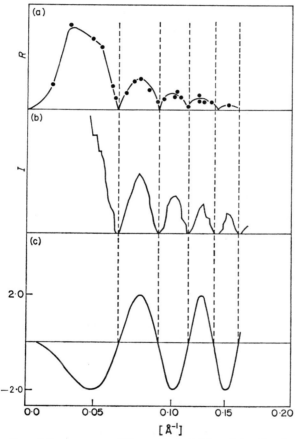

FIG. 2. Demonstration of the accuracy of the method of determining the signs of the phases by using information from a second micrograph. (a) The structure factor amplitudes calculated from a low-dose micrograph of the purple membrane as ratios (R) of their electron diffraction values, plotted against spatial frequency; they form a curve consisting of a series of maxima and minima. (b) The background-corrected intensity (I) across the optical transform of the corresponding high-dose micrograph; the positions of its maxima and minima match up almost exactly with those in (a). (c) The phase contrast transfer function appropriate to (b) (under-focus = 5750 Å; spherical aberration coefficient = 1·6 mm), illustrating how the sign assignments are made.

In principle, and as is evident from (a), a second micrograph is not strictly necessary for determining the signs of the phases. However, in practice, for a reliable determination, it is more or less essential.

that the focus does not change significantly between the low and the high-dose exposures and hence that the method involving pairs of micrographs for correcting the structure factor phases is accurate.

Figure 3 shows some typical computed matrices of amplitudes obtained by Fourier transformation of the 2048×2048 optical density arrays in the low-dose micrographs. On establishing the exact positions of a number of the strong high-resolution peaks in the lattices we were able to calculate the reciprocal space co-ordinates of all peaks in the transform with a precision of better than $1/5$ of the grid spacing in X and Y, and hence obtain objective measurements of phase at each reciprocal lattice point. However, we collected data only from those peaks whose amplitudes were equal to or higher than the r.m.s. noise level ($k = 1$ in equation (5)). Statistical considerations indicate that peaks which are appreciably weaker than the noise level will have a mean error in phase of between $45°$ and $90°$ ($90°$ being the figure for random phases). Even lower ratios of signal to noise than those used could nevertheless be accepted if one were to include data from more micrographs and extend the averaging procedure that is outlined below.

We investigated the accuracy of the phase determinations by taking account of the symmetry existing in the two structures.

In the catalase diffraction pattern, systematic absences along the $(h, 0)$ and $(0, k)$ lattice lines point to the presence of two mutually orthogonal 2_1 screw axes. Therefore the observed c-axis projection is centrosymmetric. (A related unpublished study of the same catalase crystals and single layers, negatively stained, indicates further that the crystals are orthorhombic with space group symmetry $P2_12_12_1$.) Ideally then, all the phases in the Fourier transform of the catalase micrograph, with an appropriately adjusted phase origin, should be $0°$ or $180°$. In addition, the signs, s, of symmetry related reflections in the two independently recorded quadrants are simply related $(s(h, k) = (-1)^{h+k} s(h, -k))$.

The purple membrane is composed of protein and lipid molecules which are arranged as a layer one unit cell thick in space group $P3$ (Henderson, 1965). Therefore, in this case, we can make use of the fact that the phases of each set of 3-fold related peaks should be the same.

We refined the phase origins for three catalase micrographs which were selected, by minimizing the r.m.s. phase errors, making the assumption that the nearest real values (i.e. $0°$ or $180°$) were correct. The r.m.s. error in single measurements of phase, worked out on this basis, was $39°$. On subsequently averaging the symmetry related peaks the r.m.s. error became $28°$. A final figure of $25°$ was then obtained after averaging the data common to all three micrographs. We were thus able to give a reliable phase assignment of $0°$ or $180°$ to all of the significant reflections (section (a) above) to a resolution of 9 Å.

Similarly, for three selected purple membrane micrographs, the origins were refined by minimizing the phase differences between symmetry related reflections. The r.m.s. error in single measurements of phase, estimated from the variation between the symmetry related peaks was then $18°$. Once the three symmetry related phases were averaged, the standard deviation became $10°$ to $11°$, and finally, less than $10°$ when phases from all three micrographs were averaged.

These final figures for the phase errors are low, considering that in their calculation we have accorded the strong and the weak peaks equal weight. They correspond to figures of merit (Dickerson et al., 1961) of 0.90 and 0.99, respectively, which are

```
                                    (2,0)
46  8  6 59 17 56 27 28  6      33  2 21 29 36 15  9 27 42      42 38 56 28 95 19 13 32  2
 8 12 17 15 98  4 26 38 49      45 50 52 23 31 34 12 13 14      13 42 29 45 54 69 26 29 22
51 28 18 79 68 50 27 35 38      38 15 18 41 16 28 11 15 11      21 37 23 21108 27 24 12 29
33 66 20 32186 63 15 19 21      12 17 28 58 88 33 40 25 30      26 22 22 17121 28 14 31  1
36 16 35 19186 31 39 44 40      20 44 38 58203 27 34 41 24      46 23 27 48181 70 14 48 26
14 45 23 28 46 32 65 15 50      17 42 48 66125 27 45 30 45      36 30 27 78 84 43 21 51 65
40 20 32 25 43 15 28 19 16      45 67 34 28 53 34 52 24 20      30 28  6 46 74 21 59 17  8
35 49 46 65 39  6 29 32 49      54 15 28 20 55 33 19 45 46      29 19 42 31 65 23 14 19 30
23 10 61 18 43 45 21 35 55      27 39 41 40121 31 15 25 33      45 20 30 40 23 34 16  6 28
          355°                            345°                            5°
```

```
                                    (5,0)
15 11 48 16 11 10 25 23 24       6 17 11 15 40 11 19 17 21      10 30 24 20 13 38 21 33 19
29 28 33 10 38 66 34 26 17       9 14 22 31 21  8 28 17 18      20 22 10 26 37  2 24 29  3
21 22  9 15 19 23  5 15 12      30 24  9 17 46 26 17 23  5      11 14 34 11 58 20 42 15 30
13 14 22 12 23 34  7  6 21      26 15 27 17 70 23 25 11  8      21 30 23 32 67 27  8 24 13
38 11 20 20 ⊕ 18 23 36 36      23 23 18  9 ⊕ 28 25 36  4       5 11  5 47 ⊕ 48 16 20 26
11 34 10 18 22 33 12  3 18      13  9 20 33 50 48 15 26 26       5 15 33 59 78 24 18 15 19
19 21 15 26 25  8 23 19 19      10 24  4 17 34  9 12  9 36      28 21 14 46 72  8 16 37 20
 7 29 34 20 15 32 27 19 12       4 22  6  2 11 22 19  5  9      22  1 36 11 42 39 19 13 48
14 23 16 21  3 27 25 14 23      12 25  8 15 21  2 41 39 21      37 20 22 12 29 22 36 15 11
          20°                             345°                            0°
```

```
                                    (7,0)
 9  5  5 12 10 14  3  1 14      19  4  9 18 16 22 17 22  7      18 20 25 12 14 23  6  8 17
14 18 12 14 10 11 24  2 14       9 15 18 10  9 21 15  7 24      12  5  4 18 19 16 16 20 14
16 18 11 40 27 14 10  5 14      11 14 14 14 10 30  5  8 21       8 18 24  7  4 17 18 10 10
 9  2 30 15 ⊕  8  1 16  3        5 16 22 10 ⊕ 17  7 35  3        8 19 28 13 34 12 35 12  7
21 26  7 47 ⊕ 24 10  9  6       35 21  5  8 ⊕ 11 12 13 16       14 19 31 18 18  9 14 31  1
11  6 10 22 21 18 17 14 10      11 11 22  6 19 21 12 18 27      17 20 35  5 15 13  5 20 11
14 19 11  6  2  7  4 42 25      31 10 15 29  3 21 25  9 13       9 16  9 14 27 14  5 20 16
17  3 36 20 13 13 27 18  1      13  7 14 16 18  5 18 12 14      14 11  6 12 31  2 15  4  8
18 25 20 30 18 19 28 19 16      20 11 20 13 17  9  8 15 34      21 23 16 12 26 19 29 19 18
          120°                            120°                            55°
                                     (a)
```

```
        (0,2)                            (0,4)                           (0,6)
26 30 20 38  6 22 21 16 16      13 28  8 27 34 18 18 14 26      33  8  9  7 19  6 14 12  4
42 51 50 26 29  4 41 38  9      27 13 15 11 39 13  7 12  2      12 10 17  9 22 17 19 13  7
13 32 24 22  1 34  9 14 31      23 23 29 11 31 31 12  8 19       6 14  7  9 16 26  7 16 18
17 17  8 13 42 34 15 17 15      13 16 19 11102  9 17 23 14      17 11 11 22 14 21 16 20  9
11 18 43  7116198 41 35 30      24 24 29 38300 24 11 25 14      21 12 13 26 50 30 21 19 41
26 25 17 11 33 47  2 19 25      31 52 11 50292 45 35  9 10      20 17 16 22 93 63 10 29 22
29 33 22 19 22  7 29  9  6      15  9 25 35122 32 17 15 13      21 26  6  7 66 22 40 28  3
37 10  8 14 20 20 17 22 30      25 29 16 31 90 24  7 12  9      30 16 18 37 37  5  9  9 22
 4  9 36 20 16  2 11  7 14      18 16 23 26 56 26  8 25 12       6 17 21 14 18 15 21  4  9
          5°                             225°                            155°
```

```
        (0,14)                           (0,16)                          (0,20)
 1  4  9 15  4  6 10  8 12       6  8  8 17  4  9  4  4  9       5  2  8  2  5 17  2  7  4
 8  4 14 11 16  2 11  4  1       5 17  2  3  2 15 10  8  6       2  5  4 10  2  8  5  7  6
10 10  1  9  7 13 23 12  6       7 10  6 14  7 16  2  4 11       1  6  7 13 11  8 16  7 12
 2 12  4 12  5 31  6  2  7      20  7  8  9 27  4 13  7  9      10  8 10 14 14  7 10  9  7
 7  7  9 14 29 15  8 17 13      13 11 11 20 34 10  8 13  3       2  4 10  4 ⊕  1 14 11  8
 0 11  6  2 14 22 16  8  5      18  8  3 12 46 20  7 16  4      17 16  7 10  4  8  4  8  3
 9 10  1  3 19  9  6 11 12      12  9  0  6 23 20  5  8  7      12  6  7  9 19 13 10  5 11
13  5  0 15 27 17  6  9  6      14  6 16 28  4 14 15  3 10      16  3  9  6  4 18  9  5 10
15  7  3 22 37 18  3 11  8       6 14 18 27 11  6 10 16  3       7  5  4 14  2  8  9  7  4
          355°                            220°                            195°
                                     (b)
```

FIG. 3. Fourier transform amplitudes in the region of some typical low and high-resolution peaks, computed from 2048×2048 optical density arrays in (a) a purple membrane micrograph and (b) a catalase micrograph.

The circles identify the calculated positions of each of the peaks (see text), and the phases at these positions are indicated. The strong peaks generally extended over 2 or 3 numbers, in which case they gave almost stationary values for the phase. With some of the weak peaks there were small phase gradients over the calculated positions, but in these cases the phases were able to be interpolated fairly reliably by eye.

higher (particularly in the case of the purple membrane) than those encountered when using the isomorphous replacement method in the X-ray analysis of protein crystals. An additional advantage is that the more important, stronger peaks are phased more accurately. This is not the case with the method of isomorphous replacement, which is often less accurate with stronger reflections.

The projected structures of the catalase crystals and the purple membrane sheets, obtained by Fourier synthesis of the structure factors, as determined above, are shown in Plate VI. The r.m.s. error levels in both structures estimated from the known phase and amplitude errors using the formula of Dickerson et al. (1961) are about one contour level.

Since we have already corrected for all imaging phase changes, the positive density regions in the maps correspond to the presence of high-density features in the object. In the case of the purple membrane the high-density matter is the protein ($\rho = 1.35$ g/ml) and the low-density matter, the lipid ($\rho = 1.00$ g/ml). In the case of the catalase, the mean density of the protein and the glucose fluid do not differ by significant amounts.

5. Discussion

There are several aspects concerning the projections we have just described, which warrant further comment. First, there is the important point that it is the undamaged structures we have determined and *not* radiation-altered versions. The evidence for this is provided in part by the dose-response curves (Fig. 1) which show, by extrapolation to zero dose, that the structure factor amplitudes derived from low-dose diffraction patterns can, at most, be only slightly in error. We can also be sure that the phases in the case of the centrosymmetric catalase projection do not alter during irradiation. Since they can only change by 180°, a change would imply a region of zero intensity in the dose-response curves, which is not observed†. A similar deduction about the accuracy of the phases in the purple membrane case can be made from the almost uniform fading of its diffraction pattern.

The second point to be made is that the projections given in Plate VI correspond to the same distribution of matter one would deduce by X-ray diffraction analysis. They provide information not only on the relatively gross features associated with quaternary structures (as, for example, is the case with conventional images of negatively stained specimens) but on internal detail as well. The projected potential distributions, as we have calculated, are to a good approximation equivalent to the electron density distributions that one would calculate by X-ray methods. Of course, our present structures are not yet at the resolutions we have come to expect from X-ray analysis; the analysis in Results section (c) suggests, however, that they do have the advantage of high accuracy of phase determination.

Our present limitation to resolutions of 7 and 9 Å is caused by the fall-off in contrast transfer of the medium and high spatial frequencies that is evident in the optical diffraction patterns of the images (Plates IV and V; also Fig. 2), although the effect may be influenced somewhat by the presence of non-localised inelastic scattering (Isaacson et al., 1974; Misell & Burge, 1973). The main factor causing this deterioration in contrast transfer is probably some electron optical disturbance (which may be more severe at the magnifications used here than at high magnifications). Other possible

---

† An exception is the (2, 0) reflection; it however only reappears with doses considerably greater than N', i.e. when all but the very low-resolution reflections have vanished.

contributing factors include the effect of partial coherence (Hanszen & Trepte, 1971; Frank, 1973; Reimer & Gilde, 1973) and spread of the electrons in the photographic emulsion. It can be shown, however, that because of our use of an extremely coherent incident beam and only moderate degrees of under-focus, the effect of partial coherence will be insignificant at $\sim$7 Å resolution. Experiments we have carried out to deter-mine the modulation transfer properties of the photographic plate, indicate that the reduction in contrast due to this source should be no more than 40% at this resolution (for similar results, using 60 kV electrons, see Burge & Garrard (1968)). Anticipating that the major cause of the deterioration will be overcome, it should be possible to solve the projected structures of these specimens to the resolution to which they are preserved ($\sim$3·5 Å).

Of the two projections in Plate VI, the catalase projection is the more complex. It can nevertheless be related to the projection obtained by negative staining and interpreted in terms of the superposition of single layers of molecules (Unwin, unpublished results). The possibility that the interpretation of the present projection should include consideration of dynamical interaction (Taylor & Glaeser, 1974) would seem to be discounted by the estimates given earlier. It also seems most unlikely in view of the intensities of some specific reflections in the electron diffraction pattern; the absence of a (0, 8) reflection (which is also absent or very weak in the correspond-ing X-ray powder pattern), in spite of the presence of a very intense (0, 4) reflection, is for example strong evidence that dynamical effects are unimportant.

The purple membrane, being comprised of a single layer of molecules and having only about one-tenth of the thickness of the catalase crystals, exhibits a minimum degree of superposition in projection and can accordingly be interpreted more directly. The features observed in Plate VI(a) are consistent with an analysis of the X-ray pattern of oriented purple membranes (Henderson, 1975) which showed that the protein in the membrane was composed to a considerable extent of α-helices arranged roughly perpendicular to the plane of the membrane, and which suggested that the lipids were less well-ordered in the lattice, occupying broad contiguous areas between the protein molecules. The high-density peaks, which are 10($\pm$1) Å apart in this projection, are likely to be the helices viewed end-on and in contact with one another, and the broad featureless regions near each of the 3-fold axes, the lipids. We are currently collecting three-dimensional data from the purple membrane by taking micrographs of tilted specimens.

6. Conclusion

It has been demonstrated that one can obtain *in vacuo* preservation of unstained biological specimens to resolutions close to interatomic spacings by replacing their natural aqueous environment with a medium composed of glucose or some similar compound.

Unstained specimens are sensitive to radiation damage. The detail in them can nevertheless be recorded in extremely low-dose, defocussed bright-field electron micrographs.

To determine the structures of the undamaged molecules, periodic arrays con-taining a redundancy of information and image processing methods are needed. Because of the low inherent contrast and statistically noisy nature of the images, the arrays need to be fairly large (greater than about 1000 unit cells).

Since the resolutions we have currently attained with the projected structures of the purple membrane and of catalase (7 Å and 9 Å, respectively) are not limited by the intrinsic order present (3·5 Å in both), there is a good prospect of obtaining results comparable to those obtained by using the isomorphous replacement method of protein crystallography. The electron microscopy method has the additional advantage of high accuracy of phase determination.

We thank Dr J. Pilkington (Royal Greenwich Observatory) and Dr O. Kübler (Eidg. Technische Hochschule Zurich) for help with the densitometry, Dr U. Aebi and Dr R. Smith (Biozentrum, Universitat Basel) for help with preliminary optical filtering experiments, and Chris Raeburn for construction of equipment for use with the microscope. We are also indebted to our colleagues at this Laboratory and at the Cavendish Laboratory, Cambridge, for valuable discussions and their comments on the manuscript.

REFERENCES

Blaurock, A. E. & Stoeckenius, W. (1971). *Nature (London)*, **233**, 152–154.
Bragg, W. L. & Perutz, M. F. (1952). *Acta Crystallogr.* **5**, 277–283.
Burge, R. E. & Garrard, D. F. (1968). *J. Scient. Instrum.* **1**, 715–722.
Cowley, J. M. & Moodie, A. F. (1957). *Acta Crystallogr.* **10**, 609–619.
Dickerson, R. E., Kendrew, J. C. & Strandberg, B. E. (1961). *Acta Crystallogr.* **14**, 1188–1195.
DeRosier, D. J. & Klug, A. (1968). *Nature (London)*, **217**, 130–134.
Erickson, H. P. & Klug, A. (1970). *Ber. Bunsenges. Phys. Chem.* **74**, 1129–1137.
Erickson, H. P. & Klug, A. (1971). *Phil. Trans. Roy. Soc. ser. B*, **261**, 105–118.
Frank, J. (1973). *Optik*, **38**, 519–536.
Gerchberg, R. W. (1972). *Nature (London)*, **240**, 404–406.
Glaeser, R. M. (1973). *Proc. 31st Annual Meeting EMSA*, (C. J. Arceneaux, Ed.) pp. 226–227, Claitor's Publishing Division, Baton Rouge, U.S.A.
Hamilton, J. F. & Marchant, J. C. (1967). *J. Opt. Soc. Amer.* **57**, 232–239.
Hanszen, K.-J. & Morgenstern, B. (1965). *Z. Ang. Physik.* **21**, 215–227.
Hanszen, K.-J. & Trepte, L. (1971). *Optik*, **33**, 182–198.
Harrison, S. C. (1969). *J. Mol. Biol.* **42**, 457–483.
Henderson, R. (1975). *J. Mol. Biol.* **93**, 123–138.
Hoppe, W. (1970). *Acta Crystallogr. sect. A*, **26**, 414–426.
Huxley, H. E. & Klug, A. (1971). Editors of *New Developments in Electron Microscopy. Phil. Trans. Roy. Soc. ser. B*, vol. 261.
Huxley, H. E. & Zubay, G. (1960). *J. Mol. Biol.* **2**, 10–18.
Isaacson, M., Langmore, J. & Rose, H. (1974). *Proc. 32nd Annual Meeting EMSA*, (C. J. Arceneaux, ed.), pp. 374–375, Claitor's Publishing Division, Baton Rouge, U.S.A.
Matricardi, V. R., Wray, G. & Parsons, D. F. (1972). *Micron*, **3**, 526–539.
McLachlan, D. (1958). *Proc. Nat. Acad. Sci., U.S.A.* **44**, 948–956.
Misell, D. L. & Burge, R. E. (1973). In *Image Processing and Computer-aided Design in Electron Optics* (Hawkes, P. W., ed.), pp. 168–194, Academic Press, New York.
Oesterhelt, D. & Stoeckenius, W. (1974). *Methods in Enzymology*, **31**, 667–678.
Parsons, D. F., Matricardi, V. R., Moretz, R. C. & Turner, J. N. (1974). *Advan. Biol. Med. Phys.* **15**, 162–270.
Reimer, L. & Gilde, H. (1973). In *Image Processing and Computer-aided Design in Electron Optics* (Hawkes, P. W., ed.), pp. 138–167, Academic Press, New York.
Stenn, K. & Bahr, G. R. (1970). *J. Ultrastruct. Res.* **31**, 526–550.
Sumner, J. B. & Dounce, A. L. (1955). *Methods in Enzymology*, **2**, 775–781.
Taylor, K. A. & Glaeser, R. M. (1974). *Science*, **186**, 1036–1037.
Thon, F. (1966). *Z. Naturforsch.* **219**, 476–478.
Thon, F. (1971). In *Electron Microscopy in Materials Science* (Valdrè, U., ed.), pp. 570–625, Academic Press, New York.

J. Mol. Biol. (1975) **94**, 441–448

A Rapid Method for Determining Sequences in DNA by Primed Synthesis with DNA Polymerase

F. Sanger and A. R. Coulson

Medical Research Council
Laboratory of Molecular Biology
Hills Road, Cambridge CB2 2QH, England

(Received 20 December 1974)

A simple and rapid method for determining nucleotide sequences in single-stranded DNA by primed synthesis with DNA polymerase is described. It depends on the use of *Escherichia coli* DNA polymerase I and DNA polymerase from bacteriophage T4 under conditions of different limiting nucleoside triphosphates and concurrent fractionation of the products according to size by ionophoresis on acrylamide gels. The method was used to determine two sequences in bacteriophage ϕX174 DNA using the synthetic decanucleotide A-G-A-A-A-T-A-A-A-A and a restriction enzyme digestion product as primers.

1. Introduction

In previous papers (Sanger *et al.*, 1973; Donelson, J. E., Barrell, B. G., Weith, H. L., Kössel, H. & Schott, H., unpublished data) we have described the determination of two nucleotide sequences in bacteriophage DNA using DNA polymerase primed by synthetic oligonucleotides. In this method the oligonucleotide was bound to a specific complementary region on the single-stranded DNA, and nucleoside triphosphates were added by the DNA polymerase to the 3' end of the primer. By using ^{32}P-labelled triphosphates, a radioactive complementary copy of a defined part of the template DNA was synthesized and subjected to sequence determination, which was greatly facilitated by use of the ribosubstitution technique.

This paper describes a relatively rapid and simple alternative procedure for deducing sequences by primed synthesis with DNA polymerase.

2. Principle of the Method

Figure 1 illustrates the principle of the method by considering its application to a small hypothetical sequence in a DNA chain. DNA polymerase I is first used to extend the primer oligonucleotide and copy the template in the presence of the four deoxyribotriphosphates, one of which is labelled with ^{32}P. Ideally this synthesis should be non-synchronous and as random as possible, so that the maximum number of oligonucleotides of different length, all starting from the primer, is formed. This mixture is then purified on an agarose column to remove the excess triphosphates and samples are re-treated in various ways as follows:—

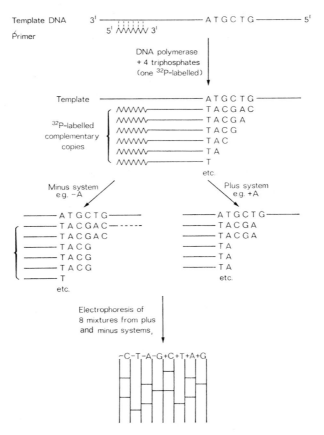

FIG. 1. The principle of the method.

(a) The "minus" system

In their original work on the "sticky ends" of phage λ DNA, Wu & Kaiser (1968) showed that if DNA polymerase acted in the absence of one triphosphate, synthesis would proceed accurately up to a position where the missing triphosphate should have been incorporated, and they used this principle to deduce sequences by assaying the relative amount of each nucleotide incorporated in the presence of different triphosphate mixtures (Wu & Taylor, 1971). The minus system described here uses a similar principle.

The random mixture of oligonucleotides, which is still hybridized to the template DNA, is reincubated with DNA polymerase I in the presence of three deoxyribo-triphosphates. Synthesis then proceeds as far as it can on each chain: thus, if dATP is the missing triphosphate (the —A† system), each chain will terminate at its 3′ end at a position before an A residue. Separate samples are incubated, with each one of the four triphosphates missing.

† Since this paper is concerned only with DNA the symbols A, C, G and T refer to the deoxy-ribonucleosides.

The four incubation mixtures are then denatured to separate the newly-synthesized strands from the template, subjected to electrophoresis on acrylamide gel in the presence of 8 M-urea, and a radioautograph prepared. In this fractionation system mobility is essentially proportional to size, so that the various synthesized oligonucleotides (which have a common 5′ end) will be arranged according to size. Ideally each oligonucleotide should be separated from its neighbour, which contains one more residue. The radioautograph from the −A system will contain bands corresponding to positions before the A residues in the synthesized chain. Thus the positions of As are located. Similarly the relative positions of the other residues may be located and, ideally, the sequence of the DNA read off from the radioautograph. This system alone is usually not sufficient to establish a sequence, so a second similar system is normally used in conjunction with it.

(b) *The "plus" system*

This system makes use of the method of Englund (1971,1972) who showed that, in the presence of a single deoxyribotriphosphate, DNA polymerase from bacteriophage T4-infected *Escherichia coli* (T4 polymerase) will degrade double-stranded DNA from its 3′ end, but that this exonuclease action will stop at residues corresponding to the one triphosphate that is present. This method is applied to the random oligonucleotide mixture obtained above. Samples are incubated with T4 polymerase and a single triphosphate and then fractionated by electrophoresis on acrylamide gel. Thus in the +A system only dATP is present and all the chains will terminate with A residues. The positions of A residues will be indicated by bands on the radioautograph. Usually these will be in products one residue larger than the corresponding bands in the −A system, but if there is more than one consecutive A residue the distance between the bands in the −A and +A systems will indicate the number of such consecutive residues. In the example illustrated in Figure 1 the smallest oligonucleotide gives a band in the −T position, indicating that the next residue after its 3′ terminus will be a T. This is confirmed by the presence of a band in the +T position in the next largest oligonucleotide. The bands in the +T and −A positions in this product show that its 3′ terminus is T and the residue following is an A, thus defining the dinucleotide sequence T-A. Similarly the next largest oligonucleotide defines the dinucleotide A-C, and so establishes the sequence T-A-C.

3. Materials and Methods

(a) *Chemicals and enzymes*

[32]P-labelled deoxyribonucleotide triphosphates were synthesized by the method of Symons (1974) or obtained from New England Nuclear at a specific activity of about 100 mCi/μmol. Phage φX DNA was a gift from H. L. Weith. Phage φX RF (replicative form) was a gift from J. W. Sedat. *E. coli* DNA polymerase (nach Klenow) was obtained from Boehringer Chemical Corporation. *Haemophilus influenzae* restriction enzyme (*Hind* II and III) was prepared by the method of Smith & Wilcox (1970), and some was a gift from T. Maniatis. T4 polymerase was prepared by the method of Goulian *et al.* (1968), and some was a gift from K. Murray and A. G. Isaksson.

Two experiments will be described here to illustrate the use of the method.

(b) *Experiment 1 (Plate I)*

The initial reaction mixture (100 μl) contained 0·02 M-Tris·HCl (pH 7·4), 0·01 M-MgCl$_2$, 0·01 M-mercaptoethanol, 0·05 mM-dATP, -dTTP and -dCTP, 10 μCi of [32P]dGTP (100

Ci/mmol), 6 μg ϕX174 DNA, 1·0 μg of the decanucleotide primer (Schott, 1974) and 16 units DNA polymerase. Incubation was at 0°C. After 2 min 50 μl were removed and added to 5 μl 0·2 M-EDTA to terminate the reaction. After 8 min 30 μl were removed and inhibited, and the remaining 20 μl incubated for a total of 30 min. The combined incubation mixtures were then applied to a column of Agarose (Bio-Gel A-0·5 m, 200 to 400 mesh, Bio-Rad Laboratories) using a 1-ml disposable plastic pipette. The column was made up and run in 2·0 mM-Tris·HCl (pH 7·4), 0·04 mM-EDTA. The fractionation was followed using a hand radiation monitor and the front band containing the synthesized DNA collected manually. It was concentrated to about 100 μl, and 5-μl samples taken for reincubation with 5 μl of the plus and minus mixtures and 1 μl of the appropriate enzyme.

The minus mixtures contained 0·04 M-Tris·HCl (pH 7·4), 0·02 M-MgCl$_2$, 0·02 M-mercaptoethanol, and the appropriate three dNTPs in 0·02 mM concentration. Incubation with DNA polymerase (0·8 unit in 1 μl) was for 30 min at 0°C.

The plus mixtures contained 0·13 M-Tris·HCl (pH 8·0), 0·013 M-MgCl$_2$, 0·02 M-mercaptoethanol, and the one triphosphate in 0·4 mM concentration. Incubation with T4 polymerase (approx. 0·02 unit in 1 μl) was for 30 min at 37°C.

The reactions were stopped by the addition of 1 μl 0·2 M-EDTA. 25 μl of freshly deionized formamide containing 0·3% xylene cyanol FF and 0·3% bromphenol blue were added and the solutions were heated at 95 to 100°C for 3 min before layering onto the acrylamide gel.

Electrophoresis was carried out on a 15% acrylamide gel (20 cm $\times$ 40 cm $\times$ 0·1 cm) at room temperature according to the method of Peacock & Dingman (1967). The buffer in the gel and in the anode compartment was Tris–glycine (3·028 g Tris-base, 14·4 g glycine/l)–8 M-urea. The cathode contained the same buffer without urea. It was run at 400 V until the bromphenol blue marker had travelled 30 cm from the origin (approx. 16 h). The gel was covered with "Saran wrap" and radioautographed.

(c) Experiment 2 (Plate II)

Hind fragment 1 was prepared by digestion of 100 μg of phage ϕX RF with *Hind* II + III and purified by ionophoresis on a 5% acrylamide gel (Edgell *et al.*, 1972). The amount obtained from 20 μg RF was mixed with 6 μg single-stranded ϕX DNA in 40 μl water and heated to 95°C for 3 min. 4 μl H $\times$ 10 buffer (66 mM-Tris·HCl (pH 7·4), 66 mM-MgCl$_2$, 100 mM-mercaptoethanol, 0·5 M-NaCl) was added and it was incubated at 67°C for 4 h. 20 μl of this annealed material was then incubated with DNA polymerase under the conditions described in section (b), above, in a volume of 50 μl. The labelled triphosphate was [^{32}P]dATP. Half of the solution was incubated for 2 min at 0°C and half for 8 min. These were combined and purified on an Agarose column and the eluate concentrated to 40 μl. 2-μl samples were then taken for reincubation with 2 μl of the plus and minus mixtures and 1 μl of the enzyme solution.

The minus mixtures were prepared by mixing equal volumes of the H $\times$ 10 buffer and 0·1 mM solutions of each of the three appropriate triphosphates. The plus mixtures contained 1 vol. H $\times$ 10 buffer, 1 vol. of a 0·2 mM solution of the triphosphate and 2 vol. of water. After incubation with enzyme as described above, 1 μl *Hind* II + III enzyme was added (this was assayed as sufficient enzyme to digest 2·5 μg bacteriophage λ DNA in 1 h at 37°C) and the solutions were reincubated at 37°C for 15 min. The reactions were terminated with EDTA and fractionated on a 12% acrylamide gel. A voltage of 1000 V was applied until the bromphenol blue marker had travelled 35 cm from the origin (4 to 5 h). The gel was allowed to become relatively hot during the run and no cooling was applied.

4. Results

(a) *The sequence primed by A-G-A-A-A-T-A-A-A-A (experiment 1)*

The method was initially tested out using the primer A-G-A-A-A-T-A-A-A-A on ϕX DNA. The sequence of the first 41 residues primed by this decamer was already known (Robertson *et al.*, 1973; Donelson *et al.*, unpublished data) and is shown in Figure 2. Plate I shows the results of one experiment. In interpreting the results it is essential to know the relative positions of the bands from the different samples on the radioautograph. This is easy where samples are run side by side but involves careful measuring for ones further apart. Thus, to show that the bands in

5   10   15   20   25
A- G-A- A- A- T- A- A- A- A- G- T- C- T- G-A- A- A- C -A- T- G- A -T – T

primer

30   35   40   45   50
A- A- A- C -T- C- C- T- A- A- G- C -A- G- A- A- A- A- C -C- T- A- C- C- G -C

Fig. 2. The sequence of the first 41 residues primed by the decanucleotide on ϕX DNA (Donelson et al., unpublished data).

the $+A$ and $-T$ system are both in position 20, the strong $-C$ band in position 28 was used as a reference. The distance between this $-C$ and the $-T$ in position 20 was the same as the sum of the distance from the $-C$ to the $+G$ in position 22 and that from this $+G$ to the $+A$ in position 20. Plate I also shows the exact lining up of the different bands, from the radioautograph.

If we consider position 18, there are two bands present, one in the $+A$ and one in the $-C$ system. This indicates that the 3' residue of the oligonucleotide in this position is an A and that the next residue in the chain will be a C. A dinucleotide, A-C, is thus defined. As would be expected, the next position contains a band in the $+C$ column, thus confirming the identification of the C. It also has a $-A$, which establishes A as the next residue. This is clearly followed by a sequence T-G; however at this stage the sequence is less easy to deduce, due to the presence of artifact bands, which appear to occur in this region and which will be discussed below. From positions 23 to 51 the sequence can be read off readily, and agrees well with the known sequence. Beyond residue 51 a partial sequence can be predicted but there are some uncertain parts, especially where "runs" of the same residue are concerned. For instance, it is clear that there are runs of T and A residues in the position labelled b, suggesting a sequence $A_{4-6} T_2 A_{3-4} T_{4-6}$, but the exact number of residues cannot be decided. This partial sequence will be discussed in a separate paper in connection with results obtained by other techniques.

(b) Priming with fragments from restriction enzyme digests

Besides using synthetic oligonucleotides as primers for DNA polymerase, it is possible to use the specific fragments obtained by the action of restriction enzymes (Maniatis et al., 1974). The method is essentially the same as that used for the oligonucleotide primer except that the products from the incubations with the plus and minus mixtures are digested with the restriction enzyme before applying to the acrylamide gel for electrophoresis. Thus the radioactive oligonucleotides all have the same 5' terminus, which corresponds to the original cleavage point of the enzyme. Plate II shows the result of an experiment in which fragment 1 from a digest prepared by the action of the restriction enzyme from H. influenzae (Hind II $+$ III) on double-stranded ϕX DNA (Edgell et al., 1972) was used as primer on the single-stranded ϕX

DNA. The results here were particularly clear-cut and it was possible to predict a sequence of 70 residues that proved to be essentially correct. This DNA sequence codes for a part of the largest coat protein (from gene F) of the bacteriophage and has been largely determined by Sedat J. W., Ziff, E. B. & Galibert, F. (unpublished data) by direct partial digestion methods (see Galibert et al., 1974), and by Blackburn (unpublished data) using transcription methods. Data were also obtained on peptides from the protein by Air (unpublished data). Although neither method gave the complete sequence, the results from the four techniques, which depend on entirely different approaches, left no doubt as to the correct sequence. These results will be discussed in a separate communication.

5. Discussion

In order that the method should give reliable results, various criteria must be satisfied. First, the products synthesized must all have the same 5′ terminus. *E. coli* DNA polymerase I normally has 5′ exonuclease activity; however, this can be removed by digestion with subtilisin and fractionation on Sephadex (Klenow & Henningsen, 1970). Thus DNA polymerase treated in this way was used in this work.

It is essential that the oligonucleotides are fractionated according to size. In preliminary experiments in which 20% acrylamide gels were used this was found not to be the case, some oligonucleotides migrating faster than corresponding smaller ones. This happened particularly in the region marked a in Plate I. The effect has not been encountered on 12% gels, though migration rate is not always exactly proportional to size, and the effect of the addition of a single residue is not exactly constant (see below). The reason for the anomalous migratory behaviour at higher gel concentrations is not clear, though it has been noted before (see Ikemura & Dahlberg, 1973). It may be connected with the secondary structure of the oligonucleotides and is certainly much worse in non-denaturing conditions.

Ideally oligonucleotides of all possible lengths should be present in the initial product of synthesis, so that all residues are represented in the plus and minus systems. In fact it is difficult to achieve this and under all conditions studied we have found that certain products are formed in relatively high yield, whereas others are absent. This suggests that the polymerase acts at different rates at different sites. It may be that this effect is partly related to the secondary structure of the template. It can also depend on the relative concentration of the triphosphates used. Thus if a low concentration of the ^{32}P-labelled triphosphate is used, "piling-up" frequently occurs before these residues. We find that the best results are obtained if synthesis is carried out for short times with a relatively high concentration of polymerase. Nevertheless it is frequently found that some expected products are missing (e.g. the +G in position 36, Plate I), and this constitutes a limitation of the method and is one reason why it is necessary to use both the plus and minus systems.

The main difficulty with the method occurs when consecutive runs of a given mononucleotide are present. Ideally if all oligonucleotides are present in the initial product of synthesis, each component of a run should appear as a band in the minus system, though the smallest component will be the strongest since it will be formed by extension of other smaller products—whereas the larger components are merely the unchanged oligonucleotides from the initial incubation. In Plate I it is possible to see the run of three As in positions 25–27 and of four As in positions 39–42. However it

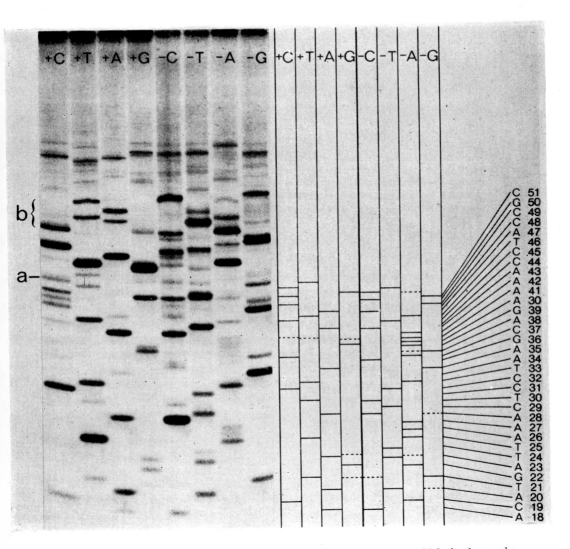

PLATE I. Radioautograph of experiment 1 (see Materials and Methods) in which the decanucleotide was used as a primer. The diagram shows the interpretation and the sequence deduced. The dashed lines represent artifact bands (see text). The bromphenol blue marker was 1 cm below the section of the gel shown in the Plate and the xylene cyanol FF marker opposite the product 50 nucleotides long.

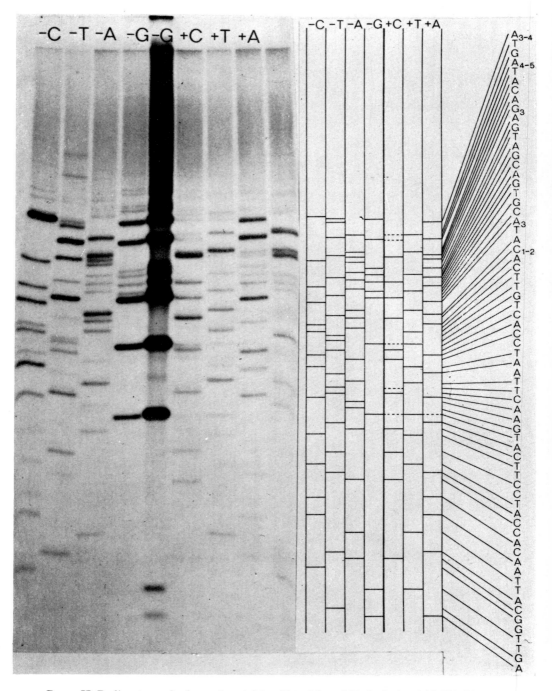

PLATE II. Radioautograph of experiment 2 (see Materials and Methods) in which *Hind* fragment 1 was used as a primer, and diagram illustrating the interpretation and sequence. The very dark centre sample labelled −G contained five times as much material as the other samples. The +G sample gave a number of artifact bands and it was not used in the interpretation. This was probably due to contamination of the +G mixture used in this experiment. A later experiment with a fresh mixture gave satisfactory results though resolution of the bands was less good.

is usually not possible to see the runs in this way and we have mainly used the distances between bands to deduce the size of a given run. In Figure 3 the change in distance travelled due to the addition of a single mononucleotide residue (the "jump") is plotted against the size of the oligonucleotides fractionated in Plates I and II. In general these values are sufficiently consistent; however they are less accurate for larger oligonucleotides and there are certain anomalies. In particular it will be seen from Plate I that position a appears to give four bands corresponding to $-T, -C,$ $+C$ and $+G$. Since the previous position contains a $-G$ and there is a $+T$ in a subsequent one, the most likely explanation is that the sequence is G-C-T and the products ending in G and C are moving at the same rate, although the expected

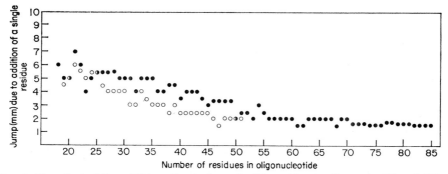

FIG. 3. The effect of the addition of a single nucleotide residue to oligonucleotides of different length on the distance travelled on ionophoresis in acrylamide gel. ($\bigcirc$) Results from experiment 1 (Plate I); ($\bullet$) results from experiment 2 (Plate II).

jump in this region should be 1·5 to 2·0 mm. Clearly in such a situation it is not possible to say how many C residues are present; however, fortunately such anomalies appear to be rare. Most of them can be avoided by using more severely denaturing conditions and low concentration gels as in the experiment shown in Plate II: thus by running the electrophoresis at 1000 V, so that the gel was at a higher temperature, some separation of the above two oligonucleotides was obtained.

One problem with methods involving primed synthesis with DNA polymerase is that one frequently finds a considerable blank incorporation of triphosphate in the absence of added primer. This is usually worse with ϕX than with phage f1 DNA and considerably worse with larger DNAs (see Maniatis et al., 1974). In general it does not affect the present method very seriously since the blank seems to represent some type of random incorporation and results in a general darkening of the radio-autograph at the top half of the gel, corresponding to oligonucleotides of over 100 residues long.

Another problem with the method, which is not fully understood, is the occasional appearance of "artifact" bands on the gel. These are usually faint bands and are marked as broken lines in Plates I and II. They can usually be recognized by the fact that they are not consistent in the plus and minus systems. The most serious one in Plate I is the $+G$ band in position 23. It is clearly anomalous since there is no corresponding $-G$ band, and two $+G$ bands should never occur in adjacent positions. These artifact bands are usually not reproducible from one experiment to another, but their presence does emphasize the need for caution in interpreting the results. They are usually found clustered in a certain region (e.g. positions 21 to 23, Plate I) and may represent positions where the polymerases react particularly slowly—perhaps due to some secondary structure of the DNA.

The above method represents a rapid and simple technique for determining sequences in specific regions of a DNA chain, if suitable primers are available. Thus, if successfully carried out, it is possible to deduce a sequence of 50 nucleotides in a few days. At present, due to the limitations discussed above, it cannot be regarded as a completely reliable method and it is necessary to have confirmatory data such as amino acid sequences or transcription results. It has nevertheless been found extremely useful and it is hoped that its reliability will increase as more experience is gained.

Some preliminary experiments in connection with this approach were done with J. E. Donelson. We wish to thank H. Kössel and H. Schott for providing the decanucleotide primer, K. Murray and A. G. Isaksson for a gift of T4 DNA polymerase, T. Maniatis for *Hind* enzyme, H. L. Weith for φX DNA, and J. W. Sedat for φX RF DNA.

REFERENCES

Edgell, M. H., Hutchison, C. A. & Sclair, M. (1972). *J. Virol.* **9**, 574–582.

Englund, P. T. (1971). *J. Biol. Chem.* **246**, 3269–3276.

Englund, P. T. (1972). *J. Mol. Biol.* **66**, 209–224.

Galibert, F., Sedat, J. & Ziff, E. (1974). *J. Mol. Biol.* **87**, 377–407.

Goulian, M., Lucas, Z. J. & Kornberg, A. (1968). *J. Biol. Chem.* **243**, 627–638.

Ikemura, T. & Dahlberg, J. E. (1973). *J. Biol. Chem.* **248**, 5024–5032.

Klenow, H. & Henningsen, I. (1970). *Proc. Nat. Acad. Sci., U.S.A.* **65**, 168–175.

Maniatis, T., Ptashne, M., Barrell, B. G. & Donelson, J. E. (1974). *Nature (London)*, **250**, 394–397.

Peacock, A. C. & Dingman, C. W. (1967). *Biochemistry*, **6**, 1818–1827.

Robertson, H. D., Barrell, B. G., Weith, H. L. & Donelson, J. E. (1973). *Nature New Biol.* **241**, 38–40.

Sanger, F., Donelson, J. E., Coulson, A. R., Kössel, H. & Fischer, D. (1973). *Proc. Nat. Acad. Sci., U.S.A.* **70**, 1209–1213.

Schott, H. (1974). *Die Makromolekulare Chemie*, **175**, 1683–1693.

Smith, H. O. & Wilcox, K. W. (1970). *J. Mol. Biol.* **51**, 379–391.

Symons, R. H. (1974). In *Methods in Enzymology* (Grossman, L. & Moldave, K., eds), vol. 29, part E, pp. 102–115, Academic Press, New York and London.

Wu, R. & Kaiser, A. D. (1968). *J. Mol. Biol.* **35**, 523–527.

Wu, R. & Taylor, E. (1971). *J. Mol. Biol.* **57**, 491–511.

J. Mol. Biol. (1975) **98**, 503–517

Detection of Specific Sequences Among DNA Fragments Separated by Gel Electrophoresis

E. M. SOUTHERN

Medical Research Council Mammalian Genome Unit
Department of Zoology
University of Edinburgh
West Mains Road, Edinburgh, Scotland

(Received 3 March 1975, and in revised form 26 June 1975)

This paper describes a method of transferring fragments of DNA from agarose gels to cellulose nitrate filters. The fragments can then be hybridized to radioactive RNA and hybrids detected by radioautography or fluorography. The method is illustrated by analyses of restriction fragments complementary to ribosomal RNAs from *Escherichia coli* and *Xenopus laevis*, and from several mammals.

1. Introduction

Since Smith and his colleagues (Smith & Wilcox, 1970; Kelly & Smith, 1970) showed that a restriction endonuclease from *Haemophilus influenzae* makes double-stranded breaks at specific sequences in DNA, this enzyme and others with similar properties have been used increasingly for studying the structure of DNA. Fragments produced by the enzymes can be separated with high resolution by electrophoresis in agarose or polyacrylamide gels. For studies of sequences in the DNA that are transcribed into RNA, it would clearly be helpful to have a method of detecting fragments in the gel that are complementary to a given RNA. This can be done by slicing the gel, eluting the DNA and hybridizing to RNA either in solution, or after binding the DNA to filters. The method is time consuming and inevitably leads to some loss in the resolving power of gel electrophoresis. This paper describes a method for transferring fragments of DNA from strips of agarose gel to strips of cellulose nitrate. After hybridization to radioactive RNA, the fragments in the DNA that contain transcribed sequences can be detected as sharp bands by radioautography or fluorography of the cellulose nitrate strip. The method has the advantages that it retains the high resolving power of the gel, it is economical of RNA and cellulose nitrate filters, and several electrophoretograms can be hybridized in one day. The main disadvantage is that fragments of 500 nucleotide pairs or less give low yields of hybrid and such fragments will be under-represented or even missing from the analysis.

2. Materials, Methods and Results

(a) Restriction endonucleases

EcoRI prepared according to the method of Yoshimuri (1971) was a gift of K. Murray. HaeIII prepared by a modification of the method of Roberts (unpublished data) was a gift of H. J. Cooke.

(b) Gel electrophoresis

Gels were cast between glass plates (de Wachter & Fiers, 1971). The plates were separated by Perspex side pieces 3 mm thick and along one edge was placed a "comb" of Perspex, which moulded the sample wells in the gel. The Perspex pieces were sealed to the glass plates with silicone grease and the plates clamped together with Bulldog clips. The assembly was stood with the comb along the lower edge. Agarose solution (Sigma electrophoresis grade agarose) was prepared by dissolving the appropriate weight in boiling electrophoresis buffer (E buffer of Loening, 1969). The solution was cooled to 60 to 70°C and poured into the assembly, where it was allowed to set for at least an hour. The assembly was then inverted, the comb removed and the wells filled with electrophoresis buffer. Samples made 5% with glycerol were loaded from a drawn-out capillary by inserting the tip below the surface and blowing gently. Electrophoresis buffer was layered carefully to fill the remaining space and a filter-paper wick inserted between the glass plates along the top edge. The lower end of the assembly was immersed in a tray of electrophoresis buffer containing the platinum anode, and the paper wick dipped into a similar cathode compartment. Electrophoresis was at 1·0 to 1·5 mA/cm width of gel for a period of about 18 h. Bromophenol blue marker travels about 3/4 the length of the gel under these conditions, but it should be noted that small DNA fragments move ahead of the bromophenol blue, especially in dilute gels. Cylindrical gels were cast in Perspex tubes 9 mm i.d. and either 12 or 24 cm long. These were run at 3 to 5 mA/tube in standard gel electrophoresis equipment.

Dr J. Spiers donated ribosomal DNA that had been purified on actinomycin/ caesium chloride gradients from DNA made from the pooled blood of several animals, and also ^{3}H-labelled 18 S and 28 S RNAs prepared from cultured *Xenopus laevis* kidney cells. *Escherichia coli* DNA was prepared by Marmur's (1961) procedure from strain MRE600. ^{32}P-labelled *E. coli* RNA was prepared from cells grown in low phosphate medium with ^{32}Pi at a concentration of 50 μCi/ml and fractionated by electrophoresis on 10% acrylamide gels. ^{32}P-labelled rat DNA was a gift of M. S. Campo. DNA from human placenta was a gift of H. J. Cooke, DNA from rat liver was a gift of A. R. Mitchell, DNA from mouse and rabbit livers were gifts of M. White. Calf thymus DNA was purchased from Sigma Biochemicals. For digestion with restriction endonucleases, the DNAs were dissolved in water to a concentration of approximately 1 mg/ml. One-tenth volume of the appropriate buffer was added and sufficient enzyme to give a complete digestion overnight at 37°C. Enzyme activity was checked on phage λ DNA and digests of this DNA were also used as size markers in gel electrophoresis, using the values given by Thomas & Davis (1975).

(c) Method of transfer

This section describes the method finally adopted: preliminary experiments and controls are described in later sections.

After electrophoresis, the gel is immersed for 1 to 2 h in electrophoresis buffer containing ethidium bromide (0·5 μg/ml), and photographed in ultraviolet light (254 nm) with a red filter on the camera. A rule laid alongside the gel aids in matching the photograph of the fluorescence of the DNA to the final radioautograph of the hybrids. Strips to be used for transfer from flat gels are cut from the gel using a flamed blade. The strips should be 0·5 cm to 1 cm wide and normally extend from the origin to the anode end of the gel. The gels used in this laboratory are 3 mm thick, and the length from the origin to the anode end is 18 cm but the method can be adapted to gels with different dimensions and to cylindrical gels. Strips of gel are then transferred to measuring cylinders containing 1·5 M-NaCl, 0·5 M-NaOH for 15 min and this solution is then replaced by 3 M-NaCl, 0·5 M-Tris·HCl (pH 7) and the gel is left for a further 15 min. The depth of liquid in the cylinders should be greater than the length of the gel strips and the cylinders should be inverted from time to time. For cylindrical gels (9 mm diam.), the times required for denaturation and neutralization are 30 and 90 min. Each gel transfer requires:

One piece of thick filter paper 20 cm × 18 cm, soaked in 20 × SSC (SSC is 0·15 M-NaCl, 0·015 M-sodium citrate).

Two pieces of thick filter paper 2 cm × 18 cm soaked in 2 × SSC.

One strip of cellulose nitrate filter (e.g. Millipore 25 HAWP), 2·2 cm × 18 cm, soaked in 2 × SSC. These strips are immersed first by floating them on the surface of the solution; otherwise air is trapped in patches, which leads to uneven transfer.

Three pieces of glass or Perspex, 5 cm × 20 cm and the same thickness as the gel.

Four or five pieces of thick, dry filter paper, 10 cm × 18 cm.

Transfer of the denatured DNA fragments is carried out as follows.

The large filter paper soaked in 20 × SSC is laid on a glass or plastic surface, care being taken to avoid trapping air bubbles below the paper. 20 × SSC is poured on so that the surface is glistening wet. One of the glass or Perspex sheets is laid on top of the wet paper. The gel strip is taken from the neutralizing solution and laid parallel to the glass or Perspex sheet, 2 to 3 mm away from it. The second glass or Perspex sheet is laid 2 to 3 mm away from the other side of the gel (Fig. 1(a)). The cellulose nitrate strip is then laid on top of the gel with its edges resting on the sheets of Perspex or glass, so that it bridges the two air spaces (Fig. 1(b)). The two narrow pieces of filter paper, moistened with 2 × SSC are laid with their edges overlapping the cellulose nitrate strip by about 5 mm (Fig. 1(c)) and the dry filter paper is then placed on top of these (Fig. 1(d)).

For cylindrical gels, the arrangement is similar, but in this case, the Perspex that supports the Millipore filter may be in contact with the gel because an air space is retained over the top of the gel. Several cylindrical gels can be transferred at the same time using the apparatus shown in Fig. 2 and similar arrangements can be used for flat gels.

20 × SSC passes through the gel drawn by the dry filter paper and carries the DNA, which becomes trapped in the cellulose nitrate. The minimum time required for complete transfer has not been measured: it depends on the size of the fragments and probably also depends on the gel concentration. A period of 3 h is enough to transfer completely all HaeIII fragments of *E. coli* DNA from 2% agarose gels 3 mm thick. But even after 20 h, transfer of large EcoRI fragments of mouse DNA from 9 mm diam. cylindrical gels is not complete. DNA remaining in the gel can be seen by the fluorescence of the ethidium bromide, which is not completely removed during treatment of the gel. During the period of the transfer, it is necessary occasionally to add more 20 × SSC to the bottom sheet of filter paper. If the paper dries too much, the gel shrinks against the cellulose nitrate strip and liquid contact is broken. The paper may be flooded, but care must be taken that liquid does not fill the air spaces between the gel and the side-pieces and soak the paper, bypassing the gel. It may be found convenient to leave the cellulose nitrate in position overnight: if the supply of

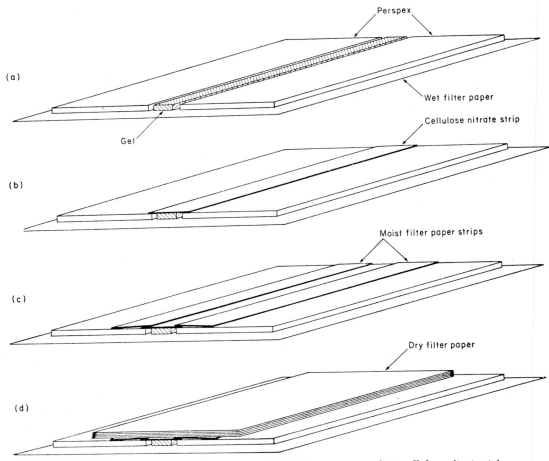

Fig. 1. Steps in the procedure for transferring DNA from agarose gels to cellulose nitrate strips.

$20 \times$ SSC has dried up it will be found that the gel has shrunk against the cellulose nitrate, but this does not impair the transfer. At the end of the transfer period the cellulose nitrate strip is lifted carefully so that the gel remains attached to its underside. It is turned over and the outline of the gel marked in pencil by a series of dots. The gel is peeled off the cellulose nitrate, the area of contact cut out with a flamed blade, and immersed in $2 \times$ SSC for 10 to 20 min. The strip is then baked in a vacuum oven at 80°C for 2 h.

(d) Hybridization

Radioactive RNAs are usually available in small quantities only and it is important to keep the volume of the solution used for hybridization as small as possible so that the RNA has a reasonable concentration. Two procedures can be used for hybridizing the cellulose nitrate strips after transferring the restriction fragments.

The procedure that uses the smallest volume is carried out by moistening the strip in hybridization mixture and then immersing it in paraffin oil. A drop of RNA solution (0·3 ml for a strip 1 cm × 18 cm) is placed on a plastic sheet. One end of the

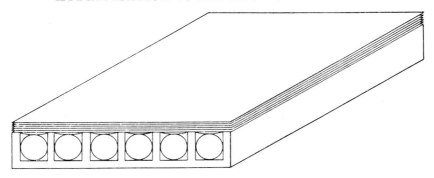

FIG. 2. Apparatus for transferring DNA from a number of cylindrical gels.

The apparatus is constructed of Perspex. The uprights which separate the gels and support the sheet of cellulose nitrate should be about 0·5 mm higher than the diameter of the gels, so that the cellulose nitrate sheet dips down to touch the gel. Thus an air gap is left between the cellulose nitrate sheet and the filter paper, above the line of contact between the gel and cellulose nitrate sheet. The apparatus is laid in a shallow tray containing 20 × SSC and the gels are then inserted into the troughs, care being taken to avoid trapping air bubbles beneath the gel. The cellulose nitrate sheet, wet with 2 × SSC, is laid over the gels and one piece of wet filter paper is laid over this. A stack of dry filter paper is then placed over the whole assembly. If necessary, a glass plate can be used to weigh down the filter papers. The depth of 20 × SSC in the tray should be enough to cover the lower part of the gels, but not so much that the air space between the Perspex and the cellulose nitrate becomes flooded.

cellulose nitrate strip is floated on the drop and when liquid is seen to soak through, the strip is drawn slowly over the surface of the drop. When it is completely wetted from one side, it is turned over and any remaining liquid is used to wet the other side. The strip is then immersed in paraffin oil saturated with the hybridization solution at the hybridization temperature. It should be borne in mind that baking the strip in 2 × SSC introduces salt, which must be taken into account when deciding on a solvent for the RNA if this method of hybridization is used. For example, if hybridization is to be carried out in 6 × SSC the RNA should be dissolved in 4 × SSC. Though this method can give good results (see Plate I) it often leads to high and uneven background. Kourilsky *et al.* (1974) found that this problem is removed if the hybridization is carried out in 2 × SSC, 40% formamide at 40°C. I have not tried this method, because this solvent removed DNA from the filters (see later section). It may well be the best method for hybridization to large fragments. I have found it convenient to carry out the hybridization in a vessel designed to hold the strip in a small volume of liquid.

The vessel (Fig. 3), which is easily made from Perspex, has internal dimensions of 0·8 mm deep by 2 cm high and about 1 cm longer than the strip to be hybridized. The vessel is filled with the solvent to be used for hybridization and the strip is fed in through the narrow opening in the top. The solvent is then drained off and the RNA solution introduced. Around 1 ml of solution is needed for a strip 1 cm × 18 cm. The wide sheets of cellulose nitrate used for transferring several gels (e.g. using the apparatus shown in Fig. 2) are too wide to be hybridized in this type of vessel. They can be hybridized in a small volume by wrapping them around a cylinder of Perspex, which is then inserted into a close-fitting tube. In this way, it is possible to hybridize a sheet 24 cm × 8 cm with about 4 ml of solution. If hybridization is carried out in a water-bath, it is not necessary to seal the top of the vessel provided the water-bath

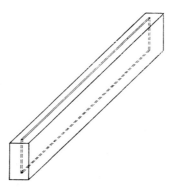

FIG. 3. Vessel used for hybridization of narrow strips.

itself is covered. The liquid in the vessel evaporates very slowly and can be replenished by small additions of water. A further advantage of this method of hybridization is that the RNA can be recovered and used again.

The period allowed for hybridization depends on the RNA concentration, its sequence complexity, its purity, and on the conditions of hybridization (see for example Bishop, 1972). After the appropriate period, strips are removed from the solution or paraffin oil, blotted between sheets of filter paper and washed, with stirring, for 20 to 30 min in a large volume of the hybridization solvent at the hybridization temperature. If the background is high, they may then be treated with a solution of RNAase A (20 μg/ml in 2 × SSC for 30 min at 20°C). After a final rinse in 2 × SSC they are dried in air.

So far the method has been tested with ^{32}P, ^{3}H, ^{35}S and ^{125}I-labelled RNAs. [^{32}P]RNAs have been detected by radioautography. For this the cellulose nitrate strips are laid on X-ray film and flattened against it with light pressure. ^{3}H, ^{125}I, ^{35}S and ^{14}C may be detected by fluorography. The cellulose nitrate strip is dipped through a solution of PPO in toluene (20%, w/v) dried in air, laid against X-ray film (Kodak RP-Royal Xomat) and kept at −70°C.

(e) Completeness of transfer and retention of DNA

Preliminary experiments showed that loading of DNA on to cellulose nitrate filters in 6 × SSC, conditions widely used in hybridization work, did not give complete retention of small fragments and a systematic study was made of the effect of salt concentration on retention. ^{3}H-labelled X. laevis DNA was sonicated to a single-strand molecular weight of 10^4 and denatured by boiling in 0·1 × SSC. Samples were made up to various salt concentrations and 0·1-ml portions of these solutions were pipetted on to cellulose nitrate filters, previously moistened with 2 × SSC, which were resting on glass-fibre filters. The solution that passed through the cellulose nitrate filter was thus collected in the glass-fibre filter. Both filters were then immersed in 5% trichloroacetic acid for 10 min, dried for 30 min in a vacuum oven at 80°C, and counted. It can be seen (Fig. 4) that the fraction of DNA retained by the cellulose nitrate increases with the salt concentration, and at concentrations above 10 × SSC the DNA is almost completely retained.

Losses of DNA at various stages of the transfer procedure were measured using ^{32}P-labelled E. coli DNA. The DNA was digested with EcoRI to give fragments in

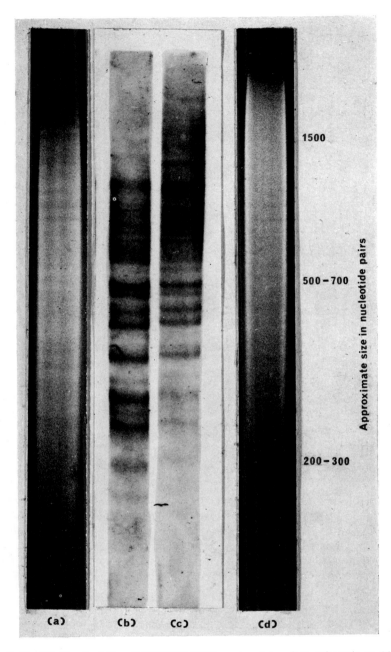

PLATE I. HaeIII digest of *E. coli* MRE600 DNA analyzed by electrophoresis on 2% agarose gel. DNA was then transferred to cellulose nitrate and hybridized with ^{32}P-labelled, high molecular weight RNA. (a) and (d) Photographs of ethidium bromide fluorescence. (b) and (c) Radioautographs of hybrids.

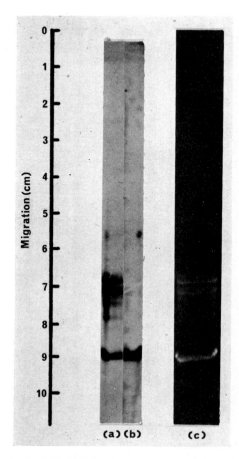

PLATE II. EcoRI digest of purified *X. laevis* ribosomal DNA analyzed by electrophoresis on 1% agarose gel. DNA was transferred to a cellulose nitrate strip, which was then cut longitudinally in two. The left-hand side was hybridized to 18 S RNA and the right-hand side to 28 S RNA (spec. act. of RNAs, 1.5×10^6 c.p.m. per μg). Hybridization was done in $1 \times$ SSC at 65°C using the vessel shown in Fig. 3. A large excess of cold 28 S RNA was added to the labelled 18 S RNA to compete out any 28 S contamination. After hybridization, the strips were washed in $1 \times$ SSC at 65°C for 1.5 h, and dried. They were then dipped through a solution of PPO in toluene (20%, w/v) dried in air and placed against Kodak RP Royal X-ray film at -70°C for 2 months. Photograph of ethidium bromide fluorescence (c). Fluorograph of 18 S hybrids (a). Fluorograph of 28 S hybrids (b).

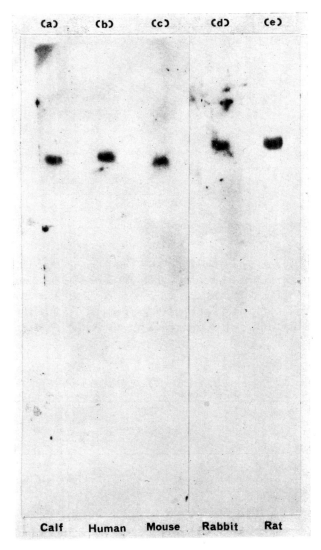

PLATE III. EcoRI digests of five mammalian DNAs, hybridized to 28 S RNA. Calf (a), human (b), mouse (c), rabbit (d) and rat (e) DNAs were digested to completion with EcoRI and separated by electrophoresis on 1% agarose gels (9mm × 12 cm, approx. 40 μg DNA per tube, 3 mA/tube for 16 h). The gels were pretreated as usual and the DNA fragments transferred to a single sheet of cellulose nitrate filter (12 cm × 8 cm) using the apparatus shown in Fig. 2. The top end of each gel was carefully aligned with one edge of the cellulose nitrate sheet. After 20 h, traces of DNA could still be seen, by ethidium bromide fluorescence, in the high molecular weight region of the gel. The filter was hybridized with 28 S RNA and radioautographed as described in the legend to Fig. 8.

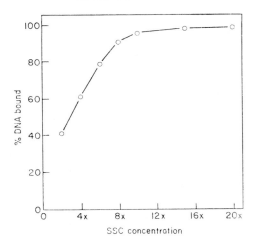

Fɪɢ. 4. Effect of salt concentration on efficiency of binding sonicated DNA to cellulose nitrate filters.

the large size range and with HaeIII to give small fragments. The fragments were then separated on a flat 1% agarose gel and transferred in the usual way. The solutions, the gel and the cellulose nitrate strip were counted. It can be seen (Table 1) that, whereas a small proportion of the DNA is leached out into the solutions during denaturation and neutralization, only traces remain in the gel after transfer.

Tᴀʙʟᴇ 1

Losses of DNA at stages of the procedure

	EcoRI fragments	HaeIII fragments
	DNA lost (%)	
Denaturing solution	2·1	4·8
Neutralizing solution	1·3	4·4
Remaining in gel after transfer	0·21	0·31

Two samples of *E. coli* DNA (0·1 μg; spec. act. approx. 10^6 c.p.m. per μg) were digested with EcoRI and HaeIII. The fragments were separated by electrophoresis on 1% gels in 1-cm wide slots, and then transferred to cellulose nitrate strips as described in Materials and Methods. The transfer was left overnight. The radioactivity leached out of the gel by the denaturing and neutralizing solutions, that remaining in the gel, and that which had been trapped on the cellulose nitrate filter were measured in a liquid scintillation counter (Cerenkov radiation).

(f) *Effect of DNA size on yield of hybrid*

Melli & Bishop (1970) have shown that hybridization by the filter method gives low yields with low molecular weight DNA. Their results were obtained using a single set of hybridization conditions and it seemed possible that losses might be reduced by using high salt concentrations. The effect of salt concentration on loss of

DNA from the filters was examined by loading filters with radioactive *X. laevis* DNA, single-strand molecular weight about 10^4, and incubating them in various salt solutions at different temperatures. Increasing the salt concentration does improve the retention of the DNA at any given temperature (Table 2) but the gain does not appear to be useful, because with increasing salt concentration it is necessary to use higher temperatures for hybridization, and this cancels the advantage of the high salt concentration. For example, the loss in $2 \times$ SSC at 65°C is the same as that in $6 \times$ SSC at 80°C and these are both typical hybridization conditions. Further experiments showed that it is disadvantageous to perform hybridization at high salt concentrations, below the optimum temperature. The optimum temperature for rate of hybridization of *X. laevis* 28 S RNA is around 80°C in $6 \times$ SSC but the rate at 70°C is still appreciable (Fig. 5). Below 70°C the rate falls rapidly. 28 S RNA was hybridized

TABLE 2

Effects of temperature and solvent on retention of sonicated DNA on cellulose nitrate filters

Solvent	50°C	Temperature 65°C DNA retained (%)	80°C	90°C
$2 \times$ SSC		77	62	48
$6 \times$ SSC		97	76	56
$10 \times$ SSC		95	83	73
$20 \times$ SSC		97	88	81
$6 \times$ SSC in 50% formamide	58	50		

[3]H-labelled *X. laevis* DNA (spec. act. approx. 5×10^5 c.p.m. per μg) was dissolved in ice-cold $0.1 \times$ SSC and sonicated in six 15-s bursts. Between each treatment the solution was cooled in ice for 1 min. The solution was boiled for 5 min, made to $20 \times$ SSC and cooled. Samples of this solution were pipetted on to 13-mm circles of cellulose nitrate, which were then washed in $2 \times$ SSC at room temperature. Approximately 650 c.p.m. were loaded on each filter, and there was no loss caused by washing in $2 \times$ SSC. The filters were dried, baked at 80°C for 2 h in a vacuum oven and immersed in 10 ml of the solvent equilibrated at the temperature used for incubation. After 90 min, the filters were removed, washed in $2 \times$ SSC at room temperature, dried under vacuum and counted in a liquid scintillation counter.

to high molecular weight and sonicated DNA in $6 \times$ SSC at 70 and 80°C (Fig. 6). As expected, the rate of hybridization at 70°C was lower than the rate at 80°C, but against expectation, both the rate and the final extent of hybridization were lower at the lower temperature, for the sonicated but not for the high molecular weight DNA. This result was unexpected because Melli & Bishop did not find an effect of DNA size on the rate of hybridization. They suggested that the decrease in yield for low molecular weight DNA is due to a loss of hybrid from the filter and it would be expected that such losses would increase with temperature. The lower yield for low molecular weight DNA at low temperature remains unexplained, but shows that there is no advantage to be gained in using high salt concentrations and low temperatures to retain small fragments of DNA during hybridization reactions. The advantage of using $6 \times$ SSC at optimum temperature is that the rate is greatly increased over the rate with, say, $2 \times$ SSC. A disadvantage is that the background of RNA that sticks to filters that have no DNA, increases with increasing salt concentration.

(g) *Methods of detecting and measuring hybrids: advantages of film detection*

Radioactive RNA may be detected and measured either by radioautography (or fluorography for weak β-emitters) or by cutting the strip into pieces, which can be counted in a scintillation counter. Film detection methods have the advantages over

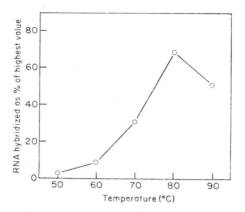

Fig. 5. Temperature dependence of hybridization of 28 S rRNA to *X. laevis* DNA.

X. laevis DNA was loaded on cellulose nitrate filters (17 μg DNA/13-mm diameter disc), which were pretreated as usual for hybridization. ^{3}H-labelled 28 S RNA from *X. laevis* kidney cells (spec. act. $1\cdot5 \times 10^6$ c.p.m./μg) was dissolved in $6 \times$ SSC (0·28 μg/ml) and warmed to the temperature used for hybridization. Two filters loaded with DNA and 2 blank filters were introduced into the solutions and left for 30 min. They were washed in 2 l of $2 \times$ SSC at room temperature, treated with 200 ml of RNAase A (20 μg/ml in $2 \times$ SSC) at room temperature for 20 min, washed in 200 ml of $2 \times$ SSC for 10 min, dried under vacuum and counted. Hybridization is expressed as a percentage of that obtained after 5 h at 80°C.

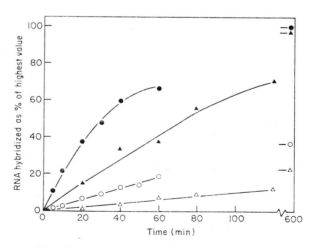

Fig. 6. Time course of hybridization of 28 S RNA to sonicated and high molecular weight DNA at 70 and 80°C.

Filters were loaded as described in the legend to Fig. 5. Two sets were loaded: one with high molecular weight DNA and one with DNA sonicated as described in the legend to Table 2. Hybridization and subsequent treatment of the filters was carried out as described in the legend to Fig. 6 and filters removed at the times indicated. $6 \times$ SSC at 80°C, high molecular weight DNA ($\bullet$); $6 \times$ SSC at 70°C, high molecular weight DNA ($\blacktriangle$): $6 \times$ SSC, 80°C sonicated DNA ($\bigcirc$): $6 \times$ SSC at 70°C, sonicated DNA ($\triangle$).

counting that they are more sensitive, give higher resolution, and can reveal artifacts not seen by counting.

The high sensitivity is illustrated by the analysis of *E. coli* rDNA (Plate I(b)). None of the bands that is clearly visible in the radioautograph contained more than 10 c.p.m. The strip of cellulose nitrate was cut into 150, 1-mm pieces and the pieces counted in a liquid scintillation counter. None of the pieces gave counts more than twice background and none of the features visible in the radioautograph was discernible from the counts. Around 100 c.p.m. of ^{32}P in a single band 1 cm wide can be detected with an overnight exposure. The radioautograph shown in Plate I was exposed for 1 week. Fluorography of ^{3}H is not so sensitive; about 3000 d.p.m. in a 1-cm band are needed to give a visible exposure overnight. The fluorograph shown in Plate II was exposed for 2 months.

The greater resolution of film detection is illustrated by a comparison of Plate II with Figure 7(c). Plate II is a fluorograph of the strip and Figure 7(c) shows the pattern of counts obtained by cutting the strip into 1-mm pieces. Many of the bands seen in the fluorograph are not discernible in the pattern of counts (compare also the tracing of the fluorograph (Fig. 7(b)) with (c)).

For ionizing radiation, blackening of the X-ray film is proportional to the amount of incident radiation, up to the limit where a high proportion of silver grains are exposed. The relative amount of radioactivity in bands can therefore be compared by tracing radioautographs in a densitometer and comparing peak areas. However, like all other photosensitivie materials, X-ray films suffer from "reciprocity failure" at low intensities of illumination by non-ionizing radiation and it is likely that bands which contain only a few counts of ^{3}H will not be detected by fluorography even after long exposures. I have not determined the lower limit of detection. Bonner & Laskey (1974) found that 500 d.p.m. of ^{3}H in a band 1 cm $\times$ 1 mm could be detected in one week and in my own experience, less than 20 d.p.m. can be detected with longer exposure. Reciprocity failure could affect quantitation of fluorographs by densitometry but comparison of Figure 7(b) and (c) suggests that the response of the film is linear within the limits of this experiment. Clearly, quantitation of ^{32}P by densitometry can be accurate and more sensitive than counting, but film response to ^{3}H may not be linear for low amounts.

An additional advantage of film detection is that non-specific binding of RNA to the cellulose nitrate is more easily distinguished from bands of hybrid. Plate III illustrates this point. In this radioautograph, non-specific binding can be seen as dots and streaks with an appearance clearly different from that of a band. Had this strip been analysed by counting, non-specific binding would not have been distinguishable from the hybrids.

(h) *Analysis of ribosomal DNA in* X. laevis

A total of 0·6 μg of purified *X. laevis* rDNA was digested with EcoRI and the fragments separated by electrophoresis in 1% agarose gels (Plate II(c)). The pattern of fragments is similar to that described by Wellauer *et al.* (1974). They compared the secondary structures of the denatured DNA fragments with those of the ribosomal RNAs and showed that the fastest running fragment (M_r approx. 3×10^6) contained most of the DNA coding for 28 S RNA, all of the transcribed spacer, and a small portion of the DNA coding for 18 S RNA. The larger fragments (M_r 4 to 6×10^6) contained most of the DNA coding for 18 S RNA, all of the non-transcribed

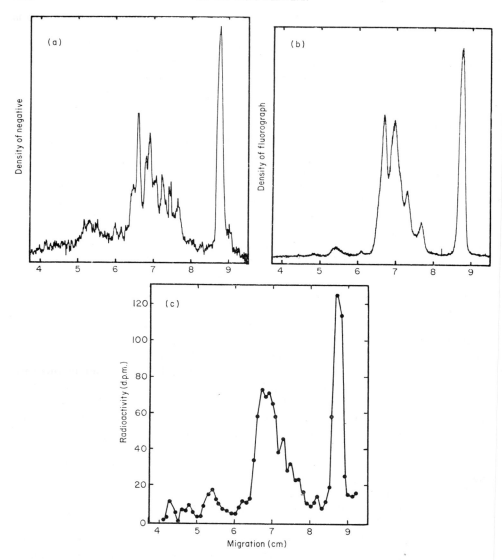

FIG. 7. (a) Microdensitometer tracing of the negative of Plate II(c). (b) Microdensitometer tracing of Plate II(a). (c) Distribution of counts in the Millipore strip which on fluorography gave Plate II(a). The strip was cut into 1 mm pieces, which were counted in a liquid scintillation counter at an efficiency of 40%.

spacer, and a small portion of the DNA coding for 28 S RNA. Different lengths of non-transcribed spacer DNA accounted for the variation in size of the longer fragments. The digest shown in Plate II(c) was transferred to cellulose nitrate as described previously. The strip was cut longitudinally into 2 parts and 1 part was hybridized with 18 S RNA and the other with 28 S RNA. Hybrids were detected by fluorography of the ^{3}H-labelled RNA (Plate II(a) and (b)). Comparison of Plate II(a) and (c) shows that the resolution of the fine bands containing the 18 S coding sequence is not as high in the fluorograph as it is in the photograph of the gel. Whereas 9 bands can be distinguished in the photograph, only 7 can be distinguished with confidence in the fluorograph. From this analysis it is possible to locate the EcoRI site within the DNA coding for 18 S RNA. As Wellauer *et al.* (1974) showed, 1 of the 2 breaks in the rDNA occurs towards one end of the 18 S region and the other is close to the distal end of the 28 S region. The 3×10^6 mol. wt fragment accounts for virtually all of the hybridization to 28 S RNA and for about 30% of the hybridization to the 18 S RNA (27% measured from the tracing of the fluorograph (Fig. 7(b)) and 31% from the counts). Only traces of 28 S RNA hybridize to the heterogeneous collection of fragments with molecular weights between 4 and 6×10^6, whereas about 70% of the 18 S hybridization is accounted for in these fragments. Thus the break in the 28 S region of the DNA is very close to the end of the coding sequence and the break in the 18 S region is about one-third of the way into the coding sequence.

(i) *Analysis of mouse and rabbit ribosomal DNAs: evidence for long, non-transcribed spacer DNA*

An EcoRI digest of total mouse DNA was separated by electrophoresis on cylindrical 1% agarose gels and transferred to strips of cellulose nitrate paper. One strip was hybridized to 18 S RNA and another to 28 S RNA prepared from rat myoblasts labelled with ^{32}P. The 28 S hybrids showed a strong, sharp band at the position of about $5 \cdot 2 \times 10^6$ daltons and a very faint, broad band in the region around 14×10^6 daltons (Fig. 8(b)). The 18 S hybrids showed corresponding bands but in this case the slower moving, broad band was relatively more intense (Fig. 8(a)). From this information, a partial structure can be derived for the ribosomal DNA in mouse. Assuming that the ribosomal genes are tandemly linked, it is clear that EcoRI makes at least 2 breaks in the sequence; one in the 18 S and one in the 28 S region. Transcription of ribosomal genes in mammals produces a precursor RNA corresponding to a DNA mol. wt of about 6×10^6, and it follows that the EcoRI fragment of about $5 \cdot 2 \times 10^6$, which contains both 28 S and 18 S sequences, must also encompass much of the transcribed spacer. The heterogeneous fragments with a mol. wt of 14×10^6 must contain a long stretch of non-transcribed spacer, and may contain some of the transcribed spacer too.

A similar analysis was carried out with rabbit DNA and gave similar results, although the size of the fragments was different from the corresponding fragments from mouse DNA. The band containing most of the 28 S sequence was larger (M_r approx. 6×10^6), whereas that containing most of the 18 S sequence was smaller (M_r approx. 12×10^6) and more homogeneous than the corresponding fragment in the mouse. The structures of mouse and rabbit ribosomal DNAs are thus rather similar to that of *X. laevis* but with longer spacer regions. The overall length of the unit in mouse is at least twice as long as that in *X. laevis*.

(j) *EcoRI sites in the rDNA of five mammals*

The analyses described above, taken with those of Wellauer *et al.* (1974) suggest that the two EcoRI sites in the ribosomal genes have been conserved since the amphibians and mammals diverged. In this case it would be expected that all

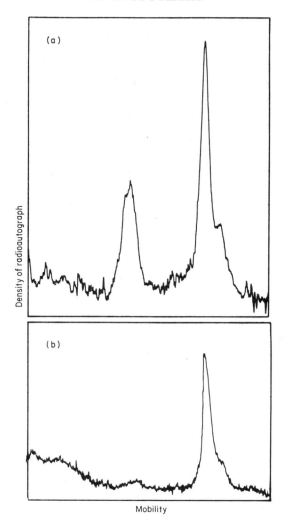

FIG. 8. EcoRI digest of mouse DNA hybridized to 18 S and 28 S RNA.

Total mouse DNA was digested to completion with EcoRI. The digest was separated by electrophoresis on 1% cylindrical agarose gels (9 mm × 24 cm, 5 mA/tube for 20 h, 40 μg of DNA/gel).

The gels were stained, photographed, and the DNA transferred to cellulose nitrate as described in Materials and Methods. One gel was hybridized to ^{32}P-labelled 18 S RNA and another to 28 S RNA. The RNA concentration was 0·1 μg/ml in 6 × SSC and hybridization was carried out at 80°C for 4 h. The filters were then washed in 2 × SSC (4 l) at 60°C for 30 min, dried and radioautographed using Kodak Blue Brand X-ray film.

(a) Densitometer tracing of the 18 S hybrids. (b) Densitometer tracing of the 28 S hybrids.

mammalian rDNAs would have equivalent EcoRI sites. Total DNAs from calf thymus, human placenta, and from livers of mouse, rabbit and rat were digested with EcoRI and the fragments separated by electrophoresis on cylindrical gels. The fragments were then transferred to a single sheet of cellulose nitrate filter and hybridized with ^{32}P-labelled rat 28 S RNA. All 5 DNAs showed a strong band in the radioautograph